Statistical Signal Processing in Engineering

Statistical Signal Processing in Engineering

Umberto Spagnolini
Politecnico di Milano
Italy

WILEY

This edition first published 2018

Registered Offices
John Wiley & Sons, Inc., 111 River Street, Hoboken, NJ 07030, USA
John Wiley & Sons Ltd, The Atrium, Southern Gate, Chichester, West Sussex, PO19 8SQ, UK

Editorial Office
The Atrium, Southern Gate, Chichester, West Sussex, PO19 8SQ, UK

For details of our global editorial offices, customer services, and more information about Wiley products visit us at www.wiley.com.

Wiley also publishes its books in a variety of electronic formats and by print-on-demand. Some content that appears in standard print versions of this book may not be available in other formats.

Library of Congress Cataloging-in-Publication Data

Names: Spagnolini, Umberto, author.
Title: Statistical signal processing in engineering / Umberto Spagnolini.
Description: Hoboken, NJ : John Wiley & Sons, 2018. | Includes bibliographical references and index. |
Identifiers: LCCN 2017021824 (print) | LCCN 2017038258 (ebook) | ISBN 9781119293958 (pdf) | ISBN 9781119293996 (ebook) | ISBN 9781119293972 (cloth)
Subjects: LCSH: Signal processing–Statistical methods.
Classification: LCC TK5102.9 (ebook) | LCC TK5102.9 .S6854 2017 (print) | DDC 621.382/23–dc23
LC record available at https://lccn.loc.gov/2017021824

Cover Design: Wiley
Cover Image: ©Vladystock/Shutterstock

Set in 10/12pt Warnock by SPi Global, Pondicherry, India

Printed in the UK

To my shining star Laura

Contents

List of Figures

List of Tables

Preface

This book is written with the intention of giving a pragmatic reference on statistical signal processing (SSP) to graduate/PhD students and engineers whose primary interest is in mixed theory and applications. It covers both traditional and more advanced SSP topics, including a brief review of algebra, signal theory, and random processes. The aim is to provide a high-level, yet easily accessible, treatment of SSP fundamental theory with some selected applications.

The book is a non-axiomatic introduction to statistical processing of signals, while still having all the rigor of SSP books. The non-axiomatic approach is purposely chosen to capture the interest of a broad audience that would otherwise be afraid to approach an axiomatic textbook due to the perceived inadequacy of their background. The intention is to stimulate the interest of readers by starting from applications from daily life, and from my personal and professional experience, I aim to demonstrate that book theory (still rigorous) is an essential tool for solving many problems. The treatment offers a unique approach to SSP: applications (somewhat simplified, but still realistic) and examples are interdisciplinary with the aim to foster interest toward the theory. The writing style is layered in order to capture the interest of different readers, offering a quick solution for field-engineers, detailed treatments to challenge the analytical skills of students, and insights for colleagues. Re-reading the same pages, one can discover more, and have a feeling of growth through seeing something not seen before.

Why a book for engineers? Engineers are pragmatic, are requested to solve problems, and use signals to "infer the world" in a way that can then be compared with the actual ground-truth. They need to quickly and reliably solve problems, and are accountable for the responsibility they take. Engineers have the attitude of looking for/finding quick-and-dirty solutions to problems, but they also need to have the skills to go deeper if necessary. Engineering students are mostly trained in this way, at graduate level up to PhD. To attract graduate/PhD engineering students, and ultimately engineers, to read another new technical book, it should contain some recipes based on solid theory, and it should convince them that the ideas therein help them to do better what they are already doing. This is a strong motivation to deal with a new theory. After delineating the solution, engineering readers can go deeper into the theory up to a level necessary to spot exceptions, limits, malfunctioning, etc. of the current solution and find that doing much better is possible, but perhaps expensive. They can then consciously make cost-benefit tradeoffs, as in the nature of engineering jobs.

Even if this book is for engineers and engineering students, all scientists can benefit from having the flavor of practical applications where SSP offers powerful problem-solving tools. The pedagogical structure for school/teachers aims to give a practical vision without losing the rigorous approach. The book is primarily for ICT engineers, these being the most conventional SSP readers, but also for mechanical, remote sensing, civil, environmental, and energy engineers. The focus is to be just deep enough in theory, and to provide the background to enable the reader to pursue books with an axiomatic approach to go deeper on theory exceptions, if necessary, or to read more on applications that are surely fascinating for their exceptions, methods, and even phenomenalism.

Typical readers will be graduate and PhD students in engineering schools at large, or in applied science (physics, geophysics, astronomy), preferably with a basic background in algebra, random processes, and signal analysis. SSP practitioners are heavily involved in software development as this is the tool to achieve solutions to many of the problems. The book contains some exercises in the form of application examples with Matlab kernel-code that can be easily adapted to solve broader problems.

I have no presumption to get all SSP knowledge into one book; rather, my focus is to give the flavor that SSP theory offers powerful tools to solve problems over broad applications, to stimulate the curiosity of readers at large, and to give guidelines on moving in depth into the SSP discipline when necessary. The book aims to stimulate the interest of readers who already have some basics to move into SSP practice. Every chapter collects into a few pages a specific professionalism, it scratches the surface of the problem and triggers the curiosity of the reader to go deeper through the essential bibliographical references provided therein. Of course, in 2017 (the time I am writing these notes), there is such easy accessibility to a broad literature, software, lecture notes about the literature, and web that my indexing to the bibliographical references would be partial and insufficient anyway. The book aims to give the reader enough critical tools to choose what is best for her/his interest among what is available.

In my professional life I have always been in the middle between applications and theory, and I have had to follow the steps illustrated in the enclosed block diagram. When facing a problem, it is important to interact with the engineers/scientists who have the deepest knowledge of the application problem itself, its approximations and bounds (stage-A). They are necessary to help to set these limits into a mathematical/statistical framework. At the start, it is preferable if one adopts the jargon of the application in order to find a good match with application people, not only for establishing (useful) personal relations, but also in order to understand the application-related literature. Once the boundary conditions of the problem have been framed (stage-A), one has to re-frame the problem into the SSP discipline. In this second stage (B), one can access the most advanced methods in algebra, statistics, and optimization. The boundary between problem definition and its solution (stage-C) is much less clearly defined than one might imagine. Except for some simple and common situations (but this happens very rarely, unfortunately!), the process is iterative with refinements, introduction of new theory-tools, or adaptations of tools developed elsewhere. No question, this stage needs experience on moving between application and theory, but it is the most stimulating one where one is continuously learning from application - experts (stage-A). Once the algorithm has been developed, it can be transferred back to the application (stage-D), and this is the concluding interaction with the application-related people.

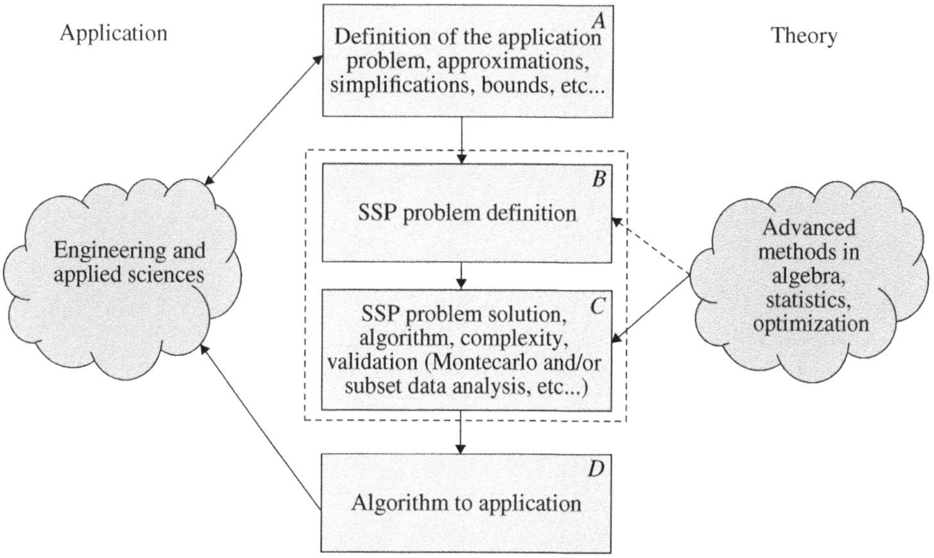

Tuning and refinement are part of the deal, and adaptation to some of the application jargon is of great help at this stage. Sometimes, in the end, the SSP-practitioner is seen as part of the application team with solid theory competences and, after many different applications, one has the impression that the SSP-practitioner knows a little of everything (but this is part of the professional experience). I hope many readers will be lured into this fascinating and diverse problem-solving loop, spanning multiple and various applications, as I have been myself. The book touches all these fields, and it contains some advice, practical rules, and warnings that stem from my personal experience. My greatest hope is to be of help to readers' professional lives.

Umberto Spagnolini, August 2017

P.S. My teaching experience led to the style of the book, and I made an effort to highlight the intuition in each page and avoid too complex a notation; the price is sometimes an awkward notation. For instance, the use of asymptotic notation that is common in many parts is replaced by "→" meaning any convenient limit indicated in the text. Errors and typos are part of the unavoidable *noise* in the text that all *SSPers* have to live with! I did my best to keep this noise as small as possible, but surely I failed somewhere.

"...all models are wrong, but some are useful"

(George E.P. Box, 1919–2013)

List of Abbreviations

$\Rightarrow$	implies or follows used to simplify the equations
$\rightarrow$	variable re-assignement or asymptotic limit
$*$	convolution or complex conjugate (when superscript)
$\circledast_N$	convolution of period N
$\simeq$	is approximately, or is approximately equal to
AIC	Akaike Criterium
AR	Autoregressive
ARMA	Autoregressive Moving Average
ART	Algebraic Reconstruction Tomography
AWGN	Additive White Gaussian Noise
BLUE	Best Linear Unbiased Estimator
CML	Conditional Maximum Likelihood
CRB	Cramer-Rao Bound
CT	Computed Tomography
CW	Continuous Wave
DFE	Decision Feedback Equalizer
DFT	Discrete Fourier Transform
DoA	Direction of Arrival
DSL	Digital Subscriber Line
DToD	Differential Time of Delay
$\mathbb{E}[]$	expectation operator
Eig	Eigenvalue decomposition
EKF	Extended Kalman Filter
EM	Expectation Maximization
ESPRIT	Estimation of Signal Parameters via Rotational Invariance
$\mathcal{F}\{.\}$ or FT	Fourier transform
FIM	Fisher Information Matrix
FM	Frequency Modulation
FN	False Negative
FP	False Positive
GLRT	Generalized Likelihood Ration Test
IID	Independent and Identically Distributed
IQML	Iterative Quadratic Maximum Likelihood
KF	Kalman Filter
LDA	Linear Discriminant Analysis

LMMSE	Linear MMSE
LRT	Likelihood Ratio Test
LS	Least Squares
LTI	Linear Time Invariant
MA	Moving Average
MAP	Maximum a-posteriori
MDL	Minimum Description Length Criterum
MIMO	Multiple Input Multiple Output
MLE	Maximum Likelihood Estimate
MMSE	Minimum Mean Square Error
MODE	Mothod of Direction Estimation
MoM	Method of moments
MRI	Magnetic Resonace Imaging
MSE	Mean Square Error
MUSIC	Multiple Signal Classification
MVDR	Minimum Variance Distortionless
MVU	Minimum Variance Unbiased
(N)LMS	(Normalized) Least Mean Square
NTP	Network Time Protocol
PDE	Partial Differential Equations
pdf	Probability Density Function
PET	Photon Emission Tomography
PLL	Phase Looked Loop
pmf	Probability Mass Function
PSD	Power Spectral Density
RLS	Recursive Least Squares
ROC	Receiver Operating Characteristic
RT	Radiotheraphy
RV	Random Variable
SINR	Signa to interference + noise ratio
SPECT	Single-Photon Emission Thomography
SSP	Statistical Signal Processing
SSS	Strict-Sense Stationary
st	Subject to
SVD	Singular Value Decomposition
SVM	Support Vector Machine
TN	True Negative
ToD	Time of Delay
TP	True Positive
UML	Unconditional Maximum Likelihood
WLS	Weighted Least Squares
WOSA	Window Overlap Spectral Analysis
wrt	With Respect To
WSS	Wide-Sense Stationary
YW	Yule Walker
$\mathcal{Z}\{.\}$	Zeta transform
ZF	Zero Forcing

How to Use the Book

The book is written for a heterogeneous audience. Graduate-level students can follow the presentation order; if skilled in the preliminary parts, they can start reading from Chapter 5 or 6, depending on whether they need to be motivated by some simple examples (in Chapter 5) to be used as guidelines. Chapters 6–9 are on non-Bayesian estimation and Chapter 10 complements this with Montecarlo methods for numerical analysis. Chapters 11–13 are on Bayesian methods, either general and specialized to stationary process (Chapter 12) or Bayesian tracking (Chapter 13). The remaining chapters can be regarded as applications of the estimation theory, starting from classical ones on spectral analysis (Chapter 14), adaptive filtering for non-stationary contexts (Chapter 15) and line-spectrum analysis (Chapter 16). The most specialized applications are in estimation on communication engineering (Chapter 17), 2D signal analysis and filtering (Chapter 18), array processing (Chapter 19), advanced methods for time of delay estimation (Chapter 20), tomography (Chapter 21), application on distributed inference (Chapter 22), and classification methods (Chapter 23).

Expert readers can start from the applications (Chapters 14–23), and follow the links to specific chapters or sections to go deeper and/or find the analytical justifications. The reader skilled in one application area can read the corresponding chapter, bounce back to the specific early sections (in Chapters 1–13), and follow a personal learning path.

The curious and perhaps unskilled reader can look at the broad applications where SSP is an essential problem-solving tool, and be motivated to start from the beginning. There is no specific reading order from Chapter 14 to Chapter 23: the proposed order seems quite logical to the author, but is certainly not the only one.

Even though SSP uses standard statistical tools, the expert statistician is encouraged to start from the applications that could be of interest, preferably after a preliminary reading of Chapters 4–5 on stochastic processes, as these are SSP-specific.

Most of the application-type chapters correspond to a whole scientific community working on that specific area, with many research activities, advances, latest methods, and results. In the introductions of these chapters, or dispersed within the text there are some essential references. The reader can start from the chapter, get an overview, and move to the specific application area if going deeper into the subject is necessary.

The companion web-page of the book contains some numerical exercises—computer-based examples that mimic real-life applications (somewhat oversimplified, but realistic):

www.wiley.com/go/spagnolini/signalprocessing

About the Companion Website

Don't forget to visit the companion website for this book:

www.wiley.com/go/spagnolini/signalprocessing

There you will find valuable material designed to enhance your learning, including:

1) Repository of theory-based and Matlab exercises
2) Videos by the author (lecture-style) detailing some aspects covered by the book

Scan this QR code to visit the companion website

Prerequisites

The reader who is somewhat familiar (at undergraduate level) with algebra, matrix analysis, optimization problems, signals, and systems (time-continuous and time-discrete) can skip Chapters 1–4 where all these concepts are revised for self-consistency and to maintain a congruent notation. The only recommended prerequisite is a good knowledge of random variables and stochastic processes, and related topics. The fundamental book by A.Papoulis and S.U. Pillai, *Probability, random variables, and stochastic processes* [11] is an excellent starting point for a quick comprehension of all relevant topics.

Why are there so many matrixes in this book?

Any textbook or journal in advanced signal processing investigates methods to solve large scale problems where there are multiple signals, variables, and measurements to be manipulated. In these situations, matrix algebra offers tools that are heavily adopted to compactly manage a large set of variables and this is a necessary background.[1] An example application can justify this statement.

The epicenter in earthquakes is obtained by measuring the delays at multiple positions, and by finding the position that best explains the collected measurements (in jargon, *data*). Figure 1 illustrates an example in 2D with epicenter at coordinates $\boldsymbol{\theta} = [\theta_1, \theta_2]^T$. At the time the earthquake occurs, it generates a spherical elastic wave that propagates with a decaying amplitude from the epicenter, and hits a set of N geophysical sensing stations after propagation through a medium with velocity v (typical values are 2000–5000 m/s for shear waves, and above 4000 m/s for compressional, or primary, waves). The signals at the sensing stations are (approximately) a replica of the same waveform as in the Figure 1 with delays $x_1, x_2, \ldots, x_N$ that depend on the propagating distance from the epicenter to each sensing station. The correspondence of each delay with the distance from the epicenter depends on the physics of propagation of elastic waves in a solid, and it is called a *forward model* (from model parameters to observations). Estimation of the epicenter is a typical *inverse problem* (from observations to model parameters) that needs at first a clear definition of the forward model. This forward model can be stated as follows: Given a set of N sensing points where the kth sensor is located at coordinates $\boldsymbol{\theta}_k = [\theta_{1,k}, \theta_{2,k}]^T$, the time of delay (ToD) of the earthquake waveform is

$$x_k = \frac{1}{v}\sqrt{(\theta_{1,k} - \theta_1)^2 + (\theta_{2,k} - \theta_2)^2} + x_0 = ||\boldsymbol{\theta}_k - \boldsymbol{\theta}||/v = h_k(\boldsymbol{\theta}) + x_0.$$

This depends on the distance from the epicenter $||\boldsymbol{\theta}_k - \boldsymbol{\theta}||$ as detailed by the relationship $h_k(.)$. The absolute time x_0 is irrelevant for the epicenter estimation as ToDs are estimated as differences between ToDs (i.e., $x_k - x_\ell = h_k(\boldsymbol{\theta}) - h_\ell(\boldsymbol{\theta})$) so to avoid the need

1 In the database of scientific publications published by the IEEE as Transactions, there is an extensive use of matrix definitions and manipulations. In the examples referred to 2001, there is at least one matrix definition and use in 65% of scientific papers published in the IEEE Trans. on Communications, 89% in those in IEEE Trans. on Information Theory, and 99.9% of papers in IEEE Trans. on Signal Processing. In 2016 these percentages have largely increased and are close to 100%.

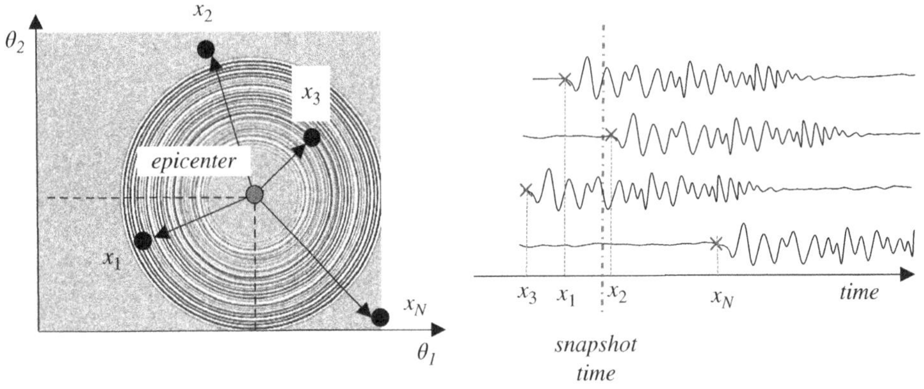

Figure 1 Example of epicenter estimation from earthquake measurements (Snapshot on the left in grey scale).

for x_0; from the reasoning here it can be assumed that $x_0 = 0$ (ToD estimation methods are in Chapter 20). All the ToDs are grouped into a vector

$$\begin{bmatrix} x_1 \\ x_2 \\ \vdots \\ x_N \end{bmatrix} = \begin{bmatrix} h_1(\boldsymbol{\theta}) \\ h_2(\boldsymbol{\theta}) \\ \vdots \\ h_N(\boldsymbol{\theta}) \end{bmatrix} \longrightarrow \mathbf{x} = \mathbf{H}(\boldsymbol{\theta}).$$

The transformation $\mathbf{H}(.)$ is $2 \mapsto N$: it maps the coordinates of the earthquake $\boldsymbol{\theta}$ onto the corresponding delays $\mathbf{x}$ at the N sensing stations. Based on the $2 \mapsto N$ transformation $\mathbf{H}(.)$, the epicenter estimation requires the $N \mapsto 2$ inverse transformation $\mathbf{G}(.)$ that now maps the N delays onto the (unknown) geographical position of the epicenter, and the estimate of the epicenter is

$$\hat{\boldsymbol{\theta}} = \mathbf{G}(\mathbf{x}),$$

so that possibly

$$\boldsymbol{\theta} = \mathbf{G}(\mathbf{H}(\boldsymbol{\theta})).$$

Measurements of ToDs are affected by errors, and this makes $\mathbf{x} \neq \mathbf{H}(\boldsymbol{\theta})$, which is conventionally accounted for by an additive perturbation model $x_k = h_k(\boldsymbol{\theta}) + w_k$ where w_k is called the "noise" affecting the kth ToD that is not known (otherwise, the problem would be trivially reduced to the original one). For the ensemble of ToDs we have

$$\mathbf{x} = \mathbf{H}(\boldsymbol{\theta}) + \mathbf{w},$$

with $\mathbf{w} = [w_1, w_2, ..., w_N]^T$ that collects noise terms of all ToDs into a vector. The estimate is $\hat{\boldsymbol{\theta}} \neq \boldsymbol{\theta}$ with an error

$$\delta\boldsymbol{\theta} = \hat{\boldsymbol{\theta}} - \boldsymbol{\theta},$$

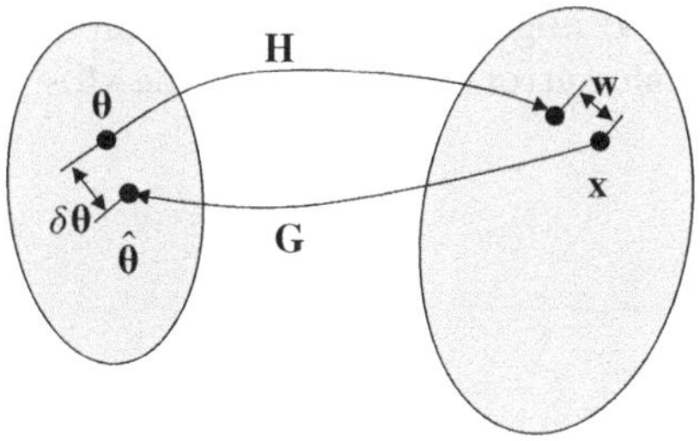

Figure 2 Mapping between parameters θ and data $\mathbf{x}$.

that should be as small as possible and/or less sensitive to noise $\mathbf{w}$ to avoid the situation where one single measurement affected by large noise might severely degrade the accuracy with large $\delta\boldsymbol{\theta}$. In general, the additive term $\mathbf{w}$ is a degree of freedom of the problem to take into account other terms such as model inaccuracies and uncertainties (e.g., propagation is not in a homogeneous 2D medium but rather it is 3D heterogeneous with depth-dependent velocity) that can be modeled as random, or deterministic (but unknown).

The concept of estimating $\boldsymbol{\theta}$ from noisy $\mathbf{x}$ can be abstracted. Let the mapping be linear:

$$\mathbf{x} = \mathbf{H}\theta + \mathbf{w}$$

where $\mathbf{H}$ is an $N \times 2$ matrix that maps every set of 2×1 parameters θ onto the N measurements ($N > 2$). Ideally, the inverse mapping is the $2 \times N$ matrix $\mathbf{G}$ such that

$$\mathbf{GH} = \mathbf{I},$$

and this means that $\mathbf{G}$ acts as the inverse of the rectangular matrix $\mathbf{H}$. The choice of the $2N$ entries $\{g_{ij}\}$ of

$$\mathbf{G} = \begin{bmatrix} g_{11} & g_{12} & \cdots & g_{1N} \\ g_{21} & g_{22} & \cdots & g_{2N} \end{bmatrix}$$

are based on the set of four equations:

$$\begin{cases} \sum_{k=1}^{N} g_{1k}h_{k1} = \sum_{k=1}^{N} g_{2k}h_{k2} = 1 \\ \sum_{k=1}^{N} g_{1k}h_{k2} = \sum_{k=1}^{N} g_{2k}h_{k1} = 0. \end{cases}$$

There are some degrees of freedom in the choice of the $2N$ entries from (only) 4 linear equations, that leaves some arbitrariness in the choice of the $2(N-2)$ entries of $\mathbf{G}$. The error of the estimate is linearly related to the noise

$$\delta\theta = \mathbf{Gw},$$

and different choices of $\mathbf{G}$ (still compliant to $\mathbf{GH} = \mathbf{I}$) might have different effects in errors. For instance, the consequence of a unitary value error from one—say the kth—observation, is:

$$\mathbf{w} = [\underbrace{0,\dots,0}_{k-1},1,\underbrace{0,\dots,0}_{N-k+1}]^T = \mathbf{e}_k$$

the corresponding error in the estimate is

$$\delta\theta_k = \mathbf{G}\mathbf{e}_k,$$

which depends on the kth column of $\mathbf{G}$ (the pair $[g_{1k}, g_{2k}]^T$), and any excessive error in estimate due to one (or more) noisy measurements can be controlled in designing the columns of $\mathbf{G}$. This highlights the fact that there is a need for an in-depth knowledge of matrix manipulation for the design of the "best" inverse transformation $\mathbf{G}(.)$: the best estimator of the parameters from the data according to some specific metrics.

1

Manipulations on Matrixes

Even if matrix algebra requires some basic analytic tools in order to handle it with confidence, the benefits are huge compared to the price paid. This chapter, and those that follow, aim to make a review of these basic operations.

1.1 Matrix Properties

$\mathbb{R}^n$ and $\mathbb{C}^n$ represent the ensemble of all the ordered tuples of n real/complex numbers $(x_1, x_2, \ldots, x_n)$. Each of these tuples can be seen as the coordinate of a point in n-dimensional (or $2n$-dimensional) space. Normally these tuples of n numbers are ordered in column vectors $\mathbf{x} = [x_1, x_2, \ldots, x_n]^T$. The notation $\mathbf{x} \in \mathbb{R}^n$ indicates that $\mathbf{x}$ is an ordered tuple (column vector) of n real numbers that define a vector in $\mathbb{R}^n$; similarly for $\mathbf{x} \in \mathbb{C}^n$. A matrix is obtained by juxtaposing a set of column vectors (tuples): $\mathbf{X} = [\mathbf{x}_1, \mathbf{x}_2, \ldots, \mathbf{x}_m]$ obtaining $\mathbf{X} \in \mathbb{R}^{n\times m}$ (or $\mathbf{X} \in \mathbb{C}^{n\times m}$ if complex-valued). The (i,j) entry of $\mathbf{X}$ (i and j indicate the row and column index, respectively) is indicated by $[\mathbf{X}]_{ij} = \mathbf{X}(i,j) = x_{ij}$, depending on the context.

Typical matrixes or vectors:

$$\mathbf{X} = \text{diag}(x_1, x_2, \ldots, x_n) = \begin{bmatrix} x_1 & 0 & \cdots & 0 \\ 0 & x_2 & \cdots & 0 \\ \vdots & \vdots & \ddots & \vdots \\ 0 & 0 & \cdots & x_n \end{bmatrix} \quad \text{(diagonal matrix)}$$

$$\mathbf{I}_n = diag(1, 1, \ldots, 1) = \mathbf{I} \quad (n \times n \text{ identity matrix})$$

$$\mathbf{e}_k \in \mathbb{R}^n \text{ such that } [\mathbf{e}_k]_{i1} = \delta_{i-k} \quad (k\text{th element selection vector})$$

$$\mathbf{1}_n = \mathbf{1} \in \mathbb{R}^n \text{ such that } [\mathbf{1}]_{i1} = 1 \text{ for } \forall i \quad \text{(one vector)}$$

$$\mathbf{X}(1:n, k) = \mathbf{X}(:, k) = \begin{bmatrix} x_{1k} \\ x_{2k} \\ \vdots \\ x_{nk} \end{bmatrix} \quad (\text{colon notation for } k\text{th column extraction})$$

Function δ_i is the Dirac delta ($\delta_0 = 1$ and $\delta_i = 0$ for $\forall i \neq 0$); for signals the equivalent notation is used: $\delta[i]$.

Statistical Signal Processing in Engineering, First Edition. Umberto Spagnolini.

Companion website: www.wiley.com/go/spagnolini/signalprocessing

1.1.1 Elementary Operations

There are a set of elementary operations for vectors and matrixes. These properties are collected for both real and complex valued matrixes, and differences are highlighted whenever necessary to distinguish operation on complex matrixes as different from real-valued ones. The reader is referred to [1, 2] or any textbook on linear algebra for extensive discussions or proofs.

- **Sum**: for $\mathbf{A}$ and $\mathbf{B} \in \mathbb{C}^{n \times m}$, sum is $\mathbf{C} = \mathbf{A} + \mathbf{B} \in \mathbb{C}^{n \times m}$ with entries: $c_{ij} = a_{ij} + b_{ij}$. Associative and commutative properties hold for the sum: $(\mathbf{A} + \mathbf{B}) + \mathbf{C} = \mathbf{A} + (\mathbf{B} + \mathbf{C})$, and $\mathbf{A} + \mathbf{B} = \mathbf{B} + \mathbf{A}$.
- **Product:** for $\mathbf{A} \in \mathbb{C}^{n \times l}$ and $\mathbf{B} \in \mathbb{C}^{l \times m}$, the matrix $\mathbf{C} = \mathbf{AB} \in \mathbb{C}^{n \times m}$ has elements $c_{ij} = \sum_{k=1}^{l} a_{ik} b_{kj}$. Distributive and associative properties hold for matrix product: ($\mathbf{A}(\mathbf{B} + \mathbf{C}) = \mathbf{AB} + \mathbf{AC}$, and $\mathbf{A}(\mathbf{BC}) = (\mathbf{AB})\mathbf{C}$), but not commutative (assuming that matrixes dimensions are appropriate): $\mathbf{AB} \neq \mathbf{BA}$, in general. Elementwise (or *Hadamard*) multiplication for $\mathbf{A}$ and $\mathbf{B} \in \mathbb{C}^{n \times m}$ is $\mathbf{C} = \mathbf{A} \odot \mathbf{B} \in \mathbb{C}^{n \times m}$ with $c_{ij} = a_{ij} b_{ij}$; other equivalent notation is pointwise multiplication $\mathbf{C} = \mathbf{A}. * \mathbf{B}$ that resembles the Matlab coding.
- **Block-partitioned matrix**: when problems are large but still with some regularity, the matrix accounting for the specific problem can be arranged into sub-matrixes. Each of these sub-matrixes (or blocks) has a predefined regularity or homogeneity such as

$$\mathbf{X} = \begin{bmatrix} \mathbf{X}_{11} & \mathbf{X}_{12} & \cdots & \mathbf{X}_{1n} \\ \mathbf{X}_{21} & \mathbf{X}_{22} & \cdots & \mathbf{X}_{2n} \\ \vdots & \vdots & \ddots & \vdots \\ \mathbf{X}_{m1} & \mathbf{X}_{m2} & \cdots & \mathbf{X}_{mn} \end{bmatrix}$$

  where the dimensions of blocks are appropriate for the problem at hand. Relevant is that block structure is preserved and all the matrix operations are equally defined on blocks. As an example, matrix-vector product for block-partitioned matrixes is

$$\mathbf{Xy} = \begin{bmatrix} \mathbf{X}_{11} & \mathbf{X}_{12} & \cdots & \mathbf{X}_{1n} \\ \mathbf{X}_{21} & \mathbf{X}_{22} & \cdots & \mathbf{X}_{2n} \\ \vdots & \vdots & \ddots & \vdots \\ \mathbf{X}_{m1} & \mathbf{X}_{m2} & \cdots & \mathbf{X}_{mn} \end{bmatrix} \begin{bmatrix} \mathbf{y}_1 \\ \mathbf{y}_2 \\ \vdots \\ \mathbf{y}_n \end{bmatrix} = \begin{bmatrix} \sum_{k=1}^{n} \mathbf{X}_{1k} \mathbf{y}_k \\ \sum_{k=1}^{n} \mathbf{X}_{2k} \mathbf{y}_k \\ \vdots \\ \sum_{k=1}^{n} \mathbf{X}_{mk} \mathbf{y}_k \end{bmatrix}.$$

  Usually, it is convenient (and *advisable*) to preserve the block-partitioned structure across all the matrix manipulation steps.
- **Ordering properties** for any $\mathbf{A} \in \mathbb{R}^{n \times m}$, or $\mathbf{A} \in \mathbb{C}^{n \times m}$ if complex-valued:

  Transpose matrix: $\mathbf{A}^T$ is $\mathbf{A}^T \in \mathbb{C}^{m \times n}$ such that $[\mathbf{A}^T]_{ij} = [\mathbf{A}]_{ji} = a_{ji}$ (row/columns are exchanged). Notice that for the product, $(\mathbf{AB})^T = \mathbf{B}^T \mathbf{A}^T$. For a *symmetric matrix* we have $\mathbf{A}^T = \mathbf{A}$.

  Conjugate matrix: $\mathbf{A}^*$ is the matrix in $\mathbb{C}^{n \times m}$ such that $[\mathbf{A}^*]_{ij} = a_{ij}^*$ (complex conjugate elementwise).

  Hermitian matrix: for a square matrix $\mathbf{A} \in \mathbb{C}^{n \times n}$ for which the following property holds: $a_{ij} = a_{ji}^*$.

Conjugate transpose (Hermitian transpose) matrix: $\mathbf{A}^H$ is $\mathbf{A}^H \in \mathbb{C}^{m \times n}$ such that $\mathbf{A}^H = (\mathbf{A}^*)^T$; therefore $[\mathbf{A}^H]_{ij} = a_{ji}^*$. As for transposition it holds that: $(\mathbf{AB})^H = \mathbf{B}^H \mathbf{A}^H$. A matrix for which $\mathbf{A}^H = \mathbf{A}$ is a *Hermitian matrix* (it has Hermitian symmetry).

Gram matrix: for any arbitrary matrix $\mathbf{A}$ the product $\mathbf{A}^T\mathbf{A}$ (or $\mathbf{A}^H\mathbf{A}$ for complex-valued matrix) is a Gram matrix, which is square, positive semidefinite (see below) and symmetric $(\mathbf{A}^T\mathbf{A})^T = \mathbf{A}^T\mathbf{A}$ (or Hermitian symmetric $(\mathbf{A}^H\mathbf{A})^H = \mathbf{A}^H\mathbf{A}$). Similarly $\mathbf{A}\mathbf{A}^T$ (or $\mathbf{A}\mathbf{A}^H$) is a Gram matrix.

- **Trace**: for a square matrix $\mathbf{A} \in \mathbb{C}^{n \times n}$ this is $\text{tr}[\mathbf{A}] = \sum_{i=1}^n a_{ii}$ (sum along main diagonal). Assuming all square matrixes, the following properties hold true: $\text{tr}[\mathbf{ABC}] = \text{tr}[\mathbf{CAB}] = \text{tr}[\mathbf{BCA}]$ (invariance for cyclic shift) that for the special case is $\text{tr}[\mathbf{AB}] = \text{tr}[\mathbf{BA}]$; for the sum $\text{tr}[\mathbf{A}+\mathbf{B}] = \text{tr}[\mathbf{A}] + \text{tr}[\mathbf{B}]$.
- **Lp norm**: for a vector $\mathbf{x} \in \mathbb{C}^n$, the *Lp* norm is a scalar ≥ 0 defined as:

$$||\mathbf{x}||_p = \left(\sum_{i=1}^n |x_i|^p\right)^{1/p}$$

there are the following special cases (or notations):

$$||\mathbf{x}||_2 = ||\mathbf{x}||_F \text{ (Euclidean or Frobenius norm, see below)}$$

$$||\mathbf{x}||_1 = \sum_{i=1}^n |x_i|$$

$$||\mathbf{x}||_\infty = \max\{|x_1|, |x_2|, \ldots, |x_n|\}$$

as limit $||\mathbf{x}||_0$ simply counts the number of non-zero elements of $\mathbf{x}$ and is used to evaluate the sparseness of a vector.
- **Euclidean (or Frobenius) norm** (special case of Lp norm):

$$\mathbf{x} \in \mathbb{C}^n \to ||\mathbf{x}||_F = \left(\sum_{i=1}^n |x_i|^2\right)^{1/2} = (\mathbf{x}^H\mathbf{x})^{1/2} = \text{tr}[\mathbf{x}\mathbf{x}^H]^{1/2}$$

$$\mathbf{X} \in \mathbb{C}^{n \times n} \to ||\mathbf{X}||_F = \left(\sum_{i=1}^n \sum_{j=1}^n |a_{ij}|^2\right)^{1/2}$$

$$||\mathbf{x}||_{\mathbf{Q}}^2 = \mathbf{x}^T\mathbf{Q}\mathbf{x} : \text{weighted norm of } \mathbf{x} \text{ for a positive definite } \mathbf{Q}.$$

The square of the Euclidean norm for vectors is typically indicated in any of the equivalent forms:

$$||\mathbf{x}||^2 = ||\mathbf{x}||_F^2 = \mathbf{x}^H\mathbf{x}.$$

- **Inner product**: for any two vectors $\mathbf{x}, \mathbf{y} \in \mathbb{C}^n$ their inner product is a *scalar* defined as $\mathbf{x}^H\mathbf{y} = \sum_{i=1}^n x_i^* y_i$. The two vectors are orthogonal with respect to each other when $\mathbf{x}^H\mathbf{y} = 0$. Due to the fact that the inner product is a scalar, the following equalities hold true: $\mathbf{x}^H\mathbf{y} = (\mathbf{x}^H\mathbf{y})^T = \mathbf{y}^T\mathbf{x}^*$ and $\mathbf{x}^H\mathbf{y} = \text{tr}[\mathbf{x}^H\mathbf{y}] = \text{tr}[\mathbf{y}\mathbf{x}^H]$. Recall that $||\mathbf{x}||_F^2 = \mathbf{x}^H\mathbf{x}$ reduces to the inner product.

- **Outer Product**: for any two vectors $\mathbf{x} \in \mathbb{C}^n$ and $\mathbf{y} \in \mathbb{C}^m$ their outer product is an $n \times m$ *matrix* defined as $\mathbf{A} = \mathbf{x}\mathbf{y}^H$, therefore, $a_{ij} = x_i y_j^*$. When $\mathbf{x} = \mathbf{y}$ the outer product is $\mathbf{A} = \mathbf{x}\mathbf{x}^H$ and it is a rank-1 Hermitian matrix. In the case where $\mathbf{x}$ is a vector of random variables (Section 3.7), the correlation matrix follows from the outer product: $\mathbf{R}_x = \mathbb{E}[\mathbf{x}\mathbf{x}^H]$.
- **Determinant** is a metric associated to any square matrix $\mathbf{A} \in \mathbb{R}^{n \times n}$. The geometric meaning of a determinant [3] denoted as $\det[\mathbf{A}]$, or $|\mathbf{A}|$, is the volume of the hypercube enclosed by the columns (or equivalently rows) of $\mathbf{A}$, and $\det[\mathbf{A}] > 0$ only when columns (or rows) enclose a non-null volume, so columns (or rows) are linearly independent. For a general context $\mathbf{A} \in \mathbb{C}^{n \times n}$ the following properties hold: $\det[\mathbf{AB}] = \det[\mathbf{A}]\det[\mathbf{B}]$; $\det[\mathbf{A}^T] = \det[\mathbf{A}]$; $\det[\mathbf{A}^H] = \det[\mathbf{A}]^*$; $\det[\alpha\mathbf{A}] = \alpha^n \det[\mathbf{A}]$; $\det[\mathbf{A} + \mathbf{x}\mathbf{y}^T] = (1 + \mathbf{y}^T\mathbf{A}^{-1}\mathbf{x})\det[\mathbf{A}]$ (matrix determinant lemma).
- **Positive definite (Positive semidefinite) symmetric square matrix**: a symmetric (or Hermitian if complex-valued) matrix $\mathbf{A}$ is positive definite if $\mathbf{z}^T\mathbf{A}\mathbf{z} > 0$ (or $\mathbf{z}^H\mathbf{A}\mathbf{z} > 0$) and positive semidefinite if $\mathbf{z}^T\mathbf{A}\mathbf{z} \geq 0$ (or $\mathbf{z}^H\mathbf{A}\mathbf{z} \geq 0$) for all non-zero vectors $\mathbf{z}$. Notice that for complex-valued matrix, the quadratic form (see Section 1.5) $\mathbf{z}^H\mathbf{A}\mathbf{z}$ is always real as $\mathbf{A}$ is Hermitian. A positive definite (or positive semidefinite) matrix is often indicated by the following notation: $\mathbf{A} \succ 0$ (or $\mathbf{A} \succeq 0$), and if $\mathbf{A} \succ 0 \Rightarrow \det[\mathbf{A}] > 0$ (or $\mathbf{A} \succeq 0 \Rightarrow \det[\mathbf{A}] \geq 0$).
- **Inverse matrix** of a square matrix $\mathbf{A} \in \mathbb{C}^{n \times n}$ is the matrix $\mathbf{A}^{-1} \in \mathbb{C}^{n \times n}$ that satisfies $\mathbf{A}\mathbf{A}^{-1} = \mathbf{A}^{-1}\mathbf{A} = \mathbf{I}$. The inverse matrix exists iff $\det[\mathbf{A}] \neq 0$ (that is equivalent to $\text{rank}[\mathbf{A}] = n$). When $\det[\mathbf{A}] = 0$ the matrix is singular and the inverse does not exist.

  The following properties hold (under the hypothesis that all the involved inverse matrixes exist):
  - $(\mathbf{A}^T)^{-1} = (\mathbf{A}^{-1})^T$ and $(\mathbf{A}^H)^{-1} = (\mathbf{A}^{-1})^H$ (or $\mathbf{A}^{-T}$ and $\mathbf{A}^{-H}$ to ease the notation)
  - $(\mathbf{AB})^{-1} = \mathbf{B}^{-1}\mathbf{A}^{-1}$
  - For a positive definite matrix it holds that:

$$[\mathbf{A}^{-1}]_{ii} \geq \frac{1}{[\mathbf{A}]_{ii}}$$

  - $\det[\mathbf{A}^{-1}] = 1/\det[\mathbf{A}]$
  - Inversion lemma:

$$(\mathbf{A} + \mathbf{BCD})^{-1} = \mathbf{A}^{-1} - \mathbf{A}^{-1}\mathbf{B}(\mathbf{D}\mathbf{A}^{-1}\mathbf{B} + \mathbf{C}^{-1})^{-1}\mathbf{D}\mathbf{A}^{-1}$$

  There are some special cases of practical interest:

$$(\mathbf{A} + \mathbf{B})^{-1} = \mathbf{A}^{-1} - \mathbf{A}^{-1}(\mathbf{A}^{-1} + \mathbf{B}^{-1})^{-1}\mathbf{A}^{-1}$$

$$(\sigma^2\mathbf{I} + \mathbf{VSV}^H)^{-1} = \frac{1}{\sigma^2}\left(\mathbf{I} - \mathbf{V}(\mathbf{V}^H\mathbf{V} + \sigma^2\mathbf{S}^{-1})^{-1}\mathbf{V}^H\right)$$

$$(\mathbf{A} + \mu\mathbf{x}\mathbf{x}^H)^{-1} = \mathbf{A}^{-1} - \frac{\mathbf{A}^{-1}\mathbf{x}\mathbf{x}^H\mathbf{A}^{-1}}{\frac{1}{\mu} + \mathbf{x}^H\mathbf{A}^{-1}\mathbf{x}} \quad \text{(Woodbury identity)}$$

$$(\mathbf{A}^{-1} + \mathbf{B}^H\mathbf{C}^{-1}\mathbf{B})^{-1}\mathbf{B}^H\mathbf{C}^{-1} = \mathbf{A}\mathbf{B}^H(\mathbf{BAB}^H + \mathbf{C})^{-1}$$

– Inversion of a block-partitioned matrix (the dimension of the blocks are arbitrary but appropriate):

$$\begin{bmatrix} \mathbf{A} & \mathbf{B} \\ \mathbf{C} & \mathbf{D} \end{bmatrix}^{-1} = \begin{bmatrix} \mathbf{E} & \mathbf{F} \\ \mathbf{G} & \mathbf{H} \end{bmatrix}$$

where (provided that all the inverse matrixes exist)

$$\begin{aligned}
\mathbf{E} &= \mathbf{A}^{-1} + \mathbf{A}^{-1}\mathbf{B}\mathbf{H}\mathbf{C}\mathbf{A}^{-1} = (\mathbf{A} - \mathbf{B}\mathbf{D}^{-1}\mathbf{C})^{-1} \\
\mathbf{F} &= -\mathbf{A}^{-1}\mathbf{B}\mathbf{H} = -\mathbf{E}\mathbf{B}\mathbf{D}^{-1} \\
\mathbf{G} &= -\mathbf{H}\mathbf{C}\mathbf{A}^{-1} = -\mathbf{D}^{-1}\mathbf{C}\mathbf{E} \\
\mathbf{H} &= \mathbf{D}^{-1} + \mathbf{D}^{-1}\mathbf{C}\mathbf{E}\mathbf{B}\mathbf{D}^{-1} = (\mathbf{D} - \mathbf{C}\mathbf{A}^{-1}\mathbf{B})^{-1}
\end{aligned}$$

– Inequality of the sum:

$$\mathbf{A}^{-1} \succeq (\mathbf{A}+\mathbf{B})^{-1} \succeq \mathbf{0}$$

for symmetric positive definite **A** and symmetric positive semidefinite **B**.

- **Rank**: for a matrix $\mathbf{A} \in \mathbb{C}^{n \times m}$ the rank $[\mathbf{A}]$ denotes the maximum number of linearly independent columns or rows of the matrix: 1.rank $[\mathbf{A}] \leq \min(n, m)$; 2.rank $[\mathbf{A}+\mathbf{B}] \leq$ rank$[\mathbf{A}]$ + rank $[\mathbf{B}] \leq \min(n, m)$; 3. rank $[\mathbf{AB}] \leq \min(\text{rank}[\mathbf{A}], \text{rank}[\mathbf{B}])$.
- **Orthogonal matrix**. A matrix $\mathbf{A} \in \mathbb{R}^{n \times n}$ is orthogonal if $\mathbf{A}^T\mathbf{A} = \mathbf{A}\mathbf{A}^T = \mathbf{I}$, or equivalently the inner product of every pair of non-identical columns (or rows) is zero. Therefore, for an orthogonal matrix, $\mathbf{A}^{-1} = \mathbf{A}^T$. Since the columns (or rows) of orthogonal matrixes have unitary norm, these are called orthonormal matrixes. Similar definitions hold for orthogonal $\mathbf{A} \in \mathbb{C}^{n \times n}$, that is, $\mathbf{A}^H\mathbf{A} = \mathbf{A}\mathbf{A}^H = \mathbf{I}$.
- **Unitary matrix**. A matrix $\mathbf{A} \in \mathbb{C}^{n \times n}$ is called unitary if $\mathbf{A}^H\mathbf{A} = \mathbf{A}\mathbf{A}^H = \mathbf{I}$. Therefore for an unitary matrix it holds that $\mathbf{A}^{-1} = \mathbf{A}^H$. Sometimes unitary matrixes are called orthogonal/orthonormal matrixes with an extension of the terms used for real matrixes.
- **Square-root** of a matrix $\mathbf{A} \in \mathbb{C}^{n \times n}$ is **B** so that $\mathbf{B}^H\mathbf{B} = \mathbf{A}$. In general, square-root is not unique as there could be any unitary matrix **Q** such that $\mathbf{B}^H\mathbf{I}\mathbf{B} = \mathbf{B}^H(\mathbf{Q}^H\mathbf{Q})\mathbf{B} = (\mathbf{Q}\mathbf{B})^H(\mathbf{Q}\mathbf{B}) = \mathbf{A}$. Common notation, with some abuse, is $\mathbf{A}^{1/2}$ and $\mathbf{A} = \mathbf{A}^{H/2}\mathbf{A}^{1/2}$.
- **Triangular matrix**. A square matrix $\mathbf{A} \in \mathbb{C}^{n \times n}$ with entry $[\mathbf{A}]_{ji}$ is upper triangular if $[\mathbf{A}]_{ij} = 0$ for $j > i$ (lower triangular if $[\mathbf{A}]_{ij} = 0$ for $i > j$), strictly upper triangular if $[\mathbf{A}]_{ij} = 0$ for $j \geq i$ (lower triangular for $i \geq j$). The inverse of an upper (lower) triangular matrix is upper (lower) triangular.
- **Kronecker** (tensor) **product** between $\mathbf{A} \in \mathbb{C}^{m \times n}$ and $\mathbf{B} \in \mathbb{C}^{k \times \ell}$ is the block-partitioned matrix

$$\mathbf{A} \otimes \mathbf{B} = \begin{bmatrix} a_{11}\mathbf{B} & a_{12}\mathbf{B} & \cdots & a_{1n}\mathbf{B} \\ a_{21}\mathbf{B} & a_{22}\mathbf{B} & \cdots & a_{2n}\mathbf{B} \\ \vdots & \vdots & \ddots & \vdots \\ a_{m1}\mathbf{B} & a_{m2}\mathbf{B} & \cdots & a_{mn}\mathbf{B} \end{bmatrix} \in \mathbb{C}^{mk \times n\ell}$$

that generalizes the matrix multiplication when handling multidimensional signals.

– *Properties:*

$$\mathbf{I}_m \otimes \mathbf{I}_n = \mathbf{I}_{mn}$$
$$\mathbf{A} \otimes (\alpha \mathbf{B}) = (\alpha \mathbf{A}) \otimes \mathbf{B} = \alpha (\mathbf{A} \otimes \mathbf{B})$$
$$(\mathbf{A} \otimes \mathbf{B})^H = \mathbf{A}^H \otimes \mathbf{B}^H$$
$$(\mathbf{A} + \mathbf{B}) \otimes \mathbf{C} = (\mathbf{A} \otimes \mathbf{C}) + (\mathbf{B} \otimes \mathbf{C})$$
$$\mathbf{A} \otimes (\mathbf{B} + \mathbf{C}) = (\mathbf{A} \otimes \mathbf{B}) + (\mathbf{A} \otimes \mathbf{C})$$
$$\mathbf{A} \otimes (\mathbf{B} \otimes \mathbf{C}) = (\mathbf{A} \otimes \mathbf{B}) \otimes \mathbf{C}$$
$$(\mathbf{A} \otimes \mathbf{B})(\mathbf{C} \otimes \mathbf{D}) = \mathbf{AC} \otimes \mathbf{BD}$$
$$(\mathbf{A} \otimes \mathbf{B})^{-1} = \mathbf{A}^{-1} \otimes \mathbf{B}^{-1}$$
$$\det[\mathbf{A} \otimes \mathbf{B}] = (\det[\mathbf{A}])^m (\det[\mathbf{B}])^n \text{ per } \mathbf{A} \in \mathbb{C}^{m\times m}, \mathbf{B} \in \mathbb{C}^{n\times n}$$
$$\mathrm{tr}[\mathbf{A} \otimes \mathbf{B}] = \mathrm{tr}[\mathbf{A}]\,\mathrm{tr}[\mathbf{B}]$$
$$\mathrm{rank}[\mathbf{A} \otimes \mathbf{B}] = \mathrm{rank}[\mathbf{A}]\,\mathrm{rank}[\mathbf{B}]$$
$$\text{if } \mathbf{A} \succ 0 \text{ and } \mathbf{B} \succ 0 \Longrightarrow (\mathbf{A} \otimes \mathbf{B}) \succ 0$$

- **Vectorization** is related to a matrix, it is based on *vec[.]* operator that for a matrix $\mathbf{A} = [\mathbf{a}_1 \mathbf{a}_2 ..., \mathbf{a}_n] \in \mathbb{C}^{m\times n}$ gives a vector with all columns ordered one after the other: $\mathrm{vec}[\mathbf{A}] = [\mathbf{a}_1^T \mathbf{a}_2^T ... \mathbf{a}_N^T]^T \in \mathbb{C}^{mn}$. There are some properties that are useful to maintain the block-partitioned structure of matrixes:

$$\mathrm{vec}[\alpha \mathbf{A} + \beta \mathbf{B}] = \alpha \mathrm{vec}[\mathbf{A}] + \beta \mathrm{vec}[\mathbf{B}]$$
$$\mathrm{vec}[\mathbf{ABC}] = (\mathbf{C}^T \otimes \mathbf{A})\mathrm{vec}[\mathbf{B}]$$
$$\mathrm{vec}[\mathbf{AB}] = (\mathbf{I} \otimes \mathbf{A})\mathrm{vec}[\mathbf{B}] = (\mathbf{B}^T \otimes \mathbf{I})\mathrm{vec}[\mathbf{A}]$$

1.2 Eigen-Decompositions

Any matrix-vector multiplication $\mathbf{Ax}$ represents a linear transformation for the vector $\mathbf{x}$. However, there are some "special" vectors $\mathbf{x}$ such that their transformation $\mathbf{Ax}$ preserves the direction of $\mathbf{x}$ up to a scaling factor; these vectors are the eigenvectors. In detail, the eigenvectors of a symmetric matrix $\mathbf{A} \in \mathbb{R}^{n\times n}$ (or a Hermitian symmetric $\mathbf{A} \in \mathbb{C}^{n\times n}$, such that $\mathbf{A}^H = \mathbf{A}$) is as a set of n vectors $\mathbf{q}_1, ..., \mathbf{q}_n$ such that:

$$\mathbf{A}\mathbf{q}_i = \lambda_i \mathbf{q}_i$$

where the $\mathbf{q}_i$ and λ_i are the (right) eigenvectors and the eigenvalues of matrix $\mathbf{A}$. Collecting all these vectors into a block-matrix

$$\mathbf{A}[\mathbf{q}_1, ..., \mathbf{q}_n] = [\mathbf{q}_1, ..., \mathbf{q}_n]\mathrm{diag}(\lambda_1, ..., \lambda_n) \Longrightarrow \mathbf{AQ} = \mathbf{Q\Lambda}$$

with $\mathbf{Q} \in \mathbb{C}^{n\times n}$ such that $\mathbf{Q}^{-1}\mathbf{AQ} = \mathbf{\Lambda}$. The eigenvalues are the solutions of the polynomial-type equation $\det(\lambda \mathbf{I} - \mathbf{A}) = 0$ and, for symmetry, are real-valued ($\lambda_i \in \mathbb{R}$). The decomposition

$$\mathbf{A} = \mathbf{Q}\boldsymbol{\Lambda}\mathbf{Q}^T = \sum_{k=1}^{n} \lambda_k \mathbf{q}_k \mathbf{q}_k^T$$

shows that $\mathbf{A}$ can be decomposed into the sum of rank-1 matrixes $\mathbf{q}_k\mathbf{q}_k^T$ weighted by λ_k (similarly for $\mathbf{A} \in \mathbb{C}^{n\times n}$: $\mathbf{A} = \mathbf{Q}\boldsymbol{\Lambda}\mathbf{Q}^H = \sum_{k=1}^{n} \lambda_k \mathbf{q}_k \mathbf{q}_k^H$). The eigenvectors are orthonormal and this is stated as:

$$\mathbf{Q}^T\mathbf{Q} = \mathbf{I} \Leftrightarrow \mathbf{q}_i^T\mathbf{q}_j = \delta_{i-j} = \begin{cases} 1 \text{ for } i = j \\ 0 \text{ for } \forall i \neq j \end{cases}$$

(and similarly $\mathbf{q}_i^H\mathbf{q}_j = \delta_{i-j}$). The inverse of a matrix has the same eigenvectors, but inverse eigenvalues:

$$\mathbf{A}^{-1} = \mathbf{Q}\boldsymbol{\Lambda}^{-1}\mathbf{Q}^T$$

The kth power of a matrix is

$$\mathbf{A}^k = (\mathbf{Q}\boldsymbol{\Lambda}\mathbf{Q}^T)(\mathbf{Q}\boldsymbol{\Lambda}\mathbf{Q}^T)\ldots(\mathbf{Q}\boldsymbol{\Lambda}\mathbf{Q}^T) = \mathbf{Q}\boldsymbol{\Lambda}^{\mathrm{k}}\mathbf{Q}^T$$

and the square-root is

$$\mathbf{A} = \mathbf{A}^{T/2}\mathbf{A}^{1/2} \rightarrow \mathbf{A}^{1/2} = \boldsymbol{\Lambda}^{1/2}\mathbf{Q}^T$$

but notice that the square-root factorization is not unique as any arbitrary $n \times n$ orthonormal matrix $\mathbf{B}$ (i.e., $\mathbf{B}^T\mathbf{B} = \mathbf{I}$) makes the square-root factorization $\mathbf{A} = \mathbf{A}^{T/2}\mathbf{B}^T\mathbf{B}\mathbf{A}^{1/2}$ and thus $\mathbf{A}^{1/2} = \mathbf{B}\boldsymbol{\Lambda}^{1/2}\mathbf{Q}^T$. The geometric mean of the eigenvalues are related to the volume of the hypercube from the columns of $\mathbf{A}$

$$\det[\mathbf{A}] = \prod_{i=1}^{n} \lambda_i \ ;$$

the volume $\det[\mathbf{A}] = 0$ when at least one eigenvalue is zero. Another very useful property is the trace:

$$\mathrm{tr}[\mathbf{A}] = \sum_{i=1}^{n} [\mathbf{A}]_{ii} = \sum_{i=1}^{n} \lambda_i$$

When a matrix is not symmetric, one can evaluate the right and left eigenvectors. For each eigenvalue λ_i, a vector $\mathbf{w}_i \in \mathbb{R}^n$ can be associated such that $\mathbf{w}_i^T\mathbf{A} = \lambda_i\mathbf{w}_i^T$; this is the left eigenvector of $\mathbf{A}$ relative to the eigenvalue λ_i. Collecting all these eigenvalues/(left) eigenvectors of $\mathbf{A}$ it follows that: $\mathbf{W}^T\mathbf{A} = \boldsymbol{\Lambda}\mathbf{W}^T$, where the columns of $\mathbf{W}$ are the left eigenvectors of $\mathbf{A}$. Notice that $(\mathbf{w}_i^T\mathbf{A})^T = \mathbf{A}^T\mathbf{w}_i = \lambda_i\mathbf{w}_i$, thus the left eigenvector is the (right) eigenvector for the matrix $\mathbf{A}^T$. For symmetric matrixes, $\mathbf{Q} = \mathbf{W}$. The same properties hold for $\mathbf{A} \in \mathbb{C}^{n\times n}$ with Hermitian transposition in place of transposition.

The most general decomposition of an arbitrary $n \times m$ matrix into eigenvectors is Singular Value Decomposition (SVD). The decomposition can be stated as follows: any $\mathbf{A} \in \mathbb{C}^{n\times m}$ can be decomposed into orthogonal matrixes

$$\underset{(m,n)}{\mathbf{A}} = \underset{(m,m)}{\mathbf{U}} \cdot \underset{(m,n)}{\mathbf{\Sigma}} \cdot \underset{(n,n)}{\mathbf{V}^H}$$

where $\mathbf{U}$ and $\mathbf{V}$ are orthonormal and (real-valued) *singular values* $\{\sigma_k\}$ usually ordered for decreasing values $\sigma_1 \geq \sigma_2 \geq \ldots \geq \sigma_r \geq \sigma_{r+1} = \ldots \sigma_p = 0$ with $r \leq p = \min(m,n)$:

$$\mathbf{\Sigma} = \begin{bmatrix} \sigma_1 & 0 & \cdots & 0 \\ 0 & \sigma_2 & & 0 \\ \vdots & & \ddots & \vdots \\ 0 & 0 & & \sigma_p \\ \vdots & & & \vdots \\ 0 & 0 & \cdots & 0 \end{bmatrix},$$

rectangular (for $m \neq n$) and, in general, non-diagonal. Decomposition highlights that a matrix can be decomposed into the sum of (up to) r rank-1 matrixes

$$\mathbf{A} = \sum_{k=1}^{r} \sigma_k \mathbf{u}_k \mathbf{v}_k^H$$

so that $\text{rank}(\mathbf{A}) = r$ (from the first r non-zero singular values). This is a relevant property exploited in many signal processing methods. The following main properties can be highlighted:

$$\begin{aligned} \mathbf{U}^H\mathbf{U} = \mathbf{U}\mathbf{U}^H &= \mathbf{I}_m \\ \mathbf{V}^H\mathbf{V} = \mathbf{V}\mathbf{V}^H &= \mathbf{I}_n \\ \mathbf{A}^H\mathbf{A}\mathbf{v}_i &= \sigma_i^2\mathbf{v}_i \\ \mathbf{A}\mathbf{A}^H\mathbf{u}_i &= \sigma_i^2\mathbf{u}_i \\ \lambda_i(\mathbf{A}\mathbf{A}^H) = \lambda_i(\mathbf{A}^H\mathbf{A}) &= \sigma_i^2 \end{aligned}$$

To summarize, columns of $\mathbf{U}$ are the eigenvectors of the Gram matrix $\mathbf{A}\mathbf{A}^H$ while columns of $\mathbf{V}$ are the eigenvectors of the Gram matrix $\mathbf{A}^H\mathbf{A}$. Based on the orthogonality property: $\mathbf{U}^H \cdot \mathbf{A} \cdot \mathbf{V} = \mathbf{\Sigma}$. Since $\text{rank}(\mathbf{A}) = r \leq \min(m,n)$, the following equivalent representation is often useful to partition the columns of $\mathbf{U}$ and $\mathbf{V}$ related to the first non-zero singular values:

$$\mathbf{A} = \mathbf{U}\mathbf{\Sigma}\mathbf{V}^H = [\underbrace{\mathbf{U}_1}_{(m,r)}, \underbrace{\mathbf{U}_2}_{(m,m-r)}] \begin{bmatrix} \mathbf{\Sigma}_1 & 0 \\ 0 & 0 \end{bmatrix} \begin{bmatrix} \mathbf{V}_1^H \\ \mathbf{V}_2^H \end{bmatrix} \begin{matrix} \}(r,n) \\ \}(n-r,n) \end{matrix} = \mathbf{U}_1\mathbf{\Sigma}_1\mathbf{V}_1^H$$

Columns are orthogonal, so $\mathbf{U}_1^H\mathbf{U}_1 = \mathbf{I}_r$ and $\mathbf{V}_1^H\mathbf{V}_1 = \mathbf{I}_r$, but $\mathbf{U}_1\mathbf{U}_1^H \neq \mathbf{U}_1^H\mathbf{U}_1$ and $\mathbf{V}_1\mathbf{V}_1^H \neq \mathbf{V}_1^H\mathbf{V}_1$. SVD highlights the properties of the linear transformation $\mathbf{A}$ as detailed in Section 2.4.

1.3 Eigenvectors in Everyday Life

The eigenvectors of a linear transformation highlight the structure of the transformation itself. However, there are engineering problems where analysis using eigenvectors computation is far more common than one might imagine; there follow some classical introductory examples.

1.3.1 Conversations in a Noisy Restaurant

In a restaurant there are N tables occupied by guests that are having conversations within each table; the mutual interference among tables induces each table to adapt the intensity of the voice to guarantee the intelligibility of the conversation and cope with the interference from the other tables. Let P_i be the power of the voice at each table with $P_i \geq 0$; the aim here is to evaluate the optimum power values $P_1, P_2, \ldots, P_N$ at every table to guarantee intelligibility without saturating the powers (i.e., *avoid the guests having to unnecessarily talk too loudly*). Tables are referred as nodes in a *weighted graph* illustrated in Figure 1.1 (see Chapter 22) where weights account for their mutual interaction.

Assuming that the gain from jth node toward the i-th node is $g_{ij} \leq 1$, the overall interference experienced by the users of the i-th group is the sum of the signal generated by all the tables scaled by the corresponding attenuation $\sum_{j \neq i} g_{ij} P_j$. What matters for intelligibility is that the voices within the ith table should be large enough compared to the overall interference. Namely, the ratio between the power of the signal and the overall interference should be

$$\frac{P_i}{\sum_{j \neq i} g_{ij} P_j} \geq \gamma_i$$

where γ_i is the minimum value for the i-th node to have an acceptable quality—usually this depends on the context and is known. Rearranging the inequality for the N nodes,

$$\gamma_i \sum_{j \neq i} g_{ij} P_j \leq P_i$$

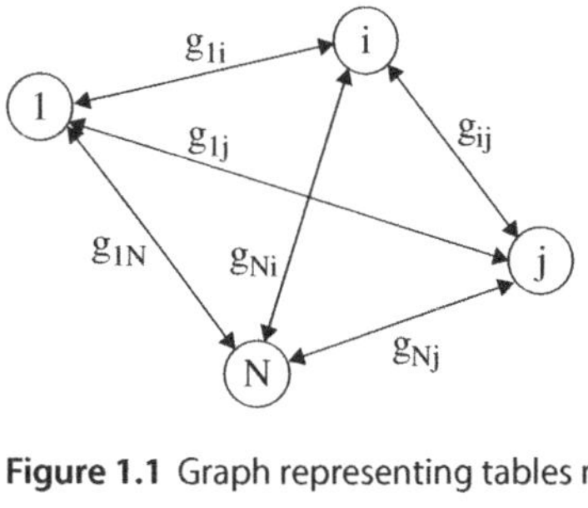

Figure 1.1 Graph representing tables mutually interfering with inter-table gain $g_{ij} \leq 1$.

that in matrix form becomes

$$\underbrace{\begin{pmatrix} \gamma_1 & 0 & \dots & 0 \\ 0 & \gamma_2 & \dots & 0 \\ \vdots & \vdots & \ddots & \vdots \\ 0 & 0 & \dots & \gamma_N \end{pmatrix}}_{\mathbf{D}} \cdot \underbrace{\begin{pmatrix} 0 & g_{12} & \dots & g_{1N} \\ g_{21} & 0 & \dots & g_{2N} \\ \vdots & \vdots & \ddots & \vdots \\ g_{N1} & g_{N2} & \dots & 0 \end{pmatrix}}_{\mathbf{G}} \cdot \underbrace{\begin{pmatrix} P_1 \\ P_2 \\ \vdots \\ P_N \end{pmatrix}}_{\mathbf{p}} \leq \underbrace{\begin{pmatrix} P_1 \\ P_2 \\ \vdots \\ P_N \end{pmatrix}}_{\mathbf{p}}$$

or equivalently

$$\mathbf{A}\mathbf{p} \leq \mathbf{p}$$

where

$$\mathbf{A} = \mathbf{D}\mathbf{G}$$

is a square matrix with known and positive entries (matrix **A** is positive) accounting for link-gains **G** and threshold signal to interference levels **D**. To solve, the inequality can be rewritten as an equality with $0 < \rho \leq 1$ such that

$$\mathbf{A}\mathbf{p} = \rho\mathbf{p}.$$

A trivial solution with all groups inactive ($P_i = 0$ for $\forall i$) should be avoided assuming that $\rho \neq 0$. The solution for the vector of powers **p** is any eigenvector of the matrix **A** associated to an eigenvalue smaller than 1. Notice that the eigenvectors are normalized to have unit-norm so that optimal power configuration across nodes is independent of any scaling factor of all nodes (called reference power). To avoid waste of power, it is advisable to choose a small reference power as this guarantees the minimum signal to interference level anyway. In practice, customers in a restaurant do not know **A** (possibly a subset or blocks of **G**) and the power configuration including the scale factor is autonomously chosen by each group to reach the global equilibrium. The computation of the eigenvector is an iterative process (see Section 2.6) *implicitly carried out by mutual respect among guests (if any)* while fulfilling the minimal intra-group constraints on quality **D** that is eased when customers are homogeneous, and all the groups have the same quality γ ($\mathbf{D} = \gamma\mathbf{I}$).

Inspection of the problem shows some minimal conditions for the solution. The **Gershgoring theorem** helps to evaluate the conditions for the eigenvalues to be bounded by unitary value. Let

$$R_i = \sum_{j=1, j\neq i}^{N} a_{ij}$$

be the sum of the ith row except the diagonal; every eigenvalue of **A** indicated as λ_k is fully contained within every Gersgorin disk

$$|\lambda_k - a_{ii}| \leq R_i = \sum_{j=1, j\neq i}^{N} a_{ij}$$

centered in a_{ii} with radius R_i. In other words, all the eigenvalues of **A** lie in the union of the Gersgorin disks. Returning to the specific problem, $a_{ii} = 0$ and thus

$$|\lambda_k| \leq \sum_{j=1, j\neq i}^{N} a_{ij} = \gamma_i \sum_{j\neq i} g_{ij}.$$

Assuming that the solution is for $\lambda_{\max} = \rho \leq 1$ as this corresponds to the strict inequality, $|\lambda_k| \leq \lambda_{\max} \leq 1$ and thus the condition

$$\gamma_i \sum_{j\neq i} g_{ij} \leq 1 \Longrightarrow \gamma_i \leq \frac{1}{\sum_{j\neq i} g_{ij}}$$

guarantees that there exists one solution. The existence of at least one eigenvector **p** with positive entries is based on the Perron–Frobenius theorem [21], fulfilled in this case.

Notice that when there are two sets of disjointed groups as shown in Figure 1.2, the gain matrix becomes

$$\mathbf{G} = \begin{pmatrix} 0 & g_{12} & 0 & 0 & g_{15} & 0 \\ g_{12} & 0 & 0 & 0 & g_{25} & 0 \\ 0 & 0 & 0 & g_{34} & 0 & g_{36} \\ 0 & 0 & g_{34} & 0 & 0 & g_{46} \\ g_{15} & g_{25} & 0 & 0 & 0 & 0 \\ 0 & 0 & g_{36} & g_{46} & 0 & 0 \end{pmatrix}$$

but since the numbering of the nodes are just conventional, these gains can be reordered by grouping connected nodes:

$$\tilde{\mathbf{G}} = \begin{pmatrix} 0 & g_{12} & g_{15} & 0 & 0 & 0 \\ g_{12} & 0 & g_{25} & 0 & 0 & 0 \\ g_{15} & g_{25} & 0 & 0 & 0 & 0 \\ 0 & 0 & 0 & 0 & g_{34} & g_{46} \\ 0 & 0 & 0 & g_{34} & 0 & g_{36} \\ 0 & 0 & 0 & g_{46} & g_{36} & 0 \end{pmatrix} = \mathbf{L} \cdot \begin{pmatrix} 0 & g_{12} & 0 & 0 & g_{15} & 0 \\ g_{12} & 0 & 0 & 0 & g_{25} & 0 \\ 0 & 0 & 0 & g_{34} & 0 & g_{36} \\ 0 & 0 & g_{34} & 0 & 0 & g_{46} \\ g_{15} & g_{25} & 0 & 0 & 0 & 0 \\ 0 & 0 & g_{36} & g_{46} & 0 & 0 \end{pmatrix} \cdot \mathbf{L}^T$$

where the ordering matrix is symmetric

$$\mathbf{L} = \begin{pmatrix} 1 & 0 & 0 & 0 & 0 & 0 \\ 0 & 1 & 0 & 0 & 0 & 0 \\ 0 & 0 & 0 & 0 & 1 & 0 \\ 0 & 0 & 0 & 1 & 0 & 0 \\ 0 & 0 & 1 & 0 & 0 & 0 \\ 0 & 0 & 0 & 0 & 0 & 1 \end{pmatrix}$$

and it swap the rows/columns $3 \leftrightarrow 5$. Gain matrix $\tilde{\mathbf{G}}$ has a block-partitioned structure

$$\tilde{\mathbf{G}} = \begin{pmatrix} \tilde{\mathbf{G}}_{11} & \mathbf{0} \\ \mathbf{0} & \tilde{\mathbf{G}}_{22} \end{pmatrix}$$

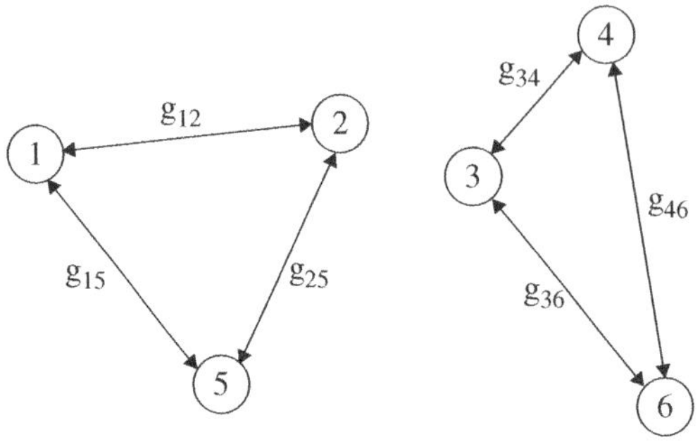

Figure 1.2 Graph for two disjoint and non-interfering sets.

this highlights that the computation of the optimal solution can be decoupled into two separated solutions for the two sets of nodes.

1.3.2 Power Control in a Cellular Communication System

In cellular systems, every mobile device (e.g., a smart phone) is connected to a base-station deployed at the center of every cell (see Figure 1.3) that in turn adjusts its transmission power to control both the received power and the interference toward the other cells (*inter-cell interference*). The interference superimposes on the background noise with power σ^2. The problem definition is the same as above except that every user with power P_i is connected to the serving base-station with intra-cell gain g_{ii} and is interfering with all the other cells. Once again, what matters is the ratio between the power of the signal reaching the serving base-station and the interference arising from all the other mobile devices (gray lines in figure), each camped in its own base-station, augmented by some background noise; the signal to interference and noise ratio (SINR) is

$$SINR_i = \frac{g_{ii}P_i}{\sum_{j\neq i} g_{ij}P_j + \sigma^2}.$$

To guarantee a reliable communication, the SINR should be above a predefined reliability value

$$SINR_i \geq \gamma_i.$$

The overall problem can be now defined as

$$\gamma_i\left(\sum_{j\neq i} G_{ij}P_j + q_i\right) \leq P_i$$

where $G_{ij} = g_{ij}/g_{ii}$ and $q_i = \sigma^2/g_{ii}$. In compact notation, this is

$$\mathbf{D}(\mathbf{GP}+\mathbf{q}) \leq \mathbf{p} \longrightarrow (\mathbf{I}-\mathbf{DG})\mathbf{p} \geq \mathbf{Dq}$$

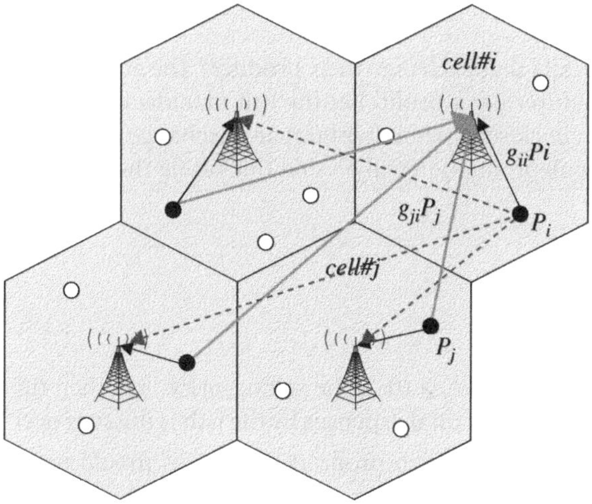

Figure 1.3 Mutual interference in a cellular communication system.

and the solution is

$$\mathbf{p} \geq (\mathbf{I} - \mathbf{DG})^{-1}\mathbf{Dq}$$

with positive values $\mathbf{p}$ provided that the inverse $(\mathbf{I} - \mathbf{DG})^{-1}$ is a positive matrix (proof now is not straightforward as Perron–Frobenius properties do not hold).

In practical systems, the solution is not obtained by directly solving the power distribution among the users within all the cells, but rather by iterative methods that assign the power to users in every cell and, based on this choice, the base-stations exchange the information on the experienced power and interference with others to change and refine the power assignments accordingly. All this must be carried out dynamically to compensate for the power losses/gains of the mobile devices g_{ij} that are changing their propagation settings. In other words, every cell measures the $SINR_i$, and the base-station, from the measurements of the ensemble of other $SINR_j$ with $j \neq i$ (typically the surrounding cells), *orders* a local correction of P_i for the devices within the serving cell (inner power control loop) while an exchange with the other base-stations defines the macro-corrections (outer power control loop). The overall system is quite complex, but the final scope is to solve the system $\mathbf{p} \geq (\mathbf{I} - \mathbf{DG})^{-1}\mathbf{Dq}$ iteratively by message exchange. Sometimes the solution cannot be attained as there are conditions in the system of equations $\mathbf{p} \geq (\mathbf{I} - \mathbf{DG})^{-1}\mathbf{Dq}$ that push some terminals with a very low SINR (being placed in a very bad propagation situation with $g_{ii} \simeq 0$—e.g., a mobile device in an elevator) to raise intolerably the power of the network to let these *bad-users* have positive values. Even if this is a solution of the system, it increases the interference for all the others and the average transmission power (and battery usage) due to a few *bad-users*. In these situations, it is convenient to drop the calls of these few users with a large benefit to all the others.

1.3.3 Price Equilibrium in the Economy

In a system with N manufacturing units that exchange their products, the equilibrium prices guarantee that every manufacturer has a profit. Let the j-th manufacturer sell a fraction a_{ij} to the i-th manufacturer; in closed system (without any exchange outside of the set of N entities), we expect that all production will be sold (including the internal-usage a_{jj}) so that

$$\sum_{i=1}^{N} a_{ij} = 1,$$

and goods are preferably delivered to others ($a_{ij} \geq 0$). If the selling-price is P_j then the i-th manufacturer spends $a_{ij}P_j$ toward j-th; the total expenses by the i-th is the sum over all the acquired goods: $\sum_{j=1}^{N} a_{ij}P_j$. In the case of zero-profit, the expenses should equal the incomes so that

$$\sum_{j=1}^{N} a_{ij}P_j = P_i$$

In compact notation, this is

$$\mathbf{A}\mathbf{p} = \mathbf{p}$$

and the price of equilibrium is the eigenvector with positive entries associated to the unit eigenvalue of matrix $\mathbf{A}$. Any scaling factor is irrelevant as the model is zero-profit, provided that all the manufacturing units adapt the pricing accordingly, possibly converging to values that correspond to the same eigenvector.

To be more realistic, the model should account for some profit

$$P_i - \sum_{j=1}^{N} a_{ij}P_j \geq \gamma_i$$

so that

$$(\mathbf{I} - \mathbf{A})\mathbf{p} \geq \begin{pmatrix} \gamma_1 \\ \gamma_2 \\ \vdots \\ \gamma_N \end{pmatrix} = \boldsymbol{\gamma}$$

with an analytical relationship similar to power control as $\mathbf{p} \geq (\mathbf{I} - \mathbf{A})^{-1}\boldsymbol{\gamma}$. Interestingly, when perturbing the solution by increasing (or decreasing) the profit in one of the manufacturers, all the others will change and the (new) equilibrium is attained iteratively by adapting prices and profits. Daily experience shows that a fast adaptation is mandatory when the market is changing, and some of the players are quick reacting at the expense of the others, pushing the profit of some to be $\gamma_i < 0$.

1.4 Derivative Rules

A set of variables $\mathbf{x} = [x_1, x_2, \cdots, x_n]^T$ ordered in a vector can be used to represent the $n \mapsto 1$ mapping that is indicated with the compact notation

$$\phi(\mathbf{x}) = \phi(x_1, x_2, \cdots, x_n),$$

and similarly for $\phi(\mathbf{X})$ if depending on the set of mn variables ordered in a matrix $\mathbf{X}$ of size $m \times n$. It is convenient to preserve the compact matrix notation in optimization problems; below are the rules for evaluating derivatives. A few definitions on notation are mandatory:

Derivative wrt a vector x (gradient wrt all the n variables)

$$\frac{\partial}{\partial \mathbf{x}}\phi(\mathbf{x}) = \begin{bmatrix} \frac{\partial\phi(\mathbf{x})}{\partial x_1} \\ \frac{\partial\phi(\mathbf{x})}{\partial x_2} \\ \vdots \\ \frac{\partial\phi(\mathbf{x})}{\partial x_n} \end{bmatrix},$$

is an $n \times 1$ vector with all the partial derivatives with respect to the n variables. Assuming that there are m multivariate functions ordered into a vector $\boldsymbol{\phi}(\mathbf{x}) = [\phi_1(\mathbf{x})\ \phi_2(\mathbf{x})\ \cdots\ \phi_m(\mathbf{x})]^T$, it is convenient to arrange the multivariate functions into a row and have the derivative wrt $\mathbf{x}$ ordered column-wise into the $n \times m$ matrix of partial derivative wrt $\mathbf{x}$:

$$\frac{\partial}{\partial \mathbf{x}}\boldsymbol{\phi}^T(\mathbf{x}) = \begin{bmatrix} \frac{\partial\phi_1(\mathbf{x})}{\partial \mathbf{x}} & \cdots & \frac{\partial\phi_m(\mathbf{x})}{\partial \mathbf{x}} \end{bmatrix} = \begin{bmatrix} \frac{\partial\phi_1(\mathbf{x})}{\partial x_1} & \cdots & \frac{\partial\phi_m(\mathbf{x})}{\partial x_1} \\ \vdots & \ddots & \vdots \\ \frac{\partial\phi_1(\mathbf{x})}{\partial x_n} & \cdots & \frac{\partial\phi_m(\mathbf{x})}{\partial x_n} \end{bmatrix}$$

Derivative wrt matrix X of size $m \times n$, in compact notation it is the matrix of derivatives

$$\frac{\partial}{\partial \mathbf{X}}\phi(\mathbf{X}) = \begin{bmatrix} \frac{\partial\phi(\mathbf{X})}{\partial x_{11}} & \cdots & \frac{\partial\phi(\mathbf{X})}{\partial x_{1n}} \\ \vdots & \ddots & \vdots \\ \frac{\partial\phi(\mathbf{X})}{\partial x_{m1}} & \cdots & \frac{\partial\phi(\mathbf{X})}{\partial x_{mn}} \end{bmatrix}$$

partial derivatives are ordered in the same way as the columns of $\mathbf{X}$.

1.4.1 Derivative with respect to $\mathbf{x} \in \mathbb{R}^n$

Two relevant cases of derivative are considered herein:

- $\phi(\mathbf{x}) = \mathbf{x}^T\mathbf{a}$ and $\phi(\mathbf{x}) = \mathbf{a}^T\mathbf{x}$

$$\frac{\partial \mathbf{x}^T\mathbf{a}}{\partial \mathbf{x}} = \begin{bmatrix} \frac{\partial \mathbf{x}^T\mathbf{a}}{\partial x_1} \\ \frac{\partial \mathbf{x}^T\mathbf{a}}{\partial x_2} \\ \vdots \\ \frac{\partial \mathbf{x}^T\mathbf{a}}{\partial x_n} \end{bmatrix} = \begin{bmatrix} a_1 \\ a_2 \\ \vdots \\ a_n \end{bmatrix} = \mathbf{a}$$

$$\frac{\partial \mathbf{a}^T\mathbf{x}}{\partial \mathbf{x}} = \begin{bmatrix} a_1 \\ a_2 \\ \vdots \\ a_n \end{bmatrix} = \mathbf{a}$$

Special cases: 1) $\frac{\partial}{\partial \mathbf{x}}\mathbf{x}^T = \frac{\partial}{\partial \mathbf{x}}\mathbf{x} = \mathbf{I}_n$; 2) $\frac{\partial}{\partial \mathbf{x}}\text{tr}[\mathbf{x}\mathbf{a}^T] = \frac{\partial}{\partial \mathbf{x}}\text{tr}[\mathbf{a}\mathbf{x}^T] = \mathbf{a}$.

- $\phi(\mathbf{x}) = \mathbf{x}^T\mathbf{A}\mathbf{x}$(*quadratic* function) with $\mathbf{A} = [\mathbf{a}_1, \mathbf{a}_2, \cdots, \mathbf{a_n}]$

$$\frac{\partial \mathbf{x}^T\mathbf{A}\mathbf{x}}{\partial \mathbf{x}} = \begin{bmatrix} \mathbf{x}^T\frac{\partial \mathbf{A}\mathbf{x}}{\partial x_1} + \frac{\partial \mathbf{x}^T\mathbf{A}}{\partial x_1}\mathbf{x} \\ \mathbf{x}^T\frac{\partial \mathbf{A}\mathbf{x}}{\partial x_2} + \frac{\partial \mathbf{x}^T\mathbf{A}}{\partial x_2}\mathbf{x} \\ \vdots \\ \mathbf{x}^T\frac{\partial \mathbf{A}\mathbf{x}}{\partial x_n} + \frac{\partial \mathbf{x}^T\mathbf{A}}{\partial x_n}\mathbf{x} \end{bmatrix} = \underbrace{\begin{bmatrix} \mathbf{x}^T\frac{\partial \mathbf{A}\mathbf{x}}{\partial x_1} \\ \frac{\partial \mathbf{A}\mathbf{x}}{\partial x_2} \\ \vdots \\ \mathbf{x}^T\frac{\partial \mathbf{A}\mathbf{x}}{\partial x_n} \end{bmatrix}}_{\mathbf{x}^T\frac{\partial \mathbf{A}\mathbf{x}}{\partial \mathbf{x}}} + \underbrace{\begin{bmatrix} \frac{\partial \mathbf{x}^T\mathbf{A}}{\partial x_1}\mathbf{x} \\ \frac{\partial \mathbf{x}^T\mathbf{A}}{\partial x_2}\mathbf{x} \\ \vdots \\ \frac{\partial \mathbf{x}^T\mathbf{A}}{\partial x_n}\mathbf{x} \end{bmatrix}}_{\frac{\partial \mathbf{x}^T\mathbf{A}}{\partial \mathbf{x}}\mathbf{x}}$$

Considering the first term, it is a gradient of multiple functions wrt the vector so that

$$\mathbf{A}\mathbf{x} = \begin{bmatrix} \mathbf{A}(1,:)\mathbf{x} \\ \mathbf{A}(2,:)\mathbf{x} \\ \vdots \\ \mathbf{A}(n,:)\mathbf{x} \end{bmatrix}$$

$$\frac{\partial \mathbf{A}\mathbf{x}}{\partial x_k} = \mathbf{a}_k = \mathbf{A}(:,k)$$

Therefore

$$\begin{bmatrix} \mathbf{x}^T\frac{\partial \mathbf{A}\mathbf{x}}{\partial x_1} \\ \mathbf{x}^T\frac{\partial \mathbf{A}\mathbf{x}}{\partial x_2} \\ \vdots \\ \mathbf{x}^T\frac{\partial \mathbf{A}\mathbf{x}}{\partial x_n} \end{bmatrix} = \begin{bmatrix} \mathbf{x}^T\mathbf{a}_1 \\ \mathbf{x}^T\mathbf{a}_2 \\ \vdots \\ \mathbf{x}^T\mathbf{a}_n \end{bmatrix} = \begin{bmatrix} \mathbf{a}_1^T\mathbf{x} \\ \mathbf{a}_2^T\mathbf{x} \\ \vdots \\ \mathbf{a}_n^T\mathbf{x} \end{bmatrix} = \begin{bmatrix} \mathbf{a}_1^T \\ \mathbf{a}_2^T \\ \vdots \\ \mathbf{a}_n^T \end{bmatrix}\mathbf{x} = \mathbf{A}^T\mathbf{x}$$

similarly for the second term

$$\frac{\partial \mathbf{x}^T \mathbf{A}}{\partial x_k} = \frac{\partial (\mathbf{A}^T \mathbf{x})^T}{\partial x_k} = \left(\frac{\partial \mathbf{A}^T \mathbf{x}}{\partial x_k}\right)^T \Rightarrow \begin{bmatrix} \frac{\partial \mathbf{x}^T}{\partial x_1}\mathbf{A}\mathbf{x} \\ \frac{\partial \mathbf{x}^T}{\partial x_2}\mathbf{A}\mathbf{x} \\ \vdots \\ \frac{\partial \mathbf{x}^T}{\partial x_n}\mathbf{A}\mathbf{x} \end{bmatrix} = \ldots = \mathbf{A}\mathbf{x}.$$

To summarize,

$$\frac{\partial \mathbf{x}^T \mathbf{A}\mathbf{x}}{\partial \mathbf{x}} = (\mathbf{A} + \mathbf{A}^T)\mathbf{x}$$

Furthermore, if $\mathbf{A}$ is symmetric ($\mathbf{A} = \mathbf{A}^T$) it reduces to

$$\frac{\partial \mathbf{x}^T \mathbf{A}\mathbf{x}}{\partial \mathbf{x}} = 2\mathbf{A}\mathbf{x}.$$

1.4.2 Derivative with respect to $\mathrm{x} \in \mathbb{C}^n$

Considering the scalar complex-valued variable $x = x_R + jx_I$, any function $\phi(x) : \mathbb{C} \mapsto \mathbb{C}$ could be managed by considering the variable as represented by a vector with two components (real and imaginary part of x) $h(.) : \mathbb{R} \times \mathbb{R} \mapsto \mathbb{C}$ such that $h(x_R, x_I) = \phi(x)$ and therefore the formalism defined for the derivative with respect to a 2×1 vector in $\mathbb{R}^2$ could be used (i.e., any complex-valued variable is represented as two real-valued variables). However, one could define a mapping $g : \mathbb{C} \times \mathbb{C} \mapsto \mathbb{C}$ such that $g(x, x^*) = \phi(x)$, where the variables x and x^* are treated as two different variables and the function is analytic (see [4]). The derivative wrt a complex variable is [1]

$$\frac{\partial \phi(x)}{\partial x} = \frac{1}{2}\left(\frac{\partial \phi(x)}{\partial x_R} - j\frac{\partial \phi(x)}{\partial x_I}\right)$$

---

1 Since

$$h(x_R, x_I) = g(x, x^*)$$

taking the partial derivative of both

$$\begin{cases} \frac{\partial h}{\partial x_R} = \frac{\partial g}{\partial x}\frac{\partial x}{\partial x_R} + \frac{\partial g}{\partial x^*}\frac{\partial x^*}{\partial x_R} \\ \frac{\partial h}{\partial x_I} = \frac{\partial g}{\partial x}\frac{\partial x}{\partial x_I} + \frac{\partial g}{\partial x^*}\frac{\partial x^*}{\partial x_I} \end{cases}$$

and the following equalities $\frac{\partial x}{\partial x_R} = \frac{\partial x^*}{\partial x_R} = 1$, $\frac{\partial x}{\partial x_I} = j$ and $\frac{\partial x^*}{\partial x_I} = -j$, solving for $\frac{\partial g}{\partial x}$ and $\frac{\partial g}{\partial x^*}$ gives:

$$\begin{cases} \frac{\partial g}{\partial x} = \frac{1}{2}\left(\frac{\partial h}{\partial x_R} - j\frac{\partial h}{\partial x_I}\right) \\ \frac{\partial g}{\partial x^*} = \frac{1}{2}\left(\frac{\partial h}{\partial x_R} + j\frac{\partial h}{\partial x_I}\right) \end{cases}$$

Derivative wrt x is equivalent to consider x^* as constant, similarly for x^*.

Below are a few simple cases of derivatives with respect a complex variable:

$$\frac{\partial x}{\partial x} = \frac{1}{2}\left(\frac{\partial x}{\partial x_R} - j\frac{\partial x}{\partial x_I}\right) = \frac{1}{2}(1+1) = 1$$

$$\frac{\partial x^*}{\partial x} = \frac{1}{2}\left(\frac{\partial x^*}{\partial x_R} - j\frac{\partial x^*}{\partial x_I}\right) = \frac{1}{2}(1-1) = 0$$

$$\frac{\partial |x|^2}{\partial x} = \frac{\partial x x^*}{\partial x} = \frac{\partial x}{\partial x}x^* + x\frac{\partial x^*}{\partial x} = x^*$$

Let us consider the case of n complex variables ($\mathbf{x} \in \mathbb{C}^n$), and following the same ordering as for real-valued multivariate variables we have

- $\phi(\mathbf{x}) = \mathbf{a}^H\mathbf{x} = \sum_{i=1}^n a_i^* x_i$

$$\frac{\partial \mathbf{a}^H\mathbf{x}}{\partial x_k} = \frac{1}{2}\left(\frac{\partial}{\partial x_{R,k}} - j\frac{\partial}{\partial x_{I,k}}\right)\sum_{i=1}^n a_i^* x_i = a_k^* \rightarrow \frac{\partial \mathbf{a}^H\mathbf{x}}{\partial \mathbf{x}} = \begin{bmatrix} a_1^* \\ a_2^* \\ \vdots \\ a_n^* \end{bmatrix} = \mathbf{a}^*$$

- $\phi(\mathbf{x}) = \mathbf{x}^H\mathbf{a} = \sum_{i=1}^n a_i x_i^*$

$$\frac{\partial \mathbf{a}^H\mathbf{x}}{\partial x_k} = \frac{1}{2}\left(\frac{\partial}{\partial x_{R,k}} - j\frac{\partial}{\partial x_{I,k}}\right)\sum_{i=1}^n a_i x_i^* = 0 \rightarrow \frac{\partial \mathbf{x}^H\mathbf{a}}{\partial \mathbf{x}} = \mathbf{0}$$

- $\phi(\mathbf{x}) = \mathbf{x}^H\mathbf{A}\mathbf{x}$ where $\mathbf{A}$ has Hermitian symmetry ($\mathbf{A}^H = \mathbf{A}$)
  Due to the fact that the kth entry is:

$$\frac{\partial \phi(\mathbf{x})}{\partial x_k} = \mathbf{x}^H\frac{\partial \mathbf{A}\mathbf{x}}{\partial x_k} + \frac{\partial \mathbf{x}^H}{\partial x_k}\mathbf{A}\mathbf{x} = \mathbf{x}^H\mathbf{a}_k + \mathbf{0}$$

  we have:

$$\frac{\partial \mathbf{x}^H\mathbf{A}\mathbf{x}}{\partial \mathbf{x}} = \begin{bmatrix} \mathbf{x}^H\frac{\partial \mathbf{A}\mathbf{x}}{\partial x_1} + \frac{\partial \mathbf{x}^H}{\partial x_1}\mathbf{A}\mathbf{x} \\ \mathbf{x}^H\frac{\partial \mathbf{A}\mathbf{x}}{\partial x_2} + \frac{\partial \mathbf{x}^H}{\partial x_2}\mathbf{A}\mathbf{x} \\ \vdots \\ \mathbf{x}^H\frac{\partial \mathbf{A}\mathbf{x}}{\partial x_1} + \frac{\partial \mathbf{x}^H}{\partial x_1}\mathbf{A}\mathbf{x} \end{bmatrix} = \begin{bmatrix} \mathbf{x}^H\mathbf{a}_1 \\ \mathbf{x}^H\mathbf{a}_2 \\ \vdots \\ \mathbf{x}^H\mathbf{a}_n \end{bmatrix} = \begin{bmatrix} \mathbf{a}_1^T\mathbf{x}^* \\ \mathbf{a}_2^T\mathbf{x}^* \\ \vdots \\ \mathbf{a}_n^T\mathbf{x}^* \end{bmatrix} = \begin{bmatrix} \mathbf{a}_1^T \\ \mathbf{a}_2^T \\ \vdots \\ \mathbf{a}_n^T \end{bmatrix}\mathbf{x}^* = \mathbf{A}^T\mathbf{x}^* = (\mathbf{A}\mathbf{x})^*$$

  the last equality holds for Hermitian symmetry.

1.4.3 Derivative with respect to the Matrix $\mathbf{X} \in \mathbb{R}^{m\times n}$

Derivation is essentially the same as for a vector as it is just a matter of ordering. However, there are some essential relationships that are often useful in optimization problems [9]:

$$\frac{\partial}{\partial \mathbf{X}}\mathrm{tr}[\mathbf{X}] = \mathbf{I}$$

$$\frac{\partial}{\partial \mathbf{X}}\mathrm{tr}[\mathbf{AX}] = \mathbf{A}^T$$

$$\frac{\partial}{\partial \mathbf{X}}\mathrm{tr}[\mathbf{AX}^{-1}] = -(\mathbf{X}^{-1}\mathbf{AX}^{-1})^T$$

$$\frac{\partial}{\partial \mathbf{X}}\mathrm{tr}[\mathbf{X}^n] = n(\mathbf{X}^{n-1})^T$$

$$\frac{\partial}{\partial \mathbf{X}}\mathrm{tr}[\exp[\mathbf{X}]] = \exp[\mathbf{X}]$$

$$\frac{\partial}{\partial \mathbf{X}}\det[\mathbf{X}] = \det[\mathbf{X}](\mathbf{X}^{-1})^T$$

$$\frac{\partial}{\partial \mathbf{X}}\ln\det[\mathbf{X}] = (\mathbf{X}^{-1})^T$$

$$\frac{\partial}{\partial \mathbf{X}}\det[\mathbf{X}^n] = n\det[\mathbf{X}]^n(\mathbf{X}^{-1})^T$$

$$\frac{\partial}{\partial \mathbf{X}}\mathrm{tr}[\mathbf{X}^T\mathbf{AXB}] = \mathbf{AXB} + \mathbf{A}^T\mathbf{XB}^T$$

$$\frac{\partial}{\partial \mathbf{X}}\mathrm{tr}[\mathbf{XX}^T] = 2\mathbf{X}$$

1.5 Quadratic Forms

A quadratic form is represented by the following expression:

$$\psi(\mathbf{x}) = \mathbf{x}^T\mathbf{Rx} = \mathrm{tr}[\mathbf{Rxx}^T] = ||\mathbf{x}||_{\mathbf{R}}^2$$

with $\mathbf{x} \in \mathbb{R}^N$ and $\mathbf{R} \in \mathbb{R}^{N\times N}$. Function $\psi(\mathbf{x})$ is a scalar and therefore $\psi(\mathbf{x})^T = (\mathbf{x}^T\mathbf{Rx})^T = \mathbf{x}^T\mathbf{R}^T\mathbf{x} = \psi(\mathbf{x}) = \mathbf{x}^T\mathbf{Rx}$: in a quadratic form $\mathbf{R}$ must be symmetric. Moreover $\mathbf{R}$ is positive definite $\mathbf{R} > \mathbf{0}$ if $\psi(\mathbf{x}) > 0$ (or positive semidefinite if $\mathbf{R} \geq \mathbf{0}$) for each $\mathbf{x} \neq \mathbf{0}$.

For any arbitrary block-partitioning of $\mathbf{R}$, it is possible to write:

$$\begin{aligned}\psi(\mathbf{x}) = \mathbf{x}^T\mathbf{Rx} &= [\mathbf{x}_1^T, \mathbf{x}_1^T]\begin{bmatrix}\mathbf{R}_{11} & \mathbf{R}_{12}\\ \mathbf{R}_{12}^T & \mathbf{R}_{22}\end{bmatrix}\begin{bmatrix}\mathbf{x}_1\\ \mathbf{x}_2\end{bmatrix}\\ &= \mathbf{x}_1^T\mathbf{R}_{11}\mathbf{x}_1 + \mathbf{x}_1^T\mathbf{R}_{12}\mathbf{x}_2 + \mathbf{x}_2^T\mathbf{R}_{12}^T\mathbf{x}_1 + \mathbf{x}_2^T\mathbf{R}_{22}\mathbf{x}_2\end{aligned}$$

If $\mathbf{R} \geq \mathbf{0}$, this implies $\mathbf{R}_{11} \geq \mathbf{0}$ and $\mathbf{R}_{22} \geq \mathbf{0}$; it is easy to verify this assertion considering $\mathbf{x}_1 = \mathbf{0}$ or $\mathbf{x}_2 = \mathbf{0}$.
When $\mathbf{x}_2$ is fixed, the constrained quadratic form is minimized (with respect to $\mathbf{x}_1$) for

$$\hat{\mathbf{x}}_1 = -\mathbf{R}_{11}^{-1}\mathbf{R}_{12}\mathbf{x}_2$$

while the value of the form for $\hat{\mathbf{x}} = [\hat{\mathbf{x}}_1^T, \mathbf{x}_2^T]^T$ is

$$\psi(\hat{\mathbf{x}}) = \mathbf{x}_2^T(\mathbf{R}_{22} - \mathbf{R}_{12}^T\mathbf{R}_{11}^{-1}\mathbf{R}_{12})\mathbf{x}_2 \geq 0$$

the quantity $\mathbf{R}_{22} - \mathbf{R}_{12}^T\mathbf{R}_{11}^{-1}\mathbf{R}_{12}$ is indicated as the Shur complement of the block-partitioned matrix $\mathbf{R}$.

1.6 Diagonalization of a Quadratic Form

Consider the generic quadratic form (see Figure 1.4 for $N = 2$)

$$\psi(\mathbf{x}) = \mathbf{x}^T\mathbf{R}\mathbf{x} - 2\mathbf{p}^T\mathbf{x} + \psi_0$$

with $\mathbf{R}^T = \mathbf{R}$ full-rank symmetric and positive definite, and $\psi_0 \geq 0$ is a constant. By nullifying the gradient of the quadratic form it is possible to identify the coordinates of its minimum:

$$\frac{\partial\psi(\mathbf{x})}{\partial\mathbf{x}} = 2\mathbf{R}\mathbf{x} - 2\mathbf{p} = 0 \Rightarrow \mathbf{x}_{\min} = \mathbf{R}^{-1}\mathbf{p}$$

there corresponds a value of ψ as

$$\psi_{\min} = \psi(\mathbf{x} = \mathbf{R}^{-1}\mathbf{p}) = \mathbf{p}^T\mathbf{R}^{-1}\mathbf{R}\mathbf{R}^{-1}\mathbf{p} - 2\mathbf{p}^T\mathbf{R}^{-1}\mathbf{p} + \psi_0 = \psi_0 - \mathbf{p}^T\mathbf{R}^{-1}\mathbf{p}.$$

Often it is useful to refer the quadratic form to its principal axis with the origin placed at the coordinates of the minimum and an appropriate rotation to have $\mathbf{A}$ replaced by a diagonal matrix.

Diagonalization is obtained as a two step procedure. First the new variable is the one wrt the translated reference

$$\mathbf{y} = \mathbf{x} - \mathbf{x}_{\min}$$

so that the quadratic form becomes

$$\psi(\mathbf{x}) = (\mathbf{x} - \mathbf{x}_{\min})^T\mathbf{R}(\mathbf{x} - \mathbf{x}_{\min}) + \psi_0 - \mathbf{p}^T\mathbf{R}^{-1}\mathbf{p}$$
$$\psi(\mathbf{y}) = \mathbf{y}^T\mathbf{R}\mathbf{y} + \psi_{\min}$$

wrt the new variable $\mathbf{y}$. As second step, the eigenvector decomposition of $\mathbf{R} = \mathbf{U}\boldsymbol{\Lambda}\mathbf{U}^T$ yields

$$\psi(\mathbf{y}) = \mathbf{y}^T\left(\mathbf{U}\boldsymbol{\Lambda}\mathbf{U}^T\right)\mathbf{y} + \psi_{\min} = \left(\mathbf{U}^T\mathbf{y}\right)^T\boldsymbol{\Lambda}\left(\mathbf{U}^T\mathbf{y}\right) + \psi_{\min}$$

The transformation:

$$\mathbf{z} = \mathbf{U}^T\mathbf{y}$$

corresponds to a rotation of the reference system $\mathbf{y}$ in order to align wrt the principal axes of the quadratic form as in Figure 1.4. In the new coordinate system, the quadratic form is diagonal:

$$\psi(\mathbf{z}) = \mathbf{z}^T\boldsymbol{\Lambda}\mathbf{z} + \psi_{\min} = \sum_{i=1}^{m}\lambda_i z_i^2 + \psi_{\min}.$$

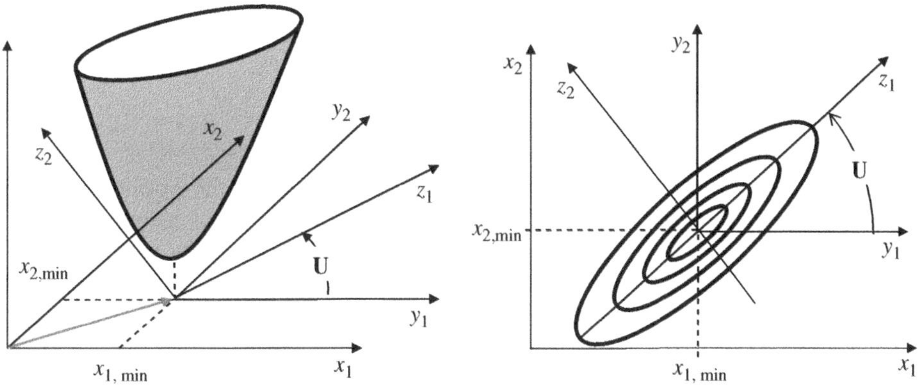

Figure 1.4 Quadratic form and its diagonalization.

This diagonalized form allows an interpretation of the eigenvalues of the quadratic form as being proportional to the curvature of the quadratic form when evaluated along the principal axes as

$$\frac{\partial^2 \psi(\mathbf{z})}{\partial z_k^2} = 2\lambda_k$$

Further deductions are possible based on this simple geometrical representation. Namely, for a positive definite matrix, all the eigenvalues are positive and the quadratic form is a paraboloid (or hyper-paraboloid for $N > 2$); the contour lines (the convex hull for $\psi(\mathbf{x}) \leq \tilde{\psi} = cost$) are ellipses (as in the Figure 1.4) that are stretched according to the ratio between the corresponding eigenvalues. When all eigenvalues are equal, these contour lines are circles.

1.7 Rayleigh Quotient

Given two quadratic forms $\psi_{\mathbf{A}}(\mathbf{x}) = \mathbf{x}^H\mathbf{A}\mathbf{x}$ and $\psi_{\mathbf{B}}(\mathbf{x}) = \mathbf{x}^H\mathbf{B}\mathbf{x}$ (for generality, $\mathbf{x} \in \mathbb{C}^N$), their ratio

$$V(\mathbf{x}) = \frac{\mathbf{x}^H\mathbf{A}\mathbf{x}}{\mathbf{x}^H\mathbf{B}\mathbf{x}}$$

is indicated as a Rayleigh quotient. The optimization (max/min) of functions like $V(\mathbf{x})$ is sometimes necessary and requires some specific considerations.

Let $\mathbf{B} \succ \mathbf{0}$ be positive definite in order to avoid meaningless conditions (i.e., $\psi_{\mathbf{B}}(\mathbf{x}) = 0$ for some value of $\mathbf{x}$); it is possible to have the square-root decomposition as $\mathbf{B} = \mathbf{B}^{H/2}\mathbf{B}^{1/2}$, and by using the transformation

$$\mathbf{y} = \mathbf{B}^{1/2}\mathbf{x}$$

(that is invertible $\mathbf{x} = \mathbf{B}^{-1/2}\mathbf{y}$) we obtain:

$$V(\mathbf{x}) = \frac{\mathbf{x}^H\mathbf{A}\mathbf{x}}{(\mathbf{B}^{1/2}\mathbf{x})^H\mathbf{B}^{1/2}\mathbf{x}} = \frac{\mathbf{y}^H(\mathbf{B}^{-H/2}\mathbf{A}\mathbf{B}^{-1/2})\mathbf{y}}{\mathbf{y}^H\mathbf{y}} = \frac{\mathbf{y}^H\mathbf{Q}\mathbf{y}}{\mathbf{y}^H\mathbf{y}}$$

The solution that maximizes/minimizes $V(\mathbf{x})$ can be found by the optimization of $\psi_\mathbf{Q}(\mathbf{y}) = \mathbf{y}^H\mathbf{Q}\mathbf{y}$ for $\mathbf{Q} = \mathbf{B}^{-H/2}\mathbf{A}\mathbf{B}^{-1/2}$ subject to the constraints of $\mathbf{y}$ with unitary norm (see also Section 1.8). Alternatively, by using the eigenvector decomposition of $\mathbf{Q} = \mathbf{V}\mathbf{\Lambda}\mathbf{V}^H$ it is possible to write:

$$V(\mathbf{x}) = \frac{\mathbf{y}^H\mathbf{Q}\mathbf{y}}{\mathbf{y}^H\mathbf{y}} = \frac{\sum_k \lambda_k |\mathbf{y}^H\mathbf{v}_k|^2}{\mathbf{y}^H\mathbf{y}}$$

moreover, the following property holds:

$$\lambda_{\min}(\mathbf{Q}) \leq V(\mathbf{x}) \leq \lambda_{\max}(\mathbf{Q})$$

for the following choices

$$\begin{aligned} V(\mathbf{x}) &= \lambda_{\min}(\mathbf{Q}) \Leftrightarrow \mathbf{y} = \mathbf{v}_{\min} \rightarrow \mathbf{x}_{\min} = \mathbf{B}^{-1/2}\mathbf{v}_{\min} \\ V(\mathbf{x}) &= \lambda_{\max}(\mathbf{Q}) \Leftrightarrow \mathbf{y} = \mathbf{v}_{\max} \rightarrow \mathbf{x}_{\max} = \mathbf{B}^{-1/2}\mathbf{v}_{\max} \end{aligned}$$

To summarize, the max/min of $V(\mathbf{x})$ is obtained from the eigenvectors corresponding to the max/min of $\mathbf{B}^{-H/2}\mathbf{A}\mathbf{B}^{-1/2}$:

$$\mathbf{x} = \mathbf{B}^{-1/2} eig_{\max/\min}\{\mathbf{B}^{-H/2}\mathbf{A}\mathbf{B}^{-1/2}\}.$$

1.8 Basics of Optimization

Optimization problems are quite common in engineering and having an essential knowledge of the main tools is vital. To exemplify, several times it is necessary to identify the coordinates of minimum (or maximum) of a function $\phi(\mathbf{x})$ wrt $\mathbf{x} \in \mathbb{R}^N$ subject to (s.t.) some conditions like $g_1(\mathbf{x}) = 0, \cdots, g_M(\mathbf{x}) = 0$. The overall minimization can be cast as follows:

$$\min_{\mathbf{x}}\{\phi(\mathbf{x})\} \text{ s.t. } \mathbf{g}(\mathbf{x}) = \begin{bmatrix} g_1(\mathbf{x}) \\ \vdots \\ g_M(\mathbf{x}) \end{bmatrix} = \mathbf{0}$$

This is a constrained optimization that is solved by using the *Lagrange multipliers* method. The objective function is redefined including the constraints

$$L(\mathbf{x}, \boldsymbol{\lambda}) = \phi(\mathbf{x}) + \sum_{k=1}^{M} \lambda_k g_k(\mathbf{x}) = \phi(\mathbf{x}) + \boldsymbol{\lambda}^T\mathbf{g}(\mathbf{x})$$

where the set of (unknown) terms λ_k are called multipliers. The optimization of $L(\mathbf{x},\lambda)$ is now wrt a total of $N+M$ unknowns and it is carried out by evaluating the gradient with respect to $\mathbf{x}$ and λ:

$$\begin{cases} \dfrac{\partial L(\mathbf{x},\lambda)}{\partial \mathbf{x}} = \mathbf{0} \\ \dfrac{\partial L(\mathbf{x},\lambda)}{\partial \lambda} = \mathbf{0} \end{cases} \Rightarrow \begin{cases} \dfrac{\partial \phi(\mathbf{x})}{\partial \mathbf{x}} + \lambda^T \dfrac{\partial \mathbf{g}(\mathbf{x})}{\partial \mathbf{x}} = \mathbf{0} \\ \mathbf{g}(\mathbf{x}) = \mathbf{0} \end{cases}$$

Notice that the second set of equations is the constraint condition. The solution of the system of equations should eliminate the multipliers and it requires some ability in manipulating equations. A comprehensive discussion on optimization methods can be found in the reference [16] with proof of a unique solution in the case of convex optimization. However, some examples widely used within the book can help to clarify the Lagrange multiplier method.

1.8.1 Quadratic Function with Simple Linear Constraint (M=1)

If $\phi(\mathbf{x}) = \mathbf{x}^T\mathbf{A}\mathbf{x}$ is quadratic function, the optimization with a linear constraint is formulated as:

$$\min_{\mathbf{x}}\{\mathbf{x}^T\mathbf{A}\mathbf{x}\} \text{ s.t. } \mathbf{c}^T\mathbf{x} = 1$$

It can be shown that this problem is convex. The formulation with the Lagrangian multiplier is:

$$L(\mathbf{x},\lambda) = \mathbf{x}^T\mathbf{A}\mathbf{x} + \lambda(\mathbf{c}^T\mathbf{x} - 1)$$

that optimized yields

$$\frac{\partial L(\mathbf{x},\lambda)}{\partial \mathbf{x}} = 0 \Rightarrow 2\mathbf{A}\mathbf{x} + \lambda\mathbf{c} = 0 \Rightarrow \mathbf{x} = -\frac{\lambda}{2}\mathbf{A}^{-1}\mathbf{c}$$

Using the constraint $\mathbf{c}^T\mathbf{x} = 1$, there is the solution for the multipliers

$$\mathbf{c}^T\mathbf{x} = 1 \Rightarrow -\frac{\lambda}{2}\mathbf{c}^T\mathbf{A}^{-1}\mathbf{c} = 1 \Rightarrow \lambda = -\frac{2}{\mathbf{c}^T\mathbf{A}^{-1}\mathbf{c}}$$

(notice that $\mathbf{c}^T\mathbf{A}^{-1}\mathbf{c}$ is a scalar) that substituted gives the solution:

$$\mathbf{x}_{\text{opt}} = \frac{\mathbf{A}^{-1}\mathbf{c}}{\mathbf{c}^T\mathbf{A}^{-1}\mathbf{c}}.$$

1.8.2 Quadratic Function with Multiple Linear Constraints

Let the quadratic form $\phi(\mathbf{x}) = \mathbf{x}^H\mathbf{A}\mathbf{x}$ with complex-valued entries and Hermitian symmetric $\mathbf{A}^H = \mathbf{A}$; the minimization is subject to M constraints $\mathbf{c}_m^H\mathbf{x} = d_m$ with $m = 1, \cdots, M$, and it is a quadratic function minimization with multiple linear constraints:

$$\min_{\mathbf{x}}\{\mathbf{x}^H\mathbf{A}\mathbf{x}\} \text{ s.t. } \mathbf{C}\mathbf{x} = \mathbf{d}.$$

Due to the fact that the optimization is generalized here to the case of complex variables, the total number of constraints is $2M$. In fact the single constraint expression can be written as:

$$\delta_m = \delta_{R,m} + j\delta_{I,m} = \mathbf{c}_m^H\mathbf{x} - d_m = 0$$

The Lagrangian function with $2M$ constraint conditions ($\boldsymbol{\lambda} \in \mathbb{C}^{M\times 1}$) is:

$$L(\mathbf{x}, \boldsymbol{\lambda}) = \mathbf{x}^T\mathbf{A}\mathbf{x} + \sum_{m=1}^{M}(\lambda_{R,m}\delta_{R,m} + \lambda_{I,m}\delta_{I,m})$$

It is useful to recall that the constraints can be written as

$$(\lambda_{R,m}\delta_{R,m} + \lambda_{I,m}\delta_{I,m}) = \frac{1}{2}(\lambda_m^*\delta_m + \lambda_m\delta_m^*)$$

to highlight the formulas of the derivatives for complex-valued variables (see Section 1.4.2). Therefore

$$L(\mathbf{x}, \boldsymbol{\lambda}) = \mathbf{x}^H\mathbf{A}\mathbf{x}+\frac{1}{2}(\boldsymbol{\lambda}^H\boldsymbol{\delta} + \boldsymbol{\delta}^H\boldsymbol{\lambda}) = \mathbf{x}^H\mathbf{A}\mathbf{x}+\frac{1}{2}\boldsymbol{\lambda}^H(\mathbf{C}\mathbf{x} - \mathbf{d}) + \frac{1}{2}(\mathbf{C}\mathbf{x} - \mathbf{d})^H\boldsymbol{\lambda}$$

and setting the gradient to zero (see Section 1.4.2):

$$\frac{\partial L(\mathbf{x}, \boldsymbol{\lambda})}{\partial \mathbf{x}} = (\mathbf{A}\mathbf{x})^* + \frac{1}{2}\mathbf{C}^T\boldsymbol{\lambda}^* = (\mathbf{A}\mathbf{x})^* + \frac{1}{2}(\mathbf{C}^H\boldsymbol{\lambda})^* = \mathbf{0}$$

whose solution is

$$\mathbf{x} = -\frac{1}{2}\mathbf{A}^{-1}\mathbf{C}^H\boldsymbol{\lambda}$$

which, inserted into the constraints $\mathbf{C}\mathbf{x} = \mathbf{d}$, gives the solution for the multipliers:

$$-\frac{1}{2}\mathbf{C}\mathbf{A}^{-1}\mathbf{C}^H\boldsymbol{\lambda} = \mathbf{d} \rightarrow \boldsymbol{\lambda} = -2(\mathbf{C}\mathbf{A}^{-1}\mathbf{C}^H)^{-1}\mathbf{d}$$

Back-substituting into the problem leads to the general solution for the constrained optimization problem:

$$\mathbf{x}_{opt} = \mathbf{A}^{-1}\mathbf{C}^H(\mathbf{C}\mathbf{A}^{-1}\mathbf{C}^H)^{-1}\mathbf{d}.$$

Appendix A: Arithmetic vs. Geometric Mean

Let $x_1, \ldots x_N$ be a set of positive values; the arithmetic mean is always larger than the geometric mean:

$$\frac{\sum_{k=1}^{N} x_k}{N} \geq \left(\prod_{k=1}^{N} x_k\right)^{1/N}$$

the equality is for $x_1 = x_2 = \ldots = x_N$.

The inequality holds for any weighted average with weighting $\lambda_k \geq 0$:

$$\sum_{k=1}^{N} \lambda_k x_k \geq \prod_{k=1}^{N} x_k^{\lambda_k},$$

the equality is still for $x_1 = x_2 = \ldots = x_N$. In some cases it is convenient (or stems from the problem itself, e.g., in information theory related problems) to adopt logarithms:

$$\log\left(\sum_{k=1}^{N} \lambda_k x_k\right) \geq \sum_{k=1}^{N} \lambda_k \log x_k$$

By extension, the following inequalities hold:

$$\sqrt{\frac{x_1^2 + \cdots + x_N^2}{N}} \geq \frac{x_1 + \cdots + x_N}{N} \geq \sqrt[N]{x_1 \cdots x_N} \geq \frac{N}{\frac{1}{x_1} + \cdots + \frac{1}{x_N}}$$

2

Linear Algebraic Systems

Linear algebraic models give a compact and powerful structure for signal modeling. Filtering, identification, and estimation can be modeled by a set of liner equations, possibly using matrix notation for the unknown $\mathbf{x}$:

$$\mathbf{A} \cdot \mathbf{x} = \mathbf{b}.$$

Its recurrence in many contexts makes it mandatory to gain insight into some essential concepts beyond the pure numerical solution, as these are necessary for an in-depth understanding of the properties of linear systems as linear transformations manipulating matrix subspaces.

2.1 Problem Definition and Vector Spaces

The linear system

$$\underset{(m,n)}{\mathbf{A}} \cdot \underset{(n,1)}{\mathbf{x}} = \underset{(m,1)}{\mathbf{b}} \tag{2.1}$$

represents the linear transformation of the vector $\mathbf{x} \in \mathbb{R}^n$ into the vector $\mathbf{b} \in \mathbb{R}^m$ through the linear transformation described by the $\mathbf{A}$ matrix where we assume that $m \geq n$ (i.e., the number of equations is larger than the number of unknowns). We can rewrite the previous equation as

$$[\mathbf{a}_1, \mathbf{a}_2, \dots, \mathbf{a}_n] \begin{bmatrix} x_1 \\ x_2 \\ \vdots \\ x_n \end{bmatrix} = \mathbf{a}_1 x_1 + \mathbf{a}_2 x_2 + \dots + \mathbf{a}_n x_n = \mathbf{b};$$

this shows that $\mathbf{b}$ is obtained as a linear combination of the columns of $\mathbf{A}$ with $\mathbf{x}$ being the coefficients. Making reference to Figure 2.1 illustrating the case for $m = 3$, every vector $\mathbf{a}_i$ is a set of m-tuples with $\mathbf{a}_i \in \mathbb{R}^m$, the span of the n columns of matrix $\mathbf{A}$ is a subspace indicated as $\mathcal{R}(\mathbf{A}) \subseteq \mathbb{R}^m$: the range of $\mathbf{A}$ or the column subspace of $\mathbf{A}$. The rank of the

Statistical Signal Processing in Engineering, First Edition. Umberto Spagnolini.

Companion website: www.wiley.com/go/spagnolini/signalprocessing

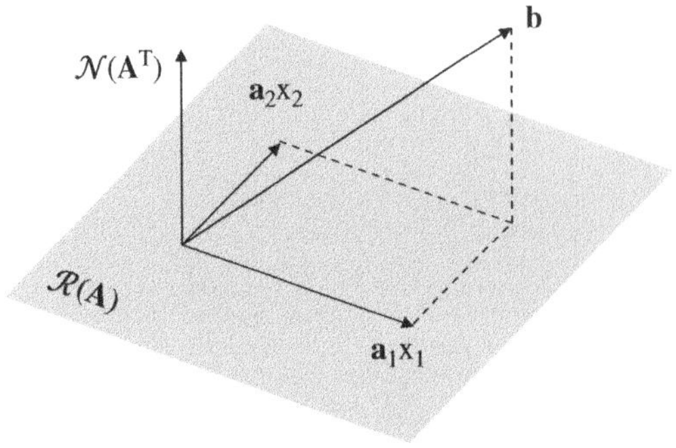

Figure 2.1 Geometric view of $\mathcal{R}(\mathbf{A})$ and $\mathcal{N}(\mathbf{A}^T)$ for $m = 3$.

matrix is defined as the dimension of this subspace: $\text{rank}(\mathbf{A}) = \dim\{\mathcal{R}(\mathbf{A})\} = r \leq m$.[1] In the case that $\mathbf{b} \in \mathcal{R}(\mathbf{A})$, the linear system has at least one solution, that is the coefficients of the columns of **A** necessary to have **b**. The easiest case is represented when $m = n$ and $\mathcal{R}(\mathbf{A}) = \mathbb{R}^m$ (i.e., all the columns of **A** exactly span $\mathbb{R}^m$); thus the rank of **A** is maximum and equal to $m = n$, every vector $\mathbf{b} \in \mathbb{R}^m$ can be described by one linear combination of columns of **A**, and the solution is given by $\mathbf{x} = \mathbf{A}^{-1}\mathbf{b}$.

The **nullspace** of A is defined as $\mathcal{N}(\mathbf{A}) = \{\mathbf{x} \in \mathbb{R}^n | \mathbf{A}\mathbf{x} = \mathbf{0}\}$ and represents the set of vectors orthogonal to all *rows* of **A**. In other words, there are combinations of columns of **A** (except the trivial case $\mathbf{x} = \mathbf{0}$) that nullify the product **Ax**, and thus the columns of **A** are not linearly independent, and $\text{rank}(\mathbf{A}) = \dim\{\mathcal{R}(\mathbf{A})\} < n$ (as $m \geq n$). On the other hand, when $\mathcal{N}(\mathbf{A}) = \{\varnothing\}$, the matrix **A** has all columns linearly independent and the mapping from **x** and **Ax** is one-to-one. It always holds that $\text{rank}(\mathbf{A}) \leq \min(m,n)$ and $\text{rank}(\mathbf{A}) + \dim \mathcal{N}(\mathbf{A}) = n$. It is also possible to define the subspaces described by the matrix rows: $\mathcal{R}(\mathbf{A}^T) = \{\boldsymbol{\alpha} \in \mathbb{R}^n | \boldsymbol{\alpha} = \mathbf{A}^T\boldsymbol{\beta} \text{ for } \boldsymbol{\beta} \in \mathbb{R}^m\} \subseteq \mathbb{R}^n$ and $\mathcal{N}(\mathbf{A}^T) = \{\boldsymbol{\beta} \in \mathbb{R}^m | \mathbf{A}^T\boldsymbol{\beta} = \mathbf{0}\}$. From these definitions, $\mathcal{R}(\mathbf{A}) \subseteq \mathbb{R}^m$ and $\mathcal{N}(\mathbf{A}^T) \subseteq \mathbb{R}^m$, or equivalently both are subspaces of $\mathbb{R}^m$. The nullspace $\mathcal{N}(\mathbf{A}^T)$ follows from the linear combination of columns of **A** as $\mathbf{A}^T\boldsymbol{\beta} = \mathbf{0} \rightarrow \boldsymbol{\beta}^T\mathbf{A} = \mathbf{0}^T$; this implies that $\boldsymbol{\beta}^T\mathbf{a}_i = 0$ for every column and thus $\boldsymbol{\beta}$ is orthogonal to every column of matrix **A** (and thus any linear combination); thus: $\mathcal{R}(\mathbf{A}) \perp \mathcal{N}(\mathbf{A}^T)$ or $\mathcal{R}(\mathbf{A}) \cup \mathcal{N}(\mathbf{A}^T) \equiv \mathbb{R}^m$. In addition, $\dim\{\mathcal{R}(\mathbf{A}) \cup \mathcal{N}(\mathbf{A}^T)\} = m$ and $\dim\{\mathcal{R}(\mathbf{A})\} + \dim\{\mathcal{N}(\mathbf{A}^T)\} = m$, or $\dim\{\mathcal{N}(\mathbf{A}^T)\} = m - r$. Notice that a similar reasoning holds for $\mathcal{R}(\mathbf{A}^T)$ and $\mathcal{N}(\mathbf{A})$, just by exchanging rows and columns (not covered here).

Reconsidering the numerical problem of solving the set of equations $\mathbf{A}\mathbf{x} = \mathbf{b}$, this can be uniquely solved only if $\mathbf{b} \in \mathcal{R}(\mathbf{A})$. All the other cases need to evaluate how to handle when $\mathbf{b} \in \mathcal{B} \supset \mathcal{R}(\mathbf{A})$. In general, it is possible to decompose **b** as $\mathbf{b} = \mathbf{b}_A + \mathbf{b}_\perp$ where $\mathbf{b}_A \in \mathcal{R}(\mathbf{A})$ and $\mathbf{b}_\perp \in \mathcal{N}(\mathbf{A}^T)$ and solve for the *equivalent* system $\mathbf{A}\mathbf{x} = \mathbf{b}_A$. This decomposition is known as *data-filtering* of **b** and the geometrical representation in Figure 2.1 is for $m = 3$, and $\text{rank}(\mathbf{A}) = n = 2$ (full-rank).

1 It holds that $\text{rank}(\mathbf{A}) = \text{rank}(\mathbf{A}^T) \leq \min(m,n)$. A matrix has full-rank in the following cases: (a) square matrixes—full-rank means non-singular, $rank(\mathbf{A}) = m = n$; (b) "tall-and-skinny" matrixes ($m \geq n$)—full-rank means that all columns are independent; (c) "fat" matrixes ($m \leq n$)—full-rank means that rows are independent.

Remark: Usually, in engineering problems, matrix **A** accounts for the model that linearly combines the minimal model parameters (the n columns of **A**) where the scaling terms are unknown (unknown vector **x**), all measurements are stack into the vector **b**. If the model fully describes the set of measurements ($m = n$), the linear system gets the scaling terms from vector **x** (or equivalently, all the measurements are exactly described by the model, and the model is complete). However, to cope with noisy measurement, the number of measurements m is purposely made redundant (larger than the number of unknown, $m > n$) so the columns of **A** span only a subspace of dimension n, that is the dimension of the true problem to be solved. In other words, the measurements should be obtained as a linear combination of columns of **A**, but not all in **A** are useful for model definition and these are usually referred as noise (noise is all that cannot be described by the model of the problem).

2.1.1 Vector Spaces in Tomographic Radiometric Inversion

The water vapor content in the atmosphere is not constant and varies over space and time—namely it depends on the the weather conditions. There is a strong interest in evaluating this water content in climatology (e.g., heat balance of the atmosphere, weather forecasts) and telecommunications (e.g., quality in satellite communications); this is carried out by measuring the microwave attenuation over predefined links and inferring from these measurements the integral of liquid and water vapor along the propagation path. See [18] for an in-depth discussion.

For the sake of reasoning here, we can assume the model in Figure 2.2, where a radiometer is based on an antenna with a narrow beamwidth pointing over a predefined elevation angle and measuring the total energy radiated by the atmosphere and attenuated by the water vapor. To further simplify, the experiment can be modeled by a constant thickness layer of height h that is divided into n cells labeled as $1,2,\ldots, n$ of equal width d with specific attenuation $x_1, x_2, \ldots, x_n$ on each of the cells and constant power (see figure). For any (elevation) angle, say θ, the attenuation is the sum of the lengths of all attenuation cells traversed, multiplied by the corresponding attenuation:

$$a(\theta) = \sum_{k=1}^{n} l_k(\theta)x_k$$

When one cell is not crossed, say $\bar{k}$, $l_{\bar{k}}(\theta) = 0$.

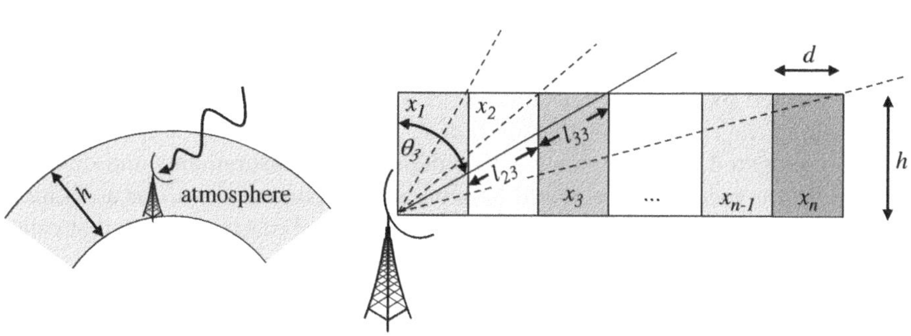

Figure 2.2 Radiometer model with constant thickness.

Angles can be chosen by the radiometer control system so that every elevation angle crosses a different number of cells, with different length in every cell, so for each angle there is one set of equations to justify that specific measurement. Assume that the kth angle θ_k crosses k cells only as in the figure

$$\tan\theta_k = k\frac{d}{h}$$

for this special case, the length in each cell is

$$l_{1k} = l_{2k} = \dots = l_{kk} = \frac{d}{\sin\theta_k}.$$

The attenuation measured by the antenna is the sum of the specific attenuation per cell times the length (Beer–Lambert law, Chapter 21)[2]

$$a(\theta_k) = l_{1k}x_1 + l_{2k}x_2 + \dots + l_{kk}x_k$$

This is commonly referred to as the model of the experiment since it relates the unknowns (attenuation on every cell) with the expected measurements (global attenuation from the antenna when pointing at angle θ_k). The ensemble of m *noisy* measurements are collected in matrix form for the linear system of equations that for the case $m = n$

$$\begin{bmatrix} l_{11} & 0 & 0 & \cdots & 0 \\ l_{12} & l_{22} & 0 & \cdots & 0 \\ l_{13} & l_{23} & l_{33} & \cdots & 0 \\ \vdots & \vdots & \vdots & \ddots & \vdots \\ l_{1n} & l_{2n} & l_{3n} & \cdots & l_{nn} \end{bmatrix} \begin{bmatrix} x_1 \\ x_2 \\ x_3 \\ \vdots \\ x_n \end{bmatrix} = \begin{bmatrix} b_1 \\ b_2 \\ b_3 \\ \vdots \\ b_n \end{bmatrix}$$

relates the model $a(\theta_k)$ with the corresponding measurement b_k for every angle. In general, for the presence of noise and uncertainties in measuring the received power, $a(\theta_k) \neq b_k$, but likely $a(\theta_k) \simeq b_k$. Recall that the choice of angles $\theta_1, \theta_2, \dots, \theta_m$ is a design parameter of the engineer that sets up the experiment, and in turn this sets the properties of matrix $\mathbf{A}$. In this special case, $\mathbf{A}$ is lower triangular and the solution of the linear system of equations follows by simple substitutions: $\hat{x}_1 = b_1/l_{11}$ from the first row, then substitute into the second row $\hat{x}_2 = (b_2 - l_{12}\hat{x}_1)/l_{22}$, etc...

Even if the choice $m = n$ guarantees a unique solution, it is very sensitive to errors (e.g., if b_1 is affected by an error, $\hat{x}_1$ is affected as well, and in turn all the other errors accumulate along the cells so that the solution $\hat{x}_n$ would be the most impaired one). Noise can be mitigated by redundancy with independent measurements and distinct experiments. More specifically, it would be advisable to choose $m > n$ for a distinct set of angles $\boldsymbol{\theta} = [\theta_1, \theta_2, \dots, \theta_m]^T$ that contribute to the model matrix $\mathbf{A}(\boldsymbol{\theta})$ so that cells

2 In engineering power is measured in decibel with respect to 1mW, or dBm, defined as $10\log_{10}(P_k/1mW)$; attenuation is in decibel, or dB, and specific attenuation x_k in decibel/m, or dB/m.

are measured by multiple raypaths, and any error in one measurement would be counterbalanced by the others. Discussion around this subject is an essential part of Chapter 21.

2.2 Rotations

Matrix multiplication represents a transformation; rotation is a special transformation in vector terminology. Figure 2.3 shows a rotation of the coordinate axis in 2D space. The original coordinates $[u_1, u_2]^T$ are transformed in the new coordinates $[v_1, v_2]^T$ through the linear transformation

$$\begin{bmatrix} v_1 \\ v_2 \end{bmatrix} = \mathbf{Q}_\theta \begin{bmatrix} u_1 \\ u_2 \end{bmatrix}$$

where the rotation matrix is given by

$$\mathbf{Q}_\theta = \begin{bmatrix} \cos\theta & \sin\theta \\ -\sin\theta & \cos\theta \end{bmatrix}$$

The following properties hold for a 2 by 2 rotation matrix:

$$\begin{aligned} &\mathbf{Q}_\theta \cdot \mathbf{Q}_{-\theta} = \mathbf{I} \longrightarrow \mathbf{Q}_\theta^{-1} = \mathbf{Q}_{-\theta} = \mathbf{Q}_\theta^T \\ &\mathbf{Q}_\theta \cdot \mathbf{Q}_\theta = \mathbf{Q}_\theta^2 = \mathbf{Q}_{2\theta} \\ &\mathbf{Q}_\theta \cdot \mathbf{Q}_\varphi = \mathbf{Q}_{\theta+\varphi} \end{aligned}$$

Therefore a rotation matrix is orthonormal (each column is a unitary norm vector and all vectors are orthogonal one another) and consequently the length (norm) of any vector is independent of any rotation.

When a vector has larger dimensions, say $n > 2$, it is possible to apply, in sequence, elementary rotations each relative to a couple of axis ($\mathbf{Q} = \prod_{i,j} \mathbf{Q}_{ij}$). For example considering the i and j axis it is possible to write:

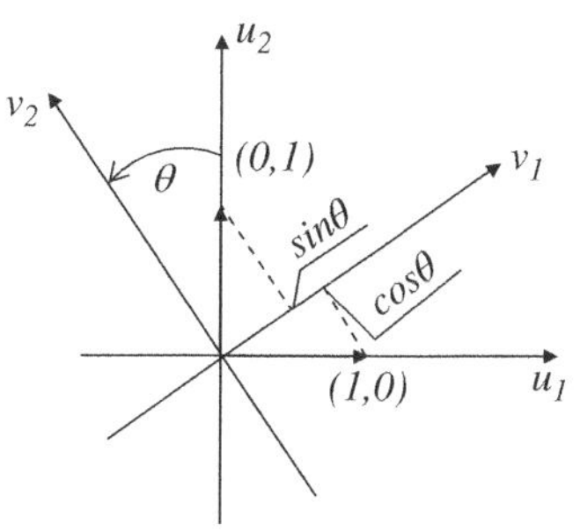

Figure 2.3 Rotation in a plane.

$$\begin{bmatrix} v_1 \\ \vdots \\ v_i \\ \vdots \\ v_j \\ \vdots \\ v_n \end{bmatrix} = \underbrace{\begin{bmatrix} 1 & \cdots & 0 & \cdots & 0 & \cdots & 0 \\ \vdots & \ddots & \vdots & & \vdots & & \vdots \\ 0 & \cdots & \cos\theta_{ij} & \cdots & \sin\theta_{ij} & \cdots & 0 \\ \vdots & & \vdots & \ddots & \vdots & & \vdots \\ 0 & \cdots & -\sin\theta_{ij} & \cdots & \cos\theta_{ij} & \cdots & 0 \\ \vdots & & \vdots & & \vdots & \ddots & \vdots \\ 0 & \cdots & 0 & \cdots & 0 & \cdots & 1 \end{bmatrix}}_{\mathbf{Q}_{ij}} \begin{bmatrix} u_1 \\ \vdots \\ u_i \\ \vdots \\ u_j \\ \vdots \\ u_n \end{bmatrix}$$

It is also possible to select the rotation angle θ_{ij} to nullify one of the components after applying a rotation. Imposing $v_j = 0$, the rotation angle θ_{ij} must be chosen to satisfy $\tan\theta_{ij} = u_j/u_i$, obtaining:

$$\sin\theta_{ij} = \frac{u_j}{\sqrt{u_i^2 + u_j^2}}, \quad \cos\theta_{ij} = \frac{u_i}{\sqrt{u_i^2 + u_j^2}}$$

$$v_i = \cos\theta_{ij} u_i + \sin\theta_{ij} u_j = \sqrt{u_i^2 + u_j^2} \text{ and } v_j = 0$$

These rotations are known as *Givens rotations* (from W. Givens who first proposed the method) and are useful to design specific rotations. Recalling that any transformation can be considered as applied column-wise on a matrix as

$$\underbrace{[\mathbf{v}_1, \mathbf{v}_2, \ldots, \mathbf{v}_n]}_{\mathbf{V}} = [\mathbf{Q}_{ij}\mathbf{u}_1, \mathbf{Q}_{ij}\mathbf{u}_2, \ldots, \mathbf{Q}_{ij}\mathbf{u}_n] = \mathbf{Q}_{ij}\underbrace{[\mathbf{u}_1, \mathbf{u}_2, \ldots, \mathbf{u}_n]}_{\mathbf{U}}$$

It is possible to iterate the application of rotation matrixes to nullify components in order to transform a generic matrix into, for example, an upper triangular matrix. The following formula shows the various iterations that generate the complete transformation. The elements of the elementary rotation are indicated by c and s to indicate $\cos\theta$ and $\sin\theta$ for a rotation angle θ specifically chosen to nullify one entry per iteration. The bold elements $\mathbf{x}$ show the modified elements of the matrix at each iteration.

$$\begin{bmatrix} c & 0 & 0 & s \\ 0 & 1 & 0 & 0 \\ 0 & 0 & 1 & 0 \\ -s & 0 & 0 & c \end{bmatrix} \begin{bmatrix} x & x & x & x \\ x & x & x & x \\ x & x & x & x \\ x & x & x & x \end{bmatrix} = \begin{bmatrix} \mathbf{x} & \mathbf{x} & \mathbf{x} & \mathbf{x} \\ x & x & x & x \\ x & x & x & x \\ \mathbf{0} & \mathbf{x} & \mathbf{x} & \mathbf{x} \end{bmatrix}$$

$$\begin{bmatrix} c & 0 & s & 0 \\ 0 & 1 & 0 & 0 \\ -s & 0 & c & 0 \\ 0 & 0 & 0 & 1 \end{bmatrix} \begin{bmatrix} x & x & x & x \\ x & x & x & x \\ x & x & x & x \\ 0 & x & x & x \end{bmatrix} = \begin{bmatrix} \mathbf{x} & \mathbf{x} & \mathbf{x} & \mathbf{x} \\ x & x & x & x \\ \mathbf{0} & \mathbf{x} & \mathbf{x} & \mathbf{x} \\ 0 & x & x & x \end{bmatrix}$$

$$\begin{bmatrix} c & s & 0 & 0 \\ -s & c & 0 & 0 \\ 0 & 0 & 1 & 0 \\ 0 & 0 & 0 & 1 \end{bmatrix} \begin{bmatrix} x & x & x & x \\ x & x & x & x \\ 0 & x & x & x \\ 0 & x & x & x \end{bmatrix} = \begin{bmatrix} \mathbf{x} & \mathbf{x} & \mathbf{x} & \mathbf{x} \\ \mathbf{0} & \mathbf{x} & \mathbf{x} & \mathbf{x} \\ 0 & x & x & x \\ 0 & x & x & x \end{bmatrix}$$

$$\begin{bmatrix} 1 & 0 & 0 & 0 \\ 0 & c & 0 & s \\ 0 & 0 & 1 & 0 \\ 0 & -s & 0 & c \end{bmatrix} \begin{bmatrix} x & x & x & x \\ 0 & x & x & x \\ 0 & x & x & x \\ 0 & x & x & x \end{bmatrix} = \begin{bmatrix} x & x & x & x \\ \mathbf{0} & \mathbf{x} & \mathbf{x} & \mathbf{x} \\ 0 & x & x & x \\ 0 & \mathbf{0} & \mathbf{x} & \mathbf{x} \end{bmatrix}$$

$$\vdots$$

$$\begin{bmatrix} 1 & 0 & 0 & 0 \\ 0 & 1 & 0 & 0 \\ 0 & 0 & c & s \\ 0 & 0 & -s & c \end{bmatrix} \begin{bmatrix} x & x & x & x \\ 0 & x & x & x \\ 0 & 0 & x & x \\ 0 & 0 & x & x \end{bmatrix} = \begin{bmatrix} x & x & x & x \\ 0 & x & x & x \\ 0 & \mathbf{0} & \mathbf{x} & \mathbf{x} \\ 0 & 0 & \mathbf{0} & \mathbf{x} \end{bmatrix}$$

Any arbitrary rotation $\mathbf{Q}$ can be written as a combination of elementary rotations, each involving only a couple of axes ($\mathbf{Q} = \prod_{i,j} \mathbf{Q}_{i,j}$). Any rotation matrix $\mathbf{Q}$ is orthonormal $\mathbf{Q}^T\mathbf{Q} = \mathbf{I} \rightarrow \mathbf{q}_i^T \mathbf{q}_j = \delta_{i-j} \rightarrow \mathbf{Q}^{-1} = \mathbf{Q}^T$, and any square orthonormal matrix is an appropriate axes rotation. By a judicious choice of these rotation matrixes it is possible to diagonalize a square matrix and thus reduce the matrix to be diagonal (e.g., for eigenvalue decomposition[3]).

2.3 Projection Matrixes and Data-Filtering

The projection of a vector onto a subspace can be represented through a linear transformation $\mathbf{P}$ (Figure 2.4):

$$\mathbf{x}_p = \mathbf{P}\mathbf{x}$$

that in this case changes the vector length (metric). The complementary matrix $\mathbf{P}^{\perp}$ projects the input vector into the complementary subspace:

$$\mathbf{x}_p^{\perp} = \mathbf{x} - \mathbf{x}_p = (\mathbf{I} - \mathbf{P})\mathbf{x} = \mathbf{P}^{\perp}\mathbf{x}.$$

The projection matrix is idempotent as $\mathbf{P}^2 = \mathbf{P}$; this is in the definition of the projection matrix as $\mathbf{P}(\mathbf{P}\mathbf{x}) = \mathbf{P}\mathbf{x}_p = \mathbf{x}_p$. Similarly, any idempotent matrix is a projection matrix. Eigenvalues of a projection matrix are $\lambda_k(\mathbf{P}) \in \{0,1\}$.

In the way it is defined, a projection matrix projects any vector $\mathbf{x}$ onto a subspace defined by the projection matrix itself that is dependent on the specific problem. For the linear system of equations in Section 2.1, the subspace is $\mathcal{R}(\mathbf{A})$ and thus the projection

3 Let $\mathbf{A}$ be a symmetric matrix; for eigenvalue computation, it is sufficient to apply pair of rotations $\mathbf{B} = \mathbf{Q}_{ij}\mathbf{A}\mathbf{Q}_{ij}^T$ to nullify b_{ij} and b_{ji} with the relationship $\cos^2\theta_{ij} + \sin^2\theta_{ij} = 1$:

$$\mathbf{Q}_{ij} \begin{bmatrix} x & x & x & x \\ x & x & x & x \\ x & x & x & x \\ x & x & x & x \end{bmatrix} \mathbf{Q}_{ij}^T = \begin{bmatrix} x & x & x & x \\ \mathbf{x} & \mathbf{x} & \mathbf{0} & \mathbf{x} \\ \mathbf{x} & \mathbf{0} & \mathbf{x} & \mathbf{x} \\ x & x & x & x \end{bmatrix}$$

For an $n \times n$ symmetric matrix, it is necessary to apply $n(n-1)/2$ rotations ordered in a way to have only non-zero terms along the diagonal, the eigenvalues, with a computation cost $O(n^2)$. For arbitrary matrixes, the rotations could be more complex, still being iterative pair-by-pair (see e.g., Jacobi rotations).

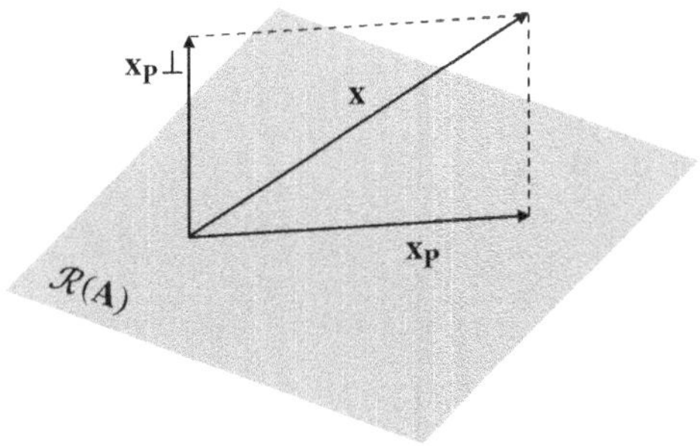

Figure 2.4 Projection onto the span of $\mathbf{A}$.

matrix is denoted as $\mathbf{P}_\mathbf{A}$ (i.e., the projection matrix onto the range space spanned by the columns of $\mathbf{A}$), its complementary as $\mathbf{P}_\mathbf{A}^\perp = \mathbf{I} - \mathbf{P}_\mathbf{A}$. Projection of measurements is a way to "clean" noisy data through a linear transformation that depends on the specific problem at hand. The projection matrix generalizes the concept of data-filtering to any arbitrary linear system of equations describing the relationship between measurements and models. It is quite common to refer to $\mathbf{P}_\mathbf{A}$ as projection onto the *signal space*, and $\mathbf{P}_\mathbf{A}^\perp$ as projection onto the *noise space* for the model $\mathbf{A}$.

2.3.1 Projections and Commercial FM Radio

Commercial (analog) FM radio stations transmit in the 88–108 MHz band; each station is allocated in one channel with a predefined bandwidth of 200 KHz. Telecommunication engineers refer to this as frequency division multiple access since multiple radios access to the same 88–108 MHz band by decoupling one another for a disjoint choice of the 200 KHz channels. To simplify, when two radio stations transmit over two adjacent channels they could interfere one another, and upon reception their effect would be deleterious. The first step carried out by any receiver is band-pass filtering of the spectral content within the channel of interest to reject all other radio stations. This is routinely employed when changing stations: the desired one is selected and others are strongly attenuated. A band-pass filter acts as a projection filter of the received signal onto the subspace spanned by the transmission on the channel of interest. The main difference is that the basis is predefined by the transmitter to be the sinusoids of Fourier transforms, rather than by an arbitrary linear transformation. There are many other examples where projection is employed as model-based filtering; this is just one that happens in everyday life.

2.4 Singular Value Decomposition (SVD) and Subspaces

The matrix that describe a linear model $\mathbf{A} \in \mathbb{C}^{m\times n}$ can be decomposed by SVD as

$$\underset{(m,n)}{\mathbf{A}} = \underset{(m,m)}{\mathbf{U}} \cdot \underset{(m,n)}{\mathbf{\Sigma}} \cdot \underset{(n,n)}{\mathbf{V}^H}$$

where $\mathbf{U}$ and $\mathbf{V}$ are unitary matrixes, and $\mathbf{\Sigma}$ collects the singular values. Let us consider the decomposition for $\text{rank}(\mathbf{A}) = r \leq \min(m, n)$ as

$$\mathbf{A} = \mathbf{U}_1\mathbf{\Sigma}_1\mathbf{V}_1^H$$

On the basis of the SVD properties given in Chapter 1, and from the property of the matrix associated to a linear system $\text{rank}(\mathbf{A}) + \dim \mathcal{N}(\mathbf{A}^H) = m$ (extended here to complex-valued matrixes), it can be shown that:

$$\mathcal{R}(\mathbf{A}) = \mathcal{R}(\mathbf{U}_1)$$
$$\mathcal{N}(\mathbf{A}^H) = \mathcal{R}(\mathbf{U}_2)$$

These are powerful relationships that justify the extensive use of SVD in statistical signal processing. Namely, $\mathbf{U}_1$ and $\mathbf{U}_2$ are the orthonormal basis for $\mathcal{R}(\mathbf{A})$ and its complementary $\mathcal{N}(\mathbf{A}^H)$. Any arbitrary vector $\mathbf{b}$ can be related to the model $\mathbf{A}$, and can be filtered to be in the span of the $\mathbf{A}$ columns:

$$\mathbf{b}_A = \mathbf{P}_\mathbf{A}\mathbf{b} \in \mathcal{R}(\mathbf{A})$$

searching for the projection matrix ($\mathbf{P}_\mathbf{A}$) that minimizes the norm $||\mathbf{b} - \mathbf{P}_\mathbf{A}\mathbf{b}||^2 = ||\mathbf{P}_\mathbf{A}^\perp\mathbf{b}||^2$, we obtain:

$$\mathbf{P}_\mathbf{A} = \mathbf{U}_1\mathbf{U}_1^H$$
$$\mathbf{P}_\mathbf{A}^\perp = \mathbf{I} - \mathbf{P}_\mathbf{A} = \mathbf{U}_2\mathbf{U}_2^H$$

(recall that $\mathbf{U}_1\mathbf{U}_1^H \neq \mathbf{U}_1^H\mathbf{U}_1 = \mathbf{I}_r$, while $\mathbf{U}^H\mathbf{U} = \mathbf{U}\mathbf{U}^H = \mathbf{I}_m$).

All properties hold for the transposed matrix and row-space:

$$\mathcal{R}(\mathbf{A}^H) = \mathcal{R}(\mathbf{V}_1)$$
$$\mathcal{N}(\mathbf{A}) = \mathcal{R}(\mathbf{V}_2)$$

with the same conclusions in terms of the corresponding orthonormal bases. SVD is a powerful tool to extract the orthonormal bases that account for $\mathcal{R}(\mathbf{A})$ and $\mathcal{R}(\mathbf{A}^H)$, or their complementary $\mathcal{N}(\mathbf{A}^H) = \mathbb{C}^m \setminus \mathcal{R}(\mathbf{A})$ and $\mathcal{N}(\mathbf{A}) = \mathbb{C}^n \setminus \mathcal{R}(\mathbf{A}^H)$.

2.4.1 How to Choose the Rank of A?

From SVD, the rank follows from the analysis of the real-valued *singular values* $\{\sigma_k\}$ that, when ordered in decreasing value, are $\sigma_1 \geq \sigma_2 \geq \ldots \geq \sigma_r \geq \sigma_{r+1} = \ldots \sigma_p = 0$ with $r = \text{rank}(\mathbf{A}) \leq p = \min(m, n)$. However, SVD is computed numerically by diagonalization methods based on rotations, similar to Givens rotations previously discussed. Numerical errors due to the finite precision of arithmetic processors unavoidably accumulate during the sequential rotations, so that computed singular values $\hat{\sigma}_k = \sigma_k + \varepsilon_k$ differ by the error ε_k. The influence of error is negligible as long as $|\varepsilon_k|/|\sigma_k| \ll 1$, so for large σ_k, $\hat{\sigma}_k = \sigma_k(1 + \varepsilon_k/\sigma_k) \simeq \sigma_k$. On the other hand, when σ_k is small, or comparable with ε_k, $\hat{\sigma}_k \simeq \varepsilon_k$ and thus singular values are never zero. The inspection of the singular values $\hat{\sigma}_k$ is still a useful tool to evaluate the rank of $\mathbf{A}$ by adopting a threshold, say $\varepsilon_{\max} = \max_k\{|\varepsilon_k|\}$, so that for $\{\hat{\sigma}_k\}$ ordered in decreasing order, $\hat{\sigma}_1 \geq \hat{\sigma}_2 \geq ... \geq \hat{\sigma}_r \geq \varepsilon_{\max} > \sigma_{r+1} > ...\sigma_p$. Since threshold $\varepsilon_{\max}$ cannot be evaluated in closed form (as errors $\{\varepsilon_k\}$ are not known), some heuristics are applied and the threshold $\varepsilon_{\max}$ is chosen to be "large enough" based on common sense (e.g., usually by visual inspection of singular values plots in contexts where rank($\mathbf{A}$) is known, or purposely forced). This problem is widely investigated when using subspace methods for estimation; see Section 16.3.

Recall that the same errors can be found in vectors of bases $\mathbf{U}_1$, $\mathbf{V}_1$ and $\mathbf{U}_2$, $\mathbf{V}_2$, but as for singular values, bases $\mathbf{U}_1$, $\mathbf{V}_1$ are less affected by numerical errors and more reliable than $\mathbf{U}_2$, $\mathbf{V}_2$. Orthogonality is always preserved, anyway.

2.5 QR and Cholesky Factorization

Given a tall matrix $\mathbf{A} \in \mathbb{C}^{m\times n}$ with $m \geq n$, it is possible to apply a sequence of rotations and reduce the matrix to a triangular matrix:

$$\mathbf{Q}^H \underbrace{\begin{bmatrix} x & x & x & x \\ x & x & x & x \\ x & x & x & x \\ x & x & x & x \\ x & x & x & x \\ \vdots & \vdots & \vdots & \vdots \\ x & x & x & x \end{bmatrix}}_{\mathbf{A}} = \underbrace{\begin{bmatrix} x & x & x & x \\ 0 & x & x & x \\ 0 & 0 & x & x \\ 0 & 0 & 0 & x \\ 0 & 0 & 0 & 0 \\ \vdots & \vdots & \vdots & \vdots \\ 0 & 0 & 0 & 0 \end{bmatrix}}_{\mathbf{R}}$$

or equivalently

$$\mathbf{A} = \mathbf{QR} = \mathbf{Q}\begin{bmatrix} \mathbf{R}_1 \\ \mathbf{0} \end{bmatrix} = \begin{bmatrix} \mathbf{Q}_1 & \mathbf{Q}_2 \end{bmatrix}\begin{bmatrix} \mathbf{R}_1 \\ \mathbf{0} \end{bmatrix} = \mathbf{Q}_1\mathbf{R}_1$$

$$= \begin{bmatrix} \mathbf{q}_1 & \mathbf{q}_2 & \cdots & \mathbf{q}_n \end{bmatrix}\begin{bmatrix} r_{11} & r_{12} & \cdots & r_{1n} \\ 0 & r_{22} & \cdots & r_{2n} \\ \vdots & \vdots & \ddots & \vdots \\ 0 & 0 & \cdots & r_{nn} \end{bmatrix}$$

where $\mathbf{R}_1$ is an $n \times n$ upper triangular matrix. $\mathbf{Q}$, $\mathbf{Q}_1$, $\mathbf{Q}_2$ have dimensions respectively of $m \times m$, $m \times n$, $m \times (m-n)$, and their columns are orthonormal. All these properties hold if $\mathbf{A}$ is full-rank (rank$[\mathbf{A}] = n$). When rank$[\mathbf{A}] = r < n$, there are only r independent columns and $\mathbf{Q}_1 \in \mathbb{C}^{m\times r}$, $\mathbf{R} = \mathbf{Q}_1^H\mathbf{A}$ is still upper triangular with dimensions $r \times r$, while $\mathbf{Q}_2^H\mathbf{A} = \mathbf{0}$.

There are several methods for QR decomposition of a matrix, and iterated rotations with progressive elimination of lower elements is one possibility. However, Gram–Schmidt orthogonalization is one of the most intuitive; highlighting that the QR over columns (i.e., $\mathbf{a}_i = \mathbf{A}(:, i)$) is

$$\mathbf{A} = \begin{bmatrix} \mathbf{q}_1 & \mathbf{q}_2 & \cdots & \mathbf{q}_n \end{bmatrix} \begin{bmatrix} \mathbf{q}_1^H\mathbf{a}_1 & \mathbf{q}_1^H\mathbf{a}_2 & \cdots & \mathbf{q}_1^H\mathbf{a}_n \\ 0 & \mathbf{q}_2^H\mathbf{a}_2 & \cdots & \mathbf{q}_2^H\mathbf{a}_n \\ \vdots & \vdots & \ddots & \vdots \\ 0 & 0 & \cdots & \mathbf{q}_n^H\mathbf{a}_n \end{bmatrix},$$

one can sequentially orthogonalize each column of $\mathbf{A}$ with respect the all the previously considered ones, and to normalize the result to have a unitary norm. Recall that projection of the vector $\mathbf{a}$ onto the vector $\mathbf{e}$ is

$$\mathbf{P}_{\mathbf{e}}\mathbf{a} = \underbrace{\left(\frac{\mathbf{e}^H}{\|\mathbf{e}\|}\mathbf{a}\right)}_{\text{inner product}} \times \underbrace{\frac{\mathbf{e}}{\|\mathbf{e}\|}}_{\text{unit-norm}} = \frac{\mathbf{e}^H\mathbf{a}}{\mathbf{e}^H\mathbf{e}}\mathbf{e}$$

Using this operator, it is possible to write the Gram–Schmidt procedure for the columns of $\mathbf{Q}_1$ starting from columns of $\mathbf{A}$ in strictly increasing order:

$$\begin{aligned}
&\mathbf{u}_1 = \mathbf{a}_1, \longrightarrow \mathbf{q}_1 = \frac{\mathbf{u}_1}{\|\mathbf{u}_1\|} \\
&\mathbf{u}_2 = \mathbf{a}_2 - \mathbf{P}_{\mathbf{q}_1}\mathbf{a}_2, \longrightarrow \mathbf{q}_2 = \frac{\mathbf{u}_2}{\|\mathbf{u}_2\|} \\
&\mathbf{u}_3 = \mathbf{a}_3 - \mathbf{P}_{\mathbf{q}_1}\mathbf{a}_3 - \mathbf{P}_{\mathbf{q}_2}\mathbf{a}_3, \longrightarrow \mathbf{q}_3 = \frac{\mathbf{u}_3}{\|\mathbf{u}_3\|} \\
&\vdots \\
&\mathbf{u}_k = \mathbf{a}_k - \sum_{j=1}^{k-1}\mathbf{P}_{\mathbf{q}_j}\mathbf{a}_k, \longrightarrow \mathbf{q}_k = \frac{\mathbf{u}_k}{\|\mathbf{u}_k\|}
\end{aligned}$$

Ordering preserves the structure of $\mathbf{R}$ as illustrated in Figure 2.5, but any other order can be considered if the desire is to have a basis for $\mathcal{R}(\mathbf{A})$. If one column appears as a linear combination of the previously considered ones, it should be skipped in the procedure to avoid unnecessary computations, and size will be $r < n$.

Given a full-rank matrix $\mathbf{A} \in \mathbb{C}^{m\times n}$ with $m > n$ and $\text{rank}(\mathbf{A}) = n$, it is possible to define a new square $(n \times n)$ matrix $\mathbf{A}^H\mathbf{A}$ (referred to as *Gram matrix* of $\mathbf{A}$) that is positive

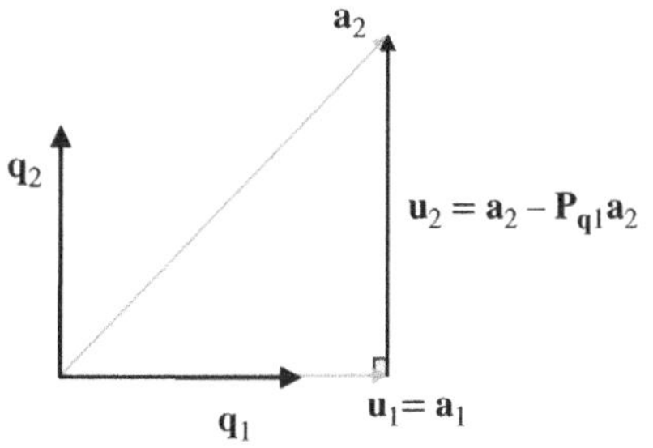

Figure 2.5 Gram-Schmidt procedure for QR decomposition ($n = 2$).

definite (as $\lambda_k(\mathbf{A}^H\mathbf{A}) > 0$, this can be easily shown from the properties of the SVD of $\mathbf{A}$) and can be factorized into a lower/upper triangular matrix:

$$\mathbf{A}^H\mathbf{A} = \mathbf{R}^H\mathbf{Q}^H\mathbf{Q}\mathbf{R} = \mathbf{R}^H\mathbf{R}.$$

In general, for any (Hermitian) symmetric square $\mathbf{X} \succ 0$, there is a unique factorization into an upper/lower triangular matrix, Cholesky factorization

$$\mathbf{X} = \mathbf{R}^H\mathbf{R}$$

with the diagonal elements of $\mathbf{R}$ being strictly positive. Cholesky factorizations can be seen as one way to have a unique definition of the square-root of a matrix, which is known not to be unique (Section 1.2). Cholesky factorization is so important in statistical signal processing that several methods and architectures (see e.g., systolic array processing [5]) have been developed by the scientific community for its efficient numerical computation.

2.6 Power Method for Leading Eigenvectors

Computation of eigenvectors is processing-intensive but there are applications where it is required to compute only one eigenvector associated with the maximum (or minimum) eigenvalue. The power method is an iterative method to compute the eigenvector associated with the largest eigenvalue for a square matrix $\mathbf{A} \in \mathbb{C}^{n\times n}$ with eigenvectors $\mathbf{u}_1, \mathbf{u}_2, ..., \mathbf{u}_n$. Let $\mathbf{q}^{(0)} \in \mathbb{C}^n$ be an arbitrary vector $\mathbf{q}^{(0)} \neq 0$; it can be decomposed into the basis (not known) for the eigenvectors of $\mathbf{A}$ ordered for decreasing eigenvalues ($\lambda_1 \geq \lambda_2 \geq ... \geq \lambda_n$):

$$\mathbf{q}^{(0)} = c_1\mathbf{u}_1 + c_2\mathbf{u}_2 + ... + c_n\mathbf{u}_n$$

multiplying by $\mathbf{A}$:

$$\mathbf{q}^{(1)} = \mathbf{A}\mathbf{q}^{(0)} = \sum_{i=1}^{n} \lambda_i \mathbf{u}_i \mathbf{u}_i^T \mathbf{q}^{(0)} = \sum_{i=1}^{n} c_i \lambda_i \mathbf{u}_i$$

where $c_i = \mathbf{u}_i^T\mathbf{q}^{(0)}$ is the norm of the projection of $\mathbf{q}^{(0)}$ onto $\mathbf{u}_i$. If this is an eigenvector, $\mathbf{A}\mathbf{q}^{(1)} = \gamma\mathbf{q}^{(1)}$ for some scalar value γ, so a normalization is necessary for $\mathbf{q}^{(1)}$. Further multiplications by $\mathbf{A}$ yield

$$\mathbf{q}^{(k)} = \mathbf{A}\mathbf{q}^{(k-1)} = \mathbf{A}^k\mathbf{q}^{(0)} = \sum_{i=1}^{N} c_i^{(k-1)} \lambda_i^k \mathbf{u}_i = c_1\lambda_1^k \left[\mathbf{u}_1 + \sum_{i=2}^{N} \frac{c_i}{c_1}\left(\frac{\lambda_i}{\lambda_1}\right)^k \mathbf{u}_i\right];$$

normalizing wrt the norm of $\mathbf{q}^{(k)}$ gives

$$\frac{\mathbf{q}^{(k)}}{\|\mathbf{q}^{(k)}\|} = \frac{\mathbf{u}_1}{\sqrt{1 + \sum_{i=2}^{N}\left(\frac{c_i}{c_1}\right)^2\left(\frac{\lambda_i}{\lambda_1}\right)^{2k}}} + \frac{\sum_{i=2}^{N}\frac{c_i}{c_1}\left(\frac{\lambda_i}{\lambda_1}\right)^k \mathbf{u}_i}{\sqrt{1 + \sum_{i=2}^{N}\left(\frac{c_i}{c_1}\right)^2\left(\frac{\lambda_i}{\lambda_1}\right)^{2k}}}.$$

It is straightforward to verify that asymptotically (for $k \to \infty$),

$$\frac{\mathbf{q}^{(k)}}{||\mathbf{q}^{(k)}||} \to \mathbf{u}_1$$

provided that $\lambda_2/\lambda_1 < 1$ (and thus $\lambda_1 \neq \lambda_2$) and $c_1 \neq 0$ (the choice of $\mathbf{q}^{(0)}$ should have a non-negligible component onto the subspace of eigenvector $\mathbf{u}_1$). The convergence speed depends on the ratio λ_2/λ_1 as the influence of the second eigenvector vanishes as $(\lambda_2/\lambda_1)^k$.

The eigenvector associated to the smallest eigenvalue can be obtained by the same iterative method, except that iterations are carried out wrt $\mathbf{A}^{-1}$ with the same conditions for the initialization $\mathbf{q}^{(0)}$. This method is referred to as the *inverse power method.*

2.7 Least Squares Solution of Overdetermined Linear Equations

Consider an overdetermined set of linear equations

$$\mathbf{A}\mathbf{x} = \mathbf{b}$$

where $\mathbf{A} \in \mathbb{C}^{m \times n}$ is a strictly tall ($m > n$) and full-rank (rank($\mathbf{A}$) $= n$) matrix. As previously discussed, for several $\mathbf{b}$, the set of equations does not have solution. It is possible to solve the problem in an "approximate" way, searching for the $\mathbf{x} = \mathbf{x}_{LS}$ that minimize the norm of the residual error $\mathbf{A}\mathbf{x} - \mathbf{b}$. This solution is called the least squares (LS) solution of $\mathbf{A}\mathbf{x} = \mathbf{b}$:

$$\mathbf{x}_{LS} = \arg\min_{\mathbf{x}} \left\{ ||\mathbf{A}\mathbf{x} - \mathbf{b}||^2 \right\}$$

In the case that $\mathbf{A} \in \mathbb{R}^{m \times n}$, we have:

$$\mathcal{E}(\mathbf{x}) = ||\mathbf{A}\mathbf{x} - \mathbf{b}||^2 = (\mathbf{A}\mathbf{x} - \mathbf{b})^T(\mathbf{A}\mathbf{x} - \mathbf{b}) = \mathbf{x}^T\mathbf{A}^T\mathbf{A}\mathbf{x} - 2\mathbf{b}^T\mathbf{A}\mathbf{x} + \mathbf{b}^T\mathbf{b}$$

This is a quadratic form as $\mathbf{A}^T\mathbf{A} = (\mathbf{A}^T\mathbf{A})^T$ is symmetric, and all derivations in Section 1.6 can be easily adapted. The solution is

$$\mathbf{x}_{LS} = \arg\min_{\mathbf{x}} \left\{ ||\mathbf{A}\mathbf{x} - \mathbf{b}||^2 \right\} = (\mathbf{A}^T\mathbf{A})^{-1}\mathbf{A}^T\mathbf{b} = \mathbf{A}^{\dagger}\mathbf{b}$$

where

$$\mathbf{A}^{\dagger} = (\mathbf{A}^T\mathbf{A})^{-1}\mathbf{A}^T$$

is the *pseudoinverse* of $\mathbf{A}$.

In the more general case of $\mathbf{A} \in \mathbb{C}^{m \times n}$, we have:

$$\mathcal{E}(\mathbf{x}) = ||\mathbf{Ax} - \mathbf{b}||^2 = (\mathbf{Ax} - \mathbf{b})^H(\mathbf{Ax} - \mathbf{b}) = \mathbf{x}^H\mathbf{A}^H\mathbf{Ax} - \mathbf{x}^H\mathbf{A}^H\mathbf{b} - \mathbf{b}^H\mathbf{Ax} + \mathbf{b}^H\mathbf{b}$$

By setting the gradient of $\mathcal{E}(\mathbf{x})$ to zero (see derivative rules for the case $\mathbf{x} \in \mathbb{C}^n$ in Section 1.4):

$$\frac{\partial \mathcal{E}(\mathbf{x})}{\partial \mathbf{x}} = ((\mathbf{A}^H\mathbf{A})^H\mathbf{x})^* - (\mathbf{A}^H\mathbf{b})^* = 0$$

Applying the conjugation operators to both sides of the equality:

$$\begin{aligned}(((\mathbf{A}^H\mathbf{A})^H\mathbf{x})^* - (\mathbf{A}^H\mathbf{b})^*)^* &= (0)^* \\ (\mathbf{A}^H\mathbf{A})^H\mathbf{x} - \mathbf{A}^H\mathbf{b} &= 0 \\ \mathbf{A}^H\mathbf{Ax} - \mathbf{A}^H\mathbf{b} &= 0\end{aligned}$$

we get

$$\mathbf{x}_{LS} = \arg\min_{\mathbf{x}} \left\{||\mathbf{Ax} - \mathbf{b}||^2\right\} = (\mathbf{A}^H\mathbf{A})^{-1}\mathbf{A}^H\mathbf{b} = \mathbf{A}^\dagger\mathbf{b}$$

that degenerates into the expression of the pseudoinverse for real matrixes.

The usage of the pseudoinverse is quite common, so it is worth highlighting some basic properties. From SVD, $\mathbf{A} = \mathbf{U}_1\mathbf{\Sigma}_1\mathbf{V}_1^H$ and the iterated usage of the orthogonality conditions yields:

$$\mathbf{A}^\dagger = (\mathbf{A}^H\mathbf{A})^{-1}\mathbf{A}^H = \left(\mathbf{V}_1^2\mathbf{\Sigma}_1\mathbf{V}_1^H\right)^{-1}\mathbf{V}_1\mathbf{\Sigma}_1\mathbf{U}_1^H = \left(\mathbf{V}_1\mathbf{\Sigma}_1^{-2}\mathbf{V}_1^H\right)\mathbf{V}_1\mathbf{\Sigma}_1\mathbf{U}_1^H = \mathbf{V}_1\mathbf{\Sigma}_1^{-1}\mathbf{U}_1^H$$

The pseudoinverse acts only on $\mathcal{R}(\mathbf{A})$. Needless to say that if $\mathbf{A} \in \mathbb{R}^{n \times n}$ and full-rank, $\mathbf{A}^\dagger = \mathbf{A}^{-1}$; this clarifies the usage of the pseudoinverse as a "generalization" of the inverse of a matrix with arbitrary rank.

The expression of the LS solution can be arranged as:

$$\mathbf{x}_{LS} = \mathbf{A}^\dagger\mathbf{b} = \left(\mathbf{V}_1\mathbf{\Sigma}_1^{-1}\mathbf{U}_1^H\right)\left(\mathbf{U}_1\mathbf{U}_1^H\right)\mathbf{b} = \left(\mathbf{V}_1\mathbf{\Sigma}_1^{-1}\mathbf{U}_1^H\right)\mathbf{b}_A,$$

where $\mathbf{b}_A = \mathbf{P}_\mathbf{A}\mathbf{b}$ is the projection of $\mathbf{b}$ on $\mathcal{R}(\mathbf{A})$, and $\mathbf{b}_\perp = \left(\mathbf{I} - \mathbf{P}_\mathbf{A}\right)\mathbf{b} = \mathbf{P}_\mathbf{A}^\perp\mathbf{b} \in \mathcal{N}(\mathbf{A}^H)$. The optimization metric $\mathcal{E}(\boldsymbol{x})$ at LS solution is

$$\mathcal{E}(\mathbf{x}_{LS}) = \|\mathbf{P}_\mathbf{A}\mathbf{b} - \mathbf{b}\|^2 = \left\|\mathbf{P}_\mathbf{A}^\perp\mathbf{b}\right\|^2 = \mathbf{b}^H\mathbf{P}_\mathbf{A}^\perp\mathbf{b} = \|\mathbf{b}_\perp\|^2$$

The norm of $\mathbf{b}_\perp$ is thus minimized as shown in Figure 2.6.

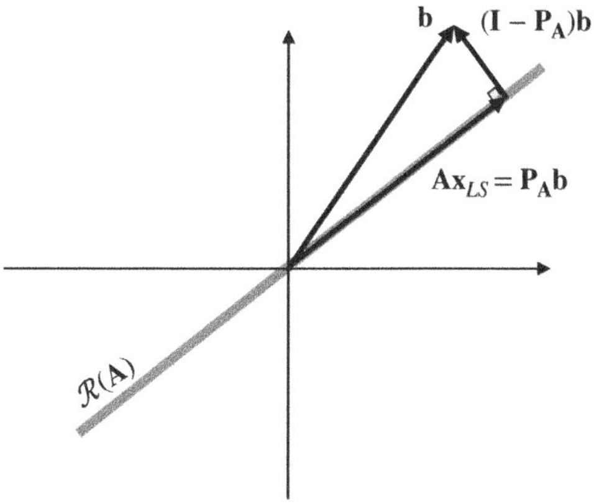

Figure 2.6 Least squares solution of linear system.

2.8 Efficient Implementation of the LS Solution

In the least squares solution of $\mathbf{Ax} = \mathbf{b}$, the inversion of $\mathbf{A}^H\mathbf{A}$ is the operation with the highest computational cost: $\mathcal{O}(n^3)$. It is possible to write the equation to solve as:

$$\underbrace{\mathbf{A}^H\mathbf{A}}_{\mathbf{C}} \cdot \mathbf{x}_{LS} = \underbrace{\mathbf{A}^H\mathbf{b}}_{\mathbf{g}}$$

Considering the Cholesky factorization of Gram matrix $\mathbf{C}$

$$\mathbf{C} = \mathbf{R}^H\mathbf{R}$$

it is possible to write

$$\mathbf{R}^H \cdot \underbrace{\mathbf{R}\mathbf{x}_{LS}}_{\mathbf{y}} = \mathbf{g}$$

so that for the first system of equations it is (forward substitutions)

$$\mathbf{R}^H\mathbf{y} = \mathbf{g}$$

$$\begin{bmatrix} r_{11} & 0 & \cdots & 0 \\ r_{12}^* & r_{22} & \cdots & 0 \\ \vdots & \vdots & \ddots & \vdots \\ r_{1n}^* & r_{2n}^* & \cdots & r_{nn} \end{bmatrix} \begin{bmatrix} y_1 \\ y_2 \\ \vdots \\ y_n \end{bmatrix} = \begin{bmatrix} g_1 \\ g_2 \\ \vdots \\ g_n \end{bmatrix}$$

$$\begin{cases} y_1 = g_1/r_{11} \\ y_2 = (g_2 - r_{12}^* y_1)/r_{22} \\ \vdots \\ y_n = (g_k - \sum_{j=1}^{k-1} r_{jk}^* y_j)/r_{kk} \text{ per } k = 3, \ldots, n \end{cases}$$

The cost is $1+2+3+...+n = n(n+1)/2 \rightarrow \mathcal{O}\{n^2/2\}$ complex multiplications. Next, the second system of equations must be solved (backward substitutions)

$$\mathbf{R}\mathbf{x}_{LS} = \mathbf{y}$$

with a cost still $\mathcal{O}\{n^2/2\}$ so that the global cost is $\mathcal{O}\{n^2\}$—lower than the direct inversion of $\mathbf{A}^H\mathbf{A}$. The cost of Cholesky factorization is of the same order ($\mathcal{O}\{n^2\}$), thus overall it is $\mathcal{O}\{2n^2\}$.

Alternatively, the most commonly adopted method in efficient systems is based on QR factorization (Section 2.5) of $\mathbf{A}$:

$$\mathbf{A} = \mathbf{Q}\mathbf{R} = \begin{bmatrix} \mathbf{Q}_1 & \mathbf{Q}_2 \end{bmatrix} \begin{bmatrix} \mathbf{R}_1 \\ \mathbf{0} \end{bmatrix}$$

where $\mathbf{Q}_1$, $\mathbf{Q}_2$ and $\mathbf{Q}$ are orthonormal matrixes while $\mathbf{R}_1$ is a $n \times n$ upper triangular matrix. Since any rotation of an orthonormal matrix does not change the norm, it is possible to write:

$$\begin{aligned}
||\mathbf{A}\mathbf{x} - \mathbf{b}||^2 &= \left\| \begin{bmatrix} \mathbf{Q}_1 & \mathbf{Q}_2 \end{bmatrix} \begin{bmatrix} \mathbf{R}_1 \\ \mathbf{0} \end{bmatrix} \mathbf{x} - \mathbf{b} \right\|^2 \\
&= \left\| \begin{bmatrix} \mathbf{Q}_1 & \mathbf{Q}_2 \end{bmatrix}^H \begin{bmatrix} \mathbf{Q}_1 & \mathbf{Q}_2 \end{bmatrix} \begin{bmatrix} \mathbf{R}_1 \\ \mathbf{0} \end{bmatrix} \mathbf{x} - \begin{bmatrix} \mathbf{Q}_1 & \mathbf{Q}_2 \end{bmatrix}^H \mathbf{b} \right\|^2 \\
&= \left\| \begin{bmatrix} \mathbf{R}_1\mathbf{x} - \mathbf{Q}_1^H\mathbf{b} \\ -\mathbf{Q}_2^H\mathbf{b} \end{bmatrix} \right\|^2 \\
&= \left\| \mathbf{R}_1\mathbf{x} - \mathbf{Q}_1^H\mathbf{b} \right\|^2 + \left\| \mathbf{Q}_2^H\mathbf{b} \right\|^2
\end{aligned}$$

This is clearly minimized by solving for

$$\mathbf{R}_1\mathbf{x} = \mathbf{Q}_1^H\mathbf{b}$$

with a complexity $\mathcal{O}(n^2/2)$ due to the fact that $\mathbf{R}_1$ is upper triangular.

2.9 Iterative Methods

Solving for $\mathbf{A}\mathbf{x} = \mathbf{b}$ with $\mathbf{A} \in \mathbb{C}^{n \times n}$ can be employed by splitting

$$\mathbf{A} = \mathbf{M} - \mathbf{N}$$

and solving iteratively

$$\mathbf{M}\mathbf{x}^{(k)} = \mathbf{N}\mathbf{x}^{(k-1)} + \mathbf{b}$$

with

$$\mathbf{x}^{(k)} = \mathbf{M}^{-1}\mathbf{N}\mathbf{x}^{(k-1)} + \mathbf{M}^{-1}\mathbf{b}$$

for some starting vector $\mathbf{x}^{(0)}$. To be of any use, the inversion of $\mathbf{M}$ should be somewhat easy or efficient. Straightforward choices are for $\mathbf{M}$ diagonal (Jacobi method: $\mathbf{M}^{-1}$ is derived elementwise) or lower triangular (Gauss–Seidel method: $\mathbf{M}^{-1}$ is obtained by ordered recursion). Let $\mathbf{x}_o = \mathbf{A}^{-1}\mathbf{b}$ be the solution; for convergence analysis the iterations can be written as

$$\mathbf{M}(\mathbf{x}^{(k)} - \mathbf{x}_o) = \mathbf{N}(\mathbf{x}^{(k-1)} - \mathbf{x}_o)$$

so that for error $\mathbf{e}^{(k)} = (\mathbf{x}^{(k)} - \mathbf{x}_o)$ wrt the solution, the iterations become

$$\mathbf{e}^{(k)} = \mathbf{M}^{-1}\mathbf{N}\mathbf{e}^{(k-1)} = \ldots = (\mathbf{M}^{-1}\mathbf{N})^k\mathbf{e}^{(0)}.$$

The convergence for $k \to \infty$ should be $\mathbf{e}^{(k)} \to 0$ for some norm. A simple proof is from the eigenvalue/eigenvector of $\mathbf{M}^{-1}\mathbf{N}$ as

$$\mathbf{M}^{-1}\mathbf{N} = \mathbf{Q}\Lambda\mathbf{Q}^H$$

so that

$$(\mathbf{M}^{-1}\mathbf{N})^k = \mathbf{Q}\Lambda^k\mathbf{Q}^H$$

and thus to guarantee convergence,

$$|\lambda_{\max}(\mathbf{M}^{-1}\mathbf{N})| < 1$$

or equivalently the spectral radius of $\mathbf{M}^{-1}\mathbf{N}$ should be strictly limited to 1.

Iterative methods for tall matrix $\mathbf{A} \in \mathbb{C}^{m\times n}$ $(m > n)$ can be similarly defined by solving for the LS problem

$$\underbrace{\mathbf{A}^H\mathbf{A}}_{\mathbf{C}} \cdot \mathbf{x}_{LS} = \underbrace{\mathbf{A}^H\mathbf{b}}_{\mathbf{g}}$$

and this time splitting the compound (square) Gram matrix $\mathbf{C} = \mathbf{M} - \mathbf{N}$. Convergence conditions are the same.

3

Random Variables in Brief

Intuitively, the probability of an outcome can be associated with its frequency of occurrence. Making an experiment with outcomes A,B,C,... and repeating the experiment N times with number of occurrences n_A, n_B, n_C, .., their relative frequencies of occurrence are given by the ratios n_A/N, n_B/N, n_C/N, ... When $N \to \infty$ these ratios approach the occurrence probability of the outcomes:

$$\Pr(A) = \lim_{N\to\infty} \frac{n_A}{N}, \ \Pr(B) = \lim_{N\to\infty} \frac{n_B}{N}, \ \Pr(C) = \lim_{N\to\infty} \frac{n_C}{N}, \ \text{etc...}$$

The set of all possible outcomes from experiments are called sample space. Outcomes can be mutually exclusive (if one occurs, another cannot) and probability is $\Pr(A \text{ or } B) = \Pr(A) + \Pr(B)$. On the other hand, two events can become a *joint* event with probability $\Pr(A \text{ and } B) = \Pr(A,B)$. In the case that two events are statistical independent, the occurrence of one is independent of the other, and thus $\Pr(A \text{ and } B) = \Pr(A)\Pr(B)$.

Starting from this notion of probability, this chapter provides some essential fundamentals of random variables, probability density for continuous and discrete variables, and an in-depth discussion of Gaussian random variables; other distributions will be detailed when necessary.

3.1 Probability Density Function (pdf), Moments, and Other Useful Properties

The *probability density function* (pdf) $p_x(x)$ of a real-valued random variable (rv) x is a non-negative function such that the probability

$$\Pr[a \le x \le b] = \int_a^b p_x(\alpha)d\alpha$$

Statistical Signal Processing in Engineering, First Edition. Umberto Spagnolini.

Companion website: www.wiley.com/go/spagnolini/signalprocessing

(the rv x is indicated as a subscript in a pdf[1]). The *cumulative density function* (cdf) is the probability of a value up to x_o:

$$\Pr[x \leq x_o] = \int_{-\infty}^{x_o} p_x(\alpha)d\alpha \leq 1$$

and the *complementary cdf* is

$$\Pr[x \geq x_o] = 1 - \Pr[x \leq x_o]$$

The *probability mass function* (pmf) gives the probability for a discrete rv that belongs to a countable set $\mathcal{X} = \{x_1, x_2, ..., x_M\}$:

$$p_k = \Pr[x = x_k]$$

so that $\sum_k p_k = 1$. To simplify most of the derivations and properties without any loss of generality, the pmf can be described in terms of pdf using Dirac delta functions

$$p_x(x) = \sum_k p_k \delta(x - x_k)$$

as $\Pr[x_k - \epsilon/2 \leq x \leq x_k + \epsilon/2] = p_k$ for ϵ arbitrary small.

The *expected value* for any transformation of the rv x is defined as

$$\mathbb{E}_x[g(x)] = \mathbb{E}[g(x)] = \int g(\alpha)p_x(\alpha)d\alpha$$

sometimes referred to as expectation of $g(x)$ wrt x. Whenever it is necessary to highlight that the expectation as operator is wrt a specific rv x, this is indicated by subscript thus: $\mathbb{E}_x[.]$; in complex contexts this notation helps to avoid confusion about what is computed with $\mathbb{E}[.]$. *Moments* are the expectation of powers of the rv (for $g(x) = x^k$)

$$\mathbb{E}[x^k] = \int \alpha^k p_x(\alpha)d\alpha \quad \text{(kth moment)}$$

The *kth central moments* are the moments about the mean:

$$\mu_k = \mathbb{E}[(x - \mathbb{E}[x])^k]$$

with properties $\mu_0 = 1$ and $\mu_1 = 0$, while the 2nd central moment μ_2 is called the variance.

The *characteristic function* is the Fourier transform of the pdf:

$$\phi_x(\omega) = \mathbb{E}[e^{-j\omega x}] = \int p_x(\alpha)e^{-j\omega\alpha}d\alpha.$$

1 Notice that $p_x(.)$ and $p_y(.)$ are two different functions; the subscript is the way to differentiate the two, and to highlight the rv each is referred to. If there is no ambiguity and it is clear from the context, this notation is redundant and subscripts are omitted.

The characteristic function is used when it is more convenient to elaborate in the Fourier transformed domain. For instance, the moments:

$$\mathbb{E}[x^k] = (-j)^{-k} \left. \frac{d^k \phi_x(\omega)}{d\omega^k} \right|_{\omega=0}$$

or when it is necessary to evaluate the pdf of the sum of N statistically independent rvs:

$$z = x_1 + x_2 + x_3 + \ldots \longrightarrow p_z(z = z_0) = p_{x_1}(\alpha) * p_{x_2}(\alpha) * p_{x_3}(\alpha) * \ldots \Big|_{\alpha=z_0}$$

here it is much simpler to use the characteristic function as convolutions become products:

$$\phi_z(\omega) = \phi_{x_1}(\omega)\phi_{x_2}(\omega)\phi_{x_3}(\omega)\ldots$$

Two (or more) rvs are characterized by the joint pdf $p_{x,y}(x, y)$ as an extension of the definition above. The joint pdf is a function of two (or more) variables such that

$$\begin{aligned} &p_{x,y}(x, y) \geq 0 \\ &\int\int p_{x,y}(x, y)dxdy = 1 \quad \text{(normalization)} \end{aligned}$$

The pdf for x or y (marginal pdf) from $p_{x,y}(x, y)$ is obtained from marginalization:

$$p_x(x) = \int p_{x,y}(x, y)dy$$

$$p_y(y) = \int p_{x,y}(x, y)dx$$

where one of the two rvs is kept constant. Notice that $\int p_{x,y}(x, y)dy$ is the *projection* of the multivariate (here 2D) function $p_{x,y}(x, y)$ along the direction y (similarly for $\int p_{x,y}(x, y)dx$ along the direction x). Two rvs are *statistical independent* when the joint pdf

$$p_{x,y}(x, y) = p_x(x)p_y(y)$$

is separable into the product of the marginal pdfs.

Conditional pdf is when one (or more) of the two rvs is kept constant (or it is not-random for the specific experiment, where it is $y = y_o$) and it becomes a pdf depending on the other (*Bayes' rule*):

$$p_{x|y}(x|y = y_o) = \frac{p_{x,y}(x, y_o)}{p_y(y_o)} = \frac{p_{x,y}(x, y_o)}{\int f_{x,y}(x, y_o)dx}$$

This is the *sectioning* of the joint pdf $p_{x,y}(x, y)$ for $y = y_o$, where the denominator is the normalization of the resulting sectioning to guarantee the normalization of the resulting pdf.

The characteristic function for a multivariate pdf is the multidimensional Fourier transform; for $p_{x,y}(x,y)$ it is

$$\phi_{x,y}(\omega_x,\omega_y)=\mathbb{E}_{x,y}[\exp(-j(\omega_x x+\omega_y y))]=\iint p_{x,y}(x,y)\exp(-j(\omega_x x+\omega_y y))dxdy$$

The characteristic function $\phi(\omega_x,\omega_y)$ along the axes (for $\omega_x=0$ and $\omega_y=0$) coincides with the characteristic function of the rvs y and x respectively:

$$\phi_{x,y}(\omega_x,0)=\iint p_{x,y}(x,y)\exp(-j\omega_x x)dxdy=\phi_x(\omega_x)$$
$$\phi_{x,y}(0,\omega_y)=\iint p_{x,y}(x,y)\exp(-j\omega_y y)dxdy=\phi_y(\omega_y)$$

Moments follow from the characteristic function by partial derivatives, for instance:

$$\left.\frac{\partial\phi_{x,y}(\omega_x,\omega_y)}{\partial\omega_x}\right|_{\omega_x=\omega_y=0}=\left.\frac{\partial}{\partial\omega_x}\iint p_{x,y}(x,y)\exp(-j(\omega_x x+\omega_y y))dxdy\right|_{\omega_x=\omega_y=0}$$
$$=-j\iint xp_{x,y}(x,y)dxdy=-j\int xp_x(x)dx=-j\mathbb{E}[x]$$

and so on for other partial derivatives.

The conditional mean depends on the conditional pdf as follows:

$$\mathbb{E}_x[x|y=y_o]=\int\alpha p_{x|y}(\alpha|y=y_o)d\alpha,$$

for the choice $y=y_o$. Since y is a rv,

$$\mathbb{E}_y[\mathbb{E}_x[x|y=y_o]]=\int\left[\int\alpha p_{x|y}(\alpha|y=y_o)d\alpha\right]p_y(y_o)dy_o=\mathbb{E}_{x,y}[x].$$

In general, for any function $g(x,y)$ of two (or more) rvs, its expectation can be partitioned into a chain of conditional expectations

$$\mathbb{E}_{x,y}[g(x,y)]=\mathbb{E}_y[\mathbb{E}_x[g(x,y)|y]]=\mathbb{E}_x[\mathbb{E}_y[g(x,y)|x]]$$

where the subscript highlights the rv in the expectation computation. It is quite common to use the conditional expectations chaining property whenever it is necessary to evaluate expectations, and these computations are easier (or more manageable) than the straight computation of $\mathbb{E}_{x,y}[g(x,y)]$. Furthermore, this property can be used when part of the computation can be evaluated analytically, and the remainder by numerical simulations.

A simple example can clarify. Let

$$z=\sum_{k=1}^{N}x_k$$

be the random sum (i.e., the number N is a rv itself) of identically distributed rvs. The function here is $z = g(x_1, .., x_N, N)$, and thus from the chain rule of conditional expectation

$$\mathbb{E}_{x,N}[z] = \mathbb{E}_N\left[\mathbb{E}_x\left[\sum_{k=1}^{N} x_k | N\right]\right] = \mathbb{E}_N[N \cdot \mathbb{E}_x[x]] = \mathbb{E}_N[N] \cdot \mathbb{E}_x[x]$$

makes the computation of the expectation trivial and straightforward.

3.2 Convexity and Jensen Inequality

Let $g(\theta)$ be any convex transformation (second derivative is $\ddot{g}(\theta) \geq 0$), then

$$g(\mathbb{E}[x]) \leq \mathbb{E}[g(x)].$$

In multidimensional space, the convexity needs a better definition. Let θ be within a line in $[\mathbf{a}, \mathbf{b}]$ arbitrarily chosen as $\theta = \mathbf{a}\gamma + \mathbf{b}(1-\gamma)$ with $\gamma \in [0,1]$; a convex transformation is one that guarantees

$$g(\mathbf{a}\gamma + \mathbf{b}(1-\gamma)) \leq \gamma g(\mathbf{a}) + (1-\gamma)g(\mathbf{b})\}.$$

The choice of $\mathbf{a}$ and $\mathbf{b}$ is arbitrary provided that $g(\mathbf{a})$ and $g(\mathbf{b})$ exist.

A very useful generalization holds for the nth power of any positive definite random matrix $\mathbf{X}$ such that [8]

$$\mathbb{E}[\mathbf{X}^n] \succeq (\mathbb{E}[\mathbf{X}])^n$$

The inequality for matrix inversion

$$\mathbb{E}[\mathbf{X}^{-1}] \succeq (\mathbb{E}[\mathbf{X}])^{-1}$$

is useful in contexts where the expectation $\mathbb{E}[\mathbf{X}]$ is more manageable than the expectation after the (cumbersome) matrix inversion $\mathbb{E}[\mathbf{X}^{-1}]$.

3.3 Uncorrelatedness and Statistical Independence

Two zero-mean random variables x, y are statistically uncorrelated if

$$\mathbb{E}[xy] = 0$$

while statistical independence implies that the joint pdf is separable: $p_{x,y}(x,y) = p_x(x)p_y(y)$. Uncorrelation is a necessary condition for independence as $\mathbb{E}[xy] = \mathbb{E}[x]\mathbb{E}[y] = 0$, but it is far more common. For a Gaussian pdf, the uncorrelation and the independence are equivalent, but this is a (useful) exception. An example can help to clarify.

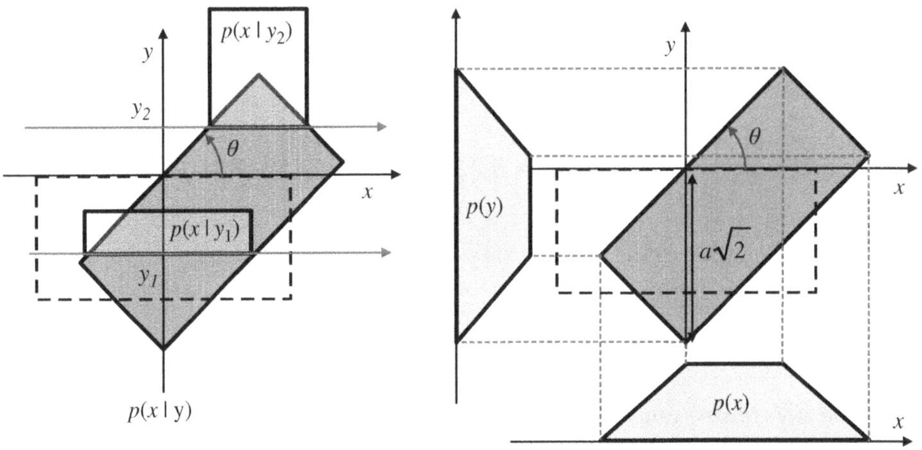

Figure 3.1 Conditional and marginal pdf from $p(x,y)$.

Let (α,β) be independent rvs with uniform pdf: $\alpha \sim \mathcal{U}(-a,a)$ and $\beta \sim \mathcal{U}(-a,0)$ so that $p_{\alpha,\beta}(\alpha,\beta) = p_\alpha(\alpha)p_\beta(\beta)$. These are linearly combined by a rotation matrix

$$\begin{bmatrix} x \\ y \end{bmatrix} = \underbrace{\begin{bmatrix} \cos\theta & -\sin\theta \\ \sin\theta & \cos\theta \end{bmatrix}}_{\mathbf{Q}_{-\theta}} \begin{bmatrix} \alpha \\ \beta \end{bmatrix}$$

along the angle $-\theta$. The marginal pdfs for any arbitrary θ are the projection of the joint pdf with a symmetric trapezoid behavior (see Figure 3.1) that degenerates into a uniform pdf for $\theta = 0, \pm\pi/2, \pi$ where the rvs are independent. Notice that the marginal pdfs are not zero-mean except for $\theta = 0, \pi$ where the pdf $p_x(x)$ (similarly for $p_y(y)$) is symmetric, while for $\mathbb{E}[y; \theta = 0] = -a/2$ and $\mathbb{E}[y; \theta = \pi] = a/2$ (the roles of x and y are symmetric for $\theta = \pm\pi/2$). The conditional pdf, say $p_{x|y}(x|y = y_0)$, can be obtained graphically by slicing the joint pdf in $y = y_o$ provided that normalization is guaranteed (unit-area)—see the figure for different values of $y = y_1, y_2$. The mean value for $\theta = \pi/4$ as in the figure can be derived from the marginal pdf $p_x(x)$ as $\mathbb{E}_x[x|\theta = \pi/4] = a\sqrt{2}/4$, or alternatively from the analytic relationship $\mathbb{E}_x[x|\theta = \pi/4] = \left(\mathbb{E}_\alpha[\alpha] - \mathbb{E}_\beta[\beta]\right)/\sqrt{2}$. The conditional mean depends on the conditional pdf, that in turn it depends on the value $y = y_o$:

$$\mathbb{E}[x|y_o, \theta = \pi/4] = \begin{cases} 0 \text{ for } -2a/\sqrt{2} \leq y_o \leq -a/\sqrt{2} \\ y + a/\sqrt{2} \text{ for } -a/\sqrt{2} \leq y_o \leq 0 \\ a/\sqrt{2} \text{ for } 0 \leq y_o \leq a/\sqrt{2} \end{cases}$$

Another example can be illustrative enough to prove that appropriate linear transformations (i.e., rotations or more generally orthonormal transformations) can manipulate rvs to be uncorrelated from being independent (and vice-versa). Let $\alpha \sim \mathcal{U}(-a,a)$ and $\beta \sim \mathcal{U}(-a,a)$ be two independent rvs; the correlation of the rvs x,y after rotation is

$\mathbb{E}[xy] = \mathbb{E}[\alpha^2 - \beta^2]\cos\theta\sin\theta + \mathbb{E}[\alpha\beta](\cos^2\theta - \sin^2\theta) = 0$, independent of any rotation θ. Any orthonormal linear transformation of zero-mean rvs preserves uncorrelation, but not independence. Conversely, rvs that are uncorrelated could be independent provided that one finds an appropriate orthonormal transformation that rotates these rvs. The relationship between uncorrelatedness and statistical independence is the basis of a class of statistical signal processing methods called independent component analysis (ICA), which is not covered here (see e.g., [29]).

3.4 Real-Valued Gaussian Random Variables

A Gaussian pdf is characterized by

$$p(x) = G(x; \mu_x, \sigma_x^2) = \frac{1}{\sqrt{2\pi\sigma_x^2}} \exp\left(-\frac{(x-\mu_x)^2}{2\sigma_x^2}\right)$$

with a "bell-shape" around the mean $\mu_x = \mathbb{E}[x]$ and variance $\sigma_x^2 = \mathbb{E}[(x-\mu_x)^2]$ that controls the "bell-width." A common notation for a Gaussian pdf is $G(x; \mu_x, \sigma_x^2)$.

The joint pdf of (x, y) is the bivariate

$$p(x,y) = \frac{1}{2\pi\sigma_x\sigma_y\sqrt{1-\rho^2}} \exp\left\{-\frac{1}{2(1-\rho^2)}\left(\frac{(x-\mu_x)^2}{\sigma_x^2} - \frac{2\rho(x-\mu_x)(y-\mu_y)}{\sigma_x\sigma_y} + \frac{(y-\mu_y)^2}{\sigma_y^2}\right)\right\}$$

where

$$\rho = \frac{\mathbb{E}[(x-\mu_x)(y-\mu_y)]}{\sigma_x\sigma_y} \in [-1, 1]$$

is the correlation between the rv $(x-\mu_x)$ and $(y-\mu_y)$. When $\rho = 0$, the rvs are uncorrelated and, since $p(x,y) = p_x(x)p_y(y)$, statistical independent. Figure 3.2 illustrates the 2-D function representing the joint pdf (here with $\rho = .6$) with the slicing for $p(x, y = 0)$ and $p(x = 0, y)$; the contouring ellipsis contains the sample data (dots) with probability $1 - \exp(-9/2) = 98.89\%$ (see discussion below).

Multivariate Gaussian

The multivariate Gaussian distribution is the joint pdf for a set of rvs $\{x_1, x_2, \cdots, x_N\}$ with Gaussian pdfs, and arbitrary correlation. Given $\mathbf{x} = [x_1, x_2, \cdots, x_N]^T$ a vector of N real Gaussian rvs, the joint pdf is

$$p(\mathbf{x}) = \frac{1}{(2\pi)^{N/2}|\mathbf{C}_{xx}|^{1/2}} \exp\left(-\frac{1}{2}(\mathbf{x}-\boldsymbol{\mu}_x)^T\mathbf{C}_{xx}^{-1}(\mathbf{x}-\boldsymbol{\mu}_x)\right) \tag{3.1}$$

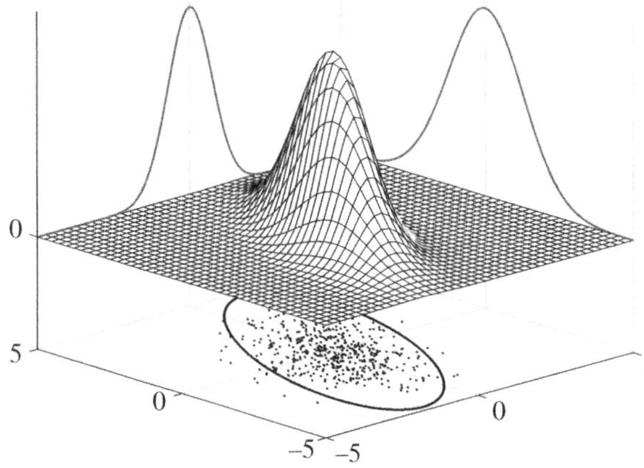

Figure 3.2 Joint Gaussian pdf with correlation $\rho = 0.6$, and sample data (bottom).

where

$$\boldsymbol{\mu}_x = \mathbb{E}[\mathbf{x}] = [\mu_{x_1}, \mu_{x2}, ..., \mu_{x_N}]^T$$

collects the means of the random variables, and $\mathbf{C}_{xx}$ is the covariance matrix of the random vector (i.e., the extension of the variance of a single random variable) defined as:

$$\mathbf{C}_{xx} = \text{cov}\{\mathbf{x}\} = \mathbb{E}[(\mathbf{x} - \boldsymbol{\mu}_x)(\mathbf{x} - \boldsymbol{\mu}_x)^T]$$

To ease the notation, the covariance of the rvs $\mathbf{x}$ is also indicated in simple subscript $\mathbf{C}_x$. Recall that for $N = 2$,

$$\mathbf{C}_{xx} = \begin{bmatrix} \mathbb{E}(x_1^2) - \mu_{x_1}^2 & \mathbb{E}(x_1 x_2) - \mu_{x_1}\mu_{x2} \\ \mathbb{E}(x_2 x_1) - \mu_{x2}\mu_{x_1} & \mathbb{E}(x_2^2) - \mu_{x2}^2 \end{bmatrix} = \begin{bmatrix} \sigma_{x_1}^2 & \rho\sigma_{x_1}\sigma_{x_2} \\ \rho\sigma_{x_1}\sigma_{x_2} & \sigma_{x_2}^2 \end{bmatrix},$$

that is the bivariate case. In a compact form it is customary to use the notation

$$\mathbf{x} \sim \mathcal{N}(\boldsymbol{\mu}_x, \mathbf{C}_{xx})$$

to denote that the rv $\mathbf{x}$ is multivariate Gaussian with mean value $\boldsymbol{\mu}_x$ and covariance $\mathbf{C}_{xx}$. The pdf (3.1) can be compactly described using the notation:

$$G(\mathbf{x}; \boldsymbol{\mu}_x, \mathbf{C}_{xx}) = |2\pi\mathbf{C}_{xx}|^{-1/2} \exp\left(-\frac{1}{2}(\mathbf{x} - \boldsymbol{\mu}_x)^T \mathbf{C}_{xx}^{-1}(\mathbf{x} - \boldsymbol{\mu}_x)\right)$$

Linear Transformation of Multivariate Gaussian

A set of Gaussian rvs $\mathbf{x}$ can be linearly transformed

$$\mathbf{y} = \mathbf{A}\mathbf{x} + \mathbf{b},$$

the resulting rvs are Gaussian

$$\mathbf{y} \sim \mathcal{N}(\boldsymbol{\mu}_y, \mathbf{C}_{yy}).$$

The parameters of the pdf are evaluated by the use of expectations:

$$\boldsymbol{\mu}_y = \mathbb{E}[\mathbf{y}] = \mathbf{A}\mathbb{E}[\mathbf{x}] + \mathbf{b} = \mathbf{A}\boldsymbol{\mu}_x + \mathbf{b}$$
$$\mathbf{C}_{yy} = \mathbb{E}[\mathbf{y}\mathbf{y}^T] - \boldsymbol{\mu}_x\boldsymbol{\mu}_{xx}^T = \mathbf{A}\mathbf{C}_{xx}\mathbf{A}^T$$

Dispersion of Multivariate Gaussian

The spread of the samples of a rv is called its dispersion (or variability or scatter) and denotes how the samples are compact or dispersed, typically from their histogram representation. There are several tools to measure around the mean such as the standard deviation, but for a multivariate Gaussian, the dispersion should be measured as

$$\Pr[\mathbf{x} \in C_x(\boldsymbol{\mu}_x)] = \bar{P}$$

to be interpreted as the convex hull $C_x(\boldsymbol{\mu}_x)$ around $\boldsymbol{\mu}_{\mathbf{x}}$ that contains with probability $\bar{P}$ the samples of $\mathbf{x}$. Setting a certain level of confidence $\bar{P}$ (say $\bar{P} = .95$, or 95th percentile), one can find $C_x(\boldsymbol{\mu}_x)$ that depends on the pdf of $\mathbf{x}$. In detail, the dispersion $C_x(\boldsymbol{\mu}_x)$ follows from the argument of the exp(.) in $p(\mathbf{x})$

$$\psi^2(\mathbf{x}) = (\mathbf{x} - \boldsymbol{\mu}_x)^T \mathbf{C}_{xx}^{-1}(\mathbf{x} - \boldsymbol{\mu}_x).$$

This is a quadratic form that can be interpreted as the weighted L2 norm of the distance between $\mathbf{x}$ and $\boldsymbol{\mu}_x$ (*Mahalanobis distance*). The contours of the pdf (level curves) are characterized by $\psi^2(\mathbf{x}) = \bar{\psi}^2 = cost$ and the convex hulls $C_x(\boldsymbol{\mu}_x)$ are hyper-ellipsoids (in N-dimensional space). The probability that $\mathbf{x}$ lies inside the hyper-ellipsoid is (from the chi-square distribution of $\psi^2(\mathbf{x})$ in Section 3.8, or see [37, p.77])

$$\Pr[\mathbf{x} \in C_x(\boldsymbol{\mu}_x)] = 1 - \exp\left(-\frac{\bar{\psi}^2}{2}\right)$$

and this links the dispersion of the pdf around the mean. To evaluate the degree of dispersion, it is useful to evaluate the volume within the hyper-ellipsoid $C_x(\boldsymbol{\mu}_x)$, that is

$$V = V_N \times |\mathbf{C}_{xx}|^{1/2}\bar{\psi}^N$$

where V_N is the volume of the (hyper)sphere of unitary radius:

$$V_N = \begin{cases} \pi^{N/2}/(N/2)! & N \text{ even} \\ 2^N \pi^{(N-1)/2}/((N-1)/2)! & N \text{ odd} \end{cases}$$

In other words, the determinant of the covariance matrix $\mathbf{C}_{xx}$ (except for the dimensions' scaling V_N) indicates the degree of compactness (or dispersion) in a similar way as the variance for a scalar rv.

Characteristic Function

The *characteristic function* for the rv $\mathbf{x}$ is the N-dimensional Fourier transform of the multivariate pdf

$$\phi(\boldsymbol{\omega}) = \mathbb{E}[e^{-j\boldsymbol{\omega}^T\mathbf{x}}]$$

where $\boldsymbol{\omega} = [\omega_1, \omega_2, ..., \omega_N]^T$. In this case it is:

$$\begin{aligned}\phi(\boldsymbol{\omega}) &= \int d\mathbf{x}(2\pi)^{-N/2}|\mathbf{C}_{xx}|^{-1/2}\exp[-j\boldsymbol{\omega}^T\mathbf{x} - \frac{1}{2}(\mathbf{x}-\boldsymbol{\mu}_x)^T\mathbf{C}_{xx}^{-1}(\mathbf{x}-\boldsymbol{\mu}_x)] \\ &\text{...adding and subtracting } j\mathbf{C}_{xx}\boldsymbol{\omega} \text{ within the bracket of the exponent...} \\ &= \exp[-j\boldsymbol{\omega}^T\boldsymbol{\mu}_x - \frac{1}{2}\boldsymbol{\omega}^T\mathbf{C}_{xx}\boldsymbol{\omega}]\end{aligned}$$

and moments can be evaluated from the gradients as

$$\begin{aligned}\mathbb{E}[\mathbf{x}] &= j\left.\frac{\partial\phi(\boldsymbol{\omega})}{\partial\boldsymbol{\omega}}\right|_{\boldsymbol{\omega}=\mathbf{0}} \\ \mathbb{E}[\mathbf{x}\mathbf{x}^T] = \mathbf{C}_{xx} + \boldsymbol{\mu}_x\boldsymbol{\mu}_x^T &= \left.\frac{\partial^2\phi(\boldsymbol{\omega})}{\partial\boldsymbol{\omega}^2}\right|_{\boldsymbol{\omega}=\mathbf{0}}\end{aligned}$$

3.5 Conditional pdf for Real-Valued Gaussian Random Variables

The conditional pdf of the Gaussian distribution is Gaussian, and conditioning (i.e., slicing the pdf) might change the mean and covariance of the resulting pdf. Derivation of conditional pdfs is particularly important in some estimation methods. Reading of this section can be postponed after Chapter 11, but it is placed here for homogeneity.

Given two sets of Gaussian rvs, $\mathbf{x} = [x_1, x_2, \cdots, x_N]^T$ and $\mathbf{y} = [y_1, y_2, \cdots, y_M]^T$

$$\begin{aligned}\mathbf{x} &\sim \mathcal{N}(\boldsymbol{\mu}_x, \mathbf{C}_{xx}) \\ \mathbf{y} &\sim \mathcal{N}(\boldsymbol{\mu}_y, \mathbf{C}_{yy})\end{aligned}$$

The conditional pdf $p(\mathbf{x}|\mathbf{y})$ (or dually $p(\mathbf{y}|\mathbf{x})$) can be evaluated from the joint pdf $p(\mathbf{x}, \mathbf{y})$. Let the vector

$$\mathbf{z} = \begin{bmatrix}\mathbf{x} \\ \mathbf{y}\end{bmatrix} \in \mathbb{R}^{(N+M)}$$

be obtained through the concatenation of $\mathbf{x}$ and $\mathbf{y}$; it is jointly Gaussian:

$$\mathbf{z} \sim \mathcal{N}(\boldsymbol{\mu}_z, \mathbf{C}_{zz})$$

where mean and covariances are block-partitioned matrixes collecting the two sets $\mathbf{x}$ and $\mathbf{y}$:

$$\boldsymbol{\mu}_z = \mathbb{E}[\mathbf{z}] = \begin{bmatrix} \boldsymbol{\mu}_x \\ \boldsymbol{\mu}_y \end{bmatrix}$$

$$\mathbf{C}_{zz} = \mathbb{E}[(\mathbf{z}-\boldsymbol{\mu}_z)(\mathbf{z}-\boldsymbol{\mu}_z)^T] = \begin{bmatrix} \mathbf{C}_{xx} & \mathbf{C}_{xy} \\ \mathbf{C}_{yx} & \mathbf{C}_{yy} \end{bmatrix}$$

The conditional pdf follows from the Bayes rule

$$p(\mathbf{x}|\mathbf{y}) = \frac{p(\mathbf{x},\mathbf{y})}{p(\mathbf{y})} = \frac{G(\mathbf{z};\boldsymbol{\mu}_z,\mathbf{C}_{zz})}{G(\mathbf{y};\boldsymbol{\mu}_y,\mathbf{C}_{yy})} = \frac{|2\pi\mathbf{C}_{zz}|^{-1/2}\exp\left(-\frac{1}{2}(\mathbf{z}-\boldsymbol{\mu}_z)^T\mathbf{C}_{zz}^{-1}(\mathbf{z}-\boldsymbol{\mu}_z)\right)}{|2\pi\mathbf{C}_{yy}|^{-1/2}\exp\left(-\frac{1}{2}(\mathbf{y}-\boldsymbol{\mu}_y)^T\mathbf{C}_{yy}^{-1}(\mathbf{y}-\boldsymbol{\mu}_y)\right)}.$$

Since the conditional pdf has the form $\exp(-\phi)$, up to a scaling factor, it is convenient to redefine all the central variables wrt their means $\delta\mathbf{x} = \mathbf{x} - \boldsymbol{\mu}_x$ and $\delta\mathbf{y} = \mathbf{y} - \boldsymbol{\mu}_y$:

$$\begin{aligned}\phi &= \begin{bmatrix} \delta\mathbf{x} \\ \delta\mathbf{y} \end{bmatrix}^T \begin{bmatrix} \mathbf{C}_{xx} & \mathbf{C}_{xy} \\ \mathbf{C}_{yx} & \mathbf{C}_{yy} \end{bmatrix}^{-1} \begin{bmatrix} \delta\mathbf{x} \\ \delta\mathbf{y} \end{bmatrix} - \delta\mathbf{y}^T\mathbf{C}_{yy}^{-1}\delta\mathbf{x} \\ &= \begin{bmatrix} \delta\mathbf{x} \\ \delta\mathbf{y} \end{bmatrix}^T \begin{bmatrix} \mathbf{T}_{xx} & \mathbf{T}_{xy} \\ \mathbf{T}_{yx} & \mathbf{T}_{yy} \end{bmatrix} \begin{bmatrix} \delta\mathbf{x} \\ \delta\mathbf{y} \end{bmatrix} - \delta\mathbf{y}^T\mathbf{C}_{yy}^{-1}\delta\mathbf{x}\end{aligned}$$

Using the inversion rules for a block-partitioned matrix (Section 1.1.1):

$$\phi = (\delta\mathbf{x} + \mathbf{T}_{xx}^{-1}\mathbf{T}_{xy}\delta\mathbf{y})^T\mathbf{T}_{xx}(\delta\mathbf{x} + \mathbf{T}_{xx}^{-1}\mathbf{T}_{xy}\delta\mathbf{y})$$

where

$$\begin{aligned}\mathbf{T}_{xx}^{-1} &= \mathbf{C}_{xx} - \mathbf{C}_{xy}\mathbf{C}_{yy}^{-1}\mathbf{C}_{yx} \\ \mathbf{T}_{xx}^{-1}\mathbf{T}_{xy} &= -\mathbf{C}_{xy}\mathbf{C}_{yy}^{-1}\end{aligned}$$

Inspection of ϕ shows that pdf $p(\mathbf{x}|\mathbf{y})$ is Gaussian with mean

$$\mathbb{E}[\delta\mathbf{x}|\delta\mathbf{y}] = \mathbf{C}_{xy}\mathbf{C}_{yy}^{-1}\delta\mathbf{y} \rightarrow \mathbb{E}[\mathbf{x}|\mathbf{y}] = \boldsymbol{\mu}_x + \mathbf{C}_{xy}\mathbf{C}_{yy}^{-1}(\mathbf{y} - \boldsymbol{\mu}_y)$$

and covariance

$$\mathrm{cov}(\mathbf{x}|\mathbf{y}) = \mathbf{T}_{xx}^{-1} = \mathbf{C}_{xx} - \mathbf{C}_{xy}\mathbf{C}_{yy}^{-1}\mathbf{C}_{yx}$$

To summarize, for a joint Gaussian pdf $p(\mathbf{x},\mathbf{y})$, its conditional rv is:

$$p(\mathbf{x}|\mathbf{y}) = G(\mathbf{x};\boldsymbol{\mu}_x + \mathbf{C}_{xy}\mathbf{C}_{yy}^{-1}(\mathbf{y}-\boldsymbol{\mu}_y), \mathbf{C}_{xx} - \mathbf{C}_{xy}\mathbf{C}_{yy}^{-1}\mathbf{C}_{yx}).$$

3.6 Conditional pdf in Additive Noise Model

It is common to consider the following model, referred to as additive noise:

$$\mathbf{y} = \mathbf{x} + \mathbf{w}$$

where $\mathbf{x}$ denotes the signal of interest, and $\mathbf{w}$ is a rv accounting for an additive noise with pdf $p_{\mathbf{w}}(\mathbf{w})$, and $\mathbf{x}$ and $\mathbf{w}$ are statistically independent. The pdf $p_{\mathbf{y}}(\mathbf{y})$ depends on the nature of signal $\mathbf{x}$; the following conditions can be distinguished:

- $\mathbf{x}$ is deterministic (non-random): the joint pdf of the observation is given by

$$p_{\mathbf{y}}(\mathbf{y}) = p_{\mathbf{w}}(\mathbf{y} - \mathbf{x})$$

- $\mathbf{x}$ is an rv with pdf $p_{\mathbf{x}}(\mathbf{x})$:

$$p_{\mathbf{y}|\mathbf{x}}(\mathbf{y}|\mathbf{x}=\mathbf{x}_o) = p_{\mathbf{w}}(\mathbf{y} - \mathbf{x}_o) \text{ for the choice } \mathbf{x} = \mathbf{x}_o$$

  but the unconditional pdf (for all the possible values of rv $\mathbf{x}$ weighted according to their respective probability) is

$$p_{\mathbf{y}}(\mathbf{y}) = \int \cdots \int p_{\mathbf{w}}(\mathbf{y} - \mathbf{x}) p_{\mathbf{x}}(\mathbf{x}) d\mathbf{x}$$

  which is the N-dimensional convolution of the pdfs: $p_{\mathbf{y}}(\mathbf{y}) = p_{\mathbf{x}}(\mathbf{y}) * p_{\mathbf{w}}(\mathbf{y})$.

In a situation with additive noise, the pdf of $p_{\mathbf{y}}(\mathbf{y})$ is very different in the two contexts.

3.7 Complex Gaussian Random Variables

A complex-valued (or simply *complex*) rv can be represented by two real rvs coupled by some predefined (and useful) operations in the complex domain. The statistical properties of these two rvs (including their mutual correlation) completely defines the properties of the complex rvs. Even if in statistical signal processing it is preferable to handle every complex rv disjointly as real and imaginary rvs, in many applications it is useful to represent the complex variable (or any set) in a compact form without explicitly indexing its real and imaginary part. The main properties for complex variables are summarized below and are extensions of the properties previously derived for real rvs. An extensive discussion of statistical signal processing for complex multivariate distribution can be found in [15], which covers in detail all the particular properties used in advanced applications.

3.7.1 Single Complex Gaussian Random Variable

Given the complex rv $x = u + jv$, the following properties hold:

$$\mathbb{E}_x[x] = \mathbb{E}_u[u] + j\mathbb{E}_v[v] = \mu_u + j\mu_v = \mu$$
$$\mathbb{E}_x[|x|^2] = \mathbb{E}_u[u^2] + \mathbb{E}_v[v^2]$$

$$\begin{aligned}
\mathbb{E}_x[x^2] &= \mathbb{E}_u[u^2] - \mathbb{E}_v[v^2] + 2j\mathbb{E}_{u,v}[uv] \neq \mathbb{E}_x[|x|^2] \\
\text{var}[x] &= \mathbb{E}_x[|x-\mu|^2] = \mathbb{E}_x[(x-\mu)(x-\mu)^*] = \mathbb{E}_x[|x|^2] - |\mu|^2 \\
&= \text{var}[u] + \text{var}[v] \\
\text{cov}[x_1, x_2] &= \mathbb{E}_{x_1,x_2}[(x_1-\mu_1)(x_2-\mu_2)^*] \\
\text{var}[x] &= \text{cov}[x, x]
\end{aligned}$$

in accordance with the joint pdf $p(x) = p(u, v)$. The rvs u and v are independent if $p(u, v) = p(u)p(v)$, and for Gaussian rvs, the uncorrelation is sufficient to guarantee statistical independence:

$$\begin{aligned}
u &\sim \mathcal{N}(\mu_u, \sigma_u^2) \\
v &\sim \mathcal{N}(\mu_v, \sigma_v^2),
\end{aligned}$$

the pdf is

$$p(x) = p(u, v) = G(u; \mu_u, \sigma_u^2)G(v; \mu_v, \sigma_v^2)$$

with

$$\begin{aligned}
\mu_x &= \mu_u + j\mu_v \\
\text{var}[x] &= \sigma_x^2 = \sigma_u^2 + \sigma_v^2
\end{aligned}$$

As stated above, it is always possible to define a real-valued 2×1 vector $\tilde{\mathbf{x}} = [u, v]^T$ so that in this case:

$$\begin{aligned}
\mathbb{E}[\tilde{\mathbf{x}}] &= \tilde{\boldsymbol{\mu}} = [\mu_u, \mu_v]^T \\
\tilde{\mathbf{C}} &= \mathbb{E}[(\tilde{\mathbf{x}} - \tilde{\boldsymbol{\mu}})(\tilde{\mathbf{x}} - \tilde{\boldsymbol{\mu}})^T] = \begin{bmatrix} \text{cov}[u, u] & \text{cov}[u, v] \\ \text{cov}[v, u] & \text{cov}[v, v] \end{bmatrix}
\end{aligned}$$

and the corresponding pdf is

$$p(\tilde{\mathbf{x}}) = \frac{1}{2\pi|\tilde{\mathbf{C}}|^{1/2}} \exp\left[-\frac{1}{2}(\tilde{\mathbf{x}} - \tilde{\boldsymbol{\mu}})^T \tilde{\mathbf{C}}^{-1}(\tilde{\mathbf{x}} - \tilde{\boldsymbol{\mu}})\right].$$

One special case is when real and imaginary variables are uncorrelated so that $\tilde{\mathbf{C}} = \text{diag}(\sigma_u^2, \sigma_v^2)$.

3.7.2 Circular Complex Gaussian Random Variable

A complex Gaussian rv x can have the same second order central moments $\sigma_u^2 = \sigma_v^2 = \sigma_x^2/2$; this condition highlights a symmetry between real and imaginary parts, and in this case x is referred to as a circular complex Gaussian rv. The pdf becomes (recall that x and μ_x are complex-valued):

$$p(x) = \frac{1}{\pi\sigma_x^2} \exp\left(-\frac{1}{\sigma_x^2}|x - \mu_x|^2\right)$$

and it is indicated by

$$x \sim \mathcal{CN}(\mu_x, \sigma_x^2),$$

in vector form $\tilde{\mathbf{C}}_{xx} = \frac{\sigma_x^2}{2}\mathbf{I}$.

The circularity condition can be easily explained by considering the independence of the statistical properties of any rotation. Let the rv $y = x \cdot e^{j\theta}$ be obtained from $x \sim \mathcal{CN}(\mu_x, \sigma_x^2)$ with any arbitrary (but deterministic) phase rotation θ. The resulting rv preserves the variance $var[y] = var[x]$ regardless of θ. This property can be generalized to any dimensions and it justifies the circularity property.

3.7.3 Multivariate Complex Gaussian Random Variables

Considering a set of N complex rvs, it is natural to write:

$$\mathbf{x} = \begin{bmatrix} x_1 \\ x_2 \\ \vdots \\ x_N \end{bmatrix} = \mathbf{u} + j\mathbf{v} \in \mathbb{C}^N$$

Alternatively, it is possible to collect in a $2N$ real-valued vector the real $\mathbf{u} = \text{Re}\{\mathbf{x}\}$ and imaginary $\mathbf{v} = \text{Im}\{\mathbf{x}\}$ parts of the rvs:

$$\tilde{\mathbf{x}} = \begin{bmatrix} \mathbf{u} \\ \mathbf{v} \end{bmatrix}$$

It is important to highlight that $\mathbf{x} \in \mathbb{C}^N$ and $\tilde{\mathbf{x}} \in \mathbb{R}^{2N}$; furthermore $\mathbf{x} \neq \tilde{\mathbf{x}}$. It holds that:

$$\mathbb{E}[\tilde{\mathbf{x}}] = \tilde{\boldsymbol{\mu}}_x = [\boldsymbol{\mu}_u^T, \boldsymbol{\mu}_v^T]$$

$$\tilde{\mathbf{C}}_{xx} = \mathbb{E}[(\tilde{\mathbf{x}} - \tilde{\boldsymbol{\mu}}_x)(\tilde{\mathbf{x}} - \tilde{\boldsymbol{\mu}}_x)^T] = \mathbb{E}\left[\begin{pmatrix} \mathbf{u} - \boldsymbol{\mu}_u \\ \mathbf{v} - \boldsymbol{\mu}_v \end{pmatrix}(\mathbf{u}^T - \boldsymbol{\mu}_u^T, \mathbf{v}^T - \boldsymbol{\mu}_v^T)\right]$$

$$= \begin{bmatrix} \mathbf{C}_{uu} & \mathbf{C}_{uv} \\ \mathbf{C}_{vu} & \mathbf{C}_{vv} \end{bmatrix}$$

From the definitions, it follows that: $\mathbf{C}_{uv} = \mathbb{E}[(\mathbf{u} - \boldsymbol{\mu}_u)(\mathbf{v}^T - \boldsymbol{\mu}_v^T)] = \mathbf{C}_{vu}^T$.

These general relationships can be specialized for the class of circularly symmetric complex Gaussian rvs (their extensive use justifies referring to them as simply *complex Gaussian,* and specializing only if necessary). Circular symmetry holds for the two properties

- $\mathbf{C}_{uv}^T = -\mathbf{C}_{uv} \Rightarrow \begin{cases} [\mathbf{C}_{uv}]_{ij} = -[\mathbf{C}_{uv}]_{ji} \\ [\mathbf{C}_{uv}]_{ii} = 0 \end{cases}$: antisymmetric matrix (or odd symmetry),
- $\mathbf{C}_{uu} = \mathbf{C}_{vv}$: real and imaginary components have the same characteristics (up to a scaling factor).

This implies that:

$$\tilde{\mathbf{C}}_{xx} = \begin{bmatrix} \mathbf{C}_{uu} & \mathbf{C}_{uv} \\ \mathbf{C}_{vu} & \mathbf{C}_{vv} \end{bmatrix} = \begin{bmatrix} \mathbf{C}_{uu} & \mathbf{C}_{uv} \\ \mathbf{C}_{uv}^T & \mathbf{C}_{vv} \end{bmatrix} = \begin{bmatrix} \mathbf{C}_{uu} & \mathbf{C}_{uv} \\ -\mathbf{C}_{uv} & \mathbf{C}_{vv} \end{bmatrix}$$

Notice that a vector of two complex rvs with covariance matrix $\tilde{\mathbf{C}}_{xx} = \text{diag}\{\sigma_1^2, \sigma_1^2, \sigma_2^2, \sigma_2^2\}$ is circularly symmetric.

The covariance matrix for complex rvs is defined as

$$\mathbf{C}_{xx} = \mathbb{E}[(\mathbf{x} - \boldsymbol{\mu}_x)(\mathbf{x} - \boldsymbol{\mu}_x)^H] = \mathbb{E}[\mathbf{x}\mathbf{x}^H] - \boldsymbol{\mu}_x\boldsymbol{\mu}_x^H$$

using the definitions of the covariance for real rvs, this is

$$\mathbf{C}_{xx} = \mathbb{E}[(\mathbf{u} + j\mathbf{v})(\mathbf{u}^T - j\mathbf{v}^T)] - (\boldsymbol{\mu}_u + j\boldsymbol{\mu}_v)(\boldsymbol{\mu}_u^T - j\boldsymbol{\mu}_v^T) = \mathbf{C}_{uu} + \mathbf{C}_{vv} + j(\mathbf{C}_{vu} - \mathbf{C}_{uv})$$

and applying the properties for the circular symmetry condition:

$$\mathbf{C}_{xx} = 2(\mathbf{C}_{uu} + j\mathbf{C}_{uv}^T)$$

Moreover, it can be verified that under the circularity condition, it holds that:

$$\mathbb{E}[(\mathbf{x} - \boldsymbol{\mu}_x)(\mathbf{x} - \boldsymbol{\mu}_x)^T] = 0$$

(sometimes, this latter expression is adopted to implicitly state that the circularity condition holds).

The pdf of a vector of circular complex Gaussian rvs is given by:

$$p(\mathbf{x}) = |\pi\mathbf{C}_{xx}|^{-1}\exp[-(\mathbf{x} - \boldsymbol{\mu}_x)^H\mathbf{C}_{xx}^{-1}(\mathbf{x} - \boldsymbol{\mu}_x)]$$

or in a more synthetic way:

$$\mathbf{x} \sim \mathcal{CN}(\boldsymbol{\mu}_x, \mathbf{C}_{xx})$$

Notice that when the real and imaginary parts of $\mathbf{x}$ are uncorrelated with each other, $\mathbf{C}_{uv} = \mathbf{0}$ and $\mathbf{C}_{xx} = 2\mathbf{C}_{uu}$.

Remark: Definitions of the moments for complex rvs resemble the same definitions for real rvs, except for replacing the transpose $(.)^T$ with the Hermitian transpose $(.)^H$. Replacement of $(.)^T \to (.)^H$ into formulas is quite common when extending operations from real-valued into complex-valued vectors; it is often correct (but not always!)—see Section 9.3.1.

3.8 Sum of Square of Gaussians: Chi-Square

Let $\mathbf{x} \sim \mathcal{N}(\mathbf{0}, \mathbf{I})$, the quadratic form

$$\psi = \sum_{i=1}^{N} x_i^2 \sim \chi_N^2$$

is a central chi-square with N degrees of freedom having a pdf

$$p(\psi) = \frac{1}{\Gamma(N/2)2^{N/2}}\psi^{(N/2)-1}e^{-\psi/2}$$

In general, for $\mathbf{x} \sim \mathcal{N}(\boldsymbol{\mu}_x, \mathbf{C}_{xx})$, the quadratic form $\psi(\mathbf{x}) = (\mathbf{x} - \boldsymbol{\mu}_x)^T\mathbf{R}(\mathbf{x} - \boldsymbol{\mu}_x)$ is the most general one having the property that, for $\mathbf{R} = \mathbf{C}_{xx}^{-1}$, $\psi(\mathbf{x}) \sim \chi^2_N$. Even if in general it is quite complex to derive the corresponding pdf, the moments are easily determined. Let

$$\psi = \text{tr}[\mathbf{R}\mathbf{x}\mathbf{x}^T] \text{ with } \mathbf{x} \sim \mathcal{N}(\mathbf{0}, \mathbf{C}_{xx})$$

then

$$\begin{aligned}
\mathbb{E}[\psi] &= \text{tr}[\mathbf{R}\mathbf{C}_{xx}] \\
\mathbb{E}[\psi^2] &= \mathbb{E}[\mathbf{x}^T\mathbf{R}\mathbf{x}\mathbf{x}^T\mathbf{R}\mathbf{x}] = \text{tr}[\mathbf{R}\mathbf{C}_{xx}]^2 + 2\text{tr}[(\mathbf{R}\mathbf{C}_{xx})^2] \\
\text{var}[\psi] &= 2\text{tr}[(\mathbf{R}\mathbf{C}_{xx})^2]
\end{aligned}$$

by using the property $\mathbb{E}[\mathbf{x}^T\mathbf{A}\mathbf{x}\mathbf{x}^T\mathbf{B}\mathbf{x}] = \text{tr}[\mathbf{A}\mathbf{C}_{xx}]\text{tr}[\mathbf{B}\mathbf{C}_{xx}] + 2\text{tr}[\mathbf{A}\mathbf{C}_{xx}\mathbf{B}\mathbf{C}_{xx}]$ derived in reference [17, Appendix].

3.9 Order Statistics for N rvs

The kth order statistic of a random sample $x_1, ..., x_N$ is its kth-smallest value $x_{(k)}$ after ordering for decreasing values:

$$x_{(1)} \leq x_{(2)} \leq ... \leq x_{(N)}$$

where the new labeling is only to highlight the position of the rv. There are two privileged positions:

- first (or minimum) order statistics, is the minimum of the sample:

$$x_{(1)} = x_{\min} = \min\{x_1, ..., x_N\}$$

- Nth (or maximum) order statistics out of the set of N samples is the maximum of the sample:

$$x_{(N)} = x_{\max} = \max\{x_1, ..., x_N\}$$

  The density of the maximum and the minimum for independent and identically distributed random variables can be evaluated analytically, and are sketched herein as they are relevant in many practical problems (see the excellent book by David and Nagaraja [10] for further details).

Let the rv x_i have pdf $p_i(x_i)$ and cdf $F_i(x_i)$, the maximum order statistics

$$x_{\max} = \max\{x_1, ..., x_N\}$$

is an rv itself with cdf

$$F_{\max}(\xi) = \Pr(x_{\max} \le \xi) = \Pr(x_1 \le \xi, x_2 \le \xi, \ldots, x_N \le \xi) = \prod_{n=1}^{N} \Pr(x_n \le \xi) = \prod_{n=1}^{N} F_n(\xi)$$

after employing the property of the independence of the rvs. The pdf of maximum order statistics $x_{\max}$ is the derivative of the cdf

$$f_{\max}(\xi) = \frac{d}{d\xi} \prod_{n=1}^{N} F_n(\xi).$$

If all the rvs are identically distributed, $F_i(x) = F(x)$, thus

$$\begin{aligned} &\Pr(x_{\max} \le \xi) = F(\xi)^N \\ &f_{\max}(\xi) = Nf(\xi)F(\xi)^{N-1} \end{aligned}$$

For the minimum order statistics the computations are simply the complementary ones:

$$F_{\min}(\xi) = \Pr(x_{\min} \ge \xi) = 1 - \prod_{n=1}^{N} (1 - F_n(\xi))$$

and the pdf is the derivative of the cdf. For identically distributed rvs:

$$\begin{aligned} &\Pr(x_{\min} \ge \xi) = 1 - (1 - F(\xi))^N \\ &f_{\min}(\xi) = Nf(\xi)(1 - F(\xi))^{N-1} \end{aligned}$$

These properties can be extended to the case of the cdf for the kth order statistics for identically distributed rvs [10]

$$\begin{aligned} F_{x_{(k)}}(\xi) &= \Pr(x_{(k)} \le \xi) = \Pr(\text{at least } k \text{ of the set } \{x_i\} \text{ are less than or equal to } \xi) \\ &= \sum_{\ell=k}^{N} \Pr(\ell \text{ of the } x_i \text{ are less than or equal to } \xi) \\ &= \sum_{\ell=k}^{N} \binom{N}{\ell} F(\xi)^{\ell} [1 - F(\xi)]^{N-\ell} \end{aligned}$$

where the term in the sum is the binomial probability.

4

Random Processes and Linear Systems

There are sequences and signals that are random, and these are referred to as random (or stochastic) processes. These random processes can be either discrete-time or continuous-time, depending on whether the random function of time is over a discrete set or a continuous one. In order to frame the random processes in the preceding chapters, the discrete-time random process can be viewed as a sequence of random variables sequentially ordered by the time variable as indexing parameter. This means that the stochastic processes are completely characterized by the Nth oder joint pdf $p(\mathbf{x})$ for the set of rvs $\{x_1, x_2, \cdots, x_N\}$ obtained by continuous-time random process $x(t)$ for any choice of the time instant $\{t_1, t_2, \cdots, t_N\}$ to have $\mathbf{x} = [x(t_1), x(t_2), \cdots, x(t_N)]^T$. Similarly, for discrete-time random process $x[n]$ where $n \in \mathbb{Z}$, one can choose the discrete-time time sampling $\{n_1, n_2, \cdots, n_N\}$ to have $\mathbf{x} = [x[n_1], x[n_2], \cdots, x[n_N]]^T$. The commonality between the continuous/discrete-time processes is that the random variables are obtained by selecting the samples indexed by the time t or n specified by the notation $x(t)$ and $x[n]$. Sometime samples of a discrete-time random process can be obtained from a continuous-time signal $x(t)$ after sampling with an appropriately chosen sampling interval T (Appendix A) as

$$x[n] = x(nT).$$

The distinction between continuous-time and discrete-time processes will be omitted in the discussion as it will be clear from the context. The notation is not differentiated for the two cases and; this could confuse beginners, but different notations would be even worse. With some abuse, random processes are called signals and are studied by their statistical properties. Here we handle both continuous-time and discrete-time random processes under the same frameworks by inspecting their central moments and correlation properties. Detailed and extensive treatments are in the milestone book by Papoulis and Pillay [11].

Needless to say, the independent variable t or n can be time (and this provides a natural and intuitive ordering of samples), but also space, as in an image. Furthermore, the independent parameters can be more than one, as for multiple spaces in images, or space and time when monitoring the time evolution of a phenomenon, etc. These generalizations will not be covered in this summary, and further extensions are left to specialized textbooks.

Statistical Signal Processing in Engineering, First Edition. Umberto Spagnolini.

Companion website: www.wiley.com/go/spagnolini/signalprocessing

4.1 Moment Characterizations and Stationarity

The moments of a random process are the mean, the (auto)covariance and the autocorrelation defined in Table 4.1.

Covariances and autocorrelations are dependent on two temporal variables, and it is straightforward to prove that are related one another and coincide for zero-mean processes:

$$r_{xx}(t,s) = c_{xx}(t,s) + \mu_x(t)\mu_x(s)$$

$$r_{xx}[m,n] = c_{xx}[m,n] + \mu_x[m]\mu_x[n]$$

The properties of covariance

$$\text{var}[x(t)] = c_{xx}(t,t)$$
$$c^2_{xx}(t,s) \leq c_{xx}(t,t)c_{xx}(s,s) = \text{var}[x(t)]\text{var}[x(s)] \quad \text{(Schwarz inequality)}$$

highlight how variables $x(t)$ and $x(s)$ are two distinct rvs (the same holds for discrete-time random processes). A *white process* is when samples of a zero-mean random process can be correlated only with themselves:

$$\text{white process:} \begin{cases} \mathbb{E}[x(t)x(s)] = \sigma^2\delta(s-t) & \text{continuous-time} \\ \mathbb{E}[x[m]x[n]] = \sigma^2\delta[m-n] & \text{discrete-time} \end{cases}$$

where σ^2 is the power of the random process (continuous/discrete-time use the same notation, but different units). Another very common process extensively discussed later is when samples are less correlated when more spaced apart such as

$$c_{xx}[m,n] = \sigma^2\rho^{|m-n|}$$

where the parameter ρ ($|\rho| < 1$) controls the degree of correlation of the samples: when $\rho \simeq 0$, the samples are quickly uncorrelated (as $\rho \rightarrow 0$, this degenerates into the white process), and for $\rho \simeq 1$, samples remain correlated even for large separations in time (for $\rho \simeq -1$ samples are anti-correlated). The continuous-time counterpart is

$$c_{xx}(t,s) = \sigma^2 e^{-\lambda|t-s|}$$

where $\lambda > 0$.

Table 4.1 Moments of a random process.

Continuous-time $x(t)$		**Discrete-time $x[n]$**	
mean function	$\mu_x(t) = \mathbb{E}[x(t)]$	mean sequence	$\mu_x[n] = \mathbb{E}[x[n]]$
(auto)covariance function	$c_{xx}(t,s) = \text{cov}[x(t),x(s)]$	(auto)covariance sequence	$c_{xx}[m,n] = \text{cov}[x[m],x[n]]$
autocorrelation function	$r_{xx}(t,s) = \mathbb{E}[x(t)x(s)]$	autocorrelation sequence	$r_{xx}[m,n] = \mathbb{E}[x[m]x[n]]$

Notice that Gaussianity of random processes is a widely adopted assumption in several (not all) signal processing problems; this justifies the in-depth analysis of the properties of the first two central moments. Furthermore, in many contexts it is understood that processes are zero-mean (or the mean value is removed during any signal manipulations to reduce to the zero-mean assumption), and all derivations are for the autocorrelations. This routinely employed assumption is avoided in this chapter, but is adopted later.

Stationarity

Samples in random processes depend onto the choice of time, but there is a set of processes where the notion of absolute time is not of interest and their properties are invariant wrt any time-shift. A strict-sense stationary (SSS) process is when the joint pdf is time-invariant for any order N. More specifically, let $\mathbf{x}(t) = [x(t+t_1), x(t+t_2), \cdots, x(t+t_N)]^T$, the Nth order stationarity is guaranteed when

$$p(\mathbf{x}(t)) = p(\mathbf{x}(\tau)) \text{ for any } \tau \neq t.$$

The 2nd order stationarity is usually referred as a wide-sense stationary (WSS) process and it is a weaker condition compared to SSS. WSS involves just two samples, and thus two random variables. The first/second order central moments for WSS are independent of absolute time:

$$\text{continuous time: } \begin{cases} \mu_x(t) = \mu_x \\ c_{xx}(\tau) = \mathbb{E}[x(t)x(t+\tau)] - \mu_x^2 \end{cases}$$

$$\text{discrete-time: } \begin{cases} \mu_x[n] = \mu_x \\ c_{xx}[n] = \mathbb{E}[x[k]x[k+n]] - \mu_x^2 \end{cases}$$

Covariance is even symmetric:

$$c_{xx}(\tau) = c_{xx}(-\tau) \text{ with } |c_{xx}(\tau)| \leq c_{xx}(0) = \text{var}[x(t)]$$

and equally for a discrete-time random process:

$$c_{xx}[n] = c_{xx}[-n] \text{ with } |c_{xx}[n]| \leq c_{xx}[0] = \text{var}[x[n]]$$

As an example, consider a sinusoid

$$x(t) = a \cdot \cos(\omega_o t + \varphi)$$

with deterministic amplitude a and frequency ω_o but random phase $\varphi \sim \mathcal{U}(-\pi, \pi)$, the shifted copy $x(t-\tau) = a \cdot \cos(\omega_o t + \varphi - \omega_o t)$ has phase modulo-2π $\bar{\varphi} = [\varphi - \omega_o t]_{2\pi} \sim \mathcal{U}(-\pi, \pi)$, and thus it is SSS with mean

$$\mu_x = \mathbb{E}_\varphi[x(t)] = 0$$

and its correlation (or covariance as it is zero-mean)

$$r_{xx}(\tau) = \mathbb{E}_\varphi[x(t)x(t+\tau)] = \frac{a^2}{2}\mathbb{E}_\varphi[\cos(\omega_o \tau) + \cos(2\omega_o t + \omega_o \tau + 2\varphi)] = \frac{a^2}{2}\cos(\omega_o \tau)$$

is also periodic. There is a similar result for discrete-time random processes involving a discrete-time sinusoid:

$$x[n] = a \cdot \cos(\omega_o n + \varphi) \rightarrow r_{xx}[n] = \mathbb{E}_\varphi[x[k+n]x[k]] = \frac{a^2}{2}\cos(\omega_o k).$$

4.2 Random Processes and Linear Systems

A linear system linearly manipulates the input signals to yield another signal as output. There are powerful mathematical tools to handle signals revised in Appendix A–B that can be extended to random processes, and enable compact but effective random process analysis.

Continuous-Time Random Process

A continuous-time random process $x(t)$ that is linearly filtered with impulse response $h(t, \tau)$ is

$$y(t) = \int h(t, \tau)x(\tau)d\tau,$$

the output $y(t)$ is also a stochastic process. The mean value of $y(t)$ is obtained by taking the expectations

$$\mu_y(t) = \int h(t, \tau)\mu_y(\tau)d\tau$$

and covariances are by definition (similarly for discrete-time process). A linear time-invariant (LTI) system is for time-invariant impulse response

$$h(t, \tau) = h(t - \tau)$$

and it yields

$$y(t) = \int h(t - \tau)x(\tau)d\tau = h(t) * x(t)$$

where "*" denotes the convolution integral (or convolution, in short). Assuming that x(t) is WSS, the output is WSS with mean

$$\mu_y(t) = \int h(t)dt \cdot \mu_x = H(0) \cdot \mu_x$$

where $H(\omega) = \mathcal{F}\{h(t)\}$ is the Fourier transform of $h(t)$ with properties given in Appendix A. The covariance is

$$c_{yy}(\tau) = \int\int h(t)h(s)c_{xx}(\tau + s - t)dtds = r_{hh}(\tau) * c_{xx}(\tau)$$

where

$$r_{hh}(\tau) = h(\tau) * h(-\tau)$$

is the autocorrelation of the impulse response $h(t)$. Taking the Fourier transform both sides

$$C_{yy}(\omega) = |H(\omega)|^2 C_{xx}(\omega),$$

the *power spectral density* (PSD) of random process $x(t)$ (*Wiener–Khinchine theorem*) is

$$C_{xx}(\omega) = \mathcal{F}\{c_{xx}(\tau)\}.$$

The name PSD is justified by the computation of the power of $x(t)$ from the power spectral density as the sum over all frequencies of the power per unit frequency:

$$\text{var}[x(t)] = c_{xx}(0) = \int C_{xx}(\omega)\frac{d\omega}{2\pi}.$$

An interesting case is for a white process, where $c_{xx}(\tau) = \sigma_x^2\delta(\tau)$. Here, the PSD is constant over frequency (this motivates the name, as white light is the combination of all colors that are all the optical frequencies)

$$C_{xx}(\omega) = \sigma_x^2$$

and the output is a correlated WSS process with

$$c_{yy}(t) = \sigma_x^2 r_{hh}(t) \rightarrow \text{var}[y(t)] = \sigma_x^2 \int h^2(t)dt$$

This proves that an LTI system should have $\int h^2(t)dt < \infty$ in order to have the variance well defined.

Discrete-Time Random Process

Discrete-time random processes have similar properties to the above. For an LTI discrete-time system:

$$y[n] = \sum_k h[n-k]x[k] = h[n] * x[n]$$

and again for WSS

$$\mu_y = \mu_x \sum_k h[k] = \mu_x H(z)|_{z=e^{j0}}$$

where $H(z) = \mathcal{Z}\{h[n]\}$ is the z-transform of the sequence $h[n]$ that is related to the Fourier transform by the identity (Appendix B):

$$\mathcal{F}\{h[n]\} = H(z)|_{z=e^{j\omega}}$$

The covariance is

$$c_{yy}[n] = r_{hh}[n] * c_{xx}[n]$$

and the PSD for a discrete-time random process is its Fourier transform (Wiener–Khinchine theorem):

$$C_{xx}(e^{j\omega}) = \sum_k c_{xx}[k]e^{-j\omega k} = \mathcal{F}\{c_{xx}[k]\}.$$

In the remaining chapters, discrete-time random processes are far more common, and this is the main focus. Sometime, it is more convenient to use the z-transform for PSD that is paired with the Fourier representation of PSD:

$$C_{yy}(z) = \mathcal{Z}\{c_{yy}[k]\} = (H(z)H^*(1/z)) \cdot C_{xx}(z) \rightarrow C_{yy}(e^{j\omega}) = |H(e^{j\omega})|^2 \cdot C_{xx}(e^{j\omega}).$$

A remarkable property is that, since the covariance sequence is even symmetric (i.e., $c_{xx}[k] = c_{xx}[-k]$), the PSD is also symmetric across the unit circle in the z-plane:

$$C_{xx}(z) = C_{xx}(1/z).$$

A simple proof of this is that any arbitrary PSD can be obtained as filtering of a white process, say with a filter $g[n] \leftrightarrow G(z)$, and thus $C_{xx}(z) = G(z)G^*(1/z)$. This is the basis of spectral factorization below.

4.3 Complex-Valued Random Processes

A random process can be represented with complex-valued variables, and this is a convenient representation in several signal processing applications. One example on the usefulness of this *representation* (as in nature signals are not complex) is narrowband signals[1]

$$x(t) = \alpha(t) \cdot \cos(\omega_o t) + \beta(t) \cdot \sin(\omega_o t) = a(t) \cdot \cos(\omega_o t + \varphi(t))$$

where ω_o is much larger that the spectral components of the pair $\{\alpha(t), \beta(t)\}$ and

$$\begin{cases} a(t) = & \sqrt{\alpha^2(t) + \beta^2(t)} \\ \varphi(t) = & \arctan(\beta(t)/\alpha(t)) \end{cases}$$

1 Narrowband modulated signals are when their bandwidth is much smaller than the center frequency. A narrowband random process $x(t)$ with PSD $C_{xx}(\omega)$ is supported in $\omega \in [\omega_o - B_x/2, \omega_o + B_x/2]$, provided that the bandwidth B_x is $B_x \ll \omega_o$.

One can define a complex-valued narrowband sinusoid (equivalent *baseband signal*)

$$\tilde{x}(t) = a(t)e^{j\varphi(t)}e^{j\omega_o t}$$

such that

$$x(t) = \text{Re}\{\tilde{x}(t)\} = \frac{\tilde{x}(t) + \tilde{x}^*(t)}{2}.$$

The baseband model is very common in electrical and telecommunication engineering as it simplifies several signal processing activities such as filtering and spectral translations (modulation/demodulation) [37, 57], mostly for the trigonometric manipulations involved. The complex-valued (and, by extension, any) periodic signal

$$\tilde{a} \cdot e^{j\omega_o t}$$

is a complex sinusoid, or a sinusoid in short (even if it would be a cosine and sine over the two components), with $\tilde{a} = ae^{j\varphi}$ complex envelope.

SSS and WSS definitions apply in the same way for both continuous/discrete-time random processes, and their moments are

$$\text{continuous time:} \begin{cases} \mu_x = \mathbb{E}[\tilde{x}(t)] \\ r_{xx}(\tau) = \mathbb{E}[\tilde{x}^*(t)\tilde{x}(t+\tau)] \\ c_{xx}(\tau) = r_{xx}(\tau) - |\mu_x|^2 \end{cases}$$

$$\text{discrete-time:} \begin{cases} \mu_x = \mathbb{E}[\tilde{x}[t]] \\ r_{xx}[n] = \mathbb{E}[\tilde{x}^*[k]\tilde{x}[k+n]] \\ c_{xx}[n] = r_{xx}[n] - |\mu_x|^2 \end{cases}$$

The covariance is the complex conjugate symmetric:

$$\text{continuous time: } c_{xx}(-\tau) = c^*_{xx}(\tau) \quad \text{discrete-time: } c_{xx}[-n] = c^*_{xx}[n]$$

and since this symmetry implies the one for real-valued signals, the definition for complex-valued signals is commonly adopted even for real-valued random processes as both coincide.

4.4 Pole-Zero and Rational Spectra (Discrete-Time)

Filtering by an LTI system with impulse response $h[n] \leftrightarrow H(z)$ can be modeled by a finite difference equation that relates the output samples ($y[n]$) with the input ($x[n]$):

$$y[n] + \sum_{k \neq 0} a[k] \cdot y[n-k] = \sum_{\ell} b[\ell] \cdot x[n-\ell],$$

Taking the z-transform of both terms

$$A(z)Y(z) = B(z)X(z)$$

where

$$A(z) = 1 + \sum_{k \neq 0} a[k]z^{-k}$$
$$B(z) = \sum_{k} b[k]z^{-k}$$

The filtering becomes

$$Y(z) = H(z)X(z)$$

and the filter transfer function

$$H(z) = \frac{B(z)}{A(z)}$$

is the ratio of two polynomials, where $A(z)$ is *monic* (see Appendix C of chapter 14 for properties). The roots of a polynomial define some symmetry properties of the polynomial coefficients: a minimum phase polynomial is when all roots are inside the unit circle and maximum phase when all are outside. Roots of $B(z)$ are called *zeros*, and roots of $A(z)$ are *poles*. A property of a filter is: if $h[n] = 0$ for $n < 0$, the filter is causal and sample $y[n]$ depends on any linear combination of some of the past samples $x[n-1], x[n-2], \ldots$, and if anti-causal, $h[n] = 0$ for $n > 0$. The recursive components $\{a[n]\}$ (i.e., those that use some output samples $k \neq 0$ to compute the output in n) play a key role in filtering. If $a[n] = 0$ for $n < 0$, the recursive component is causal, and the output depends only on the past outputs $y[n-1], y[n-2], \ldots$, and is anti-causal if $a[n] = 0$ for $n > 0$.

4.4.1 Stability of LTI Systems

Stability of an LTI system is a condition for any physically plausible filter. For the stability of a causal system it is necessary (and sufficient if $\{b[\ell]\}$ is limited) that $A(z)$ is a causal and *minimum phase* polynomial. For an anti-causal system, all the conditions are reversed and $A(z)$ should be a *maximum phase* polynomial to guarantee stability. An LTI filter with mixed phase polynomial $A(z)$ (i.e., poles inside and outside of the unit-circe) is stable: simply it is causal and anti-causal, and the impulse response contains both a causal and an anti-causal component. To illustrate, one has to factorize the polynomial into minimum $A_{\min}(z)$ and maximum $A_{\max}(z)$ phase such that $A(z) = A_{\min}(z)A_{\max}(z)$, and the z-transform of the filter with similar decomposition for $B(z)$ (not strictly necessary):

$$H(z) = \underbrace{\frac{B_{\min}(z)}{A_{\min}(z)}}_{H_c(z)} \cdot \underbrace{\frac{B_{\max}(z)}{A_{\max}(z)}}_{H_a(z)}$$

is decomposed into the convolution of two causal ($h_c[n]$) and anti-causal ($h_a[n]$) terms:

$$h[n] = h_c[n] * h_a[n].$$

Filtering with mixed phase polynomials usually trips up beginners, who conclude it to be impossible; but convolution should be carried out twice for causal and anti-causal elements, following the order of brackets:

$$Y(z) = H_a(z)\left(H_c(z)X(z)\right)$$

Filtering is first applied as causal on the input sequence $X(z)$

$$C(z) = H_c(z)X(z)$$

Then there is a second causal filtering after reversing the sequence $C(z)$

$$Y(1/z) = H_a(1/z)C(1/z)$$

The final result is obtained simply by reversing back the output $Y(1/z) = \mathcal{Z}\{y[-n]\}$. Recall that the double filtering lacks initial conditions and thus some edge effects at the beginning/end of the sequence are likely to be experienced in practice.

4.4.2 Rational PSD

After this quick overview, the rational PSD is the rational function of $e^{-j\omega}$ that can be obtained as the ratio of two polynomials

$$C_{xx}(\omega) = \sigma^2 \left. \frac{B(z)B^*(1/z)}{A(z)A^*(1/z)} \right|_{z=e^{j\omega}}$$

with scaling factor σ^2. This rational PSD can be equivalently obtained by filtering a white random process with power σ^2 with a filter with transfer function $H(z) = B(z)/A(z)$.

Assuming that polynomials $A_{N_a}(z)$ and $B_{N_b}(z)$ have degrees N_a and N_b (or roots N_a and N_b), the random process $x[n]$ is classified as being AR, MA, or ARMA with the constraints in Table 4.2, and the corresponding PSDs have different peculiarities. Usually systems are causal and stable (or are assumed to be so), so $A(z)$ is a minimum phase polynomial. The order of the filters characterizes the random process as being AR(N_a), MA(N_b), or ARMA(N_a,N_b).

Table 4.2 Classification of a random process.

Autoregressive	$AR(N_a)$	$A_{N_a}(z)$	$B_0(z) = b[0]$
Moving average	$MA(N_b)$	$A_0(z) = 1$	$B_{N_b}(z)$
Autoregressive moving average	$ARMA(N_a,N_b)$	$A_{N_a}(z)$	$B_{N_b}(z)$

4.4.3 Paley–Wiener Theorem

Spectral factorizations of PSD refer to any causal filter $G(z)$ such that the PSD of a WSS can be factorized as

$$C_{xx}(z) = G(z)G^*(1/z)$$

This holds if and only if

$$\int |\ln C_{xx}(e^{j\omega})| d\omega < \infty$$

and for a rational PSD this is always satisfied (Paley–Wiener theorem). The filter $G(z)$ can be referred to as a shaping filter for $C_{xx}(z)$ from the white process. If $G(z)$ is causal, the inverse $1/G(z)$ is causal too and it is called a *whitening filter* as the random process $x[n]$ filtered by $1/G(z)$ yields a white process.

The spectral factor $G(z)$ in PSD factorization is not unique as the PSD is constrained by the magnitude of $G(e^{j\omega})$

$$C_{xx}(e^{j\omega}) = |G(e^{j\omega})|^2$$

but not for the phase. A remarkable property is that the symmetry $C_{xx}(z) = C_{xx}(1/z)$ for a rational PSD implies that poles and zeros appear in a symmetric pattern: if $\bar{z}$ is a pole (or zero), then $\bar{z}^*$, $1/\bar{z}$, and $1/\bar{z}^*$ are also poles (or zeros). To validate this, let the elementary sequence $\{1, -\bar{z}\}$ correspond to one zero in $\bar{z}$, the following identity holds:

$$\left|(1-\bar{z}z^{-1})|_{z=e^{j\omega}}\right| = \left|(1-\bar{z}^*z)|_{z=e^{j\omega}}\right|$$

The amplitude of the Fourier transform does not change if zeros have reciprocal conjugate positions with respect to the unit-radius circle (i.e., for $\bar{z}$ or $1/\bar{z}^* = (1/\bar{z})^*$ as radial positions are respectively $|\bar{z}|$ and $|1/\bar{z}^*| = 1/|\bar{z}^*| = 1/|\bar{z}|$ but angularly the same). There are no poles/zeros on the unit circle with odd multiplicity, but only with even multiplicity.

Minimum/maximum phase factorization is a very convenient factorization for the PSD (but is not unique). Due to the symmetry of PSD, one can factor all poles/zeros inside the unit circle to $G(z)$ as this one is unique, and this is the minimum phase spectral factor:

$$G(z) = C_{xx,\min}(z)$$

The maximum phase spectral factor is

$$C_{xx,\max}(z) = C^*_{xx,\min}(1/z)$$

such that

$$C_{xx}(z) = C_{xx,\min}(z) \cdot C_{xx,\max}(z) = C_{xx,\min}(z) \cdot C^*_{xx,\min}(1/z)$$

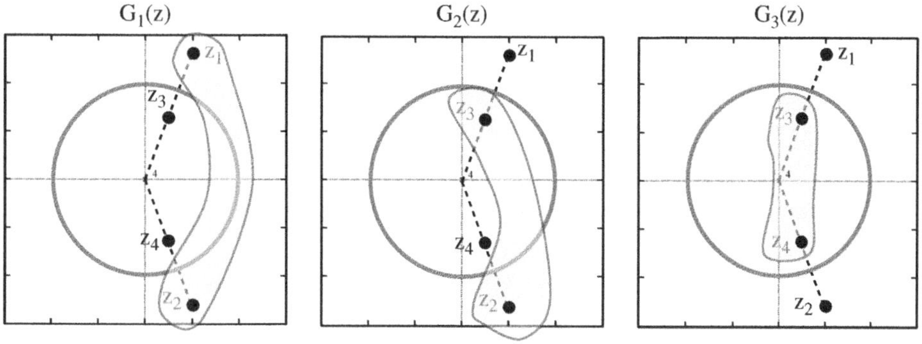

Figure 4.1 Factorization of autocorrelation sequences.

The minimum phase factorization has the remarkable benefit that all poles/zeros are inside the unit circle; filter $G(z) = C_{xx,\min}(z)$ is causal (obviously stable) and its whitening filter $1/G(z)$ is causal with poles/zeros exchanged.

In spectral analysis, the covariance sequence could have a limited length of $2N+1$ samples (say, $c_{xx}[n] = 0$ for $n \notin [-N, N]$); the $C_{xx}(z)$ has $2N$ zeros and there are $2^N - 1$ possible factorizations such as

$$C_{xx}(z) = G(z) \cdot G^*(1/z)$$

where $G(z)$ is any arbitrary complex-valued sequence obtained by selecting a subset of roots such that the sequence $G^*(1/z)$ complements all these roots. Among all the $2^N - 1$ factorizations, there is only one where all the roots are inside the unit-radius circle and thus $G(z)$ is minimum phase (and $G^*(1/z)$ is maximum phase) according to the Paley–Wiener theorem. An example can clarify this. Let $X(z) = 1 - z/2 + z^2/2$ be a polynomial with 2 roots $z_1 = (1 + j)$ and $z_2 = (1 - j)$; its autocorrelation is $C_{xx}(z) = X(z)X^*(1/z) = (3/2 - 3/2(z^{-1} + z) + (z^{-2} + z^2)$ with four roots (Figure 4.1): $z_1 = (1 + j)$, $z_2 = (1 - j)$, $z_3 = 1/2(1 + j)$, $z_4 = 1/2(1 - j)$; notice that $z_3 = 1/z_1^*$ and $z_4 = 1/z_2^*$. Factorizations are $G_1(z) = (1 - z_1 z^{-1})(1 - z_2 z^{-1})$, $G_2(z) = (1 - z_3 z^{-1})(1 - z_2 z^{-1})$, and $G_3(z) = (1 - z_3 z^{-1})(1 - z_4 z^{-1})$, which all give the same autocorrelation: $C_{xx}(z) = G_1(z)G_1^*(1/z) = G_2(z)G_2^*(1/z) = G_3(z)G_3^*(1/z)$, with $G_1(z)$ being max phase, and $G_3(z)$ min-phase. Incidentally, $G_3(z) = X(z)$.

4.5 Gaussian Random Process (Discrete-Time)

Random processes filtered by an LTI system preserve the Gaussian property, and this is a strong argument in favor of their widespread usage. Furthermore, a non-Gaussian random process $x[n]$ filtered by a filter $h[n]$ can be approximated by a Gaussian one provided that the filter is long enough and filter's coefficients are of comparable values (central limit theorem).

In statistical signal processing, the random variables of a random process are often ordered sequentially into a vector at time n such as

$$\mathbf{x}[n] = [x[n], x[n+1], \cdots, x[n+N-1]]^T.$$

The WSS guarantees that the mean is independent of the time-reference and thus

$$\mathbb{E}[\mathbf{x}[n]] = \mu_x \mathbf{1}_N$$

and the covariance is

$$\mathbf{C}_{xx} = \mathbb{E}[\mathbf{x}[n]\mathbf{x}^T[n]] - \mu_x^2 \mathbf{1}_N$$

that for the stationarity

$$\mathbf{C}_{xx} = \begin{bmatrix} c_{xx}[0] & c_{xx}[1] & c_{xx}[2] & \cdots & c_{xx}[N-1] \\ c_{xx}[-1] & c_{xx}[0] & c_{xx}[1] & \cdots & c_{xx}[N-2] \\ c_{xx}[-2] & c_{xx}[-1] & c_{xx}[0] & \cdots & c_{xx}[N-3] \\ \vdots & \vdots & \vdots & \ddots & \vdots \\ c_{xx}[1-N] & c_{xx}[2-N] & c_{xx}[3-N] & \cdots & c_{xx}[0] \end{bmatrix}$$

The covariance is independent of n (for WSS) and it is symmetric (or Hermitian symmetric, if complex-valued) characterized by being diagonal-constant; this is called a *Toeplitz structured matrix* (or Toeplitz matrix). This structure is a consequence of the WSS and it is a way of inferring this property from experiment. If the random process is Gaussian, the two central moments fully characterize the samples and WSS is also SSS, but the opposite is not true.

Recall that for any zero-mean WSS process:

$$\mathbf{C}_{xx} = \mathbf{R}_{xx} = \mathbb{E}[\mathbf{x}[n]\mathbf{x}^T[n]]$$

This situation is very common in statistical signal processing, and the exception is to have $\mu_x \neq 0$, which will always be specified when the case. Another case widely discussed in later chapters is the zero-mean AR(1) WSS with correlation (or covariance) $r_{xx}[n] = c_{xx}[n] = \sigma^2 \rho^{|n|}$, with $|\rho| < 1$; the correlation matrix is

$$\mathbf{R}_{xx} = \sigma^2 \begin{bmatrix} 1 & \rho & \rho^2 & \cdots & \rho^{N-1} \\ \rho & 1 & \rho & \cdots & \rho^{N-2} \\ \rho^2 & \rho & 1 & \cdots & \rho^{N-3} \\ \vdots & \vdots & \vdots & \ddots & \vdots \\ \rho^{N-1} & \rho^{N-2} & \rho^{N-3} & \cdots & 1 \end{bmatrix},$$

with a well-defined Toeplitz structure, that for a white random process ($\rho = 0$) is $\mathbf{R}_{xx} = \sigma^2\mathbf{I}$. This correlation matrix has explicit expressions for inverse and determinant:

$$\mathbf{R}_{xx}^{-1} = \frac{1}{\sigma^2(1-\rho^2)}\begin{bmatrix} 1 & -\rho & 0 & \cdots & 0 \\ -\rho & 1+\rho^2 & -\rho & \ddots & \vdots \\ 0 & \ddots & \ddots & \ddots & 0 \\ \vdots & \ddots & -\rho & 1+\rho^2 & -\rho \\ 0 & \cdots & 0 & -\rho & 1 \end{bmatrix}$$

$$|\mathbf{R}_{xx}| = \sigma^{2N}(1-\rho^2)^{N-1}$$

that are useful for many analytical derivations later in the text.

4.6 Measuring Moments in Stochastic Processes

From a limited set of samples of a random process, there are some sample metrics that can be conveniently evaluated. Let the WSS process $x[n]$ be zero-mean (covariance and correlation coincide), the *sample autocorrelation* from N samples $\{x[1], x[2], ..., x[N]\}$ is

$$\hat{r}_{xx}[n] = \frac{1}{2N+1}\sum_{k=1}^{N} x^*[k]x[n+k]$$

and from WSS (and ergodicity) it is

$$\mathbb{E}[\hat{r}_{xx}[n]] = r_{xx}[n]$$
$$\lim_{N\to\infty} \hat{r}_{xx}[n] = r_{xx}[n]$$

so that the z-transform

$$\hat{r}_{xx}[n] \leftrightarrow \hat{R}_{xx}(z)$$

for a limited set of observations attains the autocorrelation of the random process $R_{xx}(z)$ for $N \to \infty$.

Collecting the N samples into a vector $\mathbf{x}[n] = [x[n], x[n+1], ..., x[n+N-1]]^T$, the sample moments are dependent on the centering position (here the nth sample):

$$\hat{\mu}_x[n] = \frac{1}{N}\sum_{k=n}^{n+N-1} x[k] = \mathbf{1}_N^T\mathbf{x}[n]/\mathbf{1}_N^T\mathbf{1}_N$$

$$\hat{r}_{xx}[n] = \frac{1}{2N+1}\sum_{k=-N}^{N} x^*[k]x[n+k]$$

These denote the sample mean and autocorrelation of the sequence $\mathbf{x}[n]$ that, if WSS, coincide with the ensemble values for $N \to \infty$.

Samples can be arranged in successive blocks as

$$\mathbf{x}[n], \mathbf{x}[n+1], \ldots, \mathbf{x}[n+M]$$

so that the sample correlation matrix is

$$\hat{\mathbf{R}}_{xx} = \frac{1}{M} \sum_{k=n}^{n+M} \mathbf{x}[k]\mathbf{x}[k]^H$$

that is full-rank for $M \geq N$ (a *necessary condition*). Arrangements of the sample statistical values depend on the specific context and will be specified case-by-case.

Appendix A: Transforms for Continuous-Time Signals

The Fourier transform is an elegant tool for analysis and representation of signals. There are some basic operations on continuous-time signals that should be defined first:

- **Periodicity:** this refers to invariance vs. time; a signal $x(t)$ is periodic of period T if for a certain $T > 0$:

$$x(t) = x(t + kT), \qquad n = 0, \pm 1, \pm 2, \ldots$$

  and the smallest value of period T for the property to hold is the fundamental period. The most intuitive signal is the sinusoid: $x(t) = \cos(\omega_o t)$ with period $T = 2\pi/\omega_o$, but the exponential (or complex-valued sinusoid) $x(t) = \exp(j\omega_o t)$ is similarly periodic.
- **Convolution:**

$$x(t) * y(t) = \int x(\tau)y(t-\tau)d\tau = \int y(\tau)x(t-\tau)d\tau = y(t) * x(t)$$

$$x(t) * y(-t) = \int x(\tau)y(t+\tau)d\tau = \int y(\tau)x(t+\tau)d\tau = x(-t) * y(t)$$

(cross-correlation $x(t)$ and $y(t)$)

- **Dirac delta** $\delta(t)$ is an abstraction very in common signal analysis, it is a "function" that is zero everywhere except for $t = 0$ where it is infinite, and it has unitary area: $\int \delta(t)dt = 1$. It has some properties listed below:

scaling	$\int \delta(at)dt = \frac{1}{\lvert a \rvert}$
unitary area	$\int x(t)\delta(t)dt = x(0)$
sampling	$x(t)\delta(t-\tau) = x(\tau)\delta(t-\tau)$ as $\int x(t)\delta(t-\tau)dt = x(\tau)$
translation	$x(t) * \delta(t-\tau) = x(t-\tau)$
comb of Dirac deltas	$\sum_k \delta(t-kT)$

The comb-delta is the basis for getting periodic signals and uniform sampling from a signal $x(t)$ (Figure 4.2):

$$x(t) * \sum_k \delta(t-kT) = \sum_k x(t-kT) \text{ periodic signal with period } T$$

$$x(t) \times \sum_k \delta(t-kT) = \sum_k x(kT)\delta(t-kT) \text{ sampling with period } T$$

The continuous-time Fourier transform for angular frequency ω is

$$X(\omega) = \int x(t)e^{-j\omega t}dt = \mathcal{F}\{x(t)\}$$

while the inverse transform is

$$x(t) = \int X(\omega)e^{j\omega t}\frac{d\omega}{2\pi} = \mathcal{F}^{-1}\{X(\omega)\}$$

Sometimes the Fourier transform is defined for frequency

$$f = \frac{\omega}{2\pi}$$

as it is readily measured in engineering. Most of the time, "frequency" is short for angular frequency, and difference is clear from the context. $\mathcal{F}\{.\}$ and $\mathcal{F}^{-1}\{.\}$ denote

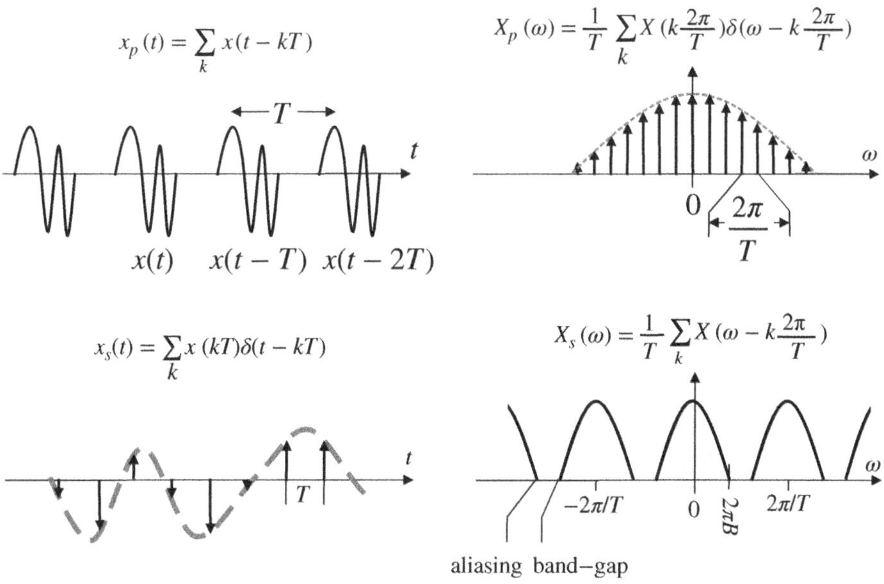

Figure 4.2 Periodic and sampled continuous-time signals.

transforms or analysis (decomposition of a signal into a sum of sinusoids) and synthesis (composition of a signal from multiple sinusoids, properly scaled). The mapping is denoted as

$$x(t) \longleftrightarrow X(\omega).$$

Some common properties (see e.g., [12]) are summarized in Table 4.3.

Sampling refers to the representation of a continuous-time signal $x(t)$ with discrete-time values of the original signal at sampling interval (or period) T (or equivalently, sampling frequency $1/T$ measured in samples per second, or Hz). The sampled signal $x_s(t)$ in Figure 4.2 looses all content everywhere except at the sampling time, where it is the value itself, and this can be conveniently represented by Dirac deltas:

$$x_s(t) = \sum_k x(kT)\delta(t - kT).$$

The sampling interval T can be properly chosen to guarantee that sampling can be reversed and there is no loss of information. The Nyquist–Shannon sampling theorem for a bandlimited signal $x(t)$ with frequency support $f \in [-B, B]$ of its Fourier transform $X(f) = \mathcal{F}[x(t)]$ (i.e., $X(f) = 0$ for $f \notin [-B, B]$) guarantees that the information content is preserved for the choice $2BT \leq 1$, or equivalently from the samples in $x_s(t)$ it is always possible to recover $x(t)$ with appropriate processing and filtering. In practice, the chosen sampling interval should be sufficient to leave a non-empty guard-band after periodic repetition of $X(\omega)$, to avoid frequency aliasing.

Table 4.3 Fourier transform properties.

$x(t) = \mathcal{F}^{-1}\{X(\omega)\}$	$X(\omega) = \mathcal{F}\{x(t)\}$	
$x(t)$ real	$X(-\omega) = X^*(\omega)$	symmetry
$\delta(t)$	1	Dirac delta
1	$\delta(\omega)$	constant
$x(t-\tau)$	$X(\omega)e^{-j\omega\tau}$	delay
$a_1x_1(t) + a_2x_2(t)$	$a_1X_1(\omega) + a_2X_2(\omega)$	linearity
$x(t) * h(t)$	$X(\omega)H(\omega)$	convolution (or filtering)
$x(t) \times h(t)$	$X(\omega) * H(\omega)$	modulation
$x(t) \times e^{-j\bar{\omega}t}$	$X(\omega - \bar{\omega})$	modulation with a complex sinusoid
$\int \lvert x(t)\rvert^2 dt = \int \lvert X(\omega)\rvert^2 \frac{d\omega}{2\pi}$		Parseval's relation (energy conservation)
$\frac{d}{dt}x(t)$	$j\omega X(\omega)$	differentiation
$\int_{-\infty}^{t} x(\tau)d\tau$	$\frac{1}{j\omega}X(\omega) + \pi X(0)\delta(\omega)$	integration
$\sum_k \delta(t - kT)$	$\frac{1}{T}\sum_\ell \delta(\omega - \ell\frac{2\pi}{T})$	Dirac-comb
$x_p(t) = \sum_k x(t - kT)$	$\frac{1}{T}\sum_\ell X(\frac{2\pi}{T})\delta(\omega - \ell\frac{2\pi}{T})$	periodic signal maps onto a comb
$x_s(t) = \sum_k x(kT)\delta(t - kT)$	$\frac{1}{T}\sum_\ell X(\omega - \ell\frac{2\pi}{T})$	sampling maps onto a periodic $X(\omega)$

Appendix B: Transforms for Discrete-Time Signals

Sequences (real or complex-valued) are conveniently represented by the coefficients of polynomials called *z-transforms*

$$X(z) = \mathcal{Z}\{x[n]\} = \sum_n x[n] \cdot z^{-n}$$

or in compact notation

$$x[n] \leftrightarrow X(z)$$

The inverse z-transform is

$$x[n] = \mathcal{Z}^{-1}\{X(z)\} = \frac{1}{j2\pi} \oint X(z) z^{n-1} dz$$

where the integral is around a circle centered at the origin and including the convergence region (see below). The roots $\{z_k\}$ of polynomial $X(z)$ are *zeros* of the sequence, and from the fundamental theorems of algebra, the Nth order polynomial for a sequence of N+1 samples can be rewritten as the product of N terms:

$$X(z) = x_0 \prod_k (1 - z_k z^{-1})$$

$X(z)$ is called a *minimum phase* polynomial if $|z_k| < 1$ for $\forall k$ (all zeros inside the unit circle), and *maximum phase* if $|z_k| > 1$. The z-transform for the sequence after the complex conjugate of all the samples is

$$X^*(z) = \sum_n x^*[n] \cdot z^{-n}$$

while

$$(X(z))^* = \left(\sum_n x[n] \cdot z^{-n} \right)^* \neq X^*(z)$$

Convolution is the product of polynomials

$$y[n] = x[n] * h[n] = \sum_k x[n-k]h[k] = \sum_k x[k]h[n-k] \leftrightarrow Y(z) = X(z) \cdot H(z)$$

and this makes the z-transform representation particularly important in handling sequences. Furthermore, since any sequence is factored into a products of monomials

$$X(z) = x_0(1 - z_1 z^{-1})(1 - z_2 z^{-1})... \leftrightarrow x_0\{1, -z_1\} * \{1, -z_2\} * ...$$

it follows that any arbitrary-length sequence (say of N+1 samples) can be written as the convolution of N elementary sequences of two samples $\{1, -z_k\}$, where the sample at delay zero is unitary (alternative equivalent notation using a discrete-time Dirac delta function is $\delta[n]-z_k\delta[n-1]$). Any property for any sequence can be validated for a smaller elementary 2-sample sequence, and then by induction to any sequence of arbitrary length.

It is customary to use z^{-1} as unit-delay operator so that $z^{-1}x[n]=x[n-1]$. It is straightforward to generalize this to an arbitrary convolution:

$$y[n]=H(z)\cdot x[n]=\left(\sum_k h[k]z^{-k}\right)\cdot x[n]=\sum_k h[k]\left(z^{-k}x[n]\right)=\sum_k h[k]x[n-k]$$

which justifies the block-diagrams that are usually adopted to account for linear filtering. Notice that the sequence $X(1/z)$ is the same as $X(z)$ except for mirror reflections wrt the origin.

The z-transform can be evaluated as the sum of polynomials, and there is a unique relationship between the sequence and its sum, given the convergence condition that establishes the *region of convergence*. To illustrate with a common z-transform, consider the infinite sample set

$$x[n]=\{1,a,a^2,a^3,...\}=a^n u[n] \text{ for } n\geq 0 \text{ (causal exponential)}$$

The z-transform converges to the sum

$$X(z)=\sum_{n\geq 0} a^n\cdot z^{-n}=\frac{1}{1-az^{-1}}$$

provided that

$$|az^{-1}|<1 \longrightarrow \text{Region of convergence: } |z|>|a|$$

and thus if $|a|<1$ there exists a Fourier transform as the region of convergence includes the unit circle (see below). Each root of the denominator polynomial is called a *pole* of the z-transform of the sequence, and in this case the single pole is at $z=a$. The sequence is causal and stable (i.e., not diverging: $\sum_{n\geq 0}|x[n]|<\infty$) if there exists a Fourier transform, or equivalently, (all) the pole(s) is (are) inside the unit circle. Conversely, for another sequence that differs from the previous one for being sequence reversed in time:

$$x[n]=\{...,b^{-3},b^{-2},b^{-1}\}=b^n u[-n-1] \text{ for } n<0 \text{ (anti-causal exponential)}$$

the z-transform converges to

$$X(z)=\sum_{n\leq -1} b^n\cdot z^{-n}=\sum_{n\geq 0} b^{-n}\cdot z^n-1=\frac{-1}{1-bz^{-1}}$$

and

$$\text{Region of convergence: } |z|<|b|$$

Table 4.4 Properties of z-transform.

$x[n] = \mathcal{Z}^{-1}\{X(z)\}$	$X(z) = \mathcal{Z}\{x[n]\}$	
$x^*[n]$	$X^*(z)$	complex conjugate
$x[-n]$	$X(1/z)$	reverse time
$\delta[n]$	1	Dirac delta
$x[n-n_o]$	$z^{-n_o}X(z)$	delay
$a_1x_1[n] + a_2x_2[n]$	$a_1X_1(z) + a_2X_2(z)$	linearity
$x[n] * h[n]$	$X(z)H(z)$	convolution (or filtering)
$x[n] \times a^n$	$X(\frac{z}{a})$	scaling in z
$nx[n]$	$-z\frac{dX(z)}{dz}$	differentiation
$x[n] \times h[n]$	$\frac{1}{j2\pi}\oint X(v)H(z/v)v^{-1}dv$	modulation
$x[0] = \lim_{z\to\infty} X(z)$		initial value theorem
$x[\infty] = \lim_{z\to 1}(z-1)X(z)$		final value theorem
$x[n] = a^n\cos(\omega_o n)u[n]$	$\frac{1-az^{-1}\cos(\omega_o)}{1-2az^{-1}\cos(\omega_o)+a^2z^{-2}}$	dumped causal sinusoid $\|z\| > a$

The pole is at $\bar{z} = b$, and the Fourier transform exists for the anti-causal sequence if $|b| > 1$. These systems with simple pole in/out of the unit circle are the basis general decomposition as a or b, respectively can be complex-valued and each represents a dumped causal or anti-causal complex sinusoid. For a real dumped sinusoid it is enough to pair two sequences: $(a^n + a^{*n})/2$. If $b = 1/a^*$, the regions of convergence are complementary one another and poles are in symmetric position wrt the unit circle. To generalize, any z-transform

$$X(z) = \frac{1}{A(z)} = \frac{1}{A_{\min}(z)} \cdot \frac{1}{A_{\max}(z)}$$

with arbitrary poles in $A(z)$ can be partitioned into min/max phase (poles inside/outside the unit circle), and since one is only interested in stable sequences, the minimum phase corresponds to the causal sequence, and the maximum phase is for the anti-causal sequence mutually convolved to get $x[n]$. Some common properties in Table 4.4 complete this brief summary (see e.g., [13, 14] for extensive treatments).

The Fourier transform of the sequence

$$X(\omega) = \mathcal{F}\{x[n]\} = \sum_n x[n]e^{-j\omega n} = \mathcal{Z}\{x[n]\}|_{z=e^{j\omega}}$$

follows from the z-transform when evaluated along the unit-radius circle

$$X(\omega) = X(z)|_{z=e^{j\omega}}$$

and this justifies the equivalent notation $X(\omega)$ or $X(e^{j\omega})$ for the Fourier transform of the sequence. It can easily be shown that for the minimal sequence $\{1, -z_k\}$, the following identity holds true:

$$\left|(1-z_k z^{-1})|_{z=e^{j\omega}}\right| = \left|(1-z_k^* z)\Big|_{z=e^{j\omega}}\right|$$

this implies that the amplitude of the Fourier transform does not change if zeros have reciprocal conjugate positions with respect to the unit-radius circle. This property holds for any arbitrary sequence and thus

$$X^*(1/z) = \sum_n x^*[n] \cdot z^n = x_0^* \prod_k (1 - z_k^* z)$$

so that, in general:

$$\left|X(z)|_{z=e^{j\omega}}\right| = \left|X^*(1/z)|_{z=e^{j\omega}}\right|.$$

The discrete Fourier transform (DFT) is the Fourier transform evaluated along N frequency-points uniformly spaced apart:

$$X_k = X(\omega = \frac{2\pi}{N}k) = X(z)\Big|_{z=e^{j\frac{2\pi}{N}k}} \text{ for } k = 0,1,2,...,N-1$$

Notice that the link between sampling of continuous-time signals and the discrete-time sequence follows from the sampling interval T as:

$$x[n] = x(nT)$$

The angular frequency for discrete-time signals ω is limited to the range $0, 2\pi$ due to the Fourier transform's representation that is periodic, but the relationship between the two is that

$$\omega = 2\pi f T = 2\pi \frac{f}{1/T}$$

where fT is called *normalized frequency*, as it is normalized to the frequency repetition of the FT after sampling (see Appendix A).

5

Models and Applications

This chapter introduces some simple models and applications that will provide a set of concrete examples for the dissertation of the estimation methods. Any data[1] should be modeled by a set of parameters, and the estimation problem reduces to getting the "best set of parameters" describing these data. But a first step is the understanding of the generation mechanism of the data and their representation; this is referred as data (or experiment) modeling, or simply model definition.

In signal processing, applications where the data is affected by additive noise is very common, and intuitive. The ith observation

$$x[i] = s[i;\theta] + w[i]$$

is the sum of a term of interest $s[i;\theta]$ (*signal*) that depends on a set of parameters $\theta \in \mathbb{R}^p$ (or in general $\theta \in \mathbb{C}^p$) such that $\theta \in \Theta$ (admissible set), and an additive random term $w(i)$ that is commonly referred as *noise*. Multiple observations $\{x[i]\}_{i=1}^N$ are used to estimate the set of parameters θ, these are ordered in a vector/matrix form (except for time series where the ordering is usually time-increasing, in general the ordering is arbitrary):

$$\begin{bmatrix} x[1] \\ \vdots \\ x[N] \end{bmatrix} = \begin{bmatrix} s[1;\theta] \\ \vdots \\ s[N;\theta] \end{bmatrix} + \begin{bmatrix} w[1] \\ \vdots \\ w[N-1] \end{bmatrix}$$

or

$$\mathbf{x} = \mathbf{s}(\theta) + \mathbf{w}$$

where $\mathbf{s}(\theta) = [s[1;\theta], s[2;\theta], ..., s[N;\theta]]^T$ is the mapping between parameters and signals.

The mapping $\mathbf{s}(.)$ is arbitrary and non-linear, this simplifies for linear regression models

$$\mathbf{x} = \mathbf{H}\theta + \mathbf{w} \tag{5.1}$$

1 Depending on the context, the observation $x[i]$ is referred to as *measurement* or gathered signal (to highlight the fact that $x[i]$ is collected in some way), *data* (to highlight that $x[i]$ are those only available in estimation), or *process* (to highlight the randomness in the observations). Here we prefer to use the term *observation* (or data, when related to applications) as this is referred to what can be observed and made available.

Statistical Signal Processing in Engineering, First Edition. Umberto Spagnolini.

Companion website: www.wiley.com/go/spagnolini/signalprocessing

where the *regressor* (or *observation) matrix*

$$\mathbf{H} = \begin{bmatrix} h_{11} & \cdots & h_{1p} \\ \vdots & \ddots & \vdots \\ h_{N1} & \cdots & h_{Np} \end{bmatrix}$$

is $\mathbf{H} \in \mathbb{R}^{N\times p}$ (or in general $\mathbf{H} \in \mathbb{C}^{N\times p}$) and is usually dependent on the specific problem. The noise $\mathbf{w}$ is random and it is characterized by the corresponding pdf.

An additive Gaussian noise model is when

$$\mathbf{w} \sim \mathcal{N}(\boldsymbol{\mu}_w, \mathbf{C}_w)$$

where $\boldsymbol{\mu}_w$ and $\mathbf{C}_w$ could be known (or not) depending on the specific application. For a linear model with deterministic parameters $\boldsymbol{\theta}$,

$$\mathbf{x} \sim \mathcal{N}(\boldsymbol{\mu}_w + \mathbf{H}\boldsymbol{\theta}, \mathbf{C}_w)$$

Some applications are discussed below to clarify the correct interpretation of each term.

5.1 Linear Regression Model

Continuous-time observations $x(t)$ are assumed to be generated by a linear model that depends on two parameters:

$$s(t) = A + Bt$$

so one expects data to be aligned along a straight line with some deviations as shown in Figure 5.1. The noisy data is modeled as $x(t) = A + Bt + w(t)$; when sampling at positions $\{t_1, t_2, ..., t_N\}$ it yields

$$x(t_n) = A + Bt_n + w(t_n) \quad \text{for } n = 1, 2, \cdots, N$$

or in matrix form

$$\begin{bmatrix} x(t_1) \\ x(t_2) \\ \vdots \\ x(t_N) \end{bmatrix} = \begin{bmatrix} 1 & t_1 \\ 1 & t_2 \\ \vdots & \vdots \\ 1 & t_N \end{bmatrix} \begin{bmatrix} A \\ B \end{bmatrix} + \begin{bmatrix} w(t_1) \\ w(t_2) \\ \vdots \\ w(t_N) \end{bmatrix}$$

This reduces to the linear model (5.1) for the choice

$$\boldsymbol{\theta} = [A, B]^T$$

$$\mathbf{H} = \begin{bmatrix} 1 & t_1 \\ 1 & t_2 \\ \vdots & \vdots \\ 1 & t_n \end{bmatrix}$$

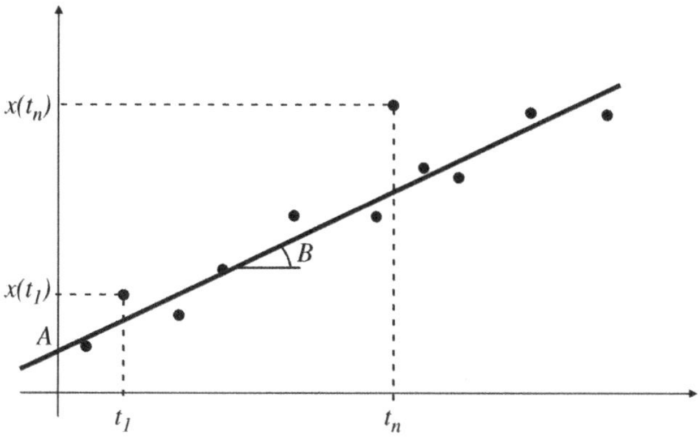

Figure 5.1 Linear regression model.

For a uniform sampling $t_n = \Delta t \cdot (n-1)$, the regression matrix simplifies to

$$\mathbf{H} = \begin{bmatrix} 1 & 0 \\ 1 & \Delta t \\ \vdots & \vdots \\ 1 & (N-1)\Delta t \end{bmatrix}$$

The estimation of the parameters A and B given a set of observations is known as a *linear regression* problem. In spite of the simplicity, linear regression occurs quite frequently in estimation problems such as frequency estimation of sinusoidal signals (Section 9.2 and Chapter 16), delay estimation in radar systems (Chapter 20), direction of arrivals for a plane wavefield impinging in a linear array of sensors (Chapter 19), and linear prediction (e.g., in stocks and finance) (Section 12.3). All these problems make use of the principles embedded in the linear regression model.

The polynomial regression model is an extension of linear regression:

$$x(t_n) = \theta_1 + \theta_2 t_n + \theta_3 t_n^2 + \cdots + \theta_p t_n^{p-1} + w(t_n)$$

In this case

$$\mathbf{H} = \begin{bmatrix} 1 & t_1 & t_1^2 & \cdots & t_1^{p-1} \\ 1 & t_2 & t_2^2 & \cdots & t_2^{p-1} \\ \vdots & \vdots & \ddots & \vdots & \vdots \\ 1 & t_N & t_N^2 & \cdots & t_N^{p-1} \end{bmatrix}$$

and the parameters are $\boldsymbol{\theta} = [\theta_1, \theta_2, ..., \theta_p]^T$ with $N \geq p$; the data model is still linear wrt the parameters. Estimation of coefficients $\theta_1, \theta_2, ..., \theta_p$ in a polynomial regression enables fitting the data and possibly predicting a value from other existing values. Let the samples be ordered to represent time $(t_1 < t_2 < ... < t_N)$, and $\hat{\boldsymbol{\theta}}_N$ be the estimate

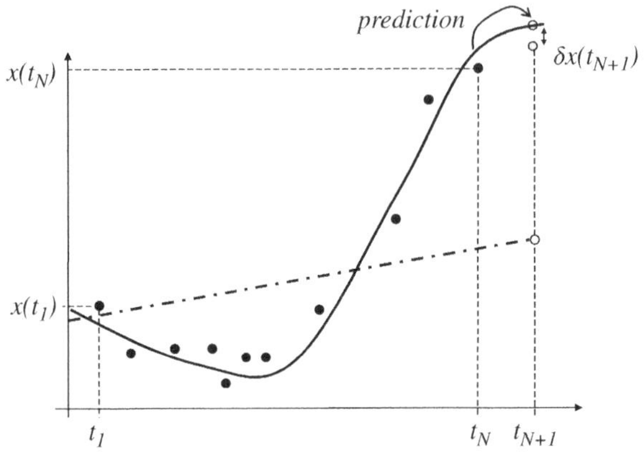

Figure 5.2 Polynomial regression and sample prediction $\hat{x}(t_{N+1})$.

from these N observations (indicated by the subscript); it is possible to predict the observations at time t_{N+1} as an extrapolation based on the current set of parameters $\hat{\theta}_N$:

$$\hat{x}(t_{N+1}) = [t_{N+1}, t_{N+1}^2, ..., t_{N+1}^{p-1}]\hat{\theta}_N$$

with a prediction error $\delta x(t_{N+1}) = x(t_{N+1}) - \hat{x}(t_{N+1})$ that depends on inaccuracies on $\hat{\theta}_N$ and the noise. For the example in Figure 5.2, if one uses a linear regression (dash-dot line) for data that needs a polynomial regression (solid line), the predicted value (empty-dot) differs from the true value (gray-dot) for a future time t_{N+1}. The prediction error $\delta x(t_{N+1})$ as shown in figure is smaller when considering a higher polynomial than for a linear regression (dash-dot line). Notice that all the N samples $(t_1, x_1), (t_2, x_2), ..., (t_N, x_N)$ contribute to estimating the parameters $\hat{\theta}_N$ to be used for a better prediction.

Remark: The observation matrix $\mathbf{H}$ is expected to be full-rank with rank$(\mathbf{H}) = p$ to guarantee that data are in the span of the model. Whenever rank$[\mathbf{H}] < p$, the regression matrix $\mathbf{H}$ is *ill conditioned* and it could be advisable to reduce the degree of the regression polynomial. Furthermore, there could be two (or more) conflicting observations carried out at the same time (i.e., $t_k = t_n$ but $x(t_k) \neq x(t_n)$), then two (or more) rows in $\mathbf{H}$ are identical.

5.2 Linear Filtering Model

Linear filtering models LTI systems and, for a filter with causal (for sake of simplicity) pulse response $\{g[0], g[1], ..., g[M-1]\}$, the signal $u[n]$ after filtering is the convolution

$$x[n] = g[n] * u[n] = \sum_{k=0}^{M-1} g[k]u[n-k]$$

Matrix notation becomes convenient to handle block-wise filtering—that is when signals are is arranged into finite length blocks. Let

$$\mathbf{u}_N = \begin{bmatrix} u[0] \\ u[1] \\ \vdots \\ u[N-1] \end{bmatrix}$$

$$\mathbf{g}_M = \begin{bmatrix} g[0] \\ g[1] \\ \vdots \\ g[M-1] \end{bmatrix}$$

be the input signal and the channel response arranged in vectors (vector lengths are as per the subscripts); the convolution has support of $M+N-1$ samples (i.e., $x[n]=0$ for $\forall n \notin [0, N+M-2]$) and the vector is

$$\mathbf{x}_{N+M-1} = \begin{bmatrix} x[0] \\ x[1] \\ \vdots \\ x[N+M-2] \end{bmatrix} = \begin{bmatrix} \sum_{k=0}^{M-1} g[k]u[0-k] \\ \sum_{k=0}^{M-1} g[k]u[1-k] \\ \vdots \\ \sum_{k=0}^{M-1} g[k]u[N+M-(k+1)] \end{bmatrix}$$
$$= \begin{bmatrix} g[0]u[0] \\ \sum_{k=0}^{1} g[k]u[1-k] \\ \vdots \\ g[M-1]u[N] \end{bmatrix}$$

where the last equality is specialized for this situation and highlights that some of the terms are zeros. Block-wise convolution is indicated by the following concise notation:

$$\mathbf{x}_{N+M-1} = \mathbf{u}_N * \mathbf{g}_M = \mathbf{g}_M * \mathbf{u}_N$$

Convolution can be arranged into a linear transformation as a matrix-vector product:

$$\mathbf{x}_{N+M-1} = \mathbf{u}_N * \mathbf{g}_M = \mathbf{U}\mathbf{g}_M = \mathbf{g}_M * \mathbf{u}_N = \mathbf{G}\mathbf{u}_N.$$

The form typically used when vector $\mathbf{g}_M$ collects the parameters is

$$\mathbf{U}\mathbf{g}_M = \underbrace{\begin{bmatrix} u[0] & 0 & \cdots & 0 & 0 \\ u[1] & g[0] & \cdots & 0 & 0 \\ \vdots & \vdots & \ddots & \vdots & \vdots \\ u[N-1] & u[N-2] & \cdots & u[0] & 0 \\ 0 & u[N-1] & \cdots & u[1] & u[0] \\ \vdots & \vdots & \ddots & \vdots & \vdots \\ 0 & 0 & \cdots & u[N-1] & u[N-2] \\ 0 & 0 & \cdots & 0 & u[N-1] \end{bmatrix}}_{\mathbf{U}\text{ dimension: }(N+M-1\times M)} \cdot \underbrace{\begin{bmatrix} g[0] \\ g[1] \\ \vdots \\ g[M-1] \end{bmatrix}}_{\mathbf{g}_M},$$

but when vector $\mathbf{u}_N$ collects the parameters is

$$\mathbf{G}\mathbf{u}_N = \underbrace{\begin{bmatrix} g[0] & 0 & \cdots & 0 & 0 \\ g[1] & g[0] & \cdots & 0 & 0 \\ \vdots & \vdots & \ddots & \vdots & \vdots \\ g[M-1] & g[M-2] & \cdots & g[0] & 0 \\ 0 & g[M-1] & \cdots & g[1] & g[0] \\ \vdots & \vdots & \ddots & \vdots & \vdots \\ 0 & 0 & \cdots & g[M-1] & g[M-2] \\ 0 & 0 & \cdots & 0 & g[M-1] \end{bmatrix}}_{\mathbf{G} \text{ dimension: } (N+M-1 \times N)} \cdot \underbrace{\begin{bmatrix} u[0] \\ u[1] \\ \vdots \\ u[N-1] \end{bmatrix}}_{\mathbf{u}_N}$$

The matrixes **U** and **G** are the convolution matrix as every column is the downward shifted copy of the previous one, and the last sample is cancelled as soon as the last sample exceeds the size of the output block. Both notations are equivalent and are used as needed as specified later in Section 5.2.3.

5.2.1 Block-Wise Circular Convolution

In the case of periodic signals the convolution is periodic and the block-size has the length of one period. So, for an input sequence with period N samples, the output sequence will have the same period N. In order to highlight this aspect, let

$$\tilde{u}[m] = u[m] * \sum_k \delta[m - kN]$$

be the periodic signal (indicated by symbol $\tilde{(\cdot)}$) obtained by the repetition of an aperiodic signal with period N; the filtering is

$$\tilde{x}[m] = g[m] * \underbrace{\left(u[m] * \sum_k \delta[m - kN] \right)}_{\tilde{u}[m]} = u[m] * \underbrace{\left(g[m] * \sum_k \delta[m - kN] \right)}_{\tilde{g}[m]}$$

or equivalently

$$\tilde{x}[m] = x[m] * \sum_k \delta[m - kN]$$

where

$$x[m] = g[m] * u[n].$$

Using the compact notation to highlight the cyclic structure of the convolution over a period N, the convolution over one block is

$$\mathbf{x}_N = \mathbf{g}_M \circledast_N \mathbf{u}_N = \tilde{\mathbf{G}}\mathbf{u}_N$$

and the $M \times M$ convolution matrix (for $M < N$ and causal filter) is

$$\tilde{\mathbf{G}} = \begin{bmatrix} g[0] & 0 & \cdots & g[M-1] & \cdots & g[2] & g[1] \\ g[1] & g[0] & \cdots & 0 & \cdots & g[3] & g[2] \\ \vdots & \vdots & \ddots & \ddots & \ddots & \vdots & \vdots \\ g[M-1] & g[M-2] & \cdots & g[0] & \cdots & 0 & 0 \\ 0 & g[M-1] & \cdots & g[1] & \cdots & 0 & 0 \\ \vdots & \vdots & \ddots & \ddots & \ddots & \vdots & \vdots \\ 0 & 0 & \cdots & g[M-3] & \cdots & g[0] & 0 \\ 0 & 0 & \cdots & g[M-2] & \cdots & g[1] & g[0] \end{bmatrix}$$

$\tilde{\mathbf{G}}$ is a circulant matrix in which each column on the right is the circularly down-shifted (by one step) version of the column on the left, and along the rows the shift moves on the right (and re-enters on the left). Notice that the structure of the upper-right corner denotes the periodic convolution matrix for causal filtering. However, for anti-causal filter the lower-left corner is non-zero, and if the filter contains both causal and anti-causal responses, the upper-right and lower-left blocks are non-zero.

5.2.2 Discrete Fourier Transform and Circular Convolution Matrixes

The discrete Fourier transform (DFT) operator of length N (DFT_N) is a linear transformation $\mathbf{W}_N$ with a Vandermonde structure:

$$\mathbf{W}_N = \frac{1}{\sqrt{N}} \begin{bmatrix} W_N^{0\cdot 0} & W_N^{0\cdot 1} & \cdots & W_N^{0\cdot m} & \cdots & W_N^{0\cdot (N-1)} \\ W_N^{1\cdot 0} & W_N^{1\cdot 1} & \cdots & W_N^{1\cdot m} & \cdots & W_N^{1\cdot (N-1)} \\ \vdots & \vdots & \ddots & \vdots & & \vdots \\ W_N^{n\cdot 0} & W_N^{n\cdot 1} & \cdots & W_N^{n\cdot m} & \cdots & W_N^{n\cdot (N-1)} \\ \vdots & \vdots & & \vdots & \ddots & \vdots \\ W_N^{(N-1)\cdot 0} & W_N^{(N-1)\cdot 1} & \cdots & W_N^{(N-1)\cdot m} & \cdots & W_N^{(N-1)\cdot (N-1)} \end{bmatrix}$$

where $W_N = \exp(-j2\pi/N)$ is the Nth root of one. The DFT is the Fourier transform $\mathcal{F}\{\mathbf{g}_M\}$ sampled regularly at intervals $2\pi/N$, and since

$$G[k] = [DFT_N(\mathbf{g}_M)]_k = \mathcal{F}\{\mathbf{g}_M\}\big|_{\omega=\frac{2\pi}{N}k}$$

is the kth DFT entry of $\mathbf{g}_M = [g[0], g[1], ..., g[M-1]]^T$, after padding with N-M zeroes,

$$\begin{bmatrix} G[1] \\ G[2] \\ \vdots \\ G[N] \end{bmatrix} = \mathbf{W}_N \begin{bmatrix} \mathbf{g}_M \\ 0 \\ \vdots \\ 0 \end{bmatrix} \updownarrow N-M$$

Rearranging gives:

$$\mathbf{W}_N \tilde{\mathbf{G}} \mathbf{W}_N^H = \mathbf{\Lambda}_G = \text{diag}\{G[1], G[2], \cdots, G[N]\}$$

The DFT transformation $\mathbf{W}_N$ is an orthonormal basis for every circulant matrix of dimension $(N \times N)$, and the corresponding DFT values are their eigenvalues. This is

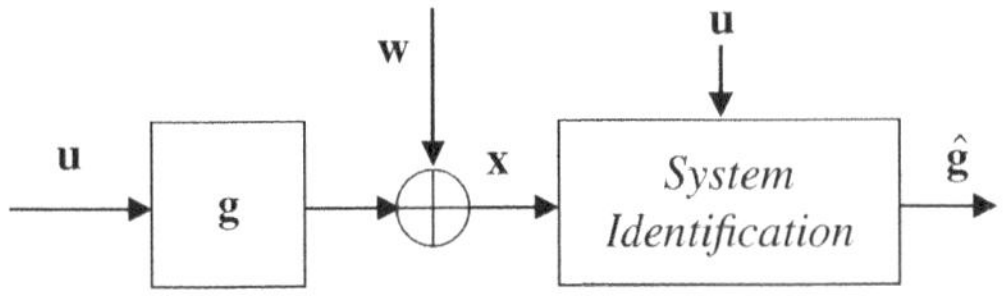

Figure 5.3 Identification problem.

an important property as it states that complex sinusoids with angular frequencies regularly spaced by $2\pi/N$ are the eigenvectors of signals when arranged as circular convolution, and the DFT outcomes are their eigenvalues. By extension, this property holds also for aperiodic signals provided that the period is large enough to avoid edge effects (as at the edges the circular and conventional outcomes of the convolution would clearly differ from one another). This property is more rigorous that the intuition given here, and is discussed later in Chapter 14.

5.2.3 Identification and Deconvolution

In linear filtering, the observations are affected by noise, and these can be reduced to the canonical model $\mathbf{x} = \mathbf{H}\boldsymbol{\theta} + \mathbf{w}$, depending on the specific problem that contributes to define the regression matrix and the parameters. Let the noisy observation model be defined as

$$\mathbf{x} = \mathbf{u} * \mathbf{g} + \mathbf{w}$$

with dimensions clear from the context. Two different types of problems can be investigated.

In **system identification** (Figure 5.3), a known signal **u** excites a linear system and the objective is to estimate the response **g** from the observations **x**. To solve this problem, the formulation to be used is

$$\mathbf{x} = \mathbf{U}\mathbf{g} + \mathbf{w}$$

where $\mathbf{U} = \mathbf{H}$ plays the role of known observation matrix, and the filter response $\mathbf{g} = \boldsymbol{\theta}$ is the unknown parameters arranged in a vector form. There are many applications of this case. Some examples follow. In wireless communication systems where the received signal propagates over an unknown link, this can be modeled as a linear system with an unknown **g**. Periodically, a known sequence **u** (called training sequence) is transmitted to estimate the propagation channel response, and the receiver employs the same sequence for the estimation (Section 15.2). In oil exploration, a known sequence excites the ground with artificially created seismic waves; the observations are the signals recorded at the surface that sense the subsurface, and the estimation of the response **g** gives information on the layered structure of the medium (Chapter 18).

Deconvolution is the inverse of convolution where the filter response **g** is known, but not the input **u**, which should be estimated (Figure 5.4). The formulation

$$\mathbf{x} = \mathbf{G}\mathbf{u} + \mathbf{w}$$

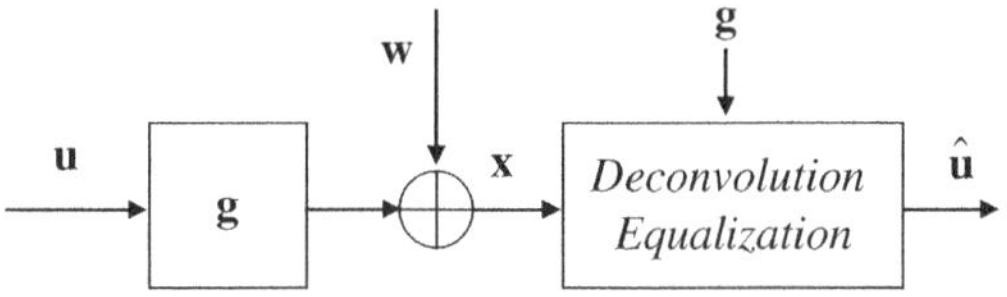

Figure 5.4 Deconvolution problem.

arranges the filter response as $\mathbf{G} = \mathbf{H}$, and the excitation $\mathbf{u} = \boldsymbol{\theta}$ now represents the unknown parameters. The problem is also referred to as equalization since the estimation $\hat{\mathbf{u}}$ of the input reconstructs the signal as if all linear distortion that occurred in the filtering is compensated for. Equalization is a crucial step in *all* communication systems to compensate for the effects of propagation over a dispersive medium (Section 15.2).

5.3 MIMO systems and Interference Models

Interference happens any time one signal is affected by the superposition of another one. A simple example of a 2×2 conversation in Figure 5.5 serves to better frame the problem. Let $\theta_1(t)$ and $\theta_2(t)$ be the two signals generated by two sources; these are received mixed by the two destinations as

$$\begin{cases} x_1(t) = h_{11}\theta_1(t) + h_{12}\theta_2(t) + w_1(t) \\ x_2(t) = h_{21}\theta_1(t) + h_{22}\theta_2(t) + w_2(t) \end{cases}$$

Each signal over the links $1 \to 1$ and $2 \to 2$ is interfered with by the other source. When receiving $x_1(t)$ (or $x_2(t)$ respectively) the destination can handle the interference as augmented noise $\bar{w}_1(t) = h_{12}\theta_2(t) + w_1(t)$ (or $\bar{w}_2(t) = h_{21}\theta_1(t) + w_2(t)$) and resume the models previously discussed, but this is clearly inefficient. The most effective approach

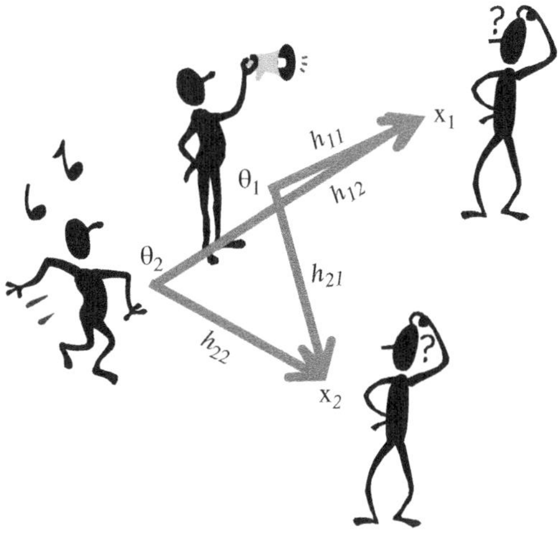

Figure 5.5 2×2 MIMO system.

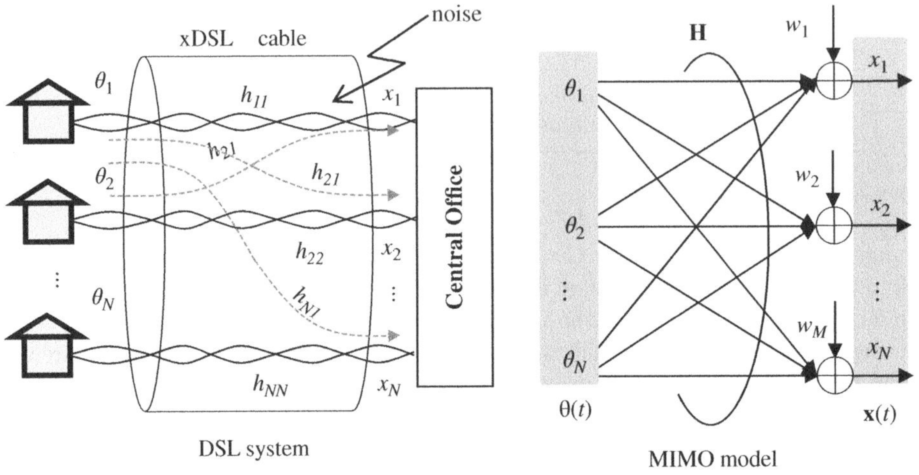

Figure 5.6 DSL system and $N \times N$ MIMO system.

is to handle the interference by appropriate modeling and exchange the information across destinations (i.e., a minimal cooperation among destinations offers a large benefit for all).

5.3.1 DSL System

Interference models are very common in statistical signal processing and are very peculiar in all communication systems. In xDSL systems, home equipment is connected to the same central office over twisted-pair copper cables. These cables are electrically coupled to one another so that the signal corresponding to the bit-stream generated by one piece of home equipment interferes with all the others within the same cable binder.

The overall problem of the transmission of N signals $\theta_1(t), \theta_2(t), ..., \theta_N(t)$ related to N pieces of home equipment toward the central office is shown in Figure 5.6. Signals can be arranged into $\theta(t) = [\theta_1(t), \theta_2(t), ..., \theta_N(t)]^T$ and the N signals at the central office can be modeled as a multiple-input-multiple-output (MIMO) system

$$\mathbf{x}(t) = \mathbf{H}\theta(t) + \mathbf{w}(t)$$

where the gain matrix $\mathbf{H} \in \mathbb{C}^{N\times N}$ is usually diagonal dominant $|\mathbf{H}(i,i)| \gg |\mathbf{H}(i,j)|$ for $\forall j \neq i$ but it could be $|\mathbf{H}(i,i)| < \sum_{j\neq i} |\mathbf{H}(i,j)|$, which for the problem at hand means that each interfering communication is smaller than the intended one if considered individually, but not the ensemble of all the interferers.

5.3.2 MIMO in Wireless Communication

In communication engineering, wireless devices (e.g., smartphones, tablets, etc.) act as mobile transmitters deployed within a certain coverage area (also called a coverage cell, or just a *cell*), each transmitting a signal $\theta_k(t)$ when active. It is very common that the fixed receiving unit (e.g., access point) is equipped with multiple antennas to enable multiple coexisting communications from multiple mobile devices. To simplify,

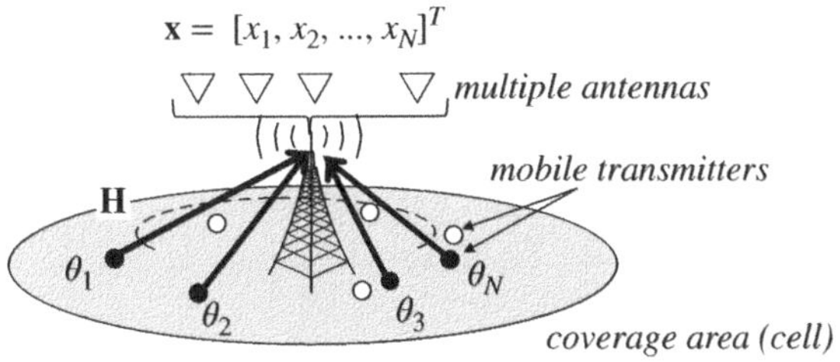

Figure 5.7 Wireless $N \times N$ MIMO system from N mobile transmitters (e.g., smartphones, tablets, etc...) to N antennas.

the number of antennas (N) is the same as the number of active users (filled-dots in Figure 5.7). The signal received by the ensemble of N transmitters over the air $\mathbf{x}(t) = \mathbf{H}\theta(t) + \mathbf{w}(t)$ is a combination of the signals by each, and the gain matrix $\mathbf{H} \in \mathbb{C}^{N\times N}$ is unstructured as the entry values depend on several complex factors not covered here (see [58]).

Moving to a realistic situation, cells are not isolated, but instead many receiving units are geographically deployed to guarantee that each wireless device can operate without interruption while moving. The scenario is illustrated in Figure 5.8 with multiple cells and multiple mobile devices; active transmitters (filled-dots) are served by the corresponding cell over orthogonal communication links.[2] To simplify, the fixed receiving unit is single antenna and can serve one mobile device at time, the kth received signal

$$x_k(t) = h_{kk}\theta_k(t) + \sum_{l\neq k} h_{kl}\theta_l(t) + w_k(t)$$

is the sum of the signals of transmitters within a cell, and the interference of the other cells (light-gray dashed lines in figure). The signals of multiple receivers is still a MIMO (also called multicell or network MIMO) model $\mathbf{x} = \mathbf{H}\boldsymbol{\theta} + \mathbf{w}$, and estimation of the transmitted signals requires cooperation among different equipment so as to handle multiple cells as one entity in a network MIMO approach.

A more realistic model of a multicell system is shown in Figure 5.9 and is based on access points with multiple antennas each, to guarantee better handling of interference by cancellation. Let the kth receiving unit be composed of M antennas; the ensemble of the M received signals $\mathbf{x}_k(t) = [x_{1,k}(t), x_{2,k}(t), ..., x_{M,k}(t)]^T$ is

$$\mathbf{x}_k(t) = \mathbf{H}_{kk}\boldsymbol{\theta}_k(t) + \sum_{l\neq k} \mathbf{H}_{kl}\boldsymbol{\theta}_l(t) + \mathbf{w}_k(t)$$

where $\mathbf{H}_{kk}\boldsymbol{\theta}_k(t)$ represents the N signals transmitted by the in-cell devices that follows the same model as for an isolated cell in Figure 5.7, and the second term is the sum of all interferers active in the other cells at the same time (light-gray dashed lines in

2 To avoid the interference among users within the same cell (intra-cell interference) each device in the same cell transmits on different time intervals (time division multiple access—TDMA), and/or different frequency (frequency division multiple acccess—FDMA), and/or different signatures (code division multiple access—CDMA).

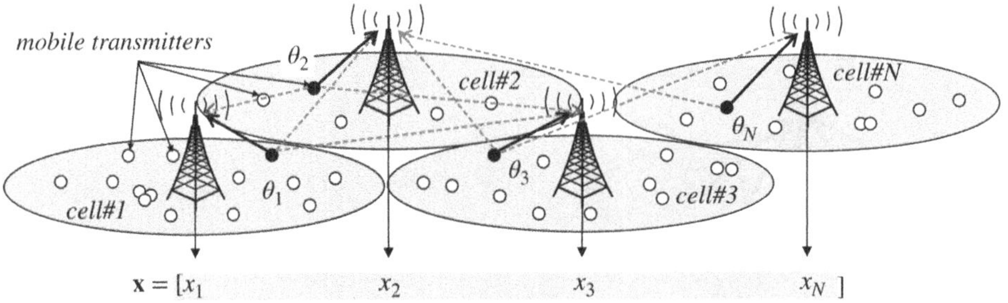

Figure 5.8 Multiple cells $N \times N$ MIMO systems.

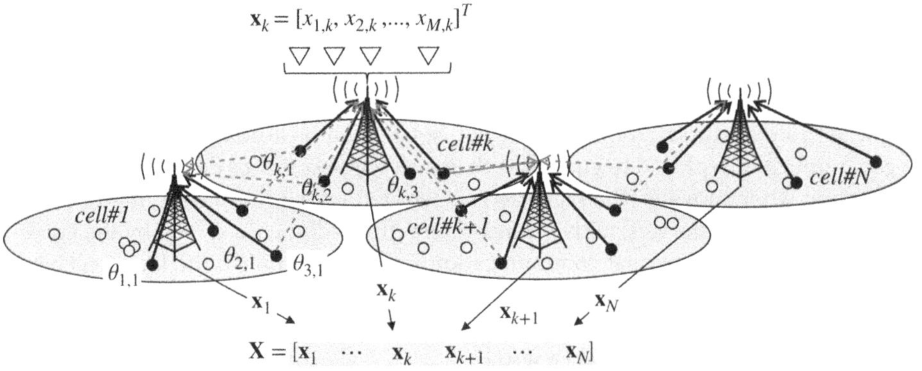

Figure 5.9 Multiple cells and multiple antennas MIMO system.

Figure 5.9). The ensemble of signals $\mathbf{X}(t) = [\mathbf{x}_1(t), \mathbf{x}_2(t), ..., \mathbf{x}_N(t)]$ is used to infer the transmitted ones using the linearity of the model. There is not just one arrangement of signals that yields a linear model, but the most intuitive one is

$$\underbrace{\begin{bmatrix} \mathbf{x}_1(t) \\ \mathbf{x}_2(t) \\ \vdots \\ \mathbf{x}_N(t) \end{bmatrix}}_{\mathbf{x}(t)} = \underbrace{\begin{bmatrix} \mathbf{H}_{11} & \mathbf{H}_{12} & \cdots & \mathbf{H}_{1N} \\ \mathbf{H}_{21} & \mathbf{H}_{22} & & \mathbf{H}_{2N} \\ \vdots & & \ddots & \vdots \\ \mathbf{H}_{N1} & \mathbf{H}_{N2} & \cdots & \mathbf{H}_{NN} \end{bmatrix}}_{\mathbf{H}} \underbrace{\begin{bmatrix} \theta_1(t) \\ \theta_2(t) \\ \vdots \\ \theta_N(t) \end{bmatrix}}_{\theta(t)} + \underbrace{\begin{bmatrix} \mathbf{w}_1(t) \\ \mathbf{w}_2(t) \\ \vdots \\ \mathbf{w}_N(t) \end{bmatrix}}_{\mathbf{w}(t)}$$

where one can easily recognize in the (large) block-partitioned vectors/matrix the conceptual counterpart of intra-cell/inter-cell communication systems. This is the key idea exploited in cooperative multicell communication systems of modern communication engineering; it is still based on the linear model, with high complexity in terms of processing capability.

MIMO Convolution

Up to now, MIMO systems have involved the instantaneous mixing of the signals $\boldsymbol{\theta}(t)$ to give $\mathbf{x}(t)$. However, the transfer matrix $\mathbf{H}$ is not necessarily instantaneous, but affected by a matrix impulse response. The original example of 2×2 can be adapted to this case, for block-wise convolution, indicated in discrete-time as:

$$\begin{cases} x_1[n] = h_{11}[n] * \theta_1[n] + h_{12}[n] * \theta_2[n] + w_1[n] \\ x_2[n] = h_{21}[n] * \theta_1[n] + h_{22}[n] * \theta_2[n] + w_2[n] \end{cases}$$

The block-wise notation is (blocks are of proper size):

$$\begin{cases} \mathbf{x}_1 = \mathbf{h}_{11} * \theta_1 + \mathbf{h}_{12} * \theta_2 + \mathbf{w}_1 \\ \mathbf{x}_2 = \mathbf{h}_{21} * \theta_1 + \mathbf{h}_{22} * \theta_2 + \mathbf{w}_2 \end{cases}$$

that in terms of convolution matrixes $\mathbf{H}_{ij}$ for the impulse response $\mathbf{h}_{ij}$ yields

$$\underbrace{\begin{bmatrix} \mathbf{x}_1 \\ \mathbf{x}_2 \end{bmatrix}}_{\mathbf{x}} = \underbrace{\begin{bmatrix} \mathbf{H}_{11} & \mathbf{H}_{12} \\ \mathbf{H}_{21} & \mathbf{H}_{22} \end{bmatrix}}_{\mathbf{H}} \underbrace{\begin{bmatrix} \theta_1 \\ \theta_2 \end{bmatrix}}_{\theta} + \underbrace{\begin{bmatrix} \mathbf{w}_1 \\ \mathbf{w}_2 \end{bmatrix}}_{\mathbf{w}}$$

which is again the linear model used up to now. Extension to larger MIMO systems is straightforward.

All developments so far isolate the relationship between the input (unknown) signal over a known MIMO system $\mathbf{H}$, and this is the basis for MIMO deconvolution as a generalization of the deconvolution in Section 5.2.3. However, one might reverse the models for MIMO identification by isolating the known excitation into a matrix $\mathbf{\Theta}$, and the unknown MIMO filters into a (tall) vector $\mathbf{h}$ such as:

$$\mathbf{x} = \mathbf{\Theta}\mathbf{h} + \mathbf{w}$$

The detail of how to make this arrangement are left to the reader.

5.4 Sinusoidal Signal

The model for a sinusoidal signal affected by noise is

$$x[n] = a_o \cos(\omega_0 n + \phi_o) + w[n] \quad \text{for } n = 0, 1, \ldots, N-1$$

The parameters are the amplitude (a_o) and/or phase (ϕ_o) and/or frequency (ω_0) of the sinusoid used to generate the signal. Using trigonometry to give a new set of variables $\alpha_1 = a_o \cos\phi_o$ and $\alpha_2 = -a_o \sin\phi_o$ for amplitude and phase, the model is

$$\underbrace{\begin{bmatrix} x[0] \\ x[1] \\ \vdots \\ x[N-1] \end{bmatrix}}_{\mathbf{x}} = \underbrace{\begin{bmatrix} \underbrace{\begin{matrix} 1 \\ \cos(\omega_0) \\ \vdots \\ \cos((N-1)\omega_0) \end{matrix}}_{\mathbf{c}(\omega_0)} & \underbrace{\begin{matrix} 0 \\ \sin(\omega_0) \\ \vdots \\ \sin((N-1)\omega_0) \end{matrix}}_{\mathbf{s}(\omega_0)} \end{bmatrix}}_{\mathbf{H}(\omega_0)} \cdot \underbrace{\begin{bmatrix} \alpha_1 \\ \alpha_2 \end{bmatrix}}_{\alpha} + \underbrace{\begin{bmatrix} w[0] \\ w[1] \\ \vdots \\ w[N-1] \end{bmatrix}}_{\mathbf{w}}$$

Inspection of this model shows that there is a linear dependence on $\boldsymbol{\alpha}$, but that it is non-linear wrt frequency (ω_0); this is highlighted in the notation as the argument of the observation matrix $\mathbf{H}(\omega_0)$.

Notice that the following properties hold (for $\omega_0 \neq 0, \pi$):

$$\mathbf{c}(\omega_0)^T\mathbf{c}(\omega_0) \simeq \frac{N}{2}$$
$$\mathbf{s}(\omega_0)^T\mathbf{s}(\omega_0) \simeq \frac{N}{2}$$
$$\mathbf{c}(\omega_0)^T\mathbf{s}(\omega_0) \simeq 0$$

as a special case (for $i \neq 0, N/2$):

$$\mathbf{c}(\frac{2\pi i}{N})^T\mathbf{c}(\frac{2\pi j}{N}) = \frac{N}{2}\delta_{i-j}$$
$$\mathbf{s}(\frac{2\pi i}{N})^T\mathbf{s}(\frac{2\pi j}{N}) = \frac{N}{2}\delta_{i-j}$$
$$\mathbf{c}(\frac{2\pi i}{N})^T\mathbf{s}(\frac{2\pi j}{N}) = 0$$

The proof is simple. Recall that the power of a discrete-time sinusoid is $\sum_{k=1}^{N}\cos(\omega_0 n)^2/N \simeq 1/2$, and this extends to the orthogonality $\mathbf{c}(\omega_0)^T\mathbf{s}(\omega_0)$. The approximation is exact for $\omega_0 = 2\pi k/N$ (i.e., when the frequency of the sinusoid completes "exactly" k cycles within N samples as for a DFT, see Section 5.2.2) and asymptotically for $N \to \infty$. Interestingly, the asymptotic orthogonality holds for any pair of frequencies $|\omega_1 - \omega_2| > 2\pi/N$ when neglecting minor terms.

The model defined here can be generalized to the sum of p sinusoids as:

$$x[n] = \sum_{\ell=1}^{p} a_\ell \cos(\omega_\ell n + \phi_\ell) + w[n]$$

which can be rewritten compactly as

$$\mathbf{x} = \left[\mathbf{c}(\omega_1), \mathbf{s}(\omega_1), ..., \mathbf{c}(\omega_p), \mathbf{s}(\omega_p)\right] \cdot \boldsymbol{\alpha} + \mathbf{w} = \mathbf{H}(\boldsymbol{\omega})\boldsymbol{\alpha} + \mathbf{w}$$

where the $N \times 2p$ matrix $\mathbf{H}(\boldsymbol{\omega})$ (with $N > 2p$) collects in columns the bases of sinusoids with arbitrary phases (as cosine and sine basis for each sinusoid) and it is full-rank if there is no pair of sinusoids with the same frequency. However, in general, $\mathbf{c}(\omega_i)^T\mathbf{c}(\omega_j) \neq 0$ for $\forall i \neq j$ (or equivalently $\mathbf{s}(\omega_i)^T\mathbf{s}(\omega_j) \neq 0$, and $\mathbf{s}(\omega_i)^T\mathbf{c}(\omega_j) \neq 0$) as the sinusoids are not orthogonal one another and thus coupled (except if $\omega_i \in \{0, 2\pi/N, 2 \times 2\pi/N, ..., (N-1) \times 2\pi/N\}$).

The model for p sinusoids depends on the set of p frequencies $\boldsymbol{\omega} = [\omega_1, \omega_2, ..., \omega_p]^T$ and $2p$ amplitudes $\boldsymbol{\alpha} = [a_1 \cos\phi_1, a_1 \sin\phi_1, ..., a_p \cos\phi_p, a_p \sin\phi_p]^T$, or equivalently p amplitudes $(a_1, ..., a_p)$ and p phases $(\phi_1, ..., \phi_p)$. The main advantage of the linear regression model is the decoupling of the terms in the model that are linear from those that are unavoidably non-linear, as for the frequencies $\boldsymbol{\omega}$.

5.5 Irregular Sampling and Interpolation

In real life, irregular sampling is far more common than one might imagine. Sampling can be irregular because regular sampling is affected by random sampling errors (*sampling jitter*), or through the impossibility of having regular sampling, or even intentionally. To visualize the problem, one can imagine a network of sensing devices as in Figure 5.10 (measuring temperature, or pressure, or precipitation [22], or the ocean's salinity and bio-geochemical data from the sea surface [23], or densities in petroleum reservoirs from oil-wells, just to mention a few) that are deployed whenever/wherever possible but in known positions, and it is of interest to interpolate the observations over a continuous set, or even on a regular set as a resampling from the interpolated data (this step is called *data re-gridding*). The model for interpolation from irregular sampling is considered first for a 1D signal; the effect of sampling jitter is investigated later and extended to 2D sampling.

Let x_i be the observation at position (say time) t_i, so the observations in Figure 5.11 are N pairs

$$(t_i, x_i)$$

We need to interpolate $x(t|\theta)$ for any t so that interpolation *honors the data* $x(t_i|\theta) \simeq x_i$. Since interpolation is the processing to estimate data in a position where data is missing, the most appropriate way would be to model the way observations are generated, but

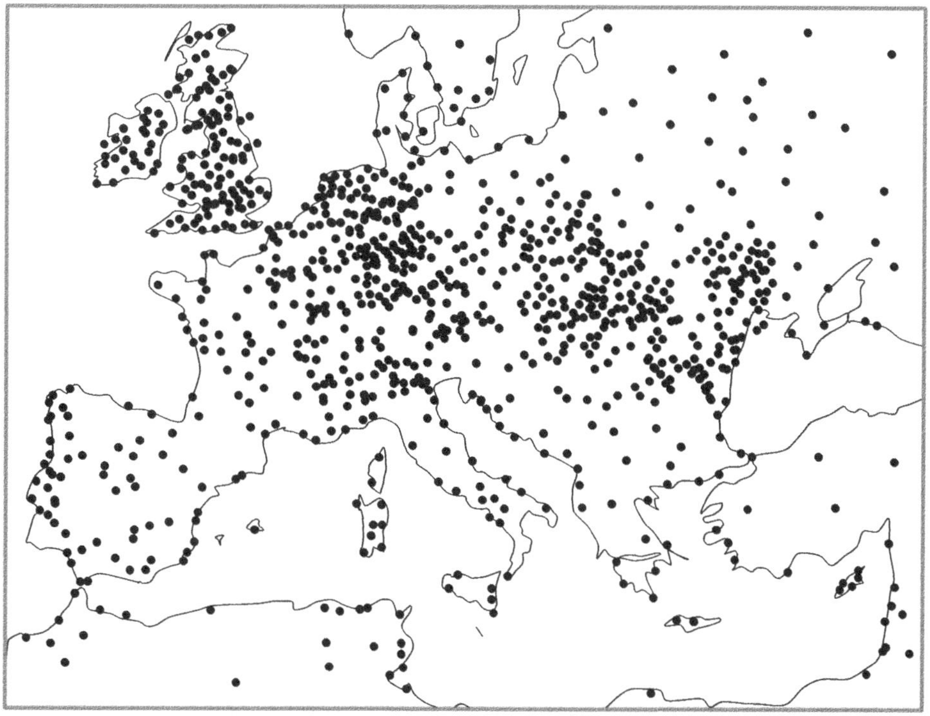

Figure 5.10 Irregular sampling of weather measuring stations (yellow dots) [*image from Google Maps*].

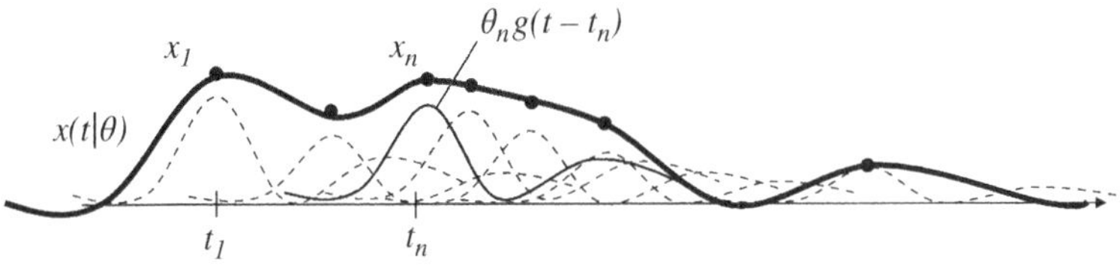

Figure 5.11 Interpolation from irregular sampling.

this is not always so trivial (imagine the annual precipitation that depends on a very complex model!) and so a way should be found to handle the problem. That is to say that the interpolation method is not unique, and below are some of the most common methods, all referred to the canonical linear model.

Irregular Basis

Interpolation is expected to honor the data while still smoothing across the available observations. An arbitrary continuous function $g(t)$ can be translated over the positions on the available observations, with a set of amplitudes $\{\theta_k\}_{k=1}^N$ to be defined

$$x(t|\theta) = \sum_{k=1}^{N} \theta_k g(t - t_k)$$

This interpolated values should honor the data and this is instrumental in order to have a set of enough equations for the unknowns $\{\theta_k\}_{k=1}^N$

$$x_i = \sum_{k=1}^{N} \theta_k g(t_i - t_k) \rightarrow \underbrace{\begin{bmatrix} x_1 \\ x_2 \\ \vdots \\ x_N \end{bmatrix}}_{\mathbf{x}} = \underbrace{\begin{bmatrix} g(0) & g(t_1 - t_2) & \dots & g(t_1 - t_N) \\ g(t_2 - t_1) & g(0) & \dots & g(t_2 - t_N) \\ \vdots & \vdots & \ddots & \vdots \\ g(t_N - t_1) & g(t_N - t_2) & \dots & g(0) \end{bmatrix}}_{\mathbf{H}} \times \underbrace{\begin{bmatrix} \theta_1 \\ \theta_2 \\ \vdots \\ \theta_N \end{bmatrix}}_{\theta},$$

The amplitudes

$$\hat{\theta} = \mathbf{H}^{-1}\mathbf{x}$$

are obtained as the solution of a full-rank linear system (if there are no conflicting observations $x_i \neq x_j$ related to the same time values $t_i = t_j$). If observations are noisy, the model becomes $\mathbf{x} = \mathbf{H}\theta + \mathbf{w}$. Choice of $g(t)$ sets the continuity and the characteristics of the interpolated values within and outside of the support of the samples. A routinely employed choice for $g(t)$ is the Gaussian function, or spline functions,[3] where the

3 A spline is an instrument for hand-drafting to smoothly connect dots. It is a flexible thin plastic tool used to pass through the given data points to yield a smooth curve in between. In signal processing, spline bases are obtained as convolution of rectangles. Let $g_1(t) = rect(t)$ be the rectangle over the unitary support as first basis, the basis of order 2 is $g_2(t) = g_1(t) * g_1(t)$ (triangle) that contributes to linear interpolation. By extention the n-th order spline is:

$$g_n(t) = \underbrace{g_1(t) * g_1(t) * \dots * g_1(t)}_{\text{convolution } n \text{ terms}} = g_{n-1}(t) * g_1(t).$$

duration should be adapted to the specific problem and sets the degree of smoothness of the continuous-time function.

Regular Basis

Alternatively, the basis function $g(t)$ can be translated over a regular grid on M positions $(M \leq N)$ with spacing Δt so that

$$x(t|\theta) = \sum_{k=1}^{M} \theta_k g(t - k\Delta t)$$

is the linear system to get the set of $\{\theta_k\}_{k=1}^{M}$ values

$$\underbrace{\begin{bmatrix} x_1 \\ x_2 \\ \vdots \\ x_N \end{bmatrix}}_{\mathbf{x}} = \underbrace{\begin{bmatrix} g(t_1 - \Delta t) & g(t_1 - 2\Delta t) & \dots & g(t_1 - M\Delta t) \\ g(t_2 - \Delta t) & g(t_2 - 2\Delta t) & \dots & g(t_2 - M\Delta t) \\ \vdots & \vdots & \ddots & \vdots \\ g(t_N - \Delta t) & g(t_N - 2\Delta t) & \dots & g(t_N - M\Delta t) \end{bmatrix}}_{\mathbf{H}} \times \underbrace{\begin{bmatrix} \theta_1 \\ \theta_2 \\ \vdots \\ \theta_M \end{bmatrix}}_{\boldsymbol{\theta}} + \mathbf{w}$$

where $\mathbf{w}$ accounts for the accuracy of the observations according to a statistical model, if available. This method is different from the irregular basis deployment one as here the bases $g(t)$ are uniformly translated and can be mutually orthogonal, say

$$g(k\Delta t) = \delta_k.$$

A common choice that resembles the bandlimited interpolation is

$$g(t) = \frac{\sin(\pi t/L\Delta t)}{\pi t/L\Delta t}$$

where $L \in \{1, 2, 3, \dots\}$ rules the degree of smoothness (larger L gives smoother interpolation).

2D Interpolation

Figure 5.12 shows an example of interpolation in 2D from the tuple

$$(u_i, v_i, x_i)$$

The problem is the same as above—to have a continuous 2D function $x(u, v|\theta)$ depending on the set of observations. Adapting the the first model to 2D gives

$$x(u, v|\theta) = \sum_{k=1}^{N} \theta_k g(u - u_k, v - v_k)$$

where the choice of the basis $g(u, v)$ constrains the interpolation in those regions without any measurements (or with coarser density of measurements). The model is loosely defined as it depends on the choice of the basis, but there is a wide literature on random sampling that can be recommended to the interested reader, such as [24].

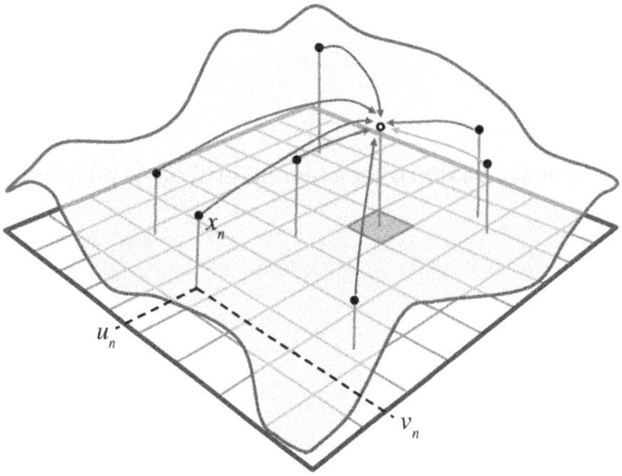

Figure 5.12 Interpolation in 2D.

5.5.1 Sampling With Jitter

Consider a zero-mean bandlimited Gaussian continuous-time process $x(t)$ that is sampled with sampling interval T, but sampling times are affected by a zero-mean random term (jitter) δt_k:

$$x(t_k) = x(kT + \delta t_k)$$

Therefore, each sample is erroneously assigned (and is considered assigned) to the variable $x(kT)$ and thus it should be considered affected by random impairments

$$\varepsilon_k = x(t_k) - x(kT)$$

due to random sampling. Properties of the equivalent additive noise ε_k can be derived assuming that $|\delta t_k| \ll T$, the linear approximation of the process around the sampling time is

$$x(kT + \delta t_k) \simeq x(kT) + \dot{x}(kT)\delta t_k$$

so that

$$\varepsilon_k \simeq \dot{x}(kT)\delta t_k$$

The process ε_k is multiplicative: it depends on signal's derivative $\dot{x}(kT)$ and sampling jitter δt_k as two statistically independent WSS processes. Being zero-mean, the autocorrelation and the power spectral density of this additive noise is

$$r_\varepsilon(\ell) = r_{\dot{x}}(\ell) r_{\delta t}(\ell) \leftrightarrow |2\pi f|^2 S_x(f) * S_{\delta t}(f)$$

and is characterized by high-frequency content.

Even if the statistical properties of sampling jitter depend on the oscillators' stability, we can further simplify to gain insight into the jitter effects. Assuming that the sampling jitter itself is an uncorrelated process with power $\sigma_{\delta t}^2$, and that the signal is bandlimited to $|f| \leq 1/2T$ with power spectral density $S_x(f) = T\sigma_x^2$, the power of jitter-equivalent additive noise

$$\sigma_\varepsilon^2 = \mathbb{E}[\varepsilon_k^2] = 2\sigma_{\delta t}^2 T\sigma_x^2 \int_0^{1/2T} (2\pi f)^2 df = \frac{8\sigma_{\delta t}^2 T\sigma_x^2}{3(2T)^3} = \frac{1}{3}\frac{\sigma_{\delta t}^2}{T^2}\sigma_x^2$$

depends on the fraction of the sampling error compared to the sampling period $\sigma_{\delta t}/T$, and it dominates the higher frequency portion. For a ratio $\sigma_{\delta t}/T = 10^{-5}$ (e.g., a good piezoelectric oscillator), the signal to sampling jitter ratio is 105 dB, but any further degradation (aging, temperature, etc.) could easily raise this to $\sigma_{\delta t}/T = 10^{-4}$, thus resulting in a loss of 20 dB. In some applications, this value is considered too large and it cannot be tolerated.

5.6 Wavefield Sensing System

In remote sensing systems, the environment is sensed from the backscattering of wavefields generated in a controlled manner. There are many situations where a wavefield is exploited to give a remotely sensed image, such as in satellite or subsurface radar imaging using electromagnetic waves, ultrasound medical or non-destructive diagnostic imaging using acoustic waves, and geophysical oil/gas/water prospecting using elastic waves. All these applications share several commonalities that are introduced below by taking radar sensing as an exemplary wavefield sensing system. An in-depth discussion of this area is in Chapters 18 and 19 for the specific processing methods.

Figure 5.13 illustrates an example of a radar system where a known and carefully designed waveform $g(t)$ is generated at transmitter (Tx). The backscattered echoes related to the targets are characterized by delays $\{\tau_\ell\}$ (or *two-way traveltimes*) that depend on the distance (*range* in jargon) and propagation velocity ($v = c = 3 \times 10^8$ m/s for electromagnetic waves in air), and amplitudes $\{\alpha_\ell\}$ that depend on the target size and its reflectivity. In general the measured signal is a superposition of L delayed/scaled echoes ($L = 2$ in Figure 5.13):

$$x(t) = \sum_{\ell=1}^{L} \alpha_\ell g(t - \tau_\ell) + w(t)$$

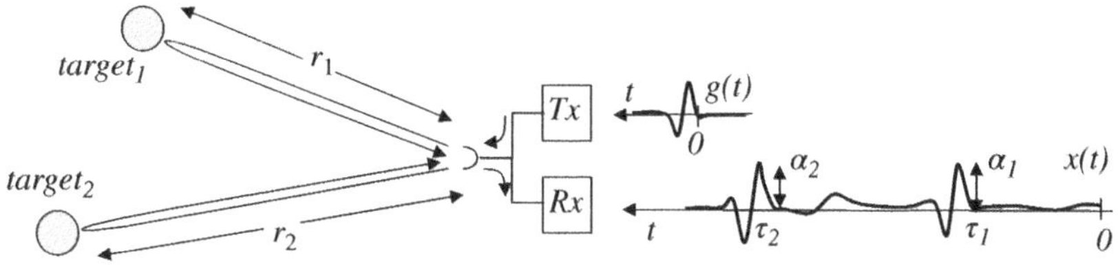

Figure 5.13 Radar system with backscattered waveforms from remote targets.

with some additive random noise $w(t)$ superimposed. The sampled signals with sampling interval Δt is

$$\begin{bmatrix} x(\Delta t) \\ x(2\Delta t) \\ \vdots \\ x(N\Delta t) \end{bmatrix} = \sum_{\ell=1}^{L} \alpha_\ell \underbrace{\begin{bmatrix} g(\Delta t - \tau_\ell) \\ g(2\Delta t - \tau_\ell) \\ \vdots \\ g(N\Delta t - \tau_\ell) \end{bmatrix}}_{\mathbf{g}(\tau_\ell)} + \begin{bmatrix} w(\Delta t) \\ w(2\Delta t) \\ \vdots \\ w(N\Delta t) \end{bmatrix}$$

where $\mathbf{g}(\tau_k)$ is the vector of the known waveform delayed by the unknown delay τ_k. The model becomes

$$\mathbf{x} = [\mathbf{g}(\tau_1), \mathbf{g}(\tau_2), ..., \mathbf{g}(\tau_L)]\boldsymbol{\alpha} + \mathbf{w} = \mathbf{H}(\boldsymbol{\tau})\boldsymbol{\alpha} + \mathbf{w}$$

where the parameters of interest $\boldsymbol{\theta} = [\boldsymbol{\tau}^T, \boldsymbol{\alpha}^T]^T$ are both the set of delays $\boldsymbol{\tau} = [\tau_1, \tau_2, ..., \tau_L]^T$ and the amplitudes $\boldsymbol{\alpha} = [\alpha_1, \alpha_2, ..., \alpha_L]^T$. The model is linear in amplitudes but non-linear wrt the delays, and this motivates special consideration of delay estimation methods in a dedicated chapter (Chapter 20).

In many situations, the only parameters of interest are the delays $\boldsymbol{\tau}$ as these give the exact distance r_ℓ from the transmitter accounting for the two-way traveltime

$$r_\ell = \frac{c}{2}\tau_\ell$$

Notice that the two-way traveltime is equivalent to one-way traveltime in a medium with half propagating velocity; this remark is crucial for imaging methods (Section 18.5).

The choice of the waveform $g(t)$ depends on the specific application and it could be narrow-pulse (or wideband) to guarantee that two closely spaced targets can be distinguished from their isolated waveforms. On the other hand, a sinusoidal waveform $g(t) = \cos(\omega_o t)$ (possibly in the form of modulated pulses $g(t) = a(t)\cos(\omega_o t)$ as detailed below) has the benefit of estimating the movement of the target. Let the distance of the target $r_\ell(t)$ be time-varying with a radial departing speed

$$v_\ell = \frac{d}{dt} r_\ell(t)$$

so that the distance variation can be approximated at first order $r_\ell(t) \simeq r_\ell(0) + v_\ell t$, and the delay is

$$\tau_\ell(t) \simeq \tau_\ell(0) + \frac{2v_\ell}{c} t$$

The ℓth backscattered echo becomes

$$g(t - \tau_\ell(t)) = \cos[\omega_{o,\ell} t - \omega_o \tau_\ell(0)]$$

where the frequency (Doppler effect)

$$\omega_{o,\ell} = \omega_o(1 - 2v_\ell/c)$$

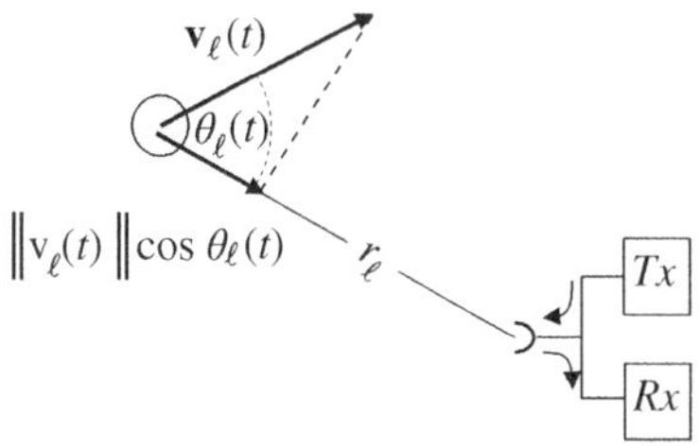

Figure 5.14 Doppler effect.

is shifted depending on the target speed v_ℓ—this is the Doppler shift (frequency increases when speed $v_\ell < 0$, i.e., the target is moving toward the radar system). In this case the additional estimation from $x(t)$ of the frequency deviation $\omega_{o,\ell} - \omega_o$ of the transmitted $g(t)$ lets one evaluate the speed of the target (from frequency estimation), in addition to its distance (from delay estimation).

Note that Doppler shift depends on the speed component along the radar direction Figure 5.14. A target with an arbitrary speed $\mathbf{v}_\ell(t)$ having an angle $\theta_\ell(t)$ wrt to the propagation direction exhibits a Doppler frequency shift

$$\omega_{o,\ell} = \omega_o(1 + 2\|\mathbf{v}_\ell(t)\|\cos\theta_\ell(t)/c)$$

that is time-varying.

Range and Doppler Frequency Estimation

In a radar system, both range and speed are equally important, and could be estimated separately. However, in many situations these two parameters need to be estimated jointly. This implies that one has to use a waveform $g(t)$ that gives the possibility to estimate both delay and Doppler shift. For the general case of $g(t) = a(t)\cos(\omega_o t)$, the ℓth backscattered echo

$$x(t) = a(t - \tau_\ell)\cos(\omega_o t + \Delta\omega_\ell t + \varphi_\ell) + w(t)$$

is delayed by the range τ_ℓ and affected by the Doppler frequency shift $\Delta\omega_\ell$; the phase is $\varphi_\ell \sim \mathcal{U}(0, 2\pi)$. The problem reduces to the estimation of the range (or delay) and Doppler shift (or frequency) for the pair $(\tau_\ell, \Delta\omega_\ell)$ from the backscattered echo $x(t)$.

Chirp-waveforms are often used for their optimal delay-frequency resolution [128]:

$$g(t) = \begin{cases} \frac{1}{\sqrt{T}}\cos(\omega_o t + \mu t^2), & -\frac{T}{2} \le t \le \frac{T}{2} \\ 0 & \text{elsewhere} \end{cases}$$

This is basically a sinusoid with $a(t) = rect(t/T)/\sqrt{T}$ and linear frequency variation around ω_o. In the presence of a moving target, the delay affects the whole waveform as above, while the Doppler shift adds a constant term to the linear frequency variation that makes the frequency estimation more accurate.

6

Estimation Theory

In statistical signal processing, the observations from each experiment are modeled as function of parameters $\boldsymbol{\theta}$ that can be either deterministic, or random, or any combination:

$$\mathbf{x}[i] = \mathbf{s}[i, \boldsymbol{\theta}]$$

where each observation of the set $\{\mathbf{x}[i]\}$ can have arbitrary size $\mathbf{x}[i] \in \mathbb{C}^M$. Another common situation is the presence of additive noise in observations

$$\mathbf{x}[i] = \mathbf{s}[i, \boldsymbol{\theta}] + \mathbf{w}[i]$$

The purpose of estimation theory is to infer the value of a limited number of p parameters $\boldsymbol{\theta}$ that model the signal component from a judicious analysis of the set $\mathbf{x}_N = \{\mathbf{x}[i]\}_{i=1}^N$, where subscript N is only to highlight the number of independent experiments, but not the way these are ordered into $\mathbf{x}_N$ (column-wise, row-wise, matrix, etc.) The estimator is any transformation (in general, non-linear and/or implicitly defined)

$$\hat{\boldsymbol{\theta}}(\mathbf{x}_N)$$

that manipulates the observations $\mathbf{x}_N$ to have an estimate of $\boldsymbol{\theta}$ according to some criteria. In addition, since $\mathbf{x}_N$ is an rv (usually multivariate), the estimate itself is an rv and metrics evaluate how good/bad an estimator $\hat{\boldsymbol{\theta}}(\mathbf{x}_N)$ is.

6.1 Historical Notes

The history of the evolution of estimation methods into what is used today involves so many scientists that making a summary would be too complex. Furthermore, it would not give the proper credit to the many statisticians who throughout history have contributed in various ways. The review here is intended to be instrumental to the methodological discussion of interest.

Statisticians can be grouped into *frequentist* and *Bayesian*, depending on their approach to modeling the parameters $\boldsymbol{\theta}$, and the history of statistics to some extent follows these different visions. For the frequentists, experiments are repeatable, data is

Statistical Signal Processing in Engineering, First Edition. Umberto Spagnolini.

Companion website: www.wiley.com/go/spagnolini/signalprocessing

just one random sample, and parameters remain constant over these repeatable processes. For the Bayesians, the parameters are unknown and described probabilistically while data (or experiments) are fixed for the realized sample.

There are many discussions about which one is better/worse, but essentially the difference is in what probability theory is, and is not, used in model definition. In a Bayesian framework, data ($\mathbf{x}$) and parameters ($\boldsymbol{\theta}$) are modeled probabilistically, including the parameters that govern the distribution of the data. In a frequentist framework, only the data is modeled probabilistically, not the parameters.

In term of estimation methods, the frequentist view will be referred to here as non-Bayesian and is based on the concept of likelihood. It seems that the first person using the term "*likelihood*" was R.A. Fisher (1921), and the process that has led up to today's vision was not at all straightforward (see e.g., [26, 27]). The frequentist view is far more intuitive if considering the model where an experiment is repeatable subject to some external impairments such as noise.

T. Bayes (1701–1761) was a Presbyterian pastor who formulated his celebrated formula of *inverse probability* (i.e., the pdf of the unobserved variable), which was published in 1763 after his death. The theory on inverse probability was ignored for years before being used for inference [25]. Bayesian approaches need skill in application and a lot of computation. Since application-practitioners seek a (possibly automated) "cookbook-type procedure," in the past the frequentists won. However, since 1950 there has been a rebirth of the Bayesian approach due to the availability of fast computing power, a better appreciation of the power of a probabilistic vision for solving dynamic systems from R.E. Kalman (1930–2016), and later the ability to solve practical problems with complex models [28]. In Bayesian inference, the jargon sets the *a-priori* distribution as the distribution of the parameters $p(\boldsymbol{\theta})$ before any experiment has been done, and the *a-posteriori* distribution as the distribution of the parameters after conducting the experiment, usually denoted as $p(\boldsymbol{\theta}|\mathbf{x})$.

6.2 Non-Bayesian vs. Bayesian

Depending on the properties of $\boldsymbol{\theta}$, there are different approaches that can be used for its estimation.

Non-Bayesian approach: $\boldsymbol{\theta}$ is as a deterministic but unknown quantity, so that one can derive the pdf of observations that are parameterized with respect to $\boldsymbol{\theta}$ accounted as a distribution parameter: $p(\mathbf{x}|\boldsymbol{\theta})$. The conditional pdf $p(\mathbf{x}|\boldsymbol{\theta})$ is the *likelihood* function of the observations and the value $\boldsymbol{\theta}$ that maximizes this parametric pdf for the current observation is the estimate. The previously indicated pdf viewed as a function of $\boldsymbol{\theta}$ is called the likelihood function of the considered parameter, and the estimator is the one that maximizes the corresponding likelihood. The estimators based on this approach are called Maximum Likelihood Estimators (MLEs).

Bayesian approach: $\boldsymbol{\theta}$ is a multivariate random variable. In this case, the *a-priori* pdf of the parameter $p(\boldsymbol{\theta})$ is known, or it is available in some way and needs to be encoded into a distribution. $p(\boldsymbol{\theta})$ is known *before* making any observation. Using Bayes' rule, it is possible to obtain the a-posteriori pdf $p(\boldsymbol{\theta}|\mathbf{x})$ as the pdf of $\boldsymbol{\theta}$ after collecting the data, which is manipulated from the conditional pdf $p(\mathbf{x}|\boldsymbol{\theta})$ by Bayes' rule:

$$p(\boldsymbol{\theta}|\mathbf{x}) = \frac{p(\mathbf{x}|\boldsymbol{\theta})p(\boldsymbol{\theta})}{p(\mathbf{x})} = \gamma\, p(\mathbf{x}|\boldsymbol{\theta})p(\boldsymbol{\theta})$$

where γ can be seen as a scaling factor to normalize the a-posteriori pdf.

The a-posteriori pdf $p(\boldsymbol{\theta}|\mathbf{x})$ contains all the statistical information about the distribution of $\boldsymbol{\theta}$ after having done the experiment and all data being available, it is a powerful probabilistic model on the parameter to be used for Bayesian estimates (Chapter 11). One can search for those values of $\boldsymbol{\theta}$ that maximize the pdf $p(\boldsymbol{\theta}|\mathbf{x})$ obtaining the so called *maximum a-posteriori* (MAP) estimator. On the other hand, the mean value of the a-posteriori pdf, called the *minimum mean square error* (MMSE) estimator, yields the estimate with minimum deviation from the true value (unknown but with a-priori distribution $p(\boldsymbol{\theta})$).

Regardless of the methods and the assumptions, the estimators can be either explicit or implicit; when explicit, the dependency on data is arbitrarily non-linear. There is a widely adopted class of estimators in which the estimate is a linear transformation on the observation (input) data:

$$\hat{\boldsymbol{\theta}} = \mathbf{A}\mathbf{x}_N$$

where the estimator is given by the matrix $\mathbf{A} \in \mathbb{C}^{p\times N}$ to be derived. Even if these linear estimators are not always optimal, their simplicity and straightforward derivation justify their widespread usage as detailed later.

6.3 Performance Metrics and Bounds

The quality of an estimator is evaluated from the analysis of the error

$$\delta\hat{\boldsymbol{\theta}}(\mathbf{x}_N) = \hat{\boldsymbol{\theta}}(\mathbf{x}_N) - \boldsymbol{\theta}_o$$

that highlights how close the estimate is to the true value $\boldsymbol{\theta}_o$. Even if $\boldsymbol{\theta}_o$ is deterministic, the estimate $\hat{\boldsymbol{\theta}}$ is a multivariate rv that depends on the data $\{\mathbf{x}[i]\}_{i=1}^N$. To exemplify, one can make multiple experiments with estimates and represent the set of estimates in the form of histogram as in Figure 6.1, being a representative behavior of the pdf $p(\hat{\boldsymbol{\theta}}|\boldsymbol{\theta}_o)$ for the value $\boldsymbol{\theta}_o$.

Even if an empirical histogram could be very meaningful to appreciate some properties of the estimate, common performance metrics are based on the first and second moments of error $\delta\hat{\boldsymbol{\theta}}(\mathbf{x}_N)$. An important property of an estimator is *consistency*, which guarantees that accuracy improves when considering more measurements; in other words, for $N \to \infty$, the histogram shrinks to $\boldsymbol{\theta}_o$, or equivalently $\delta\hat{\boldsymbol{\theta}}(\mathbf{x}_N) \to 0$ according to the performance metrics.

6.3.1 Bias

Bias is a measure of how much the mean estimate deviates from the true value

$$\mathbf{b}(\hat{\boldsymbol{\theta}}) = \mathbb{E}_x[\delta\hat{\boldsymbol{\theta}}(\mathbf{x}_N)] = \mathbb{E}_x[\hat{\boldsymbol{\theta}}(\mathbf{x}_N)] - \boldsymbol{\theta}_o$$

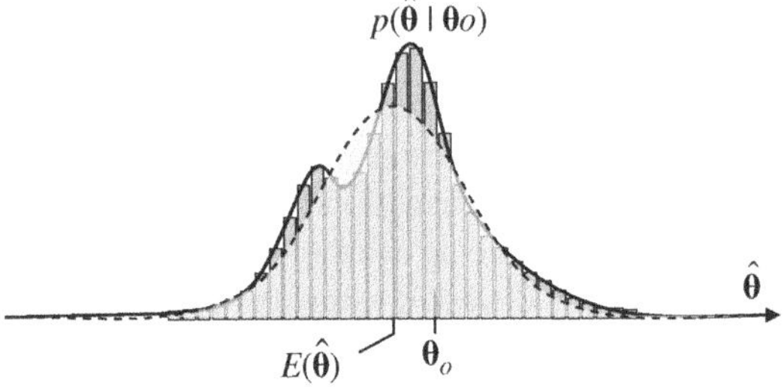

Figure 6.1 Histogram of pdf $p(\hat{\theta}|\theta_o)$ and the approximating Gaussian pdf from moments (dashed line).

for a specific value of the parameter θ_o (omitted here to ease the notation). Bias highlights the properties of the estimator in terms of mean accuracy for each value of the parameter θ_o. When $\mathbf{b}(\hat{\theta}) = 0$, the estimator is unbiased, and this is by far the most preferred condition as it highlights that the estimator is not affected by systematic errors. Even if an unbiased estimator is mostly preferred to a biased one, there are situations that tolerate a small bias. An estimator can be biased, but at least it is advisable that $\mathbb{E}[\delta\hat{\theta}(\mathbf{x}_N)] \to 0$ for $N \to \infty$ (asymptotically unbiased estimator), so that the bias can be neglected provided that the number of samples N is *large enough*.[1]

Bias can be established for the Bayesian approach as

$$\mathbf{b} = \mathbb{E}_{x,\theta}[\delta\hat{\theta}(\mathbf{x}_N)] = \mathbb{E}_{x,\theta}[\hat{\theta}(\mathbf{x}_N)] - E_\theta[\theta]$$

It is a mean property of the estimator for any possible value of θ according to its distribution $p(\theta)$. Notice that this definition is more general as it accounts for the bias in non-Bayesian problems provided that $p(\theta) = \delta(\theta - \theta_o)$.

6.3.2 Mean Square Error (MSE)

Mean square error (MSE) is a common metric for a specific estimator $\hat{\theta}(\mathbf{x}_N)$ and it condenses into one value the overall dispersion of the estimate wrt to the true value θ_o. *Non-Bayesian MSE* is defined as

$$MSE(\hat{\theta}) = \mathbb{E}_x[||\hat{\theta}(\mathbf{x}_N) - \theta_o||^2]$$

and it depends on the specific value θ_o of the parameter (the dependency on θ_o is sometimes taken as read). To clarify the importance of MSE as a metric, let us consider the estimation of a single *deterministic* parameter:

$$MSE(\hat{\theta}) = \mathbb{E}[|\hat{\theta} - \theta_o|^2] = \mathbb{E}[|(\hat{\theta} - \mathbb{E}[\hat{\theta}]) + (\mathbb{E}[\hat{\theta}] - \theta_o)|^2] = \text{var}[\hat{\theta}] + |b(\hat{\theta})|^2 \quad (6.1)$$

1 The sentence "large enough" is quite common within the engineering community as it states with pragmatism that since the asymptotic analysis from the theory guarantees some performance behaviors, a detailed analysis can prove that there will be some value N_* such that the choice $N > N_*$ fulfills the design constraints (e.g., a certain bias can be tolerated). In many cases, the quantitative value N_* is not necessary in the preliminary stage of an estimator's design and it can be established later, either analytically or numerically.

It is important to note that MSE is the sum (variance+bias^2), therefore the MSE alone cannot be considered as a complete descriptor of estimator accuracy, but is surely the most important metric that collapses both. In addition, in searching for an optimum estimator among several options, an intuitive choice is to look for the one that minimizes the MSE, as being the sum of two terms, one can minimize each of the terms in sum and start with the unbiased ($b(\hat{\theta}) = 0$) estimate. This is detailed in Section 6.5.

Even if for p parameters, the MSE can be evaluated on each parameter as $MSE(\hat{\theta}_k)$, it is far more general to evaluate the *MSE matrix:*

$$\mathbf{MSE}(\hat{\boldsymbol{\theta}}) = \mathbb{E}_x[(\hat{\boldsymbol{\theta}}(\mathbf{x}_N) - \boldsymbol{\theta}_o)(\hat{\boldsymbol{\theta}}(\mathbf{x}_N) - \boldsymbol{\theta}_o)^H] = \mathrm{cov}[\hat{\boldsymbol{\theta}}(\mathbf{x}_N)] + \mathbf{b}(\hat{\boldsymbol{\theta}}(\mathbf{x}_N)) \cdot \mathbf{b}^H(\hat{\boldsymbol{\theta}}(\mathbf{x}_N))$$

where

$$\mathrm{cov}[\hat{\boldsymbol{\theta}}(\mathbf{x}_N)] = \mathbb{E}_x[(\hat{\boldsymbol{\theta}}(\mathbf{x}_N) - \mathbb{E}_x[\hat{\boldsymbol{\theta}}(\mathbf{x}_N)])(\hat{\boldsymbol{\theta}}(\mathbf{x}_N) - \mathbb{E}_x[\hat{\boldsymbol{\theta}}(\mathbf{x}_N)])^H]$$

or equivalently, the entries are

$$[\mathbf{MSE}(\hat{\boldsymbol{\theta}})]_{ij} = \mathbb{E}[(\hat{\theta}_i - \theta_{o,i})(\hat{\theta}_j^* - \theta_{o,j}^*)] = \mathrm{cov}[\hat{\theta}_i, \hat{\theta}_j] + b(\theta_i) \cdot b^*(\theta_j)$$

and the diagonal entries are

$$[\mathbf{MSE}(\hat{\boldsymbol{\theta}})]_{kk} = MSE(\hat{\theta}_k) = \mathrm{var}[\hat{\theta}_k] + |b(\hat{\theta}_k)|^2$$

When the p parameters are homogeneous, the total MSE is the sum of the MSE for all the parameters:

$$MSE(\hat{\boldsymbol{\theta}}) = \sum_{k=1}^{p} MSE(\hat{\theta}_k) = \mathrm{tr}\{\mathbf{MSE}(\hat{\boldsymbol{\theta}})\}$$

The MSE for random parameters (*Bayesian MSE*) is somewhat less intuitive. It states the MSE for all possible values of the parameters $\boldsymbol{\theta} = \boldsymbol{\theta}_o$ depending on their distribution ($p(\boldsymbol{\theta})$):

$$\mathbf{MSE} = \mathbb{E}_{x,\theta}[(\hat{\boldsymbol{\theta}}(\mathbf{x}_N) - \boldsymbol{\theta})(\hat{\boldsymbol{\theta}}(\mathbf{x}_N) - \boldsymbol{\theta})^H] = \mathbb{E}_{\theta_o}[\mathbb{E}_x[\hat{\boldsymbol{\theta}}(\mathbf{x}_N)\text{-}\boldsymbol{\theta})(\hat{\boldsymbol{\theta}}(\mathbf{x}_N)\text{-}\boldsymbol{\theta})^H | \boldsymbol{\theta} = \boldsymbol{\theta}_o]]$$

In other words, the term $\mathbb{E}_x[\hat{\boldsymbol{\theta}}(\mathbf{x}_N) - \boldsymbol{\theta}_o)(\hat{\boldsymbol{\theta}}(\mathbf{x}_N) - \boldsymbol{\theta}_o)^H] = MSE(\hat{\boldsymbol{\theta}}(\mathbf{x}_N))$ coincides with the MSE for the specific choice of $\boldsymbol{\theta} = \boldsymbol{\theta}_o$, and then this is averaged over all possible values $\boldsymbol{\theta}_o$ according to the pdf $p(\boldsymbol{\theta})$. Further discussion will be related to the Bayesian estimators (Chapter 11 and later) once the reader is more familiar with the Bayesian approach.

6.3.3 Performance Bounds

Performance bounds play an essential role in defining the properties of every estimator with respect to an optimum estimator, provided that one exists. Pragmatically, two estimators can be compared to one another by evaluating the MSE for the same number of observations $\mathbf{x}_N$ (in a statistical sense) and selecting the estimator that has the smaller MSE as this one is more *efficient* (as it uses better the available data). If both these estimators are unbiased, the one with the minimum MSE is also the one with minimum variance according to the property (6.1). Even if the choice of the estimator that has the minimum MSE optimizes the performance among a set of choices, one might wonder how good is the estimator at hand, or if there is any better? All these questions are relevant when struggling to search for better estimators with improved performance.

For specific problems and models, the theory sets the performance bound on MSE (MSE_{opt}) such that any estimator is

$$\mathbb{E}[||\delta\hat{\boldsymbol{\theta}}(\mathbf{x}_N)||^2] \geq MSE_{opt}.$$

Any estimator that $\mathbb{E}[||\delta\hat{\boldsymbol{\theta}}(\mathbf{x}_N)||^2] \simeq MSE_{opt}$ (or better $\mathbb{E}[||\delta\hat{\boldsymbol{\theta}}(\mathbf{x}_N)||^2] = MSE_{opt}$) is considered *efficient* (or asymptotically efficient if $\mathbb{E}[||\delta\hat{\boldsymbol{\theta}}(\mathbf{x}_N)||^2] \simeq MSE_{opt}$ for $N \to \infty$) and thus it is not necessary to search for a better estimator (unless it is mandatory to reduce complexity, as is commonly the case). On the other hand, if $\mathbb{E}[||\delta\hat{\boldsymbol{\theta}}(\mathbf{x}_N)||^2] \gg MSE_{opt}$, there is room for improvement and one can make better use of the available observations, or can define a better estimator. However, if $\mathbb{E}[||\delta\hat{\boldsymbol{\theta}}(\mathbf{x}_N)||^2] < MSE_{opt}$, rather than saying that this is theoretically impossible, as is the case, one can state that there could be some errors in the estimator, or in the MSE_{opt}, or the estimator itself is using some information that it is not supposed to use, or *anyone who is claiming to have this exceptionally efficient estimator is just kidding.*

Fundamental estimation theory provides a powerful performance bound that holds for unbiased estimators: the Cramér–Rao lower bound (CRB), which will be detailed in Chapter 8 for the non-Bayesian approach and in Section 11.4 for the Bayesian case.

6.4 Statistics and Sufficient Statistics

Given a set of observations $\mathbf{x} \in \mathbb{C}^N$, the term *statistic* is any transformation

$$\mathbf{y} = T(\mathbf{x}) \in \mathbb{C}^M \text{ with } M \leq N$$

that does not depend on the parameter vector $\boldsymbol{\theta}$ to be estimated. In the case that $\mathbf{y}$ represents a set of data that carry all the information on $\boldsymbol{\theta}$, and it is possible to solve the estimation problem at hand, the set $T(\mathbf{x})$ is called a *sufficient statistic* for $\boldsymbol{\theta}$ estimation. Therefore a sufficient statistic transforms a set of N measurements into a new set of data with reduced dimension ($M \leq N$) from which it is possible to estimate $\boldsymbol{\theta}$. A sufficient statistic is *minimal* if $T(\mathbf{x})$ has the minimal dimension to allow estimation of the multi-parameters $\boldsymbol{\theta}$.

To exemplify the concept of a sufficient statistic, assume that the problem is to estimate the mean value of an rv from the set of independent and identically distributed (i.i.d.) measurements $x[1], \cdots, x[N]$. Examples of statistics are the following:

$$y_1 = \sum_{k=1}^{N} x[k] \Rightarrow \mathcal{Y}_1 = \{y_1\}$$

$$y_2 = \prod_{k=1}^{N} x[k] \Rightarrow \mathcal{Y}_2 = \{y_2\}$$

$$\mathbf{y}_3 = \mathbf{Q}\mathbf{x} \text{ where } \mathbf{Q} \in \mathbb{R}^{N \times N} \text{ is a permutation matrix} \Rightarrow \mathcal{Y}_3 = \{\mathbf{y}_3\}$$

The estimate of the mean value is the sample mean $\frac{1}{N}\sum_{n=1}^{N} x[n]$ as detailed later in terms of the optimum estimator. Clearly, $\mathcal{Y}_1$ is a sufficient statistic due to fact that it reduces the set to one value, and in turn it is the estimator of the mean value as y_1/N, and hence is minimal. $\mathcal{Y}_2$ is not a sufficient statistic as it reduces the observations to a minimal set but it is not useful for the estimate of the mean value. Since $\mathbf{Q}$ reorders the observations, $\mathcal{Y}_3$ can be considered as a sufficient statistic (trivial for $\mathbf{Q} = \mathbf{I}$) that is an ordering statistic (Section 3.9) if $\mathbf{Q}$ is chosen to order the samples for decreasing, or increasing, values.

Analytic tools and conditions to identify a family of sufficient statistics as a basis to search for a minimal one are reported in the statistical methods literature. Since conditions do not highlight estimators, here we follow a pragmatic approach to specify the estimators for a wide class of applications as meaningful enough for daily practice. The Neyman–Fisher Theorem is enunciated anyway.

Given the pdf $p(\mathbf{x}|\boldsymbol{\theta})$, if this can be factorized into the product

$$p(\mathbf{x}|\boldsymbol{\theta}) = g(T(\mathbf{x}), \boldsymbol{\theta}) \cdot h(\mathbf{x}) \tag{6.2}$$

where $g(.)$ depends on $\mathbf{x}$ through the transformation $T(.)$, then $T(\mathbf{x})$ is a sufficient statistic for $\boldsymbol{\theta}$. Conversely, if $T(\mathbf{x})$ is a sufficient statistic for $\boldsymbol{\theta}$ then $p(\mathbf{x}|\boldsymbol{\theta})$ can be factorized as in (6.2). Intuitively, the factorization implies that $p(\mathbf{x}|\boldsymbol{\theta})$ depends on $\boldsymbol{\theta}$ only through $T(\mathbf{x})$, and therefore the estimation can be reduced to the inspection of only $g(T(\mathbf{x}), \boldsymbol{\theta})$ (for example by maximizing $g(T(\mathbf{x}), \boldsymbol{\theta})$ as in methods discussed in Chapter 7). It is important to note that Gaussian pdfs, and in general any exponential-shaped pdfs, naturally allow the Neyman–Fisher factorization.

6.5 MVU and BLU Estimators

The minimization of the MSE is one of the most used approaches for setting up estimators. Since

$$MSE(\hat{\theta}) = \text{var}[\hat{\theta}] + |b(\hat{\theta})|^2$$

a common assumption is that the estimator is unbiased ($b(\hat{\theta}) = 0$) and thus the minimization of the MSE implies to the minimization of the variance. There might be different estimators based on different statistics for the wide class of unbiased estimators, and out of this class there could be one estimator that is uniformly the best with minimum variance for the whole set of parameters; this is referred to as the *minimum variance unbiased* (MVU) estimator. However, there is no guarantee that the MVU estimator is unique (and the MVU solution could be non-existent). A pragmatic rule is to compare the MSE of one candidate for being the MVU estimator with the CRB, and if the estimator attains the CRB, we can argue that the MVU estimator is optimal in the absolute sense. Figure 6.2 illustrates some typical cases of unbiased estimators for

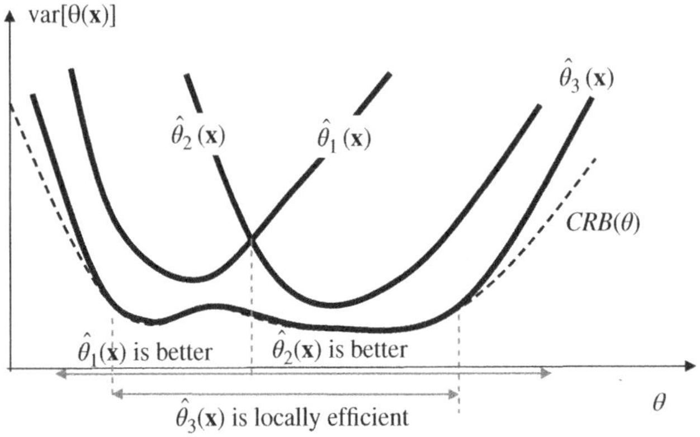

Figure 6.2 Variance of different unbiased estimators, and the Cramér–Rao bound $CRB(\theta)$.

$p = 1$. Variance (or MSE) can change depending of the parameter value, and one can find a set of values θ where one estimator (e.g., $\hat{\theta}_1(\mathbf{x})$ or $\hat{\theta}_2(\mathbf{x})$) is more efficient than another (i.e., $\hat{\theta}_2(\mathbf{x})$ or $\hat{\theta}_1(\mathbf{x})$, respectively). On the other hand, estimator $\hat{\theta}_3(\mathbf{x})$ is uniformly more efficient than $\hat{\theta}_1(\mathbf{x})$ and $\hat{\theta}_2(\mathbf{x})$, and locally is the most efficient as it attains the CRB. If there are no other estimators, $\hat{\theta}_3(\mathbf{x})$ is the MVU.

6.6 BLUE for Linear Models

BLUE stands for *best linear unbiased estimator* and it is the MVU when the estimator is constrained to be linear. One example below for the case of a linear model illustrates the method and, in spite of the linearity of the model itself, is very representative of the behavior of linear estimators in several common setups.

Let us assume that we have a linear model that describes the observations

$$\mathbf{x} = \underbrace{[\mathbf{h}_1, \mathbf{h}_2, \ldots, \mathbf{h}_p]}_{\mathbf{H}}\theta + \mathbf{w}$$

with noise model

$$\mathbf{w} \sim \mathcal{N}(\boldsymbol{\mu}_w, \mathbf{C}_w)$$

and deterministic but unknown $\boldsymbol{\theta}$. The linear estimator can be written as:

$$\hat{\theta} = \mathbf{A}\mathbf{x} + \mathbf{b}$$

where $\mathbf{A}$ and $\mathbf{b}$ should be defined by imposing unbiasedness and minimum MSE. To simplify, all vectors and matrixes are real-valued.

The bias condition is

$$\mathbb{E}[\hat{\boldsymbol{\theta}}] = \mathbf{A}\mathbb{E}[\mathbf{x}] + \mathbf{b} = \boldsymbol{\theta}$$

where

$$\mathbb{E}[\mathbf{x}] = \mathbf{H}\boldsymbol{\theta} + \boldsymbol{\mu}_w$$

It is possible to find the conditions for zero-bias by constraining

$$\mathbf{A}\mathbf{H}\boldsymbol{\theta} + \mathbf{A}\boldsymbol{\mu}_w + \mathbf{b} = \boldsymbol{\theta}$$

that in turn it becomes:

$$\begin{aligned} \mathbf{A}\mathbf{H} = \mathbf{I} &\leftrightarrow \mathbf{a}_k^T \mathbf{h}_\ell = \delta_{k-\ell} \\ \mathbf{A}\boldsymbol{\mu}_w + \mathbf{b} = \mathbf{0} & \end{aligned}$$

where $\mathbf{a}_k^T = \mathbf{A}(k,:)$ is the kth row of $\mathbf{A}$ as $\mathbf{A}^T = [\mathbf{a}_1, \mathbf{a}_2, ..., \mathbf{a}_p] \in \mathbb{R}^{N \times p}$. The constraint yields $\mathbf{b}$ as function of $\mathbf{A}$, but this corresponds to saying that the estimator can be rewritten by stripping the mean $\boldsymbol{\mu}_w$ from $\mathbf{x}$:

$$\hat{\boldsymbol{\theta}} = \mathbf{A}(\mathbf{x} - \boldsymbol{\mu}_w)$$

or equivalently, for the reasoning below, one can set $\boldsymbol{\mu}_w = \mathbf{0}$ and thus $\mathbf{b} = \mathbf{0}$.

The minimization of the variances is obtained by a simultaneous minimization of all the MSE_k relative to each entry of $\hat{\boldsymbol{\theta}}$

$$\hat{\theta}_k = \mathbf{a}_k^T \mathbf{x}$$

The MSE is

$$MSE_k = \text{var}[\hat{\theta}_k] = [\text{cov}[\hat{\boldsymbol{\theta}}]]_{kk}$$

where

$$\text{cov}[\hat{\boldsymbol{\theta}}] = \mathbb{E}[(\mathbf{A}\mathbf{x} - \boldsymbol{\theta})(\mathbf{A}\mathbf{x} - \boldsymbol{\theta})^T] = \mathbf{A}\mathbf{C}_w\mathbf{A}^T \Rightarrow [\text{cov}[\hat{\boldsymbol{\theta}}]]_{ij} = \mathbf{a}_i^T \mathbf{C}_w \mathbf{a}_j$$

The minimization is subject to the constraint on the bias $\mathbf{A}\mathbf{H} = \mathbf{I}$ as:

$$\min_{\mathbf{a}_k} \{ \underbrace{\mathbf{a}_k^T \mathbf{C}_w \mathbf{a}_k}_{\text{var}[\hat{\theta}_k]} \} \quad \text{s.t.} \underbrace{\begin{bmatrix} \mathbf{h}_1^T \\ \vdots \\ \mathbf{h}_k^T \\ \vdots \\ \mathbf{h}_p^T \end{bmatrix}}_{\mathbf{H}^T} \mathbf{a}_k = \begin{bmatrix} 0 \\ \vdots \\ 1 \\ \vdots \\ 0 \end{bmatrix} = \mathbf{e}_k$$

from the Lagrange multiplier method (see Section 1.8) it follows that

$$\mathbf{a}_{k,opt} = \mathbf{C}_w^{-1}\mathbf{H}(\mathbf{H}^T\mathbf{C}_w^{-1}\mathbf{H})^{-1}\mathbf{e}_k$$

whose variance is

$$\text{var}[\hat{\theta}_k] = \mathbf{a}_{k,opt}^T \mathbf{C}_w \mathbf{a}_{k,opt} = \mathbf{e}_k^T (\mathbf{H}^T \mathbf{C}_w^{-1} \mathbf{H})^{-1} \mathbf{e}_k = [(\mathbf{H}^T \mathbf{C}_w^{-1} \mathbf{H})^{-1}]_{kk}$$

Extending the solution to all the p vectors that compose the $\mathbf{A}$ matrix yields

$$\mathbf{A}_{opt} = \begin{bmatrix} \mathbf{a}_{1,opt}^T \\ \mathbf{a}_{2,opt}^T \\ \vdots \\ \mathbf{a}_{p,opt}^T \end{bmatrix} = \underbrace{\begin{bmatrix} \mathbf{e}_1^T \\ \mathbf{e}_2^T \\ \vdots \\ \mathbf{e}_p^T \end{bmatrix}}_{\mathbf{I}} (\mathbf{H}^T \mathbf{C}_w^{-1} \mathbf{H})^{-1} \mathbf{H}^T \mathbf{C}_w^{-1}$$

The complete expression of the estimator from data is very compact:

$$\hat{\theta} = (\mathbf{H}^T \mathbf{C}_w^{-1} \mathbf{H})^{-1} \mathbf{H}^T \mathbf{C}_w^{-1} \mathbf{x}$$

whose covariance matrix is obtained recalling that $\hat{\theta} - \theta = \mathbf{A}(\mathbf{H}\theta + \mathbf{w}) - \theta = \mathbf{A}\mathbf{w}$, and thus

$$\text{cov}[\hat{\theta}] = (\mathbf{H}^T \mathbf{C}_w^{-1} \mathbf{H})^{-1}$$

The comparison with CRB will prove that the BLUE is the optimum estimator for any linear models with Gaussian noise.

6.7 Example: BLUE of the Mean Value of Gaussian rvs

Let the observation vector be represented by N samples that have the same mean and are mutually correlated:

$$\mathbf{x} \sim N(a\mathbf{1}, \mathbf{C}_x)$$

The mean is estimated through a linear sum

$$\hat{a} = \boldsymbol{\alpha}^T \mathbf{x} = \sum_{k=1}^{N} \alpha_k x_k$$

It is a linear combination of the observations with unknown weights α_k to be determined from the BLUE conditions. The bias condition is

$$\mathbb{E}[\hat{a}] = \sum_{k=1}^{N} \alpha_k \mathbb{E}[x_k] = a \sum_{k=1}^{N} \alpha_k \Rightarrow \sum_{k=1}^{N} \alpha_k = 1$$

while the variance is

$$\text{var}[\hat{a}] = \boldsymbol{\alpha}^T cov(\mathbf{x}) \boldsymbol{\alpha} = \boldsymbol{\alpha}^T \mathbf{C}_x \boldsymbol{\alpha}$$

Minimizing the quadratic form representing the variance with the constraint $\boldsymbol{\alpha}^T\mathbf{1}=1$ is solved in Section 1.8.1; the optimum weights are

$$\boldsymbol{\alpha}_{BLUE} = \frac{\mathbf{C}_x^{-1}\mathbf{1}}{\mathbf{1}^T\mathbf{C}_x^{-1}\mathbf{1}} \Rightarrow \hat{a} = \frac{\mathbf{1}^T\mathbf{C}_x^{-1}\mathbf{x}}{\mathbf{1}^T\mathbf{C}_x^{-1}\mathbf{1}}.$$

The minimum value of the variance for $\boldsymbol{\alpha}_{BLUE}$ is

$$\text{var}[\hat{a}] = \left(\mathbf{1}^T\mathbf{C}_x^{-1}\mathbf{1}\right)^{-1} = \frac{1}{\sum_{n=1}^{N}[\mathbf{C}_x^{-1}]_{nn}}$$

which depends on the diagonal entries of the inverse of covariance $\mathbf{C}_x$.

One can use as an estimate the sample mean

$$\bar{a} = \frac{1}{N}\sum_{k=1}^{N} x_k = \frac{\mathbf{1}^T}{N}\mathbf{x}$$

This is unbiased and the variance is

$$\text{var}[\bar{a}] = \frac{\mathbf{1}^T\mathbf{C}_x\mathbf{1}}{N^2} = \frac{\sum_{n=1}^{N}[\mathbf{C}_x]_{nn}}{N^2} \geq \left(\mathbf{1}^T\mathbf{C}_x^{-1}\mathbf{1}\right)^{-1} = \text{var}[\hat{a}]$$

There are some special cases worth considering as illustrative to gain insight into the BLUE solution:

- $\mathbf{C}_x = \sigma^2\mathbf{I}$ (all the samples have the same confidence and are mutually uncorrelated):

$$\boldsymbol{\alpha}_{BLUE} = \frac{1}{N}\mathbf{1} \Rightarrow \hat{a} = \bar{a} = \frac{1}{N}\sum_{k=1}^{N} x_k$$

  the optimal estimator is the *sample mean*; there is no reason to privilege one sample wrt another in weighting definitions. The variance

$$\text{var}[\bar{a}] = \frac{\sigma^2}{N}$$

  scales as $1/N$, thus showing a remarkable benefit in increasing the data-size to reduce the variance in the so called *variance reduction by the sample mean.*
- $\mathbf{C}_x = diag\{\sigma_1^2, \sigma_2^2, \cdots, \sigma_N^2\}$ (the samples have different variances, but they are mutually uncorrelated):

$$\mathbf{1}^T\mathbf{C}_x^{-1}\mathbf{1} = \sum_{n=1}^{N}\frac{1}{\sigma_n^2}$$

$$\alpha_{k,BLUE} = \frac{1/\sigma_k^2}{\sum_{n=1}^{N} 1/\sigma_n^2}$$

and thus the BLUE reduces to:

$$\hat{a} = \frac{1}{\sum_{n=1}^{N} 1/\sigma_n^2} \sum_{k=1}^{N} \frac{x_k}{\sigma_k^2}$$

with

$$\text{var}[\hat{a}] = (\sum_{n=1}^{N} 1/\sigma_n^2)^{-1} \le \frac{1}{N^2} \sum_{n=1}^{N} \sigma_n^2 = \text{var}[\bar{a}]$$

The optimal estimator is the weighted sum of the samples where each sample has a weight that is inversely proportional to its variance. This is quite intuitive as when weighting multiple observations with different accuracy (i.e., variance), it would be better to give more credit (or larger weight) to those observations with better accuracy, and down-weight those with lower reliability. Since to a great extent many linear algorithms can be reduced to the weighted average of observations with different noise (or accuracy), weighting proportionally to the inverse of the variance is quite common provided that normalization $\sum_{n=1}^{N} 1/\sigma_n^2$ is accounted for.

- AR(1) correlation model $[\mathbf{C}_x]_{ij} = \sigma^2 \rho^{|i-j|}$ with properties detailed in Section 4.5; from the inverse one gets:

$$\mathbf{1}^T \mathbf{C}_x^{-1} = \frac{1}{\sigma^2(1-\rho^2)}[(1-\rho),(1-\rho)^2,(1-\rho)^2,\ldots,(1-\rho)^2,(1-\rho)]$$

and for BLUE the variables are uniformly weighted, except for x_1 and x_N. The variance is

$$\text{var}[\hat{a}] = \sigma^2 \frac{(1-\rho^2)/(1-\rho)}{2+(N-2)(1-\rho)}$$

that, compared to the sample mean $\text{var}[\bar{a}] = \sigma^2/N$, shows that for $\rho \to 0$, $\hat{a} \to \bar{a}$ and both estimators coincide. However, when $\rho \to 1$, there is no advantage in making N fully correlated measurements (i.e., N copies of the same data) and $\text{var}[\hat{a}] = \text{var}[\bar{a}] = \sigma^2$.

7
Parameter Estimation

7.1 Maximum Likelihood Estimation (MLE)

ML estimation is based on the maximization of the pdf associated to the observations with respect to the model parameters:

$$\hat{\boldsymbol{\theta}} = \boldsymbol{\theta}_{ML} = \arg\max_{\boldsymbol{\theta}} p(\mathbf{x}|\boldsymbol{\theta})$$

as is graphically shown in Figure 7.1 illustrating (for a simple case) the different pdfs $p(x|\theta)$ obtained by changing the parameters $\theta = \theta_1, \theta_2, \dots$, or the likelihood function $p(x = x_o|\theta)$ for the observation $x = x_o$ and the corresponding $\theta = \theta_{ML}$ that maximizes the likelihood $p(x = x_o|\theta)$ as the value of the parameter in the set $\{\theta_k\}$ that makes the specific observation $x = x_o$ maximally likely for the parametric pdf $p(x|\theta)$.

Starting from the observations $\mathbf{x}$, the MLE is the value of the parametric pdf that maximizes its probability by selecting the more likely model parameter value. The principal problems with ML estimation are connected to the fact that it requires, at first, writing the pdf of the observations (indicated as *likelihood function*) and then its maximization. Moreover, it must be taken into account that often the likelihood function is non-linear with respect to the parameters, and the optimization is carried out by conventional numerical optimization tools that do not always guarantee global optimality when the likelihood function is not convex.

Let us consider the case, very common in engineering problems, where the relationship between observations and parameters can be expressed by the additive stochastic term called noise (Chapter 5)

$$\mathbf{x} = \mathbf{s}(\boldsymbol{\theta}) + \mathbf{w}$$

where $\mathbf{s}(\cdot)$ is in principle a non-linear function, and $\mathbf{w}$ is the noise term. Assuming that the quantity $\mathbf{s}(\boldsymbol{\theta})$ is purely deterministic and that the noise pdf $p_w(\mathbf{w})$ is known, it is possible to write for the additive noise model the likelihood function as (Section 3.6)

$$p(\mathbf{x}|\boldsymbol{\theta}) = p_w(\mathbf{x} - \mathbf{s}(\boldsymbol{\theta}))$$

Statistical Signal Processing in Engineering, First Edition. Umberto Spagnolini.

Companion website: www.wiley.com/go/spagnolini/signalprocessing

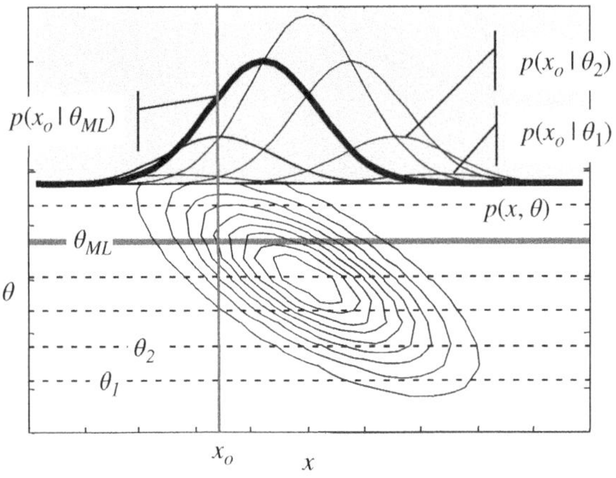

Figure 7.1 Illustration of MLE for $p = 1$ and $N = 1$.

which is equivalent to saying that the conditional pdf is the noise pdf except for a term $\mathbf{s}(\boldsymbol{\theta})$ added to its mean $\boldsymbol{\mu}_w$. In general, the likelihood $p(\mathbf{x}|\boldsymbol{\theta})$ should be written directly as the pdf of the observations where it has clearly isolated the dependency on $\boldsymbol{\theta}$.

Usually it is customary to compute log(.) of the likelihood function as it is monotonic and thus it preserves the position of its maximum (and thus the estimate). In addition, for exponential pdfs the log(.) removes some non-linearities irrelevant in numerical computations. The log-likelihood function is defined as:

$$\mathcal{L}(\mathbf{x}|\boldsymbol{\theta}) = \ln p(\mathbf{x}|\boldsymbol{\theta})$$

and therefore the MLE is

$$\boldsymbol{\theta}_{ML}(\mathbf{x}) = \arg\max_{\boldsymbol{\theta}} \mathcal{L}(\mathbf{x}|\boldsymbol{\theta})$$

Provided that the likelihood functions can always be defined, most of the challenges are on the capability to solve their maximization in closed form. However, except for some *toy problems* mostly for classroom presentations that have a closed form solution for MLE, most practical MLE problems can only be solved numerically. In spite of this limitation common to many engineering problems, it can be shown that MLE has some asymptotic properties that make it a preferred method in many contexts. The MLE is unbiased (at least asymptotically), efficient, and asymptotically Gaussian (at least, for a large enough number of observations N), with covariance matrix $\mathbf{C}_{CRB}$ defined by the CRB (Chapter 8):

$$\boldsymbol{\theta}_{ML}(\mathbf{x}) \underset{N\to\infty}{\sim} \mathcal{N}(\boldsymbol{\theta}, \mathbf{C}_{CRB})$$

This optimality condition is reached if $p(\mathbf{x}|\boldsymbol{\theta})$ fulfills some, not so strict, regularity conditions.

7.2 MLE for Gaussian Model $\mathbf{x} \sim \mathcal{N}(\boldsymbol{\mu}(\boldsymbol{\theta}); \mathbf{C}(\boldsymbol{\theta}))$

The general case of the Gaussian model is

$$\mathbf{x} \sim \mathcal{N}(\boldsymbol{\mu}(\boldsymbol{\theta}), \mathbf{C}(\boldsymbol{\theta}))$$

where the central moments $\boldsymbol{\mu}(\boldsymbol{\theta})$ and $\mathbf{C}(\boldsymbol{\theta})$ depend on parameters $\boldsymbol{\theta}$. Initially, we assume all variables/matrixes are real-valued. Some remarks on the extension for complex-valued problems will be considered later.

7.2.1 Additive Noise Model $\mathbf{x} = \mathbf{s}(\theta) + \mathbf{w}$ with $\mathbf{w} \sim \mathcal{N}(\mathbf{0}, \mathbf{C}_w)$

This models the additive noise with known covariance $\mathbf{C}_w$; the pdf of the observation can written as:

$$p(\mathbf{x}|\boldsymbol{\theta}) = p_w(\mathbf{x} - \mathbf{s}(\boldsymbol{\theta})) = |2\pi\mathbf{C}_w|^{-1/2} \exp\left\{-\frac{1}{2}[\mathbf{x} - \mathbf{s}(\boldsymbol{\theta})]^T \mathbf{C}_w^{-1} [\mathbf{x} - \mathbf{s}(\boldsymbol{\theta})]\right\}$$

the log-likelihood function is

$$\mathcal{L}(\mathbf{x}|\boldsymbol{\theta}) = -\frac{N}{2}\ln 2\pi - \frac{1}{2}\ln|\mathbf{C}_w| - \frac{1}{2}[\mathbf{x} - \mathbf{s}(\boldsymbol{\theta})]^T \mathbf{C}_w^{-1} [\mathbf{x} - \mathbf{s}(\boldsymbol{\theta})]$$

and thus $\arg\max_{\boldsymbol{\theta}} p(\mathbf{x}|\boldsymbol{\theta}) = \arg\max_{\boldsymbol{\theta}} \mathcal{L}(\mathbf{x}|\boldsymbol{\theta})$. The MLE is the minimization of the weighted norm

$$\boldsymbol{\theta}_{ML} = \arg\min_{\boldsymbol{\theta}} \left\{[\mathbf{x} - \mathbf{s}(\boldsymbol{\theta})]^T \mathbf{C}_w^{-1} [\mathbf{x} - \mathbf{s}(\boldsymbol{\theta})]\right\} = \arg\min_{\boldsymbol{\theta}} ||\mathbf{x} - \mathbf{s}(\boldsymbol{\theta})||^2_{\mathbf{C}_w^{-1}}$$

In general $||\mathbf{x} - \mathbf{s}(\boldsymbol{\theta})||^2_{\mathbf{C}_w^{-1}}$ is not quadratic (except for linear models) and therefore its minimization is far from being trivial.

In the case of a *linear model* $\mathbf{x} = \mathbf{H}\boldsymbol{\theta} + \mathbf{w}$ with $\mathbf{w} \sim \mathcal{N}(\mathbf{0}, \mathbf{C}_w)$, the MLE is

$$\boldsymbol{\theta}_{ML} = \arg\min_{\boldsymbol{\theta}} \left\{[\mathbf{x} - \mathbf{H}\boldsymbol{\theta}]^T \mathbf{C}_w^{-1} [\mathbf{x} - \mathbf{H}\boldsymbol{\theta}]\right\}$$

The minimization of the quadratic form wrt $\boldsymbol{\theta}$ leads to (provided that the inverse exists and $N \geq p$)

$$\boldsymbol{\theta}_{ML} = \left(\mathbf{H}^T \mathbf{C}_w^{-1} \mathbf{H}\right)^{-1} \mathbf{H}^T \mathbf{C}_w^{-1} \mathbf{x}$$

The MLE is unbiased and $\text{cov}[\boldsymbol{\theta}_{ML}]$ can be obtained from the definition:

$$\text{cov}[\boldsymbol{\theta}_{ML}] = (\mathbf{H}^T \mathbf{C}_w^{-1} \mathbf{H})^{-1}$$

This proves that the MLE coincides with the BLUE (Section 6.6).

A simple illustrative, and useful, example is $x[n] = \theta + w[n]$ where the N observations are affected by uncorrelated Gaussian noise $\mathbf{w} \sim \mathcal{N}(\mathbf{0}, \sigma_w^2 \mathbf{I})$. The model is

$$\mathbf{x} = \mathbf{1}\theta + \mathbf{w} \sim \mathcal{N}(\mathbf{1}\theta, \sigma_w^2 \mathbf{I})$$

and the MLE coincides with the sample mean when $\mathbf{C}_w = \sigma_w^2\mathbf{I}$:

$$\theta_{ML} = \left(\mathbf{1}^T\mathbf{1}\right)^{-1}\mathbf{1}^T\mathbf{x} = \frac{\sum_{n=1}^{N} x[n]}{N}$$

$\text{var}[\theta_{ML}] = \sigma_w^2\left(\mathbf{1}^T\mathbf{1}\right)^{-1} = \sigma_w^2/N$ scales with the number of observations N and noise is reduced by averaging (see Section 6.7 for a more complete set of cases). The degenerate case is for $N = 1$: the MLE is the sample itself $\theta_{ML} = x[1]$ with $\text{var}[\theta_{ML}] = \sigma_w^2$.

7.2.2 Additive Noise Model $\mathbf{x} = \mathbf{H}(\boldsymbol{\omega}) \cdot \boldsymbol{\alpha} + \mathbf{w}$ with $\mathbf{w} \sim \mathcal{N}(\mathbf{0}, \mathbf{C}_w)$

In some applications such as the estimation of frequency of multiple sinusoids in noise (Section 5.4), or the estimation of the direction of arrivals in uniform linear arrays (Chapter 19), the model is $\mathbf{x} = \mathbf{H}(\boldsymbol{\omega}) \cdot \boldsymbol{\alpha} + \mathbf{w}$. The parameters can be separated into into a linear ($\boldsymbol{\alpha}$) and non-linear ($\boldsymbol{\omega}$) dependence set of the signal model

$$\mathbf{s}(\boldsymbol{\omega}, \boldsymbol{\alpha}) = \mathbf{H}(\boldsymbol{\omega}) \cdot \boldsymbol{\alpha}$$

The whole set of parameters to be estimated is $\boldsymbol{\theta} = [\boldsymbol{\alpha}^T, \boldsymbol{\omega}^T]^T$. The MLE follows the same settings as above:

$$\begin{aligned}\theta_{ML} &= \arg\min_{\boldsymbol{\alpha},\boldsymbol{\omega}} \left\{[\mathbf{x} - \mathbf{H}(\boldsymbol{\omega})\boldsymbol{\alpha}]^T\mathbf{C}_w^{-1}[\mathbf{x} - \mathbf{H}(\boldsymbol{\omega})\boldsymbol{\alpha}]\right\} \\ &= \arg\min_{\boldsymbol{\omega}} \left\{ \min_{\boldsymbol{\alpha}}\{[\mathbf{x} - \mathbf{H}(\boldsymbol{\omega})\boldsymbol{\alpha}]^T\mathbf{C}_w^{-1}[\mathbf{x} - \mathbf{H}(\boldsymbol{\omega})\boldsymbol{\alpha}]\} \right\}\end{aligned}$$

where one should first optimize the quadratic objective function wrt the linear parameters $\boldsymbol{\alpha}$ (assuming that $\boldsymbol{\omega}$ are known or constant). More specifically, optimization wrt $\boldsymbol{\alpha}$ for a given $\mathbf{H}(\boldsymbol{\omega})$ gives:

$$\hat{\boldsymbol{\alpha}}(\boldsymbol{\omega}) = \left(\mathbf{H}^T(\boldsymbol{\omega})\mathbf{C}_w^{-1}\mathbf{H}(\boldsymbol{\omega})\right)^{-1}\mathbf{H}^T(\boldsymbol{\omega})\mathbf{C}_w^{-1}\mathbf{x}$$

Substituting into the quadratic form yields (apart from constant terms irrelevant for the MLE, but not for CRB in Chapter 8):

$$\mathcal{L}(\mathbf{x}|\hat{\boldsymbol{\alpha}}(\boldsymbol{\omega}), \boldsymbol{\omega}) = -\frac{1}{2}[\mathbf{x} - \mathbf{P}(\boldsymbol{\omega})\mathbf{x}]^T\mathbf{C}_w^{-1}[\mathbf{x} - \mathbf{P}(\boldsymbol{\omega})\mathbf{x}]$$

where

$$\mathbf{P}(\boldsymbol{\omega}) = \mathbf{H}(\boldsymbol{\omega})\left(\mathbf{H}^T(\boldsymbol{\omega})\mathbf{C}_w^{-1}\mathbf{H}(\boldsymbol{\omega})\right)^{-1}\mathbf{H}^T(\boldsymbol{\omega})\mathbf{C}_w^{-1} \tag{7.1}$$

is a projection matrix as $\mathbf{P}^2(\boldsymbol{\omega}) = \mathbf{P}(\boldsymbol{\omega})$. Optimization wrt $\boldsymbol{\omega}$ after substituting into the likelihood yields

$$\boldsymbol{\omega}_{ML} = \arg\min_{\boldsymbol{\omega}}\{\mathbf{x}^T[\mathbf{I} - \mathbf{P}(\boldsymbol{\omega})]^T\mathbf{C}_w^{-1}[\mathbf{I} - \mathbf{P}(\boldsymbol{\omega})]\mathbf{x}\}$$

In general, the optimization wrt $\boldsymbol{\omega}$ is a complex problem as the parameters are embedded into the projection matrix (7.1) that contains the inverse of the covariance $\mathbf{C}_w^{-1}$, and there are no closed form relationships that could be of any help to have a closed form solution.

Uncorrelated noise simplifies the problem as $\mathbf{C}_w = \sigma^2\mathbf{I}$ and thus

$$\omega_{ML} = \arg\min_{\omega}\{\mathbf{x}^T[\mathbf{I}-\mathbf{P}(\omega)]\mathbf{x}\} = \arg\max_{\omega}\{\mathbf{x}^T\mathbf{P}(\omega)\mathbf{x}\} = \arg\max_{\omega}\mathrm{tr}\{\mathbf{P}(\omega)\mathbf{x}\mathbf{x}^T\}$$

with $\mathbf{P}(\omega) = \mathbf{H}(\omega)\left(\mathbf{H}^T(\omega)\mathbf{H}(\omega)\right)^{-1}\mathbf{H}^T(\omega)$. This relationship has a relevant algebraic interpretation. When changing the set of parameters ω, the matrix $\mathbf{H}(\omega)$ changes its structure and thus the subspace spanned by its columns. $\mathbf{P}(\omega)$ is the projection matrix onto this subspace and the estimate is the combination of ω that maximizes the norm of the projection of the observation $\mathbf{x}$ onto the span of the columns of $\mathbf{H}(\omega)$, or alternatively it minimizes the norm of the component onto the complementary subspace. In spite of the algebraic interpretation, the optimization is non-linear and non-trivial. The frequency estimation problem coincides with the estimation of ω regardless of α and it is crucial in some statistical signal processing contexts to re-interpret this as an algebraic problem; this is the basis for a class of specific methods detailed in Chapter 16.

7.2.3 Additive Noise Model with Multiple Observations $\mathbf{x}_\ell = \mathbf{s}_\ell(\theta) + \mathbf{w}_\ell$ with $\mathbf{w}_\ell \sim \mathcal{N}(\mathbf{0}, \mathbf{C}_w)$, $\mathbf{C}_w$ Known

In the case of multiple independent observations with different signal model but the same noise characteristic, the model of the data is

$$\mathbf{x}_\ell = \mathbf{s}_\ell(\theta) + \mathbf{w}_\ell \text{ for } \ell = 1,2,...,L \text{ with } \mathbf{w}_\ell \sim \mathcal{N}(\mathbf{0}, \mathbf{C}_w)$$

and the same (unknown) set of parameters θ. In this case the pdf of all the independent observations is:

$$p(\{\mathbf{x}_\ell\}|\theta) = \prod_\ell p(\mathbf{x}_\ell|\theta) = |2\pi\mathbf{C}_w|^{-L/2}\exp\left\{-\frac{1}{2}\sum_\ell[\mathbf{x}_\ell - \mathbf{s}_\ell(\theta)]^T\mathbf{C}_w^{-1}[\mathbf{x}_\ell - \mathbf{s}_\ell(\theta)]\right\}$$

while the MLE is

$$\theta_{ML} = \arg\min_{\theta}\left\{\sum_\ell[\mathbf{x}_\ell - \mathbf{s}_\ell(\theta)]^T\mathbf{C}_w^{-1}[\mathbf{x}_\ell - \mathbf{s}_\ell(\theta)]\right\}$$

Some useful simplifications are possible in some special cases.

7.2.3.1 Linear Model $\mathbf{x}_\ell = \mathbf{H}\cdot\theta + \mathbf{w}_\ell$

One special case is when there are multiple independent observations with a persistent model and varying noise:

$$\mathbf{x}_\ell = \mathbf{H}\cdot\theta + \mathbf{w}_\ell \quad \text{for } \ell = 1,2,...,L$$

It can be shown that the MLE is

$$\theta_{ML} = \left(\mathbf{H}^T\mathbf{C}_w^{-1}\mathbf{H}\right)^{-1}\mathbf{H}^T\mathbf{C}_w^{-1}\bar{\mathbf{x}}$$

that is an extension of the MLE for the linear model above when using the sample mean of the observations

$$\bar{\mathbf{x}} = \frac{\sum_{\ell=1}^{L} \mathbf{x}_\ell}{L}$$

The covariance scales with the number of observations:

$$\text{cov}[\theta_{ML}] = \frac{1}{L}\left(\mathbf{H}^T\mathbf{C}_w^{-1}\mathbf{H}\right)^{-1}$$

and this motivates the common practice of repeating multiple experiments to increase the accuracy.

A special situation is when the model involves both linear and non-linear terms: $\mathbf{s}(\boldsymbol{\omega},\boldsymbol{\alpha}) = \mathbf{H}(\boldsymbol{\omega}) \cdot \boldsymbol{\alpha}$. The steps are the same as above to estimate the ensemble $\boldsymbol{\theta} = [\boldsymbol{\alpha}^T, \boldsymbol{\omega}^T]^T$. Following the same steps and considering at first an optimization with respect to $\boldsymbol{\alpha}$ for $\mathbf{H}(\boldsymbol{\omega})$ as known:

$$\boldsymbol{\alpha}_* = \boldsymbol{\alpha}(\boldsymbol{\omega}) = \left(\mathbf{H}^T\mathbf{C}_w^{-1}\mathbf{H}\right)^{-1}\mathbf{H}^T\mathbf{C}_w^{-1}\bar{\mathbf{x}}$$

where the dependency on $\boldsymbol{\omega}$ of $\mathbf{H}(\boldsymbol{\omega})$ is taken as read in the notation. Plugging this term into the general expression of the log-likelihood function leads to (apart from a constant):

$$\mathcal{L}(\{\mathbf{x}_\ell\}|\boldsymbol{\omega}) = -\frac{1}{2}\sum_\ell [\mathbf{x}_\ell - \mathbf{P}(\boldsymbol{\omega})\bar{\mathbf{x}}]^T\mathbf{C}_w^{-1}[\mathbf{x}_\ell - \mathbf{P}(\boldsymbol{\omega})\bar{\mathbf{x}}]$$

where $\mathbf{P}(\boldsymbol{\omega})$ is from (7.1). In the case of white noise ($\mathbf{C}_w = \sigma^2\mathbf{I}$), the expression of $\mathbf{P}(\boldsymbol{\omega})$ simplifies, and thus

$$\boldsymbol{\omega}_{ML} = \arg\min_{\omega}\{\sum_\ell \mathbf{x}_\ell^T\mathbf{x}_\ell/L - \bar{\mathbf{x}}^T\mathbf{P}(\boldsymbol{\omega})\bar{\mathbf{x}}\} = \arg\min_{\omega}\text{tr}\{\hat{\mathbf{R}}_x - \mathbf{P}(\boldsymbol{\omega})\bar{\mathbf{x}}\bar{\mathbf{x}}^T\}$$

where

$$\hat{\mathbf{R}}_x = \frac{\sum_{\ell=1}^{L}\mathbf{x}_\ell\mathbf{x}_\ell^T}{L}$$

is the sample correlation matrix of the data samples.

7.2.3.2 Model $\mathrm{x}_\ell = \mathrm{H}_\ell \cdot \theta + \mathrm{w}_\ell$

The models are linear with known covariance $\mathbf{C}_w$ and varying regressor $\mathbf{s}_\ell(\boldsymbol{\theta}) = \mathbf{H}_\ell \cdot \boldsymbol{\theta}$; the optimization is quadratic as for a single observation and therefore

$$\theta_{ML} = \left(\sum_\ell \mathbf{H}_\ell^T\mathbf{C}_w^{-1}\mathbf{H}_\ell\right)^{-1}\sum_\ell \mathbf{H}_\ell^T\mathbf{C}_w^{-1}\mathbf{x}_\ell$$

After some algebra one gets the covariance of the MLE:

$$\text{cov}[\boldsymbol{\theta}_{ML}|\{\mathbf{H}_\ell\}] = \left(\sum_\ell \mathbf{H}_\ell^T \mathbf{C}_w^{-1} \mathbf{H}_\ell\right)^{-1} \tag{7.2}$$

In some cases, the regressor $\mathbf{H}_\ell$ at each observation is drawn from a random (but known) set and (7.2) is the conditional covariance to the specific realizations $\{\mathbf{H}_\ell\}$. To evaluate the covariance independent of the realizations, one has to compute

$$\text{cov}[\boldsymbol{\theta}_{ML}] = \mathbb{E}_{\mathbf{H}}[\text{cov}[\boldsymbol{\theta}_{ML}|\{\mathbf{H}_\ell\}]]$$

for the joint pdf of the entries of $\mathbf{H}$, but this is usually very complex. This is bounded by Jensen's inequality for positive definite matrixes (Section 3.2):

$$\text{cov}[\boldsymbol{\theta}_{ML}] \geq \frac{1}{L}\left(\mathbb{E}_{\mathbf{H}}[\mathbf{H}^T \mathbf{C}_w^{-1} \mathbf{H}]\right)^{-1} \tag{7.3}$$

that is far more manageable, especially if entries of $\mathbf{H}$ are statistically independent. In any case, since for $L \to \infty$ the bounds are tight, and the bound (7.3) is called the asymptotic approximation to the covariance.

7.2.3.3 Model $\mathbf{x}_\ell = \mathbf{H}(\omega) \cdot \alpha_\ell + \mathbf{w}_\ell$

There are multiple observations with a constant regressor $\mathbf{H}(\boldsymbol{\omega})$, and varying scale $\boldsymbol{\alpha}_\ell$. One might estimate on each observation, say $\ell = \bar{\ell}$, both sets of parameters $\{\hat{\boldsymbol{\alpha}}_{\bar{\ell}}, \hat{\boldsymbol{\omega}}_{\bar{\ell}}\}$, but this approach does not constrain the regressor to be constant vs. ℓ. Alternatively, following the same steps in Section 7.2.2, the MLE for the sample $\mathbf{x}_\ell$ is

$$\hat{\boldsymbol{\alpha}}_\ell(\boldsymbol{\omega}) = \left(\mathbf{H}(\boldsymbol{\omega})^T \mathbf{C}_w^{-1} \mathbf{H}(\boldsymbol{\omega})\right)^{-1} \mathbf{H}(\boldsymbol{\omega})^T \mathbf{C}_w^{-1} \mathbf{x}_\ell$$

and, substituting into the log-likelihood, one gets the estimate over the whole set:

$$\boldsymbol{\omega}_{ML} = \arg\min_{\boldsymbol{\omega}} tr\{[\mathbf{I} - \mathbf{P}(\boldsymbol{\omega})]^T \mathbf{C}_w^{-1} [\mathbf{I} - \mathbf{P}(\boldsymbol{\omega})]\hat{\mathbf{R}}_x\}$$

which has no straightforward solution.

Even for $\mathbf{C}_w = \sigma^2 \mathbf{I}$, the MLE is

$$\boldsymbol{\omega}_{ML} = \arg\max_{\boldsymbol{\omega}} \text{tr}\{\mathbf{P}(\boldsymbol{\omega})\hat{\mathbf{R}}_x\}$$

which is a general case of Section 7.2.2 widely investigated in line spectra analysis (Chapter 16).

7.2.4 Model $\mathbf{x}_\ell \sim \mathcal{N}(\mathbf{0}, \mathbf{C}(\boldsymbol{\theta}))$

Each single observation is modeled as drawn from a zero-mean Gaussian rv with covariance $\mathbf{C}(\boldsymbol{\theta})$ that depends on the parameters $\boldsymbol{\theta}$:

$$\mathbf{x}_\ell \sim \mathcal{N}(\mathbf{0}, \mathbf{C}(\boldsymbol{\theta}))$$

and we have multiple independent observations $\mathbf{x}_1, \mathbf{x}_2, ..., \mathbf{x}_L$. The log-likelihood is very general

$$\mathcal{L}(\{\mathbf{x}_\ell\}|\theta) = -\left\{\ln|\mathbf{C}(\theta)| + \text{tr}[\mathbf{C}(\theta)^{-1}\hat{\mathbf{R}}_x]\right\}$$

(up to a scale-term independent on θ), it depends on the sample correlation $\hat{\mathbf{R}}_x = \sum_{\ell=1}^{L} \mathbf{x}_\ell \mathbf{x}_\ell^T / L$ from the L observations. The MLE is the value that maximizes $\mathcal{L}(\{\mathbf{x}_\ell\}|\theta)$ using any of the known numerical methods over the p-dimensional space.

For the special case that $\mathbf{x}_\ell \sim \mathcal{N}(\mathbf{0}, \mathbf{C}_x)$ where $\mathbf{C}_x$ is not structured, one should set the gradient wrt each entry of $\mathbf{C}_x$ to zero, yielding

$$[\mathbf{C}_x^T - \hat{\mathbf{R}}_x^T]_{\mathbf{C}_x = \hat{\mathbf{C}}_x} = \mathbf{0}$$

Since there is no predefined structure on $\mathbf{C}_x$:

$$\hat{\mathbf{C}}_x = \hat{\mathbf{R}}_x$$

that is, the MLE of the covariance is the sample covariance itself (recall that $\mathbf{x}_\ell$ is zero-mean).

7.2.5 Additive Noise Model with Multiple Observations $\mathbf{x}_\ell = \mathbf{s}_\ell(\theta) + \mathbf{w}_\ell$ with $\mathbf{w}_\ell \sim \mathcal{N}(\mathbf{0}, \mathbf{C}_w)$, $\mathbf{C}_w$ Unknown

When the noise covariance matrix $\mathbf{C}_w$ is unknown, its entries become part of the model parameters to be estimated. The likelihood is the same as Section 7.2.3; for a given (but unknown) $\mathbf{C}_w$ the MLE of θ is

$$\theta_{ML} = \arg\min_{\theta} \left\{\ln|\mathbf{C}_w| + \text{tr}[\mathbf{C}_w^{-1}\mathbf{Q}(\theta)]\right\}$$

where

$$\mathbf{Q}(\theta) = \frac{1}{L}\sum_{\ell=1}^{L}[\mathbf{x}_\ell - \mathbf{s}_\ell(\theta)][\mathbf{x}_\ell - \mathbf{s}_\ell(\theta)]^T$$

The optimization carried out at first with respect to $\mathbf{C}_w$ (assuming $\mathbf{s}_\ell(\theta)$ is known, as above) leads to $\hat{\mathbf{C}}_w = \mathbf{Q}(\theta)$ as this choice yields $\text{tr}[\hat{\mathbf{C}}_w^{-1}\mathbf{Q}(\theta)] = \text{tr}[\mathbf{Q}(\theta)^{-1}\mathbf{Q}(\theta)] = N$. Substituting into the likelihood function, the MLE follows from the optimization of the determinant

$$\theta_{ML} = \arg\min_{\theta} \{\ln|\mathbf{Q}(\theta)|\}$$

This is non-trivial, in general, even when setting to zero the gradients from rules in Appendix B of Chapter 8.

One special case is when the model can be represented as the product of two terms $\mathbf{s}_\ell(\theta) = \mathbf{H}_\ell \cdot \theta$ (or also $\mathbf{s}_\ell(\theta) = \mathbf{H} \cdot \theta$). One can partition $\mathbf{Q}(\theta) = \mathbf{A} + \mathbf{B}(\theta)$ and one can use the following approximations

$$|\mathbf{Q}(\theta)| = |\mathbf{A}+\mathbf{B}(\theta)| = |\mathbf{A}|\cdot|\mathbf{I}+\underbrace{\mathbf{A}^{-1}\mathbf{B}(\theta)}_{\mathbf{G}(\theta)}|$$

$$= |\mathbf{A}|\cdot\left(\mathrm{tr}[\mathbf{G}(\theta)] - \frac{1}{2}\mathrm{tr}[\mathbf{G}(\theta)^2] + \ldots\right) \simeq |\mathbf{A}|\cdot\mathrm{tr}[\mathbf{G}(\theta)]$$

which allows us to solve the equivalent problem

$$\theta_{ML} \simeq \arg\min_{\theta}\{\mathrm{tr}[\mathbf{G}(\theta)]\}.$$

7.3 Other Noise Models

Generalizing the additive noise to any independent noise model, the conditional pdf is

$$p(\mathbf{x}|\theta) = \prod_{n=1}^{N} p_w(x_n - s_n(\theta))$$

and the log-likelihood function becomes

$$\mathcal{L}(\mathbf{x}|\theta) = \sum_{n=1}^{N} \ln p_w(x_n - s_n(\theta))$$

Noise with uniform pdf. In the case that $w_n \sim \mathcal{U}(-a,a)$, the noise components have uniform pdf within $[-a,a]$ so that

$$p_w(w) = \frac{1}{2a} \quad \text{per } w \in [-a,a]$$

The ML estimator searches the parameter values in order to have a maximum error that does not exceed a:

$$\max_{\theta}\{|x_1 - s_1(\theta)|, |x_2 - s_2(\theta)|, \ldots, |x_N - s_N(\theta)|\} \leq a$$

or in another equivalent terms, the parameter values for which (Section 1.1):

$$||\mathbf{x} - \mathbf{s}(\theta)||_\infty \leq a.$$

Noise with Laplacian pdf. When the noise can be described by a Laplacian pdf:

$$p_w(w) = \frac{1}{2\sigma}\exp(-|w|/\sigma)$$

where $\mathbb{E}[w^2] = 2\sigma^2$. Therefore

$$\mathcal{L}(\mathbf{x}|\theta) = -N\ln 2\sigma - \sum_{n=1}^{N}|x_n - s_n(\theta)|/\sigma$$

and the MLE is

$$\theta_{ML} = \arg\min_{\theta} \sum_{n=1}^{N} |x_n - s_n(\theta)|$$

Note that in this case what is minimized is the sum of the L1 norm (absolute value) of the terms $x_n - s_n(\theta)$. This is the practical case when the noise contains some large amplitude noise samples that occur far more than for Gaussian noise. These are the *robust estimators* discussed later in Section 7.9.

Noise with exponential pdf. When noise samples have an arbitrary exponential pdf (up to a normalizing factor)

$$p_w(w) = \exp(-\phi(w))$$

the MLE becomes

$$\theta_{ML} = \arg\min_{\theta} \sum_{n=1}^{N} \phi(x_n - s_n(\theta))$$

which degenerates to L2 or L1 norm minimization for Gaussian or Laplace pdf. Note that the reverse is also true: when an estimator is defined as the minimization of a sum of functions $\phi(.)$ applied to estimation errors, this corresponds to the MLE applied to exponential noise terms if the noise is effectively exponential-shaped, or otherwise it is just a convenient approximation.

7.4 MLE and Nuisance Parameters

In an estimation problem, the set of parameters can be partitioned into two subsets $\boldsymbol{\theta} = [\boldsymbol{\theta}_1^T, \boldsymbol{\theta}_2^T]^T$ where $\boldsymbol{\theta}_1$ are the parameters of interest and $\boldsymbol{\theta}_2$ are *nuisance* parameters that are not strictly of interest for the application at hand, but are still necessary to tackle the estimation problem. To exemplify from Section 5.4, in the estimation of the frequency of one sinusoid in noise where amplitude and phase are unknown, the MLE needs to account for the parameters frequency, amplitude, and phase, even if the latters are not of interest for the problem at hand. Assuming that nuisance can be modeled as random with known pdf $p(\boldsymbol{\theta}_2)$ (this is a-priori information as detailed later in Chapter 11), the conditional pdf

$$p(\mathbf{x}|\boldsymbol{\theta}_1, \boldsymbol{\theta}_2)$$

can be reshaped to define the likelihood only for the parameters of interest by the marginalization

$$p(\mathbf{x}|\boldsymbol{\theta}_1) = \int p(\mathbf{x}|\boldsymbol{\theta}_1, \boldsymbol{\theta}_2) p(\boldsymbol{\theta}_2) d\boldsymbol{\theta}_2 = \mathbb{E}_{\boldsymbol{\theta}_2}[p(\mathbf{x}|\boldsymbol{\theta}_1, \boldsymbol{\theta}_2)]$$

This step removes all dependency on nuisance by using expectation, and the MLE is the minimizer of $p(\mathbf{x}|\theta_1)$. However, the expectation $\mathbb{E}_{\theta_2}[.]$ is not always so trivial, or leads to a very complicated cost function, so there are trade-offs between what is nuisance to be averaged, and what is better to keep as nuisance to be estimated. An example can make this clearer.

In frequency estimation, the complex-valued sinusoid model

$$x[n] = a \cdot \exp[j(\omega_o n + \varphi)] + w[n]$$

can account for phase as nuisance with $\varphi \sim \mathcal{U}(-\pi, \pi)$. In this case of complex-valued signals, the likelihood is

$$p(\mathbf{x}|a,\omega,\varphi) = \frac{\exp(-(||\mathbf{x}||^2 + Na^2)/\sigma_w^2)}{\pi\sigma_w^{2N}} \exp\left(\frac{2a}{\sigma_w^2} Re\left\{\sum_{n=0}^{N-1} x[n]\exp[-j(\omega n + \varphi)]\right\}\right).$$

Marginalization wrt φ is

$$p(\mathbf{x}|a,\omega) = \int_{-\pi}^{\pi} p(\mathbf{x}|a,\omega,\varphi)\frac{d\varphi}{2\pi}$$

$$p(\mathbf{x}|a,\omega) = \frac{\exp(-(||\mathbf{x}||^2 + Na^2)/\sigma_w^2)}{\pi\sigma_w^{2N}} \int_{-\pi}^{\pi} \exp\left(\frac{2a|X(\omega)|}{\sigma^2}\cos(\varphi + \angle X(\omega))\right)\frac{d\varphi}{2\pi}$$
$$= \frac{\exp(-(||\mathbf{x}||^2 + Na^2)/\sigma^2)}{\pi\sigma^{2N}} I_0\left(\frac{2a|X(\omega)|}{\sigma^2}\right)$$

where $X(\omega) = \sum_{n=0}^{N-1} x[n]\exp(-j\omega n) = \mathcal{F}(\{x[n]\})$ and $I_0(x) = \int_{-\pi}^{\pi} \exp(x\cos(\varphi))\frac{d\varphi}{2\pi}$ is the zero-th order modified Bessel function and can be approximated by $I_0(x) \simeq e^x/\sqrt{2\pi x}$ for $x > 10$. Since $I_0(x)$ is monotonic, the MLE of frequency is

$$\omega_{MLE} = \arg\max_{\omega\in[-\pi,\pi)} |X(\omega)|$$

which coincides with results in Section 9.2, or later in spectral analysis (Chapter 16). The estimate of the amplitude is from

$$p(\mathbf{x}|a,\omega_{ML}) \simeq \left(\frac{4\pi a|X_*|}{\sigma_w^2}\right)^{-1/2} \exp\left(\frac{2a|X_*| - Na^2}{\sigma_w^2}\right)$$

with $X_* = X(\omega_{MLE})$ (apart from an irrelevant scaling term and for the asymptotic approximation of $I_0(x)$ for $x \gg 1$). The MLE is

$$a_{ML} = \frac{|X_*|}{2N}\left(1 + \sqrt{1 - \frac{N\sigma_w^2}{|X_*|^2}}\right) \simeq \frac{|X_*|}{N} - \frac{\sigma_w^2}{4|X_*|} \simeq \frac{|X_*|}{N}$$

where the last approximations are for $a^2/\sigma_w^2 \gg 1$.

7.5 MLE for Continuous-Time Signals

Given a continuous-time model where the observation is limited to a time interval T:

$$x(t) = s(t;\boldsymbol{\theta}) + w(t) \quad \text{for } t \in [0, T) \tag{7.4}$$

and assuming an Additive White Gaussian Noise (AWGN) with autocorrelation function

$$r_w(\tau) = \mathbb{E}[w(t)w(t+\tau)] = \frac{N_0}{2}\delta(\tau) \tag{7.5}$$

it is possible to define a likelihood function as a limit of the one for the corresponding discrete-time observation sequence.

Let the observation be sampled with an arbitrary step Δt after low-pass filtering within the band $\pm 1/2\Delta t$; the sequence of $M = T/\Delta t$ samples $\mathbf{x} = [x(1 \cdot \Delta t), x(2 \cdot \Delta t), \ldots, x(M \cdot \Delta t)]^T$ is the starting point for the conditional pdf

$$p(\mathbf{x}|\boldsymbol{\theta}) = \left(2\pi\sigma_w^2\right)^{-M} \exp\left(-\frac{1}{2\sigma_w^2}\sum_{k=1}^{M}(x(k \cdot \Delta t) - s(k \cdot \Delta t;\boldsymbol{\theta}))^2\right) \tag{7.6}$$

where the PSD of noise is bounded within the band $\pm 1/2\Delta t$ and the power is $\sigma_w^2 = (N_0/2)/\Delta t$. Making the substitutions in (7.6) and considering $\Delta t \to 0$ (or equivalently $M \to \infty$), one obtains

$$p(x(t)|\boldsymbol{\theta}) = \Gamma \exp\left(-\frac{1}{N_0}\int_0^T (x(t) - s(t;\boldsymbol{\theta}))^2 dt\right) \tag{7.7}$$

where Γ is a normalization factor that is not-relevant for the ML estimation. The relationship (7.7) is the likelihood function for continuous-time signals within the $[0, T)$ interval.

The MLE minimizes the square error that is the argument of the likelihood function:

$$\boldsymbol{\theta}_{ML} = \arg\min_{\boldsymbol{\theta}} \int_0^T (x(t) - s(t;\boldsymbol{\theta}))^2 dt \tag{7.8}$$

Solution is by solving the set of equations (in general non-linear) from the gradient

$$\int_0^T (x(t) - s(t;\boldsymbol{\theta}))\frac{\partial s(t;\boldsymbol{\theta})}{\partial \theta_k} dt = 0 \quad \text{for } k = 1, 2, \ldots, p \tag{7.9}$$

namely the error $x(t) - s(t;\boldsymbol{\theta})$ should be orthogonal to the derivative of the signal model. The likelihood terms of (7.8) can be expanded into

$$\int_0^T (x(t) - s(t;\boldsymbol{\theta}))^2 dt = \int_0^T x(t)^2 dt + \int_0^T s(t;\boldsymbol{\theta})^2 dt - 2\int_0^T x(t)s(t;\boldsymbol{\theta})dt$$

The first term is the energy of the observation and it is irrelevant for the optimization; the second term is the energy of the model signal $s(t;\theta)$ and in some applications it does not depend on θ (e.g., in time of delay estimation, $s(t;\theta) = s(t-\theta)$ and $\int_0^T s(t-\theta)^2 dt = \int_0^T s(t)^2 dt$). In many contexts (to be evaluated case-by-case), the estimation is based on the maximization of the third term:

$$\theta_{ML} = \arg\max_{\theta} \int_0^T x(t)s(t;\theta)dt \tag{7.10}$$

that is, the cross-correlation between the signal $x(t)$ and its replica $s(t;\theta)$, whose characteristics depend on the parameters that must be estimated.

7.5.1 Example: Amplitude Estimation

Amplitude estimation is a simple and common example. The model of the data for the estimation problem is

$$x(t) = As(t) + w(t) \quad \text{for } t \in [0, T)$$

where $w(t)$ is white Gaussian. It is required to estimate the amplitude A for a known waveform $s(t)$ for the time window T. Taking into account the general relationship for an ML estimator (eq. 7.9) it follows that

$$\int_0^T (x(t) - As(t))s(t)dt = 0 \Rightarrow A_{ML} = \frac{\int_0^T x(t)s(t)dt}{E_s}$$

The MLE is the ratio between the cross-correlation between the data $x(t)$ and the waveform $s(t)$, normalized to the energy of the waveform $E_s = \int_0^T s(t)^2 dt$. For the sake of evaluating the estimator's performance, the estimate can be rearranged by using the model for $x(t)$ to isolate the random terms:

$$A_{ML} = A + \frac{\int_0^T w(t)s(t)dt}{E_s}$$

and the second term accounts for the randomness of the estimate as it depends on the stochastic term $w(t)$. The estimator is unbiased as

$$\mathbb{E}[A_{ML}] = A + \frac{\int_0^T \mathbb{E}[w(t)]s(t)dt}{E_s} = A$$

and the variance follows from the definition

$$\text{var}[A_{ML}] = \frac{1}{E_s^2}\mathbb{E}\left[\iint_0^T w(t)w(\alpha)s(t)s(\alpha)dtd\alpha\right] = \frac{N_0}{2E_s}$$

This depends on the inverse of the signal to noise ratio $2E_s/N_0$ that depends on the energy of the waveform E_s.

7.5.2 MLE for Correlated Noise $S_w(f)$

To relax the condition of White Gaussian Noise, one can characterize zero-mean noise by the autocorrelation function or the PSD:

$$r_w(\tau) \leftrightarrow S_w(f)$$

The ML approach previously described can be extended to the case of arbitrarily correlated noise (also called *colored-noise* in contrast to white noise) by applying the *whitening* filter designed from the PSD $S_w(f)$:

$$g(t) \leftrightarrow G(f) = \frac{1}{\sqrt{S_w(f)}}$$

Filtering the data by $g(t)$, the overall model becomes

$$\underbrace{g(t) * x(t)}_{\tilde{x}(t)} = \underbrace{g(t) * s(t;\theta)}_{\tilde{s}(t;\theta)} + \underbrace{g(t) * w(t)}_{\tilde{w}(t)}$$

Note that the equivalent noise $\tilde{w}(t) = g(t) * w(t)$ is white as the autocorrelation is

$$r_{\tilde{w}}(\tau) = \mathbb{E}[\tilde{w}(t)\tilde{w}(t+\tau)] = \delta(\tau) \leftrightarrow S_{\tilde{w}}(f) = 1$$

This property gives the name "whitening filter" to $g(t)$. The model after filtering

$$\tilde{x}(t) = \tilde{s}(t;\theta) + \tilde{w}(t)$$

is the same as (7.4–7.5) and the likelihood function becomes

$$p(x(t)|\theta) = \Gamma \exp\left(-\frac{1}{2}\int_0^T (\tilde{x}(t) - \tilde{s}(t;\theta))^2 dt\right)$$

leaving the analytic infrastructure unaltered.

Remark on practical applications. The approach for handling correlated noise is conceptually simple but it has several practical drawbacks that follow from an careful inspection of the design criteria of $G(f)$. A whitening filter is a zero phase filter and its impulse response $g(t)$ is even and non-casual, with long tails if $G(f)$ containing rapid transitions to zero (or equivalently, if $S_w(f) \simeq 0$ for some frequencies, or frequency band). This filter cannot be trivially implemented in practice, and tail control is mandatory to avoid artifacts. To be more specific, since the available data is temporally limited to T, the length of the response $g(t)$ should be much smaller than T to avoid $\tilde{x}(t)$ and $\tilde{s}(t;\theta)$ being excessively affected by the filter's transition effects. In practice, this can be obtained by windowing (i.e., smoothly truncating) the "exact" whitening filter $g(t)$ within a small support so that the whitening filter is replaced by $g(t) \times a(t)$, where $a(t)$ is a window with a support smaller than T. Since $g(t) \times a(t) \leftrightarrow S_w^{-1/2}(f) * A(f)$, care should be given to the design of window $a(t)$ to trade some performance degradation of MLE.

7.6 MLE for Circular Complex Gaussian

Complex-valued signals are dealt with here with examples. The observation $\mathbf{x} \in \mathbb{C}^N$ can be represented equivalently by $\tilde{\mathbf{x}} = [\mathbf{x}_R^T, \mathbf{x}_I^T]^T$; the pdfs for these two notations are (see Section 3.7):

$$\mathbf{x} \sim \mathcal{CN}(\boldsymbol{\mu}(\boldsymbol{\theta}), \mathbf{C}(\boldsymbol{\theta})) \Longleftrightarrow \tilde{\mathbf{x}} = \begin{bmatrix} \mathbf{x}_R \\ \mathbf{x}_I \end{bmatrix} \sim \mathcal{N}(\tilde{\boldsymbol{\mu}}(\boldsymbol{\theta}), \tilde{\mathbf{C}}(\boldsymbol{\theta}))$$

where

$$\boldsymbol{\mu}(\boldsymbol{\theta}) = \boldsymbol{\mu}_R(\boldsymbol{\theta}) + j\boldsymbol{\mu}_I(\boldsymbol{\theta}) \in \mathbb{C}^N$$
$$\mathbf{C}(\boldsymbol{\theta}) = 2\left(\mathbf{C}_R(\boldsymbol{\theta}) + j\mathbf{C}_I(\boldsymbol{\theta})\right) \in \mathbb{C}^{N\times N}$$

(the second equality is based on the property of circular Gaussian pdf, see Section 3.7.3 with some adaptation)

$$\tilde{\boldsymbol{\mu}}(\boldsymbol{\theta}) = \begin{bmatrix} \boldsymbol{\mu}_R(\boldsymbol{\theta}) \\ \boldsymbol{\mu}_I(\boldsymbol{\theta}) \end{bmatrix} \in \mathbb{R}^{2N}$$
$$\tilde{\mathbf{C}}(\boldsymbol{\theta}) = \begin{bmatrix} \mathbf{C}_R(\boldsymbol{\theta}) & \mathbf{C}_I(\boldsymbol{\theta}) \\ -\mathbf{C}_I(\boldsymbol{\theta}) & \mathbf{C}_R(\boldsymbol{\theta}) \end{bmatrix} \in \mathbb{R}^{2N\times 2N}$$

these two equivalent representations are used herein for MLE. More specifically, the MLE is based on the maximization of the log-likelihood (up to irrelevant constant terms)

$$\mathcal{L}(\mathbf{x}|\boldsymbol{\theta}) = -\ln|\mathbf{C}(\boldsymbol{\theta})| - (\mathbf{x} - \boldsymbol{\mu}(\boldsymbol{\theta}))^H \mathbf{C}(\boldsymbol{\theta})^{-1}(\mathbf{x} - \boldsymbol{\mu}(\boldsymbol{\theta}))$$

by nulling the gradients for complex-valued variables (Section 1.4.2) or any other iterative methods. A good alternative is by considering the fully equivalent representation as real and imaginary entries $\tilde{\mathbf{x}} \sim \mathcal{N}(\tilde{\boldsymbol{\mu}}(\boldsymbol{\theta}), \tilde{\mathbf{C}}(\boldsymbol{\theta}))$ so that the MLE (and any other estimation method) can be reduced to a known estimation for real-valued observations. In some cases, scaling by 2 is absorbed by the complex notation for compectness so that $\mathbf{C}(\boldsymbol{\theta}) = \mathbf{C}_R(\boldsymbol{\theta}) + j\mathbf{C}_1(\boldsymbol{\theta})$, and $\tilde{\mathbf{C}}(\boldsymbol{\theta}) \Rightarrow \tilde{\mathbf{C}}(\boldsymbol{\theta})/2$.

7.7 Estimation in Phase/Frequency Modulations

In communication engineering, the phase of a sinusoidal signal $s(t) = a\cos(\omega_o t + \phi(t))$ can intentionally be changed according to the information to be transmitted. More specifically, let $m(t)$ be the signal to be transmitted (modulating signal) that can include bits, voice, music, etc. The mapping of $m(t)$ and the sinusoidal signal is referred to as phase or frequency modulator depending on the approach taken:

$$\text{phase-modulation: } \phi(t) = K_{PM}m(t)$$

$$\text{frequency modulation: } \frac{1}{2\pi}\frac{d}{dt}\phi(t) = K_{FM}m(t) \Rightarrow \phi(t) = 2\pi K_{FM}\int_{-\infty}^{t} m(\zeta)d\zeta$$

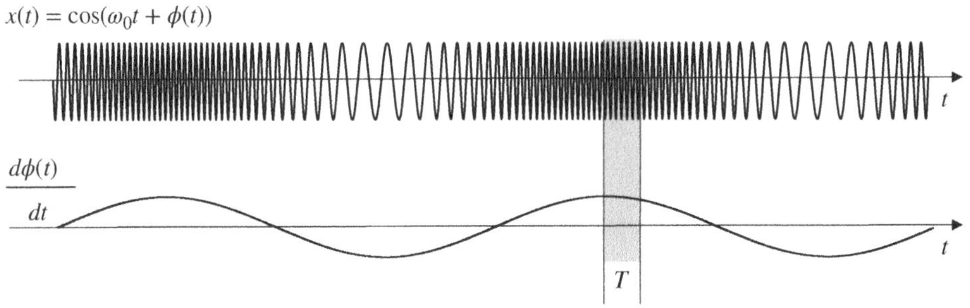

Figure 7.2 Example of frequency modulated sinusoid, and stationarity interval T.

where K_{PM} and K_{FM} are scaling terms (modulator gains). The transmitted signal $s(t)$ (see illustration in Figure 7.2) is received affected by noise

$$x(t) = a\cos(\omega_o t + \phi(t)) + w(t)$$

where the phase is continuously varying according to the modulating signal $m(t)$. The demodulator maps the estimate of the instantaneous phase at the receiver $\hat{\phi}(t)$ onto the estimate $\hat{m}(t)$ of the modulating signal for information retrieval.

7.7.1 MLE Phase Estimation

Given one sinusoid with known amplitude a and frequency ω_o (usually referred to as the carrier frequency in this application) in AWGN

$$x(t) = a\cos(\omega_o t + \phi_o) + w(t) \quad \text{for } t \in [0, T)$$

the problem here is to estimate the phase ϕ_o from a limited observation such that any phase-variation vs. time is approximately constant within T (see Figure 7.2). Truncation effects are avoided by assuming that the period of the sinusoid $2\pi/\omega_o$ is much smaller than T, or equivalently a large number of sinusoid periods is included in $[0, T)$, so that the mean value

$$\frac{1}{T}\int_0^T \cos(\omega_o t + \phi_o)dt \simeq 0$$

is independent of the phase (i.e., this relationship is strictly true only for $T \to \infty$, but errors are negligible for $\omega_o T \gg 1$).

The ML estimation can be obtained by considering the orthogonality condition (7.9)

$$\int_0^T (x(t) - a\cos(\omega_o t + \hat{\phi}))\sin(\omega_o t + \hat{\phi})dt = 0$$

which yields the implicit equation (for $\int_0^T \sin(2\omega_o t + 2\hat{\phi})) \simeq 0$ as $\omega_o T \gg 1$):

$$\int_0^T x(t)\sin(\omega_o t + \hat{\phi})dt = 0.$$

After rewriting this by using trigonometric expressions, it is possible to have

$$\hat{\phi} = -\arctan\left(\frac{\int_0^T x(t)\sin(\omega_o t)dt}{\int_0^T x(t)\cos(\omega_o t)dt}\right)$$

which is an explicit relationship for evaluating the phase. It is important to note that the arctan(.) function returns values within the range $(-\pi/2, \pi/2)$; in order to have values within the full range $(-\pi, \pi]$, it is necessary to evaluate the sign of the terms $\int_0^T x(t)\sin(\omega_o t)dt$ and $\int_0^T x(t)\cos(\omega_o t)dt$.

Accuracy of MLE can be evaluated from a sensitivity analysis assuming that the estimate includes a small (unknown) error $\hat{\phi} = \phi_o + \delta\phi$ that can be characterized from the statistical properties of the random term in the MLE that here represents the noise. Replacing this value into the orthogonality equation

$$\int_0^T [a\cos(\omega_o t + \phi_o) + w(t)]\sin(\omega_o t + \phi_o + \delta\phi)dt = 0$$

after some trigonometry

$$\frac{a}{2}T\sin\delta\phi \simeq -\int_0^T w(t)\sin(\omega_o t)dt$$

one gets the relationship between $\delta\phi$ and $w(t)$. For a small phase error $\sin\delta\phi \simeq \delta\phi$:

$$\delta\phi \simeq -\frac{2}{aT}\int_0^T w(t)\sin(\omega_o t)dt$$

One can trivially prove that the estimator is unbiased as $\mathbb{E}[\delta\phi] = 0$. The variance is evaluated by the definition as

$$\begin{aligned}\text{var}[\delta\phi] &= \frac{4}{a^2T^2}\mathbb{E}[\int_0^T w(t_1)\sin(\omega_o t_1)dt_1 \int_0^T w(t_2)\sin(\omega_o t_2)dt_2] \\ &= \frac{2N_0}{a^2T^2}\underbrace{\int_0^T \sin^2(\omega_o t_1)dt_1}_{T/2} = \frac{N_0}{a^2T}\end{aligned}$$

The accuracy increases with the observation window T as noise is reduced accordingly by the integral acting as averaging the noise over T. Needless to say, this analysis holds only when the phase error $\delta\phi$ (or equivalently, noise power $N_0/2$) is small, but it still provides a good guideline for the design of the observation window T given a degree of variance necessary for the application at hand.

7.7.2 Phase Locked Loops

Accuracy in the phase estimate calls for a large observation T, and the choice should account for a time-varying modulation signal $m(t)$ that makes the phase also vary.

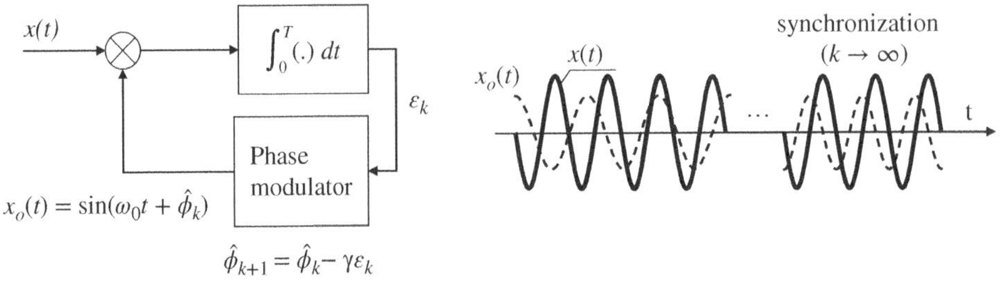

Figure 7.3 Principle of phase locked loop as iterative phase minimization between $x(t)$ and local $x_o(t)$.

Windowing T as in Figure 7.3 acts as a segmentation of the signal, and some remarks on the choice of T for time-varying signals are in order. Namely, T should be small enough to have the modulating signal approximately constant, analytically: $\phi(t_1) - \phi(t_2) \simeq 0$ for $\forall t_1, t_2 : |t_1 - t_2| \leq T$. Since phase changes smoothly from one observation window to another, the continuous time $x(t)$ can be considered as (*virtually*) segmented into a sequence of T-length observation windows with a phase estimator on each. Let ϕ_k be the phase within the kth observation window, the orthogonality equation becomes

$$\varepsilon_k(\hat{\phi}_k) = \int_0^T x(t)\sin(\omega_o t + \hat{\phi}_k)dt$$

so that ε_k is an error that depends on the instantaneous phase estimate within the window; ideally $\varepsilon_k = 0$.

The phase looked loop (PLL) is a dynamic system for phase estimation that tracks the phase-variation versus time by adaptively minimizing the error $\varepsilon_k(\hat{\phi}_k)$ according to a linear updating formula over time:

$$\hat{\phi}_{k+1} = \hat{\phi}_k - \gamma\varepsilon_k$$

for a scaling term γ (see block diagram in Figure 7.3). Assuming $\phi(t) = \phi_o = \text{const}$, the updating rule changes the phase estimate until $\varepsilon_k \to 0$ for $k \to \infty$ so that $\hat{\phi}_\infty = \phi_o$, apart from a random term due to the noise (specifically, one can guarantee only $\mathbb{E}[\hat{\phi}_\infty] = \phi_o$). When $\varepsilon_k \to 0$, the modulator embedded into the PLL generates a signal $\sin(\omega_o t + \phi_k)$ for $x(t) = \cos(\omega_o t + \phi_o)$, so that the two signals are orthogonal to one another to reach the *quadrature convergence condition* $\int_0^T \cos(\omega_o t + \phi_o)\sin(\omega_o t + \phi_k)dt = 0$.

In a dynamic system, the error ε_k is the driving term of a phase modulator that forces the signal to have a phase $\hat{\phi}_k$ that tracks the true phase-variation $\phi(t)$. The scaling γ (referred as loop-gain in PLL jargon) is a degree of freedom of the PLL design to give fast tracking capability (large γ) to follow fast varying phase signals and avoid failure-to-follow errors, but at the same time it should be small enough to smooth phase deviations to reduce the sensitivity to noise in $x(t)$ (small γ).

PLL design is a far more complex topic than the simple model introduced here. PLL can be considered as the core processing (either analog or digital) in all communication systems since it guarantees that both transmitter and receiver are synchronized when communicating information. However, the basic principle is that to ensure that two devices are synchronized, the oscillator of the receiving device should generate a sinusoid that is an exact replica of the frequency of the transmitter, up to a constant phase-shift. There are several architectures to enable the synchronization, either analog [34], digital [35], or mutual (Section 22.3). The reader is referred therein for an in-depth discussion.

7.8 Least Squares (LS) Estimation

There are applications where it is necessary to estimate the values of a set of parameters $\{\theta_1, \theta_2, ..., \theta_p\}$ based on the generation model of the deterministic part of the data but *without* a knowledge of the involved pdfs. In this case one can estimate the parameter values by minimizing the error between the data and the model, according to some metric. A common metric is the sum of the square of the error, and the estimation method is the *least squares* (LS) technique. The LS is used in a large variety of applications due to its simplicity in the absence of any statistical model, and for this reason it can be considered an optimization technique for data-fitting rather than an estimation approach. Furthermore, LS has no proof of optimality, except for some special cases where LS coincides with MLE as for additive Gaussian noise, and it is the first pass approach in complex problems.

Assuming the following generation model for data:

$$x[i] \approx s[i;\theta] \text{ with } i = 1, 2, ..., N$$

(in LS methods the symbol $\approx$ denotes that the equality is under some uncertainty, that is not modeled in any way), the LS estimation is based on the minimization of the sum of the square errors between the available data and the model outputs

$$\boldsymbol{\theta}_{LS} = \arg\min_{\theta} \left\{ \sum_{i=1}^{N} (x[i] - s[i;\theta])^2 \right\}$$

This corresponds to the minimization of the following objective function:

$$J(\boldsymbol{\theta}) = \sum_{i=1}^{N} (x[i] - s[i;\boldsymbol{\theta}])^2 = (\mathbf{x} - \mathbf{s}(\boldsymbol{\theta}))^T (\mathbf{x} - \mathbf{s}(\boldsymbol{\theta}))$$

The LS estimate $\boldsymbol{\theta}_{LS}$ is the one that minimizes

$$J(\boldsymbol{\theta}) \geq J(\boldsymbol{\theta}_{LS}) \text{ for } \boldsymbol{\theta} \neq \boldsymbol{\theta}_{LS}$$

This could be derived in closed form, but often the minimization of $J(\boldsymbol{\theta})$ needs numerical optimization procedures. In several numerical methods, there is no guarantee that the minimum obtained iteratively is the global minimum, and thus the convexity of $J(\boldsymbol{\theta})$ must be evaluated case-by-case depending on the specific problem.

A more general solution is represented by the *weighted LS* (WLS) estimator in which a positive definite matrix $\mathbf{W} \in \mathbb{R}^{N \times N}$ is used to weight the errors within the objective function:

$$J(\boldsymbol{\theta}) = (\mathbf{x} - \mathbf{s}(\boldsymbol{\theta}))^T \mathbf{W} (\mathbf{x} - \mathbf{s}(\boldsymbol{\theta}))$$

It is worth highlighting that the WLS coincides with the MLE for Gaussian noise when the covariance matrix $\mathbf{C}_w$ is known, and the optimum choice is to set $\mathbf{W} = \mathbf{C}_w^{-1}$.

The LS method is simple and flexible enough to be used in situations where the model order p is not known, or it changes and one can progressively adapt the LS order to fulfill some other constraints. The problem is as follows. Let $\boldsymbol{\theta}_p \in \mathbb{R}^p$ be a set of p parameters to be estimated, and let $J(\boldsymbol{\theta}_p)$ be the corresponding objective that once optimized yields to the LS solution $\hat{\boldsymbol{\theta}}_p$. The *order recursive* LS aims to estimate on the same data an augmented set of $p+1$ parameters, where the first p parameters coincide with the previously defined set: $\boldsymbol{\theta}_{p+1} = [\boldsymbol{\theta}_p^T, \theta_{p+1}]^T \in \mathbb{R}^{p+1}$. This approach is important whenever it is computationally efficient to update the LS solution $\hat{\boldsymbol{\theta}}_p$ rather than recomputing the whole $p+1$ parameters from scratch. Namely, the new estimate $\hat{\boldsymbol{\theta}}_{p+1}$ is such that $\hat{\boldsymbol{\theta}}_{p+1}(1:p) \neq \hat{\boldsymbol{\theta}}_p$ as the model unavoidably changes by augmentation, and possibly $J(\hat{\boldsymbol{\theta}}_{p+1}) < J(\hat{\boldsymbol{\theta}}_p)$ as the augmented model usually fits the available observations better, even if increasing the model complexity is not always beneficial (see Section 7.8.3). Derivation of an order recursive LS needs to deeply manipulate the model equation $\mathbf{s}(\boldsymbol{\theta})$, and this cannot be carried out in a general sense. One exception is for linear models $\mathbf{s}(\boldsymbol{\theta}) = \mathbf{H} \cdot \boldsymbol{\theta}$ as the order recursive form can be derived from the algebraic properties of $\mathbf{H}$ after block-partitioning into the different orders. The derivation is quite cumbersome and it is not covered here; see for example, [30] on ladder algorithms.

7.8.1 Weighted LS with $\mathbf{W} = \mathrm{diag}\{c_1, c_2, ..., c_N\}$

In the case that the weighting matrix is $\mathrm{diag}\{c_1, c_2, ..., c_N\}$, the LS objective function is

$$J(\boldsymbol{\theta}) = \sum_{i=1}^{N} c_i (x[i] - s[i; \boldsymbol{\theta}])^2$$

Each weight c_i can be used to emphasize (or de-emphasize) the contribution of each single measurement to the global metric, and thus its effect on the LS solution. Usually, more reliable measurements have larger weights than less reliable ones. It is possible to nullify the effect of one unreliable measurement by assigning a null weight $c_i = 0$, and the LS estimate does not use that measurement. Referring to the MLE with uncorrelated noise components $\mathbf{w} \sim \mathcal{N}(\mathbf{0}, \mathrm{diag}(\sigma_1^2, \sigma_2^2, ..., \sigma_N^2))$, the equivalence $c_i = 1/\sigma_i^2$ shows that unreliable measurements are those characterized by larger noise variances, and thus are down weighted in WLS.

There are applications where data is represented by the samples of a signal that slowly changes its properties over time. In this case it is useful to introduce a mechanism that weights recent observations differently from those collected in the past, as the latest observations are considered more reliable when modeling time-evolving phenomena. One simple and widely adopted solution is the exponential weight

$$J(\boldsymbol{\theta}) = \sum_{i=1}^{N} \lambda^{N-i} (x[i] - s[i; \boldsymbol{\theta}])^2$$

The *forgetting factor* λ can balance the memory as $0 < \lambda \leq 1$. For $\lambda = 1$ there is no weighting effect and all observations contribute in the same way, while choosing $\lambda < 1$ introduces a "finite memory" that fades faster for small λ.

7.8.2 LS Estimation and Linear Models

Considering a linear model

$$\mathbf{s}(\boldsymbol{\theta}) = \sum_{k=1}^{p} \theta_k \mathbf{h}_k = \mathbf{H} \cdot \boldsymbol{\theta}$$

where $\mathbf{H} \in \mathbb{R}^N$ is the combination of p columns $\{\mathbf{h}_1, \mathbf{h}_2, .., \mathbf{h}_p\}$ (with $N \geq p$) that define the algebraic structure of the model as covered in Chapter 2. In particular, the ensemble of p vectors $\{\mathbf{h}_1, \mathbf{h}_2, .., \mathbf{h}_p\}$ that are linearly independent from one another describes a subspace with p dimensions: $\mathcal{R}\{\mathbf{H}\} \subseteq \mathbb{R}^N$, or (due to the independency of columns in $\mathbf{H}$) $rank(\mathbf{H}) = \min(N, p) = p$. Therefore, the solution must be sought among those components of the observation $\mathbf{x}$ within the span of columns of $\mathbf{H}$. The observation can be decomposed as $\mathbf{x} = \mathbf{x}_{\mathbf{H}} + \mathbf{x}_{\mathbf{H}}^{\perp}$ with $\mathbf{x}_{\mathbf{H}} = \mathbf{P}_{\mathbf{H}}\mathbf{x} \in \mathcal{R}\{\mathbf{H}\}$ and $\mathbf{x}_{\mathbf{H}}^{\perp} = \mathbf{P}_{\mathbf{H}}^{\perp}\mathbf{x}$; the solution should consider only $\mathbf{x}_{\mathbf{H}}$, while the residual is $\mathbf{x}_{\mathbf{H}}^{\perp}$. The minimization of the quadratic form $J(\boldsymbol{\theta})$ yields the solution (see Section 2.7)

$$\boldsymbol{\theta}_{LS} = (\mathbf{H}^T\mathbf{H})^{-1}\mathbf{H}^T\mathbf{x}$$

Consequently, for a linear model and Gaussian uncorrelated noise, the LS, MVU, and ML estimators coincide. The value of the objective at the LS solution is

$$J_{\min} = J(\boldsymbol{\theta}_{LS}) = \mathbf{x}^T(\mathbf{I} - \underbrace{\mathbf{H}(\mathbf{H}^T\mathbf{H})^{-1}\mathbf{H}^T}_{\mathbf{P}_{\mathbf{H}}})\mathbf{x} = \mathbf{x}^T\left(\mathbf{I} - \mathbf{P}_{\mathbf{H}}\right)\mathbf{x} = \mathbf{x}^T\mathbf{P}_{\mathbf{H}}^{\perp}\mathbf{x} = \mathrm{tr}\{\mathbf{P}_{\mathbf{H}}^{\perp}\mathbf{x}\mathbf{x}^T\}$$

in other words the residual is obtained by projecting the data $\mathbf{x}$ onto the subspace that is complementary to the span of $\mathbf{H}$.

For the weighted LS estimator, the solution is obtained considering that for positive definite matrix, the Cholesky factorization holds ($\mathbf{W} = \mathbf{W}^{T/2}\mathbf{W}^{1/2}$) so that

$$J(\boldsymbol{\theta}) = (\mathbf{x} - \mathbf{H} \cdot \boldsymbol{\theta})^T \mathbf{W} (\mathbf{x} - \mathbf{H} \cdot \boldsymbol{\theta}) = \left(\tilde{\mathbf{x}} - \tilde{\mathbf{H}} \cdot \boldsymbol{\theta}\right)^T \left(\tilde{\mathbf{x}} - \tilde{\mathbf{H}} \cdot \boldsymbol{\theta}\right)$$

Once again, it is enough to redefine $\tilde{\mathbf{x}} = \mathbf{W}^{1/2}\mathbf{x}$ and $\tilde{\mathbf{H}} = \mathbf{W}^{1/2}\mathbf{H}$ so that the problem is reduced to the unweighted LS:

$$\boldsymbol{\theta}_{LS} = (\tilde{\mathbf{H}}^T\tilde{\mathbf{H}})^{-1}\tilde{\mathbf{H}}^T\tilde{\mathbf{x}} = (\mathbf{H}^T\mathbf{W}\mathbf{H})^{-1}\mathbf{H}^T\mathbf{W}\mathbf{x}$$

Note that for Gaussian noise, $\mathbf{W} = \mathbf{C}_w^{-1}$ and thus $\tilde{\mathbf{x}} = \mathbf{C}_w^{-1/2}\mathbf{x}$ and $\tilde{\mathbf{H}} = \mathbf{C}_w^{-1/2}\mathbf{H}$; the noise $\tilde{\mathbf{w}} = \mathbf{C}_w^{-1/2}\mathbf{w} \sim \mathcal{N}(\mathbf{0}, \mathbf{I})$ becomes uncorrelated and this is the *whitening* step as counterpart of the pre-whitening filter already discussed in ML for continuous-time signals (Section 7.5).

7.8.3 Under or Over-Parameterizing?

In LS, indeed in any estimation problem, the choice of the complexity of the model $\mathbf{s}(\boldsymbol{\theta})$ could be a problem itself nested into the estimation. One might wonder how to approach this, and look for some guidelines on how to choose the model complexity. Assuming that data is "truly" generated by a model $\mathbf{s}(\boldsymbol{\theta}_o)$ composed of p_o parameters, one might try a number of parameters p that are more than those really necessary $p > p_o$; this is called over-parameterization (or overfitting). Alternatively one can choose $p < p_o$, that is, under-parameterization. Even if one expects that when increasing the order, the LS metric $J(\boldsymbol{\theta})$ monotonically reduces, the reality is that practitioners are prone to under-parameterize the model and to use under-parameterization as the first cut. A general justification is not that easy at this stage and will be clearer after Chapter 8, but a simple example for a linear model can be enlightening enough.

Consider a set of data $\mathbf{x}$ generated by a linear model composed of p_o columns

$$\mathbf{x} = \sum_{k=1}^{p_o} \theta_k \mathbf{h}_k + \mathbf{w}$$

The columns are orthonormal ($\mathbf{h}_i^T \mathbf{h}_j = \delta_{i-j}$) as this simplifies the reasoning without any loss of generality, and $\mathbf{w} \sim \mathcal{N}(\mathbf{0}, \sigma^2 \mathbf{I})$. For the LS estimate, the model is assumed to be composed of p columns $\mathbf{H}_p = [\mathbf{h}_1, \mathbf{h}_2, ..., \mathbf{h}_p]$, say the first p columns with $p \leq p_o$. The LS estimate is

$$\hat{\theta}_p = \mathbf{H}_p^T \mathbf{x} = \theta_p + \mathbf{H}_p^T \mathbf{w}$$

where the second part of the equality follows from the expansion due to the true generation model. The MSE is

$$MSE(p, \sigma^2) = \sum_{k=1}^{p_o} \mathbb{E}_w[(\hat{\theta}_k - \theta_k)^2] = \sum_{k=p+1}^{p_o} \theta_k^2 + p\sigma^2$$

that is composed of two terms: the distortion due to the under-parameterization, and the projection of noise onto the span of $\mathbf{H}_p$. There is an accumulation of noise for large model order p, that is counterbalanced by a decrease in the distortion.

Figure 7.4 illustrates the distortion (black dashed line) and the noise (gray dashed line) for $\sigma^2 = \{\sigma_1^2, \sigma_2^2, \sigma_3^2\}$ (with $\sigma_1^2 > \sigma_2^2 > \sigma_3^2$) that both contribute to the $MSE(p, \sigma^2)$ evaluated here vs. model order p. The optimum model order p_* is the order p that minimizes the MSE, indicated by the dot in the figure. The best model order p_* depends on σ^2; it is always under-determined with a value p_* that depends on the noise level. For large noise (σ_1^2 in the example here) it is far better to use an under-parameterized model and tolerate some distortion, rather than attempt to increase the model order to better fit the (excessively noisy) data. The algebraic motivation is due to the excessive dimension of the span of $\mathbf{H}_p$ that collects too much noise compared to the conservative choice of setting some parameters to zero (as $\hat{\theta}_k \equiv 0$ when $k > p$). Of course, distortion would dominate if the under-parameterization is excessive.

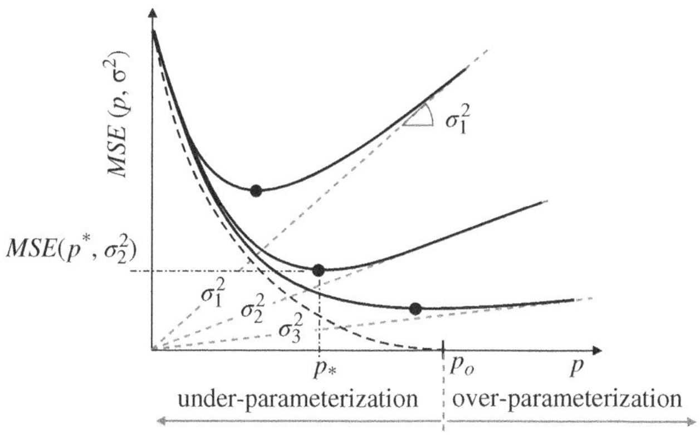

Figure 7.4 MSE versus parameterization p.

7.8.4 Constrained LS Estimation

The definition of an LS estimator can include constraints. In this case the general formulation (weighted LS estimator with constraints or *constrained LS* with linear constraints) is

$$\theta_{CLS} = \arg\min_{\theta}\{(\mathbf{x} - \mathbf{H}\cdot\theta)^T \mathbf{W}(\mathbf{x} - \mathbf{H}\cdot\theta)\} \quad \text{s.t. } \mathbf{A}\theta = \mathbf{b}$$

It is an optimization problem that must be solved by using the Lagrange multipliers for the augmented objective function (Section 1.8)

$$J(\theta) = (\mathbf{x} - \mathbf{H}\cdot\theta)^T \mathbf{W}(\mathbf{x} - \mathbf{H}\cdot\theta) + \lambda^T(\mathbf{A}\theta - \mathbf{b})$$

By setting the gradient to zero, one obtains

$$\frac{\partial J(\theta)}{\partial \theta} = \mathbf{0} \Rightarrow \theta_{CLS} = \underbrace{(\mathbf{H}^T\mathbf{W}\mathbf{H})^{-1}\mathbf{H}^T\mathbf{W}\mathbf{x}}_{\theta_{LS}} - \frac{1}{2}(\mathbf{H}^T\mathbf{W}\mathbf{H})^{-1}\mathbf{A}^T\lambda$$

where the LS solution is complemented by the constraints. Using the constraint equation for λ

$$\mathbf{A}(\theta_{LS} - \frac{1}{2}(\mathbf{H}^T\mathbf{W}\mathbf{H})^{-1}\mathbf{A}^T\lambda) = \mathbf{b} \Rightarrow \lambda = 2\left(\mathbf{A}(\mathbf{H}^T\mathbf{W}\mathbf{H})^{-1}\mathbf{A}^T\right)^{-1}\left(\mathbf{A}\theta_{LS} - \mathbf{b}\right)$$

one gets the general solution

$$\theta_{CLS} = \theta_{LS} - (\mathbf{H}^T\mathbf{W}\mathbf{H})^{-1}\mathbf{A}^T\left(\mathbf{A}(\mathbf{H}^T\mathbf{W}\mathbf{H})^{-1}\mathbf{A}^T\right)^{-1}\left(\mathbf{A}\theta_{LS} - \mathbf{b}\right)$$

that includes a correction term with respect to the unconstrained LS solution $\boldsymbol{\theta}_{LS}$. Depending on the application, more elaborate constraints can be set following the same steps as above.

7.9 Robust Estimation

The LS estimator is optimal for the case of additive Gaussian noise as it coincides with the MLE. Sometimes a subset of the available observations is not compliant with the assumption of additive Gaussian noise showing few large and anomalous values with respect to the other values: these are referred as *outliers* (Figure 7.5). In these cases, the use of the LS method leads to the estimation of model parameters that globally minimizes the estimation error when considering all the measurements (including the outliers), but the estimates are severely biased by those few outliers. Robust estimation methods aim to adopt common statistical methods that are not unduly affected by outliers, or by other deviations from the model assumptions. Literature on statistical robustness and related methods is broad; herein it is just revised, but a good introductory reference is the book by Huber [31], or the tutorial [32], which is not just on LS.

Weighted LS can offer a solution to cope with outliers, if these are preliminarly detected, by down-weighting the error elements based on the reliability of the corresponding observation components, possibly by an inspection of observations guided from a first (unweighted) LS solution. Alternatively, the LS estimation can be seen as an MLE with a quadratic penalty function (that is optimal for Gaussian noise) that can be modified to take into account the presence of outliers. This is obtained as follows. Let the objective function be

$$J(\theta) = \sum_{i=1}^{N} \phi(x[i] - s[i;\theta])$$

where $\phi(.)$ represents the penalty function; the choice of a quadratic penalty function ($\phi(\varepsilon) = \varepsilon^2$) tends to emphasize large errors that in turn corrects excessively the estimate for the presence of outliers. One robust estimation approach is based on the modification of the penalty function by "accepting" an error up to a certain value L so that

$$\phi(\varepsilon) = \begin{cases} \varepsilon^2 & \text{per } |\varepsilon| \leq L \\ L^2 & \text{per } |\varepsilon| \geq L \end{cases}$$

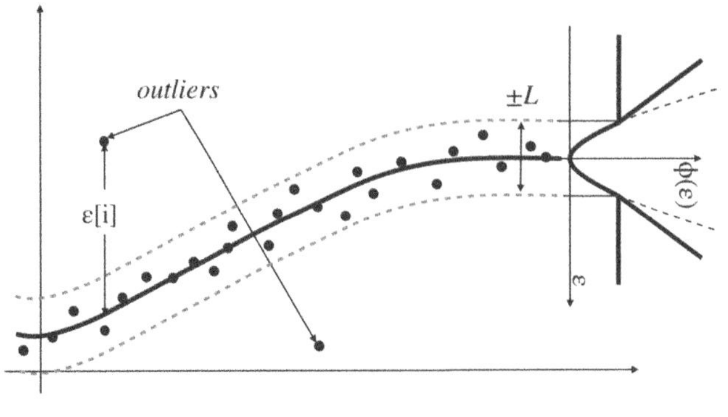

Figure 7.5 Data with outliers and the non-quadratic penalty $\phi(\varepsilon)$.

In this case, any error larger than L will not contribute proportionally to the objective function as the penalty saturates, but the optimization problem with this penalty is non-linear and it is very hard to find a closed form solution. Alternatively, the objective can nullify the influence of outliers with the choice

$$\phi(\varepsilon) = \begin{cases} \varepsilon^2 & \text{per } |\varepsilon| \leq L \\ 0 & \text{per } |\varepsilon| \geq L \end{cases}$$

as this implies stripping those values from the data.

Another robust estimation method is the use of the L1 norm as penalty function to account for a heavy-tailed pdf that could model the Laplace pdf (Section 7.3): $\phi(\varepsilon) = |\varepsilon|$. In this case, large errors (possible outliers) will contribute mildly to the global objective function if compared to the use of the L2 norm. The use of an L1 norm in the penalty function leads to the so called Huber robust estimation:

$$\phi_{Huber}(u) = \begin{cases} \varepsilon^2 & \text{per } |\varepsilon| \leq L \\ L(2|\varepsilon| - L) & \text{per } |\varepsilon| \geq L \end{cases}$$

that is optimum when the pdf is Gaussian for small values with double-exponential tails.

8

Cramér–Rao Bound

The Cramér–Rao bound (CRB) [39] [38] sets the lower value of the covariance for any unbiased estimator with parametric pdf that can be asymptotically attained by the ML estimator for a large number of observations.[1] Even if the estimator does not always exist in explicit form and needs numerical optimizations, the CRB can always be computed in closed form and its derivation needs to preserve all the scaling terms in likelihood that are usually neglected in MLE. Often the CRB computation needs care, patience, and some skill in algebra that can be eased by some software tools for symbolic calculus. Proof is provided for the case of a single parameter; it can be extended to any arbitrary set of parameters.

8.1 Cramér–Rao Bound and Fisher Information Matrix

Information indicates the degree of unpredictability of any rv, and the Fisher information matrix is a way to measure the amount of information that an rv $\mathbf{x}$ carries about the parameter $\boldsymbol{\theta}$ over the parametric pdf dependency $p(\mathbf{x}|\boldsymbol{\theta})$. The derivative with respect to the entries of $\boldsymbol{\theta}$ of the log-likelihood function (called the *score function*) plays an essential role in the derivation of the CRB.

8.1.1 CRB for Scalar Problem (P=1)

Let $\hat{\theta} = \hat{\theta}(\mathbf{x})$ be the unbiased estimator with $\theta = \theta_o$ as the true parameter; the variance is lower bounded as

$$\text{var}[\hat{\theta}(\mathbf{x})] \geq C_{CRB} = J^{-1}(\theta_o)$$

where the term

$$J(\theta_o) = -\mathbb{E}_x\left[\frac{d^2\mathcal{L}(\mathbf{x}|\theta)}{d\theta^2}\right]_{\theta=\theta_o} = \mathbb{E}_x\left[\left(\frac{d\mathcal{L}(\mathbf{x}|\theta)}{d\theta}\right)^2\right]_{\theta=\theta_o}$$

1 The importance of a lower bound in any estimator and the availability of an estimator that can reach the bound is very powerful as it can stimulate improvement of estimator design to achieve the bound even from a limited set of observations. On the other hand, this can be helpful to spot who is cheating...

Statistical Signal Processing in Engineering, First Edition. Umberto Spagnolini.

Companion website: www.wiley.com/go/spagnolini/signalprocessing

depends on the log-likelihood function and it is referred to as the *Fisher information term*. The proof is in Appendix A, but essentially the CRB is the inverse of the curvature of the log-likelihood averaged wrt the pdf of the observations $\mathbf{x}$. A simple example [85] can give insight to better visualize the conceptual meaning.

8.1.2 CRB and Local Curvature of Log-Likelihood

Let an observation x be bounded $-1 \leq x \leq 1$, and the joint pdf be

$$p(x, \theta) = \frac{1}{2}(1 + x\theta)$$

The conditional pdf $p(x|\theta)$ for likelihood is the same for the specific normalization chosen here as unitary. The log-likelihood for an ensemble of N independent measurements $\mathbf{x} = [x_1, ..., x_N]^T$ is

$$\mathcal{L}(\mathbf{x}|\theta) = \sum_{i=1}^{N} \ln\left(\frac{1}{2}(1 + x_i\theta)\right)$$

and for any set $\mathbf{x}$ one can get the corresponding log-likelihood. But each measurement for an actual value $\theta = \theta_o$ is drawn from a pdf $p[x_i|\theta_o] = \frac{1}{2}(1 + x_i\theta_o)$; hence the mean of the log-likelihood is

$$\mathbb{E}_x[\mathcal{L}(\mathbf{x}|\theta)] = N \int_{-1}^{1} \ln\left(\frac{1}{2}(1 + x\theta)\right) \frac{1}{2}(1 + x\theta_o) dx$$

The behavior of the mean likelihood and log-likelihood are illustrated in Figure 8.1 for a value $\theta_o = -0.1$ and $N = 1, 10, 10^2, 10^3$ (normalized to 1 as maximum value just for visualization purposes). The average likelihood for varying N shows that with increasing N, the behavior attains the Gaussian bell-shape that shrinks for increasing N (or equivalently, the variance scales with the number of measurements N). In addition, the insert in Figure 8.1 is the log-likelihood, which has a clear parabolic behavior with a curvature that scales with N.

The CRB follows from the analytical curvature

$$J(\theta_o) = -N \int_{-1}^{1} \frac{d^2}{d\theta^2} \ln\left(\frac{1}{2}(1 + x\theta)\right)\bigg|_{\theta=\theta_0} \frac{1}{2}(1 + x\theta_o) dx = N \int_{-1}^{1} \frac{x^2}{1 + x\theta_o} dx$$

and it is illustrated in Figure 8.2 vs. θ_o for $N = 1$, thus showing that the CRB varies wrt the actual value θ_o.

8.1.3 CRB for Multiple Parameters (p≥1)

Let $\hat{\boldsymbol{\theta}} = \hat{\boldsymbol{\theta}}(\mathbf{x})$ be the unbiased estimator for the model in $\boldsymbol{\theta} = \boldsymbol{\theta}_o$ ($\mathbb{E}[\hat{\boldsymbol{\theta}}(\mathbf{x})] = \boldsymbol{\theta}_o$); the CRB sets the bound of the covariance:

$$\text{cov}[\hat{\boldsymbol{\theta}}(\mathbf{x})] \succeq \mathbf{C}_{CRB} = \mathbf{J}^{-1}(\boldsymbol{\theta}_o)$$

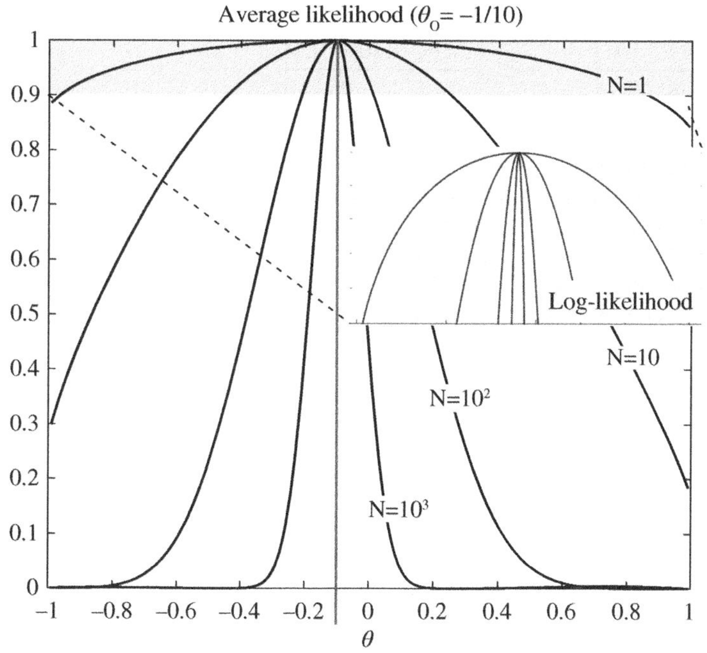

Figure 8.1 Average likelihood $p(x|\theta)$ for $\theta_o = -1/10$. Shaded insert is the log-likelihood in the neighborhood of $\theta_o = -1/10$.

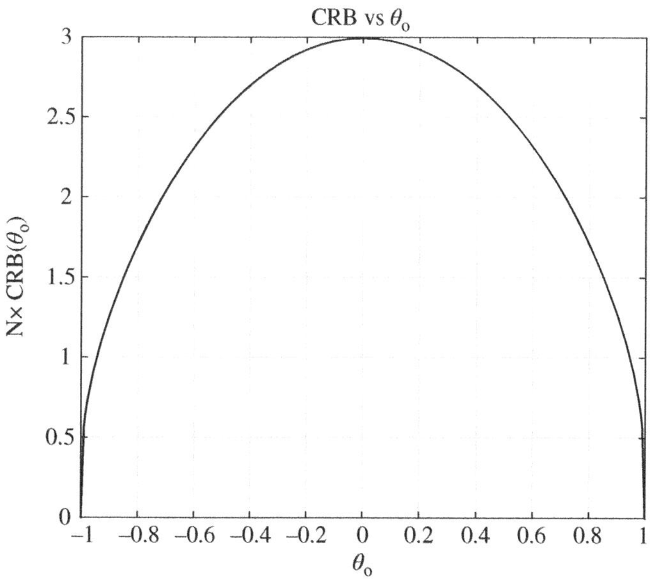

Figure 8.2 CRB vs θ_o for $N = 1$ from example in Figure 8.1.

where $\mathbf{J}(\boldsymbol{\theta}_o) \in \mathbb{R}^{p\times p}$ is the *Fisher information matrix (FIM)* with entries

$$[\mathbf{J}(\boldsymbol{\theta}_o)]_{ij} = -\mathbb{E}_x \left[\frac{\partial^2 \mathcal{L}(\mathbf{x}|\boldsymbol{\theta})}{\partial\theta_i \partial\theta_j}\right]_{\boldsymbol{\theta}=\boldsymbol{\theta}_o} = \mathbb{E}_x \left[\frac{\partial \mathcal{L}(\mathbf{x}|\boldsymbol{\theta})}{\partial\theta_i}\frac{\partial \mathcal{L}(\mathbf{x}|\boldsymbol{\theta})}{\partial\theta_j}\right]_{\boldsymbol{\theta}=\boldsymbol{\theta}_o}$$

Recall that the CRB implies that the difference $\text{cov}[\hat{\boldsymbol{\theta}}(\mathbf{x})] - \mathbf{J}^{-1}(\boldsymbol{\theta}_o) \succeq \mathbf{0}$ is positive semidefinite. The general proof follows the same steps as for $p = 1$ (see Appendix A), only is slightly more cumbersome; see for example, [7].

8.2 Interpretation of CRB and Remarks

8.2.1 Variance of Each Parameter

The CRB evaluates the bound of the covariance $\text{cov}[\hat{\boldsymbol{\theta}}(\mathbf{x})]$ that is valid for any estimator. The variance for each parameter is obtained from the diagonal entries after inversion of the FIM:

$$\text{var}[\hat{\theta}_k(\mathbf{x})] \geq [\mathbf{C}_{CRB}]_{kk} = [\mathbf{J}^{-1}(\boldsymbol{\theta}_o)]_{kk}$$

In the case of numerical simulations (see Chapter 10), it is customary to evaluate the MSE

$$MSE(\hat{\theta}_k) = \mathbb{E}[|\hat{\theta}_k - \theta_k|^2] = \text{var}[\hat{\theta}_k] + |b(\hat{\theta}_k)|^2 \geq [\mathbf{C}_{CRB}]_{kk}$$

which is strictly larger than the CRB for biased estimates ($|b(\hat{\theta}_k)| \neq 0$).

8.2.2 Compactness of the Estimates

Assume that one unbiased estimator is characterized by an estimate $\hat{\boldsymbol{\theta}} \sim \mathcal{N}(\boldsymbol{\theta}_o, \mathbf{C}_{\hat{\theta}})$ and another estimator attains the CRB $\hat{\boldsymbol{\theta}}_{CRB} \sim \mathcal{N}(\boldsymbol{\theta}_o, \mathbf{J}^{-1}(\boldsymbol{\theta}))$; it can be shown that the error of the estimator $\hat{\boldsymbol{\theta}}_{CRB}$ is maximally compact (or less dispersed) wrt $\hat{\boldsymbol{\theta}}$ for the same probability.

Since both estimators have Gaussian pdf of the estimates,

$$p(\hat{\boldsymbol{\theta}}) = |2\pi\mathbf{C}_{\hat{\theta}}|^{-1/2} \exp\left(-\frac{1}{2}(\hat{\boldsymbol{\theta}}-\boldsymbol{\theta}_o)^T \mathbf{C}_{\hat{\theta}}^{-1}(\hat{\boldsymbol{\theta}}-\boldsymbol{\theta}_o)\right)$$
$$p(\hat{\boldsymbol{\theta}}_{CRB}) = |2\pi\mathbf{C}_{CRB}|^{-1/2} \exp\left(-\frac{1}{2}(\hat{\boldsymbol{\theta}}_{CRB}-\boldsymbol{\theta}_o)^T \mathbf{C}_{CRB}^{-1}(\hat{\boldsymbol{\theta}}_{CRB}-\boldsymbol{\theta}_o)\right)$$

For any arbitrary value $\bar{\psi}^2$, the contour lines of the pdfs are hyper-ellipsoids given by $(\hat{\boldsymbol{\theta}}-\boldsymbol{\theta}_o)^T \mathbf{C}_{\hat{\theta}}^{-1}(\hat{\boldsymbol{\theta}}-\boldsymbol{\theta}_o) = \bar{\psi}^2$ and $(\hat{\boldsymbol{\theta}}_{CRB}-\boldsymbol{\theta}_o)^T \mathbf{C}_{CRB}^{-1}(\hat{\boldsymbol{\theta}}_{CRB}-\boldsymbol{\theta}_o) = \bar{\psi}^2$; their volumes increase monotonically with $\bar{\psi}^2$ and it depends on their covariance (Section 3.4). The volumes for the two hyper-ellipsoids are

$$V(\mathbf{C}_{\hat{\theta}}) = V_p \times |\mathbf{C}_{\hat{\theta}}|^{1/2} \bar{\psi}^p$$
$$V(\mathbf{C}_{CRB}) = V_p \times |\mathbf{C}_{CRB}|^{1/2} \bar{\psi}^p$$

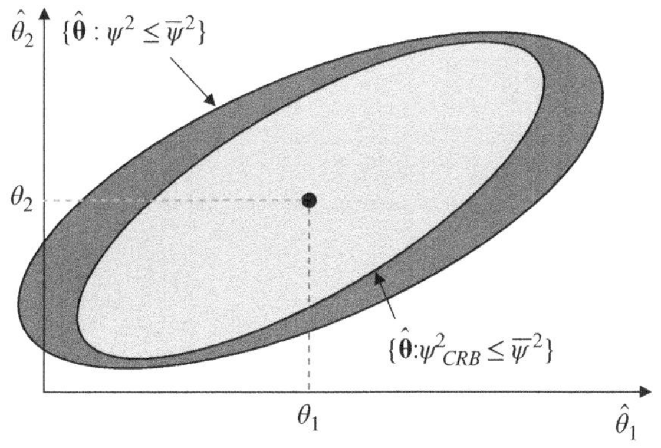

Figure 8.3 Compactness of CRB.

where the scaling term V_p depends on the dimensions p (Section 3.4), from the property $\mathbf{C}_{\hat{\theta}} - \mathbf{C}_{CRB} \succeq \mathbf{0} \Rightarrow |\mathbf{C}_{\hat{\theta}} - \mathbf{C}_{CRB}| \geq 0 \Rightarrow |\mathbf{C}_{\hat{\theta}}| - |\mathbf{C}_{CRB}| \geq 0$ it follows that

$$V(\mathbf{C}_{\hat{\theta}}) \geq V(\mathbf{C}_{CRB})$$

Since the probability $\bar{P}$ that the error vector lies inside the hyper-ellipsoid depends only on $\bar{\psi}^2$ (Section 3.4), this proves that any unbiased estimator that attains the CRB has errors that are statistically more compact (less dispersed around θ_o) than any other estimator by virtue of the smaller volumes of the dispersion ellipses, see Figure 8.3.

For a degenerate case with $p = 1$ and $\theta_o = 0$, the ellipses reduce to line intervals. Given the probability $\bar{P}$ for the two estimators:

$$\hat{\theta}^2/\sigma^2_{\hat{\theta}} \leq \bar{\psi}^2 = -2\ln(1-\bar{P}) \Rightarrow |\hat{\theta}| \leq \bar{\psi}\sigma_{\hat{\theta}}$$
$$I(\theta)\hat{\theta}^2_{CRB} = \hat{\theta}^2_{CRB}/\sigma^2_{CRB} \leq \bar{\psi}^2 \Rightarrow |\hat{\theta}_{CRB}| \leq \bar{\psi}\sigma_{CRB}$$

The estimator $\hat{\theta}_{CRB}$ has a smaller line-interval compared to any other estimator as $\sigma^2_{CRB} \leq \sigma^2_{\hat{\theta}}$.

8.2.3 FIM for Known Parameters

When one (or few) of the parameter(s) is known, the variance of the others at least reduces. Conversely, when the same number of observations is used to estimate a larger number of parameters, the variance of the parameters increases, or at least could be unchanged, but never reduces.

A simple example in Figure 8.4 illustrates the concept for $p = 2$. When estimating two parameters, the FIM is 2×2 and $\text{var}[\hat{\theta}_1] \geq [\mathbf{J}^{-1}(\theta_o)]_{11}$; this value is sketched in the figure as the maximum of the error ellipsoid as this is the result of marginalization $\text{p}(\hat{\theta}_1)$. Let θ_2 be known (or no more to be estimated); the FIM degenerates into a scalar $\mathbf{J}(\theta_o) = [\mathbf{J}(\theta_o)]_{11}$ and $\text{var}[\hat{\theta}_1|\hat{\theta}_2 = \theta_2] \geq 1/[\mathbf{J}(\theta_o)]_{11}$. This is the intersection of the error ellipsoid with axis $\hat{\theta}_2 = \theta_2$. For the special case of diagonal FIM $\mathbf{J}(\theta_o)$, the two terms coincide. In general, when the FIM is diagonal, the principal axes of the dispersion ellipses are the

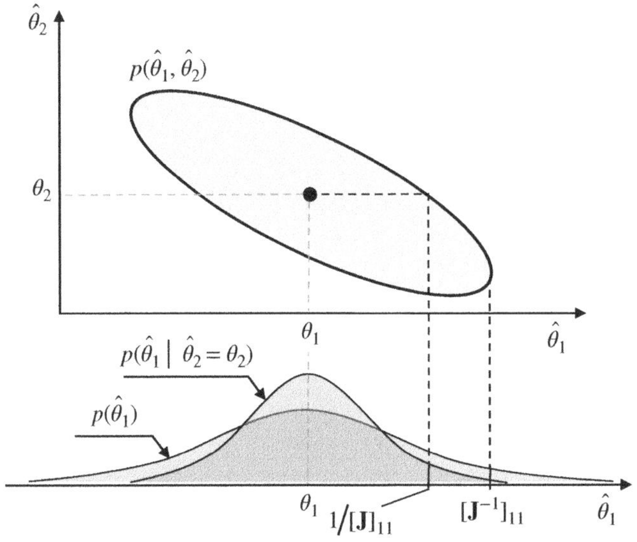

Figure 8.4 CRB and FIM for $p = 2$.

axes of $\boldsymbol{\theta}$ and the quadratic form is diagonal. In this case the estimate of each parameter is not influenced by the others and are decoupled as detailed below.

8.2.4 Approximation of the Inverse of FIM

The analytical computation of the inverse of the FIM is complex in general as it involves the inverse of a matrix $p \times p$. A convenient approximation of the CRB is the following inequality, which holds for positive definite matrixes (Section 1.1.1)

$$\text{var}[\hat{\theta}_k] \geq [\mathbf{J}^{-1}(\boldsymbol{\theta}_o)]_{kk} \geq \frac{1}{[\mathbf{J}(\boldsymbol{\theta}_o)]_{kk}}$$

The equality holds only for diagonal FIM $\mathbf{J}(\boldsymbol{\theta}_o)$.

8.2.5 Estimation Decoupled From FIM

Let the FIM be block-diagonal for a partitioning of the parameters as $\boldsymbol{\theta} = [\boldsymbol{\theta}_1^T, \boldsymbol{\theta}_2^T]^T$; then the inverse is also block-diagonal (Section 1.1.1):

$$\mathbf{J}(\boldsymbol{\theta}_o) = \begin{bmatrix} \mathbf{J}_{11} & \mathbf{0} \\ \mathbf{0} & \mathbf{J}_{22} \end{bmatrix} \Rightarrow \mathbf{J}^{-1}(\boldsymbol{\theta}_o) = \begin{bmatrix} \mathbf{J}_{11}^{-1} & \mathbf{0} \\ \mathbf{0} & \mathbf{J}_{22}^{-1} \end{bmatrix}$$

The lack of cross-terms in covariance ($\text{cov}[\hat{\boldsymbol{\theta}}_1, \hat{\boldsymbol{\theta}}_2] = 0$) means that the accuracy when estimating $\boldsymbol{\theta}_1$ is not affected by errors in the estimation of $\boldsymbol{\theta}_2$ (and vice-versa): the two sets of parameters are decoupled for an estimator that attains the CRB. This property highlights the condition of *decoupled parameters* as a partitioning of the parameters such that the accuracy of one set is independent of the accuracy of the other. Of course, if the parameters are not arranged to highlight the block-partitioning of the FIM, a reordering of $\boldsymbol{\theta}$ could be necessary. Analysis of decoupling from the FIM can help to design the estimator, as decoupled parameters guarantee that the estimator can be decoupled as well, but unfortunately gives no insight into how to do it.

8.2.6 CRB and Nuisance Parameters

Let a set of parameters be partitioned into two subsets $\boldsymbol{\theta} = [\boldsymbol{\theta}_1^T, \boldsymbol{\theta}_2^T]^T$: the parameters of interest $\boldsymbol{\theta}_1$ and *nuisance parameters* $\boldsymbol{\theta}_2$ (see Section 7.4). The FIM and its inverse are block-partitioned (Section 1.1.1):

$$\mathbf{J}(\boldsymbol{\theta}_o) = \begin{bmatrix} \mathbf{J}_{11} & \mathbf{J}_{12} \\ \mathbf{J}_{21} & \mathbf{J}_{22} \end{bmatrix} \Rightarrow \mathbf{J}^{-1}(\boldsymbol{\theta}_o) = \begin{bmatrix} \left(\mathbf{J}_{11} - \mathbf{J}_{12}\mathbf{J}_{22}^{-1}\mathbf{J}_{21}\right)^{-1} & * \\ * & * \end{bmatrix}$$

and the covariance is

$$\text{cov}[\hat{\boldsymbol{\theta}}_1] \succeq \left(\mathbf{J}_{11} - \mathbf{J}_{12}\mathbf{J}_{22}^{-1}\mathbf{J}_{21}\right)^{-1}$$

Since the second term is positive semidefinite, when increasing the number of parameters to be estimated (here nuisance parameters) the side effect is that the CRB increases always. The exception is when the parameters are decoupled ($\mathbf{J}_{12} = \mathbf{0}$).

The relationships above justify the performance degradation not only when adding nuisance parameters, but when exceeding the correct number of parameters. Assume that a model depends on p parameters (referred to as the *model order*), but the value p is unknown. One can estimate $p' > p$ parameters and expect that the estimate of the remaining $p' - p$ parameters is zero, but still affected by some uncertainty (or estimation error) that can be quantified by the CRB. The estimation of p' parameters in excess of p introduces a degradation of the estimate of all the p' parameters, including the first p parameters. Restating this differently, when the model order is not known and the estimator exceeds the true order, an increase of variance of all parameters is expected unless the excess parameters are decoupled. An intuitive view of the trade-off for the choice of p is given in Section 7.8.3.

8.2.7 CRB for Non-Gaussian rv and Gaussian Bound

The FIMs for Gaussian noise can always be evaluated in closed form, even if the derivation seems quite complex. However, if the pdf is non-Gaussian, the FIM and the CRB should be evaluated from the definition, with some patience.

Moreover, a non-Gaussian pdf can be approximated by a Gaussian one with the same first and second central moments; in this case the FIM for the Gaussian approximation ($\mathbf{J}_G(\boldsymbol{\theta}_o)$) gives a pessimistic bound wrt the non-Gaussian pdf [40]:

$$\mathbf{J}(\boldsymbol{\theta}_o) \succeq \mathbf{J}_G(\boldsymbol{\theta}_o)$$

In other words, the parameter estimation methods for the non-Gaussian case are uniformly better than the approximating Gaussian if these attain the related CRB. On the other hand, if designing methods optimized for the Gaussian approximation, these can attain the CRB for the Gaussian (CRB_G), but still be worse than CRB, given the inequality $CRB_G \succeq CRB$.

8.3 CRB and Variable Transformations

The case of a scalar variable is considered first, then the multivariate case. Let the same data be modeled by two models

$$\mathbf{x} = \mathbf{s}(\theta) + \mathbf{w}$$
$$\mathbf{x} = \mathbf{f}(\alpha) + \mathbf{w}$$

such that there exists a direct mapping of parameters that guarantees the inverse mapping too:

$$\alpha = g(\theta) \leftrightarrow \theta = g^{-1}(\alpha)$$

or equivalently $\mathbf{s}(g^{-1}(\alpha)) = \mathbf{f}(\alpha)$ and $\mathbf{f}(g(\theta)) = \mathbf{s}(\theta)$. The objective is to evaluate the variance (and thus its CRB) of $\hat{\alpha}$ when this is based on the estimate of $\hat{\theta}$ from the transformation $\hat{\alpha} = g(\hat{\theta})$. Since the estimator is unbiased (otherwise the bias should be added), $\mathbb{E}[\hat{\theta}] = \theta_o$, and from the Taylor series on θ_o (see Figure 8.5)

$$\hat{\alpha} \simeq \underbrace{g(\theta_o)}_{\alpha_o} + \left[\frac{dg(\theta)}{d\theta}\right]_{\theta=\theta_o} (\hat{\theta} - \theta_o)$$

The estimate $\hat{\alpha}$ is also unbiased. The variance is

$$\text{var}[\hat{\alpha}] = \left[\frac{dg(\theta)}{d\theta}\right]^2_{\theta=\theta_o} \text{var}[\hat{\theta}]$$

The CRB can be easily extended; since $\text{var}[\hat{\theta}] \geq 1/J(\theta)$ it is

$$\text{var}[\hat{\alpha}] \geq \frac{1}{J(\theta)} \left[\frac{dg(\theta)}{d\theta}\right]^2_{\theta=\theta_o}$$

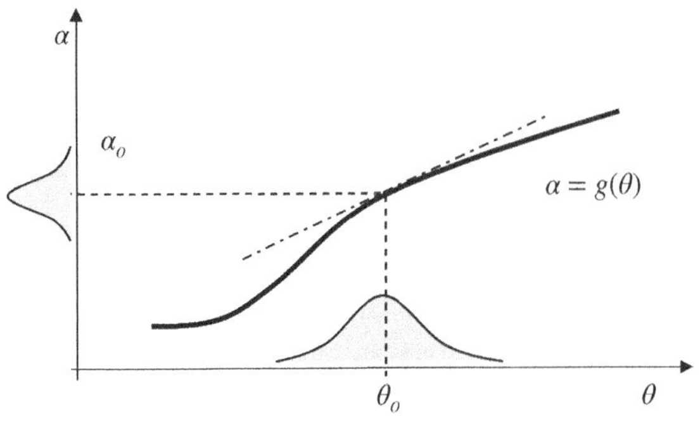

Figure 8.5 Transformation of variance and CRB ($p = 1$).

The relationship can be reversed and the same conclusions hold for the mapping $\theta = g^{-1}(\alpha)$. The choice between parameterizing as $\mathbf{s}(\theta)$ or $\mathbf{f}(\alpha)$ depends on which is computationally more convenient as the variance is the same—just a mapping.

In the case that $\boldsymbol{\theta} \in \mathbb{R}^p$ with $p > 1$, the transformation is

$$\boldsymbol{\alpha} = \mathbf{g}(\boldsymbol{\theta})$$

Once linearized from the matrix of gradients $[\partial \mathbf{g}(\boldsymbol{\theta})/\partial \boldsymbol{\theta}]_{ij} = \partial g_j(\boldsymbol{\theta})/\partial \theta_i$

$$\hat{\boldsymbol{\alpha}} = \underbrace{\mathbf{g}(\boldsymbol{\theta}_o)}_{\boldsymbol{\alpha}_o} + \frac{\partial \mathbf{g}(\boldsymbol{\theta})^T}{\partial \boldsymbol{\theta}}(\hat{\boldsymbol{\theta}} - \boldsymbol{\theta}_o)$$

This gives the relationship for transforming the covariance from one set of variables onto another

$$\text{cov}[\hat{\boldsymbol{\alpha}}] = \frac{\partial \mathbf{g}(\boldsymbol{\theta})^T}{\partial \boldsymbol{\theta}} \text{cov}[\hat{\boldsymbol{\theta}}] \frac{\partial \mathbf{g}(\boldsymbol{\theta})}{\partial \boldsymbol{\theta}}$$

For the CRB the extension is trivial since

$$\text{cov}[\hat{\boldsymbol{\theta}}] \succeq \mathbf{J}^{-1}(\boldsymbol{\theta}_o) \Rightarrow \frac{\partial \mathbf{g}(\boldsymbol{\theta})^T}{\partial \boldsymbol{\theta}} \text{cov}[\hat{\boldsymbol{\theta}}] \frac{\partial \mathbf{g}(\boldsymbol{\theta})}{\partial \boldsymbol{\theta}} \succeq \frac{\partial \mathbf{g}(\boldsymbol{\theta})^T}{\partial \boldsymbol{\theta}} \mathbf{J}^{-1}(\boldsymbol{\theta}_o) \frac{\partial \mathbf{g}(\boldsymbol{\theta})}{\partial \boldsymbol{\theta}}$$

This leads to the general relationship on CRB for any arbitrary transformation

$$\text{cov}[\hat{\boldsymbol{\alpha}}] \succeq \frac{\partial \mathbf{g}(\boldsymbol{\theta})^T}{\partial \boldsymbol{\theta}} \mathbf{J}^{-1}(\boldsymbol{\theta}_o) \frac{\partial \mathbf{g}(\boldsymbol{\theta})}{\partial \boldsymbol{\theta}}$$

8.4 FIM for Gaussian Parametric Model $\mathbf{x} \sim \mathcal{N}(\boldsymbol{\mu}(\boldsymbol{\theta}), \mathbf{C}(\boldsymbol{\theta}))$

The FIM for the general Gaussian model $\mathbf{x} \sim \mathcal{N}(\boldsymbol{\mu}(\boldsymbol{\theta}), \mathbf{C}(\boldsymbol{\theta}))$ is the following (see Appendix B for details):

$$[\mathbf{J}(\boldsymbol{\theta}_o)]_{ij} = \frac{\partial \boldsymbol{\mu}(\boldsymbol{\theta})^T}{\partial \theta_i} \mathbf{C}(\boldsymbol{\theta})^{-1} \frac{\partial \boldsymbol{\mu}(\boldsymbol{\theta})}{\partial \theta_j} + \frac{1}{2} \text{tr}\left[\mathbf{C}(\boldsymbol{\theta})^{-1} \frac{\partial \mathbf{C}(\boldsymbol{\theta})}{\partial \theta_i} \mathbf{C}(\boldsymbol{\theta})^{-1} \frac{\partial \mathbf{C}(\boldsymbol{\theta})}{\partial \theta_j}\right]\Bigg|_{\boldsymbol{\theta}=\boldsymbol{\theta}_o} \tag{8.1}$$

that can be specialized for some applications.

8.4.1 FIM for $\mathbf{x} = \mathbf{s}(\boldsymbol{\theta}) + \mathbf{w}$ with $\mathbf{w} \sim \mathcal{N}(\mathbf{0}, \mathbf{C}_w)$

For the additive Gaussian model $\mathbf{x} = \mathbf{s}(\boldsymbol{\theta}) + \mathbf{w}$ with $\mathbf{w} \sim \mathcal{N}(\mathbf{0}, \mathbf{C}_w)$, the covariance is independent of the parameters and the FIM simplifies:

$$[\mathbf{J}(\boldsymbol{\theta}_o)]_{ij} = \frac{\partial \mathbf{s}(\boldsymbol{\theta})^T}{\partial \theta_i} \mathbf{C}_w^{-1} \frac{\partial \mathbf{s}(\boldsymbol{\theta})}{\partial \theta_j}\Bigg|_{\boldsymbol{\theta}=\boldsymbol{\theta}_o}$$

as widely adopted in practice.

For the linear model $\mathbf{s}(\boldsymbol{\theta}) = \mathbf{H}\boldsymbol{\theta}$:

$$\mathbf{J}(\boldsymbol{\theta}_o) = \mathbf{H}^T\mathbf{C}_w^{-1}\mathbf{H}$$

and significantly this is independent of $\boldsymbol{\theta}_o$. The CRB is

$$\mathbf{C}_{CRB} = \mathbf{J}^{-1}(\boldsymbol{\theta}_o) = (\mathbf{H}^T\mathbf{C}_w^{-1}\mathbf{H})^{-1}$$

that coincides with the covariance of MLE, thus showing that the ML estimation for linear models in Gaussian noise always attains the CRB, regardless of the number of samples.

For uncorrelated Gaussian noise $\mathbf{C}_w = \text{diag}(\sigma_1^2, ..., \sigma_N^2)$, it is quite straightforward

$$[\mathbf{J}(\boldsymbol{\theta}_o)]_{ij} = \sum_{k=1}^{N} \frac{1}{\sigma_k^2} \frac{\partial s_k(\boldsymbol{\theta}_o)}{\partial \theta_i} \frac{\partial s_k(\boldsymbol{\theta}_o)}{\partial \theta_j}\bigg|_{\boldsymbol{\theta}=\boldsymbol{\theta}_o}$$

This further simplifies when $\mathbf{C}_w = \sigma_w^2\mathbf{I}$ as for discrete-time white noise

$$[\mathbf{J}(\boldsymbol{\theta}_o)]_{ij} = \frac{1}{\sigma_w^2} \sum_{k=1}^{N} \frac{\partial s_k(\boldsymbol{\theta}_o)}{\partial \theta_i} \frac{\partial s_k(\boldsymbol{\theta}_o)}{\partial \theta_j}\bigg|_{\boldsymbol{\theta}=\boldsymbol{\theta}_o}$$

Recall that if a model represents a sampled signal, the samples are naturally ordered for increasing time-ordering $s_k(\boldsymbol{\theta}) = s[k;\boldsymbol{\theta}]$, and this could constrain the order of $\boldsymbol{\theta}$.

8.4.2 FIM for Continuous-Time Signals in Additive White Gaussian Noise

Given the model $x(t) = s(t;\boldsymbol{\theta}) + w(t)$ over a range T, the FIM follows from the definition of the log-likelihood function as

$$\begin{aligned}
[\mathbf{J}(\boldsymbol{\theta}_o)]_{ij} &= -\mathbb{E}_x\left[\frac{\partial \log p(x|\boldsymbol{\theta})}{\partial \theta_i}\frac{\partial \log p(x|\boldsymbol{\theta})}{\partial \theta_j}\right]_{\boldsymbol{\theta}=\boldsymbol{\theta}_o} \\
&= \frac{4}{N_0^2}\mathbb{E}\left[\iint_0^T \underbrace{(x(t) - s(t;\boldsymbol{\theta}))}_{w(t)} \cdot \underbrace{(x(\alpha) - s(\alpha;\boldsymbol{\theta}))}_{w(\alpha)} \frac{\partial s(t;\boldsymbol{\theta})}{\partial \theta_i}\frac{\partial s(\alpha;\boldsymbol{\theta})}{\partial \theta_j} dt d\alpha\right] \\
&= \frac{4}{N_0^2}\iint_0^T \frac{N_0}{2}\delta(t-\alpha)\frac{\partial s(t;\boldsymbol{\theta})}{\partial \theta_i}\frac{\partial s(\alpha;\boldsymbol{\theta})}{\partial \theta_j} dt d\alpha \\
&= \frac{2}{N_0}\int_0^T \frac{\partial s(t;\boldsymbol{\theta})}{\partial \theta_i}\frac{\partial s(\alpha;\boldsymbol{\theta})}{\partial \theta_j} dt d\alpha
\end{aligned}$$

This resembles the same structure of the FIM for discrete-time processes.

8.4.3 FIM for Circular Complex Model

The MLE for the circular complex Gaussian model $\mathbf{x} \sim \mathcal{CN}(\boldsymbol{\mu}(\boldsymbol{\theta}), \mathbf{C}(\boldsymbol{\theta}))$ is in Section 7.6, and here the CRB can be evaluated by adapting the general relationships (8.1) by replacing the complex central moments

$$\boldsymbol{\mu}(\boldsymbol{\theta}) \Rightarrow \tilde{\boldsymbol{\mu}}(\boldsymbol{\theta})$$
$$\mathbf{C}(\boldsymbol{\theta}) \Rightarrow \tilde{\mathbf{C}}(\boldsymbol{\theta})$$

Even if the equivalent model $\tilde{\mathbf{x}} \sim \mathcal{N}(\tilde{\boldsymbol{\mu}}(\boldsymbol{\theta}), \tilde{\mathbf{C}}(\boldsymbol{\theta}))$ can be used, in some situations it is more convenient to transform this relationship in terms of $\boldsymbol{\mu}(\boldsymbol{\theta})$ and $\mathbf{C}(\boldsymbol{\theta})$. The second term of (8.1) can be rewritten as

$$\tilde{\mathbf{C}}(\boldsymbol{\theta})^{-1}\frac{\partial\tilde{\mathbf{C}}(\boldsymbol{\theta})}{\partial\theta_i}\tilde{\mathbf{C}}(\boldsymbol{\theta})^{-1}\frac{\partial\tilde{\mathbf{C}}(\boldsymbol{\theta})}{\partial\theta_j} = \tilde{\mathbf{D}} = \begin{bmatrix} \mathbf{D}_R & \mathbf{D}_I \\ -\mathbf{D}_I & \mathbf{D}_R \end{bmatrix}$$

and this is associated to the Hermitian matrix[2]

$$\mathbf{D} = \mathbf{D}_R + j\mathbf{D}_I = \mathbf{C}(\boldsymbol{\theta})^{-1}\frac{\partial\mathbf{C}(\boldsymbol{\theta})}{\partial\theta_i}\mathbf{C}(\boldsymbol{\theta})^{-1}\frac{\partial\mathbf{C}(\boldsymbol{\theta})}{\partial\theta_j}$$

Notice that

$$tr[\tilde{\mathbf{D}}] = 2\cdot tr[\mathbf{D}_R] = 2\cdot tr[\mathbf{D}]$$

(the second equality follows from the Hermitian symmetry as $tr[\mathbf{D}^H] = tr[\mathbf{D}]$ and thus $tr[\mathbf{D}_I] = 0$). The first term of (8.1) can be expanded into the entries

$$\frac{\partial\tilde{\boldsymbol{\mu}}(\boldsymbol{\theta})^T}{\partial\theta_i}\tilde{\mathbf{C}}(\boldsymbol{\theta})^{-1}\frac{\partial\tilde{\boldsymbol{\mu}}(\boldsymbol{\theta})}{\partial\theta_j} = 2\mathrm{Re}\left\{\frac{\partial\boldsymbol{\mu}(\boldsymbol{\theta})^H}{\partial\theta_i}\mathbf{C}(\boldsymbol{\theta})^{-1}\frac{\partial\boldsymbol{\mu}(\boldsymbol{\theta})}{\partial\theta_j}\right\}$$

where the equality is by substitution. To summarize, the general relationship for FIMs in circular symmetric complex Gaussian rv is

$$[\mathbf{J}(\boldsymbol{\theta}_o)]_{ij} = 2\mathrm{Re}\left\{\frac{\partial\boldsymbol{\mu}(\boldsymbol{\theta})^H}{\partial\theta_i}\mathbf{C}(\boldsymbol{\theta})^{-1}\frac{\partial\boldsymbol{\mu}(\boldsymbol{\theta})}{\partial\theta_j}\right\} + \mathrm{tr}\left[\mathbf{C}(\boldsymbol{\theta})^{-1}\frac{\partial\mathbf{C}(\boldsymbol{\theta})}{\partial\theta_i}\mathbf{C}(\boldsymbol{\theta})^{-1}\frac{\partial\mathbf{C}(\boldsymbol{\theta})}{\partial\theta_j}\right]\Bigg|_{\boldsymbol{\theta}=\boldsymbol{\theta}_0}$$

Needless to say, there could be adaptations depending on what is known (e.g., if $\mathbf{C}(\theta) = \mathbf{C}$ is independent on $\boldsymbol{\theta}$, the second term vanishes).

2 The two representations are fully equivalent as there is a one-to-one mapping between a complex-valued matrix $\mathbf{C} \in \mathbb{C}^{N\times N}$ and the block-partitioned real-valued $\tilde{\mathbf{C}} \in \mathbb{R}^{2N\times 2N}$

$$\mathbf{C} = \mathbf{C}_R + j\mathbf{C}_I \Leftrightarrow \tilde{\mathbf{C}} = \begin{bmatrix} \mathbf{C}_R & \mathbf{C}_I \\ -\mathbf{C}_I & \mathbf{C}_R \end{bmatrix}$$

For instance, the product is the same as

$$\mathbf{A} = \mathbf{BC} \Rightarrow \mathbf{A}_R + j\mathbf{A}_I = (\mathbf{B}_R + j\mathbf{B}_I)(\mathbf{C}_R + j\mathbf{C}_I)$$
$$\tilde{\mathbf{A}} = \tilde{\mathbf{B}}\cdot\tilde{\mathbf{C}} \Rightarrow \begin{bmatrix} \mathbf{A}_R & \mathbf{A}_I \\ -\mathbf{A}_I & \mathbf{A}_R \end{bmatrix} = \begin{bmatrix} \mathbf{B}_R & \mathbf{B}_I \\ -\mathbf{B}_I & \mathbf{B}_R \end{bmatrix}\cdot\begin{bmatrix} \mathbf{C}_R & \mathbf{C}_I \\ -\mathbf{C}_I & \mathbf{C}_R \end{bmatrix}$$

Appendix A: Proof of CRB

CRB for p=1

From the bias condition

$$\mathbb{E}[\hat{\theta}(\mathbf{x}) - \theta_o] = \int \left(\hat{\theta}(\mathbf{x}) - \theta\right) p[\mathbf{x}|\theta] d\mathbf{x} \Big|_{\theta=\theta_o} = 0$$

Taking the derivative wrt θ:

$$\frac{d}{d\theta} \int \left(\hat{\theta}(\mathbf{x}) - \theta\right) p[\mathbf{x}|\theta] d\mathbf{x} = -\underbrace{\int p[\mathbf{x}|\theta] d\mathbf{x}}_{1} + \int \left(\hat{\theta}(\mathbf{x}) - \theta\right) \frac{dp[\mathbf{x}|\theta]}{d\theta} d\mathbf{x} = 0$$

$$\Rightarrow \int \left(\hat{\theta}(\mathbf{x}) - \theta\right) \frac{dp[\mathbf{x}|\theta]}{d\theta} d\mathbf{x} = 1$$

From the property

$$\frac{d \ln p[\mathbf{x}|\theta]}{d\theta} = \frac{1}{p[\mathbf{x}|\theta]} \frac{dp[\mathbf{x}|\theta]}{d\theta} \Rightarrow \frac{dp[\mathbf{x}|\theta]}{d\theta} = p[\mathbf{x}|\theta] \frac{d \ln p[\mathbf{x}|\theta]}{d\theta}$$

we can derive

$$\int \left(\hat{\theta}(\mathbf{x}) - \theta\right) \sqrt{p[\mathbf{x}|\theta]} \times \frac{d \ln p[\mathbf{x}|\theta]}{d\theta} \sqrt{p[\mathbf{x}|\theta]} d\mathbf{x} = 1$$

The Schwartz inequality[3] can be applied for $\theta = \theta_0$:

$$\underbrace{\left(\int \left(\hat{\theta}(\mathbf{x}) - \theta_0\right)^2 p[\mathbf{x}|\theta_0] d\mathbf{x}\right)^{1/2}}_{\mathrm{var}[\hat{\theta}(\mathbf{x})]} \times \underbrace{\left(\int \left(\frac{d \ln p[\mathbf{x}|\theta]}{d\theta}\right)^2_{\theta=\theta_o} p[\mathbf{x}|\theta_o] d\mathbf{x}\right)^{1/2}}_{\mathbb{E}\left[\left(\frac{d \ln p[\mathbf{x}|\theta]}{d\theta}\right)^2\right]_{\theta=\theta_o}} \geq 1$$

so that

$$\mathrm{var}[\hat{\theta}(\mathbf{x})] \geq \left(\mathbb{E}\left[\left(\frac{d \ln p[\mathbf{x}|\theta]}{d\theta}\right)^2\right]_{\theta=\theta_o}\right)^{-1} = I^{-1}(\theta_o)$$

3 For any complex-valued functions:

$$\left|\int f_1(x) \cdot f_2(x)^* dx\right|^2 \leq \int |f_1(x)|^2 dx \cdot \int |f_2(x)|^2 dx$$

The equality holds only for $f_1(x) = c f_2(x)^*$ where c is an arbitrary constant value.

Equality holds only when $\frac{d\ln p[\mathbf{x}|\theta_o]}{d\theta} = c \times \left(\hat{\theta}(\mathbf{x}) - \theta_o\right)$, or equivalently when log-likelihood is quadratic around θ_0.

Alternative and equivalent forms of the CRB follow from the equalities

$$\int p[\mathbf{x}|\theta]d\mathbf{x} = 1 \Rightarrow \frac{d}{d\theta} \Rightarrow \int \frac{dp[\mathbf{x}|\theta]}{d\theta}d\mathbf{x} = 0 \Rightarrow \int \frac{d\ln p[\mathbf{x}|\theta]}{d\theta}p[\mathbf{x}|\theta]d\mathbf{x} = 0$$

$$\Rightarrow \frac{d}{d\theta} \Rightarrow \int \frac{d^2\ln p[\mathbf{x}|\theta]}{d\theta^2}p[\mathbf{x}|\theta]d\mathbf{x} + \int \frac{d\ln p[\mathbf{x}|\theta]}{d\theta} \underbrace{\frac{dp[\mathbf{x}|\theta]}{d\theta}}_{p[\mathbf{x}|\theta]\frac{d\ln p[\mathbf{x}|\theta]}{d\theta}} d\mathbf{x} = 0$$

$$\Rightarrow \mathbb{E}\left[\frac{d^2\ln p[\mathbf{x}|\theta]}{d\theta^2}\right] + \mathbb{E}\left[\left(\frac{d\ln p[\mathbf{x}|\theta]}{d\theta}\right)^2\right] = 0$$

CRB for p>1 (sketch)

The unbiasedness sets the basics for the derivation as

$$\int \left(\hat{\theta}(\mathbf{x}) - \theta\right)p[\mathbf{x}|\theta]d\mathbf{x}\Big|_{\theta=\theta_o} = \mathbf{0} \Rightarrow \frac{\partial}{\partial\theta} \Rightarrow$$

$$\Rightarrow \int \left(\hat{\theta}(\mathbf{x}) - \theta\right) \underbrace{\frac{\partial p[\mathbf{x}|\theta]}{\partial\theta}}_{p[\mathbf{x}|\theta]\frac{\partial\ln p[\mathbf{x}|\theta]}{\partial\theta}} d\mathbf{x}\Big|_{\theta=\theta_o}$$

$$= \mathbf{I}_p \int p[\mathbf{x}|\theta_o]d\mathbf{x} = \mathbf{I}_p$$

$$\Rightarrow \int \left(\hat{\theta}(\mathbf{x}) - \theta_o\right) \frac{\partial\ln p[\mathbf{x}|\theta]}{\partial\theta}\Big|_{\theta=\theta_o} p[\mathbf{x}|\theta_0]d\mathbf{x} = \mathbf{I}_p$$

$$\Rightarrow \mathbb{E}\left[\left(\hat{\theta}(\mathbf{x}) - \theta_o\right) \frac{\partial\ln p[\mathbf{x}|\theta]}{\partial\theta}\right]_{\theta=\theta_o} = \mathbf{I}_p$$

Rearranging into an augmented vector that depends on the rv $\mathbf{x}$:

$$\begin{bmatrix} \hat{\theta}(\mathbf{x}) - \theta \\ \frac{\partial\mathcal{L}(\mathbf{x}|\theta)}{\partial\theta} \end{bmatrix}$$

this is zero-mean with covariance (block-partitioned)

$$\mathbf{Q} = \mathbb{E}\left[\begin{bmatrix} \hat{\theta}(\mathbf{x}) - \theta \\ \frac{\partial\mathcal{L}(\mathbf{x}|\theta)}{\partial\theta} \end{bmatrix}\left[(\hat{\theta}(\mathbf{x}) - \theta)^T, (\frac{\partial\mathcal{L}(\mathbf{x}|\theta)}{\partial\theta})^T\right]\right] = \begin{bmatrix} \text{cov}[\hat{\theta}(\mathbf{x})] & \mathbf{I}_p \\ \mathbf{I}_p & \mathbf{J}(\theta) \end{bmatrix}$$

Recall that $\mathbf{Q} \succeq \mathbf{0}$. Matrix $\mathbf{Q}$ can be block-diagonalized as

$$\begin{bmatrix} \mathbf{I}_p & -\mathbf{J}^{-1}(\theta) \\ \mathbf{0}_p & \mathbf{I}_p \end{bmatrix}\begin{bmatrix} \text{cov}[\hat{\theta}(\mathbf{x})] & \mathbf{I}_p \\ \mathbf{I}_p & \mathbf{J}(\theta) \end{bmatrix}\begin{bmatrix} \mathbf{I}_p & \mathbf{0}_p \\ -\mathbf{J}^{-1}(\theta) & \mathbf{I}_p \end{bmatrix} = \begin{bmatrix} \text{cov}[\hat{\theta}(\mathbf{x})] - \mathbf{J}^{-1}(\theta) & \mathbf{0}_p \\ \mathbf{0}_p & \mathbf{J}(\theta) \end{bmatrix}$$

so that from the property $\mathbf{Q} \geq \mathbf{0}$ that is now transferred to the block-diagonals, $\text{cov}[\hat{\theta}(\mathbf{x})] - \mathbf{J}^{-1}(\theta) \geq \mathbf{0}$ that is the CRB. This concludes the sketch of the proof.

Appendix B: FIM for Gaussian Model

The FIM for the Gaussian model is based on

$$\frac{\partial \mathcal{L}(\mathbf{x}|\theta)}{\partial \theta_k} = -\frac{1}{2}\text{tr}\left[\mathbf{C}(\theta)^{-1}\frac{\partial \mathbf{C}(\theta)}{\partial \theta_k}\right] + \frac{\partial \boldsymbol{\mu}(\theta)^T}{\partial \theta_k}\mathbf{C}(\theta)^{-1}\delta\mathbf{x} - \underbrace{\frac{1}{2}\delta\mathbf{x}^T\frac{\partial \mathbf{C}(\theta)^{-1}}{\partial \theta_k}\delta\mathbf{x}}_{\text{tr}\left\{\frac{\partial \mathbf{C}(\theta)^{-1}}{\partial \theta_k}\delta\mathbf{x}\delta\mathbf{x}^T\right\}}$$

where $\delta\mathbf{x}$ is conveniently defined as $\delta\mathbf{x} = \mathbf{x} - \boldsymbol{\mu}(\theta)$. From definition of the entries of the FIM

$$[\mathbf{J}(\theta_0)]_{ij} = \mathbb{E}\left[\frac{\partial \mathcal{L}(\mathbf{x}|\theta)}{\partial \theta_i}\frac{\partial \mathcal{L}(\mathbf{x}|\theta)}{\partial \theta_j}\right]_{\theta=\theta_0}$$

we have (recall that all expectations are wrt $\mathbf{x}$ and $\mathbb{E}[\delta\mathbf{x}] = \mathbf{0}$, while $\mathbb{E}[\delta\mathbf{x}\cdot\delta\mathbf{x}^T] = \mathbf{C}(\theta)$)

$$\begin{aligned}
[\mathbf{J}(\theta_0)]_{ij} = &\ \tfrac{1}{4}\text{tr}\left[\mathbf{C}(\theta)^{-1}\tfrac{\partial \mathbf{C}(\theta)}{\partial \theta_i}\right]\text{tr}\left[\mathbf{C}(\theta)^{-1}\tfrac{\partial \mathbf{C}(\theta)}{\partial \theta_j}\right] \\
&+\tfrac{\partial \boldsymbol{\mu}(\theta)^T}{\partial \theta_i}\mathbf{C}(\theta)^{-1}\tfrac{\partial \boldsymbol{\mu}(\theta)}{\partial \theta_j} \\
&\left.\begin{aligned}
&+\tfrac{1}{4}\text{tr}\left[\mathbf{C}(\theta)^{-1}\tfrac{\partial \mathbf{C}(\theta)}{\partial \theta_i}\right]\text{tr}\left[\tfrac{\partial \mathbf{C}(\theta)^{-1}}{\partial \theta_j}\mathbf{C}(\theta)\right] \\
&+\tfrac{1}{4}\text{tr}\left[\mathbf{C}(\theta)^{-1}\tfrac{\partial \mathbf{C}(\theta)}{\partial \theta_j}\right]\text{tr}\left[\tfrac{\partial \mathbf{C}(\theta)^{-1}}{\partial \theta_i}\mathbf{C}(\theta)\right]
\end{aligned}\right\} \\
&-\tfrac{1}{2}\text{tr}\left[\mathbf{C}(\theta)^{-1}\tfrac{\partial \mathbf{C}(\theta)}{\partial \theta_i}\right]\text{tr}\left[\mathbf{C}(\theta)^{-1}\tfrac{\partial \mathbf{C}(\theta)}{\partial \theta_j}\right] \\
&+\tfrac{1}{4}\mathbb{E}\left[\delta\mathbf{x}^T\tfrac{\partial \mathbf{C}(\theta)^{-1}}{\partial \theta_i}\delta\mathbf{x}\delta\mathbf{x}^T\tfrac{\partial \mathbf{C}(\theta)^{-1}}{\partial \theta_j}\delta\mathbf{x}\right]
\end{aligned}$$

hence

$$\begin{aligned}
[\mathbf{J}(\theta_0)]_{ij} = &-\frac{1}{4}\text{tr}\left[\mathbf{C}(\theta)^{-1}\frac{\partial \mathbf{C}(\theta)}{\partial \theta_i}\right]\text{tr}\left[\mathbf{C}(\theta)^{-1}\frac{\partial \mathbf{C}(\theta)}{\partial \theta_j}\right] \\
&+\frac{\partial \boldsymbol{\mu}(\theta)^T}{\partial \theta_i}\mathbf{C}(\theta)^{-1}\frac{\partial \boldsymbol{\mu}(\theta)}{\partial \theta_j} \\
&+\frac{1}{4}\mathbb{E}\left[\delta\mathbf{x}^T\frac{\partial \mathbf{C}(\theta)^{-1}}{\partial \theta_i}\delta\mathbf{x}\delta\mathbf{x}^T\frac{\partial \mathbf{C}(\theta)^{-1}}{\partial \theta_j}\delta\mathbf{x}\right]
\end{aligned}$$

The last term is the product of two quadratic forms that after some manipulation (see Section 1.4; recall that $\frac{\partial \mathbf{C}(\theta)^{-1}}{\partial \theta_k} = -\mathbf{C}(\theta)^{-1}\frac{\partial \mathbf{C}(\theta)}{\partial \theta_k}\mathbf{C}(\theta)^{-1}$ and refer to derivations below for details)

$$\mathbb{E}\left[\delta\mathbf{x}^T\frac{\partial\mathbf{C}(\boldsymbol{\theta})^{-1}}{\partial\theta_i}\delta\mathbf{x}\delta\mathbf{x}^T\frac{\partial\mathbf{C}(\boldsymbol{\theta})^{-1}}{\partial\theta_j}\delta\mathbf{x}\right] = \mathrm{tr}\left[\mathbf{C}(\boldsymbol{\theta})^{-1}\frac{\partial\mathbf{C}(\boldsymbol{\theta})}{\partial\theta_i}\right]\times\mathrm{tr}\left[\mathbf{C}(\boldsymbol{\theta})^{-1}\frac{\partial\mathbf{C}(\boldsymbol{\theta})}{\partial\theta_j}\right] + 2\mathrm{tr}\left[\mathbf{C}(\boldsymbol{\theta})^{-1}\frac{\partial\mathbf{C}(\boldsymbol{\theta})}{\partial\theta_i}\mathbf{C}(\boldsymbol{\theta})^{-1}\frac{\partial\mathbf{C}(\boldsymbol{\theta})}{\partial\theta_j}\right]$$

Appendix C: Some Derivatives for MLE and CRB Computations

The derivative of the compound function is

$$\frac{\partial g[\mathbf{C}(\boldsymbol{\theta})]}{\partial\theta_k} = \frac{\partial g[\mathbf{C}(\boldsymbol{\theta})]}{\partial C_{11}(\boldsymbol{\theta})}\frac{\partial C_{11}(\boldsymbol{\theta})}{\partial\theta_k} + \frac{\partial g[\mathbf{C}(\boldsymbol{\theta})]}{\partial C_{12}(\boldsymbol{\theta})}\frac{\partial C_{12}(\boldsymbol{\theta})}{\partial\theta_k} + \ldots + \frac{\partial g[\mathbf{C}(\boldsymbol{\theta})]}{\partial C_{NN}(\boldsymbol{\theta})}\frac{\partial C_{NN}(\boldsymbol{\theta})}{\partial\theta_k}$$

$$\frac{\partial g[\mathbf{C}(\boldsymbol{\theta})]}{\partial\theta_k} = \sum_{i,j=1}^{N}\frac{\partial g[\mathbf{C}(\boldsymbol{\theta})]}{\partial C_{ij}(\boldsymbol{\theta})}\frac{\partial C_{ij}(\boldsymbol{\theta})}{\partial\theta_k}$$

Rearranging the terms in a matrix-matrix product of gradients:

$$\underbrace{\begin{bmatrix}\frac{\partial g[\mathbf{C}(\boldsymbol{\theta})]}{\partial C_{11}(\boldsymbol{\theta})} & \cdots & \frac{\partial g[\mathbf{C}(\boldsymbol{\theta})]}{\partial C_{1N}(\boldsymbol{\theta})}\\ \vdots & \ddots & \vdots\\ \frac{\partial g[\mathbf{C}(\boldsymbol{\theta})]}{\partial C_{N1}(\boldsymbol{\theta})} & \cdots & \frac{\partial g[\mathbf{C}(\boldsymbol{\theta})]}{\partial C_{NN}(\boldsymbol{\theta})}\end{bmatrix}}_{\frac{\partial g[\mathbf{C}(\boldsymbol{\theta})]}{\partial\mathbf{C}(\boldsymbol{\theta})}} \cdot \underbrace{\begin{bmatrix}\frac{\partial C_{11}(\boldsymbol{\theta})}{\partial\theta_k} & \cdots & \frac{\partial C_{N1}(\boldsymbol{\theta})}{\partial\theta_k}\\ \vdots & \ddots & \vdots\\ \frac{\partial C_{1N}(\boldsymbol{\theta})}{\partial\theta_k} & \cdots & \frac{\partial C_{NN}(\boldsymbol{\theta})}{\partial\theta_k}\end{bmatrix}}_{\frac{\partial\mathbf{C}(\boldsymbol{\theta})^T}{\partial\theta_k}}$$

$$= \begin{bmatrix}\sum_{j=1}^{N}\frac{\partial g[\mathbf{C}(\boldsymbol{\theta})]}{\partial C_{1j}(\boldsymbol{\theta})}\frac{\partial C_{1j}(\boldsymbol{\theta})}{\partial\theta_k} & \cdots & \sum_{j=1}^{N}\frac{\partial g[\mathbf{C}(\boldsymbol{\theta})]}{\partial C_{1j}(\boldsymbol{\theta})}\frac{\partial C_{Nj}(\boldsymbol{\theta})}{\partial\theta_k}\\ \vdots & \ddots & \vdots\\ \sum_{j=1}^{N}\frac{\partial g[\mathbf{C}(\boldsymbol{\theta})]}{\partial C_{Nj}(\boldsymbol{\theta})}\frac{\partial C_{1j}(\boldsymbol{\theta})}{\partial\theta_k} & \cdots & \sum_{j=1}^{N}\frac{\partial g[\mathbf{C}(\boldsymbol{\theta})]}{\partial C_{Nj}(\boldsymbol{\theta})}\frac{\partial C_{Nj}(\boldsymbol{\theta})}{\partial\theta_k}\end{bmatrix}$$

summing along the diagonal follows the gradient $\frac{\partial g[\mathbf{C}(\boldsymbol{\theta})]}{\partial\theta_k}$, and the general rule

$$\frac{\partial g[\mathbf{C}(\boldsymbol{\theta})]}{\partial\theta_k} = \mathrm{tr}\left[\frac{\partial g[\mathbf{C}(\boldsymbol{\theta})]}{\partial\mathbf{C}(\boldsymbol{\theta})}\frac{\partial\mathbf{C}(\boldsymbol{\theta})^T}{\partial\theta_k}\right]$$

Specializing these relationships to ML estimation, and recalling that (Section 1.4.1)

$$\frac{\partial\det[\mathbf{C}(\boldsymbol{\theta})]}{\partial\mathbf{C}(\boldsymbol{\theta})} = \det[\mathbf{C}(\boldsymbol{\theta})]\left(\mathbf{C}(\boldsymbol{\theta})^{-1}\right)^T$$

we get

$$\frac{\partial\det[\mathbf{C}(\boldsymbol{\theta})]}{\partial\theta_k} = \mathrm{tr}[\frac{\partial\det[\mathbf{C}(\boldsymbol{\theta})]}{\partial\boldsymbol{\theta}}\frac{\partial\mathbf{C}(\boldsymbol{\theta})^T}{\partial\theta_k}] = \mathrm{tr}\left[\det[\mathbf{C}(\boldsymbol{\theta})]\left(\mathbf{C}(\boldsymbol{\theta})^{-1}\right)^T\frac{\partial\mathbf{C}(\boldsymbol{\theta})^T}{\partial\theta_k}\right]$$

Another useful derivative is

$$\frac{\partial \ln\det[\mathbf{C}(\boldsymbol{\theta})]}{\partial\theta_k} = \mathrm{tr}\left[\frac{\partial \ln\det[\mathbf{C}(\boldsymbol{\theta})]}{\partial\mathbf{C}(\boldsymbol{\theta})}\frac{\partial\mathbf{C}(\boldsymbol{\theta})^T}{\partial\theta_k}\right] = \mathrm{tr}\left[\left(\mathbf{C}(\boldsymbol{\theta})^{-1}\right)^T\frac{\partial\mathbf{C}(\boldsymbol{\theta})}{\partial\theta_k}\right]$$
$$= \mathrm{tr}\left[\mathbf{C}(\boldsymbol{\theta})^{-1}\frac{\partial\mathbf{C}(\boldsymbol{\theta})}{\partial\theta_k}\right]$$

after using the property $\mathrm{tr}[\mathbf{AB}] = \mathrm{tr}[(\mathbf{AB})^T] = \mathrm{tr}[\mathbf{B}^T\mathbf{A}^T] = \mathrm{tr}[\mathbf{A}^T\mathbf{B}^T]$ and $\mathbf{C}(\boldsymbol{\theta})^T = \mathbf{C}(\boldsymbol{\theta})$. Since

$$\mathbf{C}(\boldsymbol{\theta})^{-1}\mathbf{C}(\boldsymbol{\theta}) = \mathbf{I}$$

the derivative in this case yields

$$\frac{\partial[\mathbf{C}(\boldsymbol{\theta})^{-1}\mathbf{C}(\boldsymbol{\theta})]}{\partial\theta_k} = \mathbf{C}(\boldsymbol{\theta})^{-1}\frac{\partial\mathbf{C}(\boldsymbol{\theta})}{\partial\theta_k} + \frac{\partial\mathbf{C}(\boldsymbol{\theta})^{-1}}{\partial\theta_k}\mathbf{C}(\boldsymbol{\theta}) = \frac{\partial\mathbf{I}}{\partial\theta_k} = \mathbf{0}$$

so that

$$\frac{\partial\mathbf{C}(\boldsymbol{\theta})^{-1}}{\partial\theta_k} = -\mathbf{C}(\boldsymbol{\theta})^{-1}\frac{\partial\mathbf{C}(\boldsymbol{\theta})}{\partial\theta_k}\mathbf{C}(\boldsymbol{\theta})^{-1} \in \mathbb{R}^{N\times N}$$

9

MLE and CRB for Some Selected Cases

In this chapter, MLE and the CRB are evaluated for some of the most common models and applications introduced in Chapter 5. Furthermore, from the inspection of the CRB, some methods are revised such as the common usage of numerical histograms. The numerical analysis in Chapter 10 is the natural complement of this chapter, pairing theory with numerical validations. Examples here are introductory to consolidate knowledge of the basics of estimation theory and bounds; extensive usage of these concepts can be found in applications in later chapters (Chapters 16, 17, 19, and 20).

9.1 Linear Regressions

Considering the linear regression model (Section 5.1)

$$\mathbf{x} = \underbrace{\begin{bmatrix} 1 & t_1 \\ 1 & t_2 \\ \vdots & \vdots \\ 1 & t_N \end{bmatrix}}_{\mathbf{H}=[\mathbf{1},\mathbf{t}]} \cdot \underbrace{\begin{bmatrix} A_o \\ B_o \end{bmatrix}}_{\theta_o} + \mathbf{w}$$

with $\mathbf{w} \sim \mathcal{N}(\mathbf{0}, \mathbf{C}_w)$; it is possible to obtain the MLE in a closed form. For $\mathbf{C}_w = \sigma_w^2 \mathbf{I}$ it is:

$$\begin{bmatrix} \hat{A} \\ \hat{B} \end{bmatrix} = \left(\mathbf{H}^T\mathbf{H}\right)^{-1}\mathbf{H}^T\mathbf{x} = \begin{bmatrix} \mathbf{1}^T\mathbf{1} & \mathbf{1}^T\mathbf{t} \\ \mathbf{t}^T\mathbf{1} & \mathbf{t}^T\mathbf{t} \end{bmatrix}^{-1} \begin{bmatrix} \mathbf{1}^T\mathbf{x} \\ \mathbf{t}^T\mathbf{x} \end{bmatrix}$$

$$= \begin{bmatrix} N & \sum_{k=1}^N t_k \\ \sum_{k=1}^N t_k & \sum_{k=1}^N t_k^2 \end{bmatrix}^{-1} \begin{bmatrix} \sum_{k=1}^N x(t_k) \\ \sum_{k=1}^N t_k x(t_k) \end{bmatrix}$$

Statistical Signal Processing in Engineering, First Edition. Umberto Spagnolini.

Companion website: www.wiley.com/go/spagnolini/signalprocessing

In the case of uniform sampling with $t_k = \Delta t \cdot (k-1)$ (including Δt in the variable B, or considering $\Delta t = 1$) it is[1]

$$\begin{bmatrix} \hat{A} \\ \hat{B} \end{bmatrix} = \underbrace{\begin{bmatrix} N & \sum_{k=0}^{N-1} k \\ \sum_{k=0}^{N-1} k & \sum_{k=0}^{N-1} k^2 \end{bmatrix}}_{\mathbf{H}^T\mathbf{H}}^{-1} \begin{bmatrix} \sum_{k=0}^{N-1} x[k] \\ \sum_{k=0}^{N-1} kx[k] \end{bmatrix}$$

where

$$\left(\mathbf{H}^T\mathbf{H}\right)^{-1} = \begin{bmatrix} \frac{2(2N-1)}{N(N+1)} & -\frac{6}{N(N+1)} \\ -\frac{6}{N(N+1)} & \frac{12}{N(N^2-1)} \end{bmatrix}$$

The covariance matrix is given by

$$\text{cov}[\hat{\theta}] = \sigma_w^2 \left(\mathbf{H}^T\mathbf{H}\right)^{-1} \rightarrow \begin{cases} \text{var}[\hat{A}] = \sigma_w^2 \frac{2(2N-1)}{N(N+1)} \underset{N\to\infty}{\rightarrow} \frac{4\sigma_w^2}{N} \\ \text{var}[\hat{B}] = \sigma_w^2 \frac{12}{N(N^2-1)} \underset{N\to\infty}{\rightarrow} \frac{12\sigma_w^2}{N^3} \end{cases} \tag{9.1}$$

In the case that the measurements are symmetrically distributed with respect to the origin (in this case N must be odd)

$$x[n] = \bar{A}_o + \bar{B}_o n \qquad n = -(N-1)/2, ..., -1, 0, 1, 2, ...(N-1)/2$$

such that $\bar{A}_o = A_o - (N-1)B_o/2$ but $\bar{B}_o = B_o$, the estimator is decoupled:

$$\begin{bmatrix} \hat{\bar{A}} \\ \hat{\bar{B}} \end{bmatrix} = \left(\mathbf{H}^T\mathbf{H}\right)^{-1} \begin{bmatrix} \sum_{k=-(N-1)/2}^{(N-1)/2} x[k] \\ \sum_{k=-(N-1)/2}^{(N-1)/2} kx[k] \end{bmatrix} = \frac{1}{N} \begin{bmatrix} \sum_{k=-(N-1)/2}^{(N-1)/2} x[k] \\ \frac{12}{N^2-1} \sum_{k=-(N-1)/2}^{(N-1)/2} kx[k] \end{bmatrix}$$

with

$$\left(\mathbf{H}^T\mathbf{H}\right)^{-1} = \begin{bmatrix} \frac{1}{N} & 0 \\ 0 & \frac{12}{N(N^2-1)} \end{bmatrix}$$

---

1 Some useful relationships in these contexts are $\sum_{n=0}^{N} n = \frac{1}{2}N(N+1) \simeq N^2/2$ and $\sum_{n=0}^{N} n^2 = \frac{1}{6}N(N+1)(2N+1) \simeq N^3/3$, but recall that for $N \gg 1$ the following approximation is useful:

$$\sum_{n=0}^{N} n^k \simeq \int_0^N x^k dx = \frac{N^{k+1}}{k+1}.$$

and the variance for decoupled terms is

$$\text{var}[\hat{\bar{A}}] = \frac{\sigma_w^2}{N}$$
$$\text{var}[\hat{\bar{B}}] = \frac{12\sigma_w^2}{N(N^2-1)}$$

Moreover, the variance for the symmetric arrangement $\hat{\bar{A}}$ is four times better with respect to the previously analyzed case. Note that in some applications, the numbering of the samples is just a matter of convention, and renumbering the samples by placing the origin at the center is beneficial for the estimate of the intercept (if necessary), but not of the slope. Symmetric arrangement of data should be preferred anytime is possible as the estimator is decoupled, and this does not affect the accuracy of the slope $\bar{B}_o = B_o$.

It could be possible to compute the variance var$[\hat{A}]$ for non-symmetric measurements from simple geometrical reasoning referring to Figure 9.1. The symmetric allocation decomposes line fitting into a translation ($\bar{A}_o$) and rotation ($\bar{B}_o$); the deviations of the estimate from the true values are $\delta\bar{A} = \hat{\bar{A}} - \bar{A}_o \sim \mathcal{N}(0, \frac{\sigma_w^2}{N})$ and $\delta\bar{B} = \hat{\bar{B}} - \bar{B}_o \sim \mathcal{N}(0, \frac{12\sigma_w^2}{N(N^2-1)})$, and this reflects onto the deviation for the model (A_o, B_o)

$$\delta A = \delta\bar{A} + \delta\bar{B} \times \frac{N-1}{2}$$

The variance becomes

$$\text{var}[\hat{A}] = \text{var}[\hat{\bar{A}}] + \text{var}[\hat{\bar{B}}] \times \frac{(N-1)^2}{4}$$

that is the result proved analytically in (9.1).

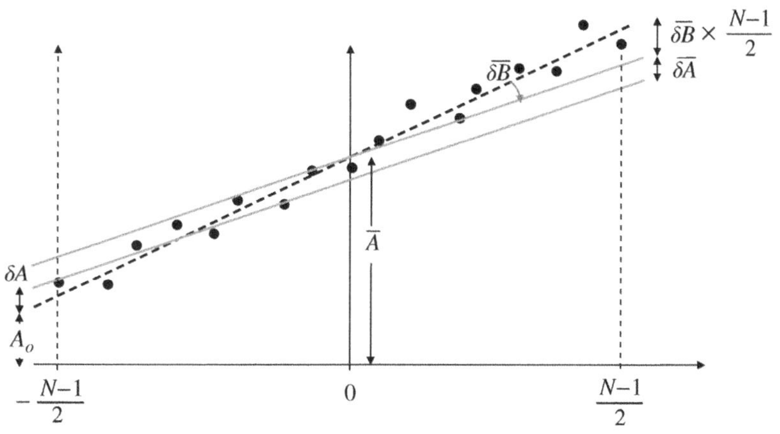

Figure 9.1 Linear regression and impact of deviations $(\delta\bar{A}, \delta\bar{B})$ on the variance.

9.2 Frequency Estimation $x[n] = a_o\cos(\omega_0 n + \phi_o) + w[n]$

Given the model defined in Section 5.4

$$\mathbf{x} = \underbrace{[\mathbf{c}(\omega), \mathbf{s}(\omega)]}_{\mathbf{H}(\omega)} \cdot \begin{bmatrix} \theta_1 \\ \theta_2 \end{bmatrix} + \mathbf{w}$$

ML estimation requires a separate optimization of the amplitudes θ_1, θ_2 and of the frequency ω. For $\mathbf{C}_w = \sigma_w^2\mathbf{I}$, the MLE reduces to the maximization of the non-linear objective $\Phi(\omega)$ (see Section 7.2.2)

$$\hat{\omega} = \arg\max_{\omega} \underbrace{\left\{\mathbf{x}^T\mathbf{H}(\omega)\left(\mathbf{H}^T(\omega)\mathbf{H}(\omega)\right)^{-1}\mathbf{H}^T(\omega)\mathbf{x}\right\}}_{\Phi(\omega)}$$

where

$$\mathbf{H}^T(\omega)\mathbf{H}(\omega) = \begin{bmatrix} \mathbf{c}^T(\omega)\mathbf{c}(\omega) & \mathbf{c}^T(\omega)\mathbf{s}(\omega) \\ \mathbf{s}^T(\omega)\mathbf{c}(\omega) & \mathbf{s}^T(\omega)\mathbf{s}(\omega) \end{bmatrix} \simeq \begin{bmatrix} N/2 & 0 \\ 0 & N/2 \end{bmatrix}$$

$$\mathbf{H}^T(\omega)\mathbf{x} = \begin{bmatrix} \mathbf{c}^T(\omega)\mathbf{x} \\ \mathbf{s}^T(\omega)\mathbf{x} \end{bmatrix}$$

and therefore

$$\Phi(\omega) = \frac{2}{N}[(\mathbf{c}^T(\omega)\mathbf{x})^2 + (\mathbf{s}^T(\omega)\mathbf{x})^2] = \frac{2}{N}|X(\omega)|^2$$

The MLE is the frequency that maximizes the amplitude of the Fourier transform of the data:

$$X(\omega) = \sum_{n=0}^{N-1} x[n]\exp(-j\omega n)$$

Different strategies for the maximization of $|X(\omega)|^2$ give different estimators. The DFT over N samples (Section 5.2.2) provides a simple (and computationally efficient) coarse estimate of $X(\omega)$ for the set of frequencies $\omega \in \{0, 2\pi/N, 2\cdot 2\pi/N, ..., (N-1)\cdot 2\pi/N\}$ over the $2\pi/N$ spaced grid. Once the bin that peaks $|X(2\pi k/N)|^2$ is identified, the frequency estimate is within that interval of dimension $2\pi/N$, which must include the maximum of $\Phi(\omega)$. Further refinements are possible by increasing the frequency resolution with DFT over $M > N$ samples after padding the sequence with zeros, or by using iterative techniques to identify the maximum of $\Phi(\omega)$. A detailed discussion on numerical methods can be found in Chapter 10.

CRB for Frequency Estimation

The computation of the CRB needs to reorganize the model and variables, and this has several implications useful for the complete understanding of the CRB computations. Let the model be defined as

$$x[n] = \underbrace{a_o \cos(\omega_0 n + \phi_o)}_{s[n;a_o,\omega_0,\phi_o]} + w[n] = s[n;\theta_o] + w[n] \quad \text{for } n = 0,1,\ldots,N-1$$

with the vector of parameters ordered as

$$\theta = \begin{bmatrix} a \\ \omega \\ \phi \end{bmatrix}$$

The FIM elements of the Gaussian model are obtained from Section 8.4.1. The derivatives of the model $s[n;\theta]$ are

$$\begin{aligned}
\frac{\partial s[n;\theta]}{\partial \theta_1} &= \left.\frac{\partial s[n;\theta]}{\partial a}\right|_{\theta=\theta_o} = \cos(\omega_0 n + \phi_o) \\
\frac{\partial s[n;\theta]}{\partial \theta_2} &= \left.\frac{\partial s[n;\theta]}{\partial \omega}\right|_{\theta=\theta_o} = -n a_o \sin(\omega_0 n + \phi_o) \\
\frac{\partial s[n;\theta]}{\partial \theta_3} &= \left.\frac{\partial s[n;\theta]}{\partial \phi}\right|_{\theta=\theta_o} = -a_o \sin(\omega_0 n + \phi_o)
\end{aligned}$$

and the elements of the FIM are (to ease notation, $\alpha = \omega_0 n + \phi_o$ and FIM entries are re-labeled by the corresponding parameters)

$$\begin{aligned}
[\mathbf{J}(\theta)]_{aa} &= \tfrac{1}{\sigma_w^2} \sum_{n=0}^{N-1} \cos^2\alpha \simeq \tfrac{N}{2\sigma_w^2} \\
[\mathbf{J}(\theta)]_{a\omega} &= -\tfrac{a_o}{\sigma_w^2} \sum_{n=0}^{N-1} n \sin\alpha \cos\alpha = -\tfrac{a_o}{2\sigma_w^2} \sum_{n=0}^{N-1} n \sin 2\alpha \simeq 0 \\
[\mathbf{J}(\theta)]_{a\phi} &= -\tfrac{a_o}{\sigma_w^2} \sum_{n=0}^{N-1} \sin\alpha \cos\alpha = -\tfrac{a_o}{2\sigma_w^2} \sum_{n=0}^{N-1} \sin 2\alpha \simeq 0 \\
[\mathbf{J}(\theta)]_{\omega\omega} &= \tfrac{a_o^2}{\sigma_w^2} \sum_{n=0}^{N-1} n^2 \sin^2\alpha = \tfrac{a_o^2}{\sigma_w^2} \sum_{n=0}^{N-1} n^2 \left(\tfrac{1}{2} - \tfrac{1}{2}\cos 2\alpha\right) \simeq \tfrac{a_o^2}{2\sigma_w^2} \sum_{n=0}^{N-1} n^2 \\
[\mathbf{J}(\theta)]_{\omega\phi} &= \tfrac{a_o^2}{\sigma_w^2} \sum_{n=0}^{N-1} n \sin^2\alpha = \tfrac{a_o^2}{\sigma_w^2} \sum_{n=0}^{N-1} n \left(\tfrac{1}{2} - \tfrac{1}{2}\cos 2\alpha\right) \simeq \tfrac{a_o^2}{2\sigma_w^2} \sum_{n=0}^{N-1} n \\
[\mathbf{J}(\theta)]_{\phi\phi} &= \tfrac{a_o^2}{\sigma_w^2} \sum_{n=0}^{N-1} \sin^2\alpha \simeq \tfrac{N a_o^2}{2\sigma_w^2}
\end{aligned}$$

where the approximations are for sufficiently large N. The FIM is

$$\mathbf{J}(\theta_o) = \frac{1}{\sigma_w^2} \begin{bmatrix} N/2 & 0 & 0 \\ 0 & a_o^2/2 \sum_{n=0}^{N-1} n^2 & a_o^2/2 \sum_{n=0}^{N-1} n \\ 0 & a_o^2/2 \sum_{n=0}^{N-1} n & N a_o^2/2 \end{bmatrix}$$

and the CRB

$$\mathbf{J}^{-1}(\theta_o) = 2\sigma_w^2 \begin{bmatrix} \frac{1}{N} & 0 & 0 \\ 0 & \frac{12/a_o^2}{N(N^2-1)} & -\frac{6/a_o^2}{N(N+1)} \\ 0 & -\frac{6/a_o^2}{N(N+1)} & \frac{2(2N-1)/a_o^2}{N(N+1)} \end{bmatrix}$$

This shows that amplitude and phase/frequency are mutually decoupled, but phase and frequency are not, and corresponding CRB depends only on amplitude a_o^2 but not on frequency or phase. More specifically

$$\text{var}[\hat{a}] \geq \frac{2\sigma_w^2}{N}$$

$$\text{var}[\hat{\omega}] \geq \frac{24\sigma_w^2/a_o^2}{N(N^2-1)} \underset{N\to\infty}{\to} \frac{24\sigma_w^2/a_o^2}{N^3}$$

$$\text{var}[\hat{\phi}] \geq \frac{4(2N-1)\sigma_w^2/a_o^2}{N(N+1)} \underset{N\to\infty}{\to} \frac{8\sigma_w^2/a_o^2}{N}$$

The decoupling guarantees that the CRB of phase/frequency is the same no matter whether amplitude a_0 is known or not. Moreover, if phase is known, the CRB follows from the FIM:

$$\text{cov}([\hat{a},\hat{\omega}]^T) \geq \sigma_w^2 \begin{bmatrix} N/2 & 0 \\ 0 & a_o^2/2\sum_{n=0}^{N-1} n^2 \end{bmatrix}^{-1} = \begin{bmatrix} \frac{2\sigma_w^2}{N} & 0 \\ 0 & \frac{12\sigma_w^2/a_o^2}{N(2N-1)(N-1)} \end{bmatrix}$$

(still decoupled), and the CRB on frequency is

$$\text{var}[\hat{\omega}] \geq \frac{12\sigma_w^2/a_o^2}{N(2N-1)(N-1)} \underset{N\to\infty}{\to} \frac{6\sigma_w^2/a_o^2}{N^3}$$

which is four times smaller (or 6dB gain) than the joint estimation of phase and frequency.

Remark: The FIM for phase and frequency are the same as for the linear regression model $A+Bn$ in Section 9.1 (phase maps onto intercept A and frequency onto slope B), but this result is not surprising as phase/frequency estimation is like considering the linear phase increase of the angle of $x[n]$:

$$\measuredangle x[n] \simeq \omega_0 n + \phi_o$$

where phase ϕ_o is just the starting point of the linear phase increase vs n, and ω_0 is the slope.

9.3 Estimation of Complex Sinusoid

The estimation of frequency for a complex sinusoid is based on the model

$$x[n] = \alpha_o \exp(j\omega_o n) + w[n] \quad \text{where } \alpha_o = |\alpha_o| \exp(j\phi_o)$$

that for an ensemble of N samples is

$$\mathbf{x} = \alpha_o \mathbf{a}(\omega_o) + \mathbf{w} \sim \mathcal{CN}(\alpha_o \mathbf{a}(\omega_o), \mathbf{C}_w)$$

where

$$\mathbf{a}(\omega_o) = \begin{bmatrix} 1 \\ \exp(j\omega_o) \\ \vdots \\ \exp(j\omega_o(N-1)) \end{bmatrix}$$

One can use this model to apply the canonical model with real and imaginary parts in Section 9.2, but further elaboration on complex notations could be beneficial to gain insight into the complex-valued notation. The frequency estimation is from minimization of the log-likelihood metric

$$\mathcal{L}(\mathbf{x}|\alpha,\omega) = (\mathbf{x} - \alpha\mathbf{a}(\omega))^H \mathbf{C}_w^{-1} (\mathbf{x} - \alpha\mathbf{a}(\omega))$$

Assuming ω is given and using the notation $\mathbf{a}_o = \mathbf{a}(\omega)$:

$$\mathcal{L}(\mathbf{x}|\alpha,\omega) = \mathbf{x}^H \mathbf{C}_w^{-1}\mathbf{x} - \alpha^* \mathbf{a}_o^H \mathbf{C}_w^{-1}\mathbf{x} - \alpha \mathbf{x}^H \mathbf{C}_w^{-1}\mathbf{a}_o + \mathbf{a}_o^H \mathbf{C}_w^{-1}\mathbf{a}_o |\alpha|^2$$

From the gradient rules of Section 1.4.1:

$$\begin{cases} \frac{\partial \mathcal{L}(\mathbf{x}|\alpha,\omega)}{\partial \alpha} = 0 \\ \frac{\partial \mathcal{L}(\mathbf{x}|\alpha,\omega)}{\partial \alpha^*} = 0 \end{cases} \Rightarrow \begin{cases} \mathbf{a}_o^H \mathbf{C}_w^{-1}\mathbf{a}_o \alpha^* = \mathbf{x}^H \mathbf{C}_w^{-1}\mathbf{a}_o \\ \mathbf{a}_o^H \mathbf{C}_w^{-1}\mathbf{a}_o \alpha = \mathbf{a}_o^H \mathbf{C}_w^{-1}\mathbf{x} \end{cases} \Rightarrow \alpha = \alpha_* = \frac{\mathbf{a}_o^H \mathbf{C}_w^{-1}}{\mathbf{a}_o^H \mathbf{C}_w^{-1}\mathbf{a}_o}\mathbf{x}$$

still dependent on ω. Plugging into the likelihood function yields the optimization wrt ω:

$$\hat{\omega} = \arg\max_{\omega} \left\{ \frac{|\mathbf{a}^H(\omega)\mathbf{C}_w^{-1}\mathbf{x}|^2}{\mathbf{a}^H(\omega)\mathbf{C}_w^{-1}\mathbf{a}(\omega)} \right\}$$

For uncorrelated Gaussian noise $\mathbf{C}_w = \sigma_w^2 \mathbf{I}$, the log-likelihood simplifies as

$$\mathcal{L}(\mathbf{x}|\alpha_*,\omega) = \frac{1}{\sigma_w^2}(||\mathbf{x}||^2 - |\mathbf{a}^H(\omega)\mathbf{x}|^2/N)$$

The MLE is equivalent to the following maximization:

$$\hat{\omega} = \arg\max_{\omega} \frac{|\mathbf{a}(\omega)^H \mathbf{x}|^2}{N} = \arg \max_{\omega\in[-\pi,\pi)} \frac{|X(\omega)|^2}{N}$$

which is once again the maximization of the Fourier transform of the sequence $\mathbf{x}$. Note that in this case, the search of the frequency is over the whole range $[-\pi, \pi)$.

The CRB can be evaluated from the FIM for the circular symmetric Gaussian in Section 8.4.3, but an in-depth discussion is deferred to Chapter 16 (in detail: Section 16.2.2) where the different assumptions on frequency estimation are relevant for the final result.

9.3.1 Proper, Improper, and Non-Circular Signals

Circular symmetric complex signals are very useful to represent narrowband signals as in communication engineering, array processing, magnetic resonance imaging, and independent component analysis (ICA), and also in wave-equation based applications such as optics, acoustics, and geophysics.

Circular symmetry is quite common in many statistical signal processing problems, and this justifies the *rule of thumb* that inference methods be "extended" to complex-valued signals by replacing the following mapping for scalars, vectors, and symmetric matrixes:

$$x^2 \longrightarrow xx^* = |x|^2$$
$$\mathbf{x}^T \longrightarrow \mathbf{x}^H$$
$$\mathbf{X} = \mathbf{X}^T \longrightarrow \mathbf{X} = \mathbf{X}^H$$

However, it is good practice to validate the mapping above at least once by making the full derivation (see the example of frequency estimation above) as these properties are mostly based on the circular symmetry of the rvs.

As soon as we depart from the most common applications, complex valued signals might have unexpected properties that appear in the implementation as errors, with several "headaches" for beginners. This implies that the properties of complex variables need to be investigated in greater detail. Recall that a complex signal can be considered as a pairing of two real-valued signals with some predefined rules of the pairing.[2]

A complex rv is *proper* if it is uncorrelated with its complex conjugate, and it is *circular* if its pdf is invariant under rotation in the complex plane. When these properties are violated, we must handle improper and/or non-circular signals. A discussion of these is beyond the scope of this book, and the reader is referred to [15]. In is worthwhile to note that the aforementioned rules of thumb do not hold for improper and non-circular signals.

9.4 Time of Delay Estimation

The signal model of a delayed waveform in the AWGN model with power spectral density $N_0/2$ is

$$x(t) = \alpha_o g(t - \tau_o) + w(t) \qquad \text{for } t \in [0, T)$$

This is representative of a wide range of remote sensing applications (see Section 5.6) where the estimation of the time of delay (ToD) is informative on the position of the target. The ML estimation of the amplitude and delay follows from the log-likelihood function (Section 7.5)

$$\mathcal{L}(x(t)|\alpha, \tau) = \int_0^T (x(t) - \alpha g(t - \tau))^2 dt$$

---

2 Let $x(t) = x_r(t) + jx_i(t)$; any multiplication by a real-valued α, $\alpha x(t) = \alpha x_r(t) + j\alpha x_i(t)$, scales both components by the same value. But if a filter with real-valued response $h(t)$ should act on one component only of $x(t)$, say $x_r(t) * h(t) + jx_i(t)$, the complex representation would be an impairment in signal handling as $x(t) * h(t) \neq x_r(t) * h(t) + jx_i(t)$.

The minimization is quadratic wrt amplitude α and non-linear wrt the ToD τ. From the expansion

$$\mathcal{L}(x(t)|\alpha,\tau) = \int_0^T x^2(t)dt - 2\alpha\int_0^T x(t)g(t-\tau)dt + \alpha^2\int_0^T g^2(t-\tau)dt$$

the estimate of the amplitude, conditioned to the delay τ is

$$\hat{\alpha}(\tau) = \frac{\int_0^T x(t)g(t-\tau)dt}{\int_0^T g(t-\tau)^2 dt} \tag{9.2}$$

and it resembles the amplitude estimation in Section 7.5.1 except for the (unknown) ToD. After substitution, the log-likelihood wrt τ

$$\mathcal{L}(x(t)|\hat{\alpha}(\tau),\tau) = \int_0^T x(t)^2 dt - \frac{\left(\int_0^T x(t)g(t-\tau)dt\right)^2}{\int_0^T g(t-\tau)^2 dt}$$

is non-linear, but there are several commonly employed simplifications to address the ML estimate of the ToD in this context. If the waveform $s(t)$ has a limited support such as $g(t) = 0$ for $t \notin [t_1, t_2]$, and the observation window T fully contains the delayed waveform, the term $\int_0^T g^2(t-\tau)dt = \int_{-\infty}^{\infty} g^2(t-\tau)dt = E_g$ is the energy of the waveform, independent of the delay, and the ML becomes[3]

$$\hat{\tau} = \arg\max_{\tau\in[0,T]}\left\{\left(\int_0^T x(t)g(t-\tau)dt\right)^2\right\}$$

The MLE for the ToD is based on the maximization of the square of the cross-correlation (or equivalently its absolute value to account for an arbitrary amplitude α_o) between the measured signal and a replica of the waveform used to maximize its match with the observations. The estimate of the amplitude (9.2) is the value of the cross-correlation at the maximum, scaled by the waveform energy.

The CRB for ToD follows from the 2×2 FIM as

$$\text{cov}([\hat{\alpha},\hat{\tau}]^T) \succeq \begin{bmatrix} J_{\alpha\alpha} & J_{\alpha\tau} \\ J_{\tau\alpha} & J_{\tau\tau} \end{bmatrix}^{-1}$$

Each entry of the FIM can be evaluated in closed form from the definitions:

$$J_{\alpha\alpha} = \frac{2}{N_0}\int_0^T g(t-\tau_o)^2 dt = \frac{2}{N_0}\int_0^T g(t)^2 dt = \frac{2E_g}{N_0}$$

3 Even if independence of the energy from the delay is necessary to guarantee that the cross-correlation estimator coincides with the ML, the method is often adopted anyway without checking whether this condition is met or not. On the other hand, the choice of the position of the window T over the observations guarantees that the delayed waveform is fully included in the T-duration window with large benefits in terms of simplicity.

$$J_{\alpha\tau} = -\frac{2\alpha_o}{N_0}\int_0^T g(t-\tau_o)\dot{g}(t-\tau_o)dt = -\frac{2\alpha_o}{N_0}\int_0^T g(t)\dot{g}(t)dt = 0$$

$$J_{\tau\tau} = \frac{2\alpha_o^2}{N_0}\int_0^T \dot{g}(t-\tau_o)^2 dt = \frac{2\alpha_o^2}{N_0}\int_0^T \dot{g}(t)^2 dt = \frac{2\alpha_o^2}{N_0}\int_{-\infty}^{\infty} \omega^2 |G(\omega)|^2 d\omega/2\pi$$

where $\dot{g}(t)$ is the derivative of the waveform wrt time, and for simplicity the delayed waveform is fully contained within the observation window and thus the delay τ_o has no influence on the FIM entries. Note that amplitude and delay are decoupled as $J_{\alpha\tau} = 0$. Furthermore, Parseval equality eases the computation of the CRB for the ToD from $J_{\tau\tau}$ as

$$\text{var}[\hat{\tau}] \geq \frac{1}{2\beta_g^2\alpha_o^2} \cdot \frac{1}{E_g/N_0}$$

where

$$\beta_g^2 = \frac{\int_{-\infty}^{+\infty} \omega^2 |G(\omega)|^2 d\omega}{\int_{-\infty}^{\infty} |G(\omega)|^2 d\omega} \qquad [\text{rad}^2/\text{s}^2]$$

is the *effective bandwidth* of the waveform. Inspection of the Fourier transform of the waveform shows that the ToD accuracy largely depends on the high portion of the spectrum $|G(\omega)|^2$, and waveforms that have the same bandwidth but energy contents unbalanced toward higher frequencies can have smaller variance of ToD. Table 9.1 summarizes a few examples of the effective bandwidth in some practical cases.

The Ricker waveform is obtained as the second order derivative of the Gaussian-pulse (for $f_o \equiv 1/\pi\sqrt{2}T_g$) and it is routinely adopted in many remote sensing application due

Table 9.1 Waveforms and effective bandwidths.

Waveforms $|G(\omega)|^2$	β_g^2
Low-pass: $g(t) = \sin(2\pi f_g t)/\pi t$ $G(\omega) = 1$ for $|\omega| \leq 2\pi f_g$	$\frac{4}{3}\pi^2 f_g^2$
Gaussian-pulse: $g(t) = \exp(-\frac{1}{2}t^2/T_g^2)$ $G(f) = \frac{1}{f_o\sqrt{\pi}}\exp(-f^2/f_o^2)$	$2\pi^2 f_o^2$
Pass-band: $|G(\omega)|^2 = 1$ for $\omega \in [\omega_0 - \Delta\omega/2, \omega_0 + \Delta\omega/2]$	$\omega_0^2 + \frac{\Delta\omega^2}{12}$
Ricker-pulse (2nd derivative of Gaussian-pulse): $g(t) = [1 - (t/T_g)^2]\exp(-\frac{1}{2}t^2/T_g^2)$ $G(f) = (\frac{2}{\sqrt{\pi}})\frac{f^2}{f_o^3}\exp(-f^2/f_o^2)$	$5\pi^2 f_o^2$

to the lack of any zero-frequency component that is intentionally or occasionally (e.g., as result of propagation effects) chosen this way. It can be seen that the effective bandwidth for the Ricker waveform is $5/2 = 2.5$ times higher than a Gaussian-pulse for the same duration T_g, and the variance of ToD is reduced accordingly with remarkable benefits in ToD accuracy and delay resolution.

Remark: When estimating the ToD, one attempts to set the observation window centered around the most likely delay. The origin is set at the center of the observation window that ranges within $\pm T/2$; in this case the upper limit of the variance is when the noise is large and $\hat{\tau} \sim \mathcal{U}(-T/2, T/2)$ so that $\text{var}[\hat{\tau}] \simeq T^2/12$.

Example of ToD for Two Overlapping Echoes with Triangular Waveform

Considering the ToD for one waveform is helpful to gain insight on ToD and set the ideal bounds on the performance limits for complex systems in Chapter 20; the case of the signal sum of two waveforms (say two echoes to mimic a remote sensing experiment) is relevant to evaluate the influence of mutual waveform interactions. The model of two echoes is

$$x(t) = \alpha_1 g(t - \tau_1) + \alpha_2 g(t - \tau_2) + w(t)$$

The amplitudes are $\alpha_1 = \alpha_2 = \alpha_o$ and are assumed known, but not the delays, which are $\tau_2 \neq \tau_1$. The waveform is triangular

$$g(t) = \begin{cases} (1 - 2|t|/T_g) & \text{for } |t| < T_g/2 \\ 0 & \text{elsewhere} \end{cases}$$

and the echoes are overlapped (and thus interfering) for $\Delta\tau = |\tau_2 - \tau_1| \leq T_g$, or equivalently when the overlapping-fraction

$$\eta = 1 - \frac{\Delta\tau}{T_g}$$

is $0 \leq \eta < 1$. The MLE becomes

$$(\hat{\tau}_1, \hat{\tau}_2) = \arg \min_{(\xi_1,\xi_2)\in[0,T]\times[0,T]} \left\{ \int_0^T (x(t) - \alpha_o g(t - \xi_1) - \alpha_o g(t - \xi_2))^2 dt \right\}$$

Neglecting all terms not influenced by the delays, the optimization kernel is

$$\psi(\xi_1, \xi_2) = \alpha_o \phi_{xg}(\xi_1) + \alpha_o \phi_{xg}(\xi_2) - 2\alpha_o^2 \phi_{gg}(\xi_1 - \xi_2)$$

to be maximized wrt the pair (ξ_1, ξ_2). It is clear that the ToD estimates are symmetric wrt the choice of ξ_1, ξ_2 as expected, and the optimization is by searching over the 2D combination (ξ_1, ξ_2) that mixes cross-correlations $\phi_{xg}(\xi)$ that estimate the ToD for the two echoes as isolated, and a correcting term $\phi_{gg}(\xi_1 - \xi_2)$ that depends only on the waveform and accounts for their mutual interference (i.e., $\phi_{gg}(\tau_1 - \tau_2) = 0$ for $\eta = 0$). Even if the correcting term $\phi_{gg}(\xi_1 - \xi_2)$ can be neglected by judicious data-windowing around each echo, the search for maxima of $\psi(\xi_1, \xi_2)$ is bidimensional in the domain $[0, T] \times [0, T]$.

The CRB depends on the terms

$$[\mathbf{J}(\theta_o)]_{\tau_1\tau_2} = \frac{2\alpha_o^2}{N_0}\int_0^T \frac{\partial g(t-\tau_1)}{\partial \tau_1}\frac{\partial g(t-\tau_2)}{\partial \tau_2}dt = \frac{2\alpha_o^2}{N_0}\int_0^T \dot{g}(t-\tau_1)\dot{g}(t-\tau_2)dt$$
$$= \dots = \frac{\alpha_o^2 E_g}{N_0/2}\frac{3}{T_g^2}\eta$$

where the computation is specialized here for the triangle waveform and its derivative $\dot{g}(t)$. The other entries of the FIM

$$[\mathbf{J}(\theta_o)]_{\tau_1\tau_1} = [\mathbf{J}(\theta_o)]_{\tau_2\tau_2} = \frac{\alpha_o^2 E_g}{N_0/2}\frac{3}{T_g^2}$$

can be easily derived as a degenerate case of FIM for full-overlapping ($\tau_1 = \tau_2$ or equivalently $\eta = 1$). The CRB follows from the FIM as

$$\mathbf{J} = \frac{\alpha_o^2 E_g}{N_0/2}\frac{3}{T_g^2}\begin{bmatrix} 1 & \eta \\ \eta & 1 \end{bmatrix} \Rightarrow \text{cov}([\hat{\tau}_1, \hat{\tau}_2]^T) \succeq \frac{T_g^2}{3\times SNR}\frac{1}{1-\eta^2}\begin{bmatrix} 1 & -\eta \\ -\eta & 1 \end{bmatrix}$$

and

$$\text{var}(\hat{\tau}_{1,2}) \geq \frac{T_g^2}{3\times SNR}\frac{1}{1-\eta^2}$$

for $SNR = 2\alpha_o^2 E_g/N_0$. The CRB depends on the mutual delays in terms of overlapping: when $\eta \neq 0$, the estimates are coupled, as expected from the mutual interference of the two waveforms, and the CRB is not diagonal. In addition, the interference degrades the variance as for $\eta < 1$ the CRB increases $1/(1-\eta^2) > 1$, and diverges for $\eta \to 1$ as in this case the two waveforms become indistinguishable.

9.5 Estimation of Max for Uniform pdf

Let

$$x[n] \sim \mathcal{U}(0, \theta)$$

be the sample from a uniform distribution. The objective is to estimate the maximum value θ of the uniform pdf from N independent observations. From definition using the step function $u(x) = \int_{-\infty}^{x} \delta(\xi)d\xi$:

$$p(x[n]|\theta) = \frac{1}{\theta}(u(x[n]) - u(x[n]-\theta))$$

The joint pdf for N independent rvs $\mathbf{x} = [x[0], x[1], ..., x[N-1]]^T$ is

$$p(\mathbf{x}|\theta) = \begin{cases} \frac{1}{\theta^N} & \text{for } x[n] \in [0, \theta], \forall n = 0, 1, ..., N-1 \\ 0 & \text{for } x[n] \notin [0, \theta], \forall n = 0, 1, ..., N-1 \end{cases}$$

From the N observations, the following relationships hold:

$$\begin{aligned} x_{\max} &= \max\{\mathbf{x}\} \leq \theta \\ x_{\min} &= \min\{\mathbf{x}\} \geq 0 \end{aligned}$$

and thus the pdf is upper and lower limited by the values of the observations. Recalling the properties of order statistics in Section 3.9, for statistically independent rvs,

$$\Pr(x[0] \leq \xi, x[1] \leq \xi, ..., x[N-1] \leq \xi) = F^N(\xi)$$

where

$$F(\xi) = \Pr(x[n] \leq \xi) = \begin{cases} 0 & \text{per } \xi \leq 0 \\ \xi/\theta & \text{per } 0 \leq \xi \leq \theta \\ 1 & \text{per } \xi \geq \theta \end{cases}$$

The pdf of the max parametric wrt θ is

$$p(\xi|\theta) = \frac{d}{d\xi} F(\xi)^N = \frac{N}{\theta^N} \xi^{N-1} \quad \text{per } 0 \leq \xi \leq \theta$$

and this is maximized for $\xi = x_{\max}$. Furthermore, the estimation of the upper value of the uniform pdf is

$$\hat{\theta} = x_{\max}$$

and this is the ML estimation of the upper value of the uniform pdf from N observations.

Bias and variance of the estimator can be evaluated from the pdf $p(\xi|\theta)$. Bias is

$$\mathbb{E}[\hat{\theta}] = \mathbb{E}[x_{\max}] = \frac{N}{\theta^N} \int_0^\theta x_{\max}^N dx_{\max} = \frac{N}{N+1} \theta$$

thus the estimator is biased. The bias can be removed by choosing as new estimator

$$\tilde{\theta} = \frac{N+1}{N} x_{\max}$$

in place of $\hat{\theta} = x_{\max}$. The variance is

$$\text{var}[\tilde{\theta}] = \frac{N}{\theta^N} \int_0^\theta \left(\frac{N+1}{N} x_{\max} - \theta \right)^2 x_{\max}^{N-1} dx_{\max} = \theta^2 \frac{1}{N(N+2)}$$

9.6 Estimation of Occurrence Probability for Binary pdf

Let the independent observations be drawn from a binary rv $x[n] \in \{0,1\}$ so that

$$p(x[n]=1)=p_o$$
$$p(x[n]=0)=1-p$$

The pdf of the ensemble $\mathbf{x}$ with K values $x[n]=1$ (regardless of the order and positions) follows the Bernulli pdf

$$\Pr(\mathbf{x} \text{ with } K \text{ values } x[n]=1)=p_B(K)=\binom{N}{K}p_o^K(1-p_o)^{N-K}$$

The estimation of the probability p_o in an experiment of N observations $\mathbf{x}$ is based on counting the number of outcomes +1. Let $\bar{K}$ be the number of $x[n]=+1$, the pdf of the ensemble of outcomes is parametric wrt p:

$$p_B(\bar{K}|p)=\binom{N}{\bar{K}}p^{\bar{K}}(1-p)^{N-\bar{K}}$$

This is the likelihood function for the parameter p depending on data $\mathbf{x}$. The ML estimation of the probability p_o follows from the maximization wrt p of

$$\mathcal{L}(\mathbf{x},\bar{K}|p)=\ln\binom{N}{\bar{K}}+\bar{K}\ln p+(N-\bar{K})\ln(1-p)$$

it is

$$\hat{p}=\frac{\bar{K}}{N}.$$

Bias and variance can be evaluated from their definitions:

$$\mathbb{E}[\hat{p}]=\frac{1}{N}\mathbb{E}_k\left[\bar{K}\right]=\sum_{\bar{K}=0}^{N}\frac{\bar{K}}{N}\binom{N}{\bar{K}}p_o^{\bar{K}}(1-p_o)^{N-\bar{K}}=p_o$$

hence the estimator is unbiased. Variance is bounded by the CRB (as unbiased). The Fisher information term is

$$\begin{aligned}
J(p_o)&=\mathbb{E}_k\left[\left(\frac{\bar{K}}{p_o}-\frac{N-\bar{K}}{1-p_o}\right)^2\right]=\mathbb{E}_k\left[\left(\frac{\bar{K}}{p_o(1-p_o)}-\frac{N}{1-p_o}\right)^2\right]\\
&=\frac{1}{p_o^2(1-p_o)^2}\mathbb{E}_k\left[\bar{K}^2\right]+\frac{N^2}{(1-p_o)^2}-\frac{2N}{p(1-p_o)^2}\mathbb{E}_k\left[\bar{K}\right]\\
&=\frac{p_oN\left(Np_o-p_o+1\right)}{p_o^2(1-p_o)^2}+\frac{N^2}{(1-p_o)^2}-\frac{2N^2p_o}{p_o(1-p_o)^2}=\frac{N}{(1-p_o)p_o}
\end{aligned}$$

To summarize:

$$\text{var}(\hat{p}) = \text{var}(\bar{K}/N) \geq \frac{p_o(1-p_o)}{N}$$

The CRB depends on the probability, and the largest value is when $p_o = 1/2$; this condition corresponds to the maximum uncertainty.

Simulation Parameters for Unlikely Events: Sample Size Definition

In some applications it is required to infer properties of events with small occurrence probabilities, and the definition of the number of observations (or *sample size*) N depends on the confidence required. Applications are covered in Chapter 23 and include the estimation of error probability in digital communications, detection/missing probability in detection theory, and error classification probability in clustering methods. These phenomena can be modeled as binary with probability of the unlikely event being p_o, with $p_o \ll 1$. The sample size N is the free parameter that should be designed to comply with the constraint that the probability of this unlikely event is estimated with due accuracy. Since these unlikely events are described on a logarithmic scale to highlight small deviations, the *normalized dispersion* (or coefficient of variation), that is the ratio of the standard deviation to the mean value, is more meaningful than just the variance. The MLE of the probability p is obtained by counting the positive hits over the total N as shown above. The normalized dispersion of the estimate is

$$\text{var}(\hat{p}/p_o) \geq \frac{1}{p_o N}$$

To exemplify, the sample size required in order to have an estimate of a probability with a minimum normalized dispersion smaller than 10% is $N \geq \sqrt{10}/p_o$, that is a sample size, at least 3 times $1/p_o$.

9.7 How to Optimize Histograms?

The histogram is recurrent in everyday life to estimate the pdf (or pmf) of phenomena by evaluating the frequency of the observations within a set of bins (i.e., non-overlapping intervals). The method to compute a histogram is quite simple: the range of values is divided into bins and the histogram follows from counting the hits in every bin. Bin-width selection (sometime referred to as *smoothing parameter*) drives the histogram as large bins unavoidably yield a biased estimate of the pdf, while small bins have fine granularity (and greater detail) but might be impaired by the small fraction of hits per bin. The focus here is to gain insight into histogram design; a comprehensive discussion can be found in [41].

From Figure 9.2, let $B_k = [t_k, t_{k+1})$ be the kth bin for an rv with pdf $p(x)$; the number of hits that fall within B_k is ν_k and for N independent samples the bin count is binomial (or Bernoulli):

$$p_B(\nu_k|p_k) = \binom{N}{\nu_k} p_k^{\nu_k}(1-p_k)^{N-\nu_k}$$

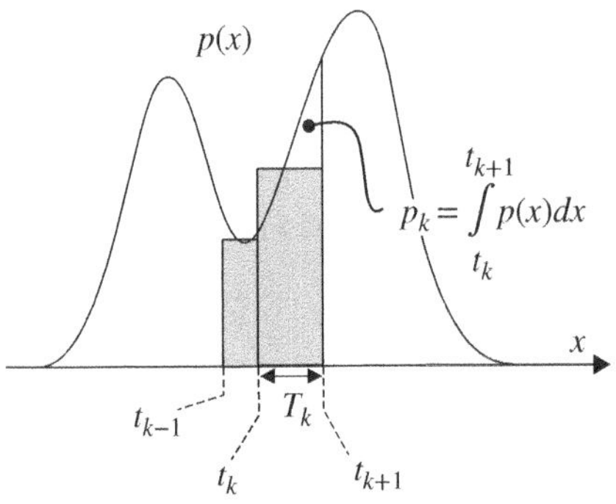

Figure 9.2 Non-uniform binning.

with probability of the kth bin

$$p_k = \Pr[x \in B_k] = \int_{B_k} p(x)dx$$

From the results in Section 9.6,

$$\mathbb{E}[\nu_k] = Np_k$$

$$\text{var}[\nu_k] = Np_k(1-p_k)$$

The histogram provides an estimate of the staircase approximation of $p(x)$ as

$$p_s(x) = \sum_{k=1}^{N} \frac{p_k}{T_k} rect\left(\frac{x-t_k}{T_k}\right)$$

that is unavoidably distorted as the bin-width $T_k = t_{k+1} - t_k$ sets the granularity of the approximating pdf. The histogram is approximating the probability by

$$\hat{p}_k = \frac{\nu_k}{N}$$

which is unbiased, with variance for the estimate of the staircase approximation $p_s(x)$ of

$$\text{var}[\hat{p}_k/T_k] = \frac{p_k(1-p_k)}{NT_k^2} \leq \frac{p_k}{NT_k^2}$$

The estimate of the approximating pdf $p_s(x)$ is

$$\hat{p}_s(x) = \sum_{k=1}^{N} \frac{\hat{p}_k}{T_k} rect\left(\frac{x-t_k}{T_k}\right)$$

The variance of the estimated pdf

$$\text{var}[\hat{p}_s(x \in B_k)] \simeq \frac{p_s(x)}{NT_k} \tag{9.3}$$

depends on the pdf value and the inverse of the bin-width T_k, so that small bin-widths are severely impaired by large uncertainties.

Uniform bin-width: The histogram with a uniform bin-width $T = t_{k+1} - t_k$ is the most common choice, and the estimate of the pdf from the histogram is the same as above:

$$\hat{p}_s(x) = \frac{1}{T}\sum_{k=1}^{N} \hat{p}_k rect\left(\frac{x - t_k}{T}\right)$$

The bins with smaller probability are those with larger fluctuations and this is annoying when visualizing the histogram's lower probability bins such as the tails of the pdf. This effect due to the paucity of the data can be appreciated from the normalized variance

$$\text{var}[\hat{p}_k/p_k] = \frac{1-p_k}{p_k NT^2} \simeq \frac{1}{p_k NT^2}$$

that shows the dependency wrt $1/p_k$.

Optimized bin-width: Bin-width can be optimized and this needs to account for a metric for pdf estimation. For a large enough number of bins, the probability is

$$p_k \simeq p(x_k)T_k$$

where $x_k = t_k + T_k/2$ is the center-bin. The error wrt the center-bin is $\delta x_k = x - x_k | x \in B_k$; this error is uniform in B_k and its spread value can be approximated by the first order derivative of pdf $\dot{p}(x_k)$, which is: $\delta x_k \sim \mathcal{U}(-\dot{p}(x_k)T_k/2, \dot{p}(x_k)T_k/2)$. The histogram can be regarded as an estimate of the staircase approximation of the pdf, the mean square distortion of the staircase approximation term depends on δx_k and is $(\dot{p}(x_k)T_k)^2/12$ for the kth bin. Introducing now the finite sample size N, the MSE of the kth bin is the sum of the variance and the distortion due to the staircase approximation (9.3):

$$MSE_k = \frac{T_k^2}{12}\dot{p}(x_k)^2 + \frac{p(x_k)}{NT_k}$$

The minimization of the MSE wrt the bin-width yields the optimum bin-width

$$T_k^{(opt)} = \left[\frac{6p(x_k)}{N\dot{p}(x_k)^2}\right]^{1/3}$$

that depends on the center-bin pdf $p(x_k)$ and its local variability from $\dot{p}(x_k)$. A flat pdf has small $\dot{p}(x_k)$ and the bin-width $T_k^{(opt)}$ becomes large, and vice-versa. This adaptive bin-spacing can better capture the details of the pdf by densifying the staircase where the pdf needs more detail, as for steep changes. Of course, this adaptive bin-spacing can be used whenever the pdf is known, even if approximately, and further refinement becomes necessary. Alternatively, optimized bin-width follows as a second step after a uniform bin-spacing histogram.

9.8 Logistic Regression

In applied statistics for data analysis, the occurrence probability of a binary event depends on another set of parameters that rule the experiments. More specifically, in the binary choice

$$Pr(u=1)=p$$
$$Pr(u=0)=1-p$$

the event $u=1$ denotes a special outcome that, to exemplify, could model disease in a large population of people, or the success of a special medical treatment, or death in other settings. The probability p depends on another set of known parameters $\mathbf{h}\in\mathbb{R}^n$ such as (still with the randomized medical trials) age, gender, geographical position, dose of a medication or radiation, or any other that could follow from an analysis of the sample values of the experiments by clustering methods (Section 23.6). The *logistic model* has the following relationship for the probability:

$$p=p(\mathbf{h};\mathbf{a},b)=\frac{\exp(\mathbf{a}^T\mathbf{h}+b)}{1+\exp(\mathbf{a}^T\mathbf{h}+b)}$$

that linearly combines the variables $\mathbf{h}$. In other cases, it is more convenient to have the logistic regression model from the log-probability ratio

$$\ln\left(\frac{Pr(u=1)}{Pr(u=0)}\right)=\ln\left(\frac{p}{1-p}\right)=\mathbf{a}^T\mathbf{h}+b$$

as being linear wrt $\mathbf{h}$. The problem at hand is to estimate the parameters of the logistic regression $(\mathbf{a},b)$ given the outcomes from a set of observations all related to the specific experiment (data-selection/classification is another part of the experiment not accounted for in the statistical inference problem considered herein).

Let a population $\mathcal{X}$ be divided into N homogeneous subsets $\mathcal{X}_1,\mathcal{X}_2,...,\mathcal{X}_N$, where the samples of the i-th population $x_i=|\mathcal{X}_i|$ are all all compliant with the set of parameters $\mathbf{h}_i$; the number of positive outcomes (corresponding to $u=1$) of the experiment are y_i, and the pdf for the binary test is still Bernoulli:

$$p(y_i|p_i)=\binom{x_i}{y_i}p_i^{y_i}(1-p_i)^{x_i-y_i}$$

with probabilities $p_i=p(\mathbf{h}_i;\mathbf{a},b)$ that are ruled by the logistic regression model

$$\ln\left(\frac{p_i}{1-p_i}\right)=\mathbf{a}^T\mathbf{h}_i+b$$

When varying the parameters $\mathbf{h}_i$, there are multiple outcomes $\{y_i\}_{i=1}^N$ of the populations $\{x_i\}_{i=1}^N$, all linearly dependent on the model parameters $(\mathbf{a}, b)$. The likelihood for multiple independent populations is

$$p(\mathbf{y}|\mathbf{a}, b) = \prod_{i=1}^{N} \binom{x_i}{y_i} p(\mathbf{h}_i; \mathbf{a}, b)^{y_i} (1 - p(\mathbf{h}_i; \mathbf{a}, b))^{x_i - y_i}$$

which can be maximized wrt $(\mathbf{a}, b)$. The estimate of the logistic regression parameters reduces to a non-linear optimization that needs to be solved by numerical methods.

10

Numerical Analysis and Montecarlo Simulations

Estimators are compared according to their performance, and numerical simulations are often the preferred method to better evaluate that performance in the range of interest, in addition to CRB or any sensitivity analysis in the neighborhood of the true solution. The benefits of performance analysis by simulation are several, such as:

- analysis of the accuracy of the estimation method compared to any analytical bound (e.g., comparison with CRB);
- evaluation of accuracy when the model is non-linear and threshold effects might appear in the estimator due to the extrema-search method (typical for small or large noise);
- analysis of robustness for any deviation from model assumptions;
- analysis of implementation method and any sub-optimal choice (e.g., signal quantization).

Montecarlo simulations are based on three main building blocks that are encoded here in MATLAB®(Figure 10.1): signal generation, estimation, and performance analysis. The signal is generated by simulating the generation mechanism with a random number generator that has a pdf that is as the one specified by the model itself (or any deviation, if this is the aspect of interest):

$$\mathbf{x}_k = \mathbf{s}_k(\boldsymbol{\theta}) + \mathbf{w}_k$$

For the kth realization of the signal, the set of parameters of interest

$$\hat{\boldsymbol{\theta}}_k = \hat{\boldsymbol{\theta}}(\mathbf{x}_k)$$

is estimated as part of the estimation step, and the performance is evaluated from the ensemble of a large number of iterations, say K, of *independent* signal generations. Analysis of the error

$$\delta\hat{\boldsymbol{\theta}}_k = \hat{\boldsymbol{\theta}}_k - \boldsymbol{\theta}$$

for the ensemble of K iterations provides useful insight into the estimation error for the signal generation model adopted. The scatter-plot, histogram, and sample moments of errors are the widely adopted performance metrics. The scatter-plot and histogram

Statistical Signal Processing in Engineering, First Edition. Umberto Spagnolini.

Companion website: www.wiley.com/go/spagnolini/signalprocessing

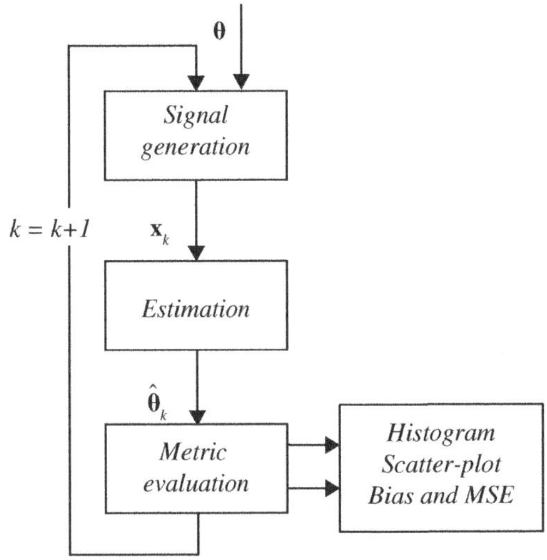

Figure 10.1 Typical Montecarlo simulations.

highlight any asymmetric behavior of the estimator that can be helpful for its refinement or optimization. The routinely employed moment metrics are bias and MSE:

$$\mathbf{b}_K(\theta) = \frac{1}{K}\sum_{k=1}^{K}\hat{\theta}_k - \theta$$

$$\mathbf{MSE}_K(\theta) = \frac{1}{K}\sum_{k=1}^{K}\begin{bmatrix} |[\hat{\theta}_k - \theta]_1|^2 \\ \vdots \\ |[\hat{\theta}_k - \theta]_p|^2 \end{bmatrix} = \frac{1}{K}\sum_{k=1}^{K}|\hat{\theta}_k - \theta|^2$$

that for

$$\lim_{K\to\infty}\mathbf{b}_K(\theta) = \mathbf{b}_\infty(\theta)$$

$$\lim_{K\to\infty}\mathbf{MSE}_K(\theta) = \mathbf{MSE}_\infty(\theta)$$

converge to the bias and MSE of the effective estimator employed so far. Analysis of the most appropriate choice of K is a problem in itself and it depends on the specific estimator and the confidence of the sample moments (a discussion is beyond the scope here). In practice, the choice of K needs care as small K is beneficial for fast simulation runs, but the bias and MSE metrics could be scattered by the specific signal realization. To get more reliable outcomes, further randomization is necessary and the number of independent signal generations K should be made larger. The efficient use of simulations for estimator tuning and optimization needs experience, and in the early stages it is better to have quick but unreliable results to tune the algorithm and thus one chooses K small. As a rule of thumb, K is in the order of few tens (up to hundreds) in the early stages and thousands for the final simulations to better simulate the asymptotic behavior of $\mathbf{b}_\infty(\theta)$ and $\mathbf{MSE}_\infty(\theta)$. In some cases, $\mathbf{MSE}_K(\theta)$ (for K large enough) is the outcome metric and it is compared with the CRB. Recall that

$$MSE_K(\theta(\ell)) = \text{var}_K(\theta(\ell)) + |b_K(\theta(\ell))|^2$$

so that

$$\lim_{K\to\infty} \text{var}_K(\theta(\ell)) = \text{var}_\infty(\theta(\ell)) \geq [\mathbf{C}_{CRB}]_{\ell\ell}$$

and thus any deviation from CRB could be due to bias, or variance, or both.

Performance can depend on the specific choice $\boldsymbol{\theta}$ and further analysis could explore those parameter choices that make the performance metric heavily deviate from the expectations. This means that a further simulation loop could explore different choices $\theta = \theta_\ell$ for $\ell = 1, 2, \ldots, L$, either being chosen to be of specific interest, or alternatively randomized.

10.1 System Identification and Channel Estimation

Notice that software implementation is an essential step in simulation, and code-architecture should be efficient and compact as iterations are repeated at least K times—even more if bias and MSE needs to be evaluated vs. another design parameter (e.g., signal to noise ratio, or the number of samples N, or a set of parameters, or any other variable). A common mistake for beginners is to write the simulation code without any concern about algorithm efficiency, or software/hardware capability, by exactly duplicating the estimator structure into the algorithm implementation as a simple mapping into software.[1]

---

1 As an example, assume the task is to segment the data $u(1), u(2), \ldots$ and evaluate the weighted sum as in filtering

$$x(k) = \sum_{i=-N}^{M} h(i)u(k-i)$$

(Bad) Matlab code that duplicates the sum could be

```
x(k)=0;
for i=-N:M
x(k)=x(k)+h(i)*u(k-i);
end
```

but the following Matlab code is far more efficient

```
x(k)=h'*u(k-N:k+M);
```

where h is the M+N+1 column vector collecting the entries of weights in the right order. The second encoding takes advantage of the Matlab kernel to efficiently employ matrix-to-matrix multiplication, and it is surely more compact (and elegant).

Let $u[1], u[2], ..., u[N_u]$ be a set of samples of a white Gaussian process with power σ_u^2; these are filtered by an unknown causal filter $\mathbf{h}$ with range $[0, N_h]$. The output

$$\mathbf{x} = \mathbf{h} * \mathbf{u} + \mathbf{w}$$

is noisy with $\mathbf{w} \sim \mathcal{N}(0, \sigma_w^2 \mathbf{I})$. This is a system identification problem: from measurement of the output $\mathbf{x}$ and knowledge of the input $\mathbf{u}$, the unknown filter response $\mathbf{h}$ can be estimated. The problem is linear and it is a good starting point, being quite simple. In performance analysis by simulation, the filter is given as

$$h[i] = \begin{cases} 1 - i/N_h & \text{for } i \in [0, N_h] \\ 0 & \text{for } i \notin [0, N_h] \end{cases}$$

The objective is to evaluate the MSE vs. the ratio σ_u^2/σ_w^2 for the estimation $\hat{\mathbf{h}}$ of the filter response $\mathbf{h}$, and compare the results with the corresponding CRB.

Before encoding the estimator and the CRB in Matlab code, it is important to review the basic theory with the software coding in mind. Convolution is more efficiently encoded in matrix-to-matrix multiplication, so the convolution becomes

$$\begin{bmatrix} x[1] \\ x[2] \\ \vdots \\ x[N_x] \end{bmatrix} = \underbrace{\begin{bmatrix} u[1] & 0 & \cdots & 0 \\ u[2] & u[1] & & \vdots \\ \vdots & \vdots & \ddots & 0 \\ u[N_u] & u[N_u - 1] & & u[1] \\ 0 & u[N_u] & & u[2] \\ 0 & 0 & \ddots & \vdots \\ \vdots & \vdots & \cdots & u[N_u] \end{bmatrix}}_{\mathbf{U}} \underbrace{\begin{bmatrix} h[1] \\ h[2] \\ \vdots \\ h[N_h] \end{bmatrix}}_{\mathbf{h}} + \begin{bmatrix} w[1] \\ w[2] \\ \vdots \\ w[N_x] \end{bmatrix}$$

Since the $N_x \times N_h$ matrix

$$\mathbf{U} = [\mathbf{u}_1, \mathbf{u}_2, ..., \mathbf{u}_{N_h}]$$

each column is the copy of the $N_u \times 1$ vector $\mathbf{u} = [u[1], u[2], ..., u[N_u]]^T$ onto portions of each column. In Matlab, this is

```
for j=1:Nh,
U(j:j+Nu-1,j)=u;
end
```

which completely fills the convolution matrix `U` with a `for`-loop over the smaller dimension (as $N_h < N_u$). The ML estimator is

$$\hat{\mathbf{h}} = (\mathbf{U}^T \mathbf{U})^{-1} \mathbf{U}^T \mathbf{x}$$

where

$$\mathbf{U}^T\mathbf{x} = \begin{bmatrix} \mathbf{u}_1^T\mathbf{x} \\ \mathbf{u}_2^T\mathbf{x} \\ \vdots \\ \mathbf{u}_{N_h}^T\mathbf{x} \end{bmatrix}$$

is the projection of the data onto the columns of $\mathbf{U}$. The CRB is

$$\mathbf{C}_{CRB} = \sigma_w^2(\mathbf{U}^T\mathbf{U})^{-1} \tag{10.1}$$

The covariance of the ML estimator coincides with the CRB as the observations are Gaussian but this statement can be checked by simulation. Careful inspection of the $N_h \times N_h$ matrix $\mathbf{U}^T\mathbf{U}$ as kernel of both estimator and CRB shows that

$$\mathbf{U}^T\mathbf{U} = \begin{bmatrix} \mathbf{u}_1^T\mathbf{u}_1 & \mathbf{u}_1^T\mathbf{u}_2 & \cdots & \mathbf{u}_1^T\mathbf{u}_{N_h} \\ \mathbf{u}_2^T\mathbf{u}_1 & \mathbf{u}_2^T\mathbf{u}_2 & \cdots & \mathbf{u}_2^T\mathbf{u}_{N_h} \\ \vdots & \vdots & \ddots & \vdots \\ \mathbf{u}_{N_h}^T\mathbf{u}_1 & \mathbf{u}_{N_h}^T\mathbf{u}_2 & \cdots & \mathbf{u}_{N_h}^T\mathbf{u}_{N_h} \end{bmatrix}$$

where each entry

$$[\mathbf{U}^T\mathbf{U}]_{ij} = \mathbf{u}_i^T\mathbf{u}_j = \sum_n u[n-i]u[n-j] = N_u \times \hat{r}_u(j-i)$$

where

$$\hat{r}_u[j-i] = \frac{1}{N_u}\sum_n u[n-i]u[n-j]$$

is the sample autocorrelation at leg $j-i$. Note that for $N_u \longrightarrow \infty$:

$$\hat{r}_u[j-i] \to r_u[j-i] = \sigma_u^2\delta[j-i]$$

as processes are WSS, and

$$\mathbf{U}^T\mathbf{U} \to N_u\sigma_u^2\mathbf{I}$$

for N_u large enough. The CRB (10.1) depends on the choice of the input sequence and it is the conditional CRB as it depends on the specific realization of $\mathbf{U}$. One might want the CRB independent on the specific realization of $\mathbf{U}$ and get $\mathbb{E}_\mathbf{U}[\mathbf{C}_{CRB}]$, but this is difficult to compute. A manageable bound is the asymptotic CRB (see also Section 7.2.3 and (7.3)) as

$$\mathbf{C}_{CRB}^a = \sigma_w^2(\mathbb{E}_\mathbf{U}[\mathbf{U}\mathbf{U}^\mathrm{T}])^{-1} = \frac{\sigma_w^2}{N_u\sigma_u^2}\mathbf{I} \preceq \mathbb{E}_\mathbf{U}[\mathbf{C}_{CRB}] \tag{10.2}$$

where the last inequality follows from Jensen's inequality (Section 3.2). Namely, the asymptotic $\mathbf{C}_{CRB}^a$ is lower than $\mathbb{E}_\mathbf{U}[\mathbf{C}_{CRB}]$ except for large N_u.

10.1.1 Matlab Code and Results

The Matlab code is the following:

```
Nh=10; Nu=40;
K=800;  % Number iterations
Nx=Nu+Nh-1;
SNR=[-10:20]; % SNR in dB
h=1-[0:Nh-1]'/Nh; % filter
U=zeros(Nx,Nh);
MSE=zeros(size(SNR));
for iSNR=1:length(SNR),
  u=10^(SNR(iSNR)/20)*randn(Nu,1);
  for j=1:Nh,
    U(j:j+Nu-1,j)=u;
  end
  invUU=inv(U'*U);
  for iter=1:K,
    x=U*h+randn(Nx,1);   % Signal generation
    hest=invUU*(U'*x);   % Estimation
    err=hest-h;  % Metric evaluation (MSE)
    MSE(iSNR)=MSE(iSNR)+(err'*err)/(K*Nh);
  end
end
CRB=10.^(-SNR/10)/Nu;   % asymptotic CRB (for Nu large)
semilogy(SNR,CRB,'-.',SNR,MSE,'.')
xlabel('SNR [dB]'); ylabel('MSE channel estimate')
title('MSE vs SNR')
```

Figure 10.2 shows the simulation results in terms of a scatter-plot of the MSE versus signal to noise ratio, with a logarithmic scale as this is more useful to highlight results when the dynamic range is large as in this case. The asymptotic CRB in logarithmic scale becomes linear as in the figure. When increasing `Nu`, the MSE are less dispersed around $\mathbf{C}^a_{CRB}$ (10.2) as should be the case; iterations are `K=800` as this is large enough to capture the effective performance of the estimator that follows the CRB (dashed line) as expected.

10.2 Frequency Estimation

Analysis of the performance of a frequency estimator is a good example of non-linear estimator where simulation is used to tune the implementation. Given the model $x[n] = A_o\cos(\omega_o n + \phi_o) + w[n]$ for $w[n] \sim N(0, \sigma_w^2)$, frequency estimation of one sinusoid in noise is from the peak of the Fourier transform $X(\omega)$ (Section 9.2, [44]). The search for the frequency value that peaks $|X(\omega)|^2/2$ involves an optimization step as the maximum

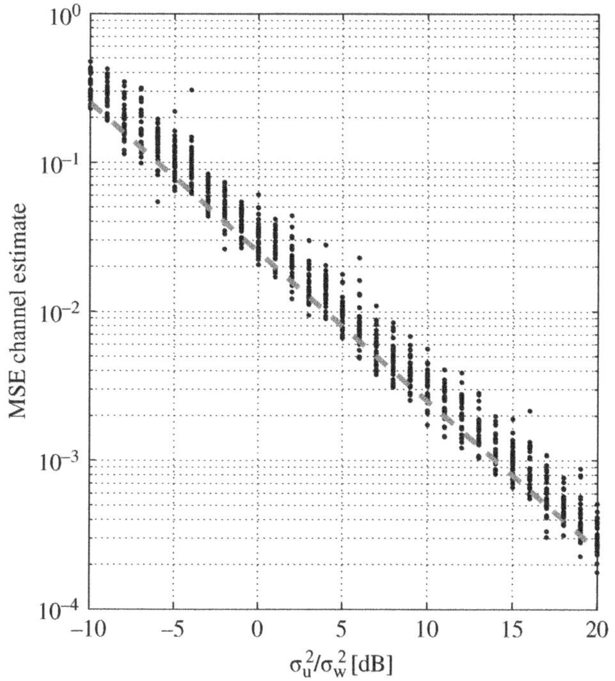

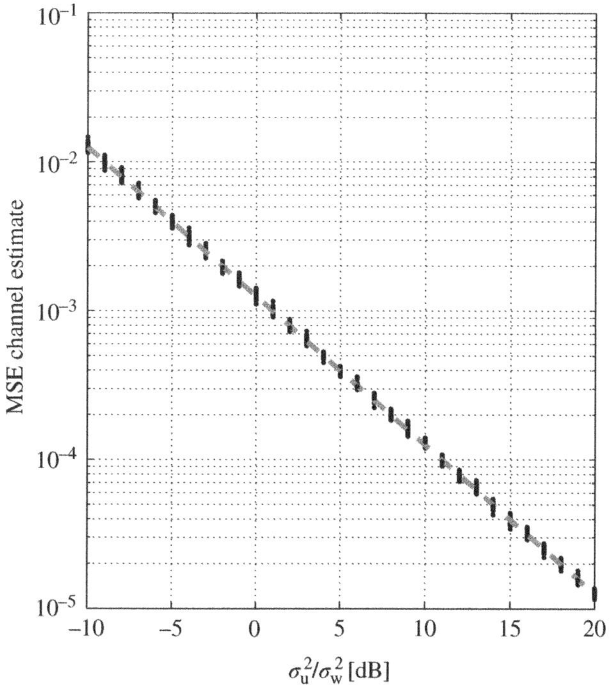

Figure 10.2 MSE vs σ_u^2/σ_w^2 for $N_u = 40$ (upper) and $N_u = 800$ (lower) for `K=800` runs (dots) of Montecarlo simulations. Asymptotic CRB (10.1) is indicated by dashed line.

can be searched for by several methods, each involving a different implementation step. These procedures are now illustrated using the following definition:

$$S(\omega) = \frac{|X(\omega)|^2}{N}$$

just to ease the notation (since the frequency that peaks $|X(\omega)|^2/N$ is the same as $|X(\omega)|$, the latter could be considered if necessary).

Given the $N \times 1$ vector $\mathbf{x}_k = \mathbf{s}(A_o, \omega_o, \phi_k) + \mathbf{w}_k$, where phase $\phi_k \sim \mathcal{U}(0, 2\pi)$ has been chosen as random to account for the asynchronicity of the time window, the computation of the metric

$$S_k(\omega) = \frac{|X_k(\omega)|^2}{N}$$

depends on the realization and the estimate is

$$\hat{\omega}_k = \arg\max_{\omega} S_k(\omega) \quad (10.3)$$

Implementation of the peak-search depends on the specific approach. To simplify the notation, the iteration subscript $k = 1, 2, ..., K$ is taken as read.

Simulation outcomes are MSE through varying some experiment parameters—the most common is the signal to noise ratio (SNR) defined as $A^2/2\sigma_w^2$. Since the frequency estimation involves the peak-search as maximization-step, the MSE vs. the SNR has a behavior that depends on the SNR value and the peak-method as shown in Figure 10.3.

The figure illustrates a common behavior of the MSE vs. SNR for an estimator that involves a non-linear step of peak-search. For low SNR, the observations are dominated by the noise, and thus the estimate is uniform within the angular frequency interval $[-\pi, \pi)$, which corresponds to the a-priori search interval for frequency estimate; the performance is that for a uniform pdf:

$$MSE_K(\hat{\omega}) \simeq \frac{(2\pi)^2}{12}$$

An opposite scenario is when the SNR is very large and one expects the estimator to perform as the CRB. However, the method for maximization of $S_k(\omega)$ can have some limitations and the MSE is saturated to a value (still larger than CRB); the corresponding SNR is referred to as *saturation region*. For an intermediate SNR interval (*CRB region*), the estimator attains the CRB. When reducing the SNR below the CRB region, the estimator starts to degrade from the CRB as the search for the maximum faces some artifacts (e.g., there are some/many local maxima and their values are dependent on the realization of noise), this is the *threshold region* of the estimator. Performance analysis for any estimator that involves a maximum (or maxima, if multiple sinusoids) search should *mandatorily* be carried out to evaluate the threshold and saturation regions for practical applications, and the algorithm modified accordingly. An example is illustrated in Section 10.2.3.

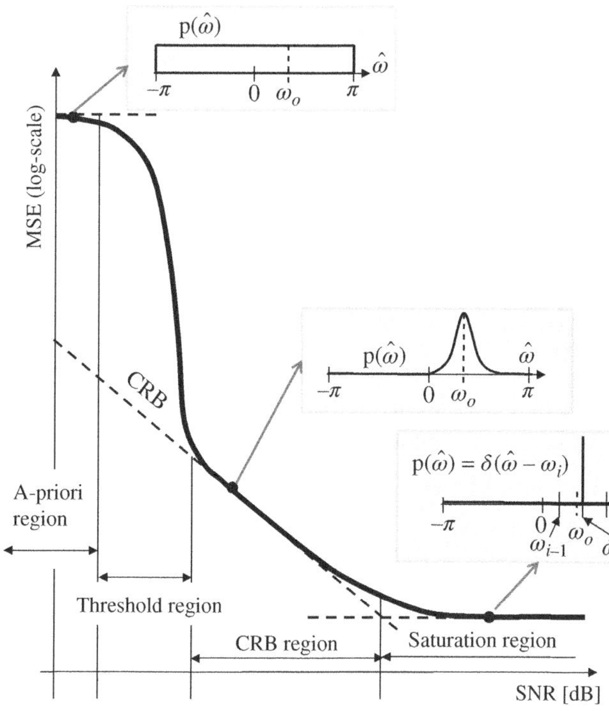

Figure 10.3 MSE vs. SNR for frequency estimation: regions and the pdf $p(\hat{\omega})$.

10.2.1 Variable (Coarse/Fine) Sampling

One simple way to develop the peak-search (10.3) is by sampling the frequency axis and extracting the value that maximizes the $S(\omega)$ as in Figure 10.4. The DFT over N samples of the sequence $\mathbf{x}$ is a way to discretize the frequency axis that has the benefit of being computationally efficient if implemented by using fast DFT algorithms (Fast Fourier Transform—FFT). From the sequence

$$\{S(2\pi i/N)\}_{i=0}^{N-1}$$

one can find the value that peaks $S(\omega)$, and around that value one can densify with a denser sampling of spacing $2\pi/M$, with $M > N$, to refine the peak-search of the continuous function $S(\omega)$ from a discrete-frequency counterpart to get $\hat{\omega}_{grid}$. The way to densify the grid spacing can be either by densifying the sampling around the frequency that peaks a coarser sampling (± 1 sample around to avoid missing the true sample), or alternatively by computing the DFT using a number of samples M with $M > N$, after padding the corresponding vector $\mathbf{x}$ with zero. Both methods to densify the sampling $S(\omega)$ are fully equivalent once M has been chosen; it is only a matter of computational effort to compare the FFT efficiency over a large number of samples, mostly zeros, with a focused sampling densification.

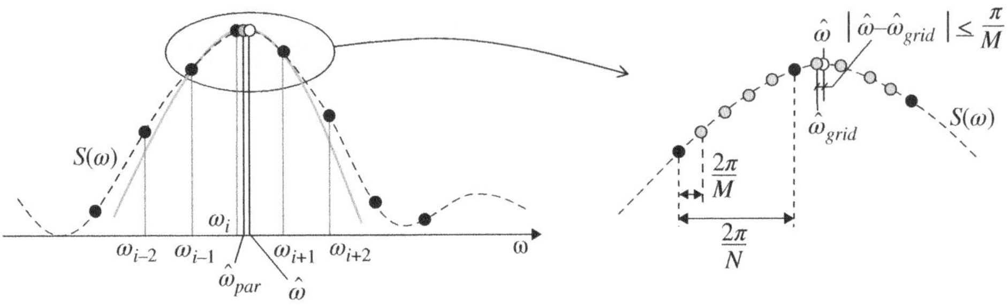

Figure 10.4 Coarse/fine search of the peak of $S(\omega)$.

Notice that sampling is affected by a granularity error $\delta\varepsilon_{grid} = \hat{\omega} - \hat{\omega}_{grid}$ due to the frequency discretization over the grid, that is (Figure 10.4)

$$|\hat{\omega} - \hat{\omega}_{grid}| \leq \frac{\pi}{M}$$

Frequency estimation with simulation codes is

$$\hat{\omega}_{grid} = \omega_o + \delta\varepsilon + \delta\varepsilon_{grid}$$

where $\delta\varepsilon$ is the estimation error, lower bounded by the CRB. The MSE metric from simulation is

$$MSE_K = \text{var}(\hat{\omega}) + \frac{1}{K}\sum_{k=1}^{K} |\delta\varepsilon_{grid,k}|^2$$

When the SNR is very large ($A^2/2\sigma^2 \to \infty$), the variance $\text{var}(\hat{\omega})$ (or the CRB) becomes negligible and the MSE is saturated by the grid spacing as

$$MSE_K \simeq \frac{1}{K}\sum_{k=1}^{K} |\delta\varepsilon_{grid,k}|^2$$

There are now a couple of cases to be considered to quantitatively describe the saturation region.

If frequency is fixed, say ω_o, the closest sample that maximizes $S(\omega)$ over the grid spacing $2\pi/M$ is by rounding the value

$$\hat{\imath} = \left\lceil \frac{\omega_o}{2\pi} M \right\rfloor$$

The error is

$$\delta\varepsilon_{grid,k} = \omega_o - \frac{2\pi}{M}\hat{\imath}$$

and the MSE is saturated by

$$MSE_K \simeq \left(\omega_o - \frac{2\pi}{M}\left[\frac{\omega_o}{2\pi}M\right]\right)^2, \text{ for } \frac{A_o^2}{2\sigma_w^2} \to \infty$$

Since

$$|\omega_o - \frac{2\pi}{M}\hat{\imath}| \leq \frac{\pi}{M}$$

the MSE is upper bounded by

$$MSE_K \leq \left(\frac{\pi}{M}\right)^2, \text{ for } \frac{A_o^2}{2\sigma_w^2} \to \infty$$

If the frequency ω_o can have any value within the range $(0,\pi)$, performance can be evaluated for any arbitrary selection of frequency (unconditional MSE); this is equivalent to considering $\omega_o \sim \mathcal{U}(0,\pi)$ and thus $\delta\varepsilon_{grid,k} \sim \mathcal{U}(-\frac{\pi}{M},\frac{\pi}{M})$. The MSE saturates to

$$MSE_K \geq \frac{1}{12}\left(\frac{2\pi}{M}\right)^2 = \frac{1}{3}\left(\frac{\pi}{M}\right)^2, \text{ for } \frac{A_o^2}{2\sigma_w^2} \to \infty$$

10.2.2 Local Parabolic Regression

The behavior of $S(\omega)$ in the neighborhood of the sample that peaks $S(\omega)$ over a grid spacing $2\pi/M$

$$\hat{\imath} = \arg\max_m S(\frac{2\pi}{M}m)$$

is very smooth and it can be approximated by a parabolic function to be used for local interpolation. One simple (but very effective) way is to find the parabolic regression over the three points (black-dots around $\hat{\omega}$ in Figure 10.4)

$$\begin{aligned} S_{-1} &= S(\tfrac{2\pi}{M}(\hat{\imath}-1)) \\ S_0 &= S(\tfrac{2\pi}{M}\hat{\imath}) \\ S_{+1} &= S(\tfrac{2\pi}{M}(\hat{\imath}+1)) \end{aligned}$$

and evaluate the correction δ with respect to the gridded maximum in $2\pi i/M$ (this is sometimes referred to as fractional correction as $|\delta| < 1$). The regression equation $\hat{S}(\delta)$ for the fractional frequency is

$$\hat{S}(\delta) = S_0(1-\delta^2) + \frac{S_{+1}-S_{-1}}{2}\delta + \frac{S_{+1}+S_{-1}}{2}\delta^2$$

The maximum is

$$\hat{\delta} = \frac{S_{+1}-S_{-1}}{2(S_{+1}+S_{-1}-2S_0)}$$

and thus the frequency estimate is

$$\hat{\omega}_{par} = \frac{2\pi}{M}(\hat{i} + \hat{\delta})$$

The benefit of this method is its simplicity as parabolic regression is less computationally expensive than dense gridding, and the performance can be traded between the choice of DFT length $M > N$ (with zero-padding) and the accuracy. The saturation region is unavoidably dependent on the choice of M, but it is much lower than for grid search over $2\pi/M$. The drawback is a moderate bias when the sample values are $S_{-1} \neq S_{+1}$ (asymmetric parabola) and "true" fractional frequency δ is not in the neighborhood of zero.

10.2.3 Matlab Code and Results

An example of Matlab code is given below for the estimation of the frequency of one sinusoid in Gaussian white noise. The code can be adapted depending on the application and the parameters to investigate (if different from MSE vs. SNR). Figure 10.5 shows the comparison of the MSE vs. SNR for frequency $f_o \sim \mathcal{U}(.2, .3)$ and $N = 128$ samples. Grid search by increasing $M = N, 8N, 64N$ is compared with local parabolic regression with $M = N, 2N, 4N$ and compared with CRB and MSE thresholds.

```
fo=.2; A=1; N=128;
Oversampling=4; M=Oversampling*N; % FFT size (zero-padding &
oversampling)
Nrun=1000; % # Montecarlo runs
snr=[-20:45]; % SNR in dB scale
t=[1:N]';
f=[-N/2:N/2-1];
for isnr=1:length(snr),
for run=1:Nrun,
 % Signal generation
  w=sqrt((A^2/2)*10^(-snr(isnr)/10))*randn(N,1);
  x=A*cos(2*pi*fo*t+2*pi*rand)+w;
  % Freq. estimation: DFT on M samples
  S=(abs(fft(x,M)).^2)/N; %DFT on M>N samples (zero-padding)
  f_est1(run)=find(S(2:M/2)==max(S(2:M/2))); % Search
f=(0:1/2)
  % Freq. estimation: quadratic interpolation
  f_cent=f_est1(run)+1;
   Num=S(f_cent-1)-S(f_cent+1);
   Den=S(f_cent-1)+S(f_cent+1)-2*S(f_cent);
   f_est2(run)=f_cent+.5*Num/Den-1;
end
MSE1(isnr)=mean((f_est1/M-fo).^2);
MSE2(isnr)=mean((f_est2/M-fo).^2);
end
```

```
CRB=12/(N*(N^2-1)).*(10.^(-snr/10)); % CRB freq. estim.
MSE_floor=(1/3)*(.5/M)^2; % quantization error +/-(.5/M)
semilogy(snr,MSE1,'-',snr,MSE2,'-o',snr,CRB/(4*pi^2),'--
',...
snr,MSE_floor*(1+0*snr),':',snr,((1/4)/12)*(1+0*snr),':')
xlabel('SNR [dB]'); ylabel('MSE_f')
```

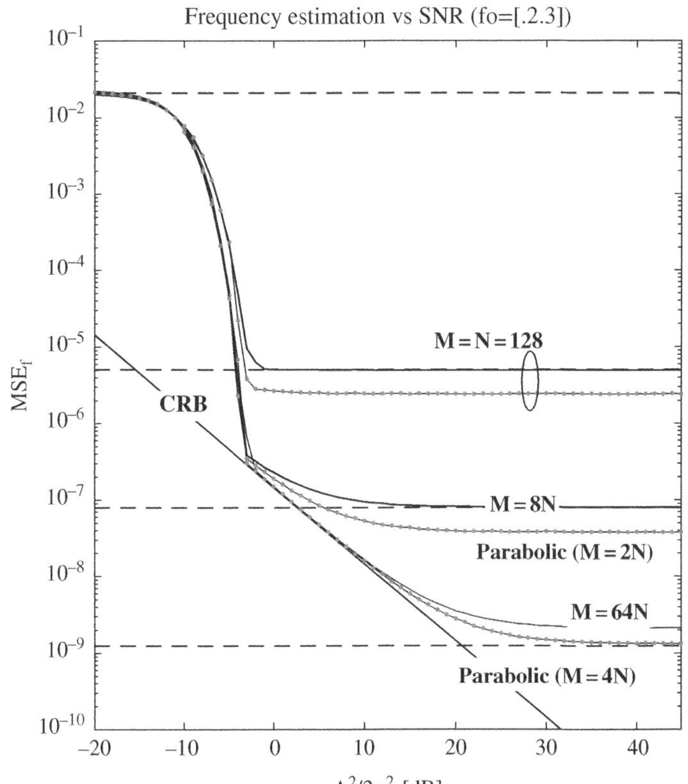

Figure 10.5 MSE vs. SNR ($A^2/2\sigma_w^2$) from Matlab code, and CRB (solid line).

The following remarks are in order:

- For $A^2/2\sigma_w^2 < -3dB$ (threshold region), the MSE degrades significantly from the CRB and the estimator becomes useless ($A^2/2\sigma_w^2 < -10dB$) (a-priori region) as the frequency estimates are random within the range [0,1/2].
- For $M = N = 128$, the MSE is dominated by the coarse spacing of the DFT and it never reaches the CRB; the parabolic regression method is slightly better.
- For $M = 2N$ and $M = 4N$, the MSE for parabolic regression outperforms the grid-method and attains the CRB for $M = 4N = 512$ samples.

- The MSE saturation equation $1/12M^2$ provides a useful approximation for grid search method, but it is largely inaccurate for the parabolic regression method (i.e., the parabolic regression method has lower MSE and it has to be preferred in many contexts).
- If an application requires a certain value of $MSE_{\max}$, the design parameters are: SNR, and number of observations N for the design of a data acquisition system based on CRB analysis, and M for the estimation design:

$$\frac{12}{N(N^2-1)}\frac{1}{A^2/2\sigma_w^2} \leq MSE_{\max}$$

$$\frac{1}{12}\left(\frac{2\pi}{M}\right)^2 \leq MSE_{\max}$$

The estimator complexity is another relevant parameter in any implementation and it depends on the architecture of the specific computer or operating system.

10.3 Time of Delay Estimation

The delay of waveforms is essentially continuous time, and the signals are sampled for the estimation, so one has to use the MLE for continuous time on discrete-time signals. Performance analysis and the Montecarlo method for time of delay estimation is another useful example where the ML estimator defined for continuous-time should be adapted to the descrete-time nature of sampled signals. Recalling the model in Section 9.4, the discrete-time model is

$$x[k] = x(k \cdot \Delta t) = \alpha_o g(k \cdot \Delta t - \tau_o) + w(k \cdot \Delta t)$$

where sampling is for a (continuous-time) waveform delayed by τ_o and noise. The noise samples are Gaussian and uncorrelated:

$$w(k \cdot \Delta t) \sim \mathcal{N}(0, N_o/2\Delta t)$$

This is according to the same assumption as in Section 7.5. The correlation-based estimator cross-correlates the waveform $g(t)$, assumed as known, with $x(t)$; In the discrete-time model this is carried out by correlating the two sequences $x[n]$ and $g[n] = g(k \cdot \Delta t)$:

$$\phi_{xg}[n] = \sum_k x[k+n]g[k]$$

(`corr_xg=xcorr(x,g)` in Matlab code) and the search for the sample n_o that maximizes is carried out by a simple search. The ToD estimation is affected by granularity due to sampling Δt that biases the estimate for large SNR E_s/N_0 with a floor in MSE. This effect is the same as granularity in frequency estimation (Section 10.2), except that in ToD the granularity is a result of the sampling interval, which depends on the experiment itself.

10.3.1 Granularity of Sampling in ToD Estimation

There are two main remedies to avoid the granularity of sampling in MLE of ToD:

- interpolate the signals before the cross-correlation to reduce the sampling interval Δt (and increase the computation complexity as N increases); or
- interpolate the cross-correlation to refine the search for the maximum.

The interpolation around the maximum with a parabolic regression is very common anyway (Figure 10.6) and similar to frequency estimation (Section 10.2.2). Let $\hat{k}$ be the sample that maximizes $\phi_{xg}^2[n]$ with $\phi_0 = |\phi_{xg}(\hat{k}\Delta t)| \geq |\phi_{xg}(\xi)|$. From the samples before and after $\phi_{-1} = |\phi_{xg}((\hat{k}-1)\Delta t)|$ and $\phi_{+1} = |\phi_{xg}((\hat{k}+1)\Delta t)|$, the maximum of the regression parabola over the 3-points wrt the variation $\delta = \tau/\Delta t - \hat{k}$ (fraction of sample) is the correcting term for the delay

$$\hat{\delta} = \frac{\phi_{-1} - \phi_{+1}}{2(\phi_{+1} + \phi_{-1} - 2\phi_0)}$$

and the ToD estimation becomes

$$\hat{\tau} = \Delta t \left(\hat{k} + \frac{\phi_{-1} - \phi_{+1}}{2(\phi_{+1} + \phi_{-1} - 2\phi_0)} \right)$$

This fractional sampling estimator $\hat{\delta}$ is biased too, but the bias is negligible when the true delay is $\delta \simeq 0$ or $\delta \simeq 1$, and it is largest for $\delta = 1/2$. The bias is reduced if the signals are oversampled, or interpolated, or the regression is over a number of samples larger than three (but odd) provided that the autocorrelation $\phi_{gg}[n]$ is smooth (or the signal is oversampled wrt its bandwidth). In any case, the maximum-search method in ToD estimation is a mixture of methods, and performance for large SNR depends on the combination of methods employed.

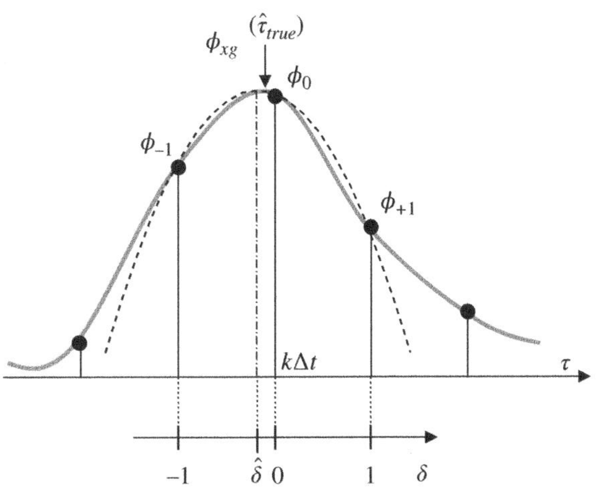

Figure 10.6 Parabolic regression over three points for ToD estimation.

10.3.2 Matlab Code and Results

The performance analysis in terms of MSE vs. SNR of ToD can be evaluated for a triangular shape waveform of duration T_g, and the CRB is (Section 9.4)

$$\text{var}(\hat{\tau}) \geq \frac{T_g^2}{3 \times SNR}$$

where the signal to noise ratio

$$SNR = \frac{\alpha_o^2 E_g}{N_0/2}$$

includes the amplitude. Remarkably, the CRB reduces on reducing the duration of the waveform T_g as cross-correlation is more spiky. For discrete-time simulations, the energy is $E_g = \Delta t \sum_k g^2[k]$ and

$$SNR = \frac{\alpha_o^2 \sum_k g^2[k]}{N_0/2\Delta t}$$

The Matlab code below exemplifies the ToD and compares the MSE with the CRB for varying SNR. Of course, the code can be easily adapted to investigate other settings and waveforms, and tailored to application-specific problems such as for radar systems. Figure 10.7 compares the case of `Tg=[10,20,40]` samples with `T=201` and MSE shows a threshold effect of the correlation-based estimator due to the search of the maxima for every noise realization for `SNR=[-20:2:50]`. When SNR is very low, the MSE gets the upper limit due to the window search of the delay, that in this case is $T^2/12$ due to the uniform pdf of $\hat{\tau}_o$ over the interval `T=201`.

From the simulation results in Figure 10.7, notice that when sampling is too coarse (here for `Tg=10`, but even lower) the discrete-time signals are not precisely replicating the continuous-time experiment, and a small loss in MSE should be accounted for if compared to the CRB.

```
tau_o=1.5; A=1;
t=-(T-1)/2:(T-1)/2;
g_o=zeros(T,1); ig=find(abs(t)<Tg); g_o(ig)=1-abs(t(ig))/Tg;
Eg=sum(g_o.^2);
% delayed waveform (triangular)
g=zeros(T,1); ig=find(abs(t-tau_o)<Tg); g(ig)=1-abs(t(ig)-
tau_o)/Tg;
Nrun=5000;
for iSNR=1:length(SNR);
 No=(2*Eg*A^2)*10^(-SNR(iSNR)/10);
 for run=1:Nrun,
  x=g+sqrt(No/2)*randn(T,1);
  corr_xg=xcorr(x,g_o); % ML estimator
```

```
  n_o=min(find(corr_xg==max(corr_xg))); %avoid dual-max
syntax-error
  tau_est(run)=min(n_o)-T;
  % Fine search by parabolic regression
   Num=corr_xg(n_o-1)-corr_xg(n_o+1);
   Den=corr_xg(n_o-1)+corr_xg(n_o+1)-2*corr_xg(n_o);
   tau_est(run)=tau_est(run)+.5*Num/Den;
 end
 MSE(iSNR)=mean((tau_est-tau_o).^2);
end
CRB=10.^-(SNR/10)*Tg^2/3; % CRB as reference
```

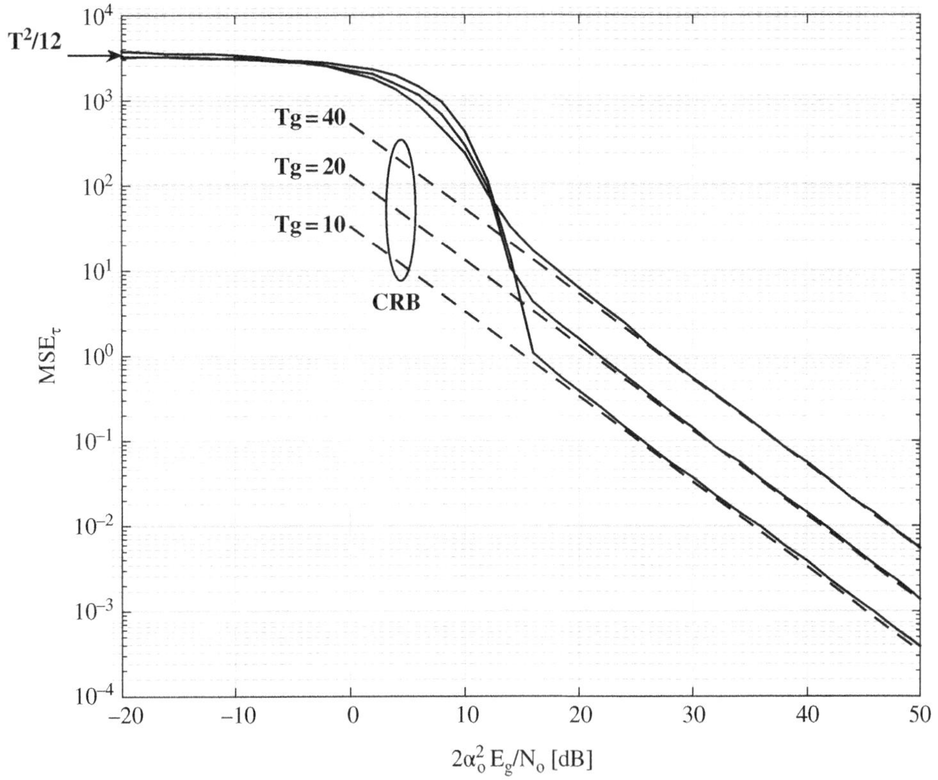

Figure 10.7 MSE vs. SNR ($2\alpha_o^2 E_g/N_0$) of ToD estimation for `Tg=[10,20,40]` samples and CRB (dashed lines).

10.4 Doppler-Radar System by Frequency Estimation

Speed affects the frequency of a sinusoid that propagates through the air according to the Doppler effect (Section 5.6). The Doppler effect is the basis of Doppler-radar systems (also known as continuous wave—CW—radar systems), as in Figure 10.8, which estimate the speed of moving objects from the frequency difference between a transmitted sinusoid (with known and predefined amplitude and frequency f_{TX}) and the backscattered sinusoid with frequency f_{RX} that depends on the speed v of the object measured along the range direction. The amplitude of the received sinusoid is attenuated by the propagation distance (L), and this presents a problem in frequency estimation accuracy. This principle is employed by (some) car-speed systems used by police to detect violators as sketched below. The frequency employed is f_{TX}=10 GHz and, according to the Doppler effect, for every 1km/h of speed of the target, the frequency shift of the return echo backscattered by the car is frequency shifted by $\Delta f_D = 2(v/c)f_{TX} = \pm 18.5\ Hz$ depending on whether the vehicle is getting closer ($\Delta f_D =$ +18.5 Hz/km/h) or receding ($\Delta f_D = -18.5$ Hz/km/h); here $c = 3\times 10^8\ m/s$. Hence, the radar receives the backscattered echoes from the static environment surrounding the car (that gives the sinusoid at the same frequency 10 GHz as this is unaffected by the static environment, known as *clutter* in radar jargon) with unknown amplitude A_c and phase φ_c, and the sinusoid of interest shifted by ±18.5 Hz for every km/h of speed. The received signal is conveniently represented in form of complex signals and it can be reduced to the superposition of two sinusoids in noise:

$$x_{RX}(t) = A_c \exp\{j(\omega_{TX}t + \varphi_c)\} + A_{RX}\exp\{j(\omega_d(v)t + \varphi_{RX})\} + w(t),$$

where the Doppler-shifted frequency is $\omega_d(v) = \omega_{TX} + \Delta\omega_D(v)$.

For technological feasibility of the radar system and estimators, the experiment can be considered as frequency-translated at much lower frequencies by the heterodyne principle, and the Doppler frequency shift remains the same regardless of any frequency translation.

To simplify the assumptions, being somewhat realistic, the transmitter transmits a sinusoid with an equivalent frequency $f_o = 10\ KHz$ and amplitude A_{TX}; the sinusoid propagating toward the car is attenuated by distance as $1/L$ (say L in meters) each path,

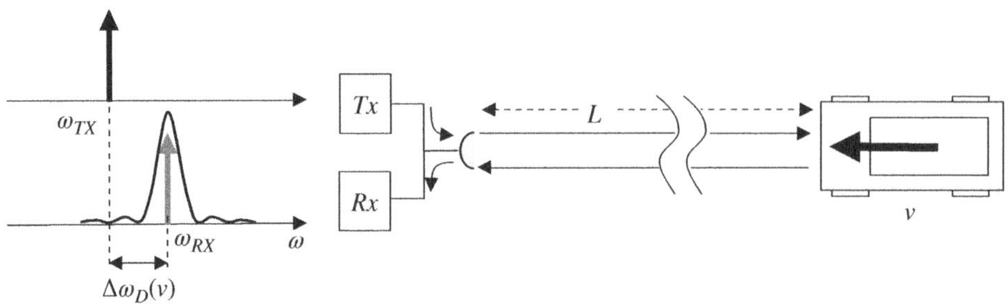

Figure 10.8 Doppler-radar system for speed estimation.

and thus the sinusoid received by reflection from the moving car has an amplitude that is scaled by the reflectivity of the car $\rho < 1$:[2]

$$A_{RX} = \frac{\rho}{L^2} A_{TX}$$

Unfortunately, reflectivity ρ and distance L are not known, and amplitude A_{RX} is a nuisance parameter in the estimation problem. Assuming that signals are sampled as at sampling frequency $f_s = 1/\Delta t = 40KHz$ with $x_{RX}[k] = x_{RX}(k\Delta t)$, the signal model becomes

$$x_{RX}[k] = A_c \exp\{j(\tilde{\omega}_o k + \tilde{\varphi}_c)\} + A_{RX} \exp\{j(\tilde{\omega}_d(v)k + \tilde{\varphi}_{RX})\} + w[k]$$

where the angular frequency of sinusoids normalized to the sampling frequency are

$$\tilde{\omega}_o = 2\pi\Delta t \cdot f_o = \frac{\pi}{2}$$

$$\tilde{\omega}_d(v) = \tilde{\omega}_o + 2\pi\Delta t \cdot f_{TX} \times \frac{2v}{c} = \frac{\pi}{2}(1 + 1.85 \cdot 10^{-3} v)$$

with v in km/h units. Once again, the problem reduces to estimating the frequency of two sinusoids closely spaced apart (as in the figure) over an observation window short enough to freeze the target regardless of vibrations etc. (say $T = 0.5$ ms, or equivalently $N = 200$ samples), with amplitudes and phases as nuisance. Typically the reflections from the backscattered environment have larger amplitudes than the sinusoid at $\tilde{\omega}_d(v)$ ($A_c \gg A_{RX}$), and the estimation of the two frequencies closely spaced apart by $2.9 \times 10^{-4} v$ could be quite cumbersome if v is too small (but this case is of least interest for detecting violations) due to the lack of resolution in distinguishing the two spectral lines (see Chapter 16).

10.4.1 EM Method

Frequency estimation for closely spaced sinusoids is even more complex when there is one strong sinusoid ($A_c \gg A_{RX}$) that severely biases the frequency estimate of the weaker one. However, the frequency of the clutter sinusoid is known except for amplitude and phase, while for the backscattered portion of the signal the unknowns are all the parameters $\{A_{RX}, \tilde{\omega}_d, \tilde{\varphi}_{RX}\}$ of the sinusoid. To exemplify for the numerical example discussed in Matlab code in the next section with $A_c = 100 A_{RX}$ and $N = 200$ samples, Figure 10.9 shows the PSD estimated by the periodogram (see later in Section 14.1) with two spectral lines for $v = 8$ km/h (black-lines) and $v = 100$ km/h (gray-lines), with the lines for the corresponding frequencies $\tilde{\omega}_d$ superimposed. The two spectral lines at frequency $\tilde{\omega}_o = \pi/2$ and $\tilde{\omega}_d = \tilde{\omega}_o(1 + 1.48 \cdot 10^{-2})$ cannot be distinguished from one another, and clutter dominates by biasing the MLE of $\tilde{\omega}_d$. However, when the Doppler shift is higher as for v=100 km/h, the two spectral lines are well separated from one another, and the MLE of frequency for estimating the speed is expected to be noise-limited only and the accuracy attains the CRB.

---

2 The reflectivity coefficient is the ratio between the intensity of the reflected wave and the intensity of the incident one. It depends on the properties of the media and the operating frequency.

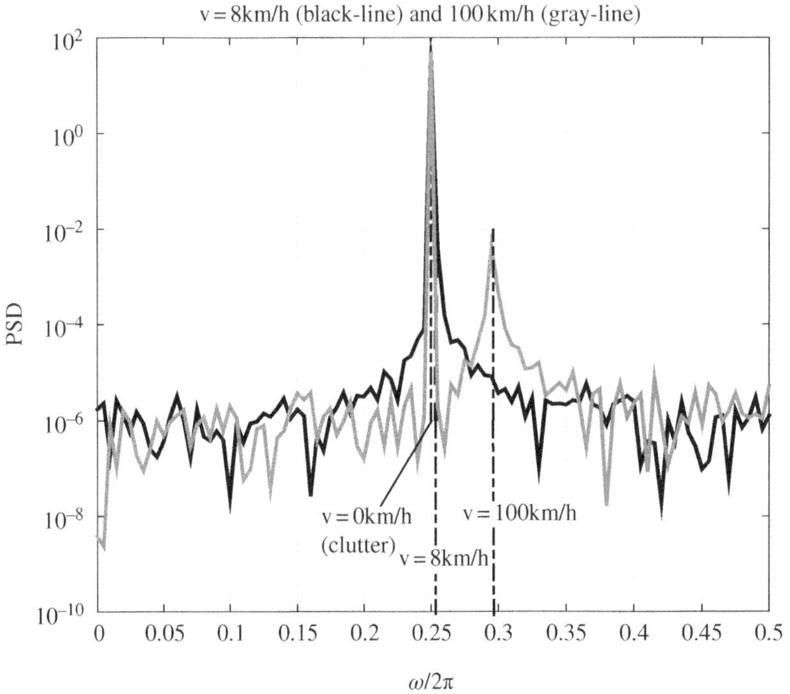

Figure 10.9 Periodogram peaks for Doppler shift for v =8 km/h (black line) and v =100 km/h (gray line) compared to the true values (dashed lines).

In this situation, the set $\{A_c, \tilde{\varphi}_c, A_{RX}, \tilde{\varphi}_{RX}\}$ are the nuisance and $\tilde{\omega}_d$ is the parameter of interest. The MLE for superimposed signal systems can be solved by the EM (expectation-maximization) method, which is an iterative MLE procedure where estimation of the parameters for the two signals are obtained one at a time, stripping the influence of the other signal from the measurements. The EM method is more rigorous than the simple intuitive description provided here, and the reader might read Section 11.6 first.

To illustrate the iterative estimate for this problem, one defines the vector model for a real-valued signal

$$\mathbf{x} = \mathbf{H}(\tilde{\omega}_o)\boldsymbol{\alpha}_c + \mathbf{H}(\tilde{\omega}_d)\boldsymbol{\alpha}_d + \mathbf{w}$$

for clutter and Doppler-shifted return, respectively. Given the estimate of the parameters at the ℓth iteration as $\{\boldsymbol{\alpha}_c^{(\ell)}, \boldsymbol{\alpha}_d^{(\ell)}, \tilde{\omega}_d^{(\ell)}\}$, one can compute the residual by stripping a local copy of the signal based on these parameters:

$$\boldsymbol{\varepsilon}^{(\ell)} = \mathbf{x} - \mathbf{H}(\tilde{\omega}_o)\boldsymbol{\alpha}_c^{(\ell)} + \mathbf{H}(\tilde{\omega}_d^{(\ell)})\boldsymbol{\alpha}_d^{(\ell)}$$

and from this one gets two signals that are (virtually) affected only by clutter and Doppler-shifted sinusoid (this is called a *complete set*—see Section 11.6)

$$\mathbf{y}_c^{(\ell)} = \mathbf{H}(\tilde{\omega}_o)\boldsymbol{\alpha}_c^{(\ell)} + \beta_c \boldsymbol{\varepsilon}^{(\ell)}$$
$$\mathbf{y}_d^{(\ell)} = \mathbf{H}(\tilde{\omega}_d^{(\ell)})\boldsymbol{\alpha}_d^{(\ell)} + \beta_d \boldsymbol{\varepsilon}^{(\ell)}$$

with scalings $\beta_c + \beta_d = 1$. Based on these new signals, one estimates the amplitudes $(\boldsymbol{\alpha}_c^{(\ell+1)}, \boldsymbol{\alpha}_d^{(\ell+1)})$ and the Doppler frequency is by using the MLE approach discussed in Section 7.2.2:

$$\boldsymbol{\alpha}_c^{(\ell+1)} = (\mathbf{H}^T(\tilde{\omega}_o)\mathbf{H}(\tilde{\omega}_o))^{-1}\mathbf{H}^T(\tilde{\omega}_o)\mathbf{y}_c^{(\ell)}$$
$$\tilde{\omega}_d^{(\ell+1)} = \arg\max_{\omega}\{\mathbf{y}_d^{(\ell)T}\mathbf{P}(\omega)\mathbf{y}_d^{(\ell)}\}$$
$$\boldsymbol{\alpha}_d^{(\ell+1)} = (\mathbf{H}^T(\tilde{\omega}_d^{(\ell+1)})\mathbf{H}(\tilde{\omega}_d^{(\ell+1)}))^{-1}\mathbf{H}^T\tilde{\omega}_d^{(\ell+1)})\mathbf{y}_d^{(\ell)}$$

The iterative procedure converges and its accuracy depends on the initialization $\{\boldsymbol{\alpha}_c^{(0)}, \boldsymbol{\alpha}_d^{(0)}, \tilde{\omega}_d^{(0)}\}$; for the specific problem it is assumed that clutter dominates, and $\mathbf{y}_c^{(0)} = \mathbf{x}$.

10.4.2 Matlab Code and Results

This numerical analysis is for the EM method, and the code is exemplary of the way that EM can be implemented. After initialization using the parameters for the problem at hand with clutter variables (`A1=1; omega_1=pi/2; phi_1=2*pi*rand`) and backscattered sinusoid variables (`A2=1/100; omega_2=pi/2*(1+1.85E-3*v); phi_2=2*pi*rand`) with scaling `beta1=beta2=5/10`, there are `N_EM=10` EM iterations. The estimation of frequency $\tilde{\omega}_d^{(\ell+1)}$ is on the vector `y2` by using the MLE method in Section 10.2.3 with the routine `freq_est(.)`.

```
sigma_n=1E-3; % noise N(0,sigma_n^2)
t=[0:Nt-1]';
x=A1*cos(omega_1*t+phi_1)+A2*cos(omega_2*t+phi_2)
+sigma_n*randn(Nt,1); %signal generation
H1=[cos(omega_1*t),sin(omega_1*t)]; % basis H(omega_1)
y1=x;
y2=beta2*(x-H1*(H1\y1)); % EM-initialization
for n_EM=1:N_EM,
  om2_est=freq_est(y2); % freq. estimation on y2 (see sect.
in book)
  H2=[cos(om2_est*t),sin(om2_est*t)]; % basis
H(omega_2_estimated)
  y1=H1*(H1\x)+beta1*(x-H1*(H1\y1)-H2*(H2\y2));
  y2=H2*(H2\y2)+beta2*(x-H2*(H2\y2)-H1*(H1\y1));
end
v_est=((2/pi)/1.85E-3)*om2_est; % transformation omega->v
```

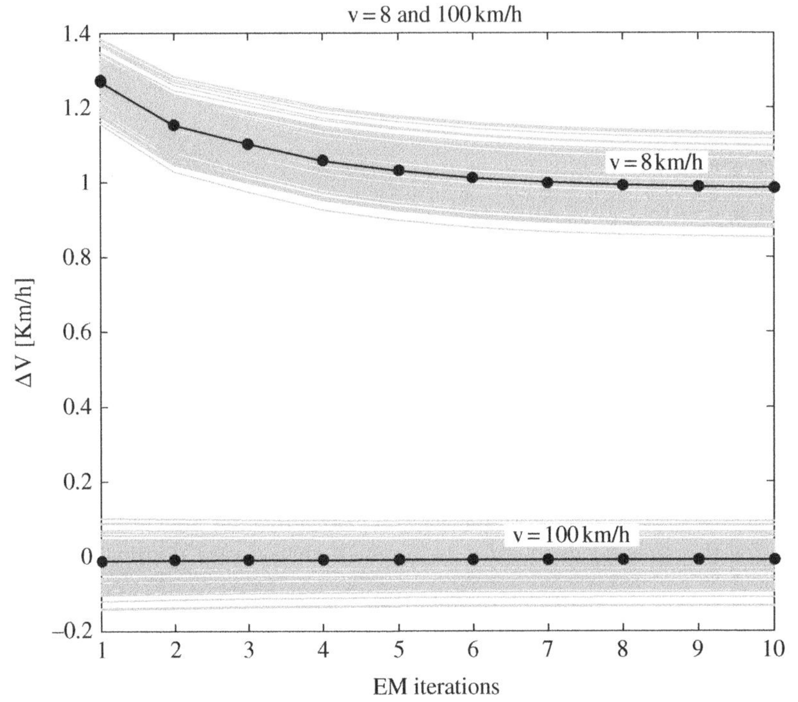

Figure 10.10 Speed-error $\Delta v = \hat{v} - v$ of `K=100` Montecarlo runs (gray lines) and their mean value (black line with markers) vs EM iterations for two choices of speed as in Figure 10.9: v =8 km/h and v =100 km/h.

Figure 10.10 shows the speed-error $\Delta v = \hat{v} - v$ with respect to the true value v vs. the EM-iteration number for the parameters of the Matlab code in the box, and v= v =8 Km/h. The convergence is monotonic and, for different Montecarlo simulations of noise and phases (here 100 independent runs), the curves for each EM iteration (gray lines) are dispersed around the mean values (black line with markers) with a notable bias. One can easily verify that on increasing the speed, the Doppler frequency would be more set apart and estimates more accurate, with error $\Delta v \leq 0.2km/h$ for a target moving at $v =$ 50km/h. Figure 10.10 shows the case of $v = 100$ km/h where the two sinusoids are mostly non-interfering and the error Δv is negligible.

11

Bayesian Estimation

In several applications there is enough (statistical) information on the most likely values of the parameters $\boldsymbol{\theta}$ to be estimated even before making any experiment, or before any data collection. The information is encoded in terms of the *a-priori pdf* $p(\boldsymbol{\theta})$ that accounts for the properties of $\boldsymbol{\theta}$ *before* any observation (Chapter 6). Bayesian methods make efficient use of the a-priori pdf to yield the *best estimate* given both the observation $\mathbf{x}$ and the a-priori knowledge on the admissible values from $p(\boldsymbol{\theta})$.

Recall that MLE is based on the conditional pdf $p(\mathbf{x} = \mathbf{x}_k|\boldsymbol{\theta} = \boldsymbol{\theta}_k)$ that sets the probability of the specific observation $\mathbf{x}_k$ for any choice $\boldsymbol{\theta} = \boldsymbol{\theta}_k$, but these choices are not all equally likely. In the Bayesian approach, the outcome of the kth experiment is part of a set of (real or conceptual) experiments with two random sets $\boldsymbol{\theta}$ and $\mathbf{x}$, that in the case of an additive noise model $\mathbf{x} = \mathbf{s}(\boldsymbol{\theta}) + \mathbf{w}$ are $\boldsymbol{\theta}$ and $\mathbf{w}$. The joint pdf is

$$p(\mathbf{x}, \boldsymbol{\theta}) = p(\mathbf{x}|\boldsymbol{\theta})p(\boldsymbol{\theta})$$

but it is meaningful to consider for each experiment (or realization of the rv $\mathbf{x}$) the pdf of the parameter $\boldsymbol{\theta}$ conditioned to the kth observation $\mathbf{x} = \mathbf{x}_k$ (here deterministic) according to Bayes' rule, which gets the pdf of the unobserved $\boldsymbol{\theta}$

$$p(\boldsymbol{\theta}|\mathbf{x} = \mathbf{x}_k) = \frac{p(\mathbf{x} = \mathbf{x}_k|\boldsymbol{\theta})p(\boldsymbol{\theta})}{p(\mathbf{x} = \mathbf{x}_k)} = \gamma p(\mathbf{x} = \mathbf{x}_k|\boldsymbol{\theta})p(\boldsymbol{\theta})$$

with γ as a scale factor to normalize the pdf. The a-posteriori pdf $p(\boldsymbol{\theta}|\mathbf{x} = \mathbf{x}_k)$ is the pdf of the unknown parameter $\boldsymbol{\theta}$ after the observation $\mathbf{x} = \mathbf{x}_k$, and reflects the statistical properties of $\boldsymbol{\theta}$ after the specific experiment.

To further stress the role of the a-priori pdf in Bayesian methods, we can make reference to Figure 11.1 to derive the a-posteriori pdf $p(\boldsymbol{\theta}|\mathbf{x})$ (as routinely employed in the notation, the specific observation $\mathbf{x} = \mathbf{x}_k$ is understood), which is obtained by multiplying the a-priori $p(\boldsymbol{\theta})$ and the conditional pdf $p(\mathbf{x}|\boldsymbol{\theta})$. In Figure 11.1, $\boldsymbol{\theta}$ is scalar (number of parameters $p = 1$) and the a-priori pdf shows some values that are more or less likely than others. After the observation, the conditional pdf $p(\mathbf{x}|\boldsymbol{\theta})$ is the one that is used in the ML method and it has a different behavior from the a-priori one. More specifically, the maximum of the conditional pdf provides the MLE $\boldsymbol{\theta}_{ML}$ without any knowledge of the probability of the values of $\boldsymbol{\theta}$. After multiplying the conditional pdf with the a-priori pdf, the a-posteriori pdf $p(\boldsymbol{\theta}|\mathbf{x})$ compounds the likelihood of

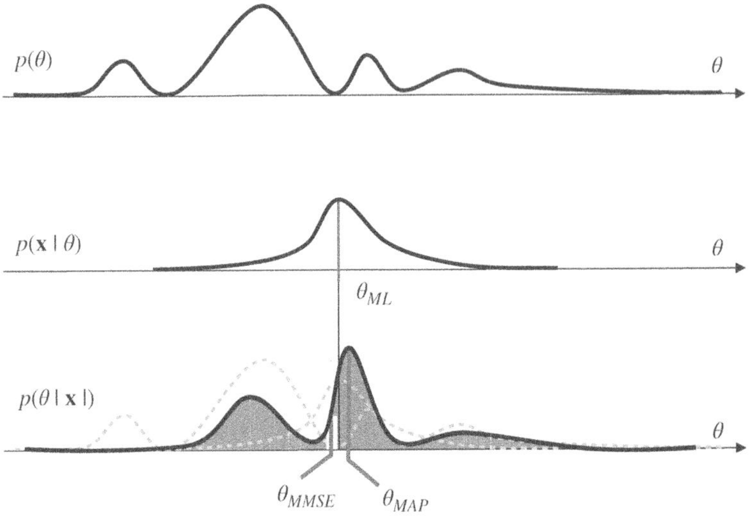

Figure 11.1 Bayesian estimation ($p = 1$).

the parameters regardless of the knowledge of their probability, and the probability associated to each value of the parameters, by the a-priori pdf $p(\boldsymbol{\theta})$.

Based on the a-posteriori pdf, Bayesian estimation can follow two different strategies:

- **Maximum a-posteriori (MAP) estimation:** the estimate is the value that maximizes the a-posteriori probability

$$\boldsymbol{\theta}_{MAP} = \arg\max_{\boldsymbol{\theta}} p(\boldsymbol{\theta}|\mathbf{x})$$

- **Minimum mean square error (MMSE) estimation**: the estimate can be in the baricentral position wrt the a-posteriori pdf:

$$\boldsymbol{\theta}_{MMSE} = \mathbb{E}[\boldsymbol{\theta}|\mathbf{x}] = \int \boldsymbol{\theta} p(\boldsymbol{\theta}|\mathbf{x}) d\boldsymbol{\theta}$$

  This is the mean value of the conditional pdf $p(\boldsymbol{\theta}|\mathbf{x})$, and it is commonly referred to as the MMSE estimate as clarified below.

Any estimator $\hat{\boldsymbol{\theta}}(\mathbf{x})$ depends on the observations, but under the Bayesian assumptions, the error $\boldsymbol{\theta} - \hat{\boldsymbol{\theta}}(\mathbf{x})$ is an rv that depends on both the data $\mathbf{x}$ and the rv $\boldsymbol{\theta}$ with pdf $p(\boldsymbol{\theta})$. The corresponding Bayesian MSE (Section 6.3.2):

$$MSE(\hat{\boldsymbol{\theta}}) = \mathbb{E}_{x,\theta}[(\boldsymbol{\theta} - \hat{\boldsymbol{\theta}}(\mathbf{x}))^H(\boldsymbol{\theta} - \hat{\boldsymbol{\theta}}(\mathbf{x}))] = \iint (\boldsymbol{\theta} - \hat{\boldsymbol{\theta}}(\mathbf{x}))^H(\boldsymbol{\theta} - \hat{\boldsymbol{\theta}}(\mathbf{x}))p(\mathbf{x}, \boldsymbol{\theta}) d\mathbf{x} d\boldsymbol{\theta}$$

is evaluated wrt the randomness of the data and $\boldsymbol{\theta}$. The MSE is thus the "mean" in the sense that it accounts for all possible choices of the rv $\boldsymbol{\theta}$ weighted by the corresponding occurrence probability $p(\boldsymbol{\theta})$ as $p(\mathbf{x},\boldsymbol{\theta}) = p(\mathbf{x}|\boldsymbol{\theta})p(\boldsymbol{\theta})$. The minimization of the MSE seeks for the parameter value that minimizes the $MSE(\hat{\boldsymbol{\theta}})$ as a function of $\hat{\boldsymbol{\theta}}$. From

$$MSE(\hat{\boldsymbol{\theta}}) = \int \left[\int (\boldsymbol{\theta} - \hat{\boldsymbol{\theta}}(\mathbf{x}))^T(\boldsymbol{\theta} - \hat{\boldsymbol{\theta}}(\mathbf{x}))p(\boldsymbol{\theta}|\mathbf{x})d\boldsymbol{\theta}\right] p(\mathbf{x})d\mathbf{x}$$

since $p(\mathbf{x})$ is semipositive definite and independent of $\hat{\boldsymbol{\theta}}$, it is enough to minimize the term within the bracket (i.e., the MSE conditioned to a specific observation $\mathbf{x}$) by nulling the gradient:

$$\frac{\partial}{\partial\hat{\boldsymbol{\theta}}}\int (\boldsymbol{\theta} - \hat{\boldsymbol{\theta}})^2 p(\boldsymbol{\theta}|\mathbf{x})d\boldsymbol{\theta} = -2\int (\boldsymbol{\theta} - \hat{\boldsymbol{\theta}})p(\boldsymbol{\theta}|\mathbf{x})d\boldsymbol{\theta} = \mathbf{0}$$

Rearranging terms yields the MMSE estimator

$$\hat{\boldsymbol{\theta}}_{MMSE} = \int \boldsymbol{\theta} p(\boldsymbol{\theta}|\mathbf{x})d\boldsymbol{\theta}$$

as the estimator that minimizes the MSE for all possible (and random) outcomes of the parameters $\boldsymbol{\theta}$.

Remark: The illustrative example at hand helps to raise some issues based on which Bayesian estimators have to be preferred. There is no unique answer, but it depends on the application. The example shows a multimodal a-posteriori pdf and MAP selects one value that occurs with the greatest probability. On the other hand, the MAP choice could experience large errors In the case that some outcomes are in other modes of the pdf. The MMSE is the best "mean solution" that could coincide with a value that is unlikely, if not admissible (e.g., $p(\theta = \hat{\theta}_{MMSE}) = 0$). In general, for multimodal a-posteriori pdf, the maximum (MAP) and the mean (MMSE) of the a-posteriori pdf do not coincide, and choosing one or another depends on the specific application problem. Some care should be taken to avoid the *Buridan's ass paradox*[1] and estimate a value that is a nonsense for the problem at hand, but still compliant with the estimator's metric—as for the MMSE estimate for binary valued parameters (Section 11.1.3). On the other hand, if the a-posteriori pdf is unimodal and even, both MAP and MMSE coincide, as in the case of Gaussian rvs—a special and favorable case as detailed below.

11.1 Additive Linear Model with Gaussian Noise

This observation is modeled by $x = \theta + w$, in which the scalar parameter θ is random and should be estimated from a single observation with additive Gaussian noise:

1 This is a paradox where an ass that is equally hungry and thirsty is placed halfway between a stack of hay and a pail of water. Since the ass will always go to whichever is closer, it will die of both hunger and thirst since it cannot make any rational decision to choose one over the other. Buridan (1295–1361?) was a French philosopher, but the paradox dates back to Aristotle: *...a man, being just as hungry as thirsty, and placed in between food and drink, must necessarily remain where he is and starve to death* (Aristotle, On the Heavens, ca.350 BCE).

$w \sim \mathcal{N}(0,\sigma_w^2)$. The Gaussian model is discussed first with $\theta \sim \mathcal{N}(\bar{\theta},\sigma_\theta^2)$, and then extended to the case of non-Gaussian rv θ. It will be shown that Bayesian estimators become non-linear for an arbitrary pdf, and this motivates the interest in linear estimators even when non-linear estimators should be adopted, at the price of some degree of sub-optimality.

11.1.1 Gaussian A-priori: $\theta \sim \mathcal{N}(\bar{\theta},\sigma_\theta^2)$

Single Observation $x = \theta + w$

In MLE, the value of the parameter θ does not change even if considering multiple experiments. For a single observation, the MLE is

$$\begin{aligned}\theta_{ML} &= x\\ \text{var}[\theta_{ML}] &= MSE_{ML} = \sigma_w^2\end{aligned}$$

Assuming now that for each observation (composed of a single sample) the parameter value θ is different and drawn from a Gaussian distribution $\theta \sim \mathcal{N}(\bar{\theta},\sigma_\theta^2)$ where $\bar{\theta}$ and σ_θ^2 are known, the joint pdf is $p(x,\theta) = p(x|\theta)p(\theta)$ while the pdf of θ conditioned to a specific observation (a-posteriori pdf) is

$$p(\theta|x) = \frac{p(x|\theta)p(\theta)}{p(x)} = \Gamma \underbrace{\exp\left(-\frac{(x-\theta)^2}{2\sigma_w^2}\right)}_{p(x|\theta)} \cdot \underbrace{\exp\left(-\frac{(\theta-\bar{\theta})^2}{2\sigma_\theta^2}\right)}_{p(\theta)}$$

up to a scale Γ. Rearranging the exponentials, the a-posteriori pdf is

$$p(\theta|x) = G(\theta; g(x), \sigma_o^2)$$

This shows the general property (see Section 11.2) that for additive linear models, the a-posteriori pdf is Gaussian if all terms are Gaussian. Returning to the specific example, the terms involved are

$$\begin{aligned}g(x) &= \frac{\sigma_w^2}{\sigma_w^2+\sigma_\theta^2}\bar{\theta} + \frac{\sigma_\theta^2}{\sigma_w^2+\sigma_\theta^2}x = \bar{\theta} + \frac{\sigma_\theta^2}{\sigma_w^2+\sigma_\theta^2}(x-\bar{\theta})\\ \sigma_o^2 &= \frac{\sigma_w^2\sigma_\theta^2}{\sigma_w^2+\sigma_\theta^2} = \text{var}(\theta|x) \le \sigma_\theta^2\end{aligned}$$

The second term shows that the variance of the a-posteriori pdf is always smaller than the a-priori (σ_θ^2) except when noise dominates the measurements ($\sigma_w^2 \gg \sigma_\theta^2$) and $\text{var}(\theta|x) \simeq \sigma_\theta^2$. The maximum of the a-posteriori $p(\theta|x)$ when Gaussian is for

$$\theta_{MAP} = g(x) = \bar{\theta} + \frac{\sigma_\theta^2}{\sigma_w^2+\sigma_\theta^2}(x-\bar{\theta}) = \theta_{MMSE}$$

Considering the symmetry of the Gaussian pdf, it is straightforward to assert that $\theta_{MAP} = \theta_{MMSE}$. It is interesting to note that for $\sigma_\theta^2 \to 0$, the a-priori information dominates and $\theta_{MAP} \to \bar{\theta}$ regardless of the specific observation, while for $\sigma_\theta^2 \to \infty$, a-priori information becomes irrelevant (in this case the a-priori information is *diffuse* or *non-informative*) and $\theta_{MAP} = x$ as in MLE. This is another general property: when the a-priori pdf $p(\theta)$ is uniform, there is no difference between one choice and the other, and the Bayesian estimator cannot rely on any a-priori pdf; in this case the MAP coincides with the MLE.

The Bayesian MSE should take into account both the parameter and noise as rvs:

$$MSE_{MAP} = \mathbb{E}_{\theta,w}\left[\left(g(x) - \theta\right)^2\right] = \frac{\sigma_w^2 \sigma_\theta^2}{\sigma_w^2 + \sigma_\theta^2} \le \sigma_w^2 = MSE_{ML}$$

and $MSE_{MAP} \to MSE_{ML}$ for diffuse a-priori ($\sigma_\theta^2 \to \infty$).

The Bayesian MSE can be evaluated in the following alternative way:

$$\mathbb{E}_{\theta,x}[(\theta_{MAP} - \theta)^2] = \mathbb{E}_\theta[\underbrace{E_{x|\theta}[(\theta_{MAP} - \theta)^2 | \theta = \theta_o]}_{MSE_{ML}(\theta_o)}]$$

Note that $MSE_{ML}(\theta_o)$ is the MSE for deterministic $\theta = \theta_o$ as for MLE. Since for the example here $\theta_{MAP} = \theta_{MMSE}$, but θ_{MMSE} optimizes the Bayesian *MSE* and not the $MSE_{ML}(\theta_o)$, the estimate θ_{MAP} cannot be optimal for the deterministic value $\theta = \theta_o$ (incidentally, it is biased). Nevertheless, when the MSE is evaluated by taking into account the variability of θ_o, the Bayesian estimation shows its superiority with respect to the ML estimation for any arbitrary (but still with pdf $p(\theta|x)$) value of θ.

To summarize, the Bayesian MSE of the MMSE estimator is always lower than the MSE of the MLE. This is due to fact that the MMSE estimator exploits the available a-priori information on the parameter value.

Multiple Observations $x[i] = \theta + w[i]$

We can consider a more complex (but somewhat more realistic) example with N observations characterized by independent noise components with the same value of the parameter θ, that in turn it can change over the realizations. The model can be written as:

$$x[i] = \theta + w[i] \Rightarrow \mathbf{x} = \theta\mathbf{1} + \mathbf{w},$$
$$\mathbf{w} \sim \mathcal{N}(0, \sigma_w^2 \mathbf{I})$$

Estimation of the parameter is the sample mean (see, e.g., Section 6.7)

$$\theta_{ML} = \sum_{i=1}^{N} x[i]/N = \bar{x}$$
$$\text{var}[\theta_{ML}] = \sigma_w^2/N$$

while the MMSE estimator is linear (it will be derived in the following section) and it is given by

$$\theta_{MMSE} = \mathbb{E}[\theta|\mathbf{x}] = \frac{\frac{N}{\sigma_w^2}\bar{x} + \frac{\bar{\theta}}{\sigma_\theta^2}}{\frac{N}{\sigma_w^2} + \frac{1}{\sigma_\theta^2}} = \frac{\sigma_\theta^2}{\sigma_w^2/N + \sigma_\theta^2}\bar{x} + \frac{\sigma_w^2/N}{\sigma_w^2/N + \sigma_\theta^2}\bar{\theta}$$

It is interesting to remark that the MMSE estimation is a linear combination between the mean parameter value $\bar{\theta}$ (known a-priori) and the sample mean $\bar{x}$. For $N \to \infty$ the a-priori information becomes irrelevant and $\theta_{MMSE} \to \bar{x} = \theta_{ML}$ while for $\sigma_\theta^2 \ll \sigma_w^2/N$ we have $\theta_{MMSE} \simeq \bar{\theta}$. From the symmetry of the Gaussian a-posteriori pdf, $\theta_{MAP} = \theta_{MMSE}$.

11.1.2 Non-Gaussian A-Priori

The observation $x = \theta + w$ is the sum of two random variables, only one of which (here noise) is Gaussian, say $w \sim \mathcal{N}(0, \sigma_w^2)$. The MMSE follows from the definition:

$$\theta_{MMSE} = \mathbb{E}[\theta|x] = \int \alpha \underbrace{\frac{p_{x|\theta}(x|\theta = \alpha)p_\theta(\alpha)}{p_x(x)}}_{p_{\theta|x}(\alpha|x)} d\alpha = \frac{\int \alpha p_w(x-\alpha)p_\theta(\alpha)d\alpha}{\int p_w(x-\alpha)p_\theta(\alpha)d\alpha}$$

where the following two equalities have been used to derive the last term: $p_{x|\theta}(x|\theta = \alpha) = p_w(x-\alpha)$ and $p_x(x) = p_\theta(x) * p_w(x) = \int p_w(x-\alpha)p_\theta(\alpha)d\alpha$. The general relationship of the MMSE estimator is[2]

$$\theta_{MMSE} = \mathbb{E}[\theta|x] = x + \frac{\sigma_w^2}{p_x(x)}\frac{d}{dx}p_x(x) = x + \sigma_w^2 \frac{d}{dx}\log p_x(x) \tag{11.1}$$

In other words, the MMSE estimator is a non-linear transformation of the observation x and its shape depends on the pdf of the observation $p_x(x) = p_\theta(x) * p_w(x)$. This example can be generalized: the MMSE estimator is non-linear except in Gaussian contexts (the proof is simple: just set $\theta \sim \mathcal{N}(\bar{\theta}, \sigma_\theta^2)$; this gives $x \sim \mathcal{N}(\bar{\theta}, \sigma_w^2 + \sigma_\theta^2)$ therefore, in this case, the MMSE estimator is linear and it coincides with the one reported previously in Section 11.1.1).

2 For a zero-mean Gaussian function, its derivative is

$$\frac{dG(u;0,\sigma^2)}{du} = -\frac{u}{\sigma^2}G(u;0,\sigma^2)$$

This can be applied to the pdf

$$\frac{d}{dx}p_x(x) = \frac{d}{dx}(p_\theta(x) * p_w(x)) = \frac{1}{\sigma_w^2}\int \alpha p_w(x-\alpha)p_\theta(\alpha)d\alpha - \frac{x}{\sigma_w^2}p_x(x)$$

where the last equality is based on the fact that w is Gaussian.

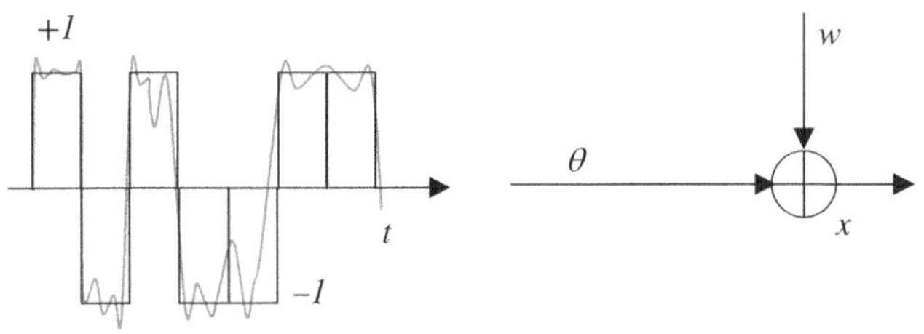

Figure 11.2 Binary communication with noise.

The additive model has the nice property that the complementary MMSE estimator (here for noise w) follows from simple reasoning. Since $\mathbb{E}[x|x] = x$, expanding this identity $\mathbb{E}[x|x] = \mathbb{E}[\theta|x] + \mathbb{E}[w|x]$ it follows that

$$w_{MMSE} = x - \theta_{MMSE} = -\sigma_w^2 \frac{d}{dx} \log p_x(x)$$

This means that the MMSE estimation of noise is complementary to the estimator of θ.

Remark: The MMSE estimator depends on the pdf of observations $p_x(x)$, but when this pdf is unknown and a large set of independent observations is available, one can use the histogram of the observations, say $\hat{p}_x(x)$, in place of $p_x(x)$, and then evaluate $\log \hat{p}_x(x)$ in a numerical (or analytical) way recalling that the histogram bins the values of x over a finite set (see Section 9.7 for the design of histogram binning). In this way an approximate MMSE estimator can be obtained. Alternatively, it is possible to fit a pdf model over the observations by matching the sample moments with the analytical ones (*method of moments*[3]) and compute the MMSE estimator from the pdf model derived so far.

11.1.3 Binary Signals: MMSE vs. MAP Estimators

In a digital communication system sketched in Figure 11.2 a binary stream $\theta(t)$ (black line) is transmitted over a noisy communication link and the received signal (grey line) can be modeled as impaired by a Gaussian noise

$$x(t) = \theta(t) + w(t)$$

The transmitted signal $\theta(t)$ is binary and its value encodes the values of a binary source to be delivered to destination. Source values are unknown, but can be modeled as a random stationary process (i.e., time t is irrelevant) with independent samples:

3 The method of moments (MoM) is fairly simple and intuitive, and thus widely adopted. Let $p_x(x|\mathbf{a})$ be a parametric pdf depending on a set of parameters $\mathbf{a}$ (unknown) with moments $\mu_k = \mathbb{E}[x^k] = \mu_k(\mathbf{a})$ that depends on parameters too, and a set of N samples $x[1], ..., x[N]$ with sample moments $\hat{\mu}_k = \sum_{i=1}^{N} x[i]^k / N$. The MoM solves for the system $\mu_k(\mathbf{a}) = \hat{\mu}_k$ wrt the parameters $\mathbf{a}$ to have the parametric pdf: $p_x(x|\hat{\mathbf{a}})$. Except for some special cases, the estimator is consistent even if it could be biased.

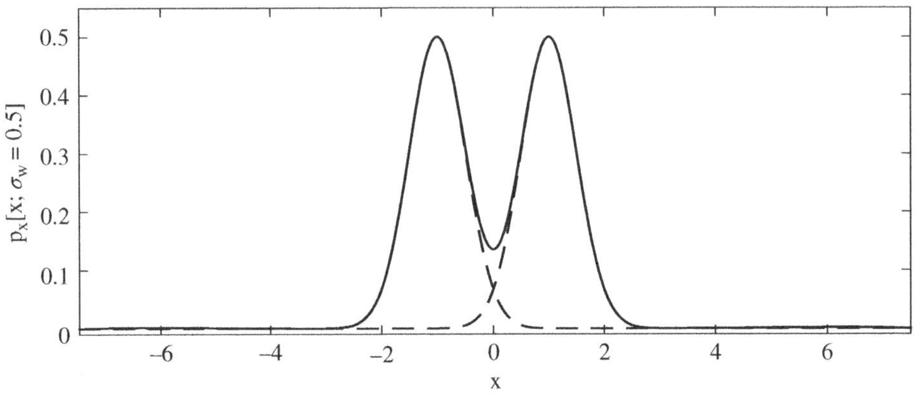

Figure 11.3 Pdf $p_x(x)$ for binary communication.

$$\theta = \begin{cases} +1 \text{ with probability } p \\ -1 \text{ with probability } 1-p \end{cases} \rightarrow p_\theta(\theta) = p \times \delta(\theta - 1) + (1-p) \times \delta(\theta + 1)$$

This represents the a-priori pdf. The unconditional pdf of the received signal $x = \theta + w$ is

$$p_x(x) = p \times p_w(x-1) + (1-p) \times p_w(x+1)$$

It is bimodal (or multimodal in the case of a multi-level signal) with the maxima positioned at ± 1 as shown in Figure 11.3 for $p = 0.5$ and $\sigma_w = 0.5$.

The MMSE approach estimates the value θ as a continuous-valued (not binary) variable by computing the derivative of the closed form (11.1). A fairly compact relationship of the MMSE estimator can be derived after some algebra (recall that $\tanh(u/2) = (e^u - 1)/(e^u + 1)$):

$$\theta_{MMSE} = \tanh\left(\frac{x}{2\sigma_w^2}\right)$$

In information theory this is referred as *soft decoding*. Figure 11.4 shows the non-linearity that represents the MMSE estimate θ_{MMSE} from the observation x. For a low noise level ($\sigma_w = 0.1$), the non-linearity has the shape of a binary detector with threshold, while for high noise level ($\sigma_w = 3$), the non-linearity attains a linear behavior with a slope of $1/\sigma_w^2$. In fact, for $\sigma_w \rightarrow \infty$ the noise dominates and the binary level that minimizes the MSE is 0 (i.e., halfway between the two choices). The MMSE estimation of w is based on the use of the non-linearity that is complementary to the previous one considered, that for $\sigma_w \rightarrow \infty$ coincides with the observation (as it is a 1:1 mapping).

The MAP estimator should decide between two source values, and the decision among a limited set is part of detection theory, discussed later in Section 23.2. However,

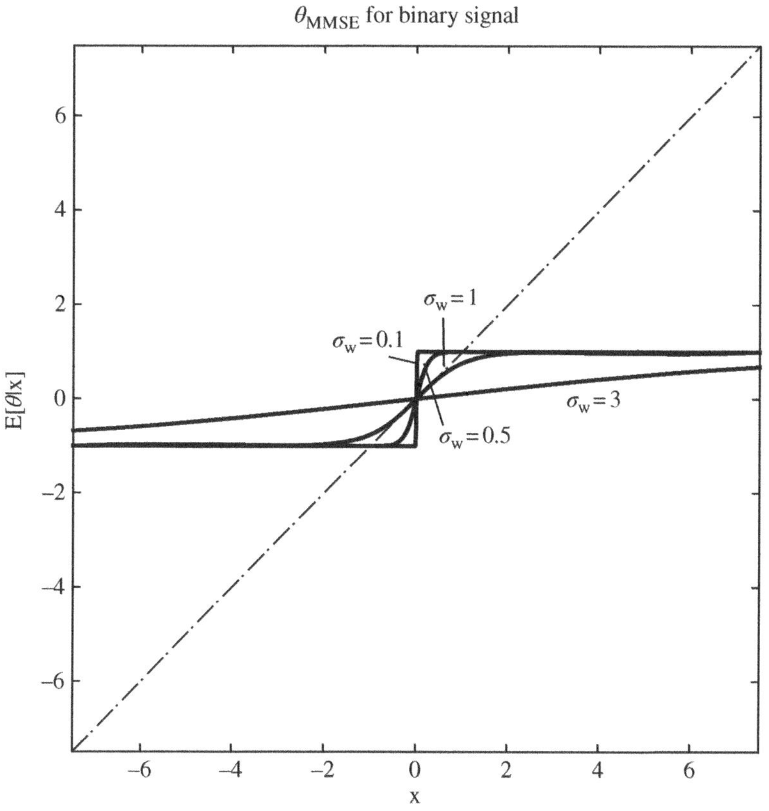

Figure 11.4 MMSE estimator for binary valued signals.

it is educational to use Bayesian inference to highlight the difference between MMSE and MAP in a non-Gaussian context. The a-posteriori pdf is:

$$p(\theta|x) = \frac{p(x|\theta)p(\theta)}{p(x)} = \begin{cases} \frac{p \times p_w(x-1)}{p \times p_w(x-1)+(1-p)\times p_w(x+1)} & \text{if } \theta = +1 \\ \frac{(1-p)\times p_w(x+1)}{p \times p_w(x-1)+(1-p)\times p_w(x+1)} & \text{if } \theta = -1 \end{cases}$$

so the MAP is

$$\theta_{MAP} = \arg\max_{\theta=\pm 1} p(\theta|x) = \begin{cases} +1 \text{ if } \frac{p_w(x-1)}{p_w(x+1)} > \frac{1-p}{p} \\ -1 \text{ if } \frac{p_w(x-1)}{p_w(x+1)} < \frac{1-p}{p} \end{cases}$$

Since

$$\frac{p_w(x-1)}{p_w(x+1)} = \exp(2x/\sigma_w^2)$$

the MAP estimator reduces to

$$\theta_{MAP} = \begin{cases} +1 \text{ if } x > \frac{\sigma_w^2}{2}\ln(\frac{1-p}{p}) \\ -1 \text{ if } x < \frac{\sigma_w^2}{2}\ln(\frac{1-p}{p}) \end{cases}$$

which for $p = 1/2$ (the most common case in digital communications) is $\theta_{MAP} = sign(x)$. It is without doubt that MMSE and MAP are clearly different from one another: the MMSE estimate returns a value that is not compatible with the generation mechanism but still has minimum MSE; the MAP estimate is simply the decision between the two alternative hypotheses $\theta = +1$ or $\theta = -1$ (see Chapter 23).

11.1.4 Example: Impulse Noise Mitigation

In real life, the stationarity assumption of Gaussian noise (or any other stochastic process) is violated by the presence of another phenomenon that occurs unexpectedly with low probability but large amplitudes: this is *impulse noise*. Gaussian mixtures can be used to model impulse noise that is characterized by long tails in the pdf (like those generated by lightning, power on/off, or other electromagnetic events) and it was modeled this way by Middleton in 1973. Still referring to the observation model $x(t) = \theta(t) + w(t)$, the rv labeled as "noise" is θ while w is the signal of interest (just to use a notation congruent with all above) with a Gaussian pdf.

The value of the random variable θ is randomly chosen between two (or more) values associated to two (or more) different random variables. Therefore the random variable θ can be expressed as

$$\theta = \begin{cases} \alpha \text{ with probability } p \\ \beta \text{ with probability } 1-p \end{cases}$$

and its pdf is the mixture

$$p_\theta(\theta) = p \times p_\alpha(\theta) + (1-p) \times p_\beta(\theta)$$

that is a linear combination of the pdfs associated to the two random variables α e β. The pdf of the observation is the convolution of the θ and w pdfs:

$$p_x(x) = p \times p_\alpha(x) * p_w(x) + (1-p) \times p_\beta(x) * p_w(x)$$

and this can be used to obtain (from (11.1)) the MMSE estimator of θ or w in many situations.

The Gaussian mixture with $\alpha \sim \mathcal{N}(\mu_\alpha, \sigma_\alpha^2)$ and $\beta \sim \mathcal{N}(\mu_\beta, \sigma_\beta^2)$ can be shaped to model the tails of the pdf; for zero-mean ($\mu_\alpha = \mu_\beta = 0$) Gaussians, the pdf is:

$$p_x(x) = p \times G(x; 0, \sigma_\alpha^2 + \sigma_w^2) + (1-p) \times G(x; 0, \sigma_\beta^2 + \sigma_w^2)$$

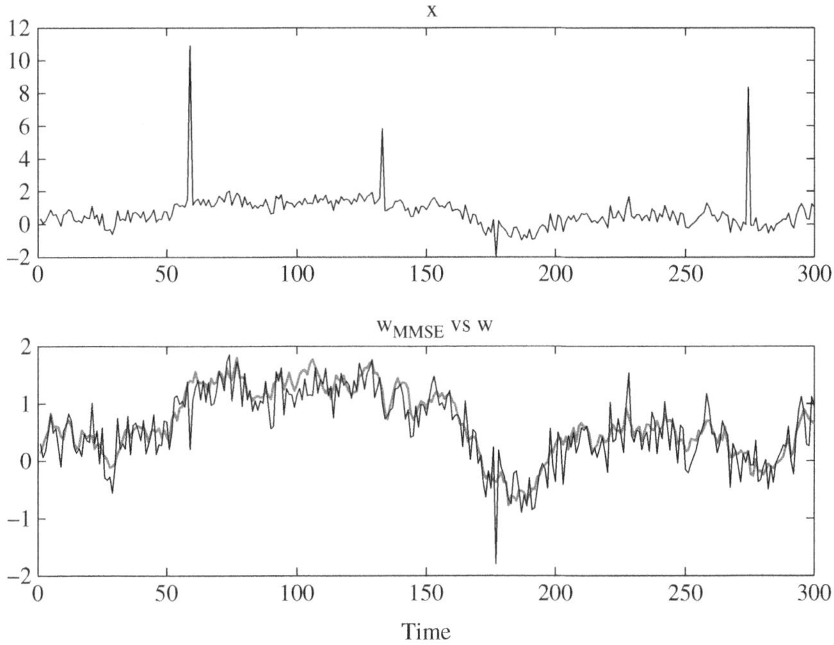

Figure 11.5 Example of data affected by impulse noise (upper figure), and after removal of impulse noise (lower figure) by MMSE estimator of the Gaussian samples (back line) compared to the true samples (gray line).

The MMSE estimator for x is

$$w_{MMSE} = x\sigma_w^2 \frac{\frac{p}{\sigma_\alpha^2+\sigma_w^2} G(x; 0, \sigma_\alpha^2 + \sigma_w^2) + \frac{1-p}{\sigma_\beta^2+\sigma_w^2} G(x; 0, \sigma_\beta^2 + \sigma_w^2)}{pG(x; 0, \sigma_\alpha^2 + \sigma_w^2) + (1-p)G(x; 0, \sigma_\beta^2 + \sigma_w^2)}$$

The effect of "spiky" data is illustrated in Figure 11.5 with x affected by Gaussian noise and some impulse noise. The estimated MMSE w_{MMSE} compared to true w (gray line) shows a strong reduction of non-Gaussian (impulse) noise but still leaves the Gaussian noise that cannot be mitigated from a non-linear MMSE estimator other than by filtering as is discussed later (simulated experiment with $\sigma_\alpha^2 = 50, \sigma_\beta^2 = 1/10, \sigma_w^2 = 1$ and $p = 10^{-2}$). Namely, the noise with large amplitude, when infrequently present ($p \ll 1/2$), is characterized by a Gaussian pdf with $\sigma_\alpha^2 \gg \sigma_\beta^2$, while normally there is little, or even no noise at all, as $\sigma_\beta^2 = 0$ ($p_\beta(\theta) \to \delta(\theta)$). In this case the estimator has a non-linearity with saturation that thresholds the observed values when too large compared to the background.

Inspection of the MMSE estimator for varying parameters in Figure 11.6 shows that the threshold effect is more sensitive to the probability $p \in \{10^{-1}, 10^{-2}, 10^{-5}\}$ of large amplitudes rather than to their variance $\sigma_\alpha^2 = 50$ (solid) and $\sigma_\alpha^2 = 10$ (dashed).

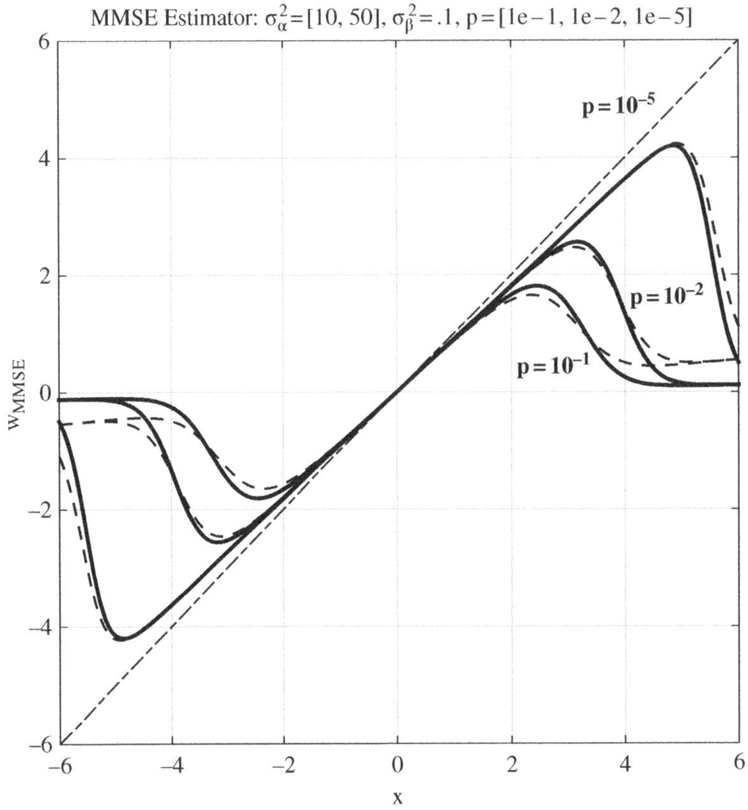

Figure 11.6 MMSE estimator for impulse noise modeled as Gaussian mixture.

11.2 Bayesian Estimation in Gaussian Settings

In general, Bayesian estimators are non-linear and their derivation could be cumbersome, or prohibitively complex. When parameters and data are jointly Gaussian, the conditional pdf $p(\theta|\mathbf{x})$ is Gaussian, and the MMSE estimator is a linear transformation of the observations. Furthermore, when the model is linear the MMSE estimator is very compact and in closed form. Even if the pdfs are non-Gaussian and the MMSE estimator would be non-linear, the use of a linear estimator in place of the true MMSE is very common and can be designed by minimizing the MSE. This is called a linear MMSE estimator (LMMSE), and in these situations the LMMSE is not optimal, but it has the benefit of being compact and pretty straightforward as it only requires knowledge of the first and second order central moments. For a Gaussian context, the LMMSE coincides with the MMSE estimator, and it is sub-optimal otherwise.

Tight lower bounds on the attainable MSE of the performance of any optimal or sub-optimal estimation scheme are useful performance analysis tools for comparisons. The Bayesian CRB in Section 11.4 is the most straightforward extension of the CRB when parameters are random.

11.2.1 MMSE Estimator

The use of a Gaussian model for parameters and observations is very common and this yields a closed form MMSE estimator. Let $p(\mathbf{x},\boldsymbol{\theta})$ be joint Gaussian; then

$$\begin{bmatrix} \mathbf{x} \\ \boldsymbol{\theta} \end{bmatrix} \sim \mathcal{N}\left(\begin{bmatrix} \boldsymbol{\mu}_x \\ \boldsymbol{\mu}_\theta \end{bmatrix}, \begin{bmatrix} \mathbf{C}_{xx} & \mathbf{C}_{x\theta} \\ \mathbf{C}_{\theta x} & \mathbf{C}_{\theta\theta} \end{bmatrix}\right)$$

The conditional pdf $p(\boldsymbol{\theta}|\mathbf{x})$ is also Gaussian (see Section 3.5 for details and derivation):

$$\boldsymbol{\theta}|\mathbf{x} \sim \mathcal{N}\left(\boldsymbol{\mu}_\theta + \mathbf{C}_{\theta x}\mathbf{C}_{xx}^{-1}(\mathbf{x}-\boldsymbol{\mu}_x), \mathbf{C}_{\theta\theta} - \mathbf{C}_{\theta x}\mathbf{C}_{xx}^{-1}\mathbf{C}_{x\theta}\right)$$

This condition is enough to prove that the MMSE estimator is

$$\boldsymbol{\theta}_{MMSE} = \mathbb{E}[\boldsymbol{\theta}|\mathbf{x}] = \boldsymbol{\mu}_\theta + \mathbf{C}_{\theta x}\mathbf{C}_{xx}^{-1}(\mathbf{x}-\boldsymbol{\mu}_x)$$

which is linear after removing the a-priori mean value $\boldsymbol{\mu}_x$ from observations, and it has an elegant closed form.

In the additive noise model

$$\mathbf{x} = \mathbf{s}(\boldsymbol{\theta}) + \mathbf{w} \text{ with } \mathbf{w} \sim \mathcal{N}(0, \mathbf{C}_{ww})$$

the covariance matrixes and the mean values lead to the following results (under the hypothesis that $\mathbf{w}$ and $\boldsymbol{\theta}$ are mutually uncorrelated):

$$\begin{aligned}
\boldsymbol{\mu}_x &= \mathbb{E}_{\theta,w}[\mathbf{x}] = \mathbb{E}_\theta[\mathbf{s}(\boldsymbol{\theta})] \\
\mathbf{C}_{xx} &= \mathbb{E}_{\theta,w}[(\mathbf{x}-\boldsymbol{\mu}_x)(\mathbf{x}-\boldsymbol{\mu}_x)^T] = \mathbb{E}_\theta[\mathbf{s}(\boldsymbol{\theta})\mathbf{s}(\boldsymbol{\theta})^T] + \mathbf{C}_{ww} - \boldsymbol{\mu}_x\boldsymbol{\mu}_x^T \\
\mathbf{C}_{\theta x} &= \mathbb{E}_{\theta,w}[(\boldsymbol{\theta}-\boldsymbol{\mu}_\theta)(\mathbf{x}-\boldsymbol{\mu}_x)^T] = \mathbb{E}_\theta[(\boldsymbol{\theta}-\boldsymbol{\mu}_\theta)(\mathbf{s}(\boldsymbol{\theta})-\boldsymbol{\mu}_x)^T]
\end{aligned}$$

Note that moments wrt the parameters $\mathbb{E}_\theta[.]$ are the only new quantities to be evaluated to derive the MMSE estimator. When the model $\mathbf{s}(\boldsymbol{\theta})$ is linear, the central moments are further simplified as shown below.

11.2.2 MMSE Estimator for Linear Models

The linear model with additive Gaussian noise is

$$\begin{aligned}
&\mathbf{x} = \mathbf{H}\cdot\boldsymbol{\theta} + \mathbf{w} \\
&\mathbf{w} \sim \mathcal{N}(0, \mathbf{C}_{ww}) \\
&\boldsymbol{\theta} \sim \mathcal{N}(\boldsymbol{\mu}_\theta, \mathbf{C}_{\theta\theta})
\end{aligned}$$

the moments can be related to the linear model $\mathbf{H}$:

$$\begin{aligned}
\boldsymbol{\mu}_x &= \mathbb{E}_{\theta,w}[\mathbf{H}\cdot\boldsymbol{\theta} + \mathbf{w}] = \mathbf{H}\boldsymbol{\mu}_\theta \\
\mathbf{C}_{\theta x} &= \mathbb{E}_{\theta,w}[(\boldsymbol{\theta}-\boldsymbol{\mu}_\theta)(\mathbf{H}\cdot(\boldsymbol{\theta}-\boldsymbol{\mu}_\theta)+\mathbf{w})^T] = \mathbf{C}_{\theta\theta}\mathbf{H}^T
\end{aligned}$$

$$\mathbf{C}_{xx} = \mathbb{E}_{\theta,w}[(\mathbf{H}\cdot(\boldsymbol{\theta}-\boldsymbol{\mu}_\theta)+\mathbf{w})(\mathbf{H}\cdot(\boldsymbol{\theta}-\boldsymbol{\mu}_\theta)+\mathbf{w})^T]$$
$$= \mathbf{H}\mathbf{C}_{\theta\theta}\mathbf{H}^T + \mathbf{C}_{ww}$$

and the MMSE estimator is now

$$\boldsymbol{\theta}_{MMSE} = \boldsymbol{\mu}_\theta + \mathbf{C}_{\theta\theta}\mathbf{H}^T(\mathbf{H}\mathbf{C}_{\theta\theta}\mathbf{H}^T + \mathbf{C}_{ww})^{-1}(\mathbf{x}-\mathbf{H}\boldsymbol{\mu}_\theta) \quad (11.2)$$

The covariance

$$\text{cov}(\boldsymbol{\theta}|\mathbf{x}) = \mathbf{C}_{\theta\theta} - \mathbf{C}_{\theta\theta}\mathbf{H}^T(\mathbf{H}\mathbf{C}_{\theta\theta}\mathbf{H}^T + \mathbf{C}_{ww})^{-1}\mathbf{H}\mathbf{C}_{\theta\theta} \quad (11.3)$$

is also useful to have the performance of the a-posteriori method. Note that the a-posteriori covariance is always an improvement over the a-priori pdf, and this results from $\mathbf{C}_{\theta\theta}\mathbf{H}^T(\mathbf{H}\mathbf{C}_{\theta\theta}\mathbf{H}^T + \mathbf{C}_{ww})^{-1}\mathbf{H}\mathbf{C}_{\theta\theta} \succeq 0$ and thus

$$\text{cov}(\boldsymbol{\theta}|\mathbf{x}) \preceq \text{cov}(\boldsymbol{\theta}) = \mathbf{C}_{\theta\theta}$$

where the equality is when $\mathbf{C}_{\theta x} = \mathbf{0}$ (i.e., data is not correlated with the parameters, and estimation of $\boldsymbol{\theta}$ from $\mathbf{x}$ is meaningless).

Computationally Efficient MMSE

There is a computational drawback in the MMSE method as the matrix $(\mathbf{H}\mathbf{C}_{\theta\theta}\mathbf{H}^T + \mathbf{C}_{ww})$ has dimension $N \times N$ that depends on the size of the observations (N) and this could be prohibitively expensive as matrix inversion costs $\mathcal{O}(N^3)$. One efficient formulation follows from the algebraic properties of matrixes. Namely, the Woodbury identity $\mathbf{A}\mathbf{B}^H(\mathbf{B}\mathbf{A}\mathbf{B}^H+\mathbf{C})^{-1} = (\mathbf{A}^{-1}+\mathbf{B}^H\mathbf{C}^{-1}\mathbf{B})^{-1}\mathbf{B}^H\mathbf{C}^{-1}$ (see Section 1.1.1) can be applied to the linear transformation in MMSE expression:

$$\mathbf{C}_{\theta\theta}\mathbf{H}^T(\mathbf{H}\mathbf{C}_{\theta\theta}\mathbf{H}^T + \mathbf{C}_{ww})^{-1} = (\mathbf{H}^T\mathbf{C}_{ww}^{-1}\mathbf{H} + \mathbf{C}_{\theta\theta}^{-1})^{-1}\mathbf{H}^T\mathbf{C}_{ww}^{-1}$$

This leads to an alternative formulation of the MMSE estimator:

$$\boldsymbol{\theta}_{MMSE} = \boldsymbol{\mu}_\theta + (\mathbf{H}^T\mathbf{C}_{ww}^{-1}\mathbf{H} + \mathbf{C}_{\theta\theta}^{-1})^{-1}\mathbf{H}^T\mathbf{C}_{ww}^{-1}(\mathbf{x}-\mathbf{H}\boldsymbol{\mu}_\theta) \quad (11.4)$$

This expression is computationally far more efficient than (11.2) as the dimension of $(\mathbf{H}^T\mathbf{C}_{ww}^{-1}\mathbf{H} + \mathbf{C}_{\theta\theta}^{-1})$ is $p \times p$, and the number of parameters is always smaller than the number of measurements ($p < N$); hence the MMSE estimator (11.4) has a cost $\mathcal{O}(p^3) \ll \mathcal{O}(N^3)$.

Moreover, the linear transformation $\mathbf{C}_{ww}^{-1/2}$ can be evaluated in advance as a Cholesky factorization of $\mathbf{C}_{ww}^{-1}$ and applied both to the regressor as $\tilde{\mathbf{H}} = \mathbf{C}_{ww}^{-1/2}\mathbf{H}$, and the observation $\mathbf{x}$ with the result of decorrelating the noise components (*whitening* filter, see Section 7.8.2) as $\tilde{\mathbf{x}} = \mathbf{C}_{ww}^{-1/2}\mathbf{x} = \tilde{\mathbf{H}}\cdot\boldsymbol{\theta} + \tilde{\mathbf{w}}$ with $\tilde{\mathbf{w}} \sim \mathcal{N}(0, \mathbf{I})$:

$$\boldsymbol{\theta}_{MMSE} = \boldsymbol{\mu}_\theta + (\tilde{\mathbf{H}}^T\tilde{\mathbf{H}} + \mathbf{C}_{\theta\theta}^{-1})^{-1}\tilde{\mathbf{H}}^T(\tilde{\mathbf{x}} - \tilde{\mathbf{H}}\boldsymbol{\mu}_\theta)$$

Example

The MMSE estimator provides a straightforward solution to the example of mean value estimate from multiple observations in Section 11.1.1 with model:

$$\mathbf{x} = \mathbf{1}\theta + \mathbf{w}$$
$$\mathbf{w} \sim \mathcal{N}(0, \sigma^2 \mathbf{I})$$
$$\theta \sim \mathcal{N}(\bar{\theta}, \sigma_\theta^2)$$

The general expression for the linear MMSE estimator is

$$\theta_{MMSE} = \bar{\theta} + \sigma_\theta^2 \mathbf{1}^T (\mathbf{1}\sigma_\theta^2 \mathbf{1}^T + \sigma_w^2 \mathbf{I})^{-1} (\mathbf{x} - \mathbf{1}\bar{\theta}) \tag{11.5}$$

After the Woodbury identity (Section 1.1.1) it is more manageable:

$$\theta_{MMSE} = \alpha \bar{\theta} + (1 - \alpha)\bar{x}$$

thus highlighting that the MMSE is the combination of the a-priori mean value $\bar{\theta}$ and the sample mean of the observations $\bar{x} = \mathbf{1}^T \mathbf{x}/N$, with

$$\alpha = \frac{\frac{\sigma_w^2}{N}}{\frac{\sigma_w^2}{N} + \sigma_\theta^2}$$

For large N, the influence of noise decreases as σ_w^2/N, and for $N \to \infty$ the importance of the a-priori information decreases as $\alpha \to 0$, and $\theta_{MMSE} \to \bar{x}$. Using the same procedure it is possible to evaluate

$$\text{cov}(\theta|\mathbf{x}) = \sigma^2_{\theta_{MMSE}} = \sigma_\theta^2 (1 - \alpha)$$

that is dependent on the weighting term α previously defined.

11.3 LMMSE Estimation and Orthogonality

The MMSE estimator $\mathbb{E}[\theta|\mathbf{x}]$ can be complex to derive as it requires a detailed statistical model and the evaluation of the conditional mean $\mathbb{E}[\theta|\mathbf{x}]$ can be quite difficult, namely when involving non-Gaussian rvs. A pragmatic approach is to derive a sub-optimal estimator that is linear with respect to the observations and is based only on the knowledge of the first and the second order moments of the involved rvs. This is the the Linear MMSE (LMMSE) estimator, where the prefix "linear" denotes the structure of the estimator itself and highlights its sub-optimality for non-Gaussian contexts.

The LMMSE is based on the linear combination of N observations

$$\hat{\theta}_k(\mathbf{x}) = a_{k,0} + \sum_{i=1}^{N} a_{k,i} x[i] = a_{k,0} + \mathbf{a}_k^T \mathbf{x}$$

while for the ensemble of p parameters it is

$$\begin{bmatrix} \hat{\theta}_1(\mathbf{x}) \\ \hat{\theta}_2(\mathbf{x}) \\ \vdots \\ \hat{\theta}_p(\mathbf{x}) \end{bmatrix} = \begin{bmatrix} \mathbf{a}_1^T \\ \mathbf{a}_2^T \\ \vdots \\ \mathbf{a}_p^T \end{bmatrix} \cdot \mathbf{x} \Rightarrow \hat{\boldsymbol{\theta}}(\mathbf{x}) = \mathbf{a}_0 + \mathbf{A}\mathbf{x} \tag{11.6}$$

The estimator is based on the $p \times N$ coefficients of matrix $\mathbf{A}$ that are obtained from the two constraints:

- **Unbiasedness**: from the Bayesian definition of bias (Section 6.3)

  $$\mathbb{E}_{\theta,x}[\theta_k - \hat{\theta}_k(\mathbf{x})] = 0 \Rightarrow \mathbb{E}[\boldsymbol{\theta}] - \mathbf{A}\mathbb{E}[\mathbf{x}] - \mathbf{a}_0 = \mathbf{0} \Rightarrow \mathbf{a}_0 = \mathbb{E}[\boldsymbol{\theta}] - \mathbf{A}\mathbb{E}[\mathbf{x}]$$

  this provides a relationship for the scaling term

  $$\mathbf{a}_0 = \boldsymbol{\mu}_\theta - \mathbf{A}\boldsymbol{\mu}_x$$

  This term can be obtained by rewriting the estimator after the subtraction of the mean from both the parameters and the observations:

  $$\underbrace{\hat{\boldsymbol{\theta}}(\mathbf{x}) - \boldsymbol{\mu}_\theta}_{\delta\hat{\boldsymbol{\theta}}(\mathbf{x})} = \mathbf{A}(\underbrace{\mathbf{x} - \boldsymbol{\mu}_x}_{\delta\mathbf{x}})$$

- **MSE minimization**: given the error

  $$\varepsilon_k(\mathbf{x}) = \theta_k - \hat{\theta}_k(\mathbf{x})$$

  that corresponds to the kth parameter, one should minimize its mean square value

  $$\min_{\mathbf{A}} \mathbb{E}_{\theta,x}[\varepsilon_k^2(\mathbf{x})], \text{ for } \forall k = 1, \ldots, p$$

  To simplify the notation it is assumed that the mean is subtracted from all the involved variables (in other words the assignment is $\delta\hat{\boldsymbol{\theta}}(\mathbf{x}) \to \hat{\boldsymbol{\theta}}(\mathbf{x})$ and $\delta\mathbf{x} \to \mathbf{x}$), or equivalently $\boldsymbol{\mu}_x = \mathbf{0}$ and $\boldsymbol{\mu}_\theta = \mathbf{0}$.
  Setting the gradients to zero:

  $$\frac{\partial}{\partial a_{k,i}} \mathbb{E}[\varepsilon_k^2(\mathbf{x})] = 0 \quad \forall k, i$$
  $$\mathbb{E}[\varepsilon_k(\mathbf{x}) \cdot x[i]] = 0 \quad \forall k, i$$

  This latter relationship is defined as the (statistical) *orthogonality condition* between the estimation error ($\varepsilon_k(\mathbf{x}) = \theta_k - \hat{\theta}_k(\mathbf{x})$) and the observations

  $$\mathbb{E}[\varepsilon_k(\mathbf{x}) \cdot \mathbf{x}^T] = \underset{1 \times N}{\mathbf{0}} = \mathbf{0}^T \quad \forall k$$

or equivalently

$$\mathbb{E}[(\theta_k - \mathbf{a}_k^T\mathbf{x})\mathbf{x}^T] = \underset{1\times N}{\mathbf{0}} = \mathbf{0}^T \quad \forall k$$

By transposing both sides of the previous equation it is possible to write:

$$\mathbb{E}[\mathbf{x}\mathbf{x}^T]\mathbf{a}_k = \mathbb{E}[\theta_k\mathbf{x}] \quad \forall k$$

that shows how the orthogonality conditions lead to a set of N linear equations (one for each parameter θ_k) with N unknowns (the elements of the $\mathbf{a}_k$ vector). The solution of this set of equation is given by (recall the assumptions here: $\boldsymbol{\mu}_\theta = \mathbf{0}$, $\boldsymbol{\mu}_x = \mathbf{0}$, or $\mathbf{a}_0 = \mathbf{0}$)

$$\mathbf{a}_k = (\mathbb{E}[\mathbf{x}\mathbf{x}^T])^{-1} \cdot \mathbb{E}[\theta_k\mathbf{x}] \tag{11.7}$$

Generalizing the orthogonality conditions between observations and estimation error:

$$\mathbb{E}[\boldsymbol{\varepsilon}(\mathbf{x}) \cdot \mathbf{x}^T] = \mathbf{0}$$

which corresponds to pN equations into the pN unknowns entries of $\mathbf{A}$. Substituting the estimator relationship $\hat{\theta}(\mathbf{x}) = \mathbf{A}\mathbf{x}$ into the previous equation yields the LMMSE estimator:

$$\mathbf{A} = \mathbf{C}_{\theta x}\mathbf{C}_{xx}^{-1} \tag{11.8}$$

Considering now also the relation to zeroing the polarization, it is possible to write the general expression for the LMMSE estimator:

$$\boldsymbol{\theta}_{LMMSE} = \boldsymbol{\mu}_\theta + \mathbf{C}_{\theta x}\mathbf{C}_{xx}^{-1}(\mathbf{x} - \boldsymbol{\mu}_x)$$

that obviously gives $\mathbb{E}[\boldsymbol{\theta}_{LMMSE}] = \boldsymbol{\mu}_\theta$.
The dispersion of the estimate is by the covariance:

$$\text{cov}[\boldsymbol{\theta}_{LMMSE}] = \mathbb{E}[(\boldsymbol{\theta} - \mathbf{C}_{\theta x}\mathbf{C}_{xx}^{-1}\mathbf{x})(\boldsymbol{\theta} - \mathbf{C}_{\theta x}\mathbf{C}_{xx}^{-1}\mathbf{x})^T] = \mathbf{C}_{\theta\theta} - \mathbf{C}_{\theta x}\mathbf{C}_{xx}^{-1}\mathbf{C}_{x\theta}$$

and the fundamental relationships of LMMSE are the same as for the MMSE for Gaussian models in Section 11.2.1.

Compared to the general MMSE, that is never simple or linear, any LMMSE estimator is based on the moments that should be known, and the cross-covariance matrix $\mathbf{C}_{\theta x}$ between the observations and the parameters. This implies that the LMMSE estimator can always be derived in closed form, but it has no proof of optimality. However, it is worthwhile to remark that the LMMSE coincides with the MMSE estimator for Gaussian variables, and it is the optimum MMSE estimator in this situation. The linear MMSE (LMMSE) estimator provides a sub-optimal estimation in the case of random variables with arbitrary pdfs even if it provides the estimation with the minimum MSE

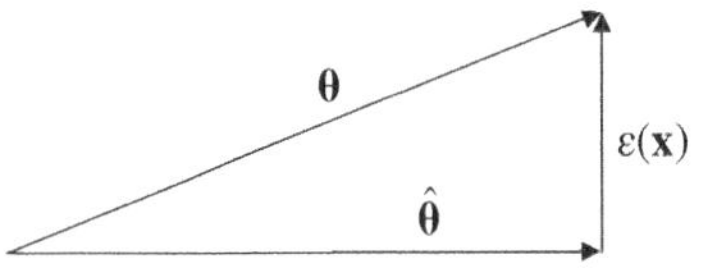

Figure 11.7 Geometric view of LMMSE orthogonality.

among all linear estimators. The degree of sub-optimality follows from the analysis of the covariance matrix cov[θ_{LMMSE}], which for the sub-optimal case (non-Gaussian pdfs) has a larger value with respect to the "true" (optimal) MMSE estimation. In summary:

$$\text{cov}[\theta_{LMMSE}] - \text{cov}[\theta_{MMSE}] \succeq \mathbf{0}$$

(the difference is semipositive definite). The simplicity of the LMMSE estimator justifies its wide adoption in science and engineering, and sometimes the optimal MMSE is so complex to derive that the LMMSE is referred to (erroneously!) as the MMSE estimator.

Orthogonality

The orthogonality condition between the observations $\mathbf{x}$ and the estimation errors $\varepsilon(\mathbf{x}) = \theta - \hat{\theta}(\mathbf{x})$ states that these are mutually orthogonal

$$\mathbb{E}[\underbrace{(\theta - \mathbf{A}\mathbf{x})}_{\varepsilon(\mathbf{x})}\mathbf{x}^T] = \mathbf{0}$$

This is the general way to derive the LMMSE estimator as the set of Np equations that are enough to derive the $p \times N$ entries of $\mathbf{A}$. The representation

$$\theta = \hat{\theta} + \varepsilon(\mathbf{x})$$

decomposes the vector θ into two orthogonal components as

$$\mathbb{E}[\varepsilon(\mathbf{x})\hat{\theta}^T] = \mathbb{E}[\varepsilon(\mathbf{x})\mathbf{x}^T\mathbf{A}] = \mathbf{0}$$

This relation is also called *Pitagora's statistical theorem*, illustrated geometrically in Figure 11.7.

11.4 Bayesian CRB

When MMSE estimators are not feasible, or practical, the comparison of any estimator with lower bounds is a useful performance index. Given any Bayesian estimator $\mathbf{g}(\mathbf{x})$, the Bayesian MSE is

$$\mathbb{E}_{x,\theta}[(\theta - \mathbf{g}(\mathbf{x}))(\theta - \mathbf{g}(\mathbf{x}))^T] \succeq \mathbb{E}_{x,\theta}[(\theta - \mathbb{E}[\theta|\mathbf{x}])(\theta - \mathbb{E}[\theta|\mathbf{x}])^T]$$

where the right hand side is the conditional mean estimator that has the minimum MSE over all possible estimators. The Bayesian CRB was introduced in the mid-1960s by Van Trees [37, 42] and it states that under some regularity conditions

$$\mathbb{E}_{x,\theta}[(\boldsymbol{\theta}-\mathbf{g}(\mathbf{x}))(\boldsymbol{\theta}-\mathbf{g}(\mathbf{x}))^T] \succeq \mathbf{C}_{BCRB} = \mathbf{J}_B^{-1}$$

where

$$[\mathbf{J}_B]_{ij} = \mathbb{E}_{x,\theta}\left[\frac{\partial \ln p(\mathbf{x},\boldsymbol{\theta})}{\partial \theta_i}\frac{\partial \ln p(\mathbf{x},\boldsymbol{\theta})}{\partial \theta_j}\right]$$

Since $\ln p(\mathbf{x},\boldsymbol{\theta}) = \ln p(\mathbf{x}|\boldsymbol{\theta}) + \ln p(\boldsymbol{\theta})$, the Bayesian CRB is the sum of two terms:

$$\mathbf{J}_B = \mathbf{J}_D + \mathbf{J}_P$$

where

$$[\mathbf{J}_D]_{ij} = \mathbb{E}_{x,\theta}\left[\frac{\partial \ln p(\mathbf{x}|\boldsymbol{\theta})}{\partial \theta_i}\frac{\partial \ln p(\mathbf{x}|\boldsymbol{\theta})}{\partial \theta_j}\right]$$
$$= \mathbb{E}_{\theta}\left[\mathbb{E}_{x|\theta}\left[\frac{\partial \ln p(\mathbf{x}|\boldsymbol{\theta})}{\partial \theta_i}\frac{\partial \ln p(\mathbf{x}|\boldsymbol{\theta})}{\partial \theta_j}\right]\right] = \mathbb{E}_{\theta}\left[[\mathbf{J}(\boldsymbol{\theta})]_{ij}\right]$$

is the contribution of data that can be interpreted as the mean of the FIM for non-Bayesian estimators (Section 8.1) over the distribution of parameters, and

$$[\mathbf{J}_P]_{ij} = \mathbb{E}_{\theta}\left[\frac{\partial \ln p(\boldsymbol{\theta})}{\partial \theta_i}\frac{\partial \ln p(\boldsymbol{\theta})}{\partial \theta_j}\right]$$

is the contribution of the a-priori information.

In detail, the MSE bound of any Bayesian estimator can be stated as follows:

$$\mathbb{E}_{x,\theta}[(\hat{\theta}_i(\mathbf{x})-\theta_i)^2] \geq \left[\mathbf{J}_B^{-1}\right]_{ii}$$

with equality only if the a-posteriori pdf $p(\boldsymbol{\theta}|\mathbf{x})$ is Gaussian. The condition for Bayesian CRB to hold requires that the estimate is unbiased according to the Bayesian context

$$\mathbb{E}_{x,\theta}\left[\boldsymbol{\theta}-\mathbf{g}(\mathbf{x})\right] = \mathbf{0}$$

which is evaluated for an average choice of parameters according to their pdf $p[\boldsymbol{\theta}]$. This unbiasedness condition for a random set of parameters is weaker than the CRB for deterministic parameters, which needs to be unbiased for *any* choice of $\boldsymbol{\theta}$. Further inspection shows that (Section 1.1.1)

$$(\mathbb{E}_{\theta}[\mathbf{J}(\boldsymbol{\theta})])^{-1} \succeq (\mathbf{J}_D+\mathbf{J}_P)^{-1} = \mathbf{C}_{BCRB}$$

where $\mathbf{J}(\boldsymbol{\theta})$ is the FIM for non-Bayesian case (Chapter 8).

11.5 Mixing Bayesian and Non-Bayesian

The Bayesian vs. non-Bayesian debate is a long-standing one. However, complex engineering problems are not always so clear— the a-priori pdf is not known, or a-priori information is not given as a pdf, and the parameters could mix nuisance, random, and deterministic ones. Making a set of cases, exceptions, and examples with special settings could take several pages, and the reader would feel disoriented faced with a taxonomy. Other than to warn the reader that one can be faced with a complex combination of parameters, the recommendation is to face the problems with a cool head using all the analytical tools discussed so far, and mix them as appropriate.

11.5.1 Linear Model with Mixed Random/Deterministic Parameters

Let the linear model with mixed parameters be:

$$\mathbf{x} = [\mathbf{H}_d, \mathbf{H}_r]\begin{bmatrix} \boldsymbol{\theta}_d \\ \boldsymbol{\theta}_r \end{bmatrix} + \mathbf{w}$$

where $\boldsymbol{\theta}_d$ is deterministic, and the other set $\boldsymbol{\theta}_r$ is random and Gaussian

$$\boldsymbol{\theta}_r \sim \mathcal{N}(\boldsymbol{\mu}_{\theta_r}, \mathbf{C}_{\theta_r\theta_r})$$

This is the called hybrid estimation problem. Assuming that the noise is $\mathbf{w} \sim \mathcal{N}(0, \mathbf{I}\sigma_w^2)$, one can write the joint pdf $p(\mathbf{x}, \boldsymbol{\theta}_r|\boldsymbol{\theta}_d) = p(\mathbf{x}|\boldsymbol{\theta}_r, \boldsymbol{\theta}_d)p(\boldsymbol{\theta}_r|\boldsymbol{\theta}_d) = p(\mathbf{x}|\boldsymbol{\theta}_r, \boldsymbol{\theta}_d)p(\boldsymbol{\theta}_r)$; under the Gaussian assumption of terms, the logarithmic of the pdf is

$$\begin{aligned}\ln p(\mathbf{x}, \boldsymbol{\theta}_r|\boldsymbol{\theta}_d) = c &- \frac{1}{2\sigma_w^2}(\mathbf{x} - \mathbf{H}_d\boldsymbol{\theta}_d - \mathbf{H}_r\boldsymbol{\theta}_r)^T(\mathbf{x} - \mathbf{H}_d\boldsymbol{\theta}_d - \mathbf{H}_r\boldsymbol{\theta}_r) \\ &- \frac{1}{2}(\boldsymbol{\theta}_r - \boldsymbol{\mu}_{\theta_r})^T\mathbf{C}_{\theta_r\theta_r}^{-1}(\boldsymbol{\theta}_r - \boldsymbol{\mu}_{\theta_r})\end{aligned}$$

and the MAP/ML estimator is obtained by setting

$$\frac{\partial}{\partial \boldsymbol{\theta}} \ln p(\mathbf{x}, \boldsymbol{\theta}_r|\boldsymbol{\theta}_d) = 0$$

for random ($\boldsymbol{\theta}_r$) and deterministic ($\boldsymbol{\theta}_d$) parameters, and every $\mathbf{x}$. The log-pdf can be rewritten more compactly as

$$\ln p(\mathbf{x}, \boldsymbol{\theta}_r|\boldsymbol{\theta}_d) = c - \frac{1}{2\sigma_w^2}(\mathbf{x} - \mathbf{H}\boldsymbol{\theta})^T(\mathbf{x} - \mathbf{H}\boldsymbol{\theta}) - \frac{1}{2}(\boldsymbol{\theta} - \boldsymbol{\mu})^T\mathbf{C}^{-1}(\boldsymbol{\theta} - \boldsymbol{\mu})$$

where for analytical convenience it is defined that

$$\mathbf{C}^{-1} = \begin{bmatrix} \mathbf{0} & \mathbf{0} \\ \mathbf{0} & \mathbf{C}_{\theta_r\theta_r}^{-1} \end{bmatrix}$$

$$\boldsymbol{\mu} = \begin{bmatrix} \mathbf{0} \\ \boldsymbol{\mu}_{\theta_r} \end{bmatrix}$$

The gradient becomes

$$\frac{\partial}{\partial \boldsymbol{\theta}} \ln p(\mathbf{x}, \boldsymbol{\theta}_r | \boldsymbol{\theta}_d) = \frac{1}{\sigma_w^2} \mathbf{H}^T (\mathbf{x} - \mathbf{H}\boldsymbol{\theta}) - \mathbf{C}^{-1}(\boldsymbol{\theta} - \boldsymbol{\mu}) = \left(\frac{1}{\sigma_w^2} \mathbf{H}^T \mathbf{H} + \mathbf{C}^{-1} \right) (\hat{\boldsymbol{\theta}}(\mathbf{x}) - \boldsymbol{\theta})$$

where the joint estimate is

$$\hat{\boldsymbol{\theta}}(\mathbf{x}) = \left(\frac{1}{\sigma_w^2} \mathbf{H}^T \mathbf{H} + \mathbf{C}^{-1} \right)^{-1} \left(\frac{1}{\sigma_w^2} \mathbf{H}^T \mathbf{x} + \mathbf{C}^{-1} \boldsymbol{\mu} \right)$$

The CRB for this hybrid system is [43](Section 11.5.2)

$$\mathbb{E}_{x,\theta_r}[(\boldsymbol{\theta} - \hat{\boldsymbol{\theta}}(\mathbf{x}))(\boldsymbol{\theta} - \hat{\boldsymbol{\theta}}(\mathbf{x}))^T] \succeq \mathbf{C}_{HCRB}(\boldsymbol{\theta}_d) = \mathbf{J}_{HB}^{-1}$$

where

$$\mathbf{J}_{HB} = \frac{1}{\sigma_w^2} \mathbf{H}^T \mathbf{H} + \mathbf{C}^{-1}$$

Example: A sinusoid has random amplitude sine/cosine components (this generalizes the case of deterministic amplitude in Section 9.2). From Section 5.2, the model is

$$\mathbf{x} = \mathbf{H}(\omega) \cdot \boldsymbol{\alpha} + \mathbf{w}$$

with

$$\mathbf{w} \sim \mathcal{N}(0, \mathbf{I}\sigma_w^2)$$
$$\boldsymbol{\alpha} \sim \mathcal{N}(\mathbf{0}, \mathbf{I}\sigma_A^2)$$

The log-pdf for random $\boldsymbol{\alpha}$ is

$$\ln p(\mathbf{x}, \boldsymbol{\alpha} | \omega) = c - \frac{1}{2\sigma_w^2} (\mathbf{x} - \mathbf{H}\boldsymbol{\alpha})^T (\mathbf{x} - \mathbf{H}\boldsymbol{\alpha}) - \frac{1}{2\sigma_A^2} \boldsymbol{\alpha}^T \boldsymbol{\alpha}$$

that minimized wrt to $\boldsymbol{\alpha}$ for a given ω yields:

$$\hat{\boldsymbol{\alpha}}(\mathbf{x}, \omega) = \left(\mathbf{H}^T \mathbf{H} + \frac{\sigma_w^2}{\sigma_A^2} \mathbf{I} \right)^{-1} \mathbf{H}^T \mathbf{x} \simeq \gamma \mathbf{H}^T \mathbf{x}$$

Since $\mathbf{H}^T \mathbf{H} = N\mathbf{I}/2$:

$$\gamma = \left(\frac{N}{2} + \frac{\sigma_w^2}{\sigma_A^2} \right)^{-1}$$

This estimate $\hat{\boldsymbol{\alpha}}(\mathbf{x},\omega)$ substituted into the log-pdf (neglecting scale-factors)

$$\ln p(\mathbf{x},\hat{\boldsymbol{\alpha}}|\omega) = c - \frac{1}{2\sigma_w^2}\mathbf{x}^T[(\mathbf{I}-\gamma\mathbf{H}\mathbf{H}^T)^2 - \frac{1}{2\sigma_A^2}\gamma^2\mathbf{H}\mathbf{H}^T]\mathbf{x} \simeq c' + \gamma'\mathbf{x}^T\mathbf{H}\mathbf{H}^T\mathbf{x}$$

proves that, regardless of whether the amplitude is deterministic or modeled as stochastic Gaussian, the frequency estimate is the value that maximizes the amplitude of the Fourier transform as for the MLE in Section 9.2, and Section 9.3.

11.5.2 Hybrid CRB

The case when the parameters are mixed non-random and random needs a separate discussion. Let

$$\boldsymbol{\theta} = \begin{bmatrix} \boldsymbol{\theta}_d \\ \boldsymbol{\theta}_r \end{bmatrix}$$

be the $p \times 1$ parameters grouped into $p_1 \times 1$ non-random $\boldsymbol{\theta}_d$, and $p_2 \times 1$ random $\boldsymbol{\theta}_r$, the MSE hybrid bound is obtained from the pdf [37]

$$p(\mathbf{x},\boldsymbol{\theta}_r|\boldsymbol{\theta}_d)$$

that decouples random and non-random parameters. The Hybrid CRB

$$\mathbb{E}_{x,\theta_r}[(\boldsymbol{\theta}-\mathbf{g}(\mathbf{x}))(\boldsymbol{\theta}-\mathbf{g}(\mathbf{x}))^T] \succeq \mathbf{C}_{HCRB}(\boldsymbol{\theta}_d) = \mathbf{J}_{HB}^{-1}$$

where

$$\mathbf{J}_{HB} = \mathbf{J}_{HD} + \mathbf{J}_{HP}$$

is now partitioned with expectations over $\boldsymbol{\theta}_r$

$$\mathbf{J}_{HD}(\boldsymbol{\theta}_d) = \mathbb{E}_{x,\theta_r}\left[\frac{\partial \ln p(\mathbf{x},\boldsymbol{\theta}_r|\boldsymbol{\theta}_d)}{\partial \boldsymbol{\theta}}\frac{\partial \ln p(\mathbf{x},\boldsymbol{\theta}_r|\boldsymbol{\theta}_d)}{\partial \boldsymbol{\theta}^T}\right] = \mathbb{E}_{\theta_r}[\mathbf{J}(\boldsymbol{\theta})]$$

$$\mathbf{J}_{HP} = \begin{bmatrix} \mathbf{0}_{p_1\times p_1} & \mathbf{0}_{p_1\times p_2} \\ \mathbf{0}_{p_2\times p_1} & \mathbb{E}_{\theta_r}\left[\frac{\partial \ln p(\boldsymbol{\theta}_r)}{\partial \boldsymbol{\theta}_r}\frac{\partial \ln p(\boldsymbol{\theta}_r)}{\partial \boldsymbol{\theta}_r^T}\right] \end{bmatrix}$$

The random parameters $\boldsymbol{\theta}_r$ can be nuisance, and the Hybrid CRB on deterministic parameters $\boldsymbol{\theta}_d$ is less tight than the CRB for the same parameters [43].

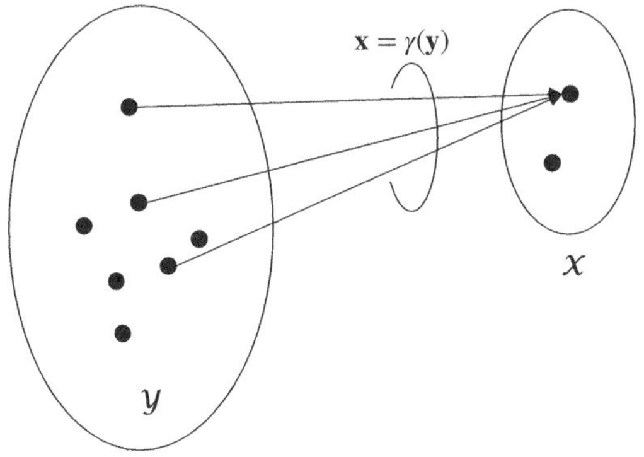

Figure 11.8 Mapping between complete $\mathcal{Y}$ and incomplete (or data) $\mathcal{X}$ set in EM method.

11.6 Expectation-Maximization (EM)

Making reference to non-Bayesian methods in Chapter 7, the MLE of the parameters $\boldsymbol{\theta}$ according to the set of observations $\mathbf{x}$ is

$$\hat{\theta}(\mathbf{x}) = \arg\max_{\theta} \ln p(\mathbf{x}|\theta)$$

However, it could happen that the optimization of the log-likelihood function is complex and/or not easily tractable, and/or it depends on too many parameters to be easily handled. The expectation maximization algorithm [45] solves the MLE by *postulating* that there exists a set of alternative observations $\mathbf{y}$ called *complete data* so that it is always possible to get the "true" observations $\mathbf{x}$ (here referred as *incomplete data*) from a not-invertible transformation (i.e., many-to-one) $\mathbf{x} = \gamma(\mathbf{y})$. The new set y is made in such a way that the MLE wrt the complete set $\ln p(\mathbf{y}|\theta)$ is better constrained and much easier to use, and/or computationally more efficient, or numerical optimization is more rapidly converging.

The choice of the set $\mathbf{y}$ (and the transformation $\gamma(.)$) is the degree of freedom of the method and is arbitrary, but in any case it cannot be inferred from the incomplete set as the transformation is not-invertible. The EM algorithm alternates the estimation (*expectation* or *E-step*) of the log-likelihood $\ln p(\mathbf{y}|\theta)$ from the observations $\mathbf{x}$ (available), and the estimate $\theta^{(n)}$ (at nth iteration) from the *maximization (M-step)* from the estimate of $\ln p(\mathbf{y}|\theta)$. More specifically, the log-likelihood $\ln p(\mathbf{y}|\theta)$ is estimated using the MMSE approach given $(\mathbf{x}, \theta^{(n)})$; this is the expected mean estimate, and the two steps of the EM algorithm are

$$Q(\theta|\theta^{(n)}) = \int \ln p(\mathbf{y}|\theta) \cdot p(\mathbf{y}|\mathbf{x}, \theta^{(n)}) d\mathbf{y} = \mathbb{E}_{\mathbf{y}} \left[\ln p(\mathbf{y}|\theta)|\mathbf{x}, \theta^{(n)}\right] \quad \text{(E-step)}$$

$$\theta^{(n+1)} = \arg\max_{\theta} Q(\theta|\theta^{(n)}) \quad \text{(M-step)}$$

The E-step is quite simple in the case that the model has additive Gaussian noise and the transformation $\gamma(.)$ is linear but still not-invertible; this case is discussed in detail below.

The EM method converges to a maximum, but not necessarily the global maximum. In any case, it can be proved that, for any choice $\theta^{(n)}$ that corresponds to the solution at the nth iteration, the inequality $Q(\theta|\theta^{(n)}) \geq Q(\theta^{(n)}|\theta^{(n)})$ holds for any choice $\theta \neq \theta^{(n)}$, and the corresponding likelihood for the MLE is always increasing: $\ln p(\mathbf{x}|\theta) - \ln p(\mathbf{x}|\theta^{(n)}) \geq 0$.

11.6.1 EM of the Sum of Signals in Gaussian Noise

The derivation here has been adapted from the original contribution of Feder and Weinstein, 1988 [46] for superimposed signals in Gaussian noise. Let the observation $\mathbf{x} \in \mathbb{R}^{N\times 1}$ be the sum of K signals each dependent on a set of parameters:

$$\mathbf{x} = \sum_{k=1}^{K} \mathbf{s}_k(\theta_k) + \mathbf{w} = \mathbf{s}(\theta) + \mathbf{w}$$

where $\mathbf{s}_k(\theta_k)$ is the kth signal that depends on the kth set of parameters θ_k and $\mathbf{w} \sim \mathcal{N}(\mathbf{0}, \mathbf{C}_w)$. The objective is to estimate all the set of parameters $\theta = [\theta_1^T, \theta_2^T, ..., \theta_K^T]^T$ from the observations by decomposing the superposition of K signals $\mathbf{s}(\theta)$ into K simple signals, all by using the EM algorithm.

The *complete* set from the *incomplete* $\mathbf{x}$ is based on the structure of the problem at hand. More specifically, one can choose as a complete set the collection of each individual signal:

$$\mathbf{y} = \begin{bmatrix} \mathbf{y}_1 \\ \mathbf{y}_2 \\ \vdots \\ \mathbf{y}_K \end{bmatrix} = \begin{bmatrix} \mathbf{s}_1(\theta_1) + \mathbf{w}_1 \\ \mathbf{s}_2(\theta_2) + \mathbf{w}_2 \\ \vdots \\ \mathbf{s}_K(\theta_K) + \mathbf{w}_K \end{bmatrix} = \underbrace{\begin{bmatrix} \mathbf{s}_1(\theta_1) \\ \mathbf{s}_2(\theta_2) \\ \vdots \\ \mathbf{s}_K(\theta_K) \end{bmatrix}}_{\tilde{\mathbf{s}}(\theta)} + \underbrace{\begin{bmatrix} \mathbf{w}_1 \\ \mathbf{w}_2 \\ \vdots \\ \mathbf{w}_K \end{bmatrix}}_{\tilde{\mathbf{w}}} \in \mathbb{R}^{KN\times 1}$$

where the incomplete set is the superposition of the signals

$$\mathbf{x} = [\mathbf{I}_N, \mathbf{I}_N, ..., \mathbf{I}_N]\mathbf{y} = \mathbf{H}\mathbf{y}$$

where $\mathbf{H} = \mathbf{1}_K^T \otimes \mathbf{I}_N \in \mathbb{R}^{N\times KN}$ is the non-invertible transformation $\gamma(\mathbf{y})$. The following conditions should hold to preserve the original MLE:

$$\mathbf{s}(\theta) = \mathbf{H}\tilde{\mathbf{s}}(\theta) = \sum_{k=1}^{K} \mathbf{s}_k(\theta_k)$$

$$\mathbf{w} = \mathbf{H}\tilde{\mathbf{w}} = \sum_{k=1}^{K} \mathbf{w}_k$$

To simplify the reasoning, each noise term in the complete set is assumed as uncorrelated wrt the others ($\mathbb{E}[\mathbf{w}_k \mathbf{w}_\ell^T] = \mathbf{0}$ for $k \neq \ell$), but with the same correlation properties of $\mathbf{w}$ except for an arbitrary scaling

$$\mathbf{w}_k \sim \mathcal{N}(\mathbf{0}, \mathbf{C}_{w,k}) \text{ with } \mathbf{C}_{w,k} = \beta_k \mathbf{C}_w$$

so that

$$\sum_{k=1}^{K} \mathbf{C}_{w,k} = \mathbf{C}_w \Rightarrow \sum_{k=1}^{K} \beta_k = 1$$

To summarize, the noise of the complete set is

$$\tilde{\mathbf{w}} \sim \mathcal{N}(\mathbf{0}, \tilde{\mathbf{C}}_w)$$

and the $KN \times KN$ covariance matrix is block-diagonal

$$\tilde{\mathbf{C}}_w = \begin{bmatrix} \mathbf{C}_{w,1} & \mathbf{0} & \cdots & \mathbf{0} \\ \mathbf{0} & \mathbf{C}_{w,2} & \cdots & \mathbf{0} \\ \vdots & \vdots & \ddots & \vdots \\ \mathbf{0} & \mathbf{0} & \cdots & \mathbf{C}_{w,K} \end{bmatrix} = \underbrace{\text{diag}\{\beta_1, \beta_2, \ldots, \beta_K\}}_{\mathbf{D}_\beta} \otimes \mathbf{C}_w$$

so it is convenient and compact (but not strictly necessary) to use the Kronecker products notation (Section 1.1.1). The inverse $\tilde{\mathbf{C}}_w^{-1} = \mathbf{D}_\beta^{-1} \otimes \mathbf{C}_w^{-1}$ is still block-diagonal. From the pdf of $\tilde{\mathbf{w}}$ for the complete set:

$$\mathbf{y} = \tilde{\mathbf{s}}(\boldsymbol{\theta}) + \tilde{\mathbf{w}} \sim \mathcal{N}(\tilde{\mathbf{s}}(\boldsymbol{\theta}), \tilde{\mathbf{C}}_w)$$

and the log-likelihood is (up to a scaling factor)

$$\begin{aligned} \ln p(\mathbf{y}|\boldsymbol{\theta}) &= c - (\mathbf{y} - \tilde{\mathbf{s}}(\boldsymbol{\theta}))^T \tilde{\mathbf{C}}_w^{-1} (\mathbf{y} - \tilde{\mathbf{s}}(\boldsymbol{\theta})) \\ &= \tilde{\mathbf{s}}(\boldsymbol{\theta})^T \tilde{\mathbf{C}}_w^{-1} \mathbf{y} + \mathbf{y}^T \tilde{\mathbf{C}}_w^{-1} \tilde{\mathbf{s}}(\boldsymbol{\theta}) - \tilde{\mathbf{s}}(\boldsymbol{\theta})^T \tilde{\mathbf{C}}_w^{-1} \tilde{\mathbf{s}}(\boldsymbol{\theta}) - \mathbf{y}^T \tilde{\mathbf{C}}_w^{-1} \mathbf{y} \end{aligned}$$

Computing the conditional mean $\mathbb{E}_{\mathbf{y}}\left[\ln p[\mathbf{y}|\boldsymbol{\theta}]|\mathbf{x}, \boldsymbol{\theta}^{(n)}\right]$ gives (apart from a scaling term $\mathbf{y}^T \tilde{\mathbf{C}}_w^{-1} \mathbf{y}$ that is independent of $\boldsymbol{\theta}$)

$$Q(\boldsymbol{\theta}|\boldsymbol{\theta}^{(n)}) = \tilde{\mathbf{s}}(\boldsymbol{\theta})^T \tilde{\mathbf{C}}_w^{-1} \hat{\mathbf{y}}(\boldsymbol{\theta}^{(n)}) + \hat{\mathbf{y}}(\boldsymbol{\theta}^{(n)})^T \tilde{\mathbf{C}}_w^{-1} \tilde{\mathbf{s}}(\boldsymbol{\theta}) - \tilde{\mathbf{s}}(\boldsymbol{\theta})^T \tilde{\mathbf{C}}_w^{-1} \tilde{\mathbf{s}}(\boldsymbol{\theta}) + d$$

where $\hat{\mathbf{y}}(\boldsymbol{\theta}^{(n)}) = \mathbb{E}\left[\mathbf{y}|\mathbf{x}, \boldsymbol{\theta}^{(n)}\right]$ is the Bayesian estimate of the complete set from the incomplete $\mathbf{x}$ and current estimate $\boldsymbol{\theta}^{(n)}$. For the optimization wrt $\boldsymbol{\theta}$ (M-step), it is convenient to group the terms of $Q(\boldsymbol{\theta}|\boldsymbol{\theta}^{(n)})$ after adding/removing the term $\hat{\mathbf{y}}(\boldsymbol{\theta}^{(n)})^T \tilde{\mathbf{C}}_w^{-1} \hat{\mathbf{y}}(\boldsymbol{\theta}^{(n)}) = e$ as independent of $\boldsymbol{\theta}$:

$$Q(\boldsymbol{\theta}|\boldsymbol{\theta}^{(n)}) = -\left(\hat{\mathbf{y}}(\boldsymbol{\theta}^{(n)}) - \tilde{\mathbf{s}}(\boldsymbol{\theta})\right)^T \tilde{\mathbf{C}}_w^{-1} \left(\hat{\mathbf{y}}(\boldsymbol{\theta}^{(n)}) - \tilde{\mathbf{s}}(\boldsymbol{\theta})\right) + e$$

where c, d, e are scaling terms independent of $\boldsymbol{\theta}$.

The estimation of the complete set $\hat{\mathbf{y}}(\boldsymbol{\theta}^{(n)}) = E\left[\mathbf{y}|\mathbf{x}, \boldsymbol{\theta}^{(n)}\right]$ is based on the linear model

$$\mathbf{y} \sim \mathcal{N}(\tilde{\mathbf{s}}(\boldsymbol{\theta}^{(n)}), \tilde{\mathbf{C}}_w)$$

and thus on the basis of the transformation $\mathbf{x} = \mathbf{H}\mathbf{y}$. The MMSE estimate of $\mathbf{y}$ from the incomplete observation set (the only one available) $\mathbf{x}$ for Gaussian noise (here equal to zero) is[4]

$$\hat{\mathbf{y}}(\theta^{(n)}) = \tilde{\mathbf{s}}(\theta^{(n)}) + \tilde{\mathbf{C}}_w \mathbf{H}^T \left(\mathbf{H}\tilde{\mathbf{C}}_w \mathbf{H}^T\right)^{-1} \left(\mathbf{x} - \mathbf{H}\cdot\tilde{\mathbf{s}}(\theta^{(n)})\right)$$

The terms of the estimate $\hat{\mathbf{y}}(\theta^{(n)})$ can be expanded by using the property of Kronecker products (Section 1.1.1)

$$\left(\mathbf{H}\tilde{\mathbf{C}}_w \mathbf{H}^T\right)^{-1} = (\underbrace{\mathbf{1}_K^T \mathbf{D}_\beta \mathbf{1}_K}_{\sum_{k=1}^K \beta_k = 1} \otimes \mathbf{C}_w)^{-1} = \mathbf{C}_w^{-1}$$

$$\tilde{\mathbf{C}}_w \mathbf{H}^T = \underbrace{\mathbf{D}_\beta \mathbf{1}_K}_{[\beta_1,\ldots,\beta_K]^T} \otimes \mathbf{C}_w = \begin{bmatrix} \beta_1 \mathbf{C}_w \\ \beta_2 \mathbf{C}_w \\ \vdots \\ \beta_K \mathbf{C}_w \end{bmatrix}$$

substituting and simplifying, it follows that the MMSE estimate of the kth signal of the complete set at the n-th iteration of the EM method is

$$\hat{\mathbf{y}}_k(\theta^{(n)}) = \mathbf{s}_k(\theta_k^{(n)}) + \beta_k \left(\mathbf{x} - \sum_{\ell=1}^{K} \mathbf{s}_\ell(\theta_\ell^{(n)}) \right)$$

which is the sum of the kth signal estimated from the parameter $\theta_k^{(n)}$ at the n-th iteration and the original observation $\mathbf{x}$ after stripping all the signals estimated up to the n-th iteration and interpreted as noise, and thus weighted for the corresponding scaling term β_k. Re-interpreting $Q(\theta|\theta^{(n)})$ now, it can be seen that

$$\begin{aligned} Q(\theta|\theta^{(n)}) &= e - \left(\hat{\mathbf{y}}(\theta^{(n)}) - \tilde{\mathbf{s}}(\theta)\right)^T \tilde{\mathbf{C}}_w^{-1} \left(\hat{\mathbf{y}}(\theta^{(n)}) - \tilde{\mathbf{s}}(\theta)\right) \\ &= e - \sum_{k=1}^{K} \frac{1}{\beta_k} \underbrace{\left(\hat{\mathbf{y}}_k(\theta^{(n)}) - \mathbf{s}_k(\theta_k)\right)^T \mathbf{C}_w^{-1} \left(\hat{\mathbf{y}}_k(\theta^{(n)}) - \mathbf{s}_k(\theta_k)\right)}_{Q_k(\theta_k|\theta_k^{(n)})} \end{aligned}$$

or equivalently, the optimization is fragmented into a set of K local optimizations, one for each set of parameters (M-step):

$$\theta_k^{(n+1)} = \arg\min_{\theta_k} \left\{ Q_k(\theta_k|\theta_k^{(n)}) \right\} \quad \text{for } k = 1, 2, \ldots, K$$

---

4 For the model $\mathbf{x} = \mathbf{H}\mathbf{y}$ with $\mathbf{y} \sim \mathcal{N}(\tilde{\mathbf{s}}(\theta^{(n)}), \tilde{\mathbf{C}}_w)$, it is (Section 11.2.1)

$$\mathbb{E}[\mathbf{y}|\mathbf{x}] = \boldsymbol{\mu}_y + \mathbf{C}_{yx}\mathbf{C}_{xx}^{-1}(\mathbf{x} - \boldsymbol{\mu}_x)$$

where $\mathbf{C}_{yx} = \tilde{\mathbf{C}}_w \mathbf{H}^T$ e $\mathbf{C}_{xx} = \mathbf{H}\tilde{\mathbf{C}}_w \mathbf{H}^T$.

The model that uses the EM method for the superposition of signals in Gaussian noise is general enough to find several applications, such as for the estimation of the parameters for a sum of sinusoids as discussed in Section 10.4 for Doppler-radar system, or a sum of waveforms (in radar or remote sensing systems), or the sum of pseudo-orthogonal modulations (in spread-spectrum and code division multiple access—CDMA—systems). An example of time of delay estimation is discussed below.

11.6.2 EM Method for the Time of Delay Estimation of Multiple Waveforms

Let the signal be the sum of K delayed echoes as in a radar system (Section 5.6); the waveform of each echo is known but not its amplitude and delay, and the ensemble of delays should be estimated from their superposition

$$\mathbf{x} = [\mathbf{g}(\tau_1), \mathbf{g}(\tau_2), ..., \mathbf{g}(\tau_K)]\boldsymbol{\alpha} + \mathbf{w} = \mathbf{H}(\boldsymbol{\tau})\boldsymbol{\alpha} + \mathbf{w}$$

Each waveform can be modeled as

$$\mathbf{s}_k(\boldsymbol{\theta}_k) = \alpha_k \begin{bmatrix} g(1-\tau_k) \\ g(2-\tau_k) \\ \vdots \\ g(N-\tau_k) \end{bmatrix} = \alpha_k \mathbf{g}(\tau_k) \Rightarrow \boldsymbol{\theta}_k = [\alpha_k, \tau_k]^T$$

and the complete set of the EM method coincides with the generation model for each individual echo as if it were non-overlapped with the others.

Given the estimate of amplitude $\alpha_k^{(n)}$ and delay $\tau_k^{(n)}$ at then nth iteration, the two steps of the EM method are as follows.

E-step (for $k = 1, 2, ..., K$):

$$\hat{\mathbf{y}}_k(\boldsymbol{\theta}^{(n)}) = \alpha_k^{(n)} \mathbf{g}(\tau_k^{(n)}) + \beta_k \left(\mathbf{x} - \sum_{\ell=1}^{K} \alpha_\ell^{(n)} \mathbf{g}(\tau_\ell^{(n)}) \right)$$

M-step (for $k = 1, 2, ..., K$):

$$(\alpha_k^{(n+1)}, \tau_k^{(n+1)}) = \arg\min_{\alpha_k, \tau_k} \left\{ \left(\hat{\mathbf{y}}_k(\boldsymbol{\theta}^{(n)}) - \alpha_k \mathbf{g}(\tau_k)\right)^T \left(\hat{\mathbf{y}}_k(\boldsymbol{\theta}^{(n)}) - \alpha_k \mathbf{g}(\tau_k)\right) \right\}$$

The M-step is equivalent to the estimation of the delay of one isolated waveform from the signal $\hat{\mathbf{y}}_k(\boldsymbol{\theta}^{(n)})$ where the effect of the interference with the other waveforms has been (approximately) removed from the E-Step (see Figure 11.9 with $K = 2$ radar-echoes). In other words, at every step the superposition of all the other waveforms (here in Figure 11.9 of the complementary waveform) is removed[5] until (at convergence) the estimator of amplitude and delay acts on one isolated waveform and thus it is optimal.

5 In signal processing it is common to use the layer-stripping (or data-stripping) as practical approach to remove the interference of multiple signals and reduce the estimate to the single echo. The performance is evaluated from the inspection of the residual left after stripping all the signals that could be estimated from the original data. Its use is widespread in geophysics and remote sensing and it resembles the EM method, in many cases just the first step of the EM is carried out as considered enough for the scope.

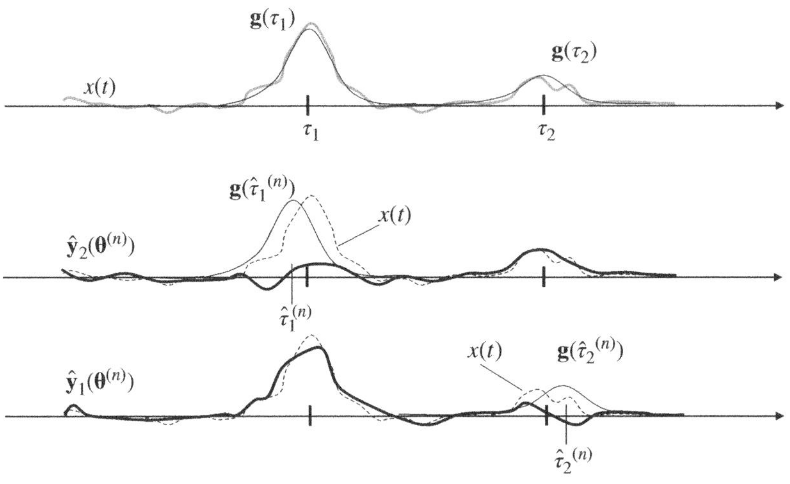

Figure 11.9 ToD estimation of multiple waveforms by EM method ($K = 2$).

Of course, any imperfect cancellation of other waveforms could cause the EM method to converge to some local maximum. This problem is not avoided at all by the EM method and should be verified case-by-case.

This example could be adapted for the estimation of frequency for sinusoids with different frequencies and amplitudes with small changes of terms, and the numerical example in Section 10.4 can suggest other applications of the EM method when trying to reduce the estimation of multiple parameters to the estimation of a few parameters carried out in parallel.

11.6.3 Remarks

The EM algorithm alternates the E-step and the M-step with the property that at every step the log-likelihood function increases. However, there is no guarantee that convergence will be obtained to the global maximum, and the ability to reach convergence to the true MLE depends on the initialization $\boldsymbol{\theta}^{(0)}$ as for any iterative optimization method. Also, in some situations EM algorithm can be painfully slow to converge.

The mapping $\gamma(.)$ that maps the incomplete set $\mathbf{x}$ from the complete (but not available) data $\mathbf{y}$ assumes that there are some *missing* measurements that complete the incomplete observations $\mathbf{x}$. In some cases, adding these missing data can simplify the maximization of the log-likelihood, but it could slow down the convergence of this EM method. Since the choice of the complete set is a free design aspect in the EM method, the convergence speed depends on how much information on the parameters are contained in the complete set.

Space-alternating EM (SAGE) has been proposed in [47] to cope with the problem of convergence speed and proved to be successful in a wide class of application problems. A detailed discussion on this matter is left to publications specialized to solve specific applications by EM method.

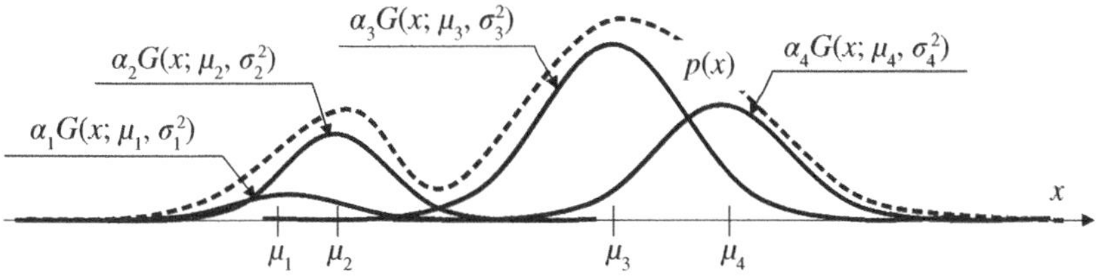

Figure 11.10 Mixture model of non-Gaussian pdfs.

Appendix A: Gaussian Mixture pdf

A non-Gaussian pdf can be represented or approximated as a mixture of pdfs, mostly when multimodal, with long (or short) tail compared to a Gaussian pdf, or any combination thereof. Given some nice properties of estimators involving Gaussian pdfs, it can be convenient to rewrite (or approximate) a non-Gaussian pdf as the weighted sum of Gaussian pdfs (*Gaussian mixture*) as follows (Figure 11.10):

$$p_x(x) = \sum_k \alpha_k G(x;\mu_k,\sigma_k^2) \tag{11.9}$$

Parameters $\{\mu_k,\sigma_k^2\}$ characterize each Gaussian function used as the basis of the Gaussian mixture, and $\{\alpha_k\}$ are the *mixing proportions* constrained by $\alpha_k \geq 0$. The problem of finding the set of parameters $\{\alpha_k,\mu_k,\sigma_k^2\}$ that better fits the non-Gaussian pdf is similar to interpolation over non-uniform sampling with one additional constraint on pdf normalization

$$\int p_x(x)dx = 1 \rightarrow \sum_k \alpha_k = 1$$

Methods to get the approximating parameters $\{\alpha_k,\mu_k,\sigma_k^2\}$ are many, and the method of moments (MoM) is a simple choice if either $p_x(x)$ is analytically known, or a set of samples $\{x_\ell\}$ are available. Alternatively, one can assume that the Gaussian mixture is the data generated from a set of independent and disjoint experiments, and one can estimate the parameters by assigning each sample to each of the experiments, as detailed in Section 23.6.

To generate a non-Gaussian process $x[n]$ with Gaussian mixture pdf based on K Gaussian functions, one can assume that there are K independent Gaussian generators $(y_1,y_2,\ldots,y_K)$ each with $y_k \sim \mathcal{N}(\mu_k,\sigma_k^2)$, and the assignment $x[n]=y_k$ is random with mixing probability α_k and it is mutually exclusive wrt the others. A simple example here can help to illustrate. Let $K=3$ be the number of normal rvs; the Gaussian mixture is obtained by selecting each sample out of three Gaussian sources, with selection probability $\{\alpha_1,\alpha_2,\alpha_3\}$ into the 3×1 vector `alfa`:

```
threshold_1=alfa(1); threshold_2=alfa(1)+alfa(2); %
probability thresholds
t=rand; % selection variable U(0,1)
if(t<=threshold_1), x=sigma(1)*randn+mu(1); end
if(t>threshold_1 & t<threshold_2), x=sigma(2)*randn+mu(2);
end
if(t>=threshold_2), x=sigma(3)*randn+mu(3); end
```

The generalization to N-dimensional Gaussian mixture pdf is straightforward:

$$p_x(\mathbf{x}) = \sum_k \alpha_k G(\mathbf{x}; \boldsymbol{\mu}_k, \mathbf{C}_k)$$

The central moments $\{\boldsymbol{\mu}_k, \mathbf{C}_k\}$ are the generalization to N rvs of the Gaussian mixture (11.9).

12

Optimal Filtering

Estimation for stochastic processes is carried out by linear filtering, and when stationary and Gaussian, the filtering can be designed to be optimal in the MMSE sense. In addition, if random processes are non-Gaussian, linear filtering is so convenient that LMMSE is a routinely employed tool.

12.1 Wiener Filter

MMSE estimators, as derived in Chapter 11, are constrained by a finite number of observations N. When removing the constraint of limited length, an MMSE estimator can be arbitrarily long in WSS Gaussian processes. This is the optimal *Wiener filter* that is based on the auto and cross-correlation sequences as detailed below.

Let $a[n] \leftrightarrow A(z)$ be the impulse response of a linear filter (estimator) that is used on a zero-mean WSS process $x[n]$ to estimate the sample $\theta[n]$ of another WSS process

$$\hat{\theta}[n] = \hat{\theta}[n; \{a[n]\}] = \sum_k a[k] \cdot x[n-k] = a[n] * x[n]$$

The estimation error

$$\varepsilon[n; \{a[n]\}] = \theta[n] - \hat{\theta}[n; \{a[n]\}]$$

depends on the impulse response $a[n]$ to be evaluated. The z-transform of the estimation error is

$$\mathcal{E}[z; \{a[n]\}] = \mathcal{Z}\{\theta[n] - a[n] * x[n]\} = \Theta(z) - A(z)X(z)$$

and the estimator is defined in terms of $A(z) = \mathcal{Z}\{a[n]\}$. Since the process is stationary, the z-transform of the autocorrelation of the estimation error is

$$R_{\varepsilon\varepsilon}(z; \{a[n]\}) = \mathbb{E}\left[\mathcal{E}[z; \{a[n]\}] \cdot \mathcal{E}^*[1/z; \{a[n]\}]\right]$$

that is the *sequence* of the MSE (i.e., the MSE for every sample of the estimator evaluated under the assumption of stationarity) that depends on $\mathcal{Z}\{a[n]\}$

$$R_{\varepsilon\varepsilon}(z; \{a[n]\}) = \mathbb{E}_{x,\theta}\left[(\Theta(z) - A(z)X(z)) \cdot (\Theta^*(1/z) - A^*(1/z)X^*(1/z))\right]$$

Statistical Signal Processing in Engineering, First Edition. Umberto Spagnolini.

Companion website: www.wiley.com/go/spagnolini/signalprocessing

The minimization of $R_{\varepsilon\varepsilon}(z;\{a[n]\})$ independently of the complex variable z is equivalent to the minimization of the MSE for every frequency, from the ensemble of gradients

$$\frac{\partial R_{\varepsilon\varepsilon}(z;\{a[n]\})}{\partial\{a[n]\}}=0$$

evaluated for every sample yields the Wiener filter:

$$\hat{A}(z)=\frac{\mathbb{E}[\Theta(z)X^*(1/z)]}{\mathbb{E}[X(z)X^*(1/z)]} \tag{12.1}$$

where samples $\hat{a}[n]\leftrightarrow\hat{A}(z)$ are obtained from the inverse z-transform.

Details of the derivation are given herein. Let the sample of the z-transform be split into real and imaginary parts

$$A(z)=\sum_n(a_R[n]+ja_I[n])z^{-n}$$

The gradients of $R_{\varepsilon\varepsilon}(z;\{a[n]\})$ for every sample n are

$$\frac{\partial}{\partial a_R[n]}R_{\varepsilon\varepsilon}(z;\{a[n]\})=0$$
$$\frac{\partial}{\partial a_I[n]}R_{\varepsilon\varepsilon}(z;\{a[n]\})=0$$

From the definitions above, the gradients

$$\frac{\partial R_{\varepsilon\varepsilon}(z;\{a[n]\})}{\partial a_R[n]}=-\mathbb{E}[[\Theta(z)-A(z)X(z)]X^*(1/z)]z^n+$$
$$-\mathbb{E}[[\Theta^*(1/z)-A^*(1/z)X^*(1/z)]X(z)]z^{-n}=0$$
$$\frac{\partial R_{\varepsilon\varepsilon}(z;\{a[n]\})}{\partial a_I[n]}=j\mathbb{E}[[\Theta(z)-A(z)X(z)]X^*(1/z)]z^n+$$
$$-j\mathbb{E}[[\Theta^*(1/z)-A^*(1/z)X^*(1/z)]X(z)]z^{-n}=0$$

lead to the solution[1]

$$\mathbb{E}[[\Theta(z)-A(z)X(z)]X^*(1/z)]=0$$

This is the orthogonality condition between the z-transform of the estimation error process $\mathcal{E}[z;\{a[n]\}]=\Theta(z)-A(z)X(z)$ (which depends on $\{a[n]\}$) and the input sequence $X(z)$ for the random processes, which easily yields the Wiener filter solutions above.

1 The trivial system of equations

$$f(z)+g(z)=0$$
$$f(z)-g(z)=0$$

has the single solution $f(z)=g(z)=0$

Remark: The Wiener filter (12.1) is the optimum MMSE estimator for WSS processes, and it is derived from the statistical properties of processes $x[n]$ and $\theta[n]$. The overall filter (12.1) is *not* in any way constrained to be causal and/or minimum phase and/or with finite range. The practical implication is that a Wiener filter can be difficult to implement (if anti-causal) and very expensive (if infinite range). Attempts to use the Wiener filter with some heuristics such as the truncation of $\hat{a}[n]$ to N_a samples are very common in practice and can make the Wiener filter more practically usable at the price of loss of MSE performance. Of course, all depends on the length of the finite-length data (N_x) compared to the (approximate) length of Wiener filter response N_a, namely if $N_x \gg N_a$ it is pretty safe to use the Wiener filter. However, if $N_x \leq N_a$ it would be advisable to adopt the LMMSE estimator for block-samples in Section 11.3 with the stationary values of the auto- and cross-covariance to build the covariance matrixes $\mathbf{C}_{xx}$ and $\mathbf{C}_{\theta x}$, respectively (see Section 4.5).

12.2 MMSE Deconvolution (or Equalization)

Deconvolution (or equalization) is very common, and the Wiener filter provides a simple and effective solution that helps to gain insight on some aspects of filtering as an estimation method. Let the observation be the output of a zero-mean white Gaussian WSS process with power σ_w^2 filtered by a filter characterized by $H(z) = \mathcal{Z}\{h[n]\}$; a zero-mean white Gaussian noise is added at the end with power σ_z^2 and the overall model is

$$x[n] = w[n] * h[n] + z[n]$$

Deconvolution seeks to estimate the excitation process $w[n]$ from the noisy observations $x[n]$ given the statistical properties of the process at hand. The Wiener filter (12.1) is based on the z-transform $R_{xx}(z)$ that follows here from the definition

$$\begin{aligned} R_{xx}(z) &= \mathbb{E}[(W(z)H(z) + Z(z))(W^*(1/z)H^*(1/z) + Z^*(1/z))] \\ &= \sigma_w^2 H(z)H^*(1/z) + \sigma_z^2 \end{aligned}$$

Here $\theta[n] \equiv w[n]$ and the numerator of the Wiener filter (12.1) follow from the definition

$$\mathbb{E}[W(z)X^*(1/z)] = \mathbb{E}[W(z)(W^*(1/z)H^*(1/z) + Z^*(1/z))] = \sigma_w^2 H^*(1/z)$$

The Wiener filter becomes

$$\hat{A}(z) = \frac{\sigma_w^2 H^*(1/z)}{\sigma_w^2 H(z)H^*(1/z) + \sigma_z^2}$$

It is useful to analyze the Wiener filter in the frequency domain by placing $z = e^{j\omega}$ so that the power spectrum densities are:

$$S_{w*h}(\omega) = \sigma_w^2 |H(\omega)|^2$$

$$S_z(\omega) = \sigma_z^2$$

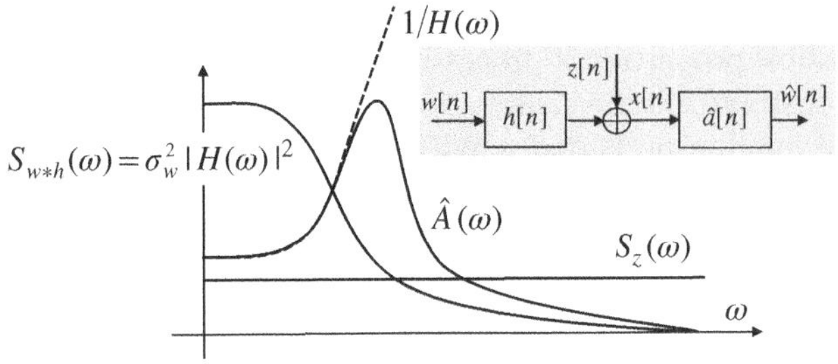

Figure 12.1 Wiener deconvolution.

The Wiener filter becomes

$$\hat{A}(\omega) = \frac{\sigma_w^2 H^*(\omega)}{\sigma_w^2 |H(\omega)|^2 + \sigma_z^2}$$

The behavior of the Wiener filter is shown in Figure 12.1. It has two prevailing behaviors depending of the characteristics of the power spectral density of Wiener filter input $S_x(\omega) = \sigma_w^2 |H(\omega)|^2 + \sigma_z^2$:

$$\hat{A}(\omega) \simeq \frac{1}{H(\omega)} \quad \text{when } S_{w*h}(\omega) \gg S_z(\omega) \text{ (large signal to noise ration in } S_x(\omega))$$

$$\hat{A}(z) \simeq \frac{\sigma_w^2}{\sigma_z^2} H^*(\omega) \quad \text{when } S_{w*h}(\omega) \ll S_z(\omega) \text{ (small signal to noise ration in } S_x(\omega))$$

The Wiener filter behaves as the inverse of $H(\omega)$ over those frequency intervals where the power spectral density of the filtered signal dominates the noise, while in other frequency intervals the Wiener attains the matched filter $H^*(\omega)$ with a small amplitude.

Example: To illustrate further, the choice $H(z) = 1 - \rho z^{-1}$ yields the Wiener filter ($\sigma^2 = \sigma_z^2 / \sigma_w^2$)

$$\hat{A}(z) = \frac{1 - \rho^* z}{(1 - \rho z^{-1})(1 - \rho^* z) + \sigma^2} = \frac{z^{-1} - \rho^*}{(1 - p_1 z^{-1})(1 - p_2 z^{-1})}$$
$$= (z^{-1} - \rho^*) \sum_k p_1^k z^{-k} \sum_n p_2^n z^{-n}$$

for the two roots $\{p_1, p_2\}$ of the denominator. Filter range is clearly infinite as the range of the two filters p_1^k and p_2^k, judiciously selected to be stable, have infinite range either causal or anti-causal.

12.3 Linear Prediction

Linear prediction is a term to denote a wide range of LMMSE estimators of a sample at time $n + k$ (with $k \geq 1$) of a signal $x[n]$ handled as a stochastic WSS process from all the observations $x[1 : n]$ up to time n; the value k is referred to as the prediction step (or interval). The usage of prediction is quite common in real life (e.g., prediction of stock exchange values, weather or rainfall, or evolution of a soccer league); it aims to predict (by estimating the corresponding value) a future sample based on all the past observations so that the error is minimized in its mean square. All the analytical tools necessary to develop prediction as an MMSE estimator are those discussed up to here, and linear prediction could be considered just a simple exercise. However, the importance of the topic has historically urged the development of specific methods that deserve to be covered herein in detail.

12.3.1 Yule–Walker Equations

Let the zero-mean WSS complex-valued random process $x[n]$ have a known autocorrelation $r_{xx}[n]$. The one-step ($k=1$) linear predictor of length p is (Figure 12.2)

$$\hat{x}_p[n] = -\sum_{k=1}^{p} a_p[k] \cdot x[n-k] = -\mathbf{a}_p^T \mathbf{x}[n]$$

(the change of sign is only to ease a later notation); subscript p indicates the length of the predictor. The set of weights $\mathbf{a}_p = [a_p[1], a_p[2], ..., a_p[p]]^T$ combines the set of observations $\mathbf{x}[n] = [x[n-1], x[n-2], ..., x[n-p]]^T$ (highlighted by a shaded area in Figure 12.2) and the objective is the minimization of the MSE of the prediction error

$$\varepsilon_p[n] = x[n] + \sum_{k=1}^{p} a_p[k] \cdot x[n-k]$$

with respect to the predictor coefficients $\{a[n]\}_{n=1}^{p}$. The MMSE is obtained from the orthogonality condition as

$$\mathbb{E}[\varepsilon_p[n] x^*[n-\ell]] = 0 \quad \text{for } \ell = 1, 2, ..., p$$

that corresponds to a set of p equations in the p unknowns $\{a[n]\}_{n=1}^{p}$. The expansion of these conditions yields

$$\underbrace{\mathbb{E}[x[n]x^*[n-\ell]]}_{r_{xx}[\ell]} + \sum_{k=1}^{p} a_p[k] \cdot \underbrace{\mathbb{E}[x[n-k]x^*[n-\ell]]}_{r_{xx}[\ell-k]} = 0 \quad \text{for } \ell = 1, 2, ..., p$$

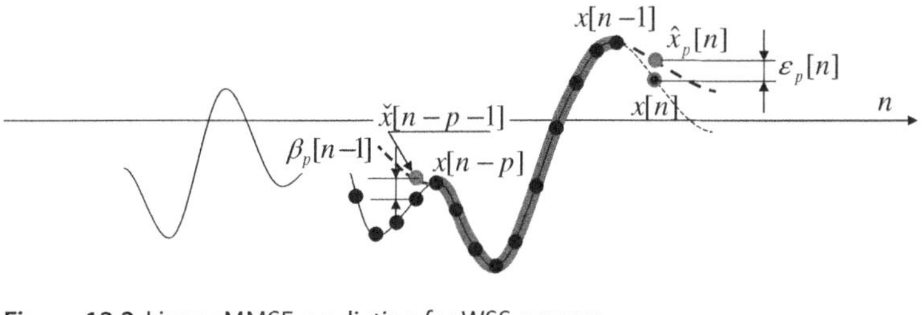

Figure 12.2 Linear MMSE prediction for WSS process.

These are the *Yule–Walker equations*

$$r_{xx}[\ell]+\sum_{k=1}^{p}a_p[k]\cdot r_{xx}[\ell-k]=0 \quad \text{for } \ell=1,2,...,p \tag{12.2}$$

for the linear prediction of length p that can be rewritten in compact notation (recall that $r_{xx}[-\ell]=r^*_{xx}[\ell]$, see Section 4.1 and 4.3) as

$$\underbrace{\begin{bmatrix} r_{xx}[0] & \cdots & r^*_{xx}[p-1] \\ \vdots & \ddots & \vdots \\ r_{xx}[p-1] & \cdots & r_{xx}[0] \end{bmatrix}}_{\mathbf{R}} \cdot \underbrace{\begin{bmatrix} a_p[1] \\ \vdots \\ a_p[p] \end{bmatrix}}_{\mathbf{a}_p} = -\underbrace{\begin{bmatrix} r_{xx}[1] \\ \vdots \\ r_{xx}[p] \end{bmatrix}}_{\mathbf{p}_1} \tag{12.3}$$

The solution of the MMSE linear predictor is

$$\hat{\mathbf{a}}_p=-\mathbf{R}^{-1}\mathbf{p}_1$$

which is fully equivalent to (11.7) except for the prediction-specific notation.

Note that the autocorrelation matrix $\mathbf{R}\in\mathbb{C}^{p\times p}$ has a Toeplitz structure (with the same entry along each diagonal) and is symmetric (or Hermitian symmetric if complex-valued) and the inversion could be quite expensive; however, there are methods tailored for this application discussed next. The vector $\mathbf{p}_1$ is the cross-correlation between the p observations $x[n-p:n-1]$ and the estimated value $x[n]$ (that is the signal itself) and because of the stationarity it depends on the autocorrelation; for a one-step predictor it is $\mathbf{p}_1$. In general, for a k-step predictor it is $\mathbf{p}_k=[r_{xx}[k],r_{xx}[k+1],...,r_{xx}[k+p-1]]^T$ and $\mathbf{R}$ remains the same. Overall, for a k-step predictor it is necessary to have the autocorrelation up to a delay of $k+p-1$ samples.

The quality of the prediction is evaluated from the MSE for the linear predictor $\hat{\mathbf{a}}$ evaluated so far; it is (notice the technicalities on how to make substitutions and use iteratively the orthogonality condition):

$$\begin{aligned}\mathbb{E}[|\varepsilon_p[n]|^2] &= \mathbb{E}\left[\varepsilon_p[n]\left(x[n]+\sum_{k=1}^{p}\hat{a}_p[k]\cdot x[n-k]\right)^*\right] \\ &= \mathbb{E}[\varepsilon_p[n]x^*[n]]+\sum_{k=1}^{p}\hat{a}^*_p[k]\cdot\underbrace{\mathbb{E}[\varepsilon_p[n]x^*[n-k]]}_{\text{ortogonality}\to 0} \\ &= \mathbb{E}\left[\left(x[n]+\sum_{k=1}^{p}\hat{a}_p[k]\cdot x[n-k]\right)x^*[n]\right]\end{aligned}$$

The mean square prediction error for the predictor of lenth p

$$\mathbb{E}[|\varepsilon_p[n]|^2]=r_{xx}[0]+\sum_{k=1}^{p}\hat{a}_p[k]\cdot r_{xx}[k] \tag{12.4}$$

depends on the predictor evaluated after solving the Yule–Walker equations (12.2). Longer predictors can make use of more information from the past and thus one expects that the MSE reduces with p as proved in next section. Since the computation complexity to invert the $p \times p$ matrix scales with the predictor length, it is important to avoid the use of excessively long predictors, unless necessary.

In practical systems, the autocorrelation is not available and it should also be estimated from all the available data. The prediction problem can be rewritten from the N observations available, following all the steps for a limited number of observations and setting the orthogonality for the sample correlation; it follows the Yule–Walker relationship:

$$\left(\frac{1}{N}\sum_{n=1}^{N}\mathbf{x}[n]\mathbf{x}[n]\right)\mathbf{a}_p = \frac{1}{N}\sum_{n=1}^{N}x[n]\mathbf{x}[n]$$

A careful inspections reveals that these Yule–Walker equations are the same as (12.2)

$$\hat{r}_{xx}[\ell] + \sum_{k=1}^{p} a_p[k]\cdot \hat{r}_{xx}[\ell - k] = 0 \quad \text{for } \ell = 1,2,\ldots,p$$

except that samples of the autocorrelation sequence are replaced by the sample estimate:

$$\hat{r}_{xx}[i] = \frac{1}{N}\sum_{k=1}^{N} x^*[k]x[i+k].$$

12.4 LS Linear Prediction

LS (least squares) prediction is an alternative method to linear prediction from the correlation properties of WSS processes. Even if the LS method needs no assumption on statistical properties, it coincides with Section 12.3.1, and it is useful to look at the steps that lead to the LS estimate. The prediction error from the process

$$\varepsilon_p[n] = x[n] + \sum_{k=1}^{p} a_p[k]\cdot x[n-k]$$

can be based on a set of N observations $\{x[n]\}_{n=1}^{N}$ and can be cast as an LS estimate

$$\hat{\mathbf{a}}_p = \arg\min_{\mathbf{a}_p}\left\{\sum_{n=1}^{N}\left(x[n] + \sum_{k=1}^{p} a_p[k]\cdot x[n-k]\right)^2\right\}$$

Some manipulations are necessary to bring the LS estimate to a manageable level. Since the observations range over N samples, $x[n] = 0$ for $n \notin [1,N]$ and thus the prediction model becomes

$$\underbrace{\begin{bmatrix} x[2] \\ x[3] \\ \vdots \\ x[p+1] \\ x[p+2] \\ \vdots \\ x[N] \\ 0 \\ 0 \\ \vdots \\ 0 \end{bmatrix}}_{\mathbf{x}} = -\underbrace{\begin{bmatrix} x[1] & 0 & \dots & 0 \\ x[2] & x[1] & \dots & 0 \\ \vdots & \vdots & \ddots & \vdots \\ x[p] & x[p-1] & \dots & 0 \\ x[p+1] & x[p] & \dots & x[1] \\ \vdots & & \ddots & \vdots \\ x[N-1] & x[N-2] & \dots & x[N-p] \\ x[N] & x[N-1] & \dots & x[N-p+1] \\ 0 & x[N] & & x[N-p+2] \\ \vdots & \vdots & \ddots & \vdots \\ 0 & 0 & \dots & x[N] \end{bmatrix}}_{\mathbf{X}} \cdot \underbrace{\begin{bmatrix} a_p[1] \\ a_p[2] \\ \vdots \\ a_p[p] \end{bmatrix}}_{\mathbf{a}_p}$$

or compactly

$$\mathbf{x} = -\mathbf{X} \cdot \mathbf{a}_p$$

The LS estimate follows from the minimization of the metric

$$J(\mathbf{a}_p) = (\mathbf{x} + \mathbf{X}\mathbf{a}_p)^T(\mathbf{x} + \mathbf{X}\mathbf{a}_p)$$

that is quadratic and thus

$$\hat{\mathbf{a}}_p = -(\mathbf{X}^T\mathbf{X})^{-1} \cdot \mathbf{X}^T\mathbf{x}$$

Inspection of the LS estimate of the linear prediction shows that this is the solution of the Yule–Walker equation (12.2) by replacing the autocorrelation matrix $\mathbf{R}$ and the vector $\mathbf{p}_1$ by the sample estimates:

$$\hat{\mathbf{R}} = \frac{1}{N}(\mathbf{X}^T\mathbf{X})$$
$$\hat{\mathbf{p}}_1 = \frac{1}{N}\mathbf{X}^T\mathbf{x}$$

and the LS predictor is

$$\hat{\mathbf{a}}_p = -\hat{\mathbf{R}}^{-1} \cdot \hat{\mathbf{p}}_1$$

Note that the choice of assigning to zero all the samples outside the interval $[1, N]$ is quite arbitrary as it reflects the lack of knowledge of the samples there. Since this (somewhat arbitrary) choice could bias the linear prediction that should comply with this constraint for samples outside $[1, N]$, an alternative solution could be to limit the estimator to use and predict only those samples that are effectively available. It is just a matter of redefining the matrixes $\mathbf{X}$ and $\mathbf{x}$ avoiding the samples outside the range $[1, N]$:

$$\underbrace{\begin{bmatrix} x[p+1] \\ x[p+2] \\ \vdots \\ x[N] \end{bmatrix}}_{\mathbf{x}} = -\underbrace{\begin{bmatrix} x[p] & x[p-1] & \dots & x[1] \\ x[p+1] & x[p] & \dots & x[2] \\ \vdots & & & \vdots \\ x[N-1] & x[N-2] & \dots & x[N-p] \end{bmatrix}}_{\mathbf{X}} \cdot \underbrace{\begin{bmatrix} a_p[1] \\ a_p[2] \\ \vdots \\ a_p[p] \end{bmatrix}}_{\mathbf{a}_p}$$

and solving accordingly. This choice coincides with the linear prediction (12.2) for $N \to \infty$.

Once again, there is no unique solution to the estimation problem when the samples are limited, it depends on the specific problem at hand. The LS approach discussed herein has the benefit that it can be adopted for non-stationary processes where it is possible to segment the observations into locally stationary portions.

12.5 Linear Prediction and AR Processes

There is a close relationship between linear prediction and an autoregressive process (Section 4.4). Let $x[n]$ be an $AR(N_a)$ process that can be modeled as generated by a filtered WSS white process $w[n]$

$$X(z) = \frac{1}{A_{N_a}(z)} W(z)$$

The prediction error $\varepsilon_p[n] \leftrightarrow E_p(z)$ from a linear prediction is the filtered signal

$$E_p(z) = A_p(z)X(z)$$

with the filter

$$A_p(z) = 1 + \sum_{k=1}^{p} a_p[k]z^{-k}$$

The error after the linear predictor (see Figure 12.3)

$$E_p(z) = A_p(z) \cdot \frac{1}{A_{N_a}(z)} \cdot W(z)$$

highlights the capability of the linear predictor to make use of the (statistical) self-similarity of the process $x[n]$ to predict at best (i.e., at minimum MSE) the future sample from the past samples until the output is an uncorrelated process, or equivalently the samples lose all correlation betwen each other. Therefore, the MSE is

$$\begin{aligned} \mathbb{E}[|\varepsilon_p[n]|^2] &> \sigma_w^2 \quad \text{for } p < N_a \\ \mathbb{E}[|\varepsilon_p[n]|^2] &= \sigma_w^2 \quad \text{for } p \geq N_a \end{aligned}$$

as shown in Figure 12.3, and thus it would be useless to increase the length of the predictor beyond the order of the AR process. In other words, for $p \geq N_a$, the linear

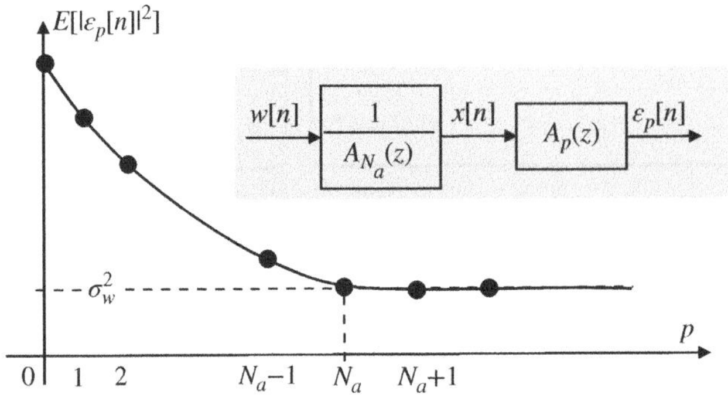

Figure 12.3 Mean square prediction error $\mathbb{E}[|\varepsilon_p[n]|^2]$ vs. predictor length p for $AR(N_a)$ random process $x[n]$.

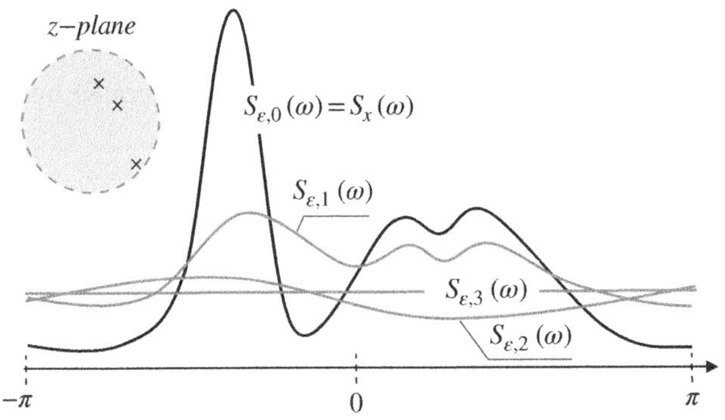

Figure 12.4 Whitening of linear prediction: PSD of $\varepsilon_p[n]$ for increasing predictor length $p = 0, 1, 2, 3$ with AR(3).

predictor deconvolves exactly the (hypothetical) filter that correlated the AR process, all that remains is the unpredictable component. By extension, the analysis of the MSE vs. the predictor length can be used to infer the order of an AR process, if truly AR.

By increasing the predictor length p, the prediction error becomes progressively more uncorrelated until $p \geq N_a$ when $\varepsilon_p[n]$ is white. Linear prediction is a simple way to decorrelate any process up to a certain degree (if $x[n]$ is $AR(N_a)$ and $p < N_a$), or in general to decorrelate any process since the objective of prediction is to orthogonalize the prediction error wrt the signal. The linear predictor acts as a *whitening filter* for the prediction error $\varepsilon_p[n]$ and its effect is sketched in Figure 12.4 for an $AR(3)$ model with some resonances corresponding to the positions of the poles of $A_3(z)$ that appear in the power spectral density of $S_x(\omega)$. The power spectral density of the prediction error $S_{\varepsilon,p}(\omega)$ for varying $p = 1, 2, 3$ shows that by increasing the prediction length, the power spectral density "flattens" (or whitens) by removing the resonances due to the poles of $AR(3)$, and by employing a pole-zero cancellation for the corresponding filters.

12.6 Levinson Recursion and Lattice Predictors

Matrix inversion (12.3) is expensive as the computational cost of the inversion of the $p \times p$ correlation matrix **R** is $\mathcal{O}(p^3)$. To reduce the computational cost, an iterative method can be used that increases the length starting from a predictor $A_p(z)$ of length p and increasing its length to $p+1$: $A_{p+1}(z)$. The method is called *Levinson recursion* and there is an associated filter structure called a *lattice filter* as sketched in Figure 12.5 with the corresponding variables and z-transforms (z^{-1} denotes the unit-delay operator) [49].

Before introducing the recursive method, it is worth recalling that since the process is WSS, there is no difference in statistical properties for the correlation whether the time axis is either forward and backward. The same coefficients of the linear predictor can be equally used to carry on *forward* prediction of sample $x[n]$ and *backward* prediction of sample $x[n-p]$ indicated by different hats:

$$\hat{x}_p[n] = -\sum_{k=1}^{p} a_p[k] \cdot x[n-k] \quad \text{(forward)}$$

$$\check{x}_p[n-p] = -\sum_{k=1}^{p} a_p[k] \cdot x[n-p-k] \quad \text{(backward)}$$

with the following prediction errors over the two branches of the lattice in Figure 12.5:

$$\varepsilon_p[n] = x[n] - \hat{x}_p[n] = x[n] + \sum_{k=1}^{p} a_p[k] \cdot x[n-k] \quad \text{(forward)}$$

$$\beta_p[n] = x[n-p] - \check{x}_p[n-p] = x[n-p] + \sum_{k=1}^{p} a_p[k] \cdot x[n-p-k] \quad \text{(backward)}$$

(subscripts refer to the length of the linear predictor). Since $\varepsilon_p[n]$ and $\beta_p[n]$ are using two different sets of samples ($\varepsilon_p[n]$ is based on samples $x[n-1], x[n-2], \ldots, x[n-p]$ while $\beta_p[n]$ depends on $x[n], x[n-1], \ldots, x[n-p+1]$) that differ by one sample only, backward prediction can be delayed by one sample (i.e., $\beta_p[n-1]$) to use the *same samples* for both forward and backward predictors, and using only the samples available at time n (notice that $x[n]$ is not available yet). The benefits of this simple delay are the following: (1) both forward $\varepsilon_p[n]$ and backward $\beta_p[n-1]$ prediction errors are based on the same samples to estimate forward prediction $\hat{x}_p[n]$ and backward prediction $\check{x}_p[n-p-1]$ (see figure); (2) the sample $x[n-p-1]$ is in the past and is available, thus the backward prediction error $\beta_p[n-1]$ can be practically measured while the forward prediction error $\varepsilon_p[n]$ needs to use the sample $x[n]$ that is not available, yet. The *measurement* of the backward prediction error $\beta_p[n-1]$ contains correlation information that has not been used in the forward predictor of length p; these can be used as this measurement adds to the p-th order predictor the sample $x[n-p-1]$ (not used in linear predictor of length p), and in this way the linear predictor can be augmented by one sample to $p+1$.

The p-th order predictor $\hat{x}_p[n]$ can be augmented by adding the backward prediction error $\beta_p[n-1]$ scaled by an appropriate coefficient c_{p+1} (to be defined):

$$\hat{x}_{p+1}[n] = \hat{x}_p[n] - c_{p+1}\beta_p[n-1]$$

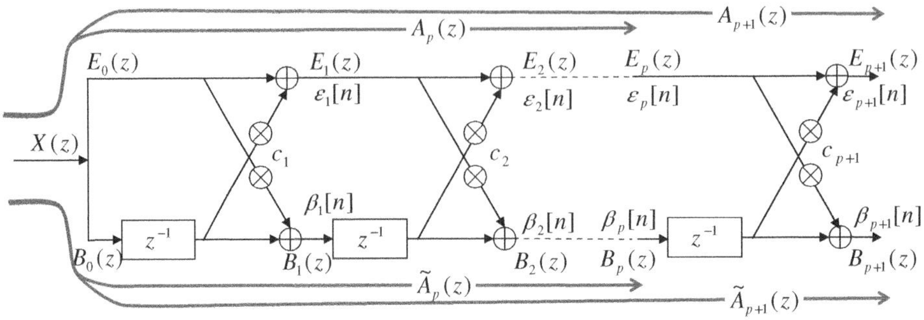

Figure 12.5 Lattice structure of linear predictors.

The forward prediction error for the predictor of length $p+1$ becomes

$$\varepsilon_{p+1}[n] = x[n] - \hat{x}_{p+1}[n] = \varepsilon_p[n] + c_{p+1}\beta_p[n-1]$$

The coefficient c_{p+1} is derived by minimizing the MSE wrt the scalar parameter c_{p+1}, using the orthogonality principle:

$$\min_{c_{p+1}}\{\mathbb{E}[|\varepsilon_{p+1}[n]|^2]\} \Rightarrow \text{(ortogonality principle)} \Rightarrow \mathbb{E}[\varepsilon_{p+1}[n]\cdot\beta_p^*[n-1]] = 0$$

The coefficient that minimizes the MSE is

$$\begin{aligned} c_{p+1} = -\frac{\mathbb{E}[\varepsilon_p[n]\cdot\beta_p^*[n-1]]}{\mathbb{E}[|\beta_p[n-1]|^2]} &= \left(\begin{array}{c}\text{for the stationarity of } x[n]\\ \mathbb{E}[|\beta_p[n]|^2] = \mathbb{E}[|\varepsilon_p[n]|^2]\end{array}\right) \\ &= -\frac{\mathbb{E}[\varepsilon_p[n]\cdot\beta_p^*[n-1]]}{\sqrt{\mathbb{E}[|\beta_p[n]|^2]\cdot\mathbb{E}[|\varepsilon_p[n]|^2]}} \end{aligned} \quad (12.5)$$

It follows that $|c_{p+1}| \leq 1$. Substituting the solution into the relationship that gives the mean square prediction error for the linear predictor of length $p+1$

$$\mathbb{E}[|\varepsilon_{p+1}[n]|^2] = \mathbb{E}[|\varepsilon_p[n]|^2](1-|c_{p+1}|^2)$$

it follows that the mean square prediction error vs. the coefficients $c_1, c_2, ..., c_{p+1}$ is

$$\mathbb{E}[|\varepsilon_{p+1}[n]|^2] = \mathbb{E}[|\varepsilon_0[n]|^2]\prod_{k=1}^{p+1}(1-|c_k|^2) = \mathbb{E}[|x[n]|^2]\prod_{k=1}^{p+1}(1-|c_k|^2) \quad (12.6)$$

where the following property was used:

$$A_0(z) = 1 \Rightarrow \varepsilon_0[n] = \beta_0[n] = x[n]$$

Since $|c_k| \leq 1$ for $\forall k$, the MSE of (12.6) reduces monotonically by increasing the predictor length.

The coefficient c_{p+1} is obtained by replacing terms in (12.5) and iterating the use of the orthogonality condition to simplify the derivations:

$$c_{p+1} = -\frac{r_{xx}[p+1] + \sum_{k=1}^{p} a_p[k] r_{xx}[p+1-k]}{\mathbb{E}[|\beta_p[n-1]|^2]} = -\frac{r_{xx}[p+1] + \sum_{k=1}^{p} a_p[k] r_{xx}[p+1-k]}{r_{xx}[0] + \sum_{k=1}^{p} a_p[k] r_{xx}[-k]}$$

the use of the first relationship highlights that its computation requires p multiplications and one division with a cost of approximately $p+1$ multiplications (by assuming the cost of division to be the same as multiplication).

The filter $A_{p+1}(z)$ follows from the z-transform for all the relationships so far (see lattice filter structure in Figure 12.5):

$$\varepsilon_p[n] \leftrightarrow E_p(z) = A_p(z) X(z)$$

$$\beta_p[n] \leftrightarrow B_p(z) = z^{-p} A_p^*(1/z) X(z) = \widetilde{A}_p(z) X(z)$$

$$\beta_p[n-1] \leftrightarrow z^{-1} B_p(z) = z^{-(p+1)} A_p^*(1/z) X(z) = z^{-1} \widetilde{A}_p(z) X(z)$$

thus

$$E_{p+1}(z) = A_{p+1}(z) X(z) = E_p(z) + c_{p+1} z^{-1} B_p(z) = \left(A_p(z) + c_{p+1} z^{-(p+1)} A_p^*(1/z)\right) X(z)$$

The recursive relationship to get the predictor of length $p+1$ from the predictor of length p is (Levinson recursion formula):

$$A_{p+1}(z) = A_p(z) + c_{p+1} z^{-(p+1)} A_p^*(1/z)$$

and the cost is p multiplications.

To evaluate the overall computational cost of obtaining the linear predictor of order p_* using Levinson recursion, the recursive formula should be iterated $p = 1, 2, \ldots, p_*$ with a cost

$$\underbrace{(2+1)}_{A_0(z)\to A_1(z)} + \underbrace{(3+2)}_{A_1(z)\to A_2(z)} + \ldots + \underbrace{(P+P-1)}_{A_{P-1}(z)\to A_P(z)} \simeq 2\left(1+2+\ldots+p_*\right) = p_*(p_*-1)$$

Thus this is $\mathcal{O}(p_*^2)$, lower than direct matrix inversion, which is $\mathcal{O}(p_*^3)$. The lattice filter related to the Levinson algorithm in Figure 12.5 is not only computationally efficient, but has several other attractive properties such as its stability (see e.g., [13, 48]).

13

Bayesian Tracking and Kalman Filter

Bayesian estimation is based on the assumption of having a-priori information on the parameter to be estimated. In some cases the pdf is known and time-invariant, but in other cases it can be variable in time. The estimators in this chapter exploit some knowledge of the mechanisms (or models) of the time evolution of the parameters (or system's *state* when referring to a time-varying quantity) of the dynamic system in order to improve the estimate by chaining the sequence of observations linked to the sequence of states that accounts for their dynamic evolution. The Bayesian estimator of the evolution of the state from an ordered sequence of observations is carried out iteratively by computing the a-priori pdf of the state (starting from the available observations before using the current observation) and the a-posteriori pdf (also adding the current observation).

The Kalman filter (KF) is a special case of linear estimator of the evolution of the state of a dynamic linear system from the sequence of observations handled as random process. As with any linear Bayesian estimator, the KF is the optimal filtering if the processes are Gaussian and it is widely investigated in the literature [51]. The linearity of the KF has the advantage of simplicity and computational efficiency, so whenever the rvs are non-Gaussian, or the evolution of the state is ruled by non-linear models, the KF can also be derived after linearizing the dynamic model and the observations in order to still use the constitutive equations of the KF. This is the Extended Kalman Filter (EKF). However, in this case there is no guarantee of global optimality (but only local).

The example of estimating the position of a point in space from a sequence of external observations (e.g., GPS navigation, radar system, mobile phones, airplane and vehicular traffic control) is quite illustrative to capture the essence of the estimation of the evolution of state represented here by the point position. This example illustrated in Figure 13.1 will be a visual guideline for the whole chapter. The target represented by a point moves in a plane with coordinates $\boldsymbol{\theta}[n] = [\theta_1[n], \theta_2[n]]^T$ (dashed line) with accelerations/decelerations and changes of direction (filled shaded gray-dots uniformly sampled in time to highlight the dynamics of the target), and the observations $\mathbf{x}[n]$ are affected by impairments. Tracking of the target is carried out by estimating its position $\hat{\boldsymbol{\theta}}[n]$ (empty dots) from the sequence of the estimated past positions $\hat{\boldsymbol{\theta}}[1:n-1]$ (solid line) combined with all the measurements up to the nth sample time $\mathbf{x}[1:n]$. The ellipses represent the confidence area for a certain probability, and in Figure 13.1 is shown the a-posteriori ellipses at time $n-1$ and n, which are more compact when compared to the prediction of the position in n. A detailed discussion is provided below in terms of covariances.

Statistical Signal Processing in Engineering, First Edition. Umberto Spagnolini.

Companion website: www.wiley.com/go/spagnolini/signalprocessing

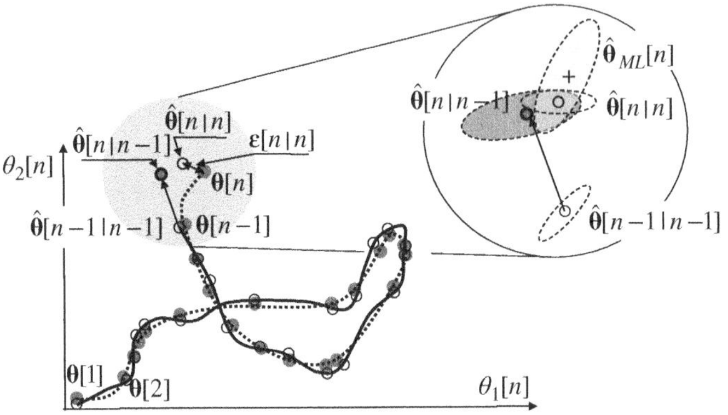

Figure 13.1 Bayesian tracking of the position of a moving point in a plane.

13.1 Bayesian Tracking of State in Dynamic Systems

The dynamic system is illustrated in Figure 13.2; the accessible data at n-th time sample $\mathbf{x}[n]$ is modeled as

$$\mathbf{x}[n] = \mathbf{h}_n(\boldsymbol{\theta}[n], \boldsymbol{\beta}[n])$$

and depends on the state $\boldsymbol{\theta}[n]$ with the transformation $\mathbf{h}_n(.)$ that is time-varying and, in general, non-linear but known. Noise in the observation model is represented by $\boldsymbol{\beta}[n]$, which is fully characterized by its statistical properties—not necessarily Gaussian. Compared to the Bayesian estimation of the state $\boldsymbol{\theta}[n]$ from the observations $\mathbf{x}[n]$ that should be carried out on a sample-by-sample basis (Chapter 11), here we assume that the constitutive equation that rules the evolution of the state is known. More specifically, the state evolves according to the recursive relationship

$$\boldsymbol{\theta}[n] = \mathbf{f}_n(\boldsymbol{\theta}[n-1], \boldsymbol{\alpha}[n])$$

that chains the evolution of the state as depending on the state at the previous time $\boldsymbol{\theta}[n-1]$ from the relationship $\mathbf{f}_n(.)$ that is arbitrarily non-linear and time-varying (but known). The excitation $\boldsymbol{\alpha}[n]$ makes the state vary vs. time in a somewhat unpredictable way, characterized by its statistical properties. In contrast to Bayesian estimation where the state is estimated from the knowledge of the a-priori pdf $p[\boldsymbol{\theta}[n]]$ and the

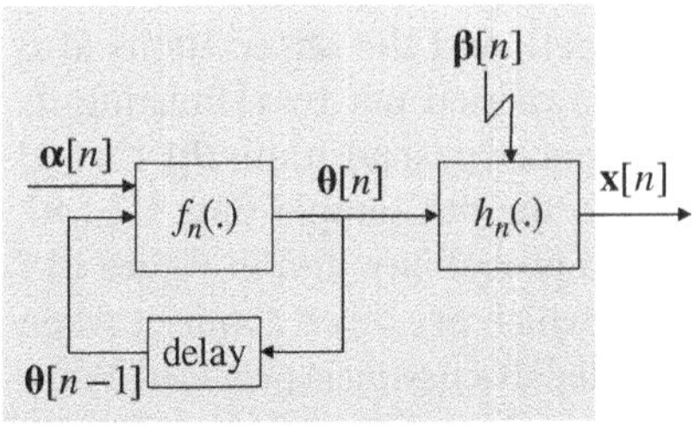

Figure 13.2 Dynamic systems model for the evolution of the state $\theta[n]$ (shaded area is not-accessible).

state-observation relationship $\mathbf{h}_n(.)$, focus of the Bayesian tracking of the state is the estimation of the *temporal evolution of the state* $\hat{\theta}[1], \hat{\theta}[2], ..., \hat{\theta}[n]$ as correlated entities from the *sequence* of observations $\mathbf{x}[1:n] = [\mathbf{x}[1], \mathbf{x}[2], ..., \mathbf{x}[n]]$ (not just the last one!) by introducing into the estimator the dynamic equation that describe the state evolution $\mathbf{f}_n(.)$. Namely, Bayesian tracking is based on the computation of the evolution of the a-posteriori pdf by accumulating all the observations available up to the current time (causality is implicitly assumed)

$$p(\theta[0]) \rightarrow p(\theta[1]|\mathbf{x}[1]) \rightarrow p(\theta[2]|\mathbf{x}[1:2]) \rightarrow ... \rightarrow p(\theta[n-1]|\mathbf{x}[1:n-1]) \\ \rightarrow p(\theta[n]|\mathbf{x}[1:n]) \rightarrow ...$$

where the initialization pdf $p(\theta[0])$ is assumed known. From the a-posteriori pdf at the nth sample $p(\theta[n]|\mathbf{x}[1:n])$ the evolution of the state can be estimated using the MAP or MMSE criteria:

$$\begin{aligned} \theta_{MAP}[n|n] &= \arg\max_{\theta[n]}\{p(\theta[n]|\mathbf{x}[1:n])\} \\ \theta_{MMSE}[n|n] &= \int \theta[n]p(\theta[n]|\mathbf{x}[1:n])d\theta[n] = \mathbb{E}[\theta[n]|\mathbf{x}[1:n]] \end{aligned}$$

where the abbreviation $\theta[n|n]$ in the notation indicates the usage of all observations $\mathbf{x}[1:n]$ up to the nth one.

Note that when $p[\theta[n]|\mathbf{x}[1:n]]$ is Gaussian

$$\theta[n|n] \sim \mathcal{N}(\hat{\theta}[n|n], \mathbf{P}[n|n])$$

it is characterized by the mean $\hat{\theta}[n|n]$ and the covariance $\mathbf{P}[n|n]$ using all the observations $\mathbf{x}[1:n]$ as highlighted in the notation. This is the case of the Kalman Filter, and the estimate is

$$\theta_{MAP}[n|n] = \theta_{MMSE}[n|n] = \hat{\theta}[n|n]$$

with a-posteriori covariance $\mathbf{P}[n|n] = \text{cov}(\theta[n|n])$.

13.1.1 Evolution of the A-Posteriori pdf

The pdfs are chained by the constitutive relationship of the dynamic of the state in a recursive relationship illustrated in Figure 13.3. The a-posteriori pdf should be considered as the update from the a-posteriori pdf from the previous time sample as $p(\theta[n-1]|\mathbf{x}[1:n-1]) \rightarrow p(\theta[n]|\mathbf{x}[1:n])$ using the evolution of the state. Assuming that the a-posteriori pdf $p(\theta[n-1]|\mathbf{x}[1:n-1])$ is known, the sequence of steps are based on the *prediction* of the state pdf (a-priori pdf at time n from $\mathbf{x}[1:n-1]$) and *update* of the state pdf from the last observation (a-posterior pdf at time n from $[\mathbf{x}[1:n-1], \mathbf{x}[n]]$).

- **Prediction:** $p(\theta[n-1]|\mathbf{x}[1:n-1]) \longrightarrow p(\theta[n]|\mathbf{x}[1:n-1])$
  Given the a-posteriori pdf at time n–1, $p(\theta[n-1]|\mathbf{x}[1:n-1])$ the prediction makes use of the dynamic model of state evolution $\mathbf{f}_n(.)$ to derive the a-priori pdf $p[\theta[n]|\mathbf{x}[1:n-1]]$ assuming that the state is not influenced by past observations:

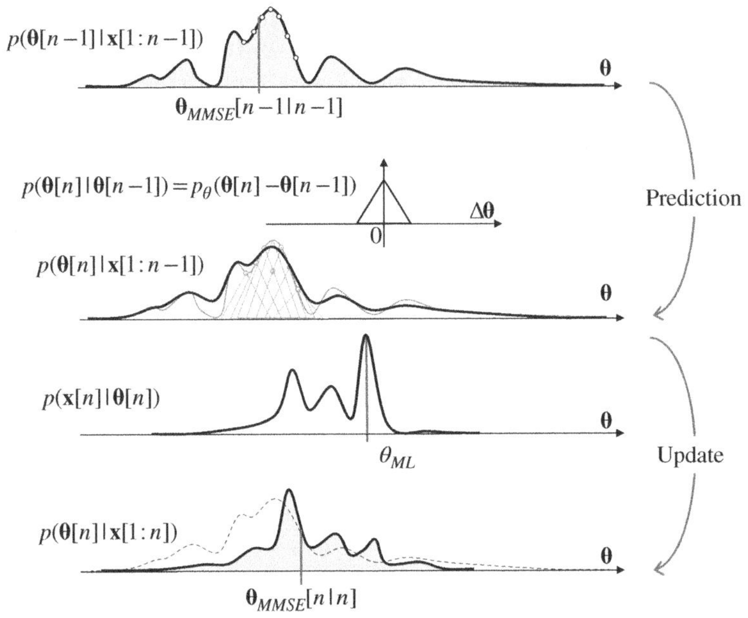

Figure 13.3 Evolution of a-posteriori pdf from a-priori pdf in Bayesian tracking.

$p(\theta[n]|\theta[n-1], \mathbf{x}[1\,:\,n-1]) = p(\theta[n]|\theta[n-1])$. The a-priori pdf at the nth step from the a-posteriori pdf at the (n–1)th step $p(\theta[n-1]|\mathbf{x}[1\,:\,n-1])$ is based on the Chapman–Kolmogorov relationship

$$p(\theta[n]|\mathbf{x}[1\,:\,n-1]) = \int p(\theta[n]|\theta[n-1]) \cdot p(\theta[n-1]|\mathbf{x}[1\,:\,n-1])d\theta[n-1]$$

The term $p[\theta[n]|\theta[n-1]]$ is referred to as the *transition pdf* as it maps the probability for every state value $\theta[n-1]$ at time n–1 onto the corresponding state value $\theta[n]$ at time n. The transition pdf depends on the state evolution $\theta[n] = \mathbf{f}_n(\theta[n-1], \boldsymbol{\alpha}[n])$ and can be time-varying. In some common contexts such as time-invariant systems, the state evolution is independent of absolute time but rather is dependent on relative time and relative state values, so that $p(\theta[n]|\theta[n-1]) = p_\theta(\theta[n] - \theta[n-1])$. Then, the Chapman–Kolmogorov formula simplifies into the convolutive relationship

$$p(\theta[n]|\mathbf{x}[1\,:\,n-1]) = \int p_\theta(\theta[n] - \theta[n-1]) \cdot p(\theta[n-1]|\mathbf{x}[1\,:\,n-1])d\theta[n-1]$$

This formula highlights that the prediction step smooths and possibly translates the a-posteriori pdf $p(\theta[n-1]|\mathbf{x}[1\,:\,n-1])$ according to the dynamic evolution of the state. The figure illustrates this step (without translation) where the transition pdf has a symmetric triangular shape implying that the probability that $\theta[n] = \theta[n-1]$ is the largest one (referring to the example of moving target, the position of the target is likely to remain still from time $n-1$ to n), but with some smearing in state according to the uncertainty of the evolution that depends on the excitation process $\boldsymbol{\alpha}[n]$.

- **Update:** $p(\theta[n]|\mathbf{x}[1:n-1]) \longrightarrow p(\theta[n]|\mathbf{x}[1:n])$
  The a-posteriori pdf can be derived when a new observation is made available to confirm or modify the a-priori pdf evaluated from all the observations up to n–1. The observation $\mathbf{x}[n]$ is accounted for using the Bayes relationship to derive the *a-posteriori* pdf:

$$p(\theta[n] \mid \mathbf{x}[1:n]) = p(\theta[n] \mid \mathbf{x}[n], \mathbf{x}[1:n-1]) = \frac{p(\theta[n], \mathbf{x}[n] \mid \mathbf{x}[1:n-1])}{p(\mathbf{x}[n]|\mathbf{x}[1:n-1])}$$
$$= \frac{p(\mathbf{x}[n] \mid \theta[n], \mathbf{x}[1:n-1]) \cdot p(\theta[n] \mid \mathbf{x}[1:n-1])}{p(\mathbf{x}[n]|\mathbf{x}[1:n-1])}$$

  Notice that the *conditional pdf* $p(\mathbf{x}[n] \mid \theta[n], \mathbf{x}[1:n-1]) = p(\mathbf{x}[n] \mid \theta[n])$ coincides with the likelihood function and depends on the relationship $\mathbf{x}[n] = \mathbf{h}_n(\theta[n], \beta[n])$ that links state and observations. The denominator is a normalization factor, and the a-posteriori pdf after update is:

$$p(\theta[n] \mid \mathbf{x}[1:n]) = \frac{1}{p(\mathbf{x}[n]|\mathbf{x}[1:n-1])} p(\mathbf{x}[n] \mid \theta[n]) \cdot p(\theta[n] \mid \mathbf{x}[1:n-1])$$

  Once again, there is no guarantee that there is a closed form relationship of the a-posteriori pdfs, except for the special case of Gaussians. From the Figure 13.3 it can be seen that the prediction step has the effect of smoothing the pdf, while the new observation in the update step sharpens some modes and downsizes others according to the likelihood of the state $\theta[n]$ for the current observation $\mathbf{x}[n]$. As sketched in the example, the likelihood can show a maximum value even far away from the predicted values as it could be by some noise realization (and the ML estimate would be excessively noisy); in this case the a-priori pdf downweights this value in favor of others more likely to fit into a sequential set of states.

The analytic relationships in Bayesian tracking estimators depend on the dynamic evolution of the state $\theta[n] = \mathbf{f}_n(\theta[n-1], \alpha[n])$, on the state-observation link $\mathbf{x}[n] = \mathbf{h}_n(\theta[n], \beta[n])$, and on the statistical properties of the processes involved. A simple and widely used tracking is the Kalman estimator (or filter) as there are closed form relationships involved for the update of the central moments.

13.2 Kalman Filter (KF)

The Kalman filter (KF) is the optimal MMSE estimate of the state for time-varying linear systems and Gaussian processes in which the evolution of the state is described by a linear dynamic system as in Figure 13.4. The main innovation introduced by R.E. Kalman in 1960 [50] was to describe the variations of signals by a finite dimension linear dynamic system. In this way there is no need to explicitly describe linear systems by their transfer functions as it maps the recursive structure of the system onto the MMSE estimator of the state of a system. The Kalman filter is derived here as a special case of Bayesian tracking by adapting the constitutive relationships to Gaussian processes and linear models.

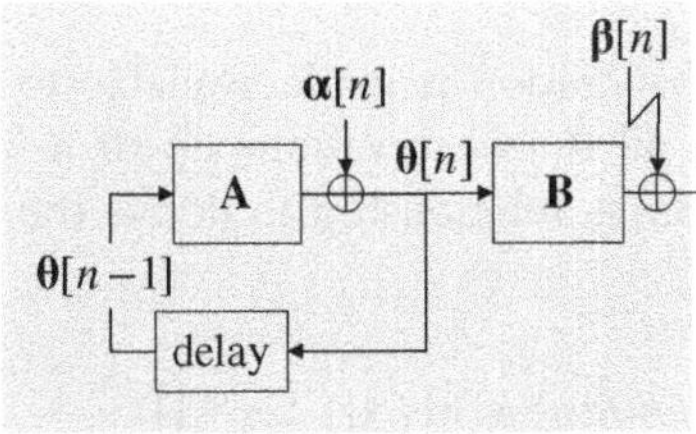

Figure 13.4 Linear dynamic model.

Let the time evolution of the state and the observations be described by the linear model (state equations)

$$\begin{aligned}\theta[n] &= \mathbf{A}\cdot\theta[n-1]+\alpha[n]\\ \mathbf{x}[n] &= \mathbf{B}\cdot\theta[n]+\beta[n]\end{aligned}$$

where the state $\theta[n]\in\mathbb{R}^p$ cannot be accessed directly, but only from noisy observations $\mathbf{x}[n]\in\mathbb{R}^N$. The excitation and noise are Gaussian and mutually uncorrelated:

$$\begin{aligned}\alpha[n] &\sim \mathcal{N}\left(\mathbf{0},\mathbf{C}_{\alpha\alpha}\right)\\ \beta[n] &\sim \mathcal{N}\left(\mathbf{0},\mathbf{C}_{\beta\beta}\right)\end{aligned}$$

In addition, the state $\theta[n]$ is uncorrelated with $\alpha[m]$ and $\beta[m]$ for $m>n$ (causality condition). The KF estimates the sequences of state variables $\theta[n]$ from all observations up to the current one $\mathbf{x}[1:n]=[\mathbf{x}[1],\mathbf{x}[2],...,\mathbf{x}[n]]$ for a known linear model $\mathbf{A}\in\mathbb{R}^{p\times p}$ and $\mathbf{B}\in\mathbb{R}^{N\times p}$, covariance of the excitation $\mathbf{C}_{\alpha\alpha}$ and noise $\mathbf{C}_{\beta\beta}$. The a-priori pdf is Gaussian $\theta[0]\sim\mathcal{N}(\hat{\theta}[0|0],\mathbf{P}[0|0])$ so that $\hat{\theta}[0|0]$ and $\mathbf{P}[0|0]$ are initialization values known from the problem at hand. Since the initialization of the pdf is Gaussian, the excitation $\alpha[n]$ is Gaussian, and the dynamic system is linear, the MMSE estimator is linear.[1]

The KF is just a special case of the previous relationships for prediction and update where the pdfs are represented by the first and second order central moments, and there is a closed form for the statistical evolution of the state from the observations. More specifically, since at every step the pdfs are Gaussian, it is enough to derive the prediction and update equations for mean and covariance. There is the following pairing of the pdfs and their properties:

$$p(\theta[n-1]|\mathbf{x}[1:n-1])=G(\theta[n-1|n-1];\hat{\theta}[n-1|n-1],\mathbf{P}[n-1|n-1])$$

1 The a-priori pdf of the state is Gaussian $\theta[n-1]\sim\mathcal{N}(\mu_\theta[n-1],\mathbf{C}_{\theta\theta}[n-1])$; the update of the pdf in prediction involves only the state evolution $\theta[n]=\mathbf{A}\cdot\theta[n-1]+\alpha[n]\sim\mathcal{N}(\mu_\theta[n],\mathbf{C}_{\theta\theta}[n])$ so it is sufficient to update the mean and covariance: $\mu_\theta[n]=\mathbf{A}\mu_\theta[n-1]$ and $\mathbf{C}_{\theta\theta}[n]=\mathbf{A}\mathbf{C}_{\theta\theta}[n-1]\mathbf{A}^T+\mathbf{C}_{\alpha\alpha}$. Therefore, the a-priori pdf of the state is still Gaussian over all its evolution. The state-observation relationship $\mathbf{x}[n]=\mathbf{B}\cdot\theta[n]+\beta[n]$ is linear with noise $\beta[n]$ (Gaussian too); all the pdfs involved are Gaussian and thus the optimal Bayes estimator (MMSE and/or MAP) is linear.

$$p(\boldsymbol{\theta}[n]|\mathbf{x}[1:n-1]) = G(\boldsymbol{\theta}[n|n-1];\hat{\boldsymbol{\theta}}[n|n-1],\mathbf{P}[n|n-1])$$
$$p(\boldsymbol{\theta}[n]\mid\mathbf{x}[1:n]) = G(\boldsymbol{\theta}[n|n];\hat{\boldsymbol{\theta}}[n|n],\mathbf{P}[n|n])$$

with the update relationships shown below. Notice that it is customary in the KF to use the following notation:

$\hat{\theta}[n|n] = \hat{\theta}[n|\mathbf{x}[1:n]]$ estimation from all the $\mathbf{x}[1:n]$ observations

$\mathbf{P}[n|n] = \mathrm{cov}[\hat{\theta}[n|n]] = \mathrm{cov}[\hat{\theta}[n|\mathbf{x}[1:n]]]$ covariance from all the $\mathbf{x}[1:n]$ observations

$\hat{\theta}[n|n-1] = \hat{\theta}[n|\mathbf{x}[1:n-1]]$ one-step prediction from the $\mathbf{x}[1:n-1]$ observations

$\mathbf{P}[n|n-1] = \mathrm{cov}[\hat{\theta}[n|n-1]]$ covariance of one-step predictor

13.2.1 KF Equations

Prediction: The pdf $\theta[n-1|n-1] \sim \mathcal{N}(\hat{\theta}[n-1|n-1],\mathbf{P}[n-1|n-1]$ at step *n-1* is the a-posteriori of the state from $\mathbf{x}[1:n-1]$. From the evolution of the state:

$$\theta[n|n-1] = \mathbf{A}\cdot\theta[n-1|n-1]+\boldsymbol{\alpha}[n] \sim \mathcal{N}(\hat{\theta}[n|n-1],\mathbf{P}[n|n-1])$$

The mean $\hat{\theta}[n|n-1]$ and covariance $\mathbf{P}[n|n-1]$ for the a-priori pdf from $\hat{\theta}[n-1|n-1]$ and $\mathbf{P}[n-1|n-1]$ are simple to derive from the linear transformation of state evolution:

$$\hat{\theta}[n|n-1] = \mathbf{A}\cdot\hat{\theta}[n-1|n-1]$$
$$\mathbf{P}[n|n-1] = \mathbf{A}\mathbf{P}[n-1|n-1]\mathbf{A}^T+\mathbf{C}_{\alpha\alpha}$$

Correction (or **update**): The equation $\mathbf{x}[n] = \mathbf{B}\cdot\theta[n]+\beta[n]$ that links state-observation is used to infer the state $\theta[n]$ from the last observation $\mathbf{x}[n]$ where all the information pertaining the set of observations $\mathbf{x}[1:n-1]$ is accumulated (through iterative updating) into the a-priori pdf at step n–1. In other words, if rewriting the observation model as

$$\mathbf{x}[n] = \mathbf{B}\cdot\boldsymbol{\vartheta}+\beta[n]$$

where the a-priori pdf for the dummy-variable $\boldsymbol{\vartheta}$ is

$$\boldsymbol{\vartheta} \sim \mathcal{N}(\hat{\theta}[n|n-1],\mathbf{P}[n|n-1])$$

the a-posteriori from all the observations $\mathbf{x}[1:n-1]$ except the last one. The model reduces to the conventional Bayes estimation for the linear Gaussian model (see Chapter 11) with Gaussian a-posteriori pdf for the new observation denoted as

$$\theta[n|n] \sim \mathcal{N}(\hat{\theta}[n|n],\mathbf{P}[n|n])$$

The mean coincides with the MMSE estimate $\hat{\theta}[n|n] = \mathbb{E}[\theta[n]|\mathbf{x}[n]]$ that can be derived by the relationships in Section 11.2:

$$\hat{\theta}[n|n] = \hat{\theta}[n|n-1] + \underbrace{\mathbf{P}[n|n-1]\mathbf{B}^T(\mathbf{B}\mathbf{P}[n|n-1]\mathbf{B}^T + \mathbf{C}_{\beta\beta})^{-1}}_{\mathbf{G}[n]} \cdot (\mathbf{x}[n] - \mathbf{B}\hat{\theta}[n|n-1]),$$

where it is highlighted that the MMSE estimate updates the prediction of the state $\hat{\theta}[n|n-1]$ from the measurements $\mathbf{x}[1:n-1]$ with the correcting term

$$\varepsilon[n|n-1] = \mathbf{x}[n] - \mathbf{B}\hat{\theta}[n|n-1]$$

which is the prediction error of the measurements, weighted appropriately by $\mathbf{G}[n]$. The KF gain matrix

$$\mathbf{G}[n] = \mathbf{P}[n|n-1]\mathbf{B}^T(\mathbf{B}\mathbf{P}[n|n-1]\mathbf{B}^T + \mathbf{C}_{\beta\beta})^{-1} \in \mathbb{R}^{p\times N}$$

plays a key role in KF and it should be recomputed at every iteration to be matched with the dynamic evolution of the variables at hand. In short, the MMSE estimate of the state (or equivalently, the first order central moment of the a-posteriori pdf) is

$$\hat{\theta}[n|n] = \hat{\theta}[n|n-1] + \mathbf{G}[n]\varepsilon[n|n-1]$$

In KF jargon, the correcting term from $\varepsilon[n|n-1]$ is the KF innovation as it refers to a term that could not be predicted in any way from the estimate of the state with the measurements available so far: $\hat{\theta}[n|n-1]$. The covariance for the a-posteriori pdf

$$\mathbf{P}[n|n] = \mathbf{P}[n|n-1] - \mathbf{G}[n](\mathbf{B}\mathbf{P}[n|n-1]\mathbf{B}^T + \mathbf{C}_{\beta\beta})\mathbf{G}[n]^T$$

can be rewritten as

$$\mathbf{P}[n|n] = \mathbf{P}[n|n-1] - \mathbf{G}[n]\mathbf{B}\mathbf{P}[n|n-1]$$

this is the *Riccati equation*, which does not involve any dependency on data.

The constitutive relationships for the KF are summarized as:

- **Prediction**

$$\hat{\theta}[n|n-1] = \mathbf{A} \cdot \hat{\theta}[n-1|n-1]$$

- **Covariance matrix of prediction**

$$\mathbf{P}[n|n-1] = \mathbf{A}\mathbf{P}[n-1|n-1]\mathbf{A}^T + \mathbf{C}_{\alpha\alpha}$$

- **Correction**

$$\hat{\theta}[n|n] = \hat{\theta}[n|n-1] + \mathbf{G}[n]\underbrace{(\mathbf{x}[n] - \mathbf{B}\hat{\theta}[n|n-1])}_{\varepsilon[n|n-1]}$$

- **Kalman Filter Gain**

$$\mathbf{G}[n] = \mathbf{P}[n|n-1]\mathbf{B}^T(\mathbf{B}\mathbf{P}[n|n-1]\mathbf{B}^T + \mathbf{C}_{\beta\beta})^{-1}$$

- **Covariance matrix of MMSE estimate** *(Riccati equation)*

$$\mathbf{P}[n|n] = \mathbf{P}[n|n-1] - \mathbf{G}[n]\mathbf{B}\mathbf{P}[n|n-1]$$

- **Initialization**

$$\hat{\theta}[0|0] = \mathbb{E}[\theta[0]] \quad \text{and} \quad \mathbf{P}[0|0] = \text{cov}[\theta[0]] \quad \text{are known}$$

13.2.2 Remarks

There are several aspects related to KF that are worthwhile to be investigated and discussed in detail. Here we restrict the discussion to those considered essential for a skilled use of KF in applications and fundamental to extrapolate other properties. For an in-depth knowledge of the topics related to KF the reader can access any of the (many) publications on the subject.

1) The KF is the optimum MMSE estimator for linear models and Gaussian processes, even if time-varying. Similarly to LMMSE, KF could be adopted as linear MMSE in contexts where a non-linear Bayesian tracking could be more appropriate (as MMSE optimum). The only things necessary to know for KF are the first and second order central moments and, if necessary, some approximations to manipulate the dynamic model.
2) The KF gain $\mathbf{G}[n]$ needs to evaluate the inverse of an $N \times N$ matrix at every iteration, and this is the computationally most expensive step in KF.
3) The KF is still the optimal MMSE tracking algorithm even if the matrixes governing the system dynamics are time-varying, provided that the variations are known. It is sufficient to make the following replacements $\mathbf{A} \to \mathbf{A}[n]$, $\mathbf{B} \to \mathbf{B}[n]$, $\mathbf{C}_{\alpha\alpha} \to \mathbf{C}_{\alpha\alpha}[n]$. These time variations can be due to the use of the KF for non-linear systems after linearization on every step. This is the *extended Kalman filter (EKF)* and will be discussed later.
4) There are some degenerate cases, such as when the parameters are constant over time and the use of the KF is to progressively refine the estimate (i.e., minimize the MSE) when increasing the observations. In this case:

$$\begin{aligned}&\boldsymbol{\alpha}[n] = \mathbf{0} \Rightarrow \mathbf{C}_{\alpha\alpha} = \mathbf{0}\\&\mathbf{A} = \mathbf{I} \Rightarrow \theta[n] = \theta[n-1] \Rightarrow \hat{\theta}[n|n-1] = \hat{\theta}[n|n] \Rightarrow \mathbf{P}[n|n-1] = \mathbf{P}[n|n]\end{aligned}$$

   and the prediction loses any meaning as stated above ($\hat{\theta}[n|n] = \hat{\theta}[n|n-1] = \hat{\theta}[n]$) and $\mathbf{P}[n|n-1] = \mathbf{P}[n|n] = \mathbf{P}[n]$. The constitutive equations

$$\begin{aligned}&\hat{\theta}[n] = \hat{\theta}[n-1] + \mathbf{G}[n](\mathbf{x}[n] - \mathbf{B}\hat{\theta}[n-1])\\&\mathbf{P}[n] = \mathbf{P}[n-1] - \mathbf{G}[n]\mathbf{B}\mathbf{P}[n-1]\\&\mathbf{G}[n] = \mathbf{P}[n-1]\mathbf{B}^T(\mathbf{B}\mathbf{P}[n-1]\mathbf{B}^T + \mathbf{C}_{\beta\beta})^{-1}\end{aligned}$$

are those for a sequential LMMSE estimator based on a KF. The initialization is still based on the mean value $\hat{\theta}[0] = \mathbb{E}[\theta[0]]$ with covariance $\mathbf{P}[0] = \mathbf{C}_{\theta\theta}$ since the a-priori value is the same regardless of time as $\theta[n] = \theta[n-1] = \ldots = \theta[1] = \theta[0] = \theta_o$ (the parameter does not change vs. time as $\mathbf{A} = \mathbf{I}$).

Another similar example is for $\boldsymbol{\alpha}[n] = \mathbf{0}$ and any deterministic and known value of the dynamic $\mathbf{A} \neq \mathbf{I}$; this models a system where state evolves in a deterministic way and it is of interest to infer some properties of this deterministic behavior (e.g., when referring to the passive positioning example, it could be the starting position of a target).

5) The update of the covariance matrixes is independent from the observations and in steady-state condition relationships can be separated:

$$\begin{aligned}
\mathbf{P}[n|n-1] &= \mathbf{A}\mathbf{P}[n-1|n-1]\mathbf{A}^T + \mathbf{C}_{\alpha\alpha} \\
\mathbf{P}[n|n] &= \mathbf{P}[n|n-1] - \mathbf{G}[n]\mathbf{B}\mathbf{P}[n|n-1] \\
\mathbf{G}[n] &= \mathbf{P}[n|n-1]\mathbf{B}^T(\mathbf{B}\mathbf{P}[n|n-1]\mathbf{B}^T + \mathbf{C}_{\beta\beta})^{-1}
\end{aligned}$$

One can expect that in steady state (or more pragmatically, for a large number of iterations), these recursive equations converge to the asymptotic covariance in prediction ($\mathbf{P}_1$) and estimation ($\mathbf{P}_0$), as well as the KF gain[2]

$$\begin{aligned}
\lim_{n\to\infty} \mathbf{P}[n|n-1] &= \mathbf{P}_1 \\
\lim_{n\to\infty} \mathbf{P}[n|n] &= \mathbf{P}_0 \\
\lim_{n\to\infty} \mathbf{G}[n] &= \mathbf{G}_\infty = \mathbf{P}_1\mathbf{B}^T(\mathbf{B}\mathbf{P}_1\mathbf{B}^T + \mathbf{C}_{\beta\beta})^{-1}
\end{aligned}$$

These values can be evaluated as (numerical) solutions of the recursive equation

$$\begin{aligned}
\mathbf{P}_1 &= \mathbf{A}\mathbf{P}_0\mathbf{A}^T + \mathbf{C}_{\alpha\alpha} \\
\mathbf{P}_0 &= \mathbf{P}_1 - \mathbf{P}_1\underbrace{\mathbf{B}^T(\mathbf{B}\mathbf{P}_1\mathbf{B}^T + \mathbf{C}_{\beta\beta})^{-1}\mathbf{B}}_{\mathbf{Q}}\mathbf{P}_1 = (\mathbf{I} - \mathbf{P}_1\mathbf{Q})\mathbf{P}_1
\end{aligned}$$

This can be solved by plugging the first expression into the second one (Riccati equation for steady-state condition); this only has a closed form solution in some cases—otherwise it is enough to simulate the recursions and get the asymptotic solution. The asymptotic MSE could be relevant in some context:

$$\begin{aligned}
MSE_1 &= \lim_{n\to\infty} \mathrm{tr}\{\mathbf{P}[n|n-1]\} = \mathrm{tr}\{\mathbf{P}_1\} \\
MSE_0 &= \lim_{n\to\infty} \mathrm{tr}\{\mathbf{P}[n|n]\} = \mathrm{tr}\{\mathbf{P}_0\}
\end{aligned}$$

to evaluate the optimality of the solution for varying time. Another benefit of using a constant gain $\mathbf{G}_\infty$ is that the gain involves no computational effort (except at

2 The proof of convergence depends on the problem at hand (technically, the dynamic model should be accessible from observations, or should be observable).

initialization) and it is asymptotically optimal, even if in the first iterations the KF $\mathbf{G}_{\infty}$ could be sub-optimal.

6) The MMSE estimate of the state $\boldsymbol{\theta}[n]$ can be derived from the set of observations $\mathbf{x}[1:n+\ell] = [\mathbf{x}[1], \mathbf{x}[2], ..., \mathbf{x}[n+\ell]]$ with $\ell \geq 0$ (the case $\ell = 0$ is the one considered here). In this case the KF is the optimal *smoothing filter*, which is based of future data and is more accurate. The practical implication is the use of anti-causal components (for $\ell > 0$) that is not always possible.
7) Initialization is by the moments $\hat{\boldsymbol{\theta}}[0|0] = \mathbb{E}[\boldsymbol{\theta}[0]]$ and $\mathbf{P}[0|0] = \text{cov}[\boldsymbol{\theta}[0]]$, but in any case the "memory" of the initial conditions is lost after some iterations. In the case of lack of any insight on moments, it is advisable to start with a diagonal $\mathbf{P}[0|0]$, or diagonal dominant or (at most) an identity with appropriate variance.

13.3 Identification of Time-Varying Filters in Wireless Communication

In wireless communication systems, the digital streams that propagate over radio links are impaired by the time-variation of the propagation environment (or *channel*), which makes the power of the received data fluctuate, and thus be "unstable" [58]. To illustrate the concept here (but the reader might also read Section 15.2 and Chapter 17), imagine a digital transmission of 1 Mb/s (each bit is 1 μs) at 1 GHz (wavelength is 0.3 m) from a moving transmitter traveling at 50 km/h (or 13.8 m/s). After the transmission of 10^4 bits, the displacement between transmitter and receiver is 0.15 m; this is exactly half a wavelength so any constructive interference (sum of waves) becomes destructive (cancellation of waves) and propagation experiences max-to-min fading. In practice, any communication channel alternates these fluctuations very rapidly and it is of utmost importance for reliable communications to estimate the channel and track (if not predict) these variations. Digital communication over a varying propagation environment is modeled as a signal $x[n]$ (transmitted stream) that is filtered by the time-varying filter $h[k|n]$ representing the propagating link condition with the convolutive relationship

$$y[n] = \sum_{k=0}^{p-1} h[k|n]x[n-k] + w[n]$$

where $w[n]$ is the additive noise and the channel response $\{h[k|n]\}_{k=0}^{p-1}$ at time n is assumed as causal. The same relationship can be rewritten to highlight the time-varying component as

$$y[n] = \mathbf{x}[n]^T \cdot \mathbf{h}[n] + w[n] \tag{13.1}$$

with $\mathbf{x}[n] = [x[n], x[n-1], ..., x[n-p+1]]^T$ and $\mathbf{h}[n] = [h[0|n], h[1|n], ..., h[p-1|n]]^T$. The topic covered here is the identification of the time-varying channel $\mathbf{h}[n]$ by tracking its variation over time.

In order to track the channel variations, we need to frame the problem as a state equation. A common assumption is that the propagating channel variations are modeled by the dynamic system as

$$\mathbf{h}[n] = \mathbf{A} \cdot \mathbf{h}[n-1] + \mathbf{v}[n] \tag{13.2}$$

where $\mathbf{A} \in \mathbb{R}^{p \times p}$ governs the time-variation of the channel that depends on the speed of the mobile terminals, and $\mathbf{v}[n] \sim \mathcal{CN}(0, \sigma_v^2 \mathbf{I}_p)$ is the random excitation that induces the channel link fluctuations. However, both equations (13.1 and 13.2) fully define the dynamic model in canonical form where the time-varying vector $\mathbf{x}[n]$ plays the role of regressor and it is a signal known both at transmitter and receiver, usually allocated in a reserved area of the signaling (*pilot* or *training* signal). The KF equations of Section 13.2.1 are adapted herein for complex-valued signals:

$$\hat{\mathbf{h}}[n|n-1] = \mathbf{A}\hat{\mathbf{h}}[n-1|n-1]$$

$$\mathbf{P}[n|n-1] = \mathbf{A}\mathbf{P}[n-1|n-1]\mathbf{A}^H + \mathbf{C}_v$$

$$\hat{\mathbf{h}}[n|n] = \hat{\mathbf{h}}[n|n-1] + \mathbf{G}[n](y[n] - \mathbf{x}[n]^T \hat{\mathbf{h}}[n|n-1])$$

$$\mathbf{G}[n] = \frac{\mathbf{P}[n|n-1]\mathbf{x}[n]}{\mathbf{x}[n]^H \mathbf{P}[n|n-1]\mathbf{x}[n] + \sigma_w^2}$$

$$\mathbf{P}[n|n] = \mathbf{P}[n|n-1] - \mathbf{G}[n]\mathbf{x}[n]^H \mathbf{P}[n|n-1]$$

$$\text{Initialization: } \hat{\mathbf{h}}[0|0] = \mathbb{E}[\hat{\mathbf{h}}[0]] \; \mathbf{P}[0|0] = \text{cov}[\mathbf{h}[0]]$$

A common model for a dynamic system is AR(1) for the channel with independent variations, so that for $\mathbf{A} = \rho \mathbf{I}_p$ and $\mathbf{C}_v = \sigma_v^2 \mathbf{I}_p$, $\mathbf{h}[n] \sim \mathcal{N}(0, \sigma_v^2/(1-\rho^2)\mathbf{I}_p)$. Training sequences are deterministic signals known at both sides of the communication link, but to derive the analytical relationships, it is convenient to consider a stochastic model where training sequences are rvs with samples uncorrelated with one another (for ideal training sequences, see the example in Section 10.1): $\mathbb{E}[\mathbf{x}[n]\mathbf{x}[n]^H] = \sigma_x^2 \mathbf{I}_p$.

Even if the time-varying channel will cause the channel estimation using KF to fluctuate, and KF is considered a preferred choice to track these variations [53], these variations are small at $n \to \infty$ and the MSE is approximated by the asymptotic relationship on the covariance [54]

$$\mathbf{P}_1 \simeq \rho^2 (\mathbf{I}_p - \mathbf{P}_1 \mathbf{Q})\mathbf{P}_1 + \sigma_v^2 \mathbf{I}_p$$

where

$$\mathbf{Q} = \mathbb{E}[\mathbf{x}[n]\mathbf{x}^H[n](\sigma_w^2 + \mathbf{x}^H[n]\mathbf{P}_1\mathbf{x}[n])^{-1}]$$

Since the channel samples are uncorrelated as $\mathbf{h}[n] \sim \mathcal{N}(0, \sigma_v^2/(1-\rho^2)\mathbf{I}_p)$, the covariance is diagonal too (as there is no term that is making the covariance deviate from a diagonal structure) and for constant amplitude training symbols with $x[n] \in \{\pm 1\}$:

$$\mathbf{Q} = \frac{\sigma_x^2}{(\sigma_w^2 + \text{tr}(\mathbf{P}_1)\sigma_x^2)} \mathbf{I}_p$$

Since the power along channel samples is evenly distributed as $\mathbf{h}[n] \sim \mathcal{N}(0, \sigma_v^2/(1-\rho^2)\mathbf{I}_p)$, the estimation error would be similarly distributed without any mutual correlation, and thus the covariance is diagonal: $\mathbf{P}_1 = \sigma^2 \mathbf{I}_p \to \text{tr}(\mathbf{P}_1) = p \times \sigma^2$. The variance of the error $\sigma^2 = [\mathbf{P}_1]_{kk}$ follows from the equation:

$$\sigma^2 = \rho^2 \left(1 - \frac{\sigma^2}{\sigma_w^2/\sigma_x^2 + p\sigma^2}\right) \sigma^2 + \sigma_v^2$$

For small σ_w^2/σ_x^2 (large signal to noise ratio), this simplifies as

$$\sigma^2 = \rho^2 \left(1 - \frac{1}{p}\right) \sigma^2 + \sigma_v^2$$

$$MSE_1 = \text{tr}(\mathbf{P}_1) = \frac{p\sigma_v^2}{1 - \rho^2\left(1 - \frac{1}{p}\right)}$$

When channel samples change unpredictably vs. time ($\rho = 0$), the KF reduces to a linear predictor with $\sigma^2 = \sigma_v^2$, while for a long filter ($p \to \infty$) it becomes $\sigma^2 = \sigma_v^2/(1-\rho^2)$, the same as the channel variance.

13.4 Extended Kalman Filter (EKF) for Non-Linear Dynamic Systems

Let us consider the non-linear dynamic system

$$\begin{aligned} \theta[n] &= \mathbf{a}(\theta[n-1]) + \boldsymbol{\alpha}[n] \\ \mathbf{x}[n] &= \mathbf{b}(\theta[n]) + \boldsymbol{\beta}[n] \end{aligned}$$

with the same conditions for the excitations and noise as for the KF (see Section 13.3). The prediction is the only step that can be retained from the KF as being dependent on the non-linear relationship

$$\hat{\theta}[n|n-1] = \mathbf{a}(\hat{\theta}[n-1|n-1])$$

However, for the remaining relationships, it is crucial to consider a linearization of the dynamic equations wrt $\hat{\theta}[n-1|n-1]$ at step n-1

$$\mathbf{a}(\theta[n-1]) \simeq \mathbf{a}(\hat{\theta}[n-1|n-1]) + \underbrace{\left.\frac{\partial \mathbf{a}(\theta[n-1])}{\partial \theta[n-1]}\right|_{\theta[n-1]=\hat{\theta}[n-1|n-1]}}_{\mathbf{A}[n|n-1]} \cdot (\theta[n-1] - \hat{\theta}[n-1|n-1])$$

$$\mathbf{b}(\theta[n]) \simeq \mathbf{b}(\hat{\theta}[n|n-1]) + \underbrace{\left.\frac{\partial \mathbf{b}(\theta[n])}{\partial \theta[n]}\right|_{\theta[n]=\hat{\theta}[n|n-1]}}_{\mathbf{B}[n|n-1]} \cdot (\theta[n] - \hat{\theta}[n|n-1])$$

The linearized relationships are

$$\theta[n] \simeq \mathbf{A}[n|n-1]\cdot\theta[n-1]+\alpha[n]+\underbrace{\left(\mathbf{a}(\hat{\theta}[n-1|n-1])-\mathbf{A}[n|n-1]\hat{\theta}[n-1|n-1]\right)}_{\Delta\hat{\theta}[n-1|n-1]}$$

$$\mathbf{x}[n] \simeq \mathbf{B}[n|n-1]\cdot\theta[n])+\beta[n]+\underbrace{\left(\mathbf{b}(\hat{\theta}[n|n-1])-\mathbf{B}[n|n-1]\hat{\theta}[n|n-1]\right)}_{\Delta\hat{\mathbf{x}}[n|n-1]}$$

Note that $\Delta\hat{\theta}[n-1|n-1]$ and $\Delta\hat{\mathbf{x}}[n|n-1]$ are constant for the linearization at the nth step and thus these have no impact onto the constitutive relationships of the EKF, except for applying a correction to the state and observation.

Based on the linearization above, the iterative relations for the EKF are easily extended from Section 13.2.1:

- **Prediction**

$$\hat{\theta}[n|n-1]=\mathbf{a}(\hat{\theta}[n-1|n-1])$$

- **Covariance matrix of prediction**

$$\mathbf{P}[n|n-1]=\mathbf{A}[n|n-1]\mathbf{P}[n-1|n-1]\mathbf{A}[n|n-1]^T+\mathbf{C}_{\alpha\alpha}$$

- **Correction**

$$\hat{\theta}[n|n]=\hat{\theta}[n|n-1]+\mathbf{G}[n]\underbrace{(\mathbf{x}[n]-\mathbf{b}(\hat{\theta}[n|n-1]))}_{\varepsilon[n|n-1]}$$

- **EKF Gain**

$$\mathbf{G}[n]=\mathbf{P}[n|n-1]\mathbf{B}[n|n-1]^T(\mathbf{B}[n|n-1]\mathbf{P}[n|n-1]\mathbf{B}[n|n-1]^T+\mathbf{C}_{\beta\beta})^{-1}$$

- **Covariance matrix of MMSE estimate** *(Riccati equation)*

$$\mathbf{P}[n|n]=\mathbf{P}[n|n-1]-\mathbf{G}[n]\mathbf{B}[n|n-1]\mathbf{P}[n|n-1]$$

- **Initialization**

$$\hat{\theta}[0|0]=\mathbb{E}[\theta[0]] \quad \text{and} \quad \mathbf{P}[0|0]=\text{cov}[\theta[0]] \quad \text{are known}$$

13.5 Position Tracking by Multi-Lateration

The introductory example of the position tracking of a moving target by multi-lateration is now reconsidered in its entire complexity. Making reference to Figure 13.5, the position of a target (person, car, smart-phone, etc. equipped with a navigation device)

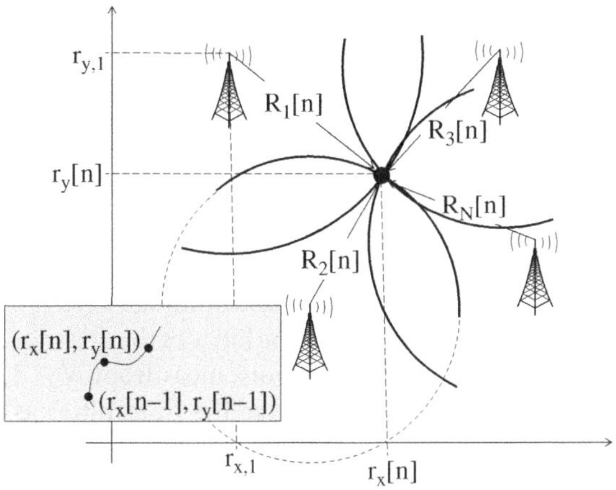

Figure 13.5 Multi-lateration positioning.

is estimated by measuring the distance (*range*) $R_k[n]$ for a set of N ($N \geq 3$) points with known positions (*anchor nodes*). Since each range defines the circle as the locus of the points from the corresponding anchor point, the set of $N = 3$ ranges (*tri-lateration*) uniquely contributes to locate the point in a plane, and if using $N > 3$ (*multi-lateration*) its position is less sensitive to unavoidable errors in $R_k[n]$. In device-based positioning, the position can be obtained at the moving device by employing multi-lateration from the position of the anchor nodes, as in GPS satellite navigation systems where the satellites have the role of anchors.[3] Each satellite transmits its instantaneous position (as each satellite is moving along the orbit, the position changes vs. time) in an absolute geographic coordinate system, and signals from multiple satellites are used by the moving receiver to estimate the ranges $R_1[n], R_2[n], \ldots$ to define the relative position on the earth using geographic longitude and latitude coordinates.

Alternatively, the target position can be estimated by the network of anchors; it is now the moving target that transmits a signal toward the anchors to let each anchor to estimate the range and still employ circular multi-laterations. This is the case of network-based positioning that is employed when the network estimates and tracks the range or the position of moving devices as in cellular communication systems. This is the operating system considered herein as more intuitive for the description of the tracking method. However, the dynamic model and the corresponding position tracking is not at all dependent on the specific positioning method that is used.

3 The Global Positioning System (GPS) is the Global Navigation Satellite System (GNSS) designed to provide position, velocity, and timing (PVT) anywhere in the world, and anytime. It was originally designed in the '70s for military use by the United States; it was released for civil use in 1983, and has been in full operation since 1995 with 24 satellites. It is maintained by US military institutions and it can be denied to civilians at any time for defense reasons. Currently (at 2011) the satellite constellation is composed of 31 satellites at an altitude of 20 200 km, and orbits are arranged to guarantee that at least four satellites are visible from any point on earth, although usually the number of satellites within visibility for positioning is 5–8.

Let us consider the experiment in Figure 13.5 where the device to be localized within an area (*service area*) transmits a signal $s(t)$ that is received simultaneously by N *base-stations* acting as anchors in a cellular communication system. The signal (*signature*) $s(t)$ is know by all base-stations and the experiment resembles the radar system (Section 5.6) where the range $R_k = c\tau_k$ of the transmitter follows from the time of delay toward the kth base-station τ_k. The propagation speed for electromagnetic waves is $c = 3 \times 10^8$ m/s, but the same concept applies for acoustic waves in air ($c = 335$ m/s) as for flying bats, or in water ($c = 1500$ m/s) as for dolphins when in fish-hunting or for sonar systems. The kth base-station has coordinate $(r_{x,k}, r_{y,k})$ while the transmitter has coordinates (r_x, r_y). From the distance measured, each listening base-station can set the locus of the distance of the transmitter on a circle with radius R_k. The collection of distances from $N \geq 3$ base-stations in non-coincident positions is instrumental to locate the transmitter as sketched in Figure 13.6.

13.5.1 Positioning and Noise

Range $\hat{R}_k = c\hat{\tau}_k$ is obtained from the estimate of the delay $\hat{\tau}_k$, that in turn has an accuracy that depends on the amplitude of the received signal as detailed in Section 9.4 and validated in Section 10.3. Signals attenuate with (true) distance R_k and signal to noise ratio degrades consequently. The loci of the estimates $\hat{R}_1, \hat{R}_2, ..., \hat{R}_N$ for N base-stations do not converge into one point, but rather they define an uncertainty region as sketched in Figure 13.6. A common (and realistic) assumption is that the ToD estimate is unbiased and the estimate of the range is modeled as

$$\hat{R}_k \sim \mathcal{N}(R_k, \sigma_k^2)$$

with variance σ_k^2 that is dependent on the distance R_k as detailed below.

To better define the problem at hand, let the moving device transmit at constant power P_T; the power received by the kth base-station decreases with distance R_k and thus also the signal to noise ratio SNR_k, which in turn makes the variance σ_k^2 increase.

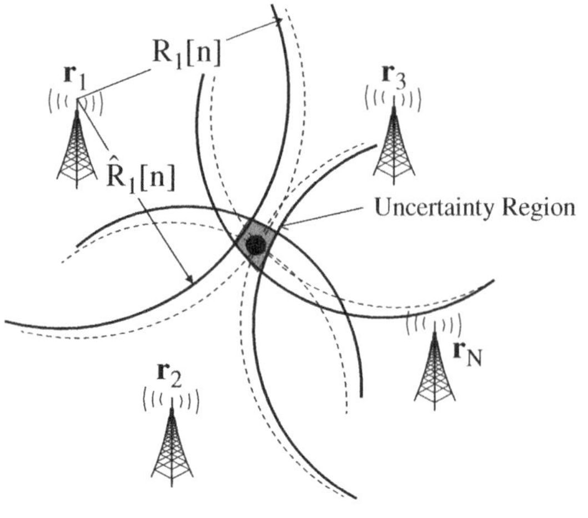

Figure 13.6 Multi-lateration from ranges with errors (solid lines).

From physics, the received power decreases with the inverse of the square of the distance (more sophisticated attenuation models could be adopted without changing the procedure to derive the impact of distance into the variance of the range)

$$P_{R,k} = \frac{P_T}{R_k^2}$$

It follows that the signal to noise ratio varies with distance as

$$SNR_k = SNR_o \left(\frac{R_o}{R_k}\right)^2$$

with respect to an arbitrary reference distance R_o (typical $R_o = 1\,m$ or $1\,km$). The variance of the delay can be evaluated from the CRB, which in turn depends on the received waveform

$$g_k(t) = \frac{1}{R_k} g(t - \tau_k)$$

with amplitude scaled by distance. For white Gaussian noise, the CRB can be evaluated following the steps in Section 8.4.2:

$$\text{var}[\hat{\tau}_k] \geq \frac{N_0}{2}\left[\int \left(\frac{dg_k(t)}{d\tau_k}\right)^2 dt\right]^{-1}$$

After some rearrangement of terms, it can be shown that

$$\text{var}[\hat{\tau}_k] \geq \frac{\beta_g^2}{SNR_o}\left(\frac{R_k}{R_o}\right)^2$$

where β_g^2 is the effective bandwidth that depends on the spectral property of the waveform $g(t)$ (Section 9.4). To simplify, let the ToD estimate attains the CRB (i.e., this is a reasonable assumption when waveforms are isolated and without multipaths or overlapping replicas), so the range estimate is thus affected by a variance

$$\sigma_k^2 = c^2 \frac{\beta_g^2}{SNR_o}\left(\frac{R_k}{R_o}\right)^2$$

This relationship is used herein to evaluate the effect of range errors that depend on the relative position of the mobile unit and the base-stations.

In practice, a moving device transmits the positioning signal periodically, say every T seconds, so the position is sampled with rate $1/T$ position/sec, and the range's variations vs. time form a set of discrete values: $R_k[1], R_k[2], ..., R_k[n]$. As a consequence of the movement, the variance is also changing vs. time (or position)

$$\sigma_k^2[n] = c^2 \frac{\beta_g^2}{SNR_o}\left(\frac{R_k[n]}{R_o}\right)^2$$

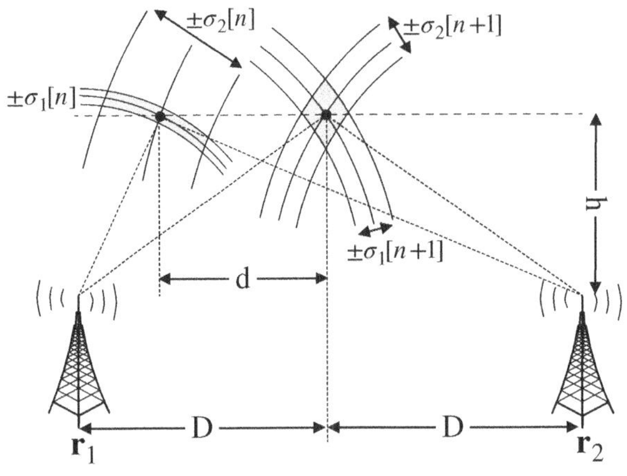

Figure 13.7 Example of uncertainty regions from range errors.

When merging all the measurements, the pdf of the position for the set of ranges is the joint pdf of all the conditional pdfs

$$p(\mathbf{r}|\hat{R}_1,...,\hat{R}_N) = \prod_{k=1}^{N} p(\mathbf{r}|\hat{R}_k)p(\hat{R}_k)$$

which is not regularly shaped and not even Gaussian; the shape of the position-ambiguity region depends on the position of the moving terminal wrt the fixed base-stations.

To exemplify, in Figure 13.7 the transmitter moves along a line at distance h from the line that connects the two base-stations (in this situation, a third base-station could avoid any ambiguity between the two solutions on the two half-spaces). By varying the distance d from the symmetry axis, the ratio between the two variances is

$$\frac{\sigma_2^2}{\sigma_1^2} = \frac{(D+d)^2 + h^2}{(D-d)^2 + h^2}$$

which coincides for $d = 0$ (see Figure 13.7). On the other hand, the ambiguity regions are approximately orthogonal but unbalanced in $d = D$ and for $D \gg h$ as $\sigma_2^2 \gg \sigma_1^2$. This example stimulates a remark: since the dimension of the position-ambiguity region increases with distance, an isotropic deployment of N receiving base-stations around the coverage area can yield an ambiguity region as intersection of all the ambiguity regions from all the base-stations that is approximately isotropic (from this example when $\sigma_i^2 \simeq \sigma_j^2$).

This simple example in Figure 13.7 highlights that in positioning, the use of accuracy as the only parameter to define performance is questionable as accuracy depends on the position of the moving transmitter and the anchors, unless the anchors are isotropic in space. This conclusion does not change even if referring to a GPS systems where each GPS receiver positions itself, except by changing some terms in the problem at hand and tracking of the positions vs. time can largely improve the overall accuracy.

13.5.2 Example of Position Tracking

The dynamic evolution of the target depends on its kinematics, when excited with time-varying accelerations $\alpha_x[n]$, $\alpha_y[n]$ modeled as a random term. The state variables are position $(r_x[n], r_y[n])$ and speed $(v_x[n], v_y[n])$

$$\underbrace{\begin{bmatrix} r_x[n] \\ r_y[n] \\ v_x[n] \\ v_y[n] \end{bmatrix}}_{\theta[n]} = \underbrace{\begin{bmatrix} 1 & 0 & T & 0 \\ 0 & 1 & 0 & T \\ 0 & 0 & 1 & 0 \\ 0 & 0 & 0 & 1 \end{bmatrix}}_{\mathbf{A}} \cdot \underbrace{\begin{bmatrix} r_x[n-1] \\ r_y[n-1] \\ v_x[n-1] \\ v_y[n-1] \end{bmatrix}}_{\theta[n-1]} + \underbrace{\begin{bmatrix} 0 \\ 0 \\ \alpha_x[n] \\ \alpha_y[n] \end{bmatrix}}_{\boldsymbol{\alpha}[n]}$$

and the relationship is linear. However, the relationship between state variables and ranges $\hat{R}_1[n], \hat{R}_2[n], ..., \hat{R}_N[n]$ estimated by the ToDs is non-linear:

$$\hat{R}_1[n] = \sqrt{(r_x[n] - r_{x,1})^2 + (r_y[n] - r_{y,1})^2} + \beta_1[n]$$
$$\hat{R}_2[n] = \sqrt{(r_x[n] - r_{x,2})^2 + (r_y[n] - r_{y,2})^2} + \beta_2[n]$$
$$\vdots$$
$$\hat{R}_N[n] = \sqrt{(r_x[n] - r_{x,N})^2 + (r_y[n] - r_{y,N})^2} + \beta_N[n]$$

and in canonical form

$$\underbrace{\begin{bmatrix} \hat{R}_1[n] \\ \hat{R}_2[n] \\ \vdots \\ \hat{R}_N[n] \end{bmatrix}}_{\hat{\mathbf{R}}[n]} = \underbrace{\begin{bmatrix} \sqrt{(r_x[n] - r_{x,1})^2 + (r_y[n] - r_{y,1})^2} \\ \sqrt{(r_x[n] - r_{x,2})^2 + (r_y[n] - r_{y,2})^2} \\ \vdots \\ \sqrt{(r_x[n] - r_{x,N})^2 + (r_y[n] - r_{y,N})^2} \end{bmatrix}}_{\mathbf{b}(\theta[n])} + \underbrace{\begin{bmatrix} \beta_1[n] \\ \beta_2[n] \\ \vdots \\ \beta_N[n] \end{bmatrix}}_{\boldsymbol{\beta}[n]}$$

where $\boldsymbol{\beta}[n]$ is the acquisition noise. The linearization is

$$\left.\frac{\partial \mathbf{b}(\theta[n])}{\partial \theta[n]}\right|_{\theta[n]=\hat{\theta}[n|n-1]}$$
$$= \underbrace{\begin{bmatrix} \frac{\hat{r}_x[n|n-1]}{\sqrt{(\hat{r}_x[n|n-1]-r_{x,1})^2+(\hat{r}_y[n|n-1]-r_{y,1})^2}} & \frac{\hat{r}_y[n|n-1]}{\sqrt{(\hat{r}_x[n|n-1]-r_{x,1})^2+(\hat{r}_y[n|n-1]-r_{y,1})^2}} & 0 & 0 \\ \vdots & \vdots & \vdots & \vdots \\ \frac{\hat{r}_x[n|n-1]}{\sqrt{(\hat{r}_x[n|n-1]-r_{x,N})^2+(\hat{r}_y[n|n-1]-r_{y,N})^2}} & \frac{\hat{r}_y[n|n-1]}{\sqrt{(\hat{r}_x[n|n-1]-r_{x,N})^2+(\hat{r}_y[n|n-1]-r_{y,N})^2}} & 0 & 0 \end{bmatrix}}_{\mathbf{B}[n|n-1] \in \mathbb{R}^{N\times 4}},$$

while target tracking follows the EKF equations in Section 13.4.

Even if position tracking can be interpreted as just a mere application of the EKF formulas, the choice of the covariances $\mathbf{C}_{\alpha\alpha}$ of the excitation, and $\mathbf{C}_{\beta\beta}$ of the noise needs some comment. Unless movement has some privileged directions (e.g., vehicles along a road follow the road with minimal off-road deviation and this can be taken into account by unbalancing the term in the covariance matrix), the acceleration can be considered uncorrelated along the two directions (for a planar problem) with the same values (isotropic movement) so that $\mathbf{C}_{\alpha\alpha} = \sigma_\alpha^2\mathbf{I}$, where the value σ_α^2 depends on the specific application (large or small depending on whether the point is rapidly varying in position or not). In this case the speed has a Rayleigh pdf with a root mean square deviation within an interval T as $\sigma_\alpha T\sqrt{2}$ [m/s]. Alternatively, the model can have an inertia to the variation of speed according to an AR(1) model, so that

$$\begin{cases} v_x[n] = \rho v_x[n-1] + \alpha_x[n] \\ v_y[n] = \rho v_y[n-1] + \alpha_y[n] \end{cases}$$

and thus the autocorrelation of the speed is $\mathbb{E}[v_x[n]v_x[n+k]] = \rho^{|k|}\sigma_\alpha^2/(1-\rho^2)$ (similarly for v_y). In this case, ρ and σ_α^2 are degrees of freedom in model definition that can be adapted to the specific problem to account for the inertia of the body.

The covariance $\mathbf{C}_{\beta\beta}$ depends on the noise in observations that is independent of the base-stations, but dependent on the distance (which is time-varying): $\mathbf{C}_{\beta\beta}[n] = \text{diag}\{\sigma_1^2[n], \sigma_2^2[n], ..., \sigma_N^2[n]\}$. A reasonable assumption is to consider that the prediction $\hat{\boldsymbol{\theta}}[n|n-1]$ is close enough to the true position for modeling of the signal attenuation so that the following approximation can be used:

$$\sigma_k^2[n] = \frac{\beta_g^2 c^2}{SNR_o}\left(\frac{R_k[n]}{R_o}\right)^2 \simeq \frac{\beta_g^2 c^2}{SNR_o} \cdot \frac{(\hat{r}_x[n|n-1] - r_{x,k})^2 + (\hat{r}_y[n|n-1] - r_{y,k})^2}{R_o^2}.$$

The EKF gain should be made position-varying as $\mathbf{C}_{\beta\beta}[n]$ and thus $\mathbf{G}[n] = \mathbf{P}[n|n-1]\mathbf{B}[n|n-1]^T(\mathbf{B}[n|n-1]\mathbf{P}[n|n-1]\mathbf{B}[n|n-1]^T + \mathbf{C}_{\beta\beta}[n])^{-1}$. This makes the use of EKF computationally expensive, but some approximations are usually employed in practice to avoid excessive costs.

In a practical GPS system, the range estimate could be biased by the effect of multipath: $\mathbb{E}[\hat{R}_k[n]] > R_k[n]$. This is mainly due to scattering when the propagation is not in line of sight as shown in Figure 13.8. The main benefit of position tracking (solid line with empty points) compared to a memoryless approach as in MLE (dashed line with gray filled points), is to track the sequence of positions according to a dynamic model that implicitly reduces the importance of those ranges excessively affected by multipath, such as $\hat{R}_k[n] \gg R_k[n]$, which otherwise would not be compatible with the motion of the point (e.g., an abrupt change of estimated position implies an acceleration that cannot be feasible for the experiment at hand, while perhaps the point is still instead as in Figure 13.8).

13.6 Non-Gaussian Pdf and Particle Filters

The EKF is based on the assumption that linearization accurately models the evolution of the state, but the a-posteriori pdf $p(\boldsymbol{\theta}[n] \mid \mathbf{x}[1:n])$ is always handled as Gaussian, as are all the pdfs involved in prediction and update. When the pdfs are non-Gaussian, the

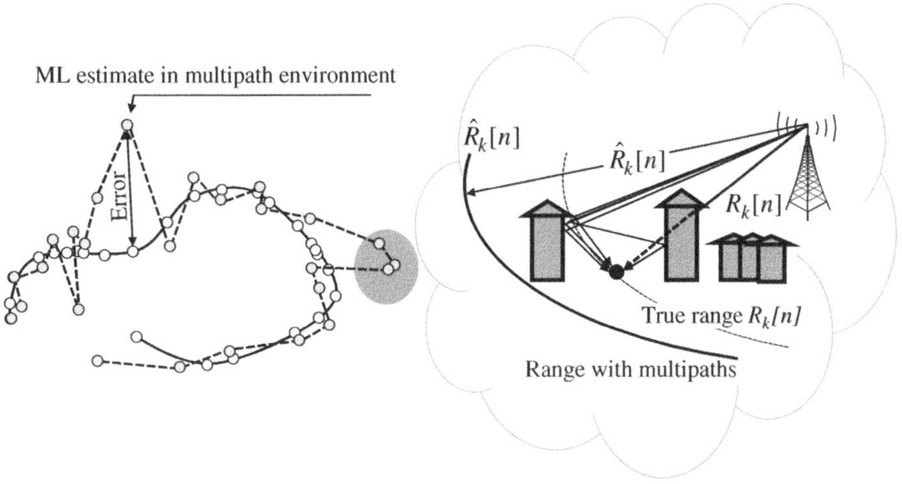

Figure 13.8 Positioning in the presence of multipaths.

tracking has to be approached in the general way following the steps in Section 13.1, but still based on prediction and update. The topic is broad; there are several solutions but all are based on a parametric description of the a-posteriori pdf $p(\boldsymbol{\theta}[n] \mid \mathbf{x}[1:n])$.

One simple and intuitive solution is to describe the arbitrary pdf $p(\boldsymbol{\theta}[n] \mid \mathbf{x}[1:n])$ with a set of K *state conditions* $\boldsymbol{\theta}_1[n], \boldsymbol{\theta}_2[n], \ldots, \boldsymbol{\theta}_K[n]$ (or *particles*) chosen to be representative enough of the true pdf. Each particle is associated to a *weight* $w_k[n]$ (note that particles and weights depend on time, made varying to better describe the a-posteriori pdfs at every step), and the pdf is approximated as

$$p(\boldsymbol{\theta}[n]|\mathbf{x}[1:n]) \simeq \sum_{k=1}^{K} w_k[n]\delta(\boldsymbol{\theta}[n] - \boldsymbol{\theta}_k[n])$$

with weights normalized $\sum_{k=1}^{K} w_k[n] = 1$. In this way the pdf is approximated by a probability mass function that provides samples of the pdf that are chosen to better fit the a-posteriori pdf. Bayesian tracking using the *particle filtering* method estimates the evolution of the state by the same prediction and updating steps discussed in Section 13.1, except that tracking steps should be adapted for discrete-valued states.

The choice of the particles is the key of these methods. Particles should be dense enough to accurately represent the pdf where variation is large, but there should be particles (perhaps less dense) even for those state values where the probability is small. Particles should change in density over time as those state values that become less likely should have a coarser particle density. Of course, increasing K makes the approximation more accurate, but much more expensive. There has been an intensive research activity on *particle filtering* (or the sequential Monte Carlo method) within the scientific community on different applications such as positioning, navigation systems, tracking of monetary flows, and inflation, just to mention a few. The reader can easily find detailed discussions from the open literature.

14

Spectral Analysis

In time series analysis, the estimation of the correlation properties of random samples is referred to as spectral analysis. If a process $x[n]$ is stationary, its correlation properties are fully defined by the autocorrelation or equivalently by its Fourier transform, in term of power spectral density (PSD):

$$r_x[k] = E\left[x^*[n]x[n+k]\right] \leftrightarrow S_x(\omega) = \mathcal{F}\{r_x[k]\}$$

Spectral analysis deals with the estimation of a PSD from a limited set of N observations $\mathbf{x} = [x[0], ..., x[N-1]]^T$:

$$\hat{S}_x(\omega) = \hat{S}_x(\omega|\mathbf{x})$$

There are two fundamental approaches to spectral analysis: parametric or non-parametric, depending if there is a parametric model for the PSD or not. Non-parametric spectral analysis makes no assumptions on the PSD and thus it is more general and flexible as it can be used in every situation when there is no a-priori knowledge on the mechanism that correlates the samples in the time series. Parametric methods model (or guess) the generation mechanism of the process and thus the spectral analysis reduces to the estimate of these parameters. Except for a few cases, spectral analysis is basically a search for the most appropriate (non-parametric/parametric) method to better estimate the PSD and some features that are of primary interest for the application.

There is no general rule to carrying out spectral analysis except for common sense related to the needs of the specific application. Therefore, the organization of this chapter aims to offer a set of tools and to establish the pros and cons of each method based on statistical performance metrics. The lack of a general method that is optimal for every application makes spectral analysis a professional expertise itself where experience particularly matters. For this reason, spectral analysis is considered "an art", and the experts are "artisans" of spectral analysis who specialize in different application areas. All the principal methods are discussed herein with the usual pragmatic approach aiming to provide the main tools to deal with the most common problems and to refine the intuition, while avoiding entering into excessive detail in an area that could easily fill a whole dedicated book, see e.g., [52].

Statistical Signal Processing in Engineering, First Edition. Umberto Spagnolini.

Companion website: www.wiley.com/go/spagnolini/signalprocessing

14.1 Periodogram

Spectral analysis estimates the PSD $\hat{S}_x(\omega) = \hat{S}_x(\omega|\mathbf{x})$ from the observations $\mathbf{x} = [x[0], ..., x[N-1]]^T$, and performance is evaluated by the bias $\mathbb{E}[\hat{S}_x(\omega|\mathbf{x})]$ and the variance $\text{var}[\hat{S}_x(\omega|\mathbf{x})]$. As for any estimator, it is desirable that the PSD estimate is unbiased (at least asymptotically as $N \to \infty$) and consistent. Non-parametric spectral analysis is the periodogram, and it is historically based on the concept of the filter-bank revised in Section 14.1.3, as every value of the PSD represents the power collected by a band-pass filter centered at the specific frequency bin.

Let

$$X(\omega_k) = \sum_{n=0}^{N-1} x[n]e^{-j\omega_k n} \tag{14.1}$$

be the discrete Fourier transform (DFT) of the sequence $\mathbf{x}$ evaluated for the kth bin of the angular frequency $\omega_k = 2\pi k/N$ (or kth frequency bin $f_k = k/N$). The periodogram is the estimator

$$\hat{S}_x(\omega_k) = \frac{1}{N}\left|\sum_{n=0}^{N-1} x[n]e^{-j\omega_k n}\right|^2 = \frac{1}{N}|X(\omega_k)|^2 \tag{14.2}$$

and it corresponds to the energy density of the sequence of N samples $|X(\omega_k)|^2$ divided by the number of samples. If considering unit-time sampling, the periodogram divides the energy density by the observation time and reflects the definition of power density (power per unit of frequency bin). The periodogram has the benefit of simplicity and computational efficiency and this motivates its widespread use. However, an in-depth knowledge of its performance is mandatory in order to establish the limits of this simple and widely adopted estimator.

14.1.1 Bias of the Periodogram

A careful inspection of the definition of the periodogram shows that the periodogram can be rewritten as an FT of the sequence of N samples $X(\omega) = \mathcal{F}\{\mathbf{x}\}$ sampled in regular frequency bins:

$$\hat{S}_x(\omega_k) = \frac{1}{N}|X(\omega)|^2\Big|_{\omega=\frac{2\pi}{N}k}$$

but

$$\frac{1}{N}|X(\omega)|^2 = \hat{R}_x(\omega) \leftrightarrow \hat{r}_x[k]$$

is the FT of the sample autocorrelation

$$\hat{r}_x[k] = \frac{1}{N}\sum_{n=0}^{N-1} x^*[n]x[n+k]$$

The properties of $\hat{R}_x(\omega)$ govern the bias, and since (see Appendix A)

$$\mathbb{E}[\hat{R}_x(\omega)] = \frac{1}{N}\left(\frac{\sin[\omega N/2]}{\sin[\omega/2]}\right)^2 \circledast_{2\pi} S_x(\omega)$$

this proves that the periodogram (a sampled version of $\hat{R}_x(\omega)$) is biased, as

$$\mathbb{E}[\hat{S}_x(\omega_k)] = \frac{1}{N}\left(\frac{\sin[\omega N/2]}{\sin[\omega/2]}\right)^2 \circledast_{2\pi} S_x(\omega)\Big|_{\omega=\frac{2\pi}{N}k} \tag{14.3}$$

where the convolution $\circledast_{2\pi}$ is periodic over 2π. The periodgram is a biased estimator of the PSD that is asymptotically unbiased as

$$\lim_{N\to\infty} \mathbb{E}[\hat{S}_x(\omega_k)] = \lim_{N\to\infty} \frac{1}{N}\left(\frac{\sin[\omega N/2]}{\sin[\omega/2]}\right)^2 \circledast_{2\pi} S_x(\omega)\Big|_{\omega=\frac{2\pi}{N}k} = S_x(\omega)\big|_{\omega=\frac{2\pi}{N}k}$$

Remark 1: The properties of the DFT makes it possible to evaluate the FT in other frequency bins, not necessarily spaced by $1/N$ (or $2\pi/N$ if angular frequency bins). This is obtained by padding with zeros the sequence $x[0],\ldots,x[N-1]$ up to a new value, say $M > N$:

$$y[n] = \begin{cases} x[n] & \text{for } n = 0,1,\ldots,N-1 \\ 0 & \text{for } n = N, N+1,\ldots,M-1 \end{cases}$$

From signal theory [12, 13], the DFT over the M samples is the FT of the sequence sampled over the new frequency bin, which in turn is correlated (i.e., not statistically independent or uncorrelated) with the DFT $X(\omega_k)$:

$$Y(\frac{2\pi}{M}m) = \sum_{\ell=0}^{N-1} X(\frac{2\pi}{N}\ell)\frac{\sin(N(\omega - 2\pi\ell/N))}{N\sin(\omega - 2\pi\ell/N)}\Big|_{\omega=\frac{2\pi}{M}m}$$

The DFT of $y[n]$ is the interpolated values of DFT of $x[n]$ (14.1) using the periodic sinc as interpolating function.

A common practice is to pad to multiple values of N, say $M/N = L$, so that the interpolating function $Y(\omega_m)$ honors the DFT $X(\omega_k)$ and thus the periodogram $\hat{S}_y(\frac{2\pi}{M}m)$ coincides with $\hat{S}_x(\frac{2\pi}{N}n)$ as

$$\hat{S}_y(\frac{2\pi}{M}m) = \hat{S}_x(\frac{2\pi}{N}n) \text{ for } m = nL$$

with $L-1$ interpolated values between the samples of the $\hat{S}_x(\frac{2\pi}{N}n)$.

Remark 2: The bias (14.3) vanishes when considering some special cases. Let the PSD of one (or more) real (or complex) sinusoids with amplitude A, random phase, and frequency ω_o be

$$S_x(\omega) = \frac{A^2}{4}\delta(\omega - \omega_o) + \frac{A^2}{4}\delta(\omega + \omega_o)$$

The bias becomes

$$\mathbb{E}[\hat{S}_x(\omega_k)] = \frac{A^2}{4N}\left(\frac{\sin[(\omega - \omega_o)N/2]}{\sin[(\omega - \omega_o)/2]}\right)^2 + \frac{A^2}{4N}\left(\frac{\sin[(\omega + \omega_o)N/2]}{\sin[(\omega + \omega_o)/2]}\right)^2\Bigg|_{\omega=\frac{2\pi}{N}k}$$

and the periodogram is biased as shown in Figure 14.1. However, if the frequency

$$\omega_o = \frac{2\pi}{N}\ell_o$$

for some integer value ℓ_o (i.e., the sinusoid has a period that makes ℓ_o cycles into the N samples of the observation window), the periodogram is unbiased, as

$$\mathbb{E}[\hat{S}_x(\omega_k)] = N\frac{A^2}{4}\delta(\frac{2\pi}{N}(k - \ell_o)) + N\frac{A^2}{4}\delta(\frac{2\pi}{N}(k + \ell_o))$$

Since the value of periodogram in ℓ_o is $N\frac{A^2}{4}$, apparently the periodogram seems to yield a wrong value for the power of the sinusoid, as it should be $A^2/2$. Recall that the periodogram is the estimator of the power spectrum "density" and thus the power is obtained by multiplying each bin by the corresponding bandwidth, which in this case it is the frequency bin spacing $1/N$, and thus

$$\mathbb{E}[|x[n]|^2] = \frac{1}{N}\sum_{k=0}^{N-1}\mathbb{E}[\hat{S}_x(\omega_k)] = \frac{1}{N}\left(N\frac{A^2}{4} + N\frac{A^2}{4}\right) = \frac{A^2}{2}$$

as expected

Another interesting case is when the PSD is for a white WSS random process:

$$S_x(\omega) = \sigma^2$$

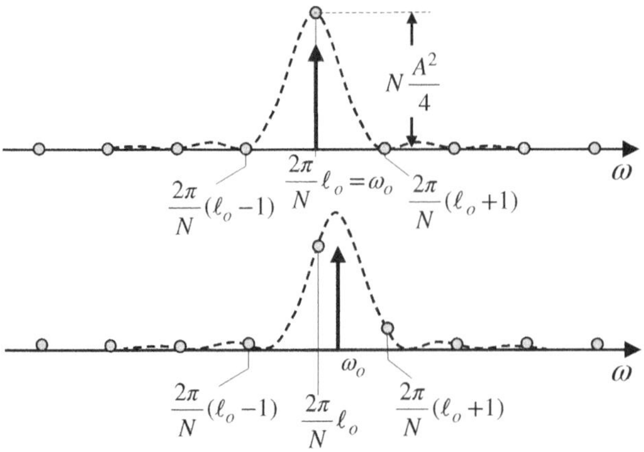

Figure 14.1 Bias of periodogram for sinusoids.

when the bias becomes

$$\mathbb{E}[\hat{S}_x(\omega_k)] = \sigma^2 \quad \text{for } \forall k = 0, 1, \ldots, N-1$$

that is, still unbiased. Once again, the power is still the same as summing N terms scaled by the bandwidth $1/N$ of the periodogram bin; the power is $\mathbb{E}[|x[n]|^2] = \frac{1}{N}\sum_{k=0}^{N-1}\mathbb{E}[\hat{S}_x(\omega_k)] = \frac{1}{N}N\sigma^2 = \sigma^2$.

14.1.2 Variance of the Periodogram

The exact evaluation of the variance of the periodogram is not straightforward, but some simplifications in the simple case can be drawn before providing the general equation. For a real-valued process, it can be written as

$$\text{var}[\hat{S}_x(\omega)] = \frac{1}{N^2}\sum_{s=0}^{N-1}\sum_{t=0}^{N-1}\sum_{u=0}^{N-1}\sum_{v=0}^{N-1}\mathbb{E}[x[s]\cdot x[t]\cdot x[u]\cdot x[v]]e^{-j\omega(s-t+u-v)} - \mathbb{E}^2[\hat{S}_x(\omega)] \tag{14.4}$$

The computation needs the 4th order moments, which are not straightforward. However, the computation becomes manageable if the process is Gaussian and white (in this way all the conclusions can be extended to every bin, regardless of the bias effect in (14.3)) so that $S_x(\omega) = \sigma_x^2 \leftrightarrow r_x[k] = \sigma_x^2\delta(k)$. For stationary processes involving stationarity up to the 4th order moment of the Gaussian process (this condition is stricter than wide-sense stationary condition, but not uncommonly seen to be fulfilled):[1]

$$\mathbb{E}[x[t]\cdot x[t+k]\cdot x[t+l]\cdot x[t+m]] = r_x[k-l]\cdot r_x[m] + r_x[k-m]\cdot r_x[l] + r_x[k]\cdot r_x[l-m]$$

and furthermore from the bias:

$$\frac{1}{N}\sum_{s=0}^{N-1}\sum_{t=0}^{N-1} r_x[s-t]e^{-j\omega(s-t)} = \sum_{n=-(N-1)}^{N-1}\frac{N-|n|}{N}r_x[n]e^{-j\omega n} = \mathbb{E}[\hat{S}_x(\omega)]. \tag{14.5}$$

These two identities let us simplify the variance (14.4) as

$$\text{var}[\hat{S}_x(\omega)] = \frac{1}{N^2}\sum_{s,t,u,v=0}^{N-1}\left(r_x[s-t]\cdot r_x[u-v] + r_x[s-v]\cdot r_x[t-u] + r_x[s-u]\cdot r_x[t-v]\right)e^{-j\omega(s-t+u-v)} - \mathbb{E}^2[\hat{S}_x(\omega)],$$

1 The 4th order cumulant for stationary processes is:

$$c_{k,l,m} = \mathbb{E}[x_t x_{t+k} x_{t+l} x_{t+m}] - r_{k-l}r_m - r_{k-m}r_l - r_k r_{l-m}$$

For zero-mean Gaussian processes, all the cumulants larger than 2nd order (i.e., the variance) are zero and thus $c_{k,l,m} \equiv 0$.

and the first term can be further simplified from (14.5) into the sum of two terms that from complex conjugate symmetry are

$$\mathrm{var}[\hat{S}_x(\omega)] = \frac{1}{N^2}\sum_{s=0}^{N-1}\sum_{t=0}^{N-1}\sum_{u=0}^{N-1}\sum_{v=0}^{N-1}\left[r_x[s-v]\cdot r_x[t-u] + r_x[s-u]\cdot r_x[t-v]\right] e^{-j\omega(s-t+u-v)} = A_1A_1^* + A_2A_2^*$$

The first term becomes (from (14.5))

$$A_1 = \frac{1}{N}\sum_{s=0}^{N-1}\sum_{v=0}^{N-1} r_x[s-v]e^{-j\omega(s-v)} \equiv \mathbb{E}[\hat{S}_x(\omega)]$$

while the second term can be evaluated in closed form from the statistical uncorrelation of the samples

$$A_2 = \frac{1}{N}\sum_{s=0}^{N-1}\sum_{u=0}^{N-1} r_x[s-u]e^{-j\omega(s+u)} = \frac{\sigma_x^2}{N}\sum_{n=0}^{N-1} e^{-j2\omega n} = \sigma_x^2 \frac{\sin\omega N}{N\sin\omega}e^{-j(N-1)\omega}.$$

To summarize, the variance of the periodogram for an observation of N samples of a white Gaussian process with power σ_x^2 is

$$\mathrm{var}[\hat{S}_x(\omega)] = \sigma_x^4\left[1 + \left(\frac{\sin\omega N}{N\sin\omega}\right)^2\right]$$

and for the samples of the periodogram it is

$$\mathrm{var}[\hat{S}_x(\omega_k)] = \sigma_x^4\left[1 + \left(\frac{\sin 2\pi k}{N\sin 2\pi k/N}\right)^2\right] = \begin{cases} \mathbb{E}^2[\hat{S}_x(\omega_k)] & \text{for } k \neq 0, N/2 \\ 2\mathbb{E}^2[\hat{S}_x(\omega_k)] & \text{for } k = 0, N/2 \end{cases}$$

The periodogram is not consistent as its variance does not vanish for $N \to \infty$ and some remedies are mandatory in order to use the periodogram as a PSD estimator, since in this simple form the variance of the periodogram is intolerably high and, surprisingly, it is a very bad estimator!

The variance derived so far can be used as a good approximation for the variance in any context (not necessarily for white Gaussian processes), and it follows the routinely adopted variance of the periodogram for real-valued discrete-time signals:

$$\mathrm{var}[\hat{S}_x(\omega_k)] \simeq S_x^2(\omega_k)(1 + \delta(k) + \delta(k - N/2)) = \begin{cases} S_x^2(\omega_k) & \text{for } k \neq 0, N/2 \\ 2S_x^2(\omega_k) & \text{for } k = 0, N/2 \end{cases} \quad (14.6)$$

without the effect of bias. Control of the variance is mandatory in order to use the periodogram effectively as an estimator of the PSD, and this will be discussed next.

Remark: The periodogram is based on the DFT and thus all the properties of symmetry of Fourier transforms hold. Namely, if the process is real, $\hat{S}_x(\omega_k) = \hat{S}_x(\omega_{-k}) =$

$\hat{S}_x(\omega_{N-k})$, and this motivates the larger variance for the samples $k = 0, N/2$. However, for a complex process the DFT is no longer symmetric: $\hat{S}_x(\omega_k) \neq \hat{S}_x(\omega_{-k}) = \hat{S}_x(\omega_{N-k})$, and the variance is the same for all frequency bins:

$$\text{var}[\hat{S}_x(\omega_k)] \simeq S_x^2(\omega_k) \quad \text{for } \forall k$$

A justification is offered in the next section.

14.1.3 Filterbank Interpretation

The relationship (14.6) is not surprising if one considers the effective measurements carried out by every frequency bin. The periodogram can be interpreted as a collection of signals from an array of band-pass filters (*filter-bank*) that separates the signal $x[n]$ into the components $y_0[n], y_1[n], ..., y_{N-1}[n]$ such that (Figure 14.2)

$$y_k[n] = h_k[n] * x[n] \leftrightarrow Y(\omega) = H_k(\omega) \times X(\omega)$$

where the kth filter is centered at the frequency $\omega_k = 2\pi k/N$ with a bandwidth $\Delta\omega = 2\pi/N$ to guarantee a uniform filter-bank (see [55] for multirate filtering and filter-banks).

The samples $y_k[n]$ are temporally correlated as the bandwidth of the process is as wide as the bandwidth of the filter $H_k(\omega)$; these samples can be extracted with a coarser sampling rate that extracts one sample every N (decimation 1:N) so that the samples after the filter-bank are a set of components $y_0[nN], y_1[nN], ..., y_{N-1}[nN]$; these components are statistically uncorrelated (as maximally decimated [55]) and can be used to extract the power of each component

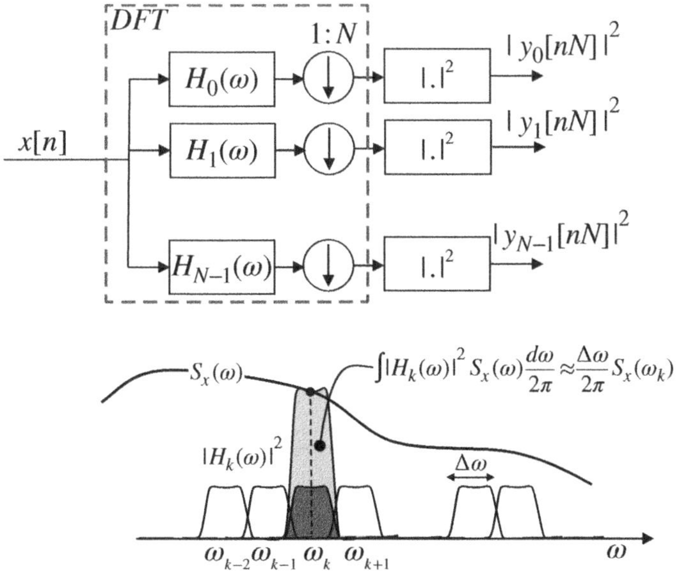

Figure 14.2 Filter-bank model of periodogram.

$$\hat{p}_k = |y_k[nN]|^2$$

as estimate of the corresponding power. Since the samples are random

$$y_k[nN] \sim \mathcal{CN}(0, S(\omega_k))$$

it follows that[2]

$$\mathbb{E}[\hat{p}_k] = S(\omega_k)$$

$$\text{var}[\hat{p}_k] = \mathbb{E}[\hat{p}_k^2] - \mathbb{E}^2[\hat{p}_k] = S^2(\omega_k)$$

thus showing that the power from every filter is the estimate of the PSD with variance $S^2(\omega_k)$. The filter-bank is not just a model, but it coincides with the periodogram where the filter-bank implements the DFT as in Figure 14.2 and the the filters are chosen as

$$h_k[n] = \begin{cases} \exp(j2\pi kn/N) & \text{for } n \in [0, N-1] \\ 0 & \text{for } n \notin [0, N-1] \end{cases} \longleftrightarrow H_k(\omega) = \frac{\sin(N(\omega - \omega_k)/2)}{\sin((\omega - \omega_k)/2)}$$

This gives an interesting interpretation to the bias of the periodogram as being due to the poor selectivity of the band-pass filtering, which can be significantly improved by choosing a different filter response as discussed below in Section 14.1.5.

14.1.4 Pdf of the Periodogram (White Gaussian Process)

As the periodogram is an rv that in every cell estimates the corresponding PSD, in order to gain insight, it is useful to evaluate the corresponding pdf for the same context of Section 14.1.2. Let the data be

$$\mathbf{x} \sim \mathcal{N}(0, \sigma_x^2 \mathbf{I}_N)$$

The DFT that is a linear transformation of $\mathbf{x}$ (see Remark 3 of Appendix B)

$$\mathbf{X} = \underbrace{\mathbf{W}_{N(R)}\mathbf{x}}_{\mathbf{X}_R} + j\underbrace{\mathbf{W}_{N(I)}\mathbf{x}}_{\mathbf{X}_I}$$

so that highlighting the bins $\mathbf{X}[0:N/2]$ as statistically independent, we have:

$$\mathbf{X}_R[0:N/2] \sim \mathcal{N}(0, \frac{N\sigma_x^2}{2}\text{diag}\{2, \underbrace{1, 1, .., 1}_{N/2-1}, 2\})$$

2 For $(A + jB) \sim \mathcal{CN}(0, \sigma^2)$ the following properties hold:

$$\mathbb{E}[|A + jB|^2] = \mathbb{E}[A^2] + \mathbb{E}[B^2] = \sigma^2 \rightarrow \mathbb{E}[A^2] = \mathbb{E}[B^2] = \sigma^2/2$$

$$\mathbb{E}[|A + jB|^4] = \mathbb{E}[(A^2 + B^2)^2] = \mathbb{E}[A^4] + \mathbb{E}[B^4] + 2\mathbb{E}[A^2B^2] = \\ = 2 \times 3[\sigma^2/2]^2 + 2[\sigma^2/2]^2 = 2\sigma^4$$

$$\mathbf{X}_I[0:N/2] \sim \mathcal{N}(0, \frac{N\sigma_x^2}{2}\text{diag}\{0,\underbrace{1,1,..,1}_{N/2-1},0\})$$

or equivalently

$$\mathbf{X}[0:N/2] \sim \mathcal{CN}(0, N\sigma_x^2\mathbf{I}_{N/2+1})$$

Each bin of the periodogram is the square of the DFT so that

$$|X[k]|^2 = \begin{cases} X_R[k]^2 & \text{for } k = 0, N/2 \\ X_R[k]^2 + X_I[k]^2 & \text{for } k \neq 0, N/2 \end{cases}$$

which is the sum of the square of two independent and Gaussian rvs, and the pdf is exponential (central Chi-Square χ_2^2 with 2 degrees of freedom):

$$|X[k]|^2 \sim \chi_2^2 \rightarrow p_{|X|^2}(|X[k]|^2) = \lambda \exp(-\lambda |X[k]|^2) \;\; \text{for } k \neq 0, N/2$$

with rate parameter $\lambda = 1/N\sigma_x^2$. From the properties of the exponential pdf:

$$\mathbb{E}[|X[k]|^2] = N\sigma_x^2$$
$$\text{var}[|X[k]|^2] = N^2\sigma_x^4\left[1 + \delta(k) + \delta(k - N/2)\right]$$

where we included the results for $k = 0, N/2$. Using the definition of the periodogram (14.2), all results derived so far should be scaled by $1/N$ and this coincides with Section 14.1.2.

For the case

$$\mathbf{x} \sim \mathcal{CN}(0, \sigma_x^2\mathbf{I}_N) \Rightarrow \mathbf{X} \sim \mathcal{CN}(0, N\sigma_x^2\mathbf{I}_N)$$

(there is no correlation between $X[k]$ and $X[-k]$ as for real valued processes) and thus

$$|X[k]|^2 = X_R[k]^2 + X_I[k]^2 \sim \chi_2^2 \rightarrow \begin{array}{c} \mathbb{E}[|X[k]|^2] = N\sigma_x^2 \\ \text{var}[|X[k]|^2] = \left(N\sigma_x^2\right)^2 \end{array}$$

that is, there is no difference for the variance among samples of the periodogram.

14.1.5 Bias and Resolution

The periodogram is biased, and the total bias depends on the length N. The main consequences of bias is loss of resolution,[3] smoothing of the PSD, and spectral leakage due to the side-lobes of the biasing periodic sinc(.) function in (14.3). To gain insight

---

3 Two sinusoids with different frequencies should appear as two lines in the PSD, but bias cannot show them as distinct unless their frequency difference $|f_2 - f_1|$ is larger than $1/N$, or N is chosen to have $|f_2 - f_1|N > 1$. Whenever f_1 and f_2 are both multiples of $1/N$, the periodogram is unbiased and the two (or multiple) lines are distinct and thus resolved.

into this, consider the periodogram to estimate the PSD in the case of a selective PSD as in Figure 14.3. In bias (14.3) the periodic sinc

$$\frac{1}{N}\left(\frac{\sin[\omega N/2]}{\sin[\omega/2]}\right)^2 = B_P(\omega) + B_L(\omega)$$

that governs the convolution with the PSD $S_x(\omega)$ can be divided into two terms over disjoint intervals (i.e., $B_P(\omega) \times B_L(\omega) = 0$): the main lobe $B_P(\omega)$ over the interval $\omega \in [-2\pi/N, 2\pi/N]$ and the side-lobes $B_L(\omega)$:

$$\mathbb{E}[\hat{S}_x(\omega)] = \underbrace{S_x(\omega) \circledast_{2\pi} B_P(\omega)}_{\text{resolution loss}} + \underbrace{S_x(\omega) \circledast_{2\pi} B_L(\omega)}_{\text{spectral leakage}}$$

as sketched in Figure 14.3. The main lobe has the effect of smoothing the PSD in the case $S_x(\omega)$ as a rapid transition with a loss of selectivity and resolution as illustrated when evaluating $\mathbb{E}[\hat{S}_x(\omega_a)]$ (solid line) compared to the PSD $S_x(\omega)$ (dashed line). The side-lobes have the effect of spreading the PSD over the frequency axis where the PSD $S_x(\omega)$ is very small (*spectral leakage*), thus creating artifacts such as $\mathbb{E}[\hat{S}_x(\omega_b)]$ that are not in the PSD. The overall effect is a "distorted" PSD (solid line) that in some cases can be unacceptable.

A detailed inspection of the reasons for bias proves that it is a consequence of the abrupt (rectangular-like) truncation of the data that can be mitigated by a smooth windowing. Re-interpreting the periodogram

$$\hat{S}_x(\omega_k) = \frac{1}{N}|\hat{X}(\omega_k)|^2$$

as the DFT of the samples

$$\hat{x}[n] = w[n] \times x[n]$$

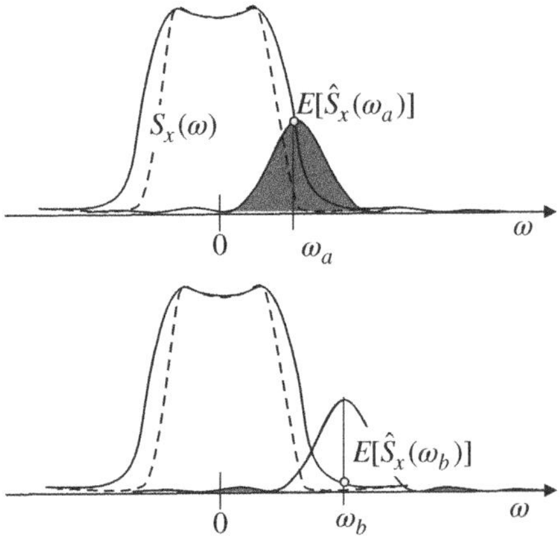

Figure 14.3 Bias and spectral leakage.

obtained after the rectangular windowing $w[n]$ (that is, $w[n]=1$ for $n\in[0,N-1]$) gives the interpretation of the bias as the consequence of the truncation. Any arbitrary window $w[n]$ on the data $x[n]$ before DFT affects the autocorrelation, as if the windowing is $w[n]*w[n]\leftrightarrow W^2(\omega)$, and the bias is

$$\mathbb{E}[\hat{S}_x(\omega)] = S_x(\omega) \circledast_{2\pi} W^2(\omega)$$

which is to be evaluated for $\omega = 2\pi k/N$. The window $w[n]$ can be smooth to reduce the side-lobes of the corresponding Fourier transform, and thus it mitigates spectral leakage. However, this is at the expense of a loss of resolution, as windows with small side-lobes have a broader main lobe. The Figure 14.4 shows an example of the most common windows and their Fourier transforms; the choice of the data-windowing is a trade-off between resolution and spectral leakage. Needless to say that windowing changes the total power of the data as the size of the effective data is smaller (edge samples are down weighted), and compensation is necessary to recover the values of the PSD. One side effect of the windowing is the increased variance due to the reduction of the effective data length.

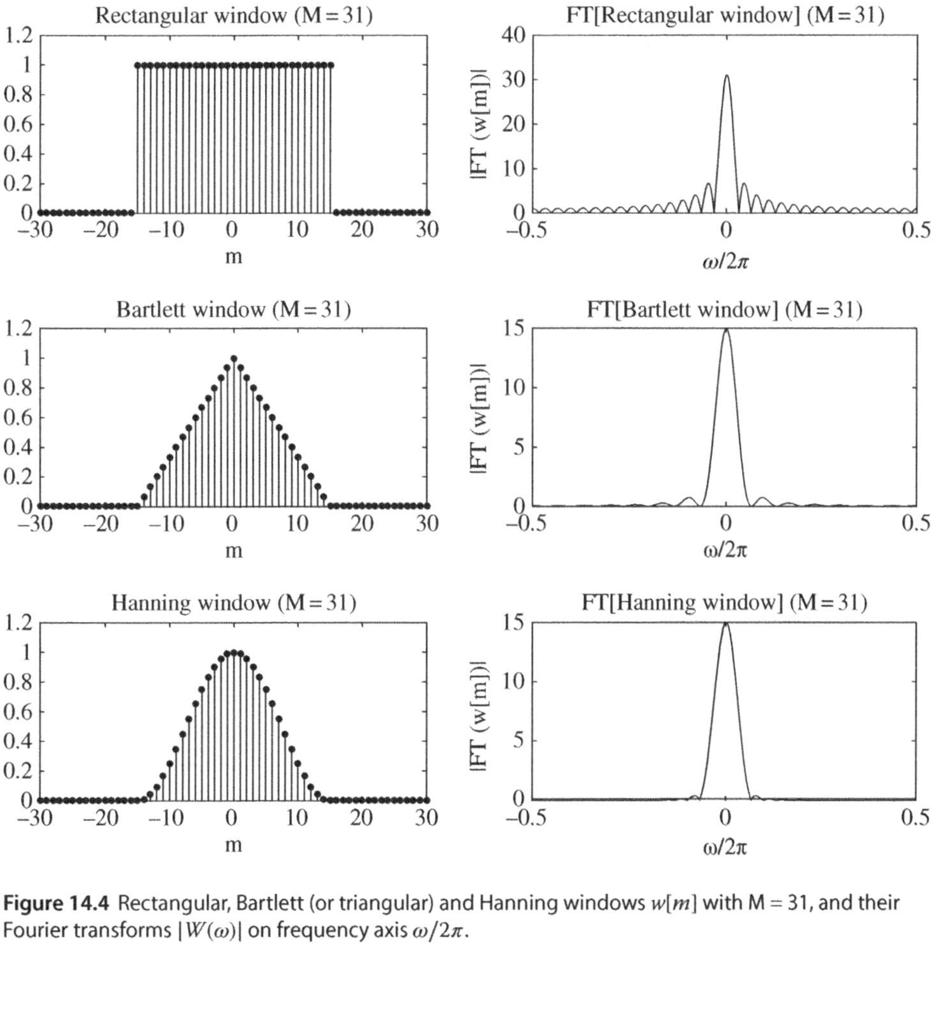

Figure 14.4 Rectangular, Bartlett (or triangular) and Hanning windows $w[m]$ with M = 31, and their Fourier transforms $|W(\omega)|$ on frequency axis $\omega/2\pi$.

14.1.6 Variance Reduction and WOSA

The variance of the periodogram is intolerably high and it prevents its reliable use as a PSD estimator. A simple method for variance reduction is by averaging independent simple periodograms to reduce the variance accordingly (see the examples in Section 6.7). The WOSA method (window overlap spectral analysis) is based on averaging of the periodograms after the segmentation of the data into smaller segments, with some overlap if windowing is adopted (to control the bias as discussed above).

The overall observations can be segmented into M blocks of N samples each as in Figure 14.5. The periodogram is the average of M simple periodograms over N samples:

$$\hat{S}_x^{(M)}(\omega_k) = \frac{1}{M}\sum_{m=0}^{M-1}\left\{\frac{1}{N}\left|\sum_{n=0}^{N-1} x[n+mN]e^{-j\omega_k n}\right|^2\right\}$$

with a requirement for $N_{tot} = MN$ observations in place of N. The variance is

$$\text{var}[\hat{S}_x^{(M)}(\omega_k)] \simeq \frac{S_x^2(\omega_k)}{M}(1+\delta(k)+\delta(k-N/2))$$

and the choice of M can control the accuracy that is needed for the specific application. In practice, the overall observations N_{tot} are given and the choice of M and N is a trade-off between the need to have small variance (with M large) and small bias (with N large). Since resolution and spectral leakage can be also controlled by data-windowing, the overall data can be windowed before computing the simple periodogram. As mentioned in the previous section, the windowing reduces the effective data used, and some overlapping can be tolerated to guarantee that the averaging is for independent periodograms. In WOSA (see Figure 14.6) the overlapping factor

$$\eta = \frac{K}{N} < 1$$

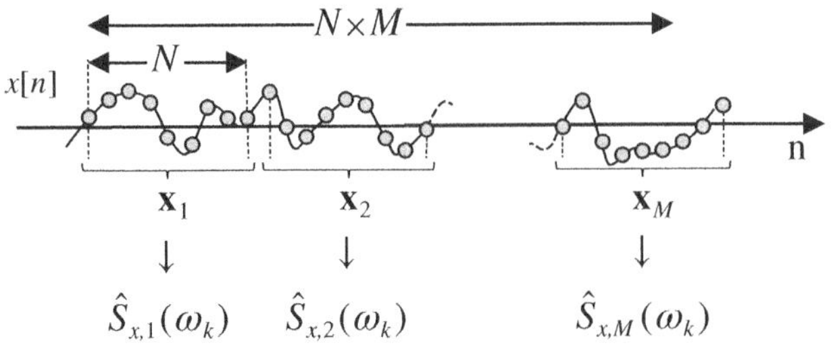

Figure 14.5 Segmentation of data into M disjoint blocks of N samples each.

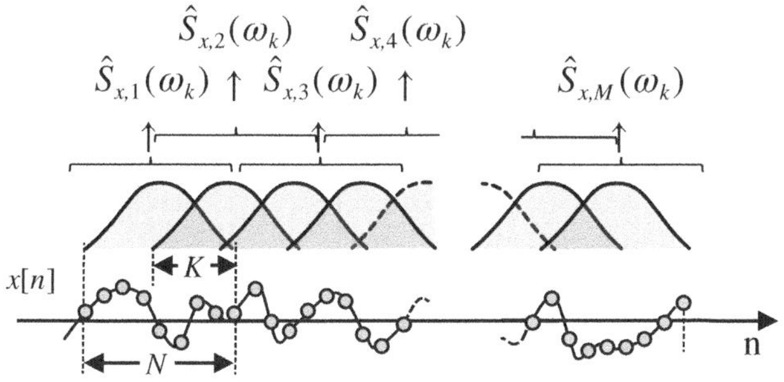

Figure 14.6 WOSA spectral analysis.

highlights the degree of overlapping in the windowed segmentation, and for a total set of samples N_{tot} the WOSA periodogram becomes

$$\hat{S}_x^{(M)}(\omega_k) = \frac{1}{M}\sum_{m=0}^{M-1}\left\{\frac{1}{N}\left|\sum_{n=0}^{N-1} x[n+m(1-\eta)N]e^{-j\omega_k n}\right|^2\right\}$$

where the number of overlapping windows is

$$M = \frac{N_{tot} - K}{N - K}$$

Once again, the choice of η should guarantee the averaging of independent periodograms. If using a Bartlett window, a reasonable choice is $\eta \simeq 50\%$; smoothed windows tolerate a larger overlapping η at the price of some spectral leakage.

Inspired by the variance mitigation of WOSA with data-windowing, the variance can be reduced by smoothing a simple periodogram over N samples with a proper smoothing $W(\omega_k)$ over the whole range of the N samples; this averaging of independent samples is by circular convolution (periodogram smoothing method)

$$\bar{S}_x(\omega_k) = \frac{1}{N}\sum_{n=0}^{N-1}\hat{S}_x(\theta_n)W(\omega_k-\theta_n) = \frac{1}{N}\times\hat{S}_x(\omega_k) \circledast_{2\pi} W(\omega_k)$$

The smoothing function $W(\omega_k)$ should be positive definite and can be chosen to be rectangular, or triangular (Bartlett shape), or any other shape. Even if the smoothing reduces the variance on the order of the smoothing interval, it introduces a bias that can be seen by the loss of resolution due to the smoothing itself. More specifically, the bias is

$$\mathbb{E}[\bar{S}_x(\omega)] = 1/N \times S_x(\omega) \circledast_{2\pi} \frac{1}{N}\left(\frac{\sin[\omega N/2]}{\sin[\omega/2]}\right)^2 \circledast_{2\pi} W(\omega) \simeq 1/N \times \sum_{n=0}^{N-1} S_x(\theta_n)W(\omega_k-\theta_n)$$

where the last approximation is a consequence of the choice of a smoothing window that is large compared to the periodogram spacing $1/N$. The choice $\sum_{k=0}^{N-1} W(\omega_k)/N = w[0] = 1$ guarantees that the smoothing does not add any bias, not even asymptotically. The variance is

$$\begin{aligned}\text{var}[\bar{S}_x(\omega_k)] &= \mathbb{E}\left[\frac{1}{N}\sum_{n=0}^{N-1}\left(\hat{S}_x(\theta_n) - \mathbb{E}[\hat{S}_x(\theta_n)]\right) W(\omega_k - \theta_n)\right]^2 \\ &= \frac{1}{N^2}\sum_{n=0}^{N-1} \text{var}[\hat{S}_x(\theta_n)] W^2(\omega_k - \theta_n)\end{aligned}$$

which depends on the variance of the other terms as consequence of the smoothing. Since the influence of the smoothing is to average in the neighborhood of the frequency ω_k, it is expected that the PSD is approximately constant within the smoothing window so that the ratio

$$\frac{\text{var}[\bar{S}_x(\omega_k)]}{\text{var}[\hat{S}_x(\omega_k)]} \simeq \frac{1}{N^2}\sum_{n=0}^{N-1} W^2(\omega_n) = \frac{1}{N}\sum_{m=-(M-1)}^{M-1} w^2[m]$$

depends approximately only on the window choice and it can be evaluated for some typical window choices that can be applied either in time (Table 14.1) or as periodogram smoothing in frequency (Table 14.2). Recall that a raised cosine with $\alpha + \beta \equiv 1$ is a general window, and it can be specialized to the Hann (or Hanning) window for $\alpha = 1/2$, and the Hamming window for $\alpha = .54$.

14.1.7 Numerical Example: Bandlimited Process and (Small) Sinusoid

The periodogram is used as the first step in spectral analysis when it would be uncertain to use any parametric method, but bias and variance should be used judiciously. Let us consider as observations

$$x[n] = z[n] + A\cos(\omega_o n + \phi)$$

modeled as the sum of a bandlimited process $z[n]$ with bandwidth B and with a small sinusoid superimposed with frequency $f_o = \omega_o/2\pi$ close to the edge of the bandlimited

Table 14.1 Time windows and variance reduction.

Time window: $w[m]$ (support $\|m\| \leq M-1$)	$\frac{\text{var}[\bar{S}_x(\omega_k)]}{\text{var}[S_x(\omega_k)]}$
rectangular: $w[m] = 1$	$= \frac{2M-1}{N}$
Bartlett: $w[m] = 1 - \|m\|/M$	$= \frac{(2M^2+1)}{3MN}$
raised cosine: $w[m] = \alpha + \beta\cos(\pi m/(M-1))$	$(2\alpha^2 + \beta^2)\frac{M}{N} = \begin{cases}\frac{3M}{4N}(\alpha = 1/2)\\ .795\frac{M}{N}(\alpha = .54)\end{cases}$

Table 14.2 Frequency smoothing and variance reduction.

Frequency smoothing: $W(\omega_k)$ (support $\lvert k\rvert \leq M-1$ with $2M-1$ % N)	$\frac{\text{var}[\bar{S}_x(\omega_k)]}{\text{var}[S_x(\omega_k)]}$
rectangular: $W(\omega_k) = \frac{N}{2M-1} \times 1$	$= \frac{1}{2M-1}$
Bartlett: $W(\omega_k) = \frac{N}{M} \times [1 - \lvert k\rvert/M]$	$= \frac{2M^2+1}{3M^3} \simeq \frac{2}{3M}$
raised cosine: $W(\omega_k) = \frac{N}{\alpha(2M-1)-\beta} \times [\alpha + \beta\cos(\pi m/(M-1))]$	$\frac{(2\alpha^2+\beta^2)M}{(\alpha(2M-1)-\beta)^2} \simeq \begin{cases} \frac{3}{4M} (\alpha = 1/2) \\ \frac{.68}{M} (\alpha = .54) \end{cases}$

process ($f_o > B$). The periodogram is used here to estimate the PSD and highlight the existence, or not, of the sinusoid (intentionally or accidentally) "hidden" on the side of the PSD $S_z(f)$. A numerical analysis can ease the discussion based on the Matlab code below.

The bandlimited process has been generated by filtering a white Gaussian process with a Chebyshev filter that is chosen to have a high order (here 15th order) to guarantee good selectivity with sharp transition between pass-band and stop-band; cutoff frequency is chosen as $B = 0.1$. The sinusoid has frequency $f_o = 0.35/2$ and amplitude $A = 10^{-4}$ that is much smaller than the in-band PSD of $S_z(f)$.

```
N=1024; % size of simple periodogram (N=128,256,512,1024)
M=20; % # windows (overlapping is not applied)
Ntot=M*N; % tot.observations

% choose the window by uncommenting
win=boxcar(N); % boxcar=no-window!
% win=bartlett(N);
% win=hamming(N);
win=sqrt(N)*win./sqrt(sum(win.^2)); % normalization

[b,a]=cheby1(15,.1,.2); % 15th order Chebychev filter
A=1E-4; omega_o=pi*.35; % sinusoid
z=filter(b,a,randn(Ntot,1));
x=z+A*cos(omega_o*[1:Ntot]');

for m=1:M,
X(m,:)=fft(win.*x(1+N*(m-1):N+N*(m-1)),N)';
X(m,:)=fftshift(X(m,:));
end

S_est=(1/N)*mean(abs(X).^2);
```

The WOSA method has been applied over $M = 20$ segments of $N = 1024$ samples/segment without overlapping by using unwindowed observations (plain periodogram), or by windowing using either Hanning or Bartlett windows. Figure 14.7 shows the superposition of 10 independent simulations of averaged periodograms for different windowing superimposed to the "true" PSD $S_z(f)$ (dashed line). Without

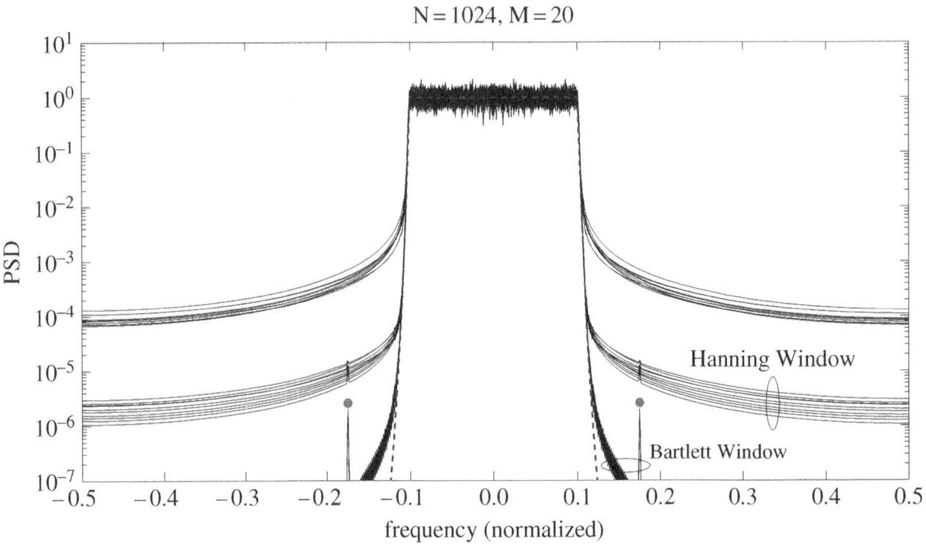

Figure 14.7 Periodogram (WOSA method) for rectangular, Bartlett, and Hamming window.

windowing, the PSD is biased with a smooth transition toward the edge of the band (f=1/2) that completely hides the PSD of the sinusoid. A similar conclusion can be drawn for the use of Hanning windowing on the data. On the other hand, the periodogram using Bartlett windows on the observations still has the effect of reducing the selectivity, but this bias does not hide the PSD of the sinusoid and the value of the periodogram at the frequency of the sinusoid restores the true value of the PSD of the sinusoid (filled dot in Figure 14.7).

The effect of the number of samples is equally important to highlight the sinusoid in the PSD as the periodogram for the sinusoid increases with N (see Section 14.1.5). This is shown in Figure 14.8 where the Bartlett window has been used on observations with increasing length thus showing that N=128 or 256 samples are not enough to avoid the bias of the periodogram hiding the line spectrum of the sinusoid, while any choice of N=512 or larger can allow the lines of the sinusoid to exceed the biased PSD of $S_z(f)$.

14.2 Parametric Spectral Analysis

In parametric spectral analysis, the PSD is modeled in terms of a set of parameters $\boldsymbol{\theta}$ so that $S(\omega|\boldsymbol{\theta})$ accounts for the PSD and the estimate of $\boldsymbol{\theta}$ yields the estimate of the PSD:

$$\hat{S}_x(\omega) = S_x(\omega|\hat{\boldsymbol{\theta}})$$

Accuracy depends on the compliance of the parametric model to the PSD of the specific process at hand. To exemplify, if the PSD contains resonances, the parametric model should account for these resonances in a number that is the same as the PSD, otherwise there could be errors due to under- or over-parameterization. Parametric models are based on AR, MA, or ARMA processes (Section 4.4) where the N samples are obtained

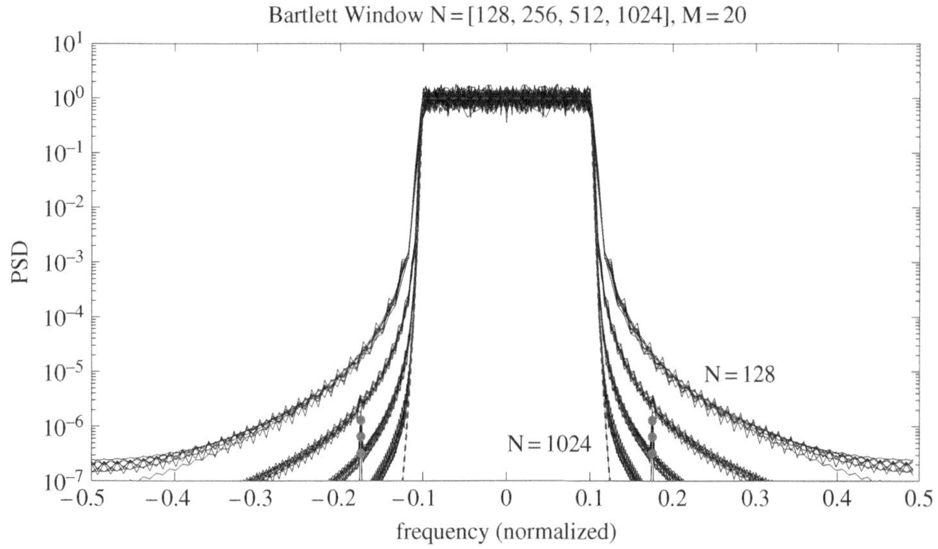

Figure 14.8 Periodogram (WOSA method) for varying $N = 128, 256, 512, 1024$ and Bartlett window.

by properly filtering an uncorrelated Gaussian process $w[n]$ with power (unknown) σ_w^2. The most general case is the parametric spectral analysis for an ARMA model (ARMA spectral analysis), where the corresponding z-transform is

$$X(z) = \frac{1 + \sum_{\ell=1}^{N_b} b[\ell] z^{-\ell}}{1 + \sum_{k=1}^{N_a} a[k] z^{-k}} W(z) = \frac{B(z|\mathbf{b})}{A(z|\mathbf{a})} W(z)$$

The PSD

$$S_x(\omega|\boldsymbol{\theta}) = \left| \frac{B(\omega|\mathbf{b})}{A(\omega|\mathbf{a})} \right|^2 \sigma_w^2$$

is parametric as it depends on the set of $N_a + N_b$ complex-valued coefficients **a** and **b**, and on the power σ_w^2 acting as a scaling term. The overall parameters

$$\boldsymbol{\theta} = \begin{bmatrix} \mathbf{a} \\ \mathbf{b} \\ \sigma_w^2 \end{bmatrix}$$

completely define the PSD, and the model can be reduced to AR (by choosing $\mathbf{b} = \mathbf{0}$) and MA (by choosing $\mathbf{a} = \mathbf{0}$) spectral analysis.

14.2.1 MLE and CRB

The MLE is based on the parametric model of the distribution of $\mathbf{x}$ that is based on the Gaussian model:

$$\mathbf{x} \sim \mathcal{N}(0, \mathbf{C}_{xx})$$

First of all, it is necessary to evaluate the relationship between the structure of the covariance $\mathbf{C}_{xx}$ and the PSD. Starting from the log-pdf of the data (see also Section 7.2.4)

$$\ln p(\mathbf{x}) = -\frac{N}{2}\ln 2\pi - \frac{1}{2}\ln|\mathbf{C}_{xx}| - \frac{1}{2}\mathbf{x}^T\mathbf{C}_{xx}^{-1}\mathbf{x}$$

each of these terms can be made dependent on the PSD by exploiting the relationships between eigenvalues/eigenvectors of the covariance matrix

$$\mathbf{Q}\cdot\mathbf{C}_{xx}\cdot\mathbf{Q}^H = \mathbf{\Lambda}$$

and its PSD (Appendix B):

$$|\mathbf{C}_{xx}| = \prod_{k=1}^{N}\lambda_k(\mathbf{C}_{xx}) \simeq \prod_{k=0}^{N-1}S_x(\frac{2\pi}{N}k)$$

$$\mathbf{C}_{xx}^{-1} = \sum_{k=1}^{N}\frac{1}{\lambda_k(\mathbf{C}_{xx})}\mathbf{q}_k\mathbf{q}_k^H \simeq \sum_{k=0}^{N-1}\frac{1}{S_x(\frac{2\pi}{N}k)}\mathbf{v}_k\mathbf{v}_k^H$$

as in spectral analysis a common assumption is to have a large value of N. The kth eigenvector can be approximated by a complex sinusoid with k cycles over N samples:

$$\mathbf{v}_k = \frac{1}{\sqrt{N}}\left[1, \exp\left(j\frac{2\pi k}{N}\right), ..., \exp\left(j\frac{2\pi k(N-1)}{N}\right)\right]^T$$

By substituting into the pdf we get:

$$\ln p(\mathbf{x}) = -\frac{N}{2}\ln 2\pi - \frac{1}{2}\sum_{k=0}^{N-1}\left(\ln S_x(\frac{2\pi}{N}k) + \frac{|\mathbf{v}_k^H\mathbf{x}|^2}{S_x(\frac{2\pi}{N}k)}\right)$$

and since $\mathbf{v}_k^H\mathbf{x} = DFT_N(\mathbf{x})|_{\omega_k} = X(\omega_k)/\sqrt{N}$, the term

$$|\mathbf{v}_k^H\mathbf{x}|^2 = \frac{1}{N}|X(\omega_k)|^2 = \hat{S}_x(\omega_k, \mathbf{x})$$

is again the periodogram. The pdf can be rewritten now in terms of the PSD values

$$\ln p(\mathbf{x}) = -\frac{N}{2}\ln 2\pi - \frac{1}{2}\sum_{k=0}^{N-1}\left(\ln S_x(\frac{2\pi}{N}k) + \frac{\hat{S}_x(\omega_k, \mathbf{x})}{S_x(\frac{2\pi}{N}k)}\right)$$

Since $1/N$ is the normalized bandwidth of the frequency bin in the periodogram (recall that $\omega = 2\pi f$), the log-likelihood can be rewritten as

$$\begin{aligned}\ln p(\mathbf{x}) &= -\frac{N}{2}\ln 2\pi - \frac{N}{2}\sum_{k=0}^{N-1}\frac{1}{N}\left(\ln S_x(\frac{2\pi}{N}k) + \frac{\hat{S}_x(\omega_k,\mathbf{x})}{S_x(\frac{2\pi}{N}k)}\right)\\ &\simeq -\frac{N}{2}\ln 2\pi - \frac{N}{2}\int_{-1/2}^{1/2}\ln S_x(f) + \frac{\hat{S}_x(f,\mathbf{x})}{S_x(f)}df\end{aligned}$$

The latter approximation holds for large enough N —in principle for $N \to \infty$, and it is the pdf of data for spectral analysis.

14.2.2 General Model for AR, MA, ARMA Spectral Analysis

For parametric spectral analysis, the PSD $S_x(f|\boldsymbol{\theta})$ depends on a set of non random parameters $\boldsymbol{\theta}$, so the log-likelihood is

$$\ln p(\mathbf{x}|\boldsymbol{\theta}) \simeq -\frac{N}{2}\ln 2\pi - \frac{N}{2}\int_{-1/2}^{1/2}\ln S_x(f|\boldsymbol{\theta}) + \frac{\hat{S}_x(f,\mathbf{x})}{S_x(f|\boldsymbol{\theta})}df \tag{14.7}$$

and the ML estimate is

$$\boldsymbol{\theta}_{ML} = \arg\min_{\boldsymbol{\theta}}\left\{\int_{-1/2}^{1/2}\ln S_x(f|\boldsymbol{\theta}) + \frac{\hat{S}_x(f,\mathbf{x})}{S_x(f|\boldsymbol{\theta})}df\right\}$$

that is specialized below for AR, MA, and ARMA models.

The asymptotic equation for the conditional pdf (14.7) is the basis for the CRB as the FIM entry is

$$[\mathbf{J}]_{ij} = -\mathbb{E}\left[\frac{\partial^2 \ln p(\mathbf{x}|\boldsymbol{\theta})}{\partial\theta_i\partial\theta_j}\right]$$

from the derivatives

$$\begin{aligned}\frac{\partial \ln p(\mathbf{x}|\boldsymbol{\theta})}{\partial\theta_i} &= -\frac{N}{2}\int_{-1/2}^{1/2}\left(\frac{1}{S_x(f|\boldsymbol{\theta})} - \frac{\hat{S}_x(f,\mathbf{x})}{S_x(f|\boldsymbol{\theta})^2}\right)\frac{\partial S_x(f|\boldsymbol{\theta})}{\partial\theta_i}df\\ \frac{\partial^2 \ln p(\mathbf{x}|\boldsymbol{\theta})}{\partial\theta_i\partial\theta_j} &= -\frac{N}{2}\int_{-1/2}^{1/2}\left(\frac{1}{S_x(f|\boldsymbol{\theta})} - \frac{\hat{S}_x(f,\mathbf{x})}{S_x(f|\boldsymbol{\theta})^2}\right)\frac{\partial^2 S_x(f|\boldsymbol{\theta})}{\partial\theta_i\partial\theta_j}df+\\ &\quad+\frac{N}{2}\int_{-1/2}^{1/2}\left(\frac{1}{S_x(f|\boldsymbol{\theta})^2} - \frac{2\hat{S}_x(f,\mathbf{x})}{S_{xx}(f|\boldsymbol{\theta})^3}\right)\frac{\partial S_x(f|\boldsymbol{\theta})}{\partial\theta_i}\frac{\partial S_x(f|\boldsymbol{\theta})}{\partial\theta_j}df\end{aligned}$$

Since the periodogram is asymptotically unbiased

$$\mathbb{E}[\hat{S}_x(f,\mathbf{x})] \underset{N\to\infty}{\to} S_x(f|\boldsymbol{\theta})$$

the first term of the FIM is zero (for $N \to \infty$) and the FIM reduces to the celebrated *Whittle formula* for the CRB [56]:

$$[\mathbf{J}]_{ij} = \frac{N}{2}\int_{-1/2}^{1/2} \frac{1}{S_x(f|\boldsymbol{\theta})^2}\frac{\partial S_x(f|\boldsymbol{\theta})}{\partial \theta_i}\frac{\partial S_x(f|\boldsymbol{\theta})}{\partial \theta_j}df = \frac{N}{2}\int_{-1/2}^{1/2} \frac{\partial \ln S_x(f|\boldsymbol{\theta})}{\partial \theta_i}\frac{\partial \ln S_x(f|\boldsymbol{\theta})}{\partial \theta_j}df \tag{14.8}$$

that is generally applicable for the CRB in a broad set of spectral analysis methods.

14.3 AR Spectral Analysis

AR spectral analysis models the PSD $S_x(\omega)$ by a generation model with poles as in Figure 14.9, so that the PSD is assumed to be

$$S_x(\omega|\{a[k]\},\sigma_{N_a}^2) = \frac{\sigma_{N_a}^2}{\left|1+\sum_{k=1}^{N_a} a[k]e^{-jk\omega}\right|^2}$$

for a certain order N_a and excitation power $\sigma_{N_a}^2$. The positions of the poles in the z-plane define the frequency and strength of the resonances, and AR is the appropriate description when the PSD shows resonances only, either because the generation mechanism is known or just assumed.

14.3.1 MLE and CRB

In N_ath order AR spectral analysis, in short $AR(N_a)$, the unknown terms are the recursive term of the filter $\mathbf{a} = [a[1], a[2], ..., a[N_a]]^T$ and the excitation power $\sigma_{N_a}^2$ that depends on the specific AR order N_a highlighted in the subscript:

$$S_x(f|\boldsymbol{\theta}) = \frac{\sigma_{N_a}^2}{|A(f|\mathbf{a})|^2} \quad \text{with } \boldsymbol{\theta} = \begin{bmatrix} \mathbf{a} \\ \sigma_{N_a}^2 \end{bmatrix}$$

The log-likelihood (14.7) is (apart from a scaling)

$$\ln p(\mathbf{x}|\boldsymbol{\theta}) \simeq -\frac{N}{2}\ln\sigma_{N_a}^2 + \frac{N}{2}\int_{-1/2}^{1/2} \ln|A(f|\mathbf{a})|^2 df - \frac{N}{2\sigma_{N_a}^2}\int_{-1/2}^{1/2} |A(f|\mathbf{a})|^2 \hat{S}_x(f,\mathbf{x})df.$$

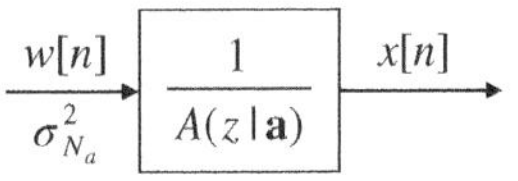

Figure 14.9 Model for AR spectral analysis.

Since for monic polynomials, $\int_{-1/2}^{1/2} \ln|A(f|\mathbf{a})|^2 df = 0$ (Appendix C), the log-likelihood simplifies to

$$\ln p(\mathbf{x}|\boldsymbol{\theta}) \simeq -\frac{N}{2}\ln\sigma_{N_a}^2 - \frac{N}{2\sigma_{N_a}^2}\int_{-1/2}^{1/2} |A(f|\mathbf{a})|^2 \hat{S}_x(f,\mathbf{x})df$$

The MLE follows by nulling the gradients in order, starting from the gradient wrt $\sigma_{N_a}^2$:

$$\frac{\partial \ln p(\mathbf{x}|\boldsymbol{\theta})}{\partial\sigma_{N_a}^2} = 0 \Rightarrow -\frac{N}{2\sigma_{N_a}^2} + \frac{N}{2\sigma_{N_a}^4}\int_{-1/2}^{1/2} |A(f|\mathbf{a})|^2 \hat{S}_x(f,\mathbf{x})df = 0$$

$$\hat{\sigma}_{N_a}^2 = \hat{\sigma}_{N_a}^2(\mathbf{a}) = \int_{-1/2}^{1/2} |A(f|\mathbf{a})|^2 \hat{S}_x(f,\mathbf{x})df$$

The ML estimate of the excitation power depends on the set of parameters $\mathbf{a}$. Plugging into the log-likelihood

$$\ln p(\mathbf{x}|\mathbf{a},\hat{\sigma}_{N_a}^2(\mathbf{a})) \simeq -\frac{N}{2}(1+\ln 2\pi) - \frac{N}{2}\ln\hat{\sigma}_{N_a}^2(\mathbf{a})$$

its maximization is equivalent to minimizing $\hat{\sigma}_{N_a}^2(\mathbf{a})$ wrt $\mathbf{a}$. From the gradients it is (to simplify $\mathbf{a} \in \mathbb{R}^{N_a}$):

$$\frac{\partial\hat{\sigma}_{N_a}^2(\mathbf{a})}{\partial a[k]} = \frac{\partial}{\partial a[k]}\int_{-1/2}^{1/2} A(f|\mathbf{a})A^*(-f|\mathbf{a})\hat{S}_x(f,\mathbf{x})df$$

Recalling that

$$\frac{\partial A(f|\mathbf{a})}{\partial a[k]} = e^{-j2\pi fk}$$

$$\frac{\partial A^*(-f|\mathbf{a})}{\partial a[k]} = e^{j2\pi fk}$$

the following condition holds:

$$\frac{\partial\hat{\sigma}_{N_a}^2(\mathbf{a})}{\partial a[k]} = \int_{-1/2}^{1/2} \underbrace{\left[A(f|\mathbf{a})e^{j2\pi fk} + A^*(-f|\mathbf{a})e^{-j2\pi fk}\right]}_{2\mathrm{Re}\{A(f|\mathbf{a})e^{j2\pi fk}\}}\hat{S}_x(f,\mathbf{x})df = 0$$

Since $\hat{S}_x(f,\mathbf{x})$ is real-valued, $\mathrm{Re}\{A(f|\mathbf{a})e^{j2\pi fk}\}\hat{S}_x(f,\mathbf{x}) = \mathrm{Re}\{A(f|\mathbf{a})\hat{S}_x(f,\mathbf{x})e^{j2\pi fk}\}$, it is sufficient to solve for

$$\int_{-1/2}^{1/2} A(f|\mathbf{a})\hat{S}_x(f,\mathbf{x})e^{j2\pi fk}df = 0$$

Substituting:

$$\int_{-1/2}^{1/2} \hat{S}_x(f,\mathbf{x})e^{j2\pi fk}df + \sum_{\ell=1}^{N_a} a[\ell]\int_{-1/2}^{1/2} \hat{S}_x(f,\mathbf{x})e^{j2\pi f(k-\ell)}df = 0$$

but recalling that the periodogram is the Fourier transform of the sample (biased) estimate of the autocorrelation, the ML estimate follows from a set of N_a equations

$$\hat{r}[k] + \sum_{\ell=1}^{N_a} a[\ell]\hat{r}[k-\ell] = 0 \quad \text{for } k = 1,2,\ldots,N_a$$

These are the Yule–Walker equations in Section 12.3.1, here rewritten wrt the sample autocorrelation that recalls the LS approach to linear prediction (Section 12.4). Furthermore

$$\begin{aligned}
\hat{\sigma}^2_{N_a}(\mathbf{a}) &= \int_{-1/2}^{1/2} A^*(-f|\mathbf{a})A(f|\mathbf{a})\hat{S}_x(f,\mathbf{x})df \\
&= \int_{-1/2}^{1/2} A(f|\mathbf{a})\hat{S}_x(f,\mathbf{x})df + \sum_{\ell=1}^{N_a} a[\ell] \underbrace{\int_{-1/2}^{1/2} A(f|\mathbf{a})\hat{S}_x(f,\mathbf{x})e^{j2\pi f\ell}df}_{=0 \text{ from the gradient } \frac{\partial\sigma^2_w(\mathbf{a})}{\partial a[\ell]}=0} \\
&= \int_{-1/2}^{1/2} \left(\hat{S}_{xx}(f,\mathbf{x}) + \sum_{k=1}^{N_a} a[k]\hat{S}_x(f,\mathbf{x})e^{-j2\pi fk}\right)df \\
&= \hat{r}[0] + \sum_{k=1}^{N_a} a[\ell]\hat{r}[k]
\end{aligned}$$

which is the power of the excitation from the sample autocorrelation as the power of the prediction error from the Yule–Walker equations (12.4) in Section 12.3.1.

Remark: The coincidence of AR spectral analysis with the Yule–Walker equations of linear prediction is not surprising. AR spectral analysis can be interpreted as a search for the linear predictor that fully uncorrelates the process under analysis. The sample autcorrelation replaces the ensemble autocorrelation as this estimates the correlation properties from a limited set of observations with positive definite Fourier transform (while still being biased).

The CRB can be evaluated from the Whittle formula (14.8) adapted for the AR model:

$$S_x(f|\boldsymbol{\theta}) = \frac{\sigma^2_{N_a}}{A(f|\mathbf{a})A^*(-f|\mathbf{a})}$$

Computing first the terms

$$\frac{\partial \ln S_x(f|\boldsymbol{\theta})}{\partial\sigma^2_{N_a}} = \frac{1}{\sigma^2_{N_a}}$$

$$\frac{\partial \ln S_x(f|\boldsymbol{\theta})}{\partial a[k]} = -\frac{e^{-j2\pi fk}}{A(f|\mathbf{a})} - \frac{e^{j2\pi fk}}{A^*(-f|\mathbf{a})}$$

and then the elements of the FIM from the Whittle formula, recalling that $A(f|\mathbf{a})$ is causal and minimum phase[4]

$$\frac{N}{2}\int_{-1/2}^{1/2}\left(\frac{\partial \ln S_x(f|\boldsymbol{\theta})}{\partial \sigma_{N_a}^2}\right)^2 df = \frac{N}{2\sigma_{N_a}^4}$$

$$\frac{N}{2}\int_{-1/2}^{1/2}\frac{\partial \ln S_x(f|\boldsymbol{\theta})}{\partial \sigma_{N_a}^2}\frac{\partial \ln S_x(f|\boldsymbol{\theta})}{\partial a[k]}df = 0$$

$$\frac{N}{2}\int_{-1/2}^{1/2}\frac{\partial \ln S_x(f|\boldsymbol{\theta})}{\partial a[k]}\frac{\partial \ln S_x(f|\boldsymbol{\theta})}{\partial a[\ell]}df = N\int_{-1/2}^{1/2}\frac{e^{j2\pi f(\ell-k)}}{A(f|\mathbf{a})A^*(-f|\mathbf{a})}df = \frac{N}{\sigma_{N_a}^2}r_x[\ell-k]$$

The FIM is block-partitioned

$$\mathbf{J}(\boldsymbol{\theta}_0) = \begin{bmatrix} \frac{N}{\sigma_{N_a}^2}\mathbf{C}_{xx} & \mathbf{0} \\ \mathbf{0} & \frac{N}{2\sigma_{N_a}^4} \end{bmatrix}$$

so the estimation of the excitation $\sigma_{N_a}^2$ is decoupled from the coefficients $\mathbf{a}$ of the AR model polynomial. The CRB follows from the inversion of the FIM with covariance

$$\text{var}\{\hat{\sigma}_{N_a}^2\} \geq \frac{2\sigma_{N_a}^4}{N}$$

$$\text{cov}\{\hat{\mathbf{a}}\} \geq \frac{\sigma_{N_a}^2}{N}\mathbf{C}_{xx}^{-1}$$

The CRB of the coefficients $\mathbf{a}$ depends on the inverse of the covariance matrix $\mathbf{C}_{xx}$. Note that all coefficients do not have the same variance, but they are unbalanced and they depend on the specific problem as discussed in the example in the next section.

14.3.2 A Good Reason to Avoid Over-Parametrization in AR

AR spectral analysis can be carried out regardless of the true model, or even when the PSD does not show resonances due to the presence of poles. Since the generation mechanism is not always known, the order N_a of the AR spectral analysis is often another unknown of the problem at hand. Even if in principle the estimates $\hat{\mathbf{a}}$ could account for the order as those terms larger that the true order should be zero, the augmentation of

4 For an arbitrary causal and minimum phase polynomial $P(z) = \sum_{k=0}^{N} p_k z^{-k}$ the sequence corresponding to $1/P(z)$ is causal too and thus the following identities hold for the inverse Fourier transform:

$$p_{-\ell} = \int_{-\pi}^{\pi}\frac{e^{-j\omega\ell}}{P(e^{j\omega})}\frac{d\omega}{2\pi} = \int_{-\pi}^{\pi}\frac{e^{j\omega\ell}}{P^*(e^{-j\omega})}\frac{d\omega}{2\pi} = 0 \text{ for } \ell \geq 1$$

the model order in AR spectral analysis has to be avoided as the parameters $\mathbf{a}$ are not at all decoupled, and the covariance $\text{cov}\{\hat{\mathbf{a}}\}$ could be unacceptably large.

A simple example can help to gain insight into the concept of model over-parameterization (set an AR order in spectral analysis larger than the true AR order). Let a set of observations be generated as an AR(1) with parameters $\{\sigma_w^2, \rho\}$ for $\rho \in \mathbb{R}$ (the complex-valued case will be considered later), the autocorrelation is

$$r_x[k] = \frac{\rho^{|k|}}{1-\rho^2}$$

and the Yule–Walker equations to estimate the parameters $\mathbf{a}$ leads to the estimates $\hat{\mathbf{a}}$ such that $\mathbb{E}[\hat{a}_1] = \rho(1-1/N)$ and $\mathbb{E}[\hat{a}_k] = 0$ for $\forall k > 1$ (the bias $(1-1/N)$ is due to the sample autocorrelation that vanishes for N large). The CRB depends on the correlation matrix $\mathbf{C}_{xx}$ that for an arbitrary order N_a is structured as

$$[\mathbf{C}_{xx}]_{i,j} = \frac{\rho^{|i-j|}}{1-\rho^2}$$

and its inverse is known (Section 4.5). The CRB for different order selection is:

$$\text{CRB for AR(1): } \text{cov}\{\hat{a}_1\} \geq \frac{\sigma_w^2(1-\rho^2)}{N}$$

$$\text{CRB for AR(2): } \text{cov}\{\hat{\mathbf{a}}\} \geq \frac{\sigma_w^2}{N}\begin{bmatrix} 1 & -\rho \\ -\rho & 1 \end{bmatrix}$$

$$\text{CRB for AR(4): } \text{cov}\{\hat{\mathbf{a}}\} \geq \frac{\sigma_w^2}{N}\begin{bmatrix} 1 & -\rho & 0 & 0 \\ -\rho & 1+\rho^2 & -\rho & 0 \\ 0 & -\rho & 1+\rho^2 & -\rho \\ 0 & 0 & -\rho & 1 \end{bmatrix}$$

thus showing that diagonal terms are not the same and that (for this specific problem) the CRB matrix is diagonal banded. Furthermore:

$$\text{var}\{\hat{a}_1 | N_a > 1\} \geq \frac{\sigma_w^2(1-\rho^2)}{N} = \text{var}\{\hat{a}_1 | N_a = 1\}$$

$$\text{var}\{\hat{a}_k | N_a > 1\} \geq \frac{\sigma_w^2}{N} > 0$$

that is, the over-parameterization increases the variance of the AR spectral analysis and the additional poles due to $\hat{a}_k$ because the choice $N_a > 1$ perturbs the PSD estimates that in turn depend on the dispersion of the poles.

The conclusion on over-parameterization is general: using more parameters than necessary reduces the accuracy of the PSD estimate so that sometimes it is better to under-parameterize AR spectral analysis. Alternatively, one can start cautiously with low order and increase the order in the hunt for a better trade-off, being aware that every augmentation step increases the variance of the PSD estimate.

14.3.3 Cramér–Rao Bound of Poles in AR Spectral Analysis

An AR model can be rewritten wrt the roots of the polynomial $A(z)$ as

$$A(z|\boldsymbol{\mu}) = \prod_{k=1}^{N_a}(1-\mu_k z^{-1})$$

where $\boldsymbol{\mu}$ collects all the roots $\mu_1, \mu_2, ..., \mu_{N_a}$ as detailed below. The z-transform of the PSD becomes

$$S_x(z|\boldsymbol{\mu},\sigma_w^2) = \frac{\sigma_w^2}{A(z|\boldsymbol{\mu})\cdot A^*(1/z|\boldsymbol{\mu})}$$

and the Whittle relationship for the CRB of an AR spectral analysis can be rewritten in terms of the roots of the z-transform. The importance of this model stems from the use of AR spectral analysis to estimate the angular position of the poles, and these in turn are related to the frequency of the resonances in PSD.

To simplify the reasoning, let the process **x** be real-valued so that poles are complex conjugates μ_k and μ_k^*; in addition poles are either real or complex but are assumed to have multiplicity one. Let the poles be decoupled into those that are real ($\mathcal{R}$) and those that are complex conjugates ($\mathcal{C}$) so that the z-transform of the PSD is

$$S_x(z|\boldsymbol{\mu},\sigma_w^2) = \sigma_w^2 \times \frac{1}{\prod_{k\in\mathcal{R}}(1-\mu_k z^{-1})(1-\mu_k z)} \times \frac{1}{\prod_{k\in\mathcal{C}}(1-\mu_k z^{-1})(1-\mu_k^* z^{-1})(1-\mu_k^* z)(1-\mu_k z)}$$

The Whittle formula needs the derivatives evaluated separately for the two sets

$$\text{real poles:} \quad \frac{\partial \ln S_x(z)}{\partial \mu_k} = -\frac{-z^{-1}}{1-\mu_k z^{-1}} - \frac{-z}{1-\mu_k z}$$

$$\text{complex conjugate poles:} \begin{cases} \frac{\partial \ln S_x(z)}{\partial \mu_k} = -\frac{-z^{-1}}{1-\mu_k z^{-1}} - \frac{-z}{1-\mu_k z} \\ \frac{\partial \ln S_x(z)}{\partial \mu_k^*} = -\frac{-z^{-1}}{1-\mu_k^* z^{-1}} - \frac{-z}{1-\mu_k^* z} \end{cases}$$

Integrating over the set for the elements of the FIM:[5]

$$k\in\mathcal{R}: \frac{N}{2}\int_{-1/2}^{1/2}\left[\left(\frac{\partial \ln S_x(z)}{\partial \mu_k}\right)^2\right]_{z=e^{j2\pi f}}$$

---

5 From Appendix B of Chapter 4 the inverse z-transform for stable system ($|\alpha| < 1$ and $|\beta| > 1$) is

$$H(z) = \frac{1}{1-\alpha z^{-1}} \times \frac{1}{1-\beta z} \leftrightarrow h[n] = \frac{1}{1-\alpha\beta} \times \begin{cases} \alpha^n & \text{for } n \geq 0 \\ \beta^{-n} & \text{for } n \leq 0 \end{cases}$$

with region of convergence $z \in (\alpha, 1/\beta)$ that contains $|z| = 1$; in addition $h[0] = (1-\alpha\beta)^{-1}$.

$$df = N\int_{-1/2}^{1/2}\left[\frac{1}{(1-\mu_k z^{-1})(1-\mu_k z)}\right]_{z=e^{j2\pi f}} df = \frac{N}{1-\mu_k^2}$$

$$k,\ell \in \mathcal{R}: \frac{N}{2}\int_{-1/2}^{1/2}\left[\frac{\partial \ln S_x(z)}{\partial \mu_k}\frac{\partial \ln S_x(z)}{\partial \mu_\ell}\right]_{z=e^{j2\pi f}}$$

$$df = N\int_{-1/2}^{1/2}\left[\frac{1}{(1-\mu_k z^{-1})(1-\mu_\ell z)}\right]_{z=e^{j2\pi f}} df = \frac{N}{1-\mu_k\mu_\ell}$$

$$k,\ell \in \mathcal{C}: \frac{N}{2}\int_{-1/2}^{1/2}\left[\frac{\partial \ln S_x(z)}{\partial \mu_k}\frac{\partial \ln S_x(z)}{\partial \mu_\ell^*}\right]_{z=e^{j2\pi f}}$$

$$df = N\int_{-1/2}^{1/2}\left[\frac{1}{(1-\mu_k z^{-1})(1-\mu_\ell^* z)}\right]_{z=e^{j2\pi f}} df = \frac{N}{1-\mu_k\mu_\ell^*}$$

and using the identity $\int_{-1/2}^{1/2}\left[H(z)z^k\right]_{z=e^{j2\pi f}} df = \int_{-1/2}^{1/2} H(f)e^{j2\pi fk}df = h[k]$ (Appendix C). The entries of the FIM partitioned either into real ($\mathcal{R}$) or complex conjugate ($\mathcal{C}$) poles have the same value for the accounted pole. To summarize, it is convenient to order the N_a poles into real ($\boldsymbol{\mu}_\mathcal{R}$) and complex ($\boldsymbol{\mu}_\mathcal{C}$) with their conjugate ($\bar{\boldsymbol{\mu}}_\mathcal{C}$)

$$\boldsymbol{\mu} = \begin{bmatrix} \boldsymbol{\mu}_\mathcal{R} \\ \boldsymbol{\mu}_\mathcal{C} \\ \bar{\boldsymbol{\mu}}_\mathcal{C} \end{bmatrix}$$

such that $[\boldsymbol{\mu}_\mathcal{C}]_k = \mu_k$ and $[\bar{\boldsymbol{\mu}}_\mathcal{C}]_k = \mu_k^*$; so the FIM is

$$[\mathbf{J}(\boldsymbol{\theta}_0)]_{k\ell} = \frac{N}{1-[\boldsymbol{\mu}_0 \cdot \boldsymbol{\mu}_0^H]_{k\ell}}$$

Example: Pole Scattering in AR(2)

Consider an AR(2) with poles $\mu_{1,2} = \rho\exp(\pm j\theta)$. From the relationship above, the FIM is

$$\mathbf{J}(\boldsymbol{\theta}_0) = N\begin{bmatrix} \frac{1}{1-\rho^2} & \frac{1}{1-\rho^2 e^{j2\theta}} \\ \frac{1}{1-\rho^2 e^{-j2\theta}} & \frac{1}{1-\rho^2} \end{bmatrix}$$

To derive the covariance for real and imaginary component of the poles it is useful to recall the following identity:

$$\begin{bmatrix} \mu_R \\ \mu_I \end{bmatrix} = \frac{1}{2}\begin{bmatrix} 1 & 1 \\ -j & j \end{bmatrix}\begin{bmatrix} \mu \\ \mu^* \end{bmatrix}$$

so that the CRB for the transformed variables (Section 8.3) is

$$\text{cov}[[\mu_R, \mu_I]^T] \geq \frac{1}{4N}\begin{bmatrix} 1 & 1 \\ -j & j \end{bmatrix}\begin{bmatrix} \frac{1}{1-\rho^2} & \frac{1}{1-\rho^2 e^{j2\theta}} \\ \frac{1}{1-\rho^2 e^{-j2\theta}} & \frac{1}{1-\rho^2} \end{bmatrix}^{-1}\begin{bmatrix} 1 & j \\ 1 & j \end{bmatrix}$$

After some algebra it follows that

$$\mathrm{var}\{\mu_R\} \geq \frac{1-\rho^4}{4N}$$

$$\mathrm{var}\{\mu_I\} \geq \frac{1}{4N}\frac{\left(1-\rho^2\right)^2}{|\tan\theta|}$$

The variance of μ_R is independent of the frequency of the poles, and the variance for the imaginary term is very large when $\theta \to 0$ (both poles are coincident) and vanishes for $\theta \to \pi/2$. The error ellipsoid is not referred to the orthogonal axes, as $\mathrm{cov}\{\mu_R, \mu_I\} \neq 0$; the following numerical example for AR(1) can clarify.

14.3.4 Example: Frequency Estimation by AR Spectral Analysis

Let a sequence be modeled as a complex sinusoid in white Gaussian noise with power $\mathbb{E}[|w[n]|^2] = \sigma^2$

$$x[n] = A\exp(j\omega_o n) + w[n]$$

The angular frequency ω and the amplitude $|A|$ are not known and should be estimated as poles in AR(1) spectral analysis. The phase of the sinusoid is denoted as $\angle A$ and $\angle A \sim \mathcal{U}(-\pi, \pi)$. AR(1) spectral analysis models observations as

$$S_x(z|z_p, \sigma^2_{AR(1)}) = \frac{\sigma^2_{AR(1)}}{1 - z_p z^{-1}}$$

such that $\angle z_p = \omega_o$ and $|z_p| \simeq 1$ as the pole should approximate at best the true PSD: $S_x(f) = |A|\delta(f - f_o) + \sigma^2_w$. AR(1) follows from the Yule–Walker equation (12.2)

$$r_x[1] - z_p r_x[0] = 0$$

For the sample estimate of the autocorrelation:

$$\hat{z}_p = \frac{\hat{r}_1}{\hat{r}_0}$$

where $\hat{r}_k = \hat{r}_x[k]$ just to ease the notation for the computations below. The objective is to evaluate the performance (bias and variance) of AR(1) spectral analysis to estimate the frequency of the sinusoid as an alternative to the conventional MLE in Section 9.3. The frequency estimate accuracy depends on the small deviations of the radial and angular position of the pole $\hat{z}_p$ due to noise.

Since the pole depends on the sample autocorrelations, $\hat{r}_k$ is decomposed into (Appendix A)

$$\hat{r}_k = \bar{r}_k + \delta r_k \quad \text{with random deviation } |\delta r_k / \bar{r}_k| \ll 1$$

and

$$\bar{r}_k = \mathbb{E}[\hat{r}_k] = \begin{cases} |A|^2 \exp(j\omega_o k)\left(1 - \frac{|k|}{N}\right) & \text{for } k \neq 0 \\ |A|^2 + \sigma^2 & \text{for } k = 0 \end{cases}$$

For large N and large SNR defined as

$$\gamma = \frac{\sigma^2}{|A|^2} = \frac{1}{SNR},$$

the pole can be approximated as

$$\hat{z}_p = \frac{\hat{r}_1}{\hat{r}_0} = \frac{\bar{r}_1}{\bar{r}_0}\frac{1 + \delta r_1/\bar{r}_1}{1 + \delta r_0/\bar{r}_0} \simeq \frac{\bar{r}_1}{\bar{r}_0}(1 + \delta r_1/\bar{r}_1)(1 - \delta r_0/\bar{r}_0) \simeq \frac{\bar{r}_1}{\bar{r}_0}(1 + \delta r_1/\bar{r}_1 - \delta r_0/\bar{r}_0).$$

The bias is

$$\mathbb{E}[\hat{z}_p] = \frac{\bar{r}_1}{\bar{r}_0} = \frac{1 - 1/N}{1 + \gamma}\exp(j\omega_o) \simeq [1 - (\gamma + 1/N)]\exp(j\omega_o) \tag{14.9}$$

This proves that the estimate of the frequency is unbiased, but the radial position of the pole $\mathbb{E}[|\hat{z}_p|]$ depends on γ—however for $\gamma \to 0$ and $N \to \infty$ the pole attains the unitary circle. The variance is evaluated from the deviations δr_k:

$$\text{var}[\hat{z}_p] = \left|\frac{\bar{r}_1}{\bar{r}_0}\right|^2 \mathbb{E}[(\delta r_1/\bar{r}_1 - \delta r_0/\bar{r}_0)(\delta r_1/\bar{r}_1 - \delta r_0/\bar{r}_0)^*]$$

Since $\mathbb{E}[\hat{z}_p] = \bar{r}_1/\bar{r}_0$ it is convenient to group the terms as follows

$$\frac{\text{var}[\hat{z}_p]}{|\mathbb{E}[\hat{z}_p]|^2} = \frac{\mathbb{E}[|\delta r_0|^2]}{|\bar{r}_0|^2} + \frac{\mathbb{E}[|\delta r_1|^2]}{|\bar{r}_1|^2} - \frac{\mathbb{E}[\delta r_1 \delta r_0^*]}{\bar{r}_1 \bar{r}_0^*} - \frac{\mathbb{E}[\delta r_0 \delta r_1^*]}{\bar{r}_0 \bar{r}_1^*}$$

From the properties in Appendix D, the variance is

$$\frac{\text{var}[\hat{z}_p]}{|\mathbb{E}[\hat{z}_p]|^2} = \frac{2\gamma}{N}\left(\frac{(2 + \frac{N-1}{N}\gamma)}{\left(1 - \frac{1}{N}\right)^2} + \frac{(2|A|^2 + \sigma^2)}{(1+\gamma)^2} - \frac{2\left(2 - \frac{1}{N}\right)}{(1+\gamma)\left(1 - \frac{1}{N}\right)}\right) \simeq \frac{2\gamma}{N}(\gamma + 1/N). \tag{14.10}$$

where the last approximation is for $\gamma \ll 1$ and $1/N \ll 1$. For $\omega_o = \pi/4$, Figure 14.10 shows the comparison between the simulated (black solid line) and the analytical (gray dashed line) bias (14.9) (upper figure) and standard deviation $\sqrt{\text{var}[\hat{z}_p]}$ from (14.10) that perfectly superimpose for AR(1) and therefore validate the approximations in the sensitivity analysis for $N = 10$ and 50 samples. Figure 14.10 also shows the dispersion for AR(2) (dash dot line) when extracting only one pole corresponding to the frequency close to $\omega_o = \pi/4$, that is less dispersed and closer to the unit circle for $|A|^2/\sigma^2 < 15dB$.

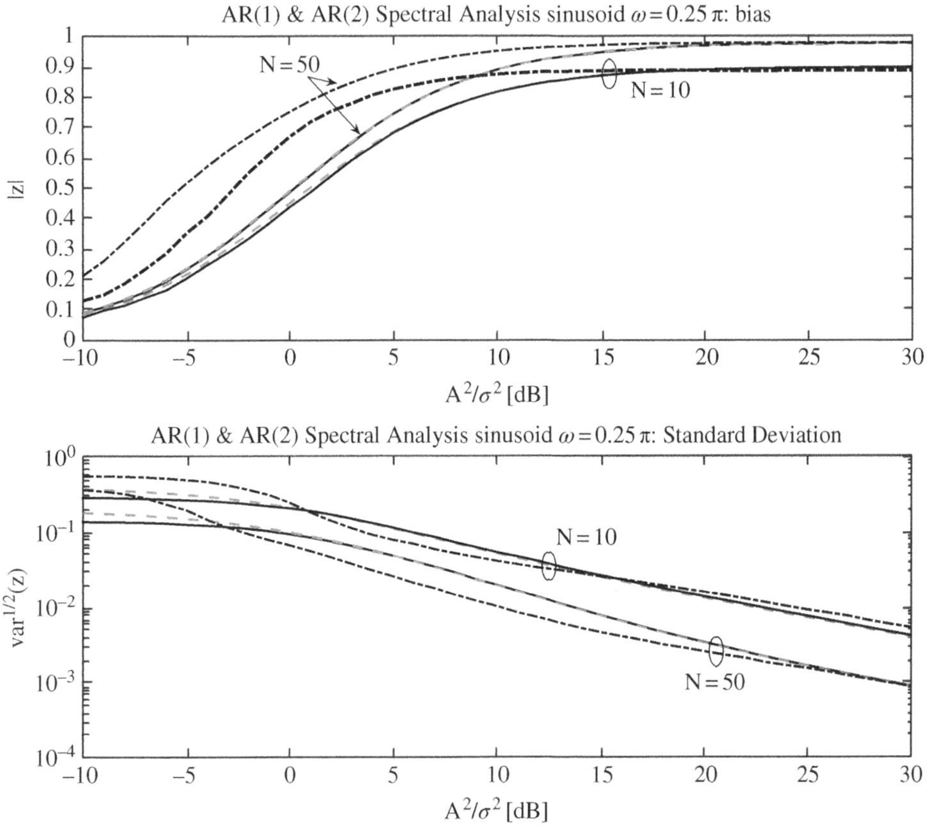

Figure 14.10 Radial position of pole (upper figure) and dispersion (lower figure) for AR(1) (solid line) and AR(2) (dash dot), compared to analytic model (14.10–14.11) (dashed gray line)

In practice, the number of sinusoids is not known and it is quite common practice to "search" for the order by trial-and-error on the poles. Figure 14.11 shows the poles from a set of independent AR spectral analyses with order $N_a = 1, 2, 3, 4$ (superposition of 50 independent trails with $N = 20$ and $\omega_o = \pi/4$) and $SNR = 1/\gamma = 5dB$. Notice that for AR(1), the poles in every trial superimpose around the true angular position with a radial location set by the bias (14.9). When increasing the order of the AR spectral analysis, the poles cluster around the angular position in $\omega_o = \pi/4$ thus providing an estimate of the frequency, while the other poles disperse within the unit circle far from the sinusoid thus allowing clear identification of the frequency of the sinusoid. However, each additional pole can be a candidate to model (and estimate) a second, or third, or forth sinusoid in observation thus providing a tool to better identify the number of sinusoids in the data (if unknown). Needless to say, ranking the poles with decreasing radial distance from the unit circle offers a powerful tool to identify the number of sinusoids, if necessary, when the number is not clear from the practical problem at hand.

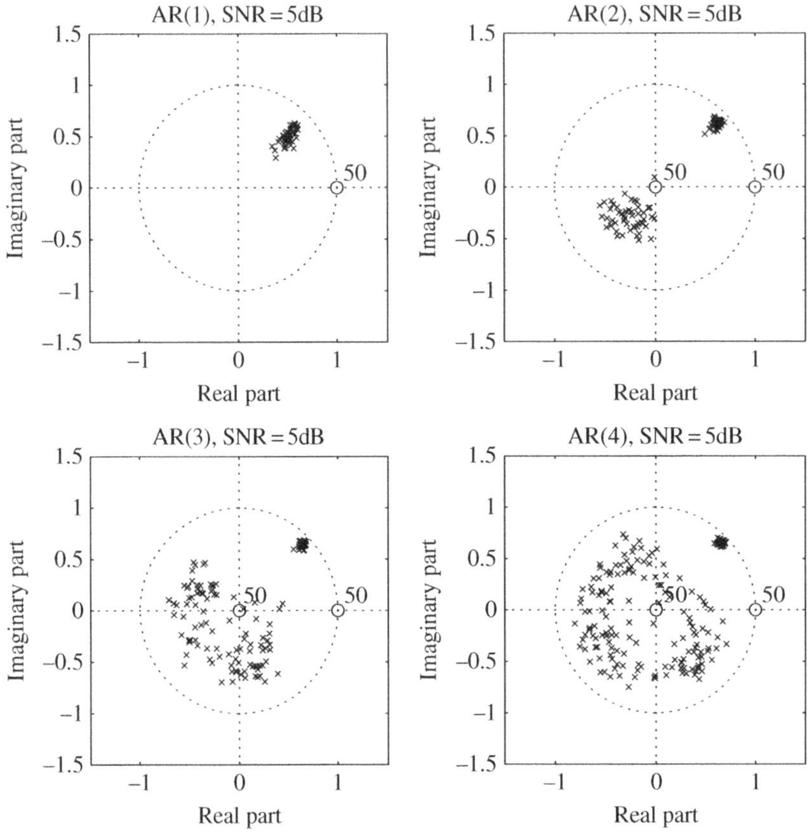

Figure 14.11 Scatter-plot of poles of Montecarlo simulations for AR(1), AR(2), AR(3), AR(4) spectral analysis.

14.4 MA Spectral Analysis

The PSD of the N_bth order moving average (MA) process is the one generated by filtering a white Gaussian noise with power $\sigma^2_{N_b}$ with a limited support filter (or averaging filter):

$$X(z) = \left(1 + \sum_{\ell=1}^{N_b} b[\ell]z^{-\ell}\right) W(z) = B(z|\mathbf{b})W(z)$$

as illustrated in Figure 14.11. The MA spectral analysis estimates the set of generation parameters $\hat{\mathbf{b}}, \hat{\sigma}^2_{N_b}$ from the limited observations so that the PSD of the $MA(N_b)$ model is

$$S_x(\omega|\hat{\mathbf{b}}, \hat{\sigma}^2_{N_b}) = \left|B(\omega|\hat{\mathbf{b}})\right|^2 \hat{\sigma}^2_{N_b}$$

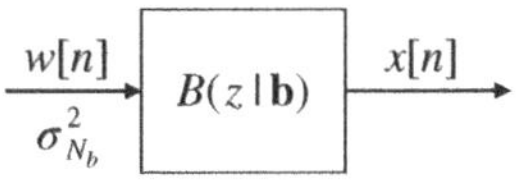

Figure 14.12 Model for MA spectral analysis.

From the sample estimate of the autocorrelation $\hat{r}_x[n]$ over the support of $2N_b+1$ samples, it is sufficient to use the identity over the z-transform

$$\hat{r}_x[n] \leftrightarrow \hat{R}_x(z) = \sigma^2_{N_b} \cdot B(z|\mathbf{b}) \cdot B^*(1/z|\mathbf{b})$$

and solve wrt $\mathbf{b}$ and $\sigma^2_{N_b}$, the latter acts as a scale factor of the estimated PSD. The overall problem cast in this way is strongly non-linear and does not have a unique solution for $\hat{\mathbf{b}}$. From the sample estimate of the autocorrelation $\hat{R}_{xx}(z)$ one can extract the $2N_b$ roots, which are in all reciprocal positions according to the Paley–Wiener theorem (Section 4.4). These roots can be paired in any arbitrary way and collected into a set $B(z|\hat{\mathbf{b}})$ provided that each selection does not contain the reciprocal set. Even if any collection of zeros in the filter $B(z|\hat{\mathbf{b}})$ is an MA estimate, the preferred one is the minimum phase arrangement $B_{\min}(z|\hat{\mathbf{b}})$ so that

$$\hat{R}_x(z) = \hat{\sigma}^2_{N_b} \cdot B_{\min}(z|\hat{\mathbf{b}}) \cdot B^*_{\min}(1/z|\hat{\mathbf{b}})$$

This is essential for the MA analysis of the sample autocorrelation, and some remarks are necessary to highlight some practical implications.

Remark 1: To reduce the variance of the sample estimate of the autocorrelation, it is customary to average the sample estimate after segmenting the set of observations into M subsets of disjoint time series of N_b+1 samples each. Assuming $x^{(m)}[n] = x[i(N_b+1)+n]$ to be the mth segment, the sample autocorrelation is

$$\hat{r}_x[n] = \frac{1}{M}\sum_{m=1}^{M}\frac{1}{N_b}\sum_{k=0}^{N_b-n} x^{(m)}[k]^* \cdot x^{(m)}[k+n]$$

and the variance for disjoint sets decreases with $1/M$. Of course, the order of the MA analysis should be traded with the variance of the sample autocorrelation.

Remark 2: Given the sample autocorrelation over an arbitrary interval, say $N > N_b$, $MA(N_b)$ cannot be carried out by simply using the first N_b samples (that would imply a truncation of the sample autocorrelation) but must rather be by windowing the autocorrelation by a window $v_{N_b}[n]$ that has a positive definite FT ($v_{N_b}[n] = 0$ for $n \notin [-N_b, N_b]$ but $\mathcal{F}\{v_{N_b}[n]\} = V_{N_b}(\omega) \geq 0$ for any ω). To justify this, let

$$\hat{r}_x[n] = \frac{1}{N}\sum_{k=0}^{N-n} x^*[k] \cdot x[k+n]$$

be the sample autocorrelation over N samples; the FT $\hat{R}_x(\omega) = \mathcal{F}\{\hat{r}_x[n]\}$ has the following property (Appendix A):

$$\mathbb{E}[\hat{R}_x(\omega)] = \frac{1}{N}\left(\frac{\sin[\omega N/2]}{\sin[\omega/2]}\right)^2 \circledast_{2\pi} S(\omega)$$

To carry out $MA(N_b)$ spectral analysis, the sample autocorrelation $\hat{r}_x[n]$ is windowed and the MA spectral analysis becomes

$$v_{N_b}[n] \times \hat{r}_x[n] \leftrightarrow S(\omega|\hat{\mathbf{b}}, \hat{\sigma}_w^2) = V_{N_b}(\omega) \circledast_{2\pi} \hat{R}_x(\omega)$$

for some choice of $\hat{\mathbf{b}}, \hat{\sigma}_w^2$. The MA PSD is biased, as

$$\mathbb{E}[S(\omega|\hat{\mathbf{b}}, \hat{\sigma}_w^2)] = V_{N_b}(\omega) \circledast_{2\pi} \mathbb{E}[\hat{R}_x(\omega)] = V_{N_b}(\omega) \circledast_{2\pi} \frac{1}{N}\left(\frac{\sin[\omega N/2]}{\sin[\omega/2]}\right)^2 \circledast_{2\pi} S(\omega),$$

and it is biased for $N \to \infty$ as $\mathbb{E}[S(\omega|\hat{\mathbf{b}}, \hat{\sigma}_w^2)] \to V_{N_b}(\omega) \circledast_{2\pi} S(\omega)$. The choice of the window should guarantee that $V_{N_b}(\omega) \geq 0$ to make the PSD positive definite. Truncation $v_{N_b}[n] = 1$ for $n \in [-N_b, N_b]$ would not satisfy this constraint, while $v_{N_b}[n] = 1 - |n|/N_b$ for $n \in [-N_b, N_b]$ (triangular window) fulfills the constraints and is the most common choice.

14.5 ARMA Spectral Analysis

$ARMA(N_a, N_b)$ spectral analysis is the composition of the $AR(N_a)$ and $MA(N_b)$ systems. Namely, the ARMA process is obtained from the causal linear system excited by white Gaussian process $w[n]$ as described by the difference equation (Figure 14.13)

$$x[n] + \sum_{i=1}^{N_a} a[i] \cdot x[n-i] = \sum_{j=0}^{N_b} b[j] \cdot w[n-j] \tag{14.11}$$

The constitutive relationship of ARMA can be derived from the definition of the autocorrelation

$$r_x[k] = \mathbb{E}[x^*[n] \cdot x[n+k]] = \mathbb{E}[x[n] \cdot x^*[n-k]]$$

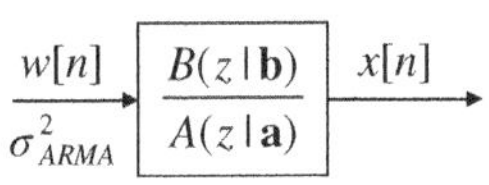

Figure 14.13 Model for ARMA spectral analysis.

using the second form of the finite difference equation (14.11) by multiplying for $x^*[n-k]$ and computing the expectation:

$$r_x[k]+\sum_{i=1}^{N_a} a[i]\cdot r_x[k-i]=\sum_{j=0}^{N_b} b[j]\cdot \mathbb{E}[w[n-j]\cdot x^*[n-k]]$$

Since the system is causal, the impulse response is $h[n]$ and the process $x[n]$ is the filtered excitation:

$$x[n]=\sum_{\ell=0}^{\infty} h[\ell]w[n-\ell]$$

Substituting into the second term of the relationship above:

$$\mathbb{E}[w[n-j]\cdot x^*[n-k]]=\mathbb{E}[w[n-j]\cdot\sum_{\ell=0}^{\infty} h^*[\ell]w^*[n-k-\ell]]$$

$$=\sum_{\ell=0}^{\infty} h^*[\ell]\cdot \underbrace{\mathbb{E}[w[n-j]\cdot w^*[n-k-\ell]]}_{\sigma_w^2\delta[k+\ell-j]}=\sigma_w^2\cdot h^*[j-k]$$

and thus

$$r_x[k]+\sum_{i=1}^{N_a} a[i]\cdot r_x[k-i]=\sigma_w^2\sum_{j=0}^{N_b} b[j]\cdot h^*[j-k]$$

In this form it is still dependent on the impulse response, but since $h[n]=0$ for $n<0$ (causal system), the second term is zero for $k>j$ and also for $k>N_b$. It follows the *modified Yule–Walker* equations:

$$r_x[k]+\sum_{i=1}^{N_a} a[i]\cdot r_x[k-i]=0 \quad \text{for } k\geq N_b+1 \tag{14.12}$$

which is a generalization of the Yule–Walker equations for the ARMA processes to derive the $AR(N_a)$ component $\hat{\mathbf{a}}$ of the $ARMA(N_a,N_b)$. Of course, for ARMA analysis from a limited set of samples $x[n]$, the autocorrelation is replaced by the sample (and biased) estimate of the autocorrelation (Appendix A).

After solving for the modified Yule–Walker equations (14.12) to get $\hat{\mathbf{a}}$, one solves for $MA(N_b)$. This is obtained by noticing that when filtering the observations with $A(z|\hat{\mathbf{a}})$ (incidentally, this is the prediction error using a linear predictor based on the modified Yule–Walker equations) obtained so far, the residual depends on the MA component only as

$$A(z|\hat{\mathbf{a}})X(z)=B(z|\mathbf{b})W(z)$$

The autocorrelation of the MA component $A(z|\hat{\mathbf{a}})X(z) = \hat{Y}(z) \longleftrightarrow \hat{y}[n] = \hat{a}[n] * x[n]$:

$$r_{\hat{y}}[n] = \sum_{i,j=0}^{N_a} a[i]a^*[j]\underbrace{\mathbb{E}[x[k+n-i]\cdot x^*[k-j]]}_{r_x[n+j-i]}$$

rewritten in the form of sample autocorrelation (recall that $a[0]=1$)

$$\hat{r}_{\hat{y}}[n] = \begin{cases} \sum_{i,j=0}^{N_a} \hat{a}[i]\cdot\hat{a}^*[j]\cdot\hat{r}_x[n+j-i] & \text{for } n = 0,1,\dots,N_b \\ \hat{r}^*_{\hat{y}}[-n] \quad \text{for } n = -1,\dots,-N_b \end{cases}$$

Therefore, the N_b+1 samples of the autocorrelation $\hat{r}_{\hat{y}}[n]$ represent the MA component of the ARMA process and can be estimated following the same rules as $MA(N_b)$ applied now on $\hat{r}_{\hat{y}}[n]$. By using the identity

$$\hat{R}_{\hat{y}}(z|\mathbf{b},\sigma^2_{ARMA}) = \sum_{n=-N_b}^{N_b} \hat{r}_{\hat{y}}[n]z^{-n} = \sigma^2_{ARMA}\cdot B(z|\mathbf{b})\cdot B^*(1/z|\mathbf{b})$$

it can be solved for $\mathbf{b}$ and the excitation power σ^2_{ARMA} that is now the scaling factor for the whole ARMA process.

Remark 1: Note that there is no guarantee that the sample autocorrelation $\hat{r}_{\hat{y}}[n]$ holds over the range $[-N_b, N_b]$, so the PSD of the MA component may not be positive semidefinite. In this case it is a good rule when handling sample estimates (that could have some non-zero samples outside of the range of the MA process) to window $\hat{r}_{\hat{y}}[n]$ with a triangular window over the range of the order of the MA component, as this guarantees that the PSD is positive definite.

Remark 2: The estimate of the AR portion by the modified Yule–Walker equation (14.12) with the sample autocorrelation could be slow, as the system of equations uses samples with lower confidence (as for $k \geq N_b+1$). Some alternative methods could be explored to make a better (and more efficient) use of the available samples.

14.5.1 Cramér–Rao Bound for ARMA Spectral Analysis

Let the PSD for the ARMA be

$$S_x(f|\boldsymbol{\theta}) = \sigma^2_w \frac{B(f|\mathbf{b})\cdot B^*(-f|\mathbf{b})}{A(f|\mathbf{a})\cdot A^*(-f|\mathbf{a})}$$

where $A(z|\mathbf{a})$ and $B(z|\mathbf{b})$ are both minimum phase factorizations ($\mathbf{a}$ and $\mathbf{b}$ are real-valued vectors). The CRB is obtained from the asymptotic Whittle formula starting from:

$$\frac{\partial \ln S_x(f|\boldsymbol{\theta})}{\partial a[k]} = -\frac{e^{-j2\pi fk}}{A(f|\mathbf{a})} - \frac{e^{j2\pi fk}}{A^*(-f|\mathbf{a})}$$
$$\frac{\partial \ln S_x(f|\boldsymbol{\theta})}{\partial b[k]} = \frac{e^{-j2\pi fk}}{B(f|\mathbf{b})} + \frac{e^{j2\pi fk}}{B^*(-f|\mathbf{b})}$$

The second order derivatives for the FIM elements are (recall the causality condition for the factorizations $A(z|\mathbf{a})$ and $B(z|\mathbf{b})$):

$$\frac{N}{2}\int_{-1/2}^{1/2}\frac{\partial \ln S_x(f|\boldsymbol{\theta})}{\partial \sigma_w^2}\frac{\partial \ln S_x(f|\boldsymbol{\theta})}{\partial a[k]}df = 0$$

$$\frac{N}{2}\int_{-1/2}^{1/2}\frac{\partial \ln S_x(f|\boldsymbol{\theta})}{\partial a[k]}\frac{\partial \ln S_x(f|\boldsymbol{\theta})}{\partial a[\ell]}df = N\int_{-1/2}^{1/2}\frac{e^{j2\pi f(\ell-k)}}{A(f|\mathbf{a})A^*(-f|\mathbf{a})}df = N\cdot r_{aa}[\ell-k]$$

$$\frac{N}{2}\int_{-1/2}^{1/2}\frac{\partial \ln S_x(f|\boldsymbol{\theta})}{\partial b[k]}\frac{\partial \ln S_x(f|\boldsymbol{\theta})}{\partial b[\ell]}df = N\int_{-1/2}^{1/2}\frac{e^{j2\pi f(\ell-k)}}{B(f|\mathbf{b})B^*(-f|\mathbf{b})}df = N\cdot r_{bb}[\ell-k]$$

$$\frac{N}{2}\int_{-1/2}^{1/2}\frac{\partial \ln S_x(f|\boldsymbol{\theta})}{\partial a[k]}\frac{\partial \ln S_x(f|\boldsymbol{\theta})}{\partial b[\ell]}df = -N\int_{-1/2}^{1/2}\frac{e^{j2\pi f(\ell-k)}}{A(f|\mathbf{a})B^*(-f|\mathbf{b})}df = -N\cdot r_{ab}[\ell-k]$$

where $r_{aa}[k]$, $r_{bb}[k]$, $r_{ab}[k]$ are the auto/cross-correlation sequences for two equivalent AR processes and a mixed one, all excited by a unit-power white process:

$$r_{aa}[k] \leftrightarrow \frac{1}{A(z|\mathbf{a})\cdot A^*(1/z|\mathbf{a})} \quad \text{AR process}$$

$$r_{bb}[k] \leftrightarrow \frac{1}{B(z|\mathbf{b})\cdot B^*(1/z|\mathbf{b})} \quad \text{AR process equivalent}$$

$$r_{ab}[k] \leftrightarrow \frac{1}{A(z|\mathbf{a})\cdot B^*(1/z|\mathbf{b})}$$

Autocorrelation $r_{aa}[k]$ and $r_{bb}[k]$ of the AR processes can be numerically evaluated from the Yule–Walker equations for the polynomials $A(z|\mathbf{a})$ and $B(z|\mathbf{b})$, respectively. The numerical computation of the cross-correlation $r_{ab}[k]$ deserves a comment as the following factorization of

$$R_{ab}(z) = \underbrace{\frac{1}{A(z|\mathbf{a})B(z|\mathbf{b})\cdot A^*(1/z|\mathbf{a})B^*(1/z|\mathbf{b})}}_{\text{AR equivalent}} \times A^*(1/z|\mathbf{a})B(z|\mathbf{b})$$

highlights the way to compute it. Namely, first the Yule–Walker equations are applied to compute the autocorrelation for the equivalent AR process $1/\{A(z|\mathbf{a})B(z|\mathbf{b})\}$ represented by the first term using the minimum phase factorization $A(z|\mathbf{a})B(z|\mathbf{b})$, and then the cross-correlation terms of interests are obtained by filtering the autocorrelation computed thus far with a finite impulse non-causal (as $A^*(1/z|\mathbf{a})$ is non-causal) filter $A^*(1/z|\mathbf{a})B(z|\mathbf{b})$.

The FIM is block-partitioned:

$$\mathbf{J}(\boldsymbol{\theta}_0) = N\begin{bmatrix} \mathbf{C}_{aa} & -\mathbf{C}_{ab} & \mathbf{0} \\ -\mathbf{C}_{ab}^T & \mathbf{C}_{bb} & \mathbf{0} \\ \mathbf{0} & \mathbf{0} & \frac{1}{2}\cdot\sigma_w^{-4} \end{bmatrix}$$

with blocks

$$[\mathbf{C}_{aa}]_{k,\ell} = r_{aa}[\ell-k]$$

$$[\mathbf{C}_{bb}]_{k,\ell} = r_{bb}[\ell - k]$$
$$[\mathbf{C}_{ab}]_{k,\ell} = r_{ab}[\ell - k]$$

The structure of the FIM confirms that for ARMA spectral analysis the AR and MA parameters are coupled, and this offers room to find different estimation procedures than in Sections 14.3–14.4, but the power of the excitation acting as a scale parameter is always decoupled. This decoupling is not surprising as the structure of the PSD in terms of resonances and nulls depends on the poles and zeros of the AR and MA components, but not on the scale factor. The CRB for ARMA spectral analysis can be considered the more general one as the CRB for AR and MA can be derived from the CRB of ARMA by removing the corresponding blocks.

Appendix A: Which Sample Estimate of the Autocorrelation to Use?

Given N samples of a random process $\{x[n]\}_{n=0}^{N-1}$, the sample estimate of the autocorrelation is:

$$\hat{r}_x[k] = \frac{1}{N}\sum_{n=0}^{N-1} x^*[n]x[n+k] \text{ for } k \in (-N, N)$$

However, this sample estimate is biased as for every value of the shift k the product terms that are summed up are $N - |k|$; analytically it is

$$\mathbb{E}\left[\hat{r}_x[k]\right] = \begin{cases} \frac{N-|k|}{N} r_x[k] & \text{for } |k| \leq (N-1) \\ 0 & \text{elsewhere} \end{cases}$$

Alternatively, one can avoid the bias by scaling the sum for the number of terms, and it follows the definition of the sample autocorrelation

$$\tilde{r}_x[k] = \frac{1}{N-|k|}\sum_{n=0}^{N-1} x^*[n]x[n+k] \text{ for } k \in (-N, N)$$

which is unbiased as

$$\mathbb{E}\left[\tilde{r}_x[k]\right] = \begin{cases} r_x[k] & \text{for } |k| \leq (N-1) \\ 0 & \text{altrove} \end{cases}$$

Herein we prove that even if the estimate $\tilde{r}_x[k]$ is apparently the preferred choice being unbiased, the biased estimate of the sample autocorrelation $\hat{r}_x[k]$ is the choice to be adopted in all spectral analysis methods.

The Fourier transform of the two sample estimates are

$$\tilde{R}_x(\omega) = \sum_{m=-(N-1)}^{N-1} \tilde{r}_x[k]e^{-j\omega k}$$
$$\hat{R}_x(\omega) = \sum_{m=-(N-1)}^{N-1} \hat{r}_x[k]e^{-j\omega k}$$

The expectations are related to the PSD $S_x(\omega)$ as follows

$$\mathbb{E}[\tilde{R}_x(\omega)] = \sum_{k=-(N-1)}^{N-1} \mathbb{E}[\tilde{r}_x[k]]e^{-j\omega k} = \sum_k w^{(R)}[k] \cdot r_x[k]e^{-j\omega k} = W^{(R)}(\omega) \circledast_{2\pi} S_x(\omega)$$
$$\mathbb{E}[\hat{R}_x(\omega)] = \sum_{k=-(N-1)}^{N-1} \mathbb{E}[\hat{r}_x[k]]e^{-j\omega k} = \sum_k w^{(B)}[k] \cdot r_x[k]e^{-j\omega k} = W^{(B)}(\omega) \circledast_{2\pi} S_x(\omega) \tag{14.13}$$

where the limited support autocorrelation has been replaced by rectangular and triangular (or Bartlett) windowing

$$w^{(R)}[k] = 1 \qquad |k| < N \quad \longleftrightarrow \quad W^{(R)}(\omega) = \frac{\sin[\omega(2N-1)/2]}{\sin[\omega/2]}$$
$$w^{(B)}[k] = \frac{N-|k|}{N} \qquad |k| < N \quad \longleftrightarrow \quad W^{(B)}(\omega) = \frac{1}{N}\left(\frac{\sin[\omega N/2]}{\sin[\omega/2]}\right)^2$$

Inspection of the relationships (14.13) proves that the FT of the unbiased sample autocorrelation $\tilde{r}_x[k]$ is not positive definite as $W^{(R)}(\omega)$ could be negative valued, for some spectral behaviors. On the other hand, the Fourier transform of the biased sample autocorrelation $\hat{r}_x[k]$ is positive definite (in the mean) as the sample estimate is implicitly windowed by a Bartlett window $w^{(B)}[k]$ that has positive definite FT. Furthermore, it can be seen that for $N \to \infty$, the Bartlett windowing becomes a constant and $W^{(B)}(\omega) \to \delta(\omega)$, therefore

$$\lim_{N\to\infty} \mathbb{E}[\hat{R}_x(\omega)] = S_x(\omega)$$

is the unbiased estimate of the PSD. Incidentally, the biased sample autocorrelation is the implicit estimator used by the periodogram in PSD estimation.

Appendix B: Eigenvectors and Eigenvalues of Correlation Matrix

The objective here is to show the equivalence between the eigenvectors of a correlation matrix with Toeplitz structure and the basis of the discrete Fourier transform (DFT). A few preliminary remarks are in order. Given a periodic sequence $\{e[n]\}$ with period of N samples, the circular (or periodic) convolution over the period N with a filter $\{h[n]\}$ is equivalent to the linear convolution $e[n] * h[n]$ made periodic with period N

$$x[n] = e[n] \circledast_N h[n] = \sum_k (e[n] * h[n]) * \delta[n - kN]$$

The DFT of the circular convolution is

$$x[n] = e[n] \circledast_N h[n] \leftrightarrow X[k] = H[k] \cdot E[k]$$

where

$$E[k] = \sum_{n=0}^{N-1} e[n] W_N^{-nk}$$

is the DFT of the sequence $\{e[n]\}_{n=0}^{N-1}$ (recall that $W_N = \exp(j2\pi/N)$ is the Nth root of unitary value). In compact notation, the DFT is

$$\underbrace{\begin{bmatrix} E[0] \\ E[1] \\ \vdots \\ E[N-1] \end{bmatrix}}_{\mathbf{E}} = \underbrace{\begin{bmatrix} 1 & 1 & \cdots & 1 \\ 1 & W_N^{-1} & \cdots & W_N^{-(N-1)} \\ \vdots & \vdots & \ddots & \vdots \\ 1 & W_N^{-1\cdot(N-1)} & \cdots & W_N^{-(N-1)(N-1)} \end{bmatrix}}_{\mathbf{W}_N} \cdot \underbrace{\begin{bmatrix} e[0] \\ e[1] \\ \vdots \\ e[N-1] \end{bmatrix}}_{\mathbf{e}}$$

where $\mathbf{W}_N$ is the matrix of forward DFT (it has a Vandermonde structure). Since the inverse DFT is

$$\mathbf{e} = \frac{1}{N}\mathbf{W}_N^H \mathbf{E}$$

it follows that

$$\mathbf{W}_N^H \mathbf{W}_N = N\mathbf{I}_N$$

Here, the columns of $\mathbf{W}_N$ are a complete basis $\mathbb{C}^N$ and each column represents a complex exponential $\exp(j2\pi nk/N)$ of the DFT transformation. The circular convolution after DFT is

$$\mathbf{X} = \mathbf{H} \cdot \mathbf{E} = \mathbf{W}_N \mathbf{x}$$

where

$$\mathbf{H} = \text{diag}\{H[0], H[1], ..., H[N-1]\}$$

The properties of an arbitrary covariance matrix $\mathbf{C}_{xx}$ are derived by assuming that random process $\mathbf{x}$ is the output of a linear system with Gaussian white input

$$\mathbf{e} \sim \mathcal{N}(0, \sigma_e^2 \mathbf{I}_N)$$

so that the following property holds for its DFT:

$$\mathbf{E} \sim \mathcal{CN}(0, N\sigma_e^2 \mathbf{I}_N)$$

Since $\mathbf{X} = \mathbf{H} \cdot \mathbf{E}$, the covariance matrix is

$$\begin{aligned} \mathbf{C}_{XX} &= \mathbb{E}[\mathbf{X}\mathbf{X}^H] = \mathbf{H} \cdot \mathbb{E}[\mathbf{E}\mathbf{E}^H] \cdot \mathbf{H}^H = N\sigma_e^2 \mathbf{H}\mathbf{H}^H \\ &= \text{diag}\{N\sigma_e^2 |H[0]|^2, N\sigma_e^2 |H[1]|^2, \ldots, N\sigma_e^2 |H[N-1]|^2\} \end{aligned}$$

but still from the definitions of covariance matrix and DFT:

$$\mathbf{C}_{XX} = \mathbb{E}[\mathbf{X}\mathbf{X}^H] = \mathbf{W}_N \cdot \underbrace{\mathbb{E}[\mathbf{x}\mathbf{x}^H]}_{\mathbf{C}_{xx}} \cdot \mathbf{W}_N^H$$

leading to the following equality:

$$\mathbf{W}_N \cdot \mathbf{C}_{xx} \cdot \mathbf{W}_N^H = \text{diag}\{N\sigma_e^2|H[0]|^2, N\sigma_e^2|H[1]|^2, ..., N\sigma_e^2|H[N-1]|^2\}$$

Each column of the DFT basis is one eigenvector of the covariance matrix $\mathbf{C}_{xx}$ obtained after circular convolution, and the entry $N\sigma_e^2|H[k]|^2$ is the corresponding eigenvalue. Resuming the definition of eigenvectors/eigenvalues with an orthonormal basis:

$$\mathbf{V} \cdot \mathbf{C}_{xx} \cdot \mathbf{V}^H = \text{diag}\{N\sigma_e^2|H[0]|^2, N\sigma_e^2|H[1]|^2, ..., N\sigma_e^2|H[N-1]|^2\}$$

where the kth eigenvector and eigenvalue are

$$\mathbf{v}_k = \frac{1}{\sqrt{N}}\left[1, \exp\left(j\frac{2\pi k}{N}\right), ..., \exp\left(j\frac{2\pi k(N-1)}{N}\right)\right]^T$$

$$\lambda_k(\mathbf{C}_{xx}) = N\sigma_e^2|H[k]|^2$$

notice that the kth eigenvalue coincides with the power spectral density

$$\lambda_k(\mathbf{C}_{xx}) = S_x(\frac{2\pi}{N}k)$$

Based on these numerical equalities, some remarks are now necessary to better focus on the importance of the results derived so far.

Remark 1 : For a stationary random process, the eigenvector decomposition of the covariance matrix $\mathbf{C}_{yy}$ is

$$\mathbf{Q} \cdot \mathbf{C}_{xx} \cdot \mathbf{Q}^H = \mathbf{\Lambda}$$

In general, the kth eigenvector $\mathbf{q}_k \neq \mathbf{v}_k$, and similarly $\lambda_k(\mathbf{C}_{xx}) \neq S_x(\frac{2\pi}{N}k)$ as the process $x[n]$ is not periodic, or it was not generated as circular convolution over a period N. However, for a very large period N, the periodic and linear convolution coincides, except for edge effects that become negligible for large N. The same conclusion holds for the covariance if $N \to \infty$ (in practice, it is enough that N is larger than the coherence interval of the process $x[n]$, or equivalently the covariance matrix $\mathbf{C}_{xx}$ is a diagonal banded matrix with a band small enough compared to the size of the covariance matrix itself):

$$\mathbf{q}_k \underset{N\to\infty}{\simeq} \mathbf{v}_k$$

$$\lambda_k(\mathbf{C}_{xx}) \underset{N\to\infty}{\simeq} S_x(\frac{2\pi}{N}k)$$

Remark 2: In software programs for numerical analysis, the eigenvectors of the covariance matrix are ordered for increasing (or decreasing) eigenvalues while power spectral density $S_x(\frac{2\pi}{N}k)$ is ordered wrt increasing frequency as the eigenvectors are predefined as the basis of the Fourier transform.

Remark 3: In the case of a complex-valued process, it is convenient to partition the DFT into real and imaginary components:

$$\mathbf{W}_N = \underbrace{\left[\mathbf{c}(0), \mathbf{c}(\frac{2\pi}{N}), ..., \mathbf{c}(\frac{2\pi(N-1)}{N})\right]}_{\mathbf{W}_{N(R)}} + j\underbrace{\left[\mathbf{s}(0), \mathbf{s}(\frac{2\pi}{N}), ..., \mathbf{s}(\frac{2\pi(N-1)}{N})\right]}_{\mathbf{W}_{N(I)}}$$

where each of the vectors are (Section 5.4)

$$\mathbf{c}(\omega) = \begin{bmatrix} 1 \\ \cos(\omega) \\ \vdots \\ \cos((N-1)\omega) \end{bmatrix}; \quad \mathbf{s}(\omega) = \begin{bmatrix} 0 \\ \sin(\omega) \\ \vdots \\ \sin((N-1)\omega) \end{bmatrix}$$

and the following properties of orthogonality hold:

$$\begin{aligned}
&\mathbf{c}(\frac{2\pi i}{N})^T \mathbf{c}(\frac{2\pi j}{N}) = \frac{N}{2}\delta_{i-j} \quad \text{for } i \neq 0, N/2 \\
&\mathbf{c}(\frac{2\pi i}{N})^T \mathbf{c}(\frac{2\pi i}{N}) = N \quad \text{for } i = 0, N/2 \\
&\mathbf{s}(\frac{2\pi i}{N})^T \mathbf{s}(\frac{2\pi j}{N}) = \frac{N}{2}\delta_{i-j} \quad \text{for } i \neq 0, N/2 \\
&\mathbf{s}(\frac{2\pi i}{N})^T \mathbf{s}(\frac{2\pi i}{N}) = 0 \quad \text{for } i = 0, N/2 \\
&\mathbf{c}(\frac{2\pi i}{N})^T \mathbf{s}(\frac{2\pi j}{N}) = 0
\end{aligned}$$

In summary

$$\begin{aligned}
&\mathbf{W}_{N(R)}^T \mathbf{W}_{N(R)} = \frac{N}{2}\text{diag}\{2, \underbrace{1, 1, .., 1}_{N/2-1}, 2, \underbrace{1, 1, .., 1}_{N/2-1}\} \\
&\mathbf{W}_{N(R)}^T \mathbf{W}_{N(I)} = \mathbf{0} \\
&\mathbf{W}_{N(I)}^T \mathbf{W}_{N(I)} = \frac{N}{2}\text{diag}\{0, \underbrace{1, 1, .., 1}_{N/2-1}, 0, \underbrace{1, 1, .., 1}_{N/2-1}\}
\end{aligned}$$

and thus

$$\mathbf{W}_N^H \mathbf{W}_N = (\mathbf{W}_{N(R)}^T - j\mathbf{W}_{N(I)}^T)(\mathbf{W}_{N(R)} + j\mathbf{W}_{N(I)}) = N\mathbf{I}_N$$

as expected.

Appendix C: Property of Monic Polynomial

This Appendix extends some properties of the z-transform in Appendix B of Chapter 4. Let $H(z)$ be the monic polynomial of a minimum phase filter ($h_o = 1$) excited by a white

Gaussian noise with unitary power ($\sigma^2 = 1$); the z-transform of the autocorrelation sequence is

$$R(z) = H(z)H^*(1/z)$$

Since along the unitary circle the Fourier transforms $R(\omega) = R(z = e^{j\omega})$ and $|H(\omega)|^2 = H(z = e^{j\omega})H^*(z = e^{-j\omega})$ are positive semidefinite, after the log as monotonic transformation:

$$\int_{-\pi}^{\pi} \ln R(\omega)d\omega = \int_{-\pi}^{\pi} \ln |H(\omega)|^2 d\omega$$

The careful inspection of

$$\ln H(z)H^*(1/z) = \ln H(z) + \ln H^*(1/z) = C(z) + C^*(1/z)$$

proves that since $H(z)$ is minimum phase, it is analytic for $|z| \geq 1$ (poles/zeros are for $|z| < 1$), and from the properties of the log it follows that poles/zeros are transformed as follows: zeros of $H(z) \to$ poles of $\ln H(z)$; poles of $H(z) \to$ poles of $\ln H(z)$. Since $H(z)$ is minimum phase, the polynomial $\ln H(z) = C(z) = \sum_{k=0}^{\infty} c_k z^{-k}$ is minimum phase also. From the initial value theorem $H(z \to \infty) = h_o = 1$ (monic polynomial) and thus $c_o = \ln H(z \to \infty) = 0$, hence[6]

$$\int_{-\pi}^{\pi} C(\omega)d\omega + \int_{-\pi}^{\pi} C^*(\omega)d\omega = 2c_o = 0$$

and thus

$$\int_{-\pi}^{\pi} \ln R(\omega)d\omega = 0$$

Appendix D: Variance of Pole in AR(1)

The biased estimate of the autocorrelation is (sample index is by subscript):

$$\hat{r}_n = |A|^2 e^{j\omega n}\left(1 - \frac{|n|}{N}\right) + Ae^{j\omega(k+n)}\frac{1}{N}\sum_{k=1}^{N} w_k^* + A^* e^{-j\omega n}\frac{1}{N}\sum_{k=1}^{N} w_{n+k} + \frac{1}{N}\sum_{k=1}^{N} w_k^* w_{k+n}$$

---

6 Alternatively, this follows from complex function theory and the Cauchy integral theorem

$$\frac{1}{2\pi j}\oint z^{-k}dz = \begin{cases} 1 \text{ for } k = 1 \\ 0 \text{ for } \forall k \neq 1 \end{cases}$$

when applied to

$$\oint \ln H(z)H^*(1/z)dz = \sum_{k=0}^{\infty} \oint (c_k z^{-k} + c_k^* z^k)dz = 0$$

and the mean value is

$$\mathbb{E}[\hat{r}_n] = |A|^2 \exp(j\omega n)\left(1-\frac{|n|}{N}\right)+\sigma^2\delta_n$$

To compute the variance it is necessary to evaluate the expectation for all mixed terms $\mathbb{E}[\hat{r}_n\hat{r}_m^*]$. After some algebra:

$$\begin{aligned}\hat{r}_n\hat{r}_m^* = &\left[|A|^2 e^{j\omega n}\left(1-\frac{|n|}{N}\right)+Ae^{j\omega n}\frac{1}{N}\sum_{k=1}^{N}e^{j\omega k}w_k^*\right.\\
&\left.+A^*e^{-j\omega n}\frac{1}{N}\sum_{k=1}^{N}w_{n+k}+\frac{1}{N}\sum_{k=1}^{N}w_k^*w_{k+n}\right]\\
&\times\left[|A|^2 e^{-j\omega m}\left(1-\frac{|m|}{N}\right)+A^*e^{-j\omega m}\frac{1}{N}\sum_{h=1}^{N}e^{-j\omega h}w_h\right.\\
&\left.+Ae^{j\omega m}\frac{1}{N}\sum_{h=1}^{N}w_{m+h}^*+\frac{1}{N}\sum_{h=1}^{N}w_hw_{h+m}^*\right]\end{aligned}$$

and the expectations are

$$\begin{aligned}\mathbb{E}[\hat{r}_n\hat{r}_m^*] = &|A|^4 e^{j\omega(n-m)}\left(1-\frac{|n|}{N}\right)\left(1-\frac{|m|}{N}\right)\\
&+\frac{1}{N}|A|^2\sigma^2\left(2-\frac{|n-m|}{N}\right)e^{j\omega(n-m)}\\
&+|A|^2\sigma^2\left[e^{j\omega n}\left(1-\frac{|n|}{N}\right)\delta_m+e^{-j\omega m}\left(1-\frac{|m|}{N}\right)\delta_n\right]\\
&+\frac{1}{N^2}\sum_{k,l}\mathbb{E}[w_k^*w_{k+n}w_lw_{l+m}^*].\end{aligned}$$

Under the assumption that $\gamma=\sigma^2/|A|^2 \ll 1$ and $1/N \ll 1$, one can specify the following moments:

$$\begin{aligned}\mathbb{E}[|\hat{r}_0|^2] &= |A|^4+(2+\frac{2}{N})|A|^2\sigma^2+\frac{1}{N^2}\sum_{k,l}\mathbb{E}[|w_k|^2|w_l|^2]=\\
&=|A|^4+(2+\frac{2}{N})|A|^2\sigma^2+\frac{(N+1)\sigma^4}{N}\\
&\simeq |A|^4+2|A|^2\sigma^2+\sigma^4\simeq|A|^4(1+2\gamma)\\
\mathbb{E}[|\hat{r}_1|^2] &= |A|^4\left(1-\frac{1}{N}\right)^2+\frac{2}{N}|A|^2\sigma^2+\frac{1}{N^2}\sum_{k,l}\mathbb{E}[w_k^*w_{k+1}w_lw_{l+1}^*]\\
&=|A|^4\left(1-\frac{1}{N}\right)^2+\frac{2}{N}|A|^2\sigma^2+\frac{N-1}{N^2}\mathbb{E}[|w_k|^2|w_l|^2]\\
&\simeq |A|^4\left(1-\frac{1}{N}\right)^2+\frac{2}{N}|A|^2\sigma^2+\frac{1}{N}\sigma^4\simeq|A|^4\left(1-\frac{2}{N}\right)\end{aligned}$$

$$\begin{aligned}\mathbb{E}[\hat{r}_0\hat{r}_1^*] &= |A|^4 e^{-j\omega}\left(1-\frac{1}{N}\right)+\frac{1}{N}|A|^2\sigma^2\left(2-\frac{1}{N}\right)e^{-j\omega}\\ &\quad+|A|^2\sigma^2 e^{-j\omega}\left(1-\frac{1}{N}\right)+\frac{1}{N^2}\sum_{k,l}\mathbb{E}[|w_k|^2 w_l w_{l+1}^*]\\ &=|A|^2\left\{\left(1-\frac{1}{N}\right)(|A|^2+\sigma^2)\right.\\ &\quad\left.+\frac{1}{N}\sigma^2\left(2-\frac{1}{N}\right)\right\}e^{-j\omega}\simeq|A|^4\left(1-\frac{1}{N}+\gamma\right)e^{-j\omega}\\ \mathbb{E}[\hat{r}_1\hat{r}_0^*] &= \left(\mathbb{E}[\hat{r}_0\hat{r}_1^*]\right)^*\end{aligned}$$

Recalling that the random term in the sample estimate is redefined as deviation wrt the mean value $\hat{r}_k = \bar{r}_k + \delta r_k$, it follows that:

$$\mathbb{E}[|\delta r_0|^2] = \mathbb{E}[|\hat{r}_0|^2] - |\bar{r}_0|^2 = \frac{\sigma^2}{N}(2|A|^2+\sigma^2)$$

$$\mathbb{E}[|\delta r_1|^2] = \mathbb{E}[|\hat{r}_1|^2] - |\bar{r}_1|^2 = \frac{2}{N}|A|^2\sigma^2 + \frac{N-1}{N^2}\sigma^4 = \frac{\sigma^2}{N}(2|A|^2+\frac{N-1}{N}\sigma^2)$$

$$\mathbb{E}[\delta r_0\cdot\delta r_1^*] = \mathbb{E}[(\hat{r}_0-\bar{r}_0)\cdot(\hat{r}_1^*-\bar{r}_1^*)] = \mathbb{E}[\hat{r}_0\hat{r}_1^*] - \bar{r}_0\bar{r}_1^* = \frac{|A|^2\sigma^2}{N}\left(2-\frac{1}{N}\right)e^{-j\omega}$$

these are useful for the derivation in the main text.

15

Adaptive Filtering

The subject of this chapter is parameter estimation in non-stationary conditions such as when the process at hand is non-stationary, or the generating mechanism changes vs. time. The context is quite general and involves a wide range of applications where stationarity is not at all guaranteed. In non-stationary situations the estimation has to be adaptive to track the varying parameters at best without any (sometimes not even partial) knowledge of the varying dynamics that would preclude the use of Bayesian tracking (e.g., Kalman filter).

The adaptive filtering is represented by a linear estimator that adapts to the context (possibly time-varying) by minimizing the mean square error (MSE) as objective function. The MSE optimization is carried out iteratively while observations are gathered, and preference criteria among different estimators are convergence speed, accuracy, and cost in terms of operation per iteration (or per sample). Convergence speed indicates the capability of the adaptive filters to cope with the non-stationary situation by tracking the variations of the parameters.

Applications of adaptive filtering are so broad that a summary is almost impossible, and we refer to the textbook by Widrow and Stearns [59] for an overview. However, a couple of representative examples will be used to introduce the topic and made reference to during the discussion of the theoretical foundations of adaptive filtering.

Adaptive filtering solves iteratively the minimization of the MSE, but a discussion of iterative methods for MSE optimization in stationary processes is given before the discussion of adaptive filtering. Performance analysis and convergence conditions in stationary environments are also provided while the convergence speed represents the only metric that addresses the tracking capability of adaptive filtering methods (i.e., fast convergence denotes the capability of the adaptive filtering to track rapid time-varying processes).

15.1 Adaptive Interference Cancellation

In many situations, the interference $i[n]$ is superimposed on the signal of interest $s[n]$ as illustrated herein. In interference cancellation, a redundant observation of the interference $x[n] = h[n] * i[n]$ is made available so that a filter $a[n]$ can be designed to cancel the interference from the noisy signal, and leave the signal of interest, possibly with some

Statistical Signal Processing in Engineering, First Edition. Umberto Spagnolini.

Companion website: www.wiley.com/go/spagnolini/signalprocessing

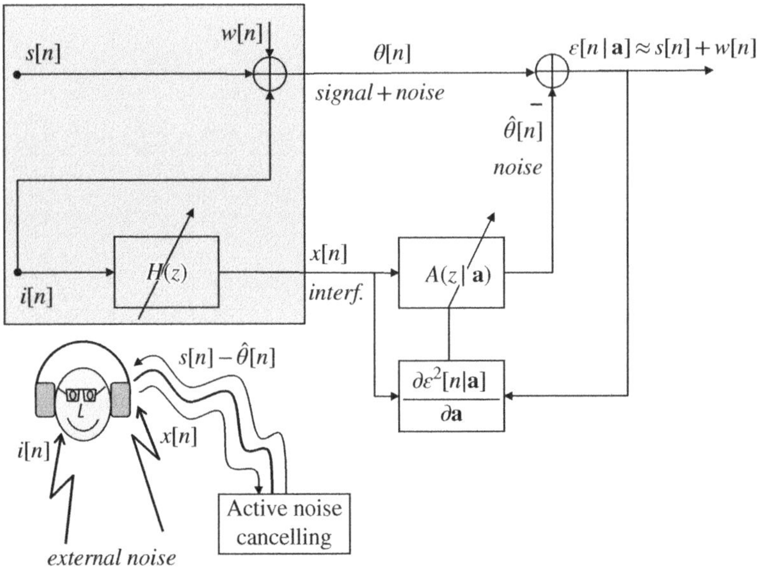

Figure 15.1 Adaptive noise cancelling.

additional noise: $s[n] + w[n]$. Since the redundant measurement of interference could be filtered by a time-varying filter $h[n]$, the filter $a[n]$ should be adaptive in order to comply with these variations. This is the principle of adaptive interference cancellation (or *active noise canceling*). There are many applications where active noise cancellation is employed, such as to reduce the noise in helicopters to allow pilots to communicate (in this case there is a second microphone to measure the interference $i[n]$), or to cancel the maternal heartbeat in fetal electrocardiography, or to cancel external noise in active headsets (see Figure 15.1), or to cancel the echo in satellite communication systems, just to mention a few.

Making reference to the figure, the signal $\theta[n]$ plays the role of the desired signal and the signal $x[n]$ is filtered by the filter $a[n]$ to provide the estimate $\hat{\theta}[n]$ of the desired signal $\theta[n]$. Let

$$\varepsilon[n|\mathbf{a}] = \theta[n] - x[n] * a[n] = \theta[n] - \mathbf{a}^T \cdot \mathbf{x}[n]$$

be the error, the filter coefficients $\mathbf{a}$ are obtained by the minimization of the mean square error (MSE)

$$J(\mathbf{a}) = \mathbb{E}[\varepsilon^2[n|\mathbf{a}]] = \mathbb{E}[(\theta[n] - \mathbf{a}^T \cdot \mathbf{x}[n])^2]$$

Even if the filter is time-varying $\mathbf{a}[n]$ the processes can be considered stationary (at least within a limited time interval) and the MSE is quadratic wrt $\mathbf{a}$:

$$J(\mathbf{a}) = \mathbb{E}[|\theta[n]|^2] - 2\mathbf{p}^T\mathbf{a} + \mathbf{a}^T\mathbf{R}\mathbf{a}$$

where

$$\mathbf{p} = \mathbb{E}[\mathbf{x}[n]\theta[n]]$$
$$\mathbf{R} = \mathbb{E}[\mathbf{x}[n]\mathbf{x}^T[n]]$$

indicates, in order, the crosscorelation between the set of observations used in filtering $\mathbf{x}[n]$ and the signal $\theta[n]$, and the autocorrelation matrix of the observations $\mathbf{x}[n]$—both are known (or estimated separately or updated iteratively as clarified below). The MSE is minimized for the choice (*Wiener–Hopf equations*)

$$\mathbf{R}\hat{\mathbf{a}}_{opt} = \mathbf{p}$$

or in explicit form if $\mathbf{R}$ is positive definite

$$\hat{\mathbf{a}}_{opt} = \mathbf{R}^{-1}\mathbf{p}$$

This is the filter that provides the interference cancellation with minimum MSE. Optimization of the MSE objective function $J(\mathbf{a})$ can be carried out iteratively to avoid direct matrix inversion, and to adapt to changes of the correlation as a consequence of changing the interference-coupling filter $h[n] \leftrightarrow H(z)$ (e.g., in an active noise canceling headset, this is due to the movement of the head, which changes the coupling with the source of external interference).

15.2 Adaptive Equalization in Communication Systems

Modern communication systems guarantee mobility to users, and this makes the propagation environment rapidly time-varying, with significant degradation of the quality of connectivity, and consequently of the service experienced by users. This is one of the most relevant fields for adaptive filtering as the communication channel is modeled as a filter, and compensation of this impairment is by adaptive equalization. To gain insight in this relevant topic, some preliminary introduction to the problem at hand is mandatory (see e.g., [57, 58] out of many).

15.2.1 Wireless Communication Systems in Brief

The communication link in Figure 15.2 contains the cascade of a filter at the transmitter *(shaping filter)* $g^{(Tx)}(t)$, the multipath propagation channel $h^{(ch)}(t)$, and a filter at the receiver $g^{(Rx)}(t)$. The overall equivalent channel is represented by a filter

$$h(t) = g^{(Tx)}(t) * h^{(ch)}(t) * g^{(Rx)}(t)$$

The response $\varphi(t) = g^{(Tx)}(t) * g^{(Rx)}(t)$ is the cascade of two filters that are independent of the multipath channel, and their design is a degree of freedom for the

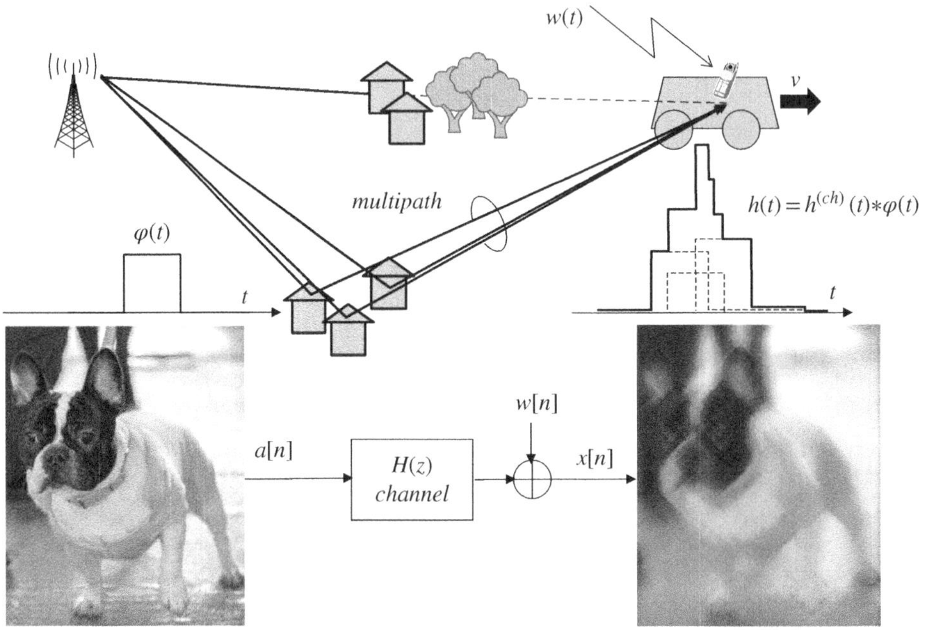

Figure 15.2 Multipath communication channel.

telecommunication engineer.[1] The source-destination filter $h(t) = h^{(ch)}(t) * \varphi(t)$ couples the propagation-dependent term ($h^{(ch)}(t)$) with the true filtering of the communication system ($\varphi(t)$).

The multipath channel is modeled as a combination of multiple (say $P \gg 1$) raypaths, each characterized by a delay τ_p (that depends on the length of the raypath) and an amplitude r_p (that depends on the reflectivity of the environment):

$$h^{(ch)}(t) = \sum_{p=1}^{P} r_p \delta(t - \tau_p) \quad \text{with } 0 \leq \tau_1 \leq \ldots \leq \tau_P$$

Absolute delay is irrelevant: what matters is the relative delays of the multipaths, and these are ordered within $[0, T_{\max}]$: $\tau_p \in [0, T_{\max}]$. The overall *communication channel* is the superposition of delayed and amplitude scaled signatures $\varphi(t)$

$$h(t) = \sum_{p=1}^{P} r_p \varphi(t - \tau_p)$$

as illustrated in the figure for $\varphi(t) = rect(t/T)$.

1 The most employed choice is $g^{(Rx)}(t) = g^{(Tx)}(-t)$ (matched filter) and $g^{(Tx)}(t)$ is designed to guarantee the *Nyquist criterion* that avoids inter-symbol interference (ISI):

$$\varphi(t)|_{t=kT} = \delta[k]$$

for the symbol-rate period T. Raised cosine waveforms for the cascade $\varphi(t) = g^{(Rx)}(t) * g^{(Tx)}(t)$ are the widely adopted solution in many contexts [58].

At the transmitter, each symbol of the set $\{a[n]\}$ encloses the *information bits* after mapping (say B bits/symbol, $B \geq 1$).[2] Symbols are generated at a signaling rate $1/T$ measured in symbols/sec (or $B/T = R_b$ bits/sec, often referred to as the data-rate); the signal received (apart from a delay) after filtering by $g^{(Rx)}(t)$ is

$$x(t) = \sum_k a[k]h(t - kT) + v(t) * g^{(Rx)}(t)$$

where $v(t)$ is the noise at the receiver (e.g., at the antenna). The receiver samples the received signal $x(t)$ at the same rate as the transmitted symbols with the purpose of estimating (*decoding*) the transmitted symbols $a[k]$:

$$x[n] = x(nT) = \sum_k a[k]h((n-k)T) + w[n] = a[n] * h[n] + w[n]$$

where $h[n] = h(nT)$ is the discrete-time communication channel response and $w[n]$ is white Gaussian noise.[3] Estimation of the transmitted symbols needs knowledge of the channel response, which is carried out as a separate estimation step as discussed below.

Remark: The length of communication channel $h[n]$ depends on $\varphi(t)$ evaluated in terms of symbol intervals T and the delay of multipath $[0, T_{\max}]$. Even if the span of $\varphi(t)$ could be very large, it is common practice to limit its value by assuming that $\varphi(t) = 0$ for $|t| > N_\varphi T$ (range within $\pm N_\varphi$ signaling intervals); it follows that $h(t) = 0$ for $t \notin [-N_\varphi T, N_\varphi T + T_{\max} + T)$ or alternatively the time span of $h[n]$ is

$$N_h = 2N_\varphi + 1 + \left\lceil \frac{T_{\max}}{T} \right\rceil$$

It is convenient to ease the notation by assuming that the communication channel $h[n]$ is causal, and thus $h[n] = 0$ for $n \notin [0, N_h - 1]$.

15.2.2 Adaptive Equalization

In modern communication systems, signals are arranged in blocks (*frames*) that can be either sequential or discontinuous (*packets*). Each block includes a set of signals: the *training* signal $a_T[n]$ is known to transmitter and receiver(s) as part of the communication setup, and the unknown data-symbols $a_D[n]$ that deliver the information bits to the destination. The communication models for these two sub-blocks follow the same models for identification and equalization problems (Section 5.2.3); in sequence:

$$\mathbf{x}_T = \mathbf{h} * \mathbf{a}_T + \mathbf{w}_T = \mathbf{A}_T\mathbf{h} + \mathbf{w}_T \Rightarrow \text{(identification)} \Rightarrow \hat{\mathbf{h}}_T$$

2 Mapping is a step carried out in any digital communication system that maps every set of information bits onto a waveform. The bits are grouped into B bits and then this group of B bits is mapped into one out of 2^B waveforms (or symbols). Since the duration of each bit is T_b (bit period), each symbol is $T = BT_B$ (symbol period). Compacting bits into symbols has several benefits, namely the bandwidth is reduced by $1/B$ and over the same spectrum multiple communications can be allocated without mutually interfering as in frequency division multiplexing (FDM).

3 If $v(t)$ is uncorrelated (at least within the bandwidth $1/T$) with power spectral density $N_0/2$, the autocorrelation sequence is $\mathbb{E}[w^*[n+k]w[n]] = (N_0/2)\varphi(kT) = (N_0/2)\delta[k]$, where the last equality follows from the Nyquist criterion for $\varphi(t)$.

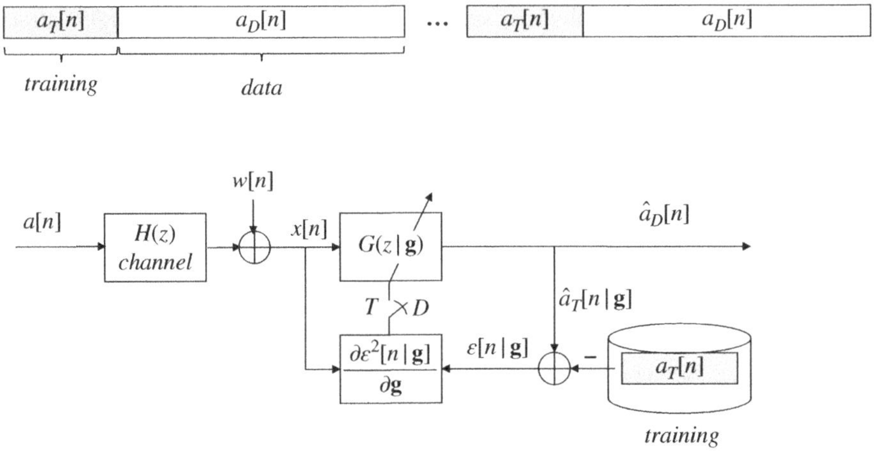

Figure 15.3 Adaptive identification and equalization in packet communication systems.

$$\mathbf{x}_D \quad = \quad \mathbf{h} * \mathbf{a}_D + \mathbf{w}_D \simeq \hat{\mathbf{H}}_T \mathbf{a}_D + \mathbf{w}_D \Rightarrow \text{(lin.equalization)} \Rightarrow \hat{\mathbf{a}}_D(\hat{\mathbf{h}}_T) = \mathbf{g} * \mathbf{x}_D$$

where training symbols $\mathbf{a}_T$ are used for identification, and data-symbols are estimated $\hat{\mathbf{a}}_D(\hat{\mathbf{h}}_T)$ by using the channel response estimated from training period $\hat{\mathbf{h}}_T$. The receiver alternates channel estimation and equalization (or more generally decoding) on every frame, possibly adaptively adjusting to the varying context as in Figure 15.3. These steps are common to many communication systems (e.g., cellular communication systems such as UMTS, LTE, and 5G); differences are in the frame structure, signals, and equalization methods.

Mobility enables a data service everywhere, but at the same time communication channel response $h[n]$ changes vs. time (e.g., due to the changes of delays $\{\tau_p\}$ and amplitudes $\{r_p\}$) and thus the linear equalization $\mathbf{g}$ should be adapted accordingly. Rather than estimating the communication channel, one can directly estimate the equalization filter $\mathbf{g}$ during the training section such that $\mathbf{g} * \mathbf{h} \simeq \mathbf{I}$. The adaptive equalization can be cast as follows: during the training sub-block, the signal $a_T[n]$ stored locally at the receiver plays the role of the desired signal, and the received signal $x_T[n]$ filtered by the filter $g[n]$ provides the estimate $\hat{a}_T[n] = x_T[n] * g[n]$ with error

$$\varepsilon[n|\mathbf{g}] = a_T[n] - x_T[n] * g[n] = a_T[n] - \mathbf{g}^H \cdot \mathbf{x}_T[n]$$

The minimization of the MSE wrt $\mathbf{g}$

$$J(\mathbf{g}) = \mathbb{E}[\varepsilon^2[n|\mathbf{g}]] = \mathbb{E}[(a_T[n] - \mathbf{g}^H \cdot \mathbf{x}_T[n])^2]$$

can be carried out by considering the training $\{a_T[n]\}$ as stationary process so that the MSE is quadratic wrt $\mathbf{g}$:

$$J(\mathbf{g}) = \mathbb{E}[|a_T[n]|^2] - \mathbf{p}^H \mathbf{g} - \mathbf{g}^H \mathbf{p} + \mathbf{g}^H \mathbf{R} \mathbf{g}$$

where

$$\mathbf{p} = \mathbb{E}[\mathbf{x}_T[n]a_T^*[n]]$$
$$\mathbf{R} = \mathbb{E}[\mathbf{x}_T[n]\mathbf{x}_T^H[n]]$$

Alternatively, if the limited training sequence cannot be considered as the realization of a random process, the auto/cross-correlations can be replaced by the sample values.

Adaptation can be carried out on a block-by-block basis (by locally minimizing the MSE during the training sub-block, and disabling afterward, as sketched by the switch in Figure 15.3) or while data is being received.

15.3 Steepest Descent MSE Minimization

Making reference to the system in Figure 15.4, the adaptive filtering is based on the iterative minimization of the quadratic MSE

$$J(\mathbf{a}) = \mathbb{E}[(\theta[n] - \mathbf{a}^T \cdot \mathbf{x}[n])^2] = \mathbb{E}[|\theta[n]|^2] - 2\mathbf{p}^T\mathbf{a} + \mathbf{a}^T\mathbf{R}\mathbf{a}$$

where $\mathbf{p} = \mathbb{E}[\mathbf{x}[n]\theta[n]]$ and $\mathbf{R} = \mathbb{E}[\mathbf{x}[n]\mathbf{x}[n]^T]$, so that the objective function is optimized iteratively at every step

$$J(\hat{\mathbf{a}}_{k+1}) \leq J(\hat{\mathbf{a}}_k)$$

such that

$$\hat{\mathbf{a}}_k \underset{k\to\infty}{\to} \hat{\mathbf{a}}_{opt} = \mathbf{R}^{-1}\mathbf{p} \tag{15.1}$$

The gradient method updates the estimator pointing in the (opposite direction of the) local gradient

$$\hat{\mathbf{a}}_{k+1} = \hat{\mathbf{a}}_k - \frac{\mu}{2} \cdot \left.\frac{\partial J(\mathbf{a})}{\partial \mathbf{a}}\right|_{\mathbf{a}=\hat{\mathbf{a}}_k} \tag{15.2}$$

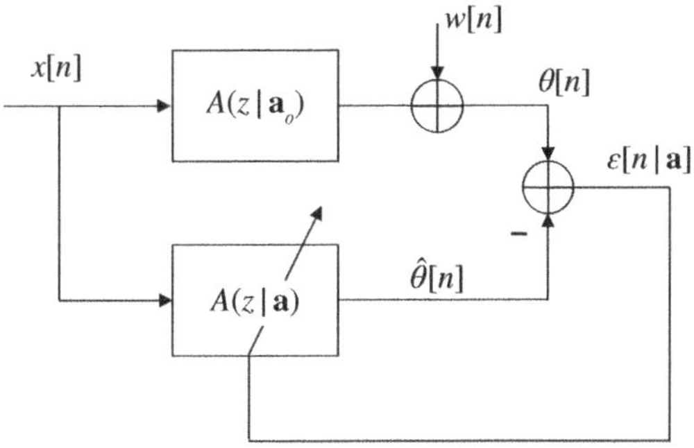

Figure 15.4 Adaptive filter identification.

where the step-size μ is chosen according to convergence criteria (scaling factor 1/2 is only for analytical convenience). The gradient can be computed analytically as

$$\left.\frac{\partial J(\mathbf{a})}{\partial \mathbf{a}}\right|_{\mathbf{a}=\hat{\mathbf{a}}_k} = -2\mathbf{p} + 2\mathbf{R}\hat{\mathbf{a}}_k \tag{15.3}$$

and the update equation (15.2) is

$$\hat{\mathbf{a}}_{k+1} = \hat{\mathbf{a}}_k + \mu \cdot \left(\mathbf{p} - \mathbf{R}\hat{\mathbf{a}}_k\right) \tag{15.4}$$

The significant advantage of this iterative method is that the minimum MSE solution (15.1) can be obtained without the inverse of $\mathbf{R}$ even if the convergence $\hat{\mathbf{a}}_k \to \hat{\mathbf{a}}_{opt}$ guarantees that the gradient method embeds the computation of the inverse by an iterative search.

15.3.1 Convergence Analysis and Step-Size μ

The convergence depends on the step-size μ. The MSE $J(\mathbf{a})$ is a quadratic form and for the analysis of the convergence it is convenient to refer the quadratic form wrt the principal axis as discussed in Section 1.6 and adapted to the problem at hand. The iterations can be referred wrt the solution $\hat{\mathbf{a}}_{opt}$ by defining the variable $\delta\hat{\mathbf{a}}_k = \hat{\mathbf{a}}_k - \hat{\mathbf{a}}_{opt}$ (translate wrt the solution), and then it can be referred to the principal axes from the eigenvector decomposition of the (positive semidefinite) correlation matrix $\mathbf{R}$. The updating relationship is

$$\delta\hat{\mathbf{a}}_{k+1} = \left(\mathbf{I} - \mu \cdot \underbrace{\mathbf{R}}_{\mathbf{U}\boldsymbol{\Lambda}\mathbf{U}^T}\right)\delta\hat{\mathbf{a}}_k$$

$$\underbrace{\mathbf{U}^T\delta\hat{\mathbf{a}}_{k+1}}_{\hat{\mathbf{z}}_{k+1}} = (\mathbf{I} - \mu\boldsymbol{\Lambda})\,\underbrace{\mathbf{U}^T\delta\hat{\mathbf{a}}_k}_{\hat{\mathbf{z}}_k}$$

which becomes

$$\hat{\mathbf{z}}_{k+1} = (\mathbf{I} - \mu\boldsymbol{\Lambda})\hat{\mathbf{z}}_k \tag{15.5}$$

where $\hat{\mathbf{z}}_k = \mathbf{U}^T\delta\hat{\mathbf{a}}_k = \mathbf{U}^T(\hat{\mathbf{a}}_k - \hat{\mathbf{a}}_{opt})$ is the deviation of the solution wrt the convergence $\hat{\mathbf{a}}_{opt}$ at the kth step; the convergence condition (for $k \to \infty$) is

$$\hat{\mathbf{z}}_k \to \mathbf{0}$$

After diagonalization, the iterative update at every iteration is decoupled into N parallel iterative problems (15.5) coupled together by the choice of the *same* step-size μ. The update of the ℓth term

$$[\hat{\mathbf{z}}_{k+1}]_\ell = (1-\mu\lambda_\ell)[\hat{\mathbf{z}}_k]_\ell = (1-\mu\lambda_\ell)^2[\hat{\mathbf{z}}_{k-1}]_\ell = \ldots = (1-\mu\lambda_\ell)^{k+1}[\hat{\mathbf{z}}_0]_\ell$$

shows that its convergence is dominated by the corresponding eigenvalue (i.e., the curvature of the quadratic form along the ℓth component as detailed in Section 1.6). To guarantee the convergence for all terms constrained to have the same step μ it should be the case that

$$-1 < 1 - \mu\lambda_\ell < 1 \quad \text{for } \forall\ell$$

and this is enough to guarantee the convergence for the component along the largest curvature:

$$0 < \mu < \frac{2}{\lambda_{\max}}$$

The geometric series in every principal direction can be approximated by the exponential (recall that $\exp(-x) \simeq 1 - x$)

$$\exp(-k/\tau_\ell) \simeq (1 - \mu\lambda_\ell)^k$$

so the *time-of-convergence* (measured in iterations) along the ℓth principal direction is

$$\tau_\ell \simeq \frac{1}{\mu\lambda_\ell} \quad \text{[iterations]}$$

Since global convergence implies convergence in every direction, this should be guaranteed for the slowest one:

$$T_{conv} = \max\{\tau_\ell\} = (1/\mu)\max\{1/\lambda_\ell\} = \frac{1/\mu}{\min\{\lambda_k\}} \quad \text{[iterations]}$$

The choice of $\mu < 1/\lambda_{\max}$ is to avoid convergence with oscillations as clarified later, and the time of convergence is

$$T_{conv} \geq \kappa(\mathbf{R}) = \frac{\lambda_{\max}}{\lambda_{\min}}$$

where $\kappa(\mathbf{R})$ is the condition number of the matrix $\mathbf{R}$. To summarize, the convergence of the whole method follows the slowest eigenvalue (or the flat portion of the quadratic MSE) according to the progression law $(1 - 1/T_{conv})^k \geq (1 - 1/\kappa(\mathbf{R}))^k$, and this gives the minimum number of iterations to reach convergence.[4] Alternatively, the choice of the

4 The scaling term $(1 - \mu\lambda_\ell)^k$ in $[\hat{\mathbf{z}}_k]_\ell = (1 - \mu\lambda_\ell)^k[\hat{\mathbf{z}}_0]_\ell$ indicates the downsizing of the initial condition $\hat{\mathbf{z}}_0$ that is tapered to zero for $k \to \infty$. It is quite common to evaluate the *minimum* number of iterations when this scaling term reaches 1% so that the initial condition becomes irrelevant (called 1% convergence); it follows by solving for the iterations

$$\left(1 - \frac{1}{\kappa(\mathbf{R})}\right)^k = 10^{-2} \Rightarrow K_{1\%} = -\frac{2}{\log_{10}\left(1 - \frac{1}{\kappa(\mathbf{R})}\right)}$$

step-size can be framed as a global optimization to choose the minimum value of μ that maximizes $|1-\mu\lambda_\ell|$ subject to $|1-\mu\lambda_\ell|<1$ for convergence; the solution of this min-max problem is the optimum choice [60]

$$\mu = \mu_{opt} = \frac{2}{\lambda_{\max} + \lambda_{\min}}$$

where some values can be fluctuating but converge quickly.

The step-size μ depends on the maximum eigenvalue, and this is not trivial to compute. A practical choice follows from the following inequality (Section 1.1.1):

$$\mathrm{tr}\{\mathbf{R}\} = \sum_{k=1}^{N} \lambda_k \geq \lambda_{\max} \Rightarrow \frac{1}{\mathrm{tr}\{\mathbf{R}\}} \leq \frac{1}{\lambda_{\max}}$$

and the (conservative) choice is

$$\mu < \frac{1}{\mathrm{tr}\{\mathbf{R}\}}$$

The drawback of this choice is the increased number of iterations needed to converge.

15.3.2 An Intuitive View of Convergence Conditions

To gain insight in convergence conditions herein we provide an interpretation of some assumptions that will be developed later. The motivation for placing these remarks here is that the framework of the iterative minimization of the quadratic MSE is simple enough to capture the essence of the iterative optimization methods embedded in any adaptive estimation.

Geometric View of the Convergence Condition

The quadratic form referred to the principal axes

$$J(\mathbf{z}) = J_{\min} + \sum_{k=1}^{N} \lambda_\ell z_\ell^2$$

is the sum of quadratic terms in each direction. Along the ith direction ($z_\ell = 0$ for $\forall \ell \neq i$) it is (see Figure 15.5)

$$J_i(z_i) = J_{\min} + \lambda_i z_i^2$$

The minimization with gradient search is (iterations are now indicated within square brackets)

$$z_i[k+1] = (1-\mu\lambda_i) z_i[k]$$

from which

$$J_i(z_i[n+1]) = J_{\min} + \lambda_i (1-\mu\lambda_i)^2 z_i^2[n]$$

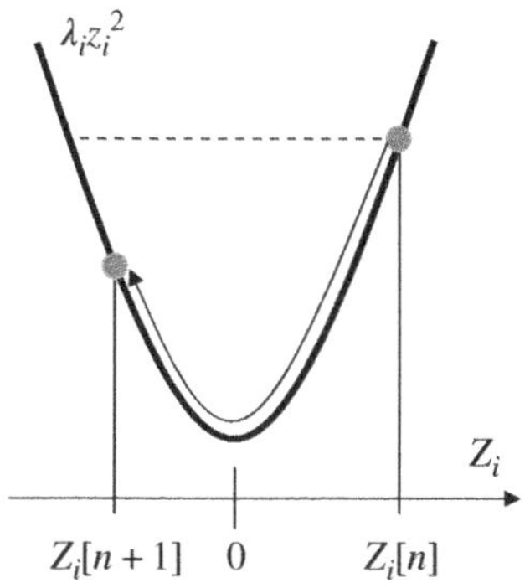

Figure 15.5 Iterations along the principal axes.

Iterative methods should use a step-size μ that avoids increasing the MSE at every step by constraining the step to be always decreasing:

$$\begin{aligned}
&J(z_i[n+1]) \le J(z_i[n]) \\
&\lambda_i(1-\mu\lambda_i)^2 z_i^2[n] \le \lambda_i z_i^2[n] \\
&\mu^2\lambda_i - 2\mu \le 0 \\
&0 < \mu \le \frac{2}{\lambda_i}
\end{aligned}$$

However, the analysis of the optimization along the ith direction shows that for $\mu = 2/\lambda_i$, $J(z_i[n+1]) = J(z_i[n])$ (*the step-size is too large and after one iteration the MSE is the same, except on the other side of a parabolic shape*) while for $\mu = 1/\lambda_i$ the minimum is attained in one step. Even if the relationship for convergence is $\mu < 2/\lambda_i$, there are two intervals for convergence: $0 < \mu \le 1/\lambda_i$ convergence without oscillations (*overdamped*); $1/\lambda_i < \mu < 2/\lambda_i$ convergence with decaying oscillations (*underdamped*). Since one step-size μ fits all, if the choice is $0 < \mu < 1/\lambda_{\max}$, the convergence is overdamped, otherwise some components could oscillate when reaching the convergence. The choice $\mu \simeq 1/\lambda_{\max}$ yields the time of convergence

$$T_{conv} \simeq \frac{\lambda_{\max}}{\lambda_{\min}} \text{ [iterations]}$$

This is the most common assumption if the correlation matrixes $\mathbf{R}, \mathbf{p}$ in (15.1) are known, or the optimization method is iterative and not adaptive.

Convergence and Power Spectral Density

The convergence speed depends on the eigenvalue spread $\lambda_{\max}/\lambda_{\min}$ of $\mathbf{R}$, and a large spread denotes a slow convergence in the direction corresponding to the smallest eigenvector (or smaller curvature) as illustrated in Figure 15.6. However, Appendix B of Chapter 14 has shown that there is a good correspondence between the power spectral density $S_x(\omega)$ of $x[n]$ and the eigenvalues of $\mathbf{R}$, so that

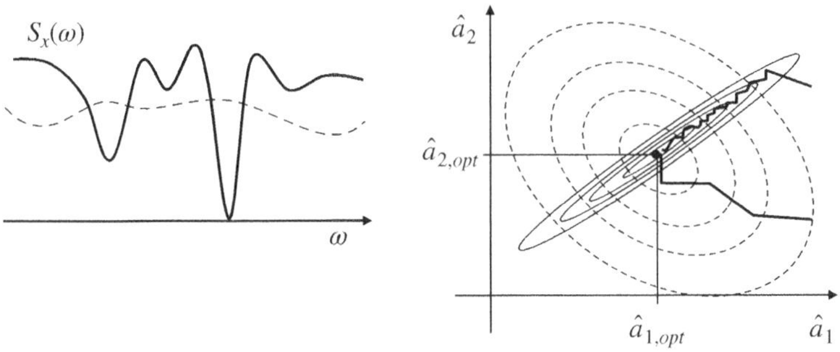

Figure 15.6 Power spectral density $S_x(\omega)$ and iterations over the MSE surface.

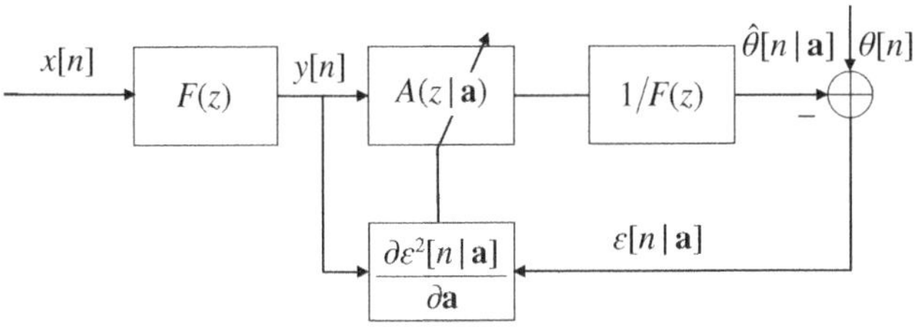

Figure 15.7 Whitening in adaptive filters.

$$\frac{\lambda_{\max}}{\lambda_{\min}} \simeq \frac{\max_\omega\{S_x(\omega)\}}{\min_\omega\{S_x(\omega)\}}$$

This relationship shows that T_{conv} depends on the PSD $S_x(\omega)$. This is somewhat intuitive as the iterative method constrains the step-size from the most energetic components of $S_x(\omega)$ but convergence depends on the least energetic components of $S_x(\omega)$ (or the least sinusoidal components of the process $x[n]$) that are the minima of $S_x(\omega)$.

The dependency on the power spectral density offers a solution to increase the convergence speed. Since convergence depends on $\mathbf{R}$, convergence can be increased if $x[n]$ is filtered to yield a pseudo-constant power spectral density. This is the *whitening* filter (as power spectral density after filtering is white), which should compensate for the spread in power spectral density. The whitening filter need not to be precise as a residual spread of $\lambda_{\max}/\lambda_{\min} \simeq 10$ is enough in many cases, and it should be fixed and known so that it can be deconvolved after the adaptive filtering. Referring to Figure 15.7 with the whitening filter $F(z)$, the pre-whitened process $y[n] = f[n] * x[n] \leftrightarrow Y(z) = F(z)X(z)$ is used to get the correlation matrix $\mathbf{R}_{yy} = \mathbb{E}[\mathbf{y}\mathbf{y}^T]$ that is used to yield the adaptive filter. After filtering, the error is evaluated by comparison with $\theta[n]$ and the effect of pre-whitening should be removed as

$$\hat{\theta}(z|\mathbf{a}) = F^{-1}(z) \times A(z|\mathbf{a}) \times Y(z)$$

More on the Acceleration of Convergence

Convergence is dominated by the slowest component as fast and slow components are tied together by the same step-size μ. In addition to pre-whitening, the step sizes could be made different and optimized on each component. Namely, iterations (15.4) could be replaced by

$$\hat{\mathbf{a}}_{k+1} = \hat{\mathbf{a}}_k + \boldsymbol{\mu} \cdot (\mathbf{p} - \mathbf{R}\hat{\mathbf{a}}_k)$$

where $\boldsymbol{\mu} \in \mathbb{R}^{N \times N}$ is a set of steps that vary between components and chosen to maximize the convergence rate. From convergence analysis:

$$\delta\hat{\mathbf{a}}_{k+1} = (\mathbf{I} - \boldsymbol{\mu} \cdot \mathbf{R})\,\delta\hat{\mathbf{a}}_k$$

The choice $\boldsymbol{\mu} = \mathbf{R}^{-1}$ guarantees convergence in one step. Evaluating $\mathbf{R}^{-1}$ is computationally expensive as it needs the inverse of the matrix $\mathbf{R}$, and this is not within the scope of this method. One alternative is to use an approximation of the inverse that need not be very accurate (this is basically the principle of pre-whitening and deconvolution); the other alternative is to get $\mathbf{R}^{-1}$ by iteratively updating its estimate from a set of rank-1 matrixes: this is the essence of the recursive least squares (RLS) method discussed later in Section 15.10.

15.4 From Iterative to Adaptive Filters

In applications, the correlations $\mathbf{R}$ and $\mathbf{p}$ are not known and should be estimated as sample estimates, possibly varying block-by-block as in adaptive equalization in Section 15.2 where training samples are used for estimating the equalizing filter. However, the benefit of adaptive filtering is from the sample-by-sample adaptation as in active noise cancellation (Section 15.1) where one gradient adaptation step coincides with one time sample. This is the case discussed here and for this reason the adaptation notation becomes sample-based: $\hat{\mathbf{a}}_k \rightarrow \mathbf{a}[k]$.

The response of the time-varying filter at the nth iteration (or nth time sample) is

$$\mathbf{a}[n] = [a[-K_1|n], a[-K_1+1|n], \ldots, a[0|n], \ldots, a[K_2-1|n], a[K_2|n]]^T$$

over the length

$$N = K_1 + K_2 + 1$$

that is generalized to account for causal (if $K_2 \neq 0$) and/or anti-causal (if $K_1 \neq 0$) components; the estimate is

$$\hat{\theta}[n] = \sum_{k=-K_1}^{K_2} a[k|n]x[n-k] = \mathbf{a}^T[n] \cdot \mathbf{x}[n]$$

and the (instantaneous) error from the comparison with $\theta[n]$ is

$$\varepsilon[n|\mathbf{a}[n]] = \theta[n] - \mathbf{a}^T[n] \cdot \mathbf{x}[n]$$

The filter $\mathbf{a}[n]$ at the nth time sample follows from the the MSE evaluated from the sample mean up to the nth sample

$$J(\mathbf{a}[n]) = \frac{1}{n}\sum_{k=1}^{n}(\theta[k] - \mathbf{a}^T[n] \cdot \mathbf{x}[k])^2 = \frac{1}{n}\sum_{k=1}^{n}\theta[k]^2 - 2\hat{\mathbf{p}}^T[n]\mathbf{a}[n] + \mathbf{a}^T[n]\hat{\mathbf{R}}[n]\mathbf{a}[n]$$

dependent on the sample correlations

$$\hat{\mathbf{p}}[n] = \frac{1}{n}\sum_{k=1}^{n}(\mathbf{x}[k] \cdot \theta[k]) \tag{15.6}$$

$$\hat{\mathbf{R}}[n] = \frac{1}{n}\sum_{k=1}^{n}\left(\mathbf{x}[k] \cdot \mathbf{x}^T[k]\right) \tag{15.7}$$

The iterative update is based on the gradient

$$\frac{\partial J(\mathbf{a}[n])}{\partial \mathbf{a}[n]} = 2\left(\hat{\mathbf{R}}[n]\mathbf{a}[n] - \hat{\mathbf{p}}[n]\right) \tag{15.8}$$

which is the same as (15.3) except with sample correlations, and the update of the time-varying filter coefficients becomes:

$$\mathbf{a}[n+1] = \mathbf{a}[n] + \mu \cdot \left(\hat{\mathbf{p}}[n] - \hat{\mathbf{R}}[n]\mathbf{a}[n]\right) \tag{15.9}$$

which is the same as (15.4). The adaptive filter (15.9) has the capability to refine the estimate in a stationary environment as the sample correlations $\hat{\mathbf{R}}[n] \to \mathbf{R}$ and $\hat{\mathbf{p}}[n] \to \mathbf{p}$ for $n \to \infty$, and thus $\mathbf{a}[n] \to \hat{\mathbf{a}}_{opt}$ in (15.1). If dumping the sample estimates (15.6–15.7) as in Section 7.8 (see also the RLS in Section 15.10 below) the updating (15.9) can let the filter adapt to the time-varying environment, but it is again computationally too expensive as every step involves outer products and matrix-vector multiplications with a cost in the order of $\mathcal{O}(N^2)$.

15.5 LMS Algorithm and Stochastic Gradient

The *Least Mean Square* (LMS) method is based on the stochastic gradient that replaces the gradient (15.8) obtained from the sample estimates (15.6–15.7) of all the n observations with the instantaneous values $\hat{\mathbf{p}}[n] \simeq \mathbf{x}[n] \cdot \theta[n]$ and $\hat{\mathbf{R}}[n] \simeq \mathbf{x}[n] \cdot \mathbf{x}[n]^T$. The gradient is based on the instantaneous (one sample) error

$$\frac{\partial J(\mathbf{a}[n])}{\partial \mathbf{a}[n]} \simeq \frac{\partial}{\partial \mathbf{a}[n]}(\theta[n] - \mathbf{a}^T[n] \cdot \mathbf{x}[n])^2 = 2\left(\theta[n] - \mathbf{x}^T[n]\mathbf{a}[n]\right)\mathbf{x}[n]$$

The updating of the filter coefficients reduces to the so called *stochastic gradient*[5]

$$\mathbf{a}[n+1] = \mathbf{a}[n] + \mu \cdot \varepsilon[n]\mathbf{x}[n] \quad (15.10)$$

where

$$\varepsilon[n] = \theta[n] - \mathbf{x}^T[n]\mathbf{a}[n]$$

is the instantaneous error. The LMS algorithm (15.10) is extremely simple and inexpensive as the scalar-vector multiplication (i.e., $\varepsilon[n]\mathbf{x}[n]$) costs only N multiplications to update (the minimum for updating a vector of length N). Below is the corresponding Matlab code that shows a very compact structure.

Crucial for the LMS algorithm is the choice of the step-size μ as it must not be too large to avoid incorrect values of the instantaneous error $\varepsilon[n]$ changing the filter response too far with a deleterious update. On the other hand, the persistence of the correcting values inclines the updates toward the true solution, and this persistency replaces the averaging that is employed by sample estimates (15.6–15.7). To simplify with n everyday experience, LMS adaptive filtering optimizes the MSE toward the minimum MSE value with few samples. *It is like skiing to the valley with limited visibility: the skier cannot take a long hop (large μ) from just a limited view of a rough surface, but rather the skier modifies the strategy at every short-hop (small μ) as this avoids the skier interpreting a temporary hump as a deep valley to ski into.*

15.6 Convergence Analysis of LMS Algorithm

Convergence analysis of LMS is similar to the iterative method (Section 15.3) except that here the residual fluctuations due to the stochastic gradient need to be accounted for when assessing the convergence conditions. Let us consider the identification problem (see Figure 15.4) to estimate the filter response $\mathbf{a}_o$ from the measurements of input and noisy output according to the following model:

$$\theta[n] = \mathbf{a}_o^T\mathbf{x}[n] + w[n]$$

with noise $w[n] \sim \mathcal{N}(0, \sigma_w^2)$. The adaptive filter estimates the (unknown) filter by comparing the output

$$\hat{\theta}[n] = \mathbf{a}^T[n] \cdot \mathbf{x}[n]$$

5 *Stochastic* is due to the term $2\mathbf{x}[n]\varepsilon[n]$, which is the gradient of the square of the instantaneous error $\varepsilon[n]^2 = (\theta[n] - \mathbf{x}^T[n]\mathbf{a}[n])^2$ (recall that the gradient should be $\mathbb{E}[2\mathbf{x}[n]\varepsilon[n]]$, that in sample form is the sum of many terms. As a consequence of the continuous update, this gradient is sensitive to instantaneous error, but the persistency of the correction and the reduced step μ (compared to the convergence conditions of iterative methods) guarantee that instantaneous gradient and solution change still preserving the correct value of the mean, and in this way an unbiased estimate is reached.

and adapting the filter according to the LMS rule (15.10) redefined here:

$$\mathbf{a}[n+1] = \mathbf{a}[n] + \mu \cdot \mathbf{x}[n](\theta[n] - \mathbf{x}[n]^T \mathbf{a}[n]) \tag{15.11}$$

In convergence analysis, updating should be referred to the true solution $\mathbf{a}_o$ and $\delta\mathbf{a}[n] = \mathbf{a}[n] - \mathbf{a}_o$ is the deviation of the filter coefficients wrt $\mathbf{a}_o$. The LMS updating (15.11) becomes

$$\delta\mathbf{a}[n+1] = (\mathbf{I} - \mu \cdot \mathbf{x}[n]\mathbf{x}[n]^T)\delta\mathbf{a}[n] + \mu \cdot \mathbf{x}[n]w[n] \tag{15.12}$$

Converge is based on the evaluation of the evolution of the moments of $\delta\mathbf{a}[n+1]$ by assuming stationary processes (analysis for the non-stationary situation would be intolerably complex, and not that useful in practice). The convergence conditions are illustrated in Figure 15.8 and are derived herein for

- **convergence in the mean**: from the evolution of $\mathbb{E}[\delta\mathbf{a}[n]]$ vs. iteration n
- **convergence in the mean square**: from the evolution of the covariance $\mathbf{C}[n] = \mathbb{E}[\delta\mathbf{a}[n]\delta\mathbf{a}[n]^T]$

The process is zero-mean, and the covariance matrix is $\mathbf{R} = \mathbb{E}[\mathbf{x}[n]\mathbf{x}^T[n]] = \mathbf{U}\boldsymbol{\Lambda}\mathbf{U}^T$; since $\mathbf{U}\mathbf{U}^T = \mathbf{I}$, the update (15.12) can be referred to the principal axes of $\mathbf{U}$ following the same steps of Section 15.3. To simplify, we will use the same relationship (15.12) after having done the following substitutions: $\mathbf{U}^T\mathbf{a}_o \rightarrow \mathbf{a}_o$, $\mathbf{U}^T\mathbf{x}[n] \rightarrow \mathbf{x}[n]$ and $\mathbb{E}[\mathbf{x}[n]\mathbf{x}^T[n]] \rightarrow \boldsymbol{\Lambda}$.

15.6.1 Convergence in the Mean

From the update relationship:

$$\begin{aligned}\mathbb{E}[\delta\mathbf{a}[n+1]] &= \mathbb{E}[(\mathbf{I} - \mu \cdot \mathbf{x}[n]\mathbf{x}^T[n])\delta\mathbf{a}[n]] + \mu\mathbb{E}[\mathbf{x}[n]w[n]] \\ &= (\mathbf{I} - \mu\boldsymbol{\Lambda})\mathbb{E}[\delta\mathbf{a}[n]] = (\mathbf{I} - \mu\boldsymbol{\Lambda})^{n+1}\mathbb{E}[\delta\mathbf{a}[0]]\end{aligned}$$

and the convergence condition

$$0 < \mu < \frac{2}{\lambda_{\max}}$$

guarantees that $\lim_{n\rightarrow\infty} \mathbb{E}[\delta\mathbf{a}[n+1]] = \mathbf{0}$ as for the convergence of the steepest descent method (Section 15.3).

15.6.2 Convergence in the Mean Square

Convergence depends on the evolution of the diagonal terms of the covariance $\mathbf{C}[n] = \mathbb{E}[\delta\mathbf{a}[n]\delta\mathbf{a}[n]^T]$ that are arranged in the vector of variances for each estimate

$$\mathbf{c}[n] = [C_{11}[n], C_{22}[n], ..., C_{NN}[n]]^T$$

After some manipulation (see Appendix A):

$$\mathbf{c}[n+1] = \mathbf{B}\mathbf{c}[n] + \mu^2\sigma_w^2\boldsymbol{\lambda}$$

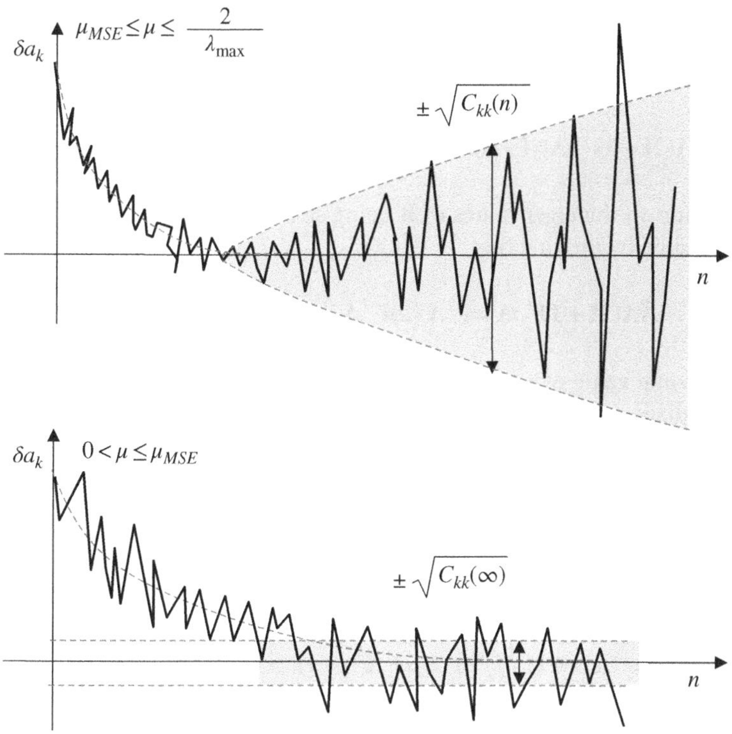

Figure 15.8 Convergence in mean and in mean square.

where

$$\boldsymbol{\lambda} = [\lambda_1, \lambda_2, \ldots, \lambda_N]^T$$
$$\mathbf{B} = \mathbf{I} + \mu^2(\mathcal{K} - 1)\boldsymbol{\Lambda}^2 - 2\mu\boldsymbol{\Lambda} + \mu^2\boldsymbol{\lambda}\boldsymbol{\lambda}^T$$

and $\mathcal{K}$ is the kurtosis[6] of the kth random variable. By chaining the iterations:

$$\mathbf{c}[n+1] = \mathbf{B}^n\mathbf{c}[0] + \mu^2\sigma_w^2\sum_{k=0}^{n-1}\mathbf{B}^k\boldsymbol{\lambda} \tag{15.13}$$

for convergence in the mean square to

$$\mathbf{c}[n] \to \mathbf{c}[\infty] < +\infty$$

There are two issues to be solved: the convergence condition $\mu \leq \mu_{MSE}$ for a certain value μ_{MSE}, and the corresponding convergence value $\mathbf{c}[\infty]$.

6 The kurtosis indicates how compact the pdf is. For a Gaussian rv, $\mathcal{K} = 3$ while for rv with larger values (e.g., Laplace pdf), $\mathcal{K} > 3$—also called super-Gaussian. A compact pdf is sub-Gaussian and has $\mathcal{K} < 3$.

Derivation of μ_{MSE}: The convergence conditions on μ are not trivial to derive—see the book by A.H. Sayed for details [60]. Let $x[n]$ be Gaussian ($\mathcal{K}=3$) and $\boldsymbol{\lambda}\boldsymbol{\lambda}^T=\boldsymbol{\Lambda}\mathbf{1}\mathbf{1}^T\boldsymbol{\Lambda}$. It follows that

$$\mathbf{B}=\mathbf{I}-2\mu\boldsymbol{\Lambda}+\mu^2\boldsymbol{\Lambda}(2\mathbf{I}+\mathbf{1}\mathbf{1}^T)\boldsymbol{\Lambda}=I-2\mu\boldsymbol{\Lambda}+\mu^2\boldsymbol{\Lambda}(2\mathbf{I}+\mathbf{1}\mathbf{1}^T)\boldsymbol{\Lambda}$$

The convergence depends on the eigenvalues of $\mathbf{B}$ being strictly smaller than 1, but all can be equivalently restated by proving that

$$\mathbf{M}=\mathbf{I}-\mathbf{B}=2\mu\boldsymbol{\Lambda}-\mu^2\boldsymbol{\Lambda}(2\mathbf{I}+\mathbf{1}\mathbf{1}^T)\boldsymbol{\Lambda}=\mu^2\boldsymbol{\Lambda}\left(2\mu^{-1}\boldsymbol{\Lambda}^{-1}-2\mathbf{I}-\mathbf{1}\mathbf{1}^T\right)\boldsymbol{\Lambda}$$

is positive definite for some value of $\mu\leq\mu_{MSE}$. Given that the structure of the matrix is mostly composed of diagonals, this can be derived from Sylvester's criterion, which states that the positive definiteness if all the leading principal minors have positive determinants. Some manipulations of the matrix in view of the determinant computation could simplify the derivation, namely:

$$\begin{aligned}|\mathbf{M}|&=|\mu^2\boldsymbol{\Lambda}\left(2\mu^{-1}\boldsymbol{\Lambda}^{-1}-2\mathbf{I}-\mathbf{1}\mathbf{1}^T\right)\boldsymbol{\Lambda}|=\mu^2|\boldsymbol{\Lambda}||2(\mu^{-1}\boldsymbol{\Lambda}^{-1}-\mathbf{I})-\mathbf{1}\mathbf{1}^T||\boldsymbol{\Lambda}|\\&=\mu^2|\boldsymbol{\Lambda}|^2|2(\mu^{-1}\boldsymbol{\Lambda}^{-1}-\mathbf{I})-\mathbf{1}\mathbf{1}^T|\\&=\mu^2|\boldsymbol{\Lambda}|^2\left(1-\mathbf{1}^T(\mu^{-1}\boldsymbol{\Lambda}^{-1}-\mathbf{I})^{-1}\mathbf{1}/2\right)|2(\mu^{-1}\boldsymbol{\Lambda}^{-1}-\mathbf{I})|>0\end{aligned}$$

Analysis of the determinant of each minor $\mathbf{M}_m$ can be decomposed into the three inequalities, in order:

$$\sum_{k=1}^{m}\lambda_k>0$$

$$2-\mathbf{1}_m^T(\mu^{-1}\boldsymbol{\Lambda}_m^{-1}-\mathbf{I}_m)^{-1}\mathbf{1}_m=2-\mu\sum_{k=1}^{m}\frac{\lambda_k}{1-\mu\lambda_k}>0$$

$$\sum_{k=1}^{m}\frac{2-\mu\lambda_k}{\mu\lambda_k}>0$$

The first is always guaranteed to be positive definite, the last is guaranteed as $\mu<2/\lambda_{\max}$ for the convergence in the mean, and the second inequality becomes

$$\frac{\mu}{2}\sum_{k=1}^{m}\frac{\lambda_k}{1-\mu\lambda_k}<1$$

which should hold for every minor $\mathbf{M}_m$, so for $m=N$, and this is the general condition for the convergence in the mean square:

$$\frac{\mu}{2}\sum_{k=1}^{N}\frac{\lambda_k}{1-\mu\lambda_k}<1$$

Solution of this condition is not straightforward, but the choice is always $\mu \ll 1/\lambda_k$ to have low-noise filters, and thus a reasonable *approximation* is

$$\mu < \frac{2}{\sum_{k=1}^{N} \lambda_k} = \frac{2}{\text{tr}\{\mathbf{R}\}} \tag{15.14}$$

with remarkable simplicity in implementation (see Matlab code below) and for this reason widely adopted in practice.

Derivation of $c[\infty]$: The asymptotic value

$$\mathbf{c}[\infty] = \mu^2\sigma_w^2 \sum_{k=0}^{\infty} \mathbf{B}^k \boldsymbol{\lambda}$$

depends on the second term in (15.13) and it can be promptly evaluated from the identity on the recursive update

$$\mathbf{c}[\infty] = \mathbf{B}\mathbf{c}[\infty] + \mu^2\sigma_w^2\boldsymbol{\lambda}$$

that solved yields

$$\mathbf{c}[\infty] = \mu^2\sigma_w^2(\mathbf{I}-\mathbf{B})^{-1}\boldsymbol{\lambda} = \mu^2\sigma_w^2(\mu^2(\mathcal{K}-1)\boldsymbol{\Lambda}^2 - 2\mu\boldsymbol{\Lambda} - \mu^2\boldsymbol{\lambda}\boldsymbol{\lambda}^T)^{-1}\boldsymbol{\lambda}$$

and therefore $\mathbf{c}[\infty] \neq \mathbf{0}$. The convergence in the mean square leaves some randomness in filter response values regardless of the choice of μ.

15.6.3 Excess MSE

The fluctuations of the adaptive filter coefficients $\delta\mathbf{a}$ are due to the LMS adaptations (Figure 15.9), and their covariance $\mathbf{c}[\infty]$ let quantify the increases the MSE at the minimum by adding an *excess MSE*. Let the fluctuations be referred to the principal axes (recall that $\mathbb{E}[\mathbf{x}[n]\mathbf{x}^T[n]] \to \boldsymbol{\Lambda}$)

$$J(\delta\mathbf{a}) = J_{\min} + \sum_{k=1}^{N} J_k(\delta a_k) \quad J_k(\delta a_k) = \lambda_k \delta a_k^2$$

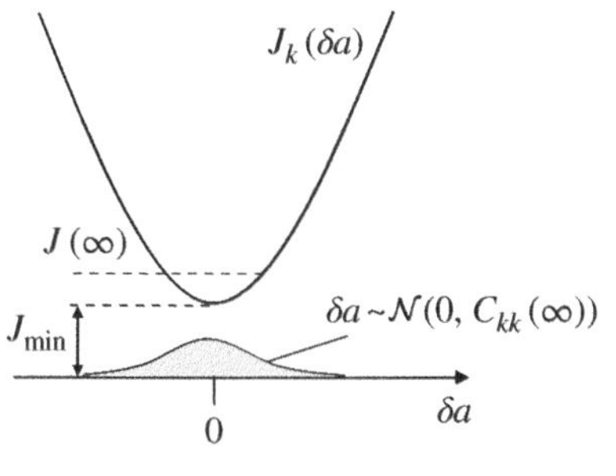

Figure 15.9 Excess MSE from fluctuations of $\delta\mathbf{a}[n]$.

Every coefficient has an asymptotic variability $\mathbb{E}[\delta a_k^2] = c_k[\infty]$ (assuming that fluctuations are uncorrelated among the coefficients) and the MSE at convergence is

$$J(\infty) = J_{\min} + \sum_{k=1}^{N} \lambda_k \mathbb{E}[\delta a_k^2] = \sum_{k=1}^{N} \lambda_k c_k[\infty]$$

Since the covariance is

$$\mathbf{c}[\infty] = \mu^2\sigma_w^2(\mathbf{D} - \mu^2\boldsymbol{\lambda}\boldsymbol{\lambda}^T)^{-1}\boldsymbol{\lambda}$$

with

$$\mathbf{D} = -\text{diag}\{\mu^2(\mathcal{K}-1)\lambda_1^2 - 2\mu\lambda_1, \ldots, \mu^2(\mathcal{K}-1)\lambda_N^2 - 2\mu\lambda_N\}$$

it follows from Woodbury's identity (Section 1.1.1) that

$$\mathbf{c}[\infty] = \mu^2\sigma_w^2\left(\mathbf{D}^{-1} + \frac{\mathbf{D}^{-1}\mu^2\boldsymbol{\lambda}\boldsymbol{\lambda}^T\mathbf{D}^{-1}}{1-\mu^2\boldsymbol{\lambda}^T\mathbf{D}^{-1}\boldsymbol{\lambda}}\right) = \boldsymbol{\lambda}\left(\frac{\mu^2\sigma_w^2}{1-\mu^2\boldsymbol{\lambda}^T\mathbf{D}^{-1}\boldsymbol{\lambda}}\right)\mathbf{D}^{-1}\boldsymbol{\lambda}$$

$$= \frac{\mu\sigma_w^2}{1+\mu\sum_{k=1}^{N}\frac{\lambda_k}{(\mathcal{K}-1)\mu\lambda_k-2}}\begin{bmatrix}\frac{1}{2-(\mathcal{K}-1)\mu\lambda_1}\\ \vdots \\ \frac{1}{2-(\mathcal{K}-1)\mu\lambda_N}\end{bmatrix}$$

Using the approximation of small step-size $\mu \leq 2/\sum\lambda_k$ from the convergence in the mean square (15.14), entries of the covariance can be simplified as follows:

$$c_k[\infty] = \frac{\mu\sigma_w^2}{1+\mu\sum_{k=1}^{N}\frac{\lambda_k}{(\mathcal{K}-1)\mu\lambda_k-2}} \times \frac{1}{2-(\mathcal{K}-1)\mu\lambda_k} \simeq \frac{\mu\sigma_w^2/2}{1-\mu\sum_{k=1}^{N}\lambda_k/2} \simeq \frac{\mu\sigma_w^2}{2}$$

so that

$$J(\infty) = J_{\min} + \frac{\mu\sigma_w^2}{2}\sum_{k=1}^{N}\lambda_k$$

Assuming that the problem at hand is referred to the system identification, the optimum MSE leaves the noise only and thus $J_{\min} = \sigma_w^2$; it follows that

$$J(\infty) = J_{\min} + \underbrace{J_{\min} \times \frac{\mu}{2}\sum_{k=1}^{N}\lambda_k}_{\text{excess MSE}} = J_{\min} + J_{\min} \times \frac{\mu}{2}\text{tr}\{\mathbf{R}\} < 2J_{\min}$$

The MSE is increased by the excess MSE, which depends on the optimum MSE $J_{\min}$ and the step-size μ.

There is another parameter called misadjustment:

$$\mathcal{M} = \frac{J(\infty) - J_{\min}}{J_{\min}}$$

as the ratio between the the excess MSE and the MSE, it indicates the degree of fluctuation of the filter coefficient around the optimum value $\mathbf{a}_o$.

15.7 Learning Curve of LMS

The learning curve represents the behavior of the MSE vs. iterations, and it is correlated to the tracking capability of the adaptive filtering. Derivation is quite complex but some remarks can be easily derived. The transition is approximated by the exponential behavior (Figure 15.10)

$$J(n) \simeq J(\infty) + J(0)\exp(-n\mu\lambda_{av})$$

when starting from some initial conditions with MSE $J(0)$. The average number of iterations for convergence

$$\tau_{av} \simeq \frac{1}{\mu\lambda_{av}}$$

depends of the mean eigenvalues

$$\lambda_{av} = \frac{1}{N}\sum_{k=1}^{N}\lambda_k = \frac{1}{N}\mathrm{tr}\{\mathbf{R}\}$$

The steady-state MSE

$$J(\infty) = J_{\min} + J_{\min} \times \frac{\mu}{2}\sum_{k=1}^{N}\lambda_k = J_{\min}\left(1 + \frac{N}{2\tau_{av}}\right)$$

depends on the time of convergence and the excess MSE, so that a large excess MSE corresponds to a faster convergence, that in turn depends on the step μ. The overall learning curve can be approximated by separating the terms

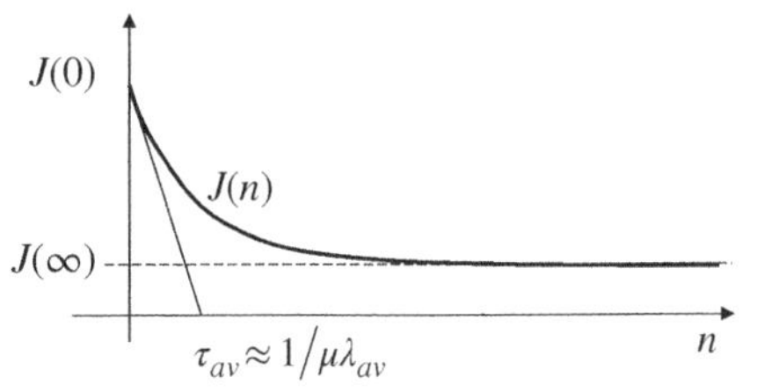

Figure 15.10 Learning curve.

$$J(n) \simeq J_{\min} + J_{\min} \times \frac{\mu}{2} N\lambda_{av} + J(0)\exp(-n\mu\lambda_{av}) \tag{15.15}$$

This is the basis for the optimization of the step-size vs. iterations.

15.7.1 Optimization of the Step-Size μ

The learning curve (15.15) depends on the step-size μ; if decreasing $\mu_2 < \mu_1$ the convergence is slower ($\tau_{av,2} > \tau_{av,1}$) but the excess MSE decreases ($J_{\min}N/2\tau_{av,2} < J_{\min}N/2\tau_{av,1}$) as illustrated Figure 15.11. There is a trade-off, and intuitively the step-size should be reduced by increasing the iterations, but how? The locus that maximizes the decreasing rate (or the learning curve speed) is found from the gradient of (15.5) wrt the step-size μ

$$\frac{\partial J(n)}{\partial \mu} = 0 \rightarrow J_{\min} \times \frac{N\lambda_{av}}{2} = J(0)n\lambda_{av}\exp(-n\mu\lambda_{av})$$

Solving for the step-size:

$$\mu(n) = \frac{1}{n\lambda_{av}} \ln\left(\frac{2nJ(0)}{NJ_{\min}}\right)$$

it follows that the decreasing rate is maximized if the step-size μ decreases with iterations as expected. The decremental rate

$$\Delta\mu \simeq -\frac{1}{n\lambda_{av}} \ln\left(\frac{2nJ(0)}{NJ_{\min}}\right)\frac{\Delta n}{n}$$

can be made independent of the other parameters and thus

$$\frac{\Delta\mu}{\mu(n)} = -\frac{\Delta n}{n}$$

the percentage reduction of $\mu(n)$ depends on the increment wrt the iteration and thus the step should decrease (call also *gear-shifting*) as $1/n$ from $n \geq N/2$ to guarantee the initial convergence.[7]

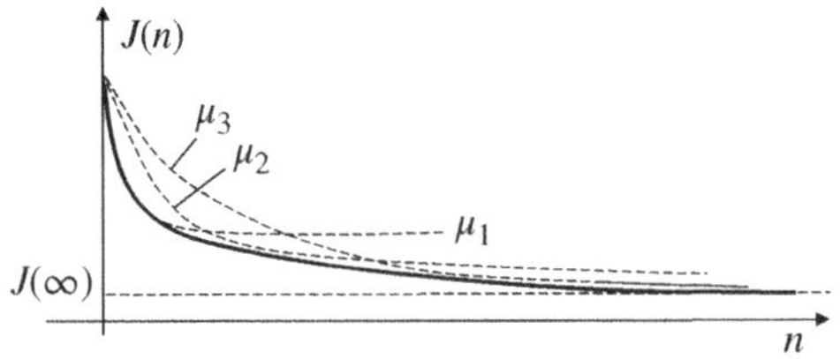

Figure 15.11 Optimization of the step-size μ.

7 Assuming $\mu \simeq 2/\mathrm{tr}\{\mathbf{R}\}$, the learning curve decreases as $\exp(-n\mu\lambda_{av}) \simeq \exp(-2n/N)$ and thus for $n \geq N/2$, the initialization error becomes negligible as $J(0)/2.7$.

15.8 NLMS Updating and Non-Stationarity

The choice of the step-size should be smaller than the maximum value to avoid the learning curve diverging as a consequence of small model mismatch, so a safe choice is $\mu < 1/\text{tr}\{\mathbf{R}\}$. It is quite common to use the normalized step-size

$$\tilde{\mu} = \mu \times tr\{\mathbf{R}\}$$

so that the convergence condition is simply

$$\tilde{\mu} < 1.$$

Alternatively, the step-size is

$$\mu = \frac{\tilde{\mu}}{\text{tr}\{\mathbf{R}\}}.$$

The LMS updating (15.10) becomes

$$\mathbf{a}[n+1] = \mathbf{a}[n] + \tilde{\mu} \cdot \varepsilon[n]\frac{\mathbf{x}[n]}{N\sigma_x^2} \tag{15.16}$$

under the condition of stationarity of the process $x[n]$, or at least within the filtering interval N.

The computation of the normalization variance σ_x^2 should be carried out from the sample estimate within the filtering window so that $\text{tr}\{\mathbf{R}\} \simeq \mathbf{x}^T[n] \cdot \mathbf{x}[n]$, and this is the *Normalized LMS* (NLMS) algorithm

$$\mathbf{a}[n+1] = \mathbf{a}[n] + \tilde{\mu}\frac{\mathbf{x}[n]\varepsilon[n]}{\mathbf{x}^T[n] \cdot \mathbf{x}[n]}$$

The step-size $\tilde{\mu}/\mathbf{x}^T[n] \cdot \mathbf{x}[n]$ changes continuosly over iterations and samples. The convergence analysis is slightly complicated due to the varying step-size according to the local samples of $\mathbf{x}[n]$, but it is faster than conventional LMS ((15.10) or (15.16)) because of its capability to scale the step-size to the instantaneous vector $\mathbf{x}[n]$, especially if time-varying.

When computation time is an issue for the specific application, there are other alternatives to LMS and NLMS by using the sign(.) functions as these avoid multiplications at the price of larger excess MSE and convergence time. These alternative solutions are listed in Table 15.1, which provides a complete overview of the main LMS methods.

Table 15.1 LMS algorithms.

Sign LMS	$\mathbf{a}[n+1] = \mathbf{a}[n] + \mu \cdot \mathbf{x}[n] \cdot \text{sign}\{\varepsilon[n]\}$
Clipped LMS	$\mathbf{a}[n+1] = \mathbf{a}[n] + \mu \cdot \text{sign}\{\mathbf{x}[n]\} \cdot \varepsilon[n]$
Sign-Sign LMS	$\mathbf{a}[n+1] = \mathbf{a}[n] + \mu \cdot \text{sign}\{\mathbf{x}[n]\} \cdot \text{sign}\{\varepsilon[n]\}$

In the case of time-varying situations where the optimum filter coefficient $\mathbf{a}_{opt}[n]$ is time-varying, the choice of the step μ should guarantee that the convergence speed is comparable with the dynamics of the filter. In this case, adaptive estimation can provide both the optimum estimate and the tracking capability. The estimation can be divided into two steps (at least conceptually): *acquisition* of the estimate to enable $\mathbf{a}[n]$ to quickly get close to the optimum solution, and *tracking* of the fluctuations (i.e., $\mathbf{a}[n] \simeq \mathbf{a}_{opt}[n]$). The step-size should be chosen large at the beginning (in acquisition-mode) and then relaxed later (in tracking-mode)—similarly to the optimum step-size in Section 15.7.1, but with the purpose of complying with acquisition and tracking in non-stationary environments.

In tracking-mode it is not possible to increase the step-size beyond the limit dictated by the convergence of the estimator. In particular, assume that the estimator has to be adapted to track the variation of a non-stationary experiment that is varying with a time constant of T_0 samples; the update must therefore be at least fast enough to chase the variations from which $\tau_{av} < T_0$ and then

$$\mu > \frac{1}{\lambda_{av} T_0} = \frac{N}{T_0 \cdot \mathrm{tr}\{\mathbf{R}\}}$$

the step can be scaled up to this limit still considering the convergence in mean square condition (15.14):

$$\frac{N}{T_0 \cdot \mathrm{tr}\{\mathbf{R}\}} < \mu < \frac{2}{\mathrm{tr}\{\mathbf{R}\}}$$

and thus $T_0 > N/2$ for the convergence conditions. The tracking capability is limited by the length of the estimator and the eigenvalues (as shown in Section 15.6), and it cannot track variations faster than $N/2$ but it can easily track variations on the order of $T_0 \simeq 2N \div 4N$.

15.9 Numerical Example: Adaptive Identification

Let us consider the identification problem with a causal filter

$$[\mathbf{a}_o]_k = \sin(\pi k/N) \text{ for } k \in (0, N]$$

of range $N = 10$ samples, and used a symmetric estimator consisting either a casual and anti-causal component over the interval $[-15, 15]$ (i.e., $K_1 = K_2 = 15$). Since the filter is causal it is expected that

$$[\mathbf{a}_{opt}]_k = \begin{cases} \sin(\pi k/N) \text{ for } k \in (0, N] \\ 0 \text{ for } k \notin (0, N] \end{cases}$$

Since the LMS iterations update the filter coefficient at every step, the filter coefficients $a[k|n]$ for $k \notin (0,N)$ should be zero, but are affected by random fluctuations as a consequence of the step μ. The Matlab code for signal generation and LMS estimate is below for the MSE evolution vs. iteration by a Montecarlo simulation with `Nrun=1000` independent runs.

```
a0=sin(pi*[0:N]/N); % filter generation
snrdB=30;
mu=.1; % normalized step
K1=15;
K2=15;
Nrun=1000;

MSE=zeros(Nx,1); %initialization
for irun=1:Nrun; % Montecarlo Runs
  w=randn(Nx,1);
  x=sqrt(10^(snrdB/10)/sum(a0.^2))*randn(Nx,1);
  theta=filter(a0,1,x)+w;
  theta_est=zeros(length(theta),1);
  a=zeros(K1+K2+1,1);
  for n=K1+1:Nx-K2,
    theta_est(n)=a'*x(n-K1:n+K2);
    eps=theta(n)-theta_est(n);
    trR=x(n-K1:n+K2)'*x(n-K1:n+K2);
    a=a+mu*eps*x(n-K1:n+K2)/trR; % NLMS updating
  end
  MSE=MSE+(theta-theta_est).^2;
end
MSE=MSE/Nrun; % MSE vs iterations (learning curve)
```

Figure 15.12 shows the true and estimated filter coefficients after $n = 500$ iterations with some fluctuations of $a[k|n]$ either for $k \notin (0,N)$ where $[\mathbf{a}_o]_k = 0$, and also for $k \in (0,N)$.

A detailed analysis of the evolution of the NLMS updates vs. iterations is shown in Figure 15.13 for $a[k = 5|n]$ where the true solution should be $[\mathbf{a}_o]_5 = 1$, and for the anti-causal part $a[k = -2|n]$ that should be zero. Smaller steps increases the time of convergence with the benefit of smaller random fluctuations when at convergence.

Figure 15.14 is the learning curve (in logarithmic scale) from the Matlab code that illustrates the excess MSE for the larger step μ that becomes almost negligible for $\mu < 0.1$ as it attains the value $J_{\min} = \sigma_w^2 = 1$. The convergence time is better highlighted by the linear scale where the gray-area denotes the anti-causal part ($K_1 = 15$).

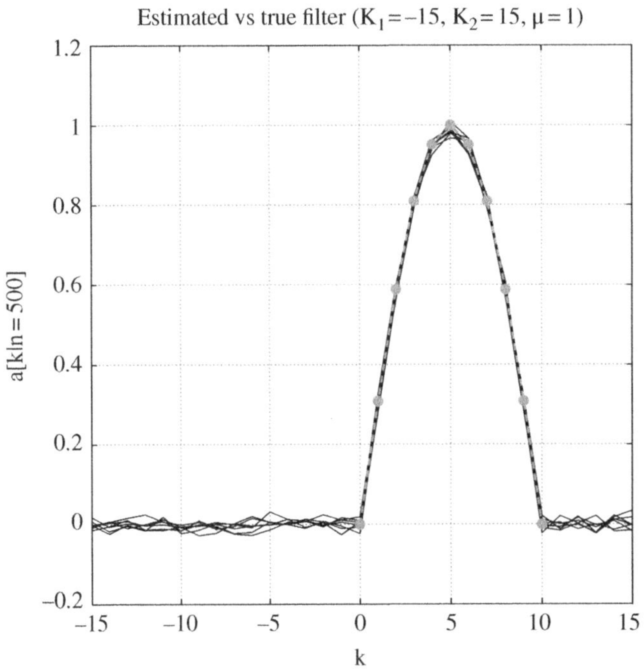

Figure 15.12 Filter response $\mathbf{a}_o$ (dots), and estimated samples over the interval $[-15, 15]$ (solid lines).

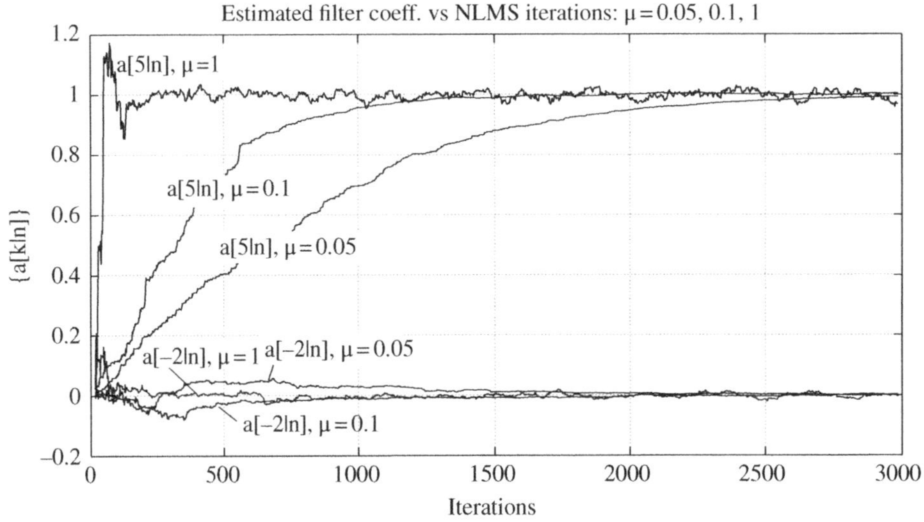

Figure 15.13 Estimated filter response for samples in $k = -2$ and $k = 5$ vs. iterations for varying normalized step-size μ.

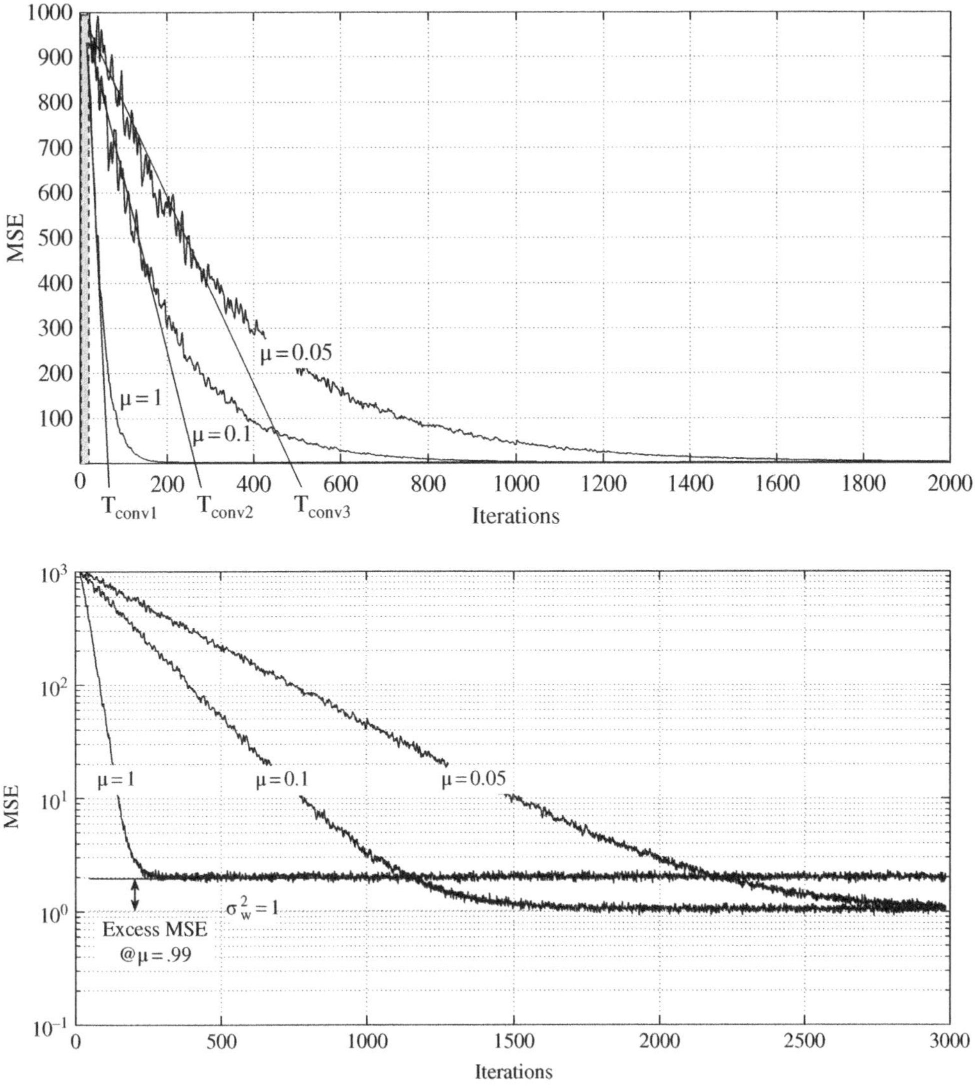

Figure 15.14 Learning curve (Top) in linear scale (to visualize the convergence iterations) and (Bottom) in logarithmic scale (to visualize the excess MSE), shaded area is the anti-causal part.

15.10 RLS Algorithm

The estimate at the nth iteration

$$\mathbf{a}[n] = \hat{\mathbf{R}}^{-1}[n] \cdot \hat{\mathbf{p}}[n]$$

follows from the sample estimate of the correlations (15.6–15.7): the recursive LS (RLS) algorithm updates the estimate $\mathbf{a}[n] \to \mathbf{a}[n+1]$ by augmenting all the observations $\{\mathbf{x}[i]\}_{i=0}^{n}$ with $\mathbf{x}[n+1]$ without recomputing the inverse of $\hat{\mathbf{R}}[n+1]$. The RLS method is a nice example of algebraic manipulation and it can be decoupled into the new sample estimate by adding the new observation $\mathbf{x}[n+1]$ and the update of the inverse $\hat{\mathbf{R}}^{-1}[n] \to \hat{\mathbf{R}}^{-1}[n+1]$ from the Woodbury identity (Section 1.1.1). Note that in the sample estimate of the correlations $\hat{\mathbf{R}}[n]$ and $\hat{\mathbf{p}}[n]$, the division by n in sample correlations is irrelevant as it is a scaling factor that does not influence the estimate $\mathbf{a}[n]$ and it will be omitted herein.

The update of the estimate is:

$$\mathbf{a}[n+1] = \hat{\mathbf{R}}^{-1}[n+1] \cdot \hat{\mathbf{p}}[n+1] \tag{15.17}$$

The two correlations can be written recursively (without the division by $n+1$):

$$\begin{aligned}
\hat{\mathbf{p}}[n+1] &= \hat{\mathbf{p}}[n] + \mathbf{x}[n+1] \cdot \theta[n+1] = \hat{\mathbf{R}}[n]\mathbf{a}[n] + \mathbf{x}[n+1] \cdot \theta[n+1] \\
\hat{\mathbf{R}}[n+1] &= \hat{\mathbf{R}}[n] + \mathbf{x}[n+1] \cdot \mathbf{x}^T[n+1]
\end{aligned}$$

The update of the inverse is

$$\hat{\mathbf{R}}^{-1}[n+1] = \hat{\mathbf{R}}^{-1}[n] - \frac{\hat{\mathbf{R}}^{-1}[n] \cdot \mathbf{x}[n+1]\mathbf{x}^T[n+1] \cdot \hat{\mathbf{R}}^{-1}[n]}{1 + \mathbf{x}^T[n+1] \cdot \hat{\mathbf{R}}^{-1}[n] \cdot \mathbf{x}[n+1]} \tag{15.18}$$

Substituting into (15.17) gives

$$\begin{aligned}
\mathbf{a}[n+1] &= \hat{\mathbf{R}}^{-1}[n+1] \cdot \left(\hat{\mathbf{R}}[n]\mathbf{a}[n] + \mathbf{x}[n+1] \cdot \theta[n+1]\right) \\
&= \hat{\mathbf{R}}^{-1}[n+1] \cdot \left(\left(\hat{\mathbf{R}}[n+1] - \mathbf{x}[n+1] \cdot \mathbf{x}^T[n+1]\right)\mathbf{a}[n] + \mathbf{x}[n+1] \cdot \theta[n+1]\right) \\
&= \mathbf{a}[n] + \underbrace{\hat{\mathbf{R}}^{-1}[n+1] \cdot \mathbf{x}[n+1]}_{\mathbf{g}[n+1]} \cdot \underbrace{\left(\theta[n+1] - \mathbf{x}^T[n+1]\mathbf{a}[n]\right)}_{\varepsilon[n+1|n]}
\end{aligned} \tag{15.19}$$

where $\varepsilon[n+1|n] = \theta[n+1] - \mathbf{x}^T[n+1]\mathbf{a}[n]$ follows the notation introduced in Kalman filtering (Section 15.4) and is the a-priori error by comparison with $\theta[n+1]$ when using the estimate at the nth iteration $\mathbf{a}[n]$. The RLS update adds to the filter coefficient at the previous iteration the *gain vector* $\mathbf{g}[n+1] = \hat{\mathbf{R}}^{-1}[n+1] \cdot \mathbf{x}[n+1]$ scaled by the error based on the previous estimate.

To summarize, the RLS update is

$$\mathbf{a}[n+1] = \mathbf{a}[n] + \mathbf{g}[n+1] \cdot \varepsilon[n+1|n] \tag{15.20}$$

with gain vector (after some algebra from (15.18))

$$\mathbf{g}[n+1] = \frac{\hat{\mathbf{R}}^{-1}[n]\mathbf{x}[n+1]}{1+\mathbf{x}^T[n+1]\cdot\hat{\mathbf{R}}^{-1}[n]\cdot\mathbf{x}[n+1]} \tag{15.21}$$

The dominant computational cost is in the computation of the gain vector $\mathbf{g}[n+1]$, which depends on the product matrix-vector $\hat{\mathbf{R}}^{-1}[n]\mathbf{x}[n+1]$ which in turn costs $\mathcal{O}(N^2)$ multiplications/iteration. Initialization is commonly $\hat{\mathbf{R}}[0] = \delta\cdot\mathbf{I}$ and $\mathbf{a}[0] = \mathbf{0}$.

Remark 1: The RLS (15.19) can be rewritten as

$$\mathbf{a}[n+1] = \mathbf{a}[n] + \hat{\mathbf{R}}^{-1}[n+1]\cdot\mathbf{x}[n+1]\cdot\varepsilon[n+1|n]$$

The comparison with the iterative steepest gradient method (Section 15.5) shows that the RLS replaces the (scalar) step μ with the matrix $\hat{\mathbf{R}}^{-1}[n+1]$ which is iteratively updated with the Woodbury identity. This justifies the intuition that RLS convergence is expected to be faster than LMS as after $n \geq N$ iterations the sample covariance $\hat{\mathbf{R}}[n]$ is full-rank—still very noisy, but it can be inverted thus providing the iterations with a update that is calibrated to guarantee the convergence into one-step (at least for the convergence in the mean).

Remark 2: The gain $\mathbf{g}[n+1]$ is based on the matrix inversion of the sample estimate $\hat{\mathbf{R}}[n]$, but numerical (e.g., rounding due to the finite arithmetic of digital signal processors) errors can make the matrix become positive semidefinite (some eigenvalues could be $\lambda(\hat{\mathbf{R}}[n]^{-1}) \simeq 0$, if not negative), with instabilities that are difficult to solve (and even to spot). A common strategy to avoid numerical errors is to update the Cholesky factorization of the inverse $\hat{\mathbf{R}}^{-1/2}[n]$; this guarantees the conditioning of the inversion. This method is called *square-root RLS*—see [Chapter 35 of 60] for details.

15.10.1 Convergence Analysis

Convergence analysis of RLS is based on the identification problem similar to LMS method (Section 15.6) the generation model is

$$\theta[n] = \mathbf{a}_o^T\mathbf{x}[n] + w[n]$$

with filter coefficient $\mathbf{a}_0$ and noise power σ_w^2. The RLS estimate

$$\mathbf{a}[n] = \hat{\mathbf{R}}^{-1}[n]\cdot\hat{\mathbf{p}}[n] \tag{15.22}$$

should consider the initialization $\hat{\mathbf{R}}[0]$ that is iteratively updated as

$$\hat{\mathbf{R}}[n] = \sum_{k=1}^{n}\mathbf{x}[k]\cdot\mathbf{x}^T[k] + \hat{\mathbf{R}}[0]$$

$$\hat{\mathbf{p}}[n] = \sum_{k=1}^{n}\mathbf{x}[k]\cdot\theta[k] = \left(\sum_{k=1}^{n}\mathbf{x}[k]\mathbf{x}^T[k]\right)\mathbf{a}_o + \sum_{k=1}^{n}\mathbf{x}[k]w[k]$$

$$= (\hat{\mathbf{R}}[n] - \hat{\mathbf{R}}[0])\mathbf{a}_o + \sum_{k=1}^{n}\mathbf{x}[k]w[k]$$

Substituting these sample correlations into the RLS estimate (15.22) yields

$$\mathbf{a}[n] = \mathbf{a}_o - \hat{\mathbf{R}}^{-1}[n]\hat{\mathbf{R}}[0]\mathbf{a}_o + \hat{\mathbf{R}}^{-1}[n]\sum_{k=1}^{n}\mathbf{x}[k]w[k] \tag{15.23}$$

In addition to a stochastic term due to the noise contribution $\hat{\mathbf{R}}^{-1}[n]\sum_{k=1}^{n}\mathbf{x}[k]w[k]$, there is one additional term $\hat{\mathbf{R}}^{-1}[n]\hat{\mathbf{R}}[0]\mathbf{a}_o$ that depends on the initialization and induces a bias (i.e., it is non-random, and not zero-mean) in the estimate. The relation (15.23) is the baseline to evaluate the convergence of the RLS algorithm.

Convergence in the Mean

The convergence in the mean is by considering the mean of the iterations (15.23). Let $n > N$ and the asymptotic ($n \to \infty$) condition of the sample correlation $\hat{\mathbf{R}}[n]/n \to \mathbf{R}$; the RLS estimate is

$$\mathbb{E}[\mathbf{a}[n]] = \mathbf{a}_o - \frac{1}{n}\mathbf{R}^{-1}\hat{\mathbf{R}}[0]\mathbf{a}_o = (\text{for } \hat{\mathbf{R}}[0] = \delta\mathbf{I}) = (\mathbf{I} - \frac{\delta}{n}\mathbf{R}^{-1})\mathbf{a}_o$$

The RLS estimate is biased due to the initialization $\hat{\mathbf{R}}[0] = \delta\mathbf{I}$, but it is asymptotically ($n \to \infty$) unbiased. Bias can be avoided if the outcomes of the RLS are delayed to $n > N$ as in this case the first $n = N$ samples $\mathbf{x}[1], \mathbf{x}[2], ..., \mathbf{x}[N]$ are solely used for the sample covariance estimate.

Convergence in the Mean Square

The convergence in the mean square neglects the bias, and in any case it is for $n \gg N$. The error is

$$\delta\mathbf{a}[n] = \mathbf{a}[n] - \mathbf{a}_o = \hat{\mathbf{R}}^{-1}[n] \cdot \sum_{k=1}^{n}\mathbf{x}[k]w[k]$$

The covariance of the filter coefficient estimate is

$$\mathbf{C}[n] = \mathbb{E}\left[\delta\mathbf{a}[n]\delta\mathbf{a}^T[n]\right] = \mathbb{E}_{x,w}\left[\hat{\mathbf{R}}^{-1}[n] \cdot \sum_{i=1}^{n}\sum_{j=1}^{n}\mathbf{x}[i]\mathbf{x}^T[j] \cdot \hat{\mathbf{R}}^{-1}[n]w[i]w[j]\right]$$

$$= \sigma_w^2\mathbb{E}_x\left[\hat{\mathbf{R}}^{-1}[n] \cdot \underbrace{\sum_{i=1}^{n}\mathbf{x}[i]\mathbf{x}^T[i]}_{\hat{\mathbf{R}}[n]} \cdot \hat{\mathbf{R}}^{-1}[n]\right] = \sigma_w^2\mathbb{E}_x\left[\hat{\mathbf{R}}^{-1}[n]\right]$$

and by assuming $\hat{\mathbf{R}}[n]/n \simeq \mathbf{R}$:

$$\mathbf{C}[n] = \frac{\sigma_w^2}{n}\mathbf{R}^{-1}$$

or alternatively the MSE is

$$MSE(\mathbf{a}[n]) = \mathbb{E}\left[\delta\mathbf{a}^T[n]\delta\mathbf{a}[n]\right] = \text{tr}\{\mathbf{C}[n]\} = \frac{\sigma_w^2}{n}\text{tr}\{\mathbf{R}^{-1}\} = \frac{\sigma_w^2}{n}\sum_{k=1}^{N}\frac{1}{\lambda_k}$$

The MSE shows that the RLS is largely sensitive to the minimum eigenvalue of $\mathbf{R}$, and thus the RLS behaves uncontrollably when $\mathbf{R}$ is ill-conditioned. Nevertheless, asymptotically the MSE decays vs. iterations with $1/n$ thus proving the convergence in mean square for large number of iterations without any excess MSE.

15.10.2 Learning Curve of RLS

Convergence speed is established from the analysis of $J(\mathbf{a}[n])$ vs. iterations: the RLS learning curve. Without any loss of generality, let $\mathbf{x}[n]$ be uncorrelated and $\mathbb{E}[\mathbf{x}[n]\mathbf{x}[n]^T] = \mathbf{R} = \mathbf{\Lambda}$ as this simplifies the derivation. The square error can be rewritten as

$$\varepsilon^2[n|\mathbf{a}] = w^2[n] + \sum_{k=1}^{N}\lambda_k \cdot \delta a_k^2[n]$$

considering only the terms along the principal axes and avoiding mixed terms (as zero-mean). The MSE follows from the expectation $J(n) = \mathbb{E}[\varepsilon^2[n|\mathbf{a}]]$ and it is

$$J(n) = \sigma_w^2 + \sum_{k=1}^{N}\lambda_k[\mathbf{C}[n]]_{kk} = \sigma_w^2 + \sum_{k=1}^{N}\lambda_k \times \frac{\sigma_w^2}{n\lambda_k}$$

since $J_{\min} = \sigma_w^2$ it follows the learning curve

$$J(n) = J_{\min} + \frac{N}{n}J_{\min} \tag{15.24}$$

that decreases with $1/n$.

To summarize:

- The RLS method converges when $n \geq 2N$ as the second term in MSE (15.24) starts to become negligible wrt the asymptotic value $J_{\min}$. The convergence into $2N$ steps is not surprising as during the first N steps the RLS algorithm gets the sample covariance $\hat{\mathbf{R}}[n]$ (if initialization is $\hat{\mathbf{R}}[0] = 0$), or downsizes any bias due to the initialization (when $\hat{\mathbf{R}}[0] = \delta\mathbf{I}$ with $\delta \neq 0$); the true iterations for filter coefficient estimation start for $n > N$.
- The convergence is independent on the eigenvalues of $\mathbf{R}$ as the RLS update can be considered as an iterative method where the steps are calibrated along all the directions of the axes of $\mathbf{R}$ (once $\hat{\mathbf{R}}[n]$ is consistently estimated, the convergence in the mean is in one step); the convergence in the mean square is only impaired by the sample estimate of the correlation $\hat{\mathbf{R}}[n]$
- For $n \to \infty$, the convergence has no excess MSE as $J(n) \to J_{\min}$, and filter coefficient estimates are stable.

15.11 Exponentially-Weighted RLS

RLS provides a stable estimate for stationary and stable situations, but for non-stationary processes it is necessary to modify the RLS method by adding a limited memory as in WLS (Section 7.8). The sample auto/cross-correlations are redefined with a forgetting factor λ and can be updated as follows

$$\hat{\mathbf{R}}[n+1] = \lambda\hat{\mathbf{R}}[n] + \mathbf{x}[n+1]\cdot\mathbf{x}^T[n+1]$$
$$\hat{\mathbf{p}}[n+1] = \lambda\hat{\mathbf{p}}[n] + \mathbf{x}[n+1]\cdot\theta[n+1]$$

where $\lambda \leq 1$ and for $\lambda = 1$ gives the RLS with infinite memory. The result is that all the fundamental updating relationships are modified for the update of the inverse with a forgetting as

$$\hat{\mathbf{R}}^{-1}[n+1] = \frac{1}{\lambda}\left(\hat{\mathbf{R}}^{-1}[n] - \frac{\hat{\mathbf{R}}^{-1}[n]\cdot\mathbf{x}[n+1]\mathbf{x}^T[n+1]\cdot\hat{\mathbf{R}}^{-1}[n]}{\lambda + \mathbf{x}^T[n+1]\cdot\hat{\mathbf{R}}^{-1}[n]\cdot\mathbf{x}[n+1]}\right)$$
$$\mathbf{g}[n+1] = \frac{\hat{\mathbf{R}}^{-1}[n]\mathbf{x}[n+1]}{\lambda + \mathbf{x}^T[n+1]\cdot\hat{\mathbf{R}}^{-1}[n]\cdot\mathbf{x}[n+1]}$$

with a comparable cost compared to RLS. All conclusions remain except for the influence of the forgetting factor that is analyzed separately below.

The forgetting factor is a moving averaging window over the observations and to analyze its effect it is better to evaluate if the forgetting introduces a bias. From the definitions

$$\hat{\mathbf{R}}[n] = \sum_{k=1}^{n}\lambda^{n-k}\mathbf{x}[k]\cdot\mathbf{x}^T[k]$$
$$\hat{\mathbf{p}}[n] = \sum_{k=1}^{n}\lambda^{n-k}\mathbf{x}[k]\cdot\theta[k] = \sum_{k=1}^{n}\lambda^{n-k}\mathbf{x}[k]\mathbf{x}^T[k]\mathbf{a}_o + \sum_{k=1}^{n}\lambda^{n-k}\mathbf{x}[k]w[k]$$
$$= \hat{\mathbf{R}}[n]\mathbf{a}_o + \sum_{k=1}^{n}\lambda^{n-k}\mathbf{x}[k]w[k] \rightarrow \mathbb{E}[\hat{\mathbf{p}}[n]] = \frac{1-\lambda^n}{1-\lambda}\mathbf{R}\mathbf{a}_o$$

it is straightforward to prove that

$$\mathbb{E}[\hat{\mathbf{R}}[n]] = \mathbf{R}\sum_{k=1}^{n}\lambda^{n-k} = L_{eff}\times\mathbf{R}$$
$$\mathbb{E}[\hat{\mathbf{p}}[n]] = L_{eff}\times\mathbf{R}\mathbf{a}_o$$

which is dependent on the effective aperture of the exponential window

$$L_{eff} = \sum_{k=1}^{n}\lambda^k = \frac{1-\lambda^n}{1-\lambda} \tag{15.25}$$

and thus the RLS is unbiased (except for the initialization, here omitted).

The analysis of the convergence in the mean square follows the same steps above:

$$\begin{aligned}\mathbf{C}[n] &= \sigma_w^2 \mathbb{E}\left[\hat{\mathbf{R}}^{-1}[n]\sum_{k=1}^{n}(\lambda^2)^{n-k}\mathbf{x}[k]\mathbf{x}[k]^T\hat{\mathbf{R}}^{-1}[n]\right] \\ &= \sigma_w^2\left(\frac{1-\lambda}{1-\lambda^n}\right)^2 \mathbf{R}^{-1}\sum_{k=1}^{n}(\lambda^2)^{n-k}\mathbb{E}[\mathbf{x}[k]\mathbf{x}[k]^T]\cdot\mathbf{R}^{-1} \\ &= \sigma_w^2\left(\frac{1-\lambda}{1-\lambda^n}\right)^2 \sum_{k=1}^{n}(\lambda^2)^{n-k}\cdot\mathbf{R}^{-1} = \frac{\sigma_w^2}{n_{eq}}\mathbf{R}^{-1}\end{aligned}$$

where the equivalent number of steps to guarantee MSE convergence is indicated as n_{eq}:

$$n_{eq} = \frac{\left(\sum_{\ell=1}^{n}\lambda^\ell\right)^2}{\sum_{\ell=1}^{n}(\lambda^2)^\ell} = \left(\frac{1-\lambda^n}{1-\lambda}\right)^2\cdot\left(\frac{1-\lambda^2}{1-\lambda^{2n}}\right) = \frac{1+\lambda}{1-\lambda}\cdot\frac{1-\lambda^n}{1+\lambda^n}$$

for $\lambda = 1$, $n_{eq} = n$. The learning curve of RLS with forgetting factor is (derivations are omitted)

$$J(n) = J_{\min} + \frac{N}{n_{eq}}J_{\min}$$

which proves again that $n_{eq} \simeq 2N$ guarantees convergence. Similarly, the equivalent steps are related to the exponential window aperture (15.25)

$$\frac{n_{eq}}{L_{eff}} = \frac{1+\lambda}{1+\lambda^n} \underset{n\to\infty}{\to} 1+\lambda \simeq 2$$

which shows that for large n, $n_{eq} \simeq L_{eff}$.

Since n_{eq} is not the effective number of iterations, the number of steps to guarantee convergence of the RLS[8] is

$$n_{conv} = \frac{2N}{1+N(1-\lambda)}$$

This relationship is useful to design the forgetting factor λ in non-stationary situations where the convergence should be guaranteed to be comparable with the dynamics of the non-stationary processes.

8 Let $\lambda = 1-\rho$ with $\rho \ll 1$. Then $\lambda^n \simeq 1 - n\rho$ and thus:

$$n_{eq} \simeq \frac{2-\rho}{\rho}\cdot\frac{n\rho}{2-n\rho} \to \frac{2n}{2-n\rho} = 2N$$

Table 15.2 Comparison between LMS and RLS.

	NLMS	RLS
Cost (multiplications)/iteration	$2N$	$N(N+2)$
Convergence (iterations) T_{conv}	$N\frac{\lambda_{max}}{\lambda_{min}}$	$\simeq 2N$
Cost up to convergence	$2N^2\frac{\lambda_{max}}{\lambda_{min}}$	$\simeq 2N^3$

15.12 LMS vs. RLS

The comparison between LMS and RLS should take into account convergence, costs (both per iteration and to reach convergence), and limitations. Table 15.2 summarizes the main properties. Recall that LMS, or NLMS, leaves an excess MSE due to the fluctuations of the filter coefficient that could be reduced by decoupling the acquisition from tracking.

In non-stationary environments, the tracking capabilities of adaptive filtering are solely based on the possibility of making the convergence faster, or comparable, to the time of convergence (in samples or iterations). Since convergence depends on filter length N, in fast time-varying situations, it is advisable to trade a small filter N for a fast convergence.

Appendix A: Convergence in Mean Square

The covariance can be recursively updated:

$$\begin{aligned}\mathbf{C}[n+1] &= \mathbb{E}\left[\delta\mathbf{a}[n+1]\delta\mathbf{a}^T[n+1]\right] = \\ &= \mathbb{E}[(\mathbf{I}-\mu\cdot\mathbf{x}[n]\mathbf{x}^T[n])\delta\mathbf{a}[n]\delta\mathbf{a}^T[n](\mathbf{I}-\mu\cdot\mathbf{x}[n]\mathbf{x}^T[n])]+ \\ &\quad +\mu^2\mathbb{E}[w[n]^2\mathbf{x}[n]\mathbf{x}^T[n]]+2\mu\mathbb{E}[w[n](\mathbf{I}-\mu\cdot\mathbf{x}[n]\mathbf{x}^T[n])\delta\mathbf{a}[n]\mathbf{x}[n]^T] \\ &= \mathbf{C}[n]-\mu(\mathbf{C}[n]\boldsymbol{\Lambda}+\boldsymbol{\Lambda}\mathbf{C}[n])+\mu^2\mathbf{A}[n]+\mu^2\sigma_w^2\boldsymbol{\Lambda}\end{aligned}$$

where

$$\mathbf{A}[n]=\mathbb{E}[\mathbf{x}[n]\mathbf{x}^T[n]\delta\mathbf{a}[n]\delta\mathbf{a}^T[n]\mathbf{x}[n]\mathbf{x}^T[n]]$$

The evolution of the terms along the diagonal are of interest for the convergence in the mean square:

$$C_{kk}[n+1]=C_{kk}[n](1-2\mu\lambda_k)+\mu^2(A_{kk}[n]+\sigma_w^2\lambda_k)$$

To evaluate $A_{kk}[n]$, all terms along the diagonal should be evaluated:

$$\text{Entry}_{\text{kk}} = \underbrace{\underbrace{\left(\mathbf{x}[n]\mathbf{x}^T[n]\delta\mathbf{a}[n]\delta\mathbf{a}^T[n]\right)}_{f_{kj}=\sum_i x_k x_i \delta a_i \delta a_j}\mathbf{x}[n]\mathbf{x}^T[n]}_{\sum_j f_{kj}x_j x_k=\sum_{i,j} x_k x_i \delta a_i \delta a_j x_j x_k}$$

in detail

$$A_{kk}[n] = \mathbb{E}[\sum_{i,j=1}^{N} x_k x_i \delta a_i \delta a_j x_j x_k] = \sum_{i,j=1}^{N} \mathbb{E}[x_k^2 x_i x_j]\mathbb{E}[\delta a_i \delta a_j] = (\mathbb{E}[x_k^2 x_i x_j] = 0 \text{ per } i \neq j)$$

$$= \sum_{\substack{i=1\\ i\neq k}}^{N} \mathbb{E}[x_k^2]\mathbb{E}[x_i^2]C_{ii}[n] + \mathbb{E}[x_k^4]C_{kk}[n] = \sum_{\substack{i=1\\ i\neq k}}^{N} \lambda_k \lambda_i C_{ii}[n] + \lambda_k^2 \mathcal{K}_k C_{kk}[n].$$

For the kurtosis of the kth entry:

$$\mathcal{K}_k = \frac{\mathbb{E}[x_k^4]}{\mathbb{E}[x_k^2]^2} = \frac{\mathbb{E}[x_k^4]}{\lambda_k^2}$$

Substitutions yields

$$C_{kk}[n+1] = C_{kk}[n](1 - 2\mu\lambda_k) + \mu^2\lambda_k(\sum_{i=1}^{N} \lambda_i C_{ii}[n] + \lambda_k(\mathcal{K}_k - 1)C_{kk}[n] + \sigma_w^2)$$

All entries along the diagonal are arranged into a vector

$$\mathbf{c}[n] = [C_{11}[n], C_{22}[n], ..., C_{NN}[n]]^T$$

so that

$$\mathbf{c}[n+1] = \left(\mathbf{I} + \mu^2(\mathcal{K} - 1)\mathbf{\Lambda}^2 - 2\mu\mathbf{\Lambda} + \mu^2\boldsymbol{\lambda}\boldsymbol{\lambda}^T\right)\mathbf{c}[n] + \mu^2\sigma_w^2\boldsymbol{\lambda}$$

where $\boldsymbol{\lambda} = [\lambda_1, \lambda_2, ..., \lambda_N]^T$

16

Line Spectrum Analysis

The estimation of the frequencies of multiple sinusoidal signals is remarkably important and common in engineering, namely for the Fourier analysis of observations or experiments that model the observations as random set of values whose statistical properties depend on the angular frequencies. Since the problem itself reduces to the estimation of the (angular) frequencies as a generalization of Section 9.2 and Section 9.3, one might debate the rationale of dedicating a whole chapter to this. The problem itself needs care in handling the assumptions behind every experiment and these were deliberately omitted in the early chapters. Crucial problems addressed here are (1) resolution (estimating two sinusoids with similar frequencies), (2) finding alternative methods to solve non-linear optimizations, and (3) evaluating accuracy wrt the CRB. All these matters triggered several frantic research efforts around the '80s in this scientific area. A strong boost was given by array processing (Chapter 19) as the estimation of the spatial properties of wavefields from multiple radiating sources in space is equivalent to the estimation of (spatial) angular frequencies, and resolution therein pertains to the possibility to spatially distinguish the emitters. Since MLE is not trivial, and closed form optimizations are very cumbersome, the scientific community investigated a set of so called *high-resolution* (or super-resolution[1]) methods based on some ingenious algebraic reasoning. All these high-resolution estimators treated here are consistent and have comparable performance wrt the CRB, so using one or another is mostly a matter of taste, but the discussion of the methodological aspects is very useful to gain insight into these challenging methods. As elsewhere, this topic is very broad and it deserves more space than herein, so readers are referred to [62, 65, 66, 101] (just to mention a few) for extensions to some of the cases not covered here, and for further details.

Why Line Spectrum Analysis?

The sinusoidal model is

$$x[n] = \sum_{\ell=1}^{L} A_\ell \cos(\omega_\ell n + \phi_\ell)$$

1 Super-resolution refers to the capability to resolve two sinusoids that have a frequency difference below the resolution of the periodogram as first sight non-parametric method, frequency spacing is $2\pi/N$.

Statistical Signal Processing in Engineering, First Edition. Umberto Spagnolini.

Companion website: www.wiley.com/go/spagnolini/signalprocessing

where $\{A_\ell, \omega_\ell, \phi_\ell\}$ are amplitudes, angular frequencies, and phases. The problem at hand is the estimation of the frequencies $\{\omega_\ell\}$ that are $\omega_\ell \in (-\pi, \pi]$ from a noisy set of observations $\{x[n]\}$. In radar, telecommunications, or acoustic signal processing, the complex-valued form

$$x[n] = \sum_{\ell=1}^{L} A_\ell e^{j(\omega_\ell n + \phi_\ell)}$$

is far more convenient even if this form is not encountered in real life as it rather stands for manipulation of real-valued measurements (Section 4.3). This is the model considered here and some of the assumptions need to be detailed.

The phases $\{\phi_\ell\}$ account for the initial value that depends on the origin (sample $n = 0$) and they are intrinsically ambiguous (i.e., the set $\{A_\ell, \omega_\ell, \phi_\ell\}$ is fully equivalent to $\{-A_\ell, \omega_\ell, \pi + \phi_\ell\}$). Phases are nuisance parameters that are conveniently assumed as random and independent so that

$$\phi_\ell \sim \mathcal{U}(0, 2\pi)$$

and thus

$$\mathbb{E}[e^{j\phi_\ell} e^{-j\phi_n}] = \delta_{\ell - n}$$

It follows that the autocorrelation is

$$r_{xx}[k] = \mathbb{E}[x^*[n]x[n+k]] = \sum_{\ell=1}^{L} A_\ell^2 e^{j\omega_\ell k}$$

and the PSD is a line spectrum

$$S_x(f) = \mathcal{F}\{r_{xx}[k]\} = \sum_{\ell=1}^{L} A_\ell^2 \delta(f - f_\ell)$$

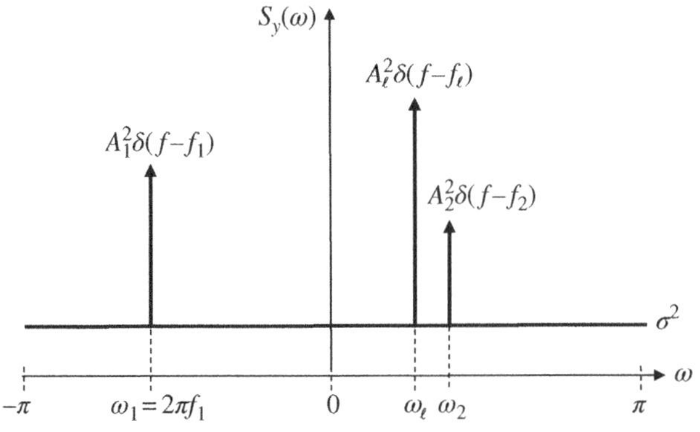

Figure 16.1 Line spectrum analysis.

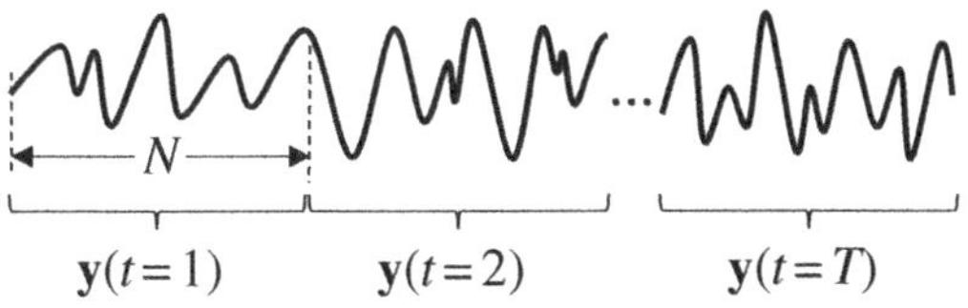

Figure 16.2 Segmentation of data into N samples.

as the sum of Dirac-deltas. The signal affected by the additive white noise $w[n] \sim \mathcal{CN}(0, \sigma^2)$ and the corresponding PSD is

$$y[n] = x[n] + w[n] \qquad S_y(f) = S_x(f) + \sigma^2$$

as illustrated in Figure 16.1. Spectrum analysis for line spectra refers to the estimation of the frequencies of $x[n]$ assuming the model above.

16.1 Model Definition

The data can be sliced into a set of N samples each (Figure 16.2), where t indexes the specific sample with $t = 1, 2, ..., T$ that is usually referred to as *time* (e.g., this is the time in data segmentation as in Section 14.1.6, or time of independent snapshots of the same process as later in Chapter 19). The general model of the observations at time t is

$$y[n;t] = \sum_{\ell=1}^{L} \alpha_\ell(t) e^{j\omega_\ell n} + w[n;t]$$

where $\alpha_\ell(t) = A_\ell(t)e^{j\phi_\ell(t)}$ collects amplitudes and phases that are arbitrarily varying (not known), and $w[n;t]$ accounts for the noise. Since the variation of $\alpha_\ell(t)$ vs. time t is usually informative of the specific phenomena, or encoded information, here this will be called *signal* (or source signal or simply amplitudes) as it refers to the informative signal at frequency ω_ℓ.

Each set of N observations, with $N \geq L$, are collected into a vector that represents the compact model used for line spectra analysis

$$\mathbf{y}(t) = \sum_{\ell=1}^{L} \alpha_\ell(t) \underbrace{\begin{bmatrix} 1 \\ e^{j\omega_\ell} \\ \vdots \\ e^{j\omega_\ell(N-1)} \end{bmatrix}}_{\mathbf{a}(\omega_\ell)} + \mathbf{w}(t) = \mathbf{A}(\omega) \cdot \mathbf{s}(t) + \mathbf{w}(t)$$

where

$$\mathbf{A}(\boldsymbol{\omega}) = [\mathbf{a}(\omega_1), \mathbf{a}(\omega_2), ..., \mathbf{a}(\omega_L)]$$

for the set of L angular frequencies

$$\boldsymbol{\omega} = [\omega_1, \omega_2, ..., \omega_L]^T$$

with signals

$$\mathbf{s}(t) = [\alpha_1(t), \alpha_2(t), ..., \alpha_L(t)]^T$$

Noise is uncorrelated

$$\mathbf{w}(t) \sim \mathcal{CN}(0, \sigma^2\mathbf{I})$$

and uncorrelated across time $\mathbb{E}[\mathbf{w}(t)\mathbf{w}^H(\tau)] = \sigma^2\delta_{t-\tau}\mathbf{I}$. This assumption does not limit the model in any way, as for correlated noise a pre-whitening can reduce the model to the one discussed here (Section 7.5.2).

16.1.1 Deterministic Signals s (t)

Signals $\mathbf{s}(t)$ are nuisance parameters in line spectrum analysis and typically change vs. time index t but, once estimated the frequencies $\boldsymbol{\omega}$, the data $\{\mathbf{y}(t)\}$ can be written as a linear regression wrt $\{\mathbf{s}(t)\}$ and these are estimated accordingly. However, the way to model the nuisance $\mathbf{s}(t)$ contributes in defining the estimators and these are essentially modeled either as deterministic and unknown, or random. If deterministic, the set $\{\mathbf{s}(t)\}$ enters into the budget of the parameters to estimate as described in Section 7.2, and the model is

$$\mathbf{y}(t) \sim \mathcal{CN}(\mathbf{A}(\boldsymbol{\omega}) \cdot \mathbf{s}(t), \sigma^2\mathbf{I})$$

and the corresponding ML will be reviewed in Section 16.2. When $\{\mathbf{s}(t)\}$ are handled as random, the model deserves further assumptions detailed below.

16.1.2 Random Signals s (t)

A reasonable (and common) assumption is that signals are random

$$\mathbf{s}(t) \sim \mathcal{CN}(0, \mathbf{R}_s)$$

where the zero-mean is a direct consequence of the uniform phases $\{\phi_\ell\}$, and the covariance $\mathbf{R}_s$ reflects the power of the signal associated with each sinusoid, and their mutual correlation. According to the random assumption of $\mathbf{s}(t)$, the data is modeled as

$$\mathbf{y}(t) \sim \mathcal{CN}(0, \mathbf{R}_y)$$

where the covariance

$$\mathbf{R}_y = \mathbf{A}(\boldsymbol{\omega})\mathbf{R}_s\mathbf{A}^H(\boldsymbol{\omega}) + \sigma^2\mathbf{I} \in \mathbb{C}^{N\times N}$$

is structured according to the structure of $\mathbf{A}(\boldsymbol{\omega})$ that in turn it depends on the (unknown) $\omega_1, \omega_2, ..., \omega_L$, and mutual correlation among signals from $\mathbf{R}_s$.

16.1.3 Properties of Structured Covariance

Let the columns of $\mathbf{A}(\boldsymbol{\omega})$ be independent, or equivalently the frequencies be distinct ($\omega_n \neq \omega_m$ for $n \neq m$); thus:

$$\text{rank}\{\mathbf{A}(\boldsymbol{\omega})\mathbf{R}_s\mathbf{A}^H(\boldsymbol{\omega})\} = \text{rank}\{\mathbf{R}_s\} \leq N, \tag{16.1}$$

or stated differently, the algebraic properties of $\mathbf{A}(\boldsymbol{\omega})\mathbf{R}_s\mathbf{A}^H(\boldsymbol{\omega})$ depend on the correlation properties of the sources with $\mathbf{R}_s$. When all the signals are uncorrelated with arbitrary powers

$$\mathbf{R}_s = \text{diag}(\sigma_1^2, \sigma_2^2, ..., \sigma_L^2),$$

the covariance of $\mathbf{s}(t)$ is full-rank $\text{rank}\{\mathbf{R}_s\} = L \leq N$ with

$$\mathcal{R}(\mathbf{A}(\boldsymbol{\omega})) \subseteq \mathbb{C}^N$$

The eigenvalue decomposition is

$$\mathbf{A}(\boldsymbol{\omega})\mathbf{R}_s\mathbf{A}^H(\boldsymbol{\omega}) = \mathbf{U}\boldsymbol{\Lambda}\mathbf{U}^H$$

but for the property (16.1), the eigenvalues in $\boldsymbol{\Lambda}$ ordered for decreasing values are

$$\lambda_1 \geq \lambda_2 \geq ... \geq \lambda_L > \lambda_{L+1} = \lambda_{L+2} = ... = \lambda_N = 0$$

Rearranging the eigenvectors into the first L leading eigenvalues and the remaining ones

$$\mathbf{U} = [\mathbf{U}_S, \mathbf{U}_N]$$

the basis

$$\mathbf{U}_S = [\mathbf{u}_1, \mathbf{u}_2, ..., \mathbf{u}_L] \in \mathbb{C}^{N \times L}$$

spans the same subspace of the columns of $\mathbf{A}(\boldsymbol{\omega})$:

$$\mathcal{R}(\mathbf{A}(\boldsymbol{\omega})) = \mathcal{R}(\mathbf{U}_S) \subseteq \mathbb{C}^N$$

and this is referred to as *signal subspace.* To stress this point, there exists an orthonormal transformation (complex set of rotations) $\mathbf{Q}$ such that it transforms the columns of $\mathbf{A}(\boldsymbol{\omega})$ into $\mathbf{U}_S$:

$$\mathbf{U}_S = \mathbf{A}(\boldsymbol{\omega})\mathbf{Q}$$

with the known properties of the othornormal transformation $\mathbf{Q}^{-1} = \mathbf{Q}^H$.

The complementary basis

$$\mathbf{U}_N = [\mathbf{u}_{L+1}, \mathbf{u}_{L+2}, ..., \mathbf{u}_N] \in \mathbb{C}^{N \times (N-L)}$$

spans the *noise subspace* as complementary to signal subspace. This is clear when considering that the eigenvector decomposition of $\mathbf{R}_y$ is

$$\mathbf{R}_y = \mathbf{U}\boldsymbol{\Lambda}\mathbf{U}^H + \sigma^2\mathbf{U}\mathbf{U}^H = \mathbf{U}(\boldsymbol{\Lambda} + \sigma^2\mathbf{I})\mathbf{U}^H$$

and thus

$$\lambda_k(\mathbf{R}_y) = \lambda_k + \sigma^2$$

The effect of the noise is only to increase the values of the eigenvalues, but not to change the subspace partitioning.

To exemplify, we can consider the following case for $L = 2$ with $\omega_1 \neq \omega_2$:

$$\mathbf{R}_s = \begin{bmatrix} 1 & \rho \\ \rho & 1 \end{bmatrix}$$

For $\rho \simeq 0$, the two signals are uncorrelated with $\lambda_1 \simeq \lambda_2$, but increasing ρ makes the two signals correlated and $\lambda_1 > \lambda_2 > 0$ up to the case $\lambda_1 > \lambda_2 = 0$ for $\rho \rightarrow 1$. But $\rho = 1$ is the case where the two signals are temporally fully correlated (the same source $\alpha_1(t) = \alpha_2(t) = \alpha_o(t)$) and the model becomes

$$\mathbf{y}(t) = \alpha_o(t)\mathbf{a}_o + \mathbf{w}(t)$$

where $\mathbf{a}_o = \mathbf{a}(\omega_1) + \mathbf{a}(\omega_2)$. In this case the column space of the two sinusoids cannot distinguish two signals even if the frequencies are distinct.

16.2 Maximum Likelihood and Cramér–Rao Bounds

MLE depends on the model and we can distinguish two methods for the two models:

- **Conditional (or deterministic) ML (CML)**: in this case the nuisance parameters are deterministic and the model is

  $$\mathbf{y}(t) \sim \mathcal{CN}(\mathbf{A}(\boldsymbol{\omega}) \cdot \mathbf{s}(t), \sigma^2\mathbf{I})$$

  The parameters are

  $$\theta = \{\boldsymbol{\omega}, \mathbf{s}(t), \sigma^2\}$$

  or more specifically the 3L+1 parameters:

  $$\theta = [A_1, ..., A_L, \phi_1, ..., \phi_L, \omega_1, ..., \omega_L, \sigma^2]^T$$

  with possibly time-varying amplitudes and phases.
- **Unconditional (or stochastic) ML (UML)**: the amplitudes $\mathbf{s}$ are nuisance parameters modeled as random (Section 7.4) and the model is

  $$\mathbf{y}(t) \sim \mathcal{CN}(0, \mathbf{A}(\boldsymbol{\omega})\mathbf{R}_s\mathbf{A}^H(\boldsymbol{\omega}) + \sigma^2\mathbf{I})$$

Table 16.1 Deterministic vs. stochastic ML in line spectrum analysis.

	Deterministic ML (CML)	Stochastic ML (UML)
Model	$\mathbf{y}(t) \sim \mathcal{CN}(\mathbf{A}(\boldsymbol{\omega}) \cdot \mathbf{s}(t), \sigma^2\mathbf{I})$ $\theta = \{\boldsymbol{\omega}, \mathbf{s}(t), \sigma^2\}$	$\mathbf{y}(t) \sim \mathcal{CN}(0, \mathbf{A}(\boldsymbol{\omega})\mathbf{R}_s\mathbf{A}^H(\boldsymbol{\omega}) + \sigma^2\mathbf{I})$ $\theta = \{\boldsymbol{\omega}, \mathbf{R}_s, \sigma^2\}$
ML	$\hat{\boldsymbol{\omega}}_c = \arg\min_{\boldsymbol{\omega}} tr[\mathbf{P}^{\perp}_{\mathbf{A}(\boldsymbol{\omega})}\hat{\mathbf{R}}_y]$ $\hat{\mathbf{R}}_y = \frac{1}{T}\sum_{t=1}^{T}\mathbf{y}(t)\mathbf{y}^H(t)$	$\hat{\boldsymbol{\omega}}_u = \arg\min_{\boldsymbol{\omega}} \ln\lvert\mathbf{A}(\boldsymbol{\omega})\hat{\mathbf{R}}_s\mathbf{A}^H(\boldsymbol{\omega}) + \hat{\sigma}^2\mathbf{I}\rvert$ $\hat{\mathbf{R}}_s = (\mathbf{A}^H\mathbf{A})^{-1}\mathbf{A}^H\hat{\mathbf{R}}_y\mathbf{A}(\mathbf{A}^H\mathbf{A})^{-1} - \hat{\sigma}^2(\mathbf{A}^H\mathbf{A})^{-1}$ $\hat{\sigma}^2 = \mathrm{tr}[\mathbf{P}^{\perp}_{\mathbf{A}(\boldsymbol{\omega})}\hat{\mathbf{R}}_y]/(N-L)$
$CRB(\boldsymbol{\omega})$	$\frac{\sigma^2}{2T}\left(\mathrm{Re}\{\dot{\mathbf{A}}^H\mathbf{P}^{\perp}_{\mathbf{A}}\dot{\mathbf{A}} \odot \hat{\mathbf{R}}_s\}\right)^{-1}$	$\frac{\sigma^2}{2T}\left(\mathrm{Re}\left[(\dot{\mathbf{A}}^H\mathbf{P}^{\perp}_{\mathbf{A}}\dot{\mathbf{A}}) \odot (\mathbf{R}_s\mathbf{A}^H\mathbf{C}_y^{-1}\mathbf{A}\mathbf{R}_s)\right]\right)^{-1}$
	$\dot{\mathbf{A}} = [\dot{\mathbf{a}}(\omega_1), \dot{\mathbf{a}}(\omega_2), ..., \dot{\mathbf{a}}(\omega_L)]$ where $\dot{\mathbf{a}}(\omega_i) = \begin{bmatrix} 0 \\ je^{j\omega_i} \\ \vdots \\ j(N-1)e^{j(N-1)\omega_i} \end{bmatrix}$	
	$\mathrm{cov}(\hat{\boldsymbol{\omega}}_c) \succeq \mathrm{cov}(\hat{\boldsymbol{\omega}}_u) = CRB_{UML} \succeq CRB_{CML}$	

The parameters are

$$\theta = \{\boldsymbol{\omega}, \mathbf{R}_s, \sigma^2\}$$

Overall the L+1 parameters given by $\{\boldsymbol{\omega}, \sigma^2\}$ are augmented by the $L(L+1)/2$ complex-valued parameters of $\mathbf{R}_s$ (or L^2 real-valued) that are fewer than the L^2 complex-valued parameters due to symmetry.

Table 16.1 summarizes taxonomically the main results of this section (from [62, 63, 101]) in terms of model, ML metrics, and CRB for the estimation of $\boldsymbol{\omega}$.

16.2.1 Conditional ML

This model has been widely investigated in Section 7.2 for multiple measurements and the CML follows from the optimization of

$$\hat{\boldsymbol{\omega}}_c = \arg\min_{\boldsymbol{\omega}} tr[\mathbf{P}^{\perp}_{\mathbf{A}(\boldsymbol{\omega})}\hat{\mathbf{R}}_y]$$

where $\mathbf{P}_{\mathbf{A}(\boldsymbol{\omega})} = \mathbf{A}(\boldsymbol{\omega})\left(\mathbf{A}^H(\boldsymbol{\omega})\mathbf{A}(\boldsymbol{\omega})\right)^{-1}\mathbf{A}^H(\boldsymbol{\omega})$ is the projection onto $\mathcal{R}\{\mathbf{A}(\boldsymbol{\omega})\}$ and

$$\hat{\mathbf{R}}_y = \frac{1}{T}\sum_{t=1}^{T}\mathbf{y}(t)\mathbf{y}^H(t)$$

is the sample correlation. For $L = 1$, the CML reduces to the frequency estimation in Section 9.2, but in general the optimization is not trivial, and various methods have been investigated in the past and are reproposed herein.

16.2.2 Cramér–Rao Bound for Conditional Model

Conditional Model for $T = 1$

For $T = 1$ (one observation), the CRB follows from the FIM for the Gaussian model $\mathbf{y} \sim CN(\mathbf{A}(\boldsymbol{\omega}) \cdot \mathbf{s}, \sigma^2\mathbf{I})$ in Section 8.4.3:

$$[\mathbf{J}(\theta_0)]_{ij} = \frac{2}{\sigma^2}\text{Re}\left\{\frac{\partial \mathbf{x}(\theta)^H}{\partial \theta_i}\frac{\partial \mathbf{x}(\theta)}{\partial \theta_j}\right\}\Bigg|_{\theta=\theta_0}$$

where

$$\mathbf{x}(\theta) = \mathbf{A}(\boldsymbol{\omega})\mathbf{s}$$

and noise power σ^2 is assumed known. Parameters can be separated into amplitudes and phase so that the derivatives are

$$\frac{\partial \mathbf{x}(\theta)}{\partial A_m} = e^{j\phi_m}\mathbf{a}(\omega_m)$$

$$\frac{\partial \mathbf{x}(\theta)}{\partial \phi_m} = jA_m e^{j\phi_m}\mathbf{a}(\omega_m)$$

$$\frac{\partial \mathbf{x}(\theta)}{\partial \omega_m} = A_m e^{j\phi_m}\underbrace{[0, je^{j\omega_m}, \ldots, j(N-1)e^{j\omega_m(N-1)}]^T}_{\dot{\mathbf{a}}(\omega_m)} = A_m e^{j\phi_m}\dot{\mathbf{a}}(\omega_m)$$

and for the L sinusoids the FIM is block-partitioned; the size of each block is $L \times L$

$$\mathbf{J}(\theta_0) = \begin{bmatrix} \mathbf{J}_{AA} & \mathbf{J}_{A\phi} & \mathbf{J}_{A\omega} \\ \mathbf{J}_{A\phi}^H & \mathbf{J}_{\phi\phi} & \mathbf{J}_{\phi\omega} \\ \mathbf{J}_{A\omega}^H & \mathbf{J}_{\phi\omega}^H & \mathbf{J}_{\omega\omega} \end{bmatrix}$$

and the entries are

$$[\mathbf{J}_{AA}]_{km} = 2\text{Re}\left\{\frac{\partial \mathbf{x}(\theta)^H}{\partial A_k}\frac{\partial \mathbf{x}(\theta)}{\partial A_m}\right\} = 2/\sigma^2\text{Re}\left\{e^{j(\phi_k-\phi_m)}\mathbf{a}(\omega_k)^H\mathbf{a}(\omega_m)\right\}$$

$$[\mathbf{J}_{A\phi}]_{km} = 2\text{Re}\left\{\frac{\partial \mathbf{x}(\theta)^H}{\partial \alpha_k}\frac{\partial \mathbf{x}(\theta)}{\partial \phi_m}\right\} = 2/\sigma^2\text{Re}\left\{jA_m e^{j(\phi_k-\phi_m)}\mathbf{a}(\omega_k)^H\mathbf{a}(\omega_m)\right\}$$

$$[\mathbf{J}_{A\omega}]_{km} = 2\text{Re}\left\{\frac{\partial \mathbf{x}(\theta)^H}{\partial \alpha_k}\frac{\partial \mathbf{x}(\theta)}{\partial \omega_m}\right\} = 2/\sigma^2\text{Re}\left\{jA_m e^{j(\phi_k-\phi_m)}\mathbf{a}(\omega_k)^H\dot{\mathbf{a}}(\omega_m)\right\}$$

$$[\mathbf{J}_{\phi\phi}]_{km} = 2\text{Re}\left\{\frac{\partial \mathbf{x}(\theta)^H}{\partial \phi_k}\frac{\partial \mathbf{x}(\theta)}{\partial \phi_m}\right\} = 2/\sigma^2\text{Re}\left\{A_k A_m e^{j(\phi_k-\phi_m)}\mathbf{a}(\omega_k)^H\mathbf{a}(\omega_m)\right\}$$

$$[\mathbf{J}_{\phi\omega}]_{km} = 2\text{Rc}\left\{\frac{\partial \mathbf{x}(\theta)^H}{\partial \phi_k}\frac{\partial \mathbf{x}(\theta)}{\partial \omega_m}\right\} = -2/\sigma^2\text{Re}\left\{A_k A_m e^{j(\phi_k-\phi_m)}\mathbf{a}(\omega_k)^H\dot{\mathbf{a}}(\omega_m)\right\}$$

$$[\mathbf{J}_{\omega\omega}]_{km} = 2\mathrm{Re}\left\{\frac{\partial \mathbf{x}(\boldsymbol{\theta})^H}{\partial \omega_k}\frac{\partial \mathbf{x}(\boldsymbol{\theta})}{\partial \omega_m}\right\} = 2/\sigma^2 \mathrm{Re}\left\{A_k A_m e^{j(\phi_k-\phi_m)}\dot{\mathbf{a}}(\omega_k)^H \dot{\mathbf{a}}(\omega_m)\right\}$$

The CRB in this way does not yield a simple closed form solution for the correlation between frequencies. Just to exemplify, the inner products are

$$\mathbf{a}(\omega_k)^H \mathbf{a}(\omega_m) = \sum_{\ell=0}^{N-1} e^{j(\omega_m-\omega_k)\ell} = e^{j\varphi_{mk}}\frac{\sin((\omega_m-\omega_k)N/2)}{\sin((\omega_m-\omega_k)/2)}$$

where φ_{mk} is a phase-term (not evaluated here). The degree of coupling among the different frequencies decreases provided that the frequency difference is $|\omega_m - \omega_k| \gg 2\pi/N$, and for increasing N as $\mathbf{a}(\omega_k)^H\mathbf{a}(\omega_m)/N \to 0$. Asymptotically with $N \to \infty$, the frequencies are decoupled and the covariance is well approximated by the covariance for one sinusoid in Gaussian noise: $\mathrm{var}[\hat{\omega}_k] \propto 1/N^3$ (See Section 9.2).

Conditional Model for $T \geq 1$

For any T, a closed form for the frequency only was derived in [62]. More specifically, the CRB for the frequency is

$$CRB_c(\boldsymbol{\omega}) = \frac{\sigma^2}{2T}\left(\mathrm{Re}\{\dot{\mathbf{A}}^H \mathbf{P}_{\mathbf{A}}^{\perp}\dot{\mathbf{A}} \odot \hat{\mathbf{R}}_s\}\right)^{-1}$$

where

$$\hat{\mathbf{R}}_s = \frac{1}{T}\sum_{t=1}^{T}\mathbf{s}(t)\mathbf{s}^H(t)$$

$$\dot{\mathbf{A}} = [\dot{\mathbf{a}}(\omega_1), \dot{\mathbf{a}}(\omega_2), \ldots, \dot{\mathbf{a}}(\omega_L)]$$

is the sample correlation of the sources, and the derivatives are $\dot{\mathbf{a}}(\omega_m) = [0, je^{j\omega_m}, \ldots, j(N-1)e^{j\omega_m(N-1)}]^T$.

From the CRB one can obtain some asymptotic properties for $N \to \infty$ and/or $T \to \infty$ that are useful to gain insight into the pros and cons of the frequency estimation methods. Once the CRB has been defined as above ($CRB_c(\boldsymbol{\omega}|N,T)$), the following properties hold when $\hat{\mathbf{R}}_s \to \mathbf{R}_s$ for $T \to \infty$ (or T large enough) [62, 63]:

$$CRB_c(\boldsymbol{\omega}|N, T\to\infty) = \frac{\sigma^2}{2T}\left(\mathrm{Re}\{\dot{\mathbf{A}}^H \mathbf{P}_{\mathbf{A}}^{\perp}\dot{\mathbf{A}} \odot \mathbf{R}_s\}\right)^{-1} = CRB_c^a(\boldsymbol{\omega}) \quad (16.2)$$

$$CRB_c(\boldsymbol{\omega}|N\to\infty, T\to\infty) = \frac{6}{TN^3}\mathrm{diag}(\sigma^2/[\mathbf{R}_s]_{11}, \ldots, \sigma^2/[\mathbf{R}_s]_{LL}) \quad (16.3)$$

where $[\mathbf{R}_s]_{\ell\ell}/\sigma^2$ is the signal to noise ratio for the ℓth source, and $CRB_c^a(\boldsymbol{\omega})$ is the asymptotic CRB. The variance for the ℓth frequency $\mathrm{var}(\hat{\omega}_\ell)$ using the CML is

$$\frac{\mathrm{var}(\hat{\omega}_\ell)}{CRB_c^a(\omega_\ell)} = 1 + \frac{\sigma^2}{N[\mathbf{R}_s]_{\ell\ell}}$$

which proves that the CML is inefficient for $N < \infty$ even though $T \to \infty$, and this makes the use of the conditional model questionable.

16.2.3 Unconditional ML

The model is

$$\mathbf{y}(t) \sim \mathcal{CN}(0, \mathbf{C}_y(\boldsymbol{\theta}))$$

and the covariance is parametric:

$$\mathbf{C}_y(\boldsymbol{\theta}) = \mathbf{A}(\boldsymbol{\omega})\mathbf{R}_s\mathbf{A}^H(\boldsymbol{\omega}) + \sigma^2\mathbf{I}$$

The MLE is the minimizer of the cost function (Section 7.2)

$$\mathcal{L}(\mathbf{y}(t)|\boldsymbol{\theta}) = -\ln|\mathbf{C}_y(\boldsymbol{\theta})| + tr[\mathbf{C}_y^{-1}(\boldsymbol{\theta}) \cdot \hat{\mathbf{R}}_y]$$

for the sample correlation

$$\hat{\mathbf{R}}_y = \frac{1}{T}\sum_{t=1}^{T} \mathbf{y}(t)\mathbf{y}^H(t)$$

that has the property

$$\lim_{T\to\infty} \hat{\mathbf{R}}_y = \mathbf{C}_y(\boldsymbol{\theta})$$

from the law of large numbers or the stationarity of the measurements.

The minimization is very complex using the properties of derivative wrt matrix entries in Section 1.4. However, the minimization is separable in two steps [61]: first the log-likelihood is optimized wrt $\mathbf{R}_s$ (i.e., for each entry of $\mathbf{R}_s$) for a fixed $\boldsymbol{\omega}$ to get a first estimate $\hat{\mathbf{R}}_s = \hat{\mathbf{R}}_s(\boldsymbol{\omega})$:

$$\hat{\mathbf{R}}_s = (\mathbf{A}^H\mathbf{A})^{-1}\mathbf{A}^H\hat{\mathbf{R}}_y\mathbf{A}(\mathbf{A}^H\mathbf{A})^{-1} - \hat{\sigma}^2(\mathbf{A}^H\mathbf{A})^{-1}$$
$$\hat{\sigma}^2 = \text{tr}[\mathbf{P}^{\perp}_{\mathbf{A}(\boldsymbol{\omega})}\hat{\mathbf{R}}_y]/(N-L)$$

and then this is plugged into the log-likelihood resulting in a function of $\boldsymbol{\omega}$ only:

$$\hat{\boldsymbol{\omega}}_u = \arg\min_{\boldsymbol{\omega}} \ln|\mathbf{A}(\boldsymbol{\omega})\hat{\mathbf{R}}_s\mathbf{A}^H(\boldsymbol{\omega}) + \hat{\sigma}^2\mathbf{I}|$$

Minimization is not straightforward and one should use numerical methods.

16.2.4 Cramér–Rao Bound for Unconditional Model

The CRB has no compact and closed form and it can be derived by following the definitions (Section 7.2). A strong argument in favor of the CRB for the unconditional model is that the CRB is attained for $N \to \infty$ as in standard ML theory. The CRB can be evaluated from the FIM (see Section 8.4.3)

$$[\mathbf{J}(\boldsymbol{\theta})]_{ij} = \text{tr}\left\{\mathbf{C}_y^{-1}\frac{\partial \mathbf{C}_y}{\partial \theta_i}\mathbf{C}_y^{-1}\frac{\partial \mathbf{C}_y}{\partial \theta_j}\right\}$$

To sketch the proof, which is very complex for the general case, let $\mathbf{R}_s$ be known (e.g., uncorrelated signals); the derivative

$$\frac{\partial \mathbf{C}_y(\boldsymbol{\theta})}{\partial \omega_i} = \dot{\mathbf{a}}(\omega_i)\mathbf{R}_s\mathbf{A}(\boldsymbol{\omega})^H + \mathbf{A}(\boldsymbol{\omega})\mathbf{R}_s\dot{\mathbf{a}}(\omega_i)^H$$

depends on the ith column of $\mathbf{A}(\boldsymbol{\omega})$ as all the others are zero. The proof can be developed further for this special case, but we propose the general formula for FIM that was derived initially in [63], and subsequently in [104], for T observations:

$$\mathbf{J}(\boldsymbol{\theta}) = \frac{2T}{\sigma^2}\text{Re}\left[(\dot{\mathbf{A}}^H\mathbf{P}_{\mathbf{A}}^{\perp}\dot{\mathbf{A}}) \odot (\mathbf{R}_s\mathbf{A}^H\mathbf{R}_y^{-1}\mathbf{A}\mathbf{R}_s)\right]$$

Evaluation of the corresponding CRB needs to evaluate each of the terms of the FIM that can be simplified in some situations such as for uncorrelated sources ($\mathbf{R}_s = \text{diag}\{\sigma_1^2, \sigma_2^2, ..., \sigma_L^2\}$) and/or orthogonal sinusoids.

16.2.5 Conditional vs. Unconditional Model & Bounds

Conditional and unconditional are just two ways to account for the nuisance of complex amplitudes. Both methods are based on the sample correlation that averages all the T observations, provided that these are drawn from independent experiments with different signals $\mathbf{s}(t)$. Performance comparison between the two methods has been subject of investigations in the past and the main results are summarized below (from [63]).

Let CRB_u and CRB_c be the CRB for unconditional and conditional models, and C_u and C_c the covariance matrix of the UML and CML; then the following inequalities hold:

$$C_c \geq C_u = CRB_u \geq CRB_c$$

This means that:

a) CML is statistically less efficient than UML;
b) UML achieves the unconditional CRB CRB_u;
c) CRB_u is a lower bound on the asymptotic statistical accuracy of any (consistent) frequency estimate based on the data sample covariance matrix;
d) CRB_c cannot be attained.

In summary, the unconditional model has to be preferred as the estimate attains the CRB_u while the estimate based on conditional model can only be worse or equal.

16.3 High-Resolution Methods

Any of the MLE methods needs to search over the L-dimensional space for frequencies, and this search could be cumbersome even if carried out as a coarse and fine search. High-resolution methods are based on the exploration of the algebraic structure of the model in Section 16.1 and attain comparable performance to MLE, and close to the CRB. All these high-resolution estimators are consistent with the number of observations T and have comparable performance.

16.3.1 Iterative Quadratic ML (IQML)

Iterative quadratic ML (IQML) was proposed by Y. Besler and A. Makovski in 1986 [64] to solve iteratively an optimization problem that is made quadratic after re-parameterizing the cost function at every iteration. IQML is the conditional ML for T=1, and generalization to any T is the algorithm MODE (Method for Direction Estimation) tailored for DoAs estimation (Chapter 19) [67].

Let $\mathbf{A}_k = \mathbf{A}(\boldsymbol{\omega}_k)$ be the linear transformation at the kth iteration; IQML solves at the kth iteration an objective where the non-linear term in the kernel is replaced by the solution $\mathbf{A}_{k-1}$ at iteration (k-1):

$$\Psi(\boldsymbol{\omega}_k) = \mathbf{y}^H \mathbf{A}_k \left(\mathbf{A}_{k-1}^H \mathbf{A}_{k-1}\right)^{-1} \mathbf{A}_k^H \mathbf{y}$$

The optimization is wrt $\boldsymbol{\omega}_k$, and then $\mathbf{A}_k$ to iterate until convergence. The advantage is that the optimization of $\Psi(\omega_k)$ can become quadratic provided that the the cost function is properly manipulated by rewriting the optimization using a polynomial representation of the minimization.

Given the signal

$$x[n|\boldsymbol{\omega}] = \sum_{\ell=1}^{L} \alpha_\ell(t) e^{j\omega_\ell n} \rightarrow \mathbf{x} = \mathbf{A}(\boldsymbol{\omega})\mathbf{s}$$

any filter represented in terms of z-transform by the polynomial

$$B(z) = b_0 \prod_{\ell=1}^{L} (1 - e^{j\omega_\ell} z^{-1}) = \sum_{m=0}^{L} b_m z^{-m}$$

is a linear predictor for $x[n|\boldsymbol{\omega}]$ and it is represented by a filter with $M+1$ coefficients $\mathbf{b} = [b_0, b_1, ..., b_M]^T$ with roots in $e^{j\omega_\ell}$: $B(e^{j\omega_\ell}) = 0$ for $\ell = 1, ..., L$. Sinusoids are exactly predictable, so from the definition of prediction, any inner product between vector $\mathbf{b}$ and any subset of length $L+1$ extracted from $x[n|\boldsymbol{\omega}]$ is zero. From the definition of the convolution matrix

$$\mathbf{B}^H = \begin{bmatrix} b_0 & \cdots & b_L & 0 & \cdots & 0 \\ 0 & b_0 & \cdots & b_L & \ddots & \vdots \\ \vdots & \ddots & \ddots & & \ddots & 0 \\ 0 & \cdots & 0 & b_0 & \cdots & b_L \end{bmatrix} \in \mathbb{C}^{(N-L)\times N}$$

it follow that

$$\mathbf{B}^H \mathbf{x} = \mathbf{0}$$

From this identity, it follows that

$$\mathbf{B}^H \mathbf{A}(\boldsymbol{\omega}) = \mathbf{0} \rightarrow \mathcal{R}(\mathbf{B}) = \mathcal{N}(\mathbf{A}^H)$$

and thus

$$\mathbf{P}^{\perp}_{\mathbf{A}(\boldsymbol{\omega})} = \mathbf{P}_{\mathbf{B}}$$

The MLE becomes the estimation of the entries of $\mathbf{b}$ such that

$$\hat{\mathbf{b}} = \arg\min_{\mathbf{b}} \mathbf{y}^H \mathbf{P}_{\mathbf{B}} \mathbf{y} \rightarrow \{\hat{\omega}_\ell\} = \text{roots}\{\hat{\mathbf{b}}\}$$

from the noisy data $\mathbf{y} = \mathbf{x}(\boldsymbol{\omega}) + \mathbf{w}$.

Recalling the idea of IQML, this optimization becomes

$$\mathbf{b}_k = \arg\min_{\mathbf{b}} \{\mathbf{y}^H \mathbf{B} \left(\mathbf{B}^H_{k-1}\mathbf{B}_{k-1}\right)^{-1} \mathbf{B}^H \mathbf{y}\}$$

where the convolution matrix $\mathbf{B}_{k-1}$ is obtained at the previous iteration $\mathbf{b}_{k-1}$. The solution does not appear quadratic, but the product $\mathbf{B}^H\mathbf{y}$ is a convolution and it can be written alternatively as

$$\mathbf{B}^H\mathbf{y} = \underbrace{\begin{bmatrix} y[0] & y[1] & \cdots & y[L] \\ y[1] & y[2] & \cdots & y[L+1] \\ \vdots & \vdots & \ddots & \vdots \\ y[N-L-1] & y[N-L] & \cdots & y[N-1] \end{bmatrix}}_{\mathbf{Y}} \cdot \mathbf{b}$$

and so the optimization becomes quadratic

$$\mathbf{b}_k = \arg\min_{\mathbf{b}} \{\mathbf{b}^H \mathbf{C}_{k-1} \mathbf{b}\}$$

where

$$\mathbf{C}_{k-1} = \mathbf{Y}^H \left(\mathbf{B}^H_{k-1}\mathbf{B}_{k-1}\right)^{-1} \mathbf{Y}$$

Notice that at every step of the minimization, there is the need to add a constraint to avoid the trivial solution $\mathbf{b} = \mathbf{0}$. By adding the constraint that $||\mathbf{b}||^2 = 1$, the optimization reduces to a Rayleight quotient (Section 1.7) and the solution is the eigenvector associated with the minimum eigenvector of $\mathbf{C}_{k-1}$; this is denoted as IQML with quadratic constraint (IQML-QC). Other constraints are $\text{Re}\{b_0\} = 1$ (provided that it is not required to estimate a zero-frequency component, or $\omega_m = 0$). Alternatively $\text{Im}\{b_0\} = 1$ as there is a control of the position of the roots along the circle that is not controlled by IQML-QC. In order to constrain roots along the unit circle of the z-transform, it should at least have complex conjugate symmetry $b_\ell = b^*_{M-\ell}$ (necessary, but not sufficient condition, see Appendix B of Chapter 4) and it follows from the overall constraints (there are L constraints if L is even) that the total number of unknowns of the quadratic problem is L. Furthermore, usually one prefers to rewrite the unknowns into a vector that collects real and imaginary components of $\mathbf{b}_k$: $\tilde{\mathbf{b}} = [\text{Re}\{b_0\}, \text{Im}\{b_0\}, ..., \text{Re}\{b_{(L-1)/2}\}, \text{Im}\{b_{(L-1)/2}\}]^T \in \mathbb{R}^{L-1}$ (assuming to constrain one sample of the complex conjugate symmetry $b_{(L-1)/2+1} = 1$ and removing all the constraints from the optimization problem). This optimization is unconstrained and quadratic and it can be solved using QR factorization [67].

IQML is not guaranteed to converge to the CML solution, and as for any iterative method, convergence depends on the initialization. The initialization $\mathbf{A}_0^H\mathbf{A}_0 = \mathbf{I}_L$ implies that all frequencies are mutually orthogonal (e.g., the sinusoids have frequencies on the grid spacing $2\pi/N$) but any other initialization that exploits the a-priori knowledge of the frequencies (even if approximate) could ease the convergence.

16.3.2 Prony Method

This method was proposed by Prony more than two centuries ago (in 1795) and it is based on the predictability of sinusoids. Similar to the IQML method, for sinusoids:

$$x[n|\boldsymbol{\omega}] + \sum_{k=1}^{L} b_k x[n-k|\boldsymbol{\omega}] = 0$$

Now the set of equations for the observations $y[n] = x[n|\boldsymbol{\omega}] + w[n]$ becomes

$$\begin{bmatrix} y[L-1] & y[L-2] & \cdots & y[0] \\ y[L] & y[L-1] & \cdots & y[1] \\ \vdots & \vdots & \ddots & \vdots \\ y[N-1] & y[N-2] & \cdots & y[N-L] \end{bmatrix} \begin{bmatrix} b_1 \\ b_2 \\ \vdots \\ b_L \end{bmatrix} = - \begin{bmatrix} y[L] \\ y[L+1] \\ \vdots \\ y[N] \end{bmatrix}$$

to estimate the frequencies from the roots: $\{\hat{\omega}_\ell\}$=roots$\{\hat{\mathbf{b}}\}$. The linear system requires that the number of observations are $N \geq 2L-1$, and for $N > 2L-1$ it is overdetermined (Section 2.7) and is appropriate when noise is high. Note that for L real-valued sinusoids, the polynomial order is $2L$ and the full predictability needs $N \geq 4L$ observations.

16.3.3 MUSIC

MUSIC stands for *multiple signal classification [65]* and it is a subspace method based on the property that the first L eigenvectors associated with the leading eigenvalues of $\mathbf{R}_y$ is a basis of the columns of $\mathbf{A}(\omega)$ detailed in Section 16.1.3. Considering the sample correlation

$$\hat{\mathbf{R}}_y = \frac{1}{T}\sum_{t=1}^{T}\mathbf{y}(t)\mathbf{y}^H(t)$$

one can extend the subspace properties as $\hat{\mathbf{R}}_y \to \mathbf{R}_y$ for $T \to \infty$. Let the eigenvector decomposition $\hat{\mathbf{R}}_y = \hat{\mathbf{Q}}\hat{\boldsymbol{\Lambda}}\hat{\mathbf{Q}}^H$, the L eigenvectors $\hat{\mathbf{Q}}_S = [\hat{\mathbf{q}}_1, \hat{\mathbf{q}}_2, ..., \hat{\mathbf{q}}_L]$ associated to the leading eigenvalues span the signal subspace:

$$\mathcal{R}(\hat{\mathbf{Q}}_S) \approx \mathcal{R}(\mathbf{A}(\boldsymbol{\omega}))$$

or alternatively, the remaining eigenvectors $\hat{\mathbf{Q}}_N = [\hat{\mathbf{q}}_{M+1}, \hat{\mathbf{q}}_{M+2}, ..., \hat{\mathbf{q}}_L]$ span the noise subspace

$$\mathcal{R}(\hat{\mathbf{Q}}_N) \approx \mathcal{N}(\mathbf{A}(\boldsymbol{\omega})^H)$$

where "$\approx$" highlights that the equality is exact only asymptotically as $\mathcal{R}(\hat{\mathbf{Q}}_S) \rightarrow \mathcal{R}(\mathbf{A}(\boldsymbol{\omega}))$ for $T \rightarrow \infty$. According to geometrical similarity, $\hat{\mathbf{Q}}_S$ is the basis of the *signal subspace* and $\hat{\mathbf{Q}}_N$ is the complementary basis of *noise subspace* drawn from the limited T observations. From the decomposition, the orthogonality condition

$$\mathbf{A}(\boldsymbol{\omega})^H \hat{\mathbf{Q}}_N \approx \mathbf{0}$$

follows, which is useful for frequency estimation.

MUSIC is based on the condition above rewritten for *each* sinusoid

$$\mathbf{a}(\omega)^H \hat{\mathbf{Q}}_N \hat{\mathbf{Q}}_N^H \mathbf{a}(\omega) \approx 0 \quad \text{for } \omega = \omega_1, ..., \omega_L$$

According to this property for the noise subspace basis, one can use a trial-sinusoid $\mathbf{a}(\omega)$ to have the *noise spectrum*

$$S_N(\omega) = \frac{1}{\sum_{k=L+1}^{N} |\hat{\mathbf{q}}_k^H \mathbf{a}(\omega)|^2} = \frac{1}{\mathbf{a}(\omega)^H \hat{\mathbf{Q}}_N \hat{\mathbf{Q}}_N^H \mathbf{a}(\omega)}$$

and the frequencies are obtained by picking the L values of ω that maximize the noise spectrum $S_N(\omega)$, as for those values the denominator approaches zero. Note that the noise spectrum is characterized by a large dynamic and coarse/fine search strategies are recommended—but still of one variable only.

Root-MUSIC

There are several variants of the MUSIC algorithm. One can look at the noise spectrum as a search for a frequency that nulls $\mathbf{a}(\omega)^H \hat{\mathbf{Q}}_N \hat{\mathbf{Q}}_N^H \mathbf{a}(\omega)$, but this can be replaced by searching for the L pairs of roots $\{z_\ell\}$ in reciprocal position wrt the unit circle over the $2N$ roots of the polynomial

$$\mathbf{a}^H(1/z)\left(\sum_{k=L+1}^{N} \hat{\mathbf{q}}_k \hat{\mathbf{q}}_k^H\right)\mathbf{a}(z) = 0 \leftrightarrow \mathbf{a}^H(1/z)\hat{\mathbf{Q}}_N \hat{\mathbf{Q}}_N^H \mathbf{a}(z) = 0$$

where

$$\mathbf{a}(z) = [1, z^{-1}, ..., z^{-L+1}]^T$$

Since

$$\mathbf{a}(\omega) = \mathbf{a}(z)|_{z=e^{j\omega}}$$

the frequency follows from the angle of these roots:

$$\{\hat{\omega}_\ell\} = \angle \text{roots}\{\mathbf{a}^H(1/z)\hat{\mathbf{Q}}_N \hat{\mathbf{Q}}_N^H \mathbf{a}(z)\}$$

Pisarenko Method

This method was proposed in 1973 by Pisarenko when solving geophysical problems [66] and it is the first example of a subspace method, which inspired all the others. It is basically MUSIC with $N = L + 1$, and thus the noise subspace is spanned by one vector only. Since $\hat{\mathbf{Q}}_N = \hat{\mathbf{q}}_{L+1} = \hat{\mathbf{q}}_N$, the noise spectrum is

$$S_N(\omega) = \frac{1}{|\hat{\mathbf{q}}_L^H \mathbf{a}(\omega)|^2}$$

The root-method reduces to the search for the L roots of the polynomial

$$\sum_{k=1}^{N} \hat{q}^*_{N,k} z^{-(k-1)} = 0$$

with frequencies being the corresponding phases.

MUSIC vs. CRB

MUSIC is a consistent estimator and performance was investigated by the seminal work of Stoica and Nehorai [62]. Once again, the closed form of the covariance reveals some properties of the estimator, but more importantly the comparison with CRB highlights the consistency. Basically, the performance for $T \to \infty$ is more simple and meaningful as this reflects the asymptotic performance. The CRB and the covariance of MUSIC are

$$\text{cov}_{CRB^a}[\hat{\omega}_\ell] = \frac{6\sigma^2}{TN^3} \cdot \frac{1}{[\mathbf{R}_s]_{\ell\ell}}$$
$$\text{cov}_{MUSIC}[\hat{\omega}_\ell] = \frac{6\sigma^2}{TN^3} [\mathbf{R}_s^{-1}]_{\ell\ell}$$

so the ratio

$$\frac{\text{cov}_{MUSIC}[\hat{\omega}_\ell]}{\text{cov}_{CRB^a}[\hat{\omega}_\ell]} = [\mathbf{R}_s]_{\ell\ell}[\mathbf{R}_s^{-1}]_{\ell\ell}$$

increases when $\mathbf{R}_s$ is singular, or sources are correlated. On the other hand, for uncorrelated sources $\text{cov}_{MUSIC}[\hat{\omega}_\ell] = \text{cov}_{CRB^a}[\hat{\omega}_\ell]$, and this motivates its widespread use in this situation. To conclude, great care should be taken when using MUSIC as to the degree of correlation of sources.

Resolution

One of the key aspect of subspace methods such as MUSIC is the large resolution capability. An example can illustrate this aspect. Let $\boldsymbol{\omega} = [-2\pi \times .25,\ 2\pi \times .1,\ 2\pi \times .105]^T$ be the frequencies of three sinusoids with unit amplitudes and statistically independent phases (MUSIC attains the CRB^a in this case); the noise has power $\sigma^2 = 2 \times 10^{-2}$. The frequencies of the two sinusoids are so closely spaced so that they can only be resolved by the periodogram if (necessary condition)

$$2\pi(.105 - .1) \geq \frac{2\pi}{N} \to N \geq 200$$

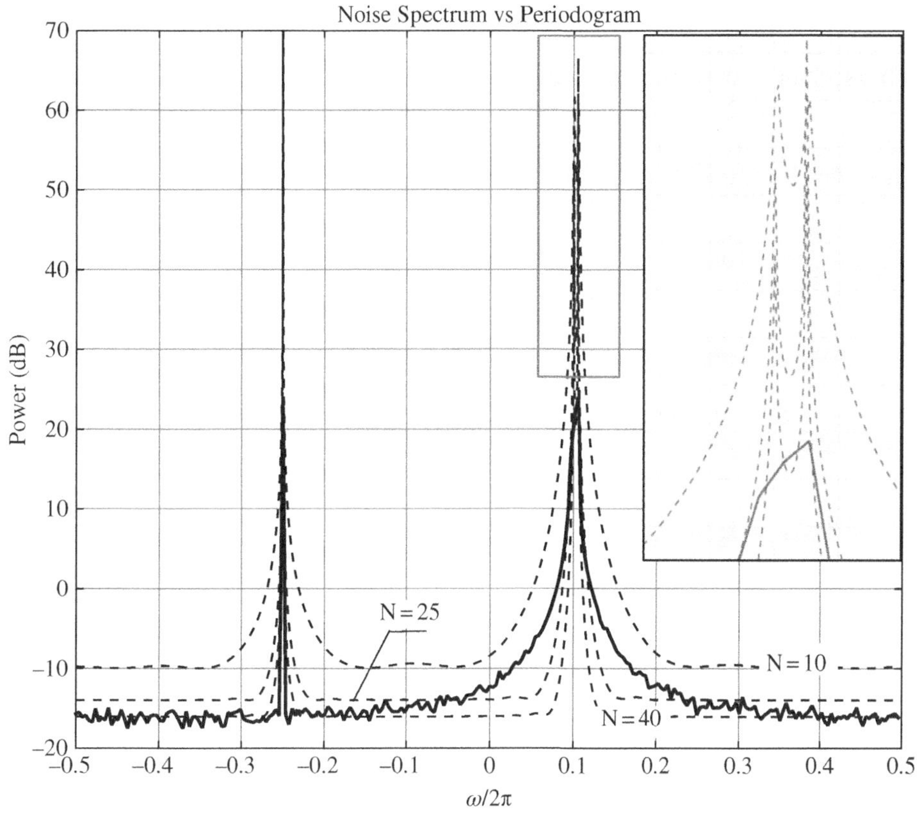

Figure 16.3 Resolution of MUSIC (dashed line) vs. periodogram (solid line).

Figure 16.3 shows the comparison between the periodogram when using a segmentation of $N = 256$ samples (solid line) and MUSIC using $N = 10, 25, 40$ samples (dashed line); details around the dashed box are magnified. The periodogram cannot distinguish two local maxima as they are too close even for $N > 200$, while MUSIC has two well-separated maxima at the two frequencies even for just $N = 10$ samples. On increasing N, the maxima are even more separated (for the sake of plot clarity the total number of samples here are 20×256, simply to average the estimates from 20 Montecarlo trials).

16.3.4 ESPRIT

Estimation of signal parameters via rotational invariance techniques (ESPRIT) was proposed by Roy and Kailath [68] and it is based on the generalization of the predictability of sinusoids. To introduce the idea, let N observations $\mathbf{x}$ be noise-free as superposition of L sinusoids with $N > L$. If partitioning these observations into two subsets of the first and the last $N-1$ observations:

$$\mathbf{x}_1 = [x[0], x[1], ..., x[N-2]]^T = [\mathbf{I}_{N-1}, 0]\mathbf{x}$$
$$\mathbf{x}_2 = [x[1], x[2], ..., x[N-1]]^T = [0, \mathbf{I}_{N-1}]\mathbf{x}$$

the subset $\mathbf{x}_2$ can be derived from $\mathbf{x}_1$ by applying to the subset $\mathbf{x}_1$ a phase-shift to each sinusoid that depends on its frequency, and thus to all sinusoids. If rewriting all to set this phase-shift as an unknown parameter, estimating the phase-shifts to match $\mathbf{x}_1$ and $\mathbf{x}_2$ is equivalent in estimating the frequency of all sinusoids. Moving to a quantitative analysis:

$$\mathbf{x}_1 = \underbrace{[\mathbf{I}_{N-1}, \mathbf{0}_{(N-1)\times 1}]\mathbf{A}(\boldsymbol{\omega})}_{\mathbf{B}_1} \cdot \mathbf{s} = \mathbf{B}_1 \cdot \mathbf{s}$$

$$\mathbf{x}_2 = \underbrace{[\mathbf{0}_{(N-1)\times 1}, \mathbf{I}_{N-1}]\mathbf{A}(\boldsymbol{\omega})}_{\mathbf{B}_2} \cdot \mathbf{s} = \mathbf{B}_2 \cdot \mathbf{s}$$

where $\mathbf{B}_1$ and $\mathbf{B}_2$ are obtained by stripping the first and the last row of

$$\mathbf{A}(\boldsymbol{\omega}) = \begin{bmatrix} 1 & \cdots & 1 \\ e^{j\omega_1} & \cdots & e^{j\omega_M} \\ \vdots & \ddots & \vdots \\ e^{j\omega_1(N-1)} & \cdots & e^{j\omega_M(N-1)} \end{bmatrix}$$

To exemplify, when removing one sample (say the last) for one sinusoid with frequency $\bar{\omega}$, the same data can be resumed by multiplying all the samples by a frequency-dependent term:

$$e^{-j\bar{\omega}}[\mathbf{I}_{N-1}, \mathbf{0}_{(N-1)\times 1}]\mathbf{a}(\bar{\omega}) = [\mathbf{0}_{(N-1)\times 1}, \mathbf{I}_{N-1}]\mathbf{a}(\bar{\omega})$$

that makes the shifted vector coincident with the signal when removing the first sample. It is straightforward to generalize this property

$$\mathbf{B}_1 = \mathbf{B}_2 \cdot \underbrace{\begin{bmatrix} e^{-j\omega_1} & \cdots & 0 \\ \vdots & \ddots & \vdots \\ 0 & \cdots & e^{-j\omega_L} \end{bmatrix}}_{\mathbf{D}(\boldsymbol{\omega})} \Rightarrow \mathbf{B}_1 = \mathbf{B}_2 \cdot \mathbf{D}(\boldsymbol{\omega})$$

where

$$\mathbf{x}_2 \xrightarrow[\mathbf{D}(\boldsymbol{\omega})]{} \mathbf{x}_1$$

maps from $\mathbf{x}_2$ onto $\mathbf{x}_1$ by a proper set of phase-shifts $\mathbf{D}(\boldsymbol{\omega})$ that depend on the unknown frequencies.

In practice, one cannot act on $\mathbf{A}(\boldsymbol{\omega})$ as it is unknown, but rather it is possible to act on the basis $\mathbf{U}_S$ of signal subspace that is obtained from the covariance matrix $\mathbf{R}_y = \mathbf{A}(\boldsymbol{\omega})\mathbf{R}_s\mathbf{A}^H(\boldsymbol{\omega}) + \sigma^2\mathbf{I}$ (in practice from $\hat{\mathbf{R}}_y$). The two matrixes are dependent on one other according to an orthonormal transformation $\mathbf{C}$: $\mathbf{U}_S = \mathbf{A}(\boldsymbol{\omega})\mathbf{C}$. Stripping the first and last row from the basis of the signal subspace $\mathbf{U}_S$ gives:

$$\mathbf{U}_{B_1} = \mathbf{B}_1\mathbf{C} = [\mathbf{I}_{N-1}, 0]\mathbf{A}(\omega)\mathbf{C} = [\mathbf{I}_{N-1}, 0]\mathbf{U}_S$$

$$\mathbf{U}_{B_2} = \mathbf{B}_2\mathbf{C} = [0, \mathbf{I}_{N-1}]\mathbf{A}(\omega)\mathbf{C} = [0, \mathbf{I}_{N-1}]\mathbf{U}_S$$

since

$$\mathbf{B}_2 = \mathbf{U}_{B_2}\mathbf{C}^{-1}$$

Multiplying both sides of $\mathbf{B}_1 = \mathbf{B}_2 \cdot \mathbf{D}$ for $\mathbf{C}$ yields

$$\mathbf{B}_1\mathbf{C} = \mathbf{B}_2 \cdot \mathbf{D} \cdot \mathbf{C}$$

and thus

$$\mathbf{U}_{B_1} = \mathbf{U}_{B_2} \cdot \left(\mathbf{C}^{-1}\mathbf{D}\mathbf{C}\right) = \mathbf{U}_{B_2} \cdot \boldsymbol{\Phi}$$

Revisiting this relationship, it follows that ESPRIT estimation of the frequencies as the phases of the eigenvalues follows from the eigenvalue/eigenvector decomposition of the (unknown) $\boldsymbol{\Phi}$ that maps (by a rotation of the components) $\mathbf{U}_{B_2}$ onto $\mathbf{U}_{B_1}$.

All that remains is the estimation of the transformation $\boldsymbol{\Phi}$, which can be obtained by the LS solution

$$\hat{\boldsymbol{\Phi}} = \left(\mathbf{U}_{B_2}^H \mathbf{U}_{B_2}\right)^{-1} \mathbf{U}_{B_2}^H \mathbf{U}_{B_1}$$

and thus from the eigenvalues of $\hat{\boldsymbol{\Phi}} \in \mathbb{C}^{(N-1)\times(N-1)}$ one gets the frequencies as those phases of the L eigenvalues closer to the unit circle. Notice that the LS estimate of $\boldsymbol{\Phi}$ searches for the combination of $\mathbf{U}_{B_2}$ that justifies (in the LS-sense) $\mathbf{U}_{B_1}$, and any mismatch is attributed to $\mathbf{U}_{B_1}$. However, for the problem at hand, both $\mathbf{U}_{B_1}$ and $\mathbf{U}_{B_2}$ are noisy as they were obtained from noisy observations, and thus any mismatch is not only due to one of the terms. The LS can be redefined in this case and the total LS (TLS) should be adopted as mandatory for ESPRIT [68].

Remark

ESPRIT is more general than only referring to the first and last row of the data vector. It is sufficient that the following property holds:

$$\mathbf{B}_1 = \mathbf{J}_1\mathbf{A}(\boldsymbol{\omega})$$
$$\mathbf{B}_2 = \mathbf{J}_2\mathbf{A}(\boldsymbol{\omega})$$

for any pair of selection matrixes $\mathbf{J}_1$ and $\mathbf{J}_2$ (with $\mathbf{J}_1 \neq \mathbf{J}_2$) of the same dimension $d \times N$ with $L \le d < N$ such that the two subsets differ only by a rigid translation that maps onto a frequency-dependent phase-shift.

16.3.5 Model Order

Many details have been omitted up to now as being simplified or neglected, but when dealing with real engineering problems one cannot avoid the details as the results might have surprises. The methods above are the most used; others have been proposed in the literature with some differences. However, all the estimators assume that the number of frequencies L is known, and this is not always true in practical problems. The number of sinusoids is called the *model order* and its knowledge is as crucial as the estimator itself. Under-estimating the model order misses some sinusoids, mostly those closely spaced apart, while over-estimating might create artifacts. Using the same example in Figure 16.3 with three sinusoids in noise of MUSIC, the noise spectrum when over-estimating the order (say estimating $L_{est} = 4$ frequencies when there are $L = 3$ sinusoids) shows the spectrum with some peaks that add to the peaks of the true sinusoids. This is illustrated in Figure 16.4 for a set of 30 Montecarlo trials with the same situations as the MUSIC example (Figure 16.3).

There are criteria for model order selection based on the analysis of the eigenvalues of the sample correlation matrix $\hat{\mathbf{R}}_y$. Namely, the eigenvalues ordered for decreasing values are

$$\lambda_1(\hat{\mathbf{R}}_y) \geq \lambda_2(\hat{\mathbf{R}}_y) \geq \dots \geq \lambda_L(\hat{\mathbf{R}}_y) > \lambda_{L+1}(\hat{\mathbf{R}}_y) \simeq \dots \simeq \lambda_N(\hat{\mathbf{R}}_y) \simeq \sigma^2$$

so that an empirical criterion to establish the model order is by inspection of the eigenvalues to identify the point when the $N - L$ eigenvalues become approximately constant. Still using the example above of MUSIC, the eigenvalues are shown in Figure 16.5. All of the first three eigenvalues are larger than the others (whose eigenvectors span the noise subspace used for the noise spectrum) and the model order can be easily established as $L_{est} = 3$.

The statistical criterion for model order selection is a classification problem among N possible choices (Section 23.2), and it is based on the likelihood ratio between two

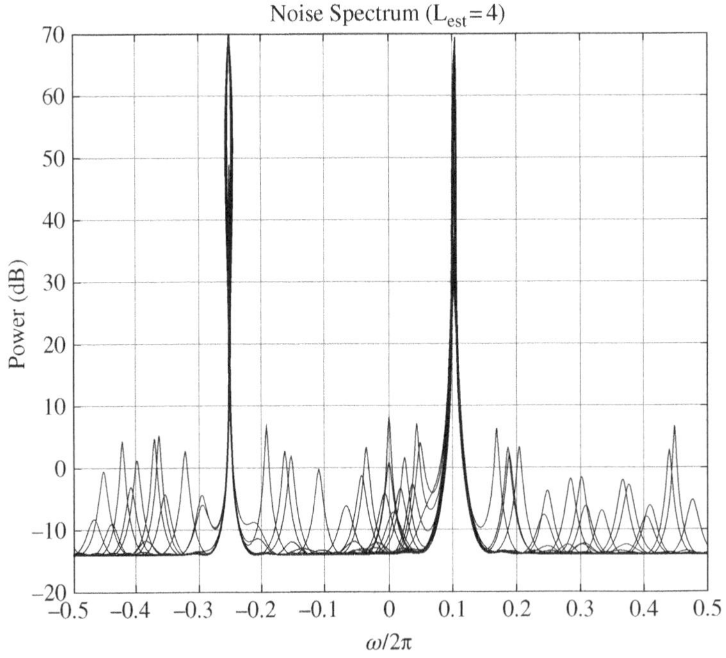

Figure 16.4 MUSIC for four lines from three sinusoids (setting of Figure 17.3).

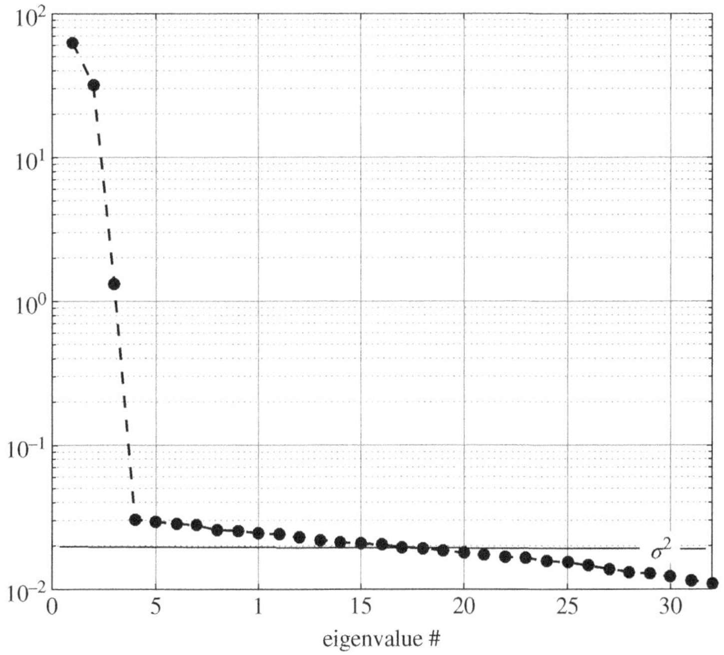

Figure 16.5 Eigenvalues for setting in Figure 17.3.

alternative hypotheses on the eigenvalues $\{\lambda_k(\hat{\mathbf{R}}_y)\}$. By setting d as the test variable, the hypothesis that the $(N-d)$ smallest eigenvalues are equal is compared with the other hypothesis that $(N-(d+1))$ of the smallest are equal. A sufficient statistic is the ratio between the arithmetic and the geometric mean [69]

$$L_d(d) = T(N-d)\cdot \ln\left\{\frac{\frac{1}{N-d}\sum_{k=d+1}^{N}\lambda_k(\hat{\mathbf{R}}_y)}{\left(\prod_{k=d+1}^{N}\lambda_k(\hat{\mathbf{R}}_y)\right)^{\frac{1}{N-d}}}\right\}$$

and when $N-d$ eigenvalues are all equal, $L_d(d)=0$.

Other approaches add to the metric $L_d(d)$ a penalty function related to the degrees of freedom

$$L_d(d)+p(d)$$

and the model order is the minimizer of this new metric. Using the information-theoretic criterion there are two main methods:

- Akaike Information Criterion (AIC) test [70]

$$AIC(d) = L_d(d) + d(2N-d)$$

- Minimum Description Length (MDL) test [71]

$$MDL(d) = L_d(d) + \frac{1}{2}[d(2N - d) + 1]\ln T$$

so that the model order is the argument that minimizes one of these two criteria as being the order that matches the degrees of freedom of the sample covariance $\hat{\mathbf{R}}_y$, which is representative of the dimension of the signal subspace in the frequency estimation problem. The MDL estimate d_{MDL} is consistent and for $T \to \infty$ with $d_{MDL} = L$; while AIC is inconsistent and tends to overestimate the model order. However, for a small number of measurements T, the AIC criterion is to be preferred.

17

Equalization in Communication Engineering

Deconvolution (Section 5.2.3 and Section 10.1) is the linear filter that compensates for a waveform distortion of convolution. In communication systems, equalization refers to the compensation of the convolutive mixing due to the signals' propagation (in this context it is called the *communication channel*), either in wireless or wired digital communications. Obviously, deconvolution is the same as equalization, even if it is somewhat more general as being independent of the specific communication system setup. In MIMO systems (Section 5.3) there are multiple simultaneous signals that are propagating from dislocated sources and cross-interfering; the equalization should separate them (*source separation*) and possibly compensate for the temporal/spatial-channel convolution. The focus of this chapter is to tailor the basic estimation methods developed so far to the communication systems where single or multiple channels are modeled as time-varying, possibly random with some degree of correlation. Signals are drawn from a finite alphabet of messages and are thus non-Gaussian by nature [57]. To limit the contribution herein to equalization methods, the estimation of the communication channel (or channel identification, see Section 5.2.3) is not considered as being estimation in a linear model with a known excitation. In common communication systems, channel estimation is part of the alternate communication of a training signal interleaved with the signal of interest as described in Section 15.2. There are *blind-estimation* methods that can estimate the channel without a deterministic knowledge of transmitted signals, relying only on their non-Gaussian nature. However, blind methods have been proved to be too slow in convergence to cope with time-variation in real systems and thus of low practical interest for most of the communication engineering community.

17.1 Linear Equalization

In linear modulation, the transmitted messages are generated by scaling the same waveform for the corresponding information message that belongs to a finite set of values (or *alphabet*) $\mathcal{A}$ with cardinality $|\mathcal{A}|$ that depends on symbol mapping (Section 15.2). It is common to consider a multi-level modulation with $\mathcal{A} = \{\pm1, \pm3, \dots\}$, or $\mathcal{A} = \{\pm1, \pm3, \dots\} \cup \{\pm j, \pm j3, \dots\}$ if complex valued. The transmitted signal is received after convolution by the channel and symbol-spaced sampling that is modeled as (see Section 15.2):

$$x[n] = h[n] * a[n] + w[n]$$

Statistical Signal Processing in Engineering, First Edition. Umberto Spagnolini.

Companion website: www.wiley.com/go/spagnolini/signalprocessing

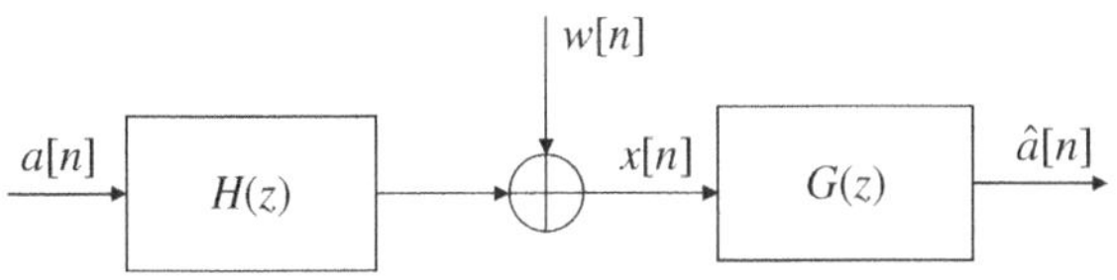

Figure 17.1 Communication system model: channel $H(z)$ and equalization $G(z)$.

with $a[n] \in \mathcal{A}$ and white Gaussian noise

$$r_{ww}[n] = \sigma_w^2 \delta[n]$$

Linear equalization (Figure 17.1) needs to compensate for the channel distortion by an appropriate linear filtering $g[n] \leftrightarrow G(z)$ as

$$\hat{a}[n] = g[n] * x[n] \leftrightarrow \hat{A}(z) = G(z)X(z)$$

so that the equalization error

$$e[n] = \hat{a}[n] - a[n]$$

is small enough, and the equalized signal coincides with the undistorted transmitted one up to a certain degree of accuracy granted by the equalizer.

Inspection of the error highlights two terms:

$$\begin{aligned} e[n] = g[n] * w[n] + (\delta[n] - g[n] * h[n]) * a(n) \leftrightarrow \\ E(z) = G(z)W(z) + (1 - H(z)G(z))A(z) \end{aligned} \tag{17.1}$$

The first is the noise filtered by the equalizer $g[n]$, and the second one is the incomplete cancellation of the channel distortions. Design of linear equalizers as linear estimators of $a[n]$ have been already discussed as deconvolution (Section 12.2); below is a review using the communication engineering jargon. In digital communications, the symbols $a[n]$ are modeled as a stochastic WSS process that is multivalued and non-Gaussian; samples are zero-mean and independent and identically distributed (iid):

$$\mathbb{E}[a[n]a^*[m]] = \sigma_a^2 \delta[m - n]$$

and the autocorrelation sequence is

$$r_{aa}[n] = \sigma_a^2 \delta[n]$$

There are two main criteria adopted for equalization based on which the error of (17.1) is accounted for in optimization.

17.1.1 Zero Forcing (ZF) Equalizer

The zero-forcing (ZF) equalizer is

$$\begin{aligned} G_{ZF}(z) = 1/H(z) \\ \hat{a}[n] = a[n] + g[n] * w[n] \end{aligned}$$

called this way as it nullifies any residual channel distortion: $g[n] * h[n] = \delta[n]$. The benefit is the simplicity but the drawbacks are the uncontrolled amplification of the noise that is filtered by the equalizer with error $e[n] = g[n] * w[n]$ (e.g., any close to zero $H(\omega)$ enhances uncontrollably the noise as $G(\omega) = 1/H(\omega) \rightarrow \infty$), and the complexity in filter design that could contain causal/anti-causal responses over an infinite range.

17.1.2 Minimum Mean Square Error (MMSE) Equalizer

The minimization of the overall error $\mathbb{E}[|e[n]|^2]$ accounts for both noise and residual channel distortion. The z-transform of the autocorrelation of the error:

$$\mathcal{Z}\{r_{ee}[n]\} = S_e(z|\{g[n]\}) = \sigma_w^2 G(z)G^*(1/z) + \sigma_a^2(1 - H(z)G(z))(1 - H^*(1/z)G^*(1/z))$$

depends on filter response $\{g[n]\}$. The minimization of $S_e(z|\{g[n]\})$ wrt the equalizer yields to the orthogonality condition as for the Wiener filter (Section 12.2):

$$\mathbb{E}[E(z)(W^*(1/z) - H^*(1/z)A^*(1/z))] = 0$$

and the (linear) MMSE equalizer is

$$G_{MMSE}(z) = \frac{\sigma_a^2 H^*(1/z)}{\sigma_a^2 H(z)H^*(1/z) + \sigma_w^2}$$

The MMSE equalizer converges to

$$\begin{aligned} G_{MMSE}(z) &\rightarrow G_{ZF}(z) && \text{for } \sigma_w^2 \rightarrow 0 \\ G_{MMSE}(z) &\simeq \tfrac{\sigma_a^2}{\sigma_w^2} H^*(1/z) && \text{for } \sigma_a^2 H(z)H^*(1/z) \gg \sigma_w^2 \end{aligned}$$

for the two extreme situations of small or large noise, respectively and $H^*(1/z)$ is called matched filter as it acts as a filter that correlates exactly with $H(z)$. Needless to say, linear MMSE is sub-optimal and some degree of non-linearity that accounts for the alphabet $\mathcal{A}$ is necessary to attain the performance of a *true* MMSE equalizer.

17.1.3 Finite-Length/Finite-Block Equalizer

This is the convolution of a finite block of symbols **a** with the finite length channel **h**, and its equalization is

$$\mathbf{x} = \mathbf{h} * \mathbf{a} + \mathbf{w} \Longrightarrow \hat{\mathbf{a}} = \mathbf{g} * \mathbf{x} = (\mathbf{g} * \mathbf{h}) * \mathbf{a} + \mathbf{g} * \mathbf{w}$$

The equalization using the ZF condition seeks for

$$\mathbf{g}_{ZF} = \arg\min_{\mathbf{g}} \mathbb{E}_w[||\mathbf{g} * \mathbf{w}||^2] \quad \text{s.t. } \mathbf{g} * \mathbf{h} = \delta[n]$$

and the MMSE

$$\mathbf{g}_{MMSE} = \arg\min_{\mathbf{g}} \mathbb{E}_{a,w}[||\mathbf{g} * \mathbf{x} - \mathbf{a}||^2]$$

To solve these minimizations, one has to replace the notation with the convolution matrix for the channel **h** as in Section 5.2 and Section 5.2.3. The optimizations reduce to the one for MIMO systems and solutions are detailed below in Section 17.3.

17.2 Non-Linear Equalization

Given the statistical properties of signals, the optimal Bayesian estimator is non-linear and resembles the one derived in Section 11.1.3. However, some clever reasoning can simplify the complexity of non-linear Bayesian estimator by the so called Decision Feedback Equalization (DFE). The channel response $h[n] \longleftrightarrow H(z)$ contains causal (post-cursor) and anti-causal (pre-cursor) terms, and the DFE exploits the a-priori information on the finite alphabet $\mathcal{A}$ for some temporary linear estimates based on the causal part of $h[n]$. The approximation of the Bayesian estimator is to use the past to predict the current degree of self-interference for its reduction, after these temporary estimates are mapped onto the finite alphabet $\mathcal{A}$.

Referring to the block diagram in Figure 17.2, the equalizer is based onto two filters $C(z)$ and $D(z)$, and one mapper onto the alphabet $\mathcal{A}$ (*decision*). The decision is a non-linear transformation that maps the residual onto the nearest-level of the alphabet (see Figure 17.3 for $\mathcal{A} = \{\pm 1, \pm 3\}$). The filter $C(z)$ is designed based on the ZF or MMSE criteria, and the filter $D(z) = \sum_{k=1}^{\infty} d_k z^{-k}$ is strictly causal as it uses all the past samples in attempt to cancel the contribution of the past onto the current sample value and leave $\epsilon[n]$ for decisions. When the decision $\hat{a}[n]$ is correct, the cancellation reduces the effect of the tails of the channel response but when noise is too large, the decisions could be wrong and these errors accumulate; the estimation is worse than linear equalizers. This is not surprising as the detector replaces a Bayes estimator with a sharp piecewise non-linearity in place of a smooth one, even when the noise is large (see Section 11.1.3). Overall, the DFE reduces to the design of two filters.

The filter $C(z)$ is decoupled into two terms

$$C(z) = T(z)(1 + D(z))$$

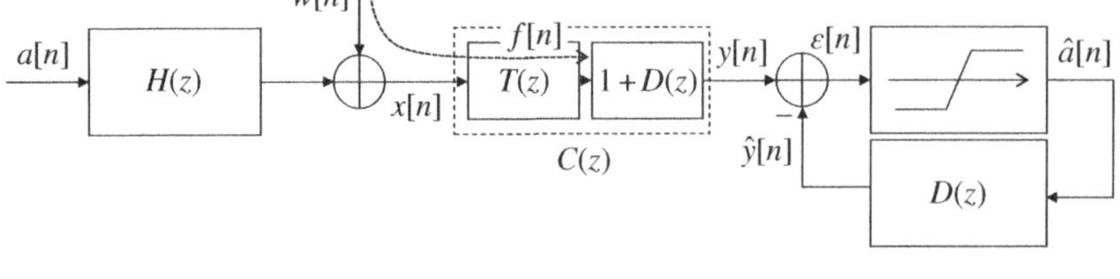

Figure 17.2 Decision feedback equalization.

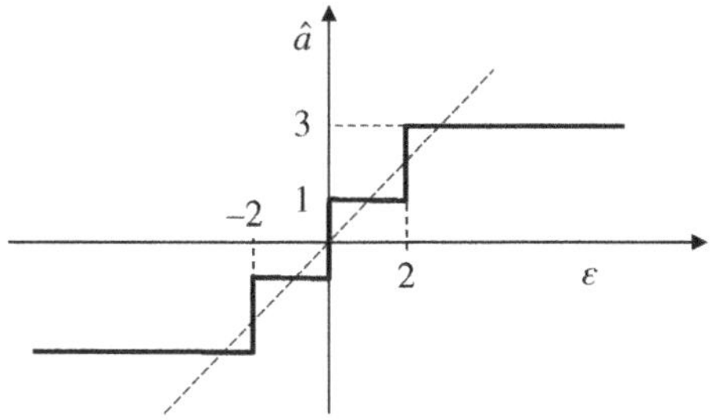

Figure 17.3 Decision levels for $\mathcal{A} = \{\pm1, \pm3\}$.

where $(1 + D(z))$ acts as a compensation of the feedback filter $D(z)$, and the filters to be designed are $T(z)$ and $D(z)$. Assuming that all decisions are correct $\hat{a}[n] = a[n]$ (almost guaranteed for noise σ_w^2 small enough), the z-transform of the error

$$E(z) = \underbrace{A(z)[1 - H(z)T(z)](1 + D(z))}_{\text{residual equalization}} + \underbrace{W(z)T(z)(1 + D(z))}_{\text{noise}} \tag{17.2}$$

contains two terms: the residual of the equalization, and the filtered noise. The design depends on how to account for these two terms.

17.2.1 ZF-DFE

The filter $T(z)$ can be designed to nullify the residual of the filter equalization (zero-forcing condition); then the design of the causal filter $D(z)$ is a free parameter to minimize the filtered noise. More specifically, the choice is

$$T_{ZF}(z) = 1/H(z)$$

and the filter $D(z)$ should be chosen to minimize the power of the filtered noise $f[n] = w[n] * t_{ZF}[n]$ with power spectral density in z-transform

$$S_f(z) = \sigma_w^2 T(z) T^*(1/z)$$

The structure of the filter $1 + D(z)$ is monic and causal, it can be considered as a linear predictor (Section 12.3) for the process $f[n]$ correlated by $T(z)$. The factorization $S_f(z) = S_{f,\min}(z) \times S_{f,\max}(z)$ into min/max phase terms according to the Paley–Wiener theorem (Section 4.4.3) (recall that $S_{f,\max}(z) = S^*_{f,\min}(1/z)$ but it is not necessarily true that $S_{f,\min}(z) = 1/T(z)$, see Appendix B of Chapter 4) yields to the choice

$$1 + D(z) = \frac{1}{S_{f,\min}(z)}$$

Alternatively, and without any change of the estimator, one can make the factorization of the filter into min/max phase $T(z) = T_{\min}(z)T_{\max}(z)$ and the filter becomes

$$1 + D(z) = \frac{1}{T_{\min}(z)T^*_{\max}(1/z)}$$

Filters are not constrained in length, but one can design $1 + D(z)$ using a limited-length linear predictor at the price of a small performance degradation.

17.2.2 MMSE–DFE

In the MMSE method, the minimization is over the ensemble of error (17.2). Namely, the error can be rewritten by decoupling the terms due to the filter $T(z)$

$$f[n] \leftrightarrow F(z) = A(z)(1 - H(z)T(z)) + W(z)T(z)$$

and the linear predictor $1 + D(z)$ as

$$E(z) = F(z)(1 + D(z))$$

The power spectral density of $f[n]$ depends on $t[n]$:

$$S_f(z|\{t[n]\}) = \sigma_a^2(1 - H(z)T(z))(1 - H^*(1/z)T^*(1/z)) + \sigma_w^2 T(z)T^*(1/z)$$

and the MMSE solution is from orthogonality, or equivalently the minimization

$$\frac{\partial S_f(z|\{t[n]\})}{\partial\{t[n]\}} = 0 \rightarrow T_{MMSE}(z) = \frac{H^*(1/z)}{\phi_H(z) + \eta}$$

where $\eta = \sigma_w^2/\sigma_a^2$, and $\phi_H(z) = H(z)H^*(1/z)$.

The linear predictor $1 + D(z)$ follows by replacing the terms. The (minimal) PSD for the filter $t_{MMSE}[n] \leftrightarrow T_{MMSE}(z)$ is

$$S_f(z|\{t_{MMSE}[n]\}) = \frac{\sigma_w^2}{\phi_H(z) + \eta}$$

and its min/max phase factorization defines the predictor

$$1 + D(z) = \frac{1}{S_{f,\min}(z|\{t_{MMSE}[n]\})} = [\phi_H(z) + \eta]_{\min}$$

The filter $C(z)$ for the MMSE–DFE equalizer is

$$C_{MMSE}(z) = T_{MMSE}(z)(1 + D(z)) = \frac{H^*(1/z)}{\phi_H(z) + \eta} \cdot [\phi_H(z) + \eta]_{\min} = \frac{H^*(1/z)}{[\phi_H(z) + \eta]_{\max}}$$

which is a filter with causal/anti-causal component.

Remark. Since the filter's length is unbounded and can have long tails, the implementation of the MMSE–DFE could be quite expensive in terms of computations. A rule of thumb is to smoothly truncate the response $c_{MMSE}[n]$ by a window over an interval that is in the order of 4–6 times the length of the channel $h[n]$. In other cases, the design could account for the limited filter length with better performance compared to a plain truncation, even if smooth.

17.2.3 Finite-Length MMSE–DFE

The MMSE–DFE can be designed by constraining the length of the filters to be limited [75]. Making reference to Figure 17.4, the $N_c \times 1$ filter $\mathbf{c}$ can have a causal and anti-causal component so that

$$y[n] = \sum_{k=-N_1}^{N_2} c[k]x[n-k] = \mathbf{c}^H\mathbf{x}[n]$$

$$\hat{y}[n] = \mathbf{d}^T\hat{\mathbf{a}}[n] \simeq \mathbf{d}^T\mathbf{a}[n]$$

where the last equality is for correct decisions, and $\mathbf{a}[n] = [a[n-1],...,a[n-N_d]]^T$. The error

$$\varepsilon[n] = y[n] - \hat{y}[n] = \mathbf{c}^H\mathbf{x}[n] - \tilde{\mathbf{d}}^T\tilde{\mathbf{a}}[n]$$

is represented in terms of the augmented terms $\tilde{\mathbf{d}} = [1, \mathbf{d}^T]^T$ and $\tilde{\mathbf{a}}[n] = [a[n], \mathbf{a}[n]^T]^T$, and the MSE becomes

$$MSE = \mathbb{E}[|\varepsilon[n]|^2] = \mathbf{c}^H\mathbf{R}_{xx}\mathbf{c} + \tilde{\mathbf{d}}^H\mathbf{R}_{\tilde{a}\tilde{a}}\tilde{\mathbf{d}} - \mathbf{c}^H\mathbf{R}_{x\tilde{a}}\tilde{\mathbf{d}} - \tilde{\mathbf{d}}^H\mathbf{R}_{\tilde{a}x}\mathbf{c}$$

The optimization wrt $\mathbf{c}$ for known $\mathbf{d}$ (the constraint condition is usually added at last step) is quadratic:

$$\mathbf{c} = \mathbf{R}_{xx}^{-1}\mathbf{R}_{x\tilde{a}}\tilde{\mathbf{d}}$$

and the MSE

$$MSE(\tilde{\mathbf{d}}) = \tilde{\mathbf{d}}^H(\mathbf{R}_{\tilde{a}\tilde{a}} - \mathbf{R}_{\tilde{a}x}\mathbf{R}_{xx}^{-1}\mathbf{R}_{x\tilde{a}})\tilde{\mathbf{d}} = \tilde{\mathbf{d}}^H\mathbf{R}\tilde{\mathbf{d}} \tag{17.3}$$

is a quadratic form, but the constraint condition $[\tilde{\mathbf{d}}]_{11} = 1$ has not been used yet.

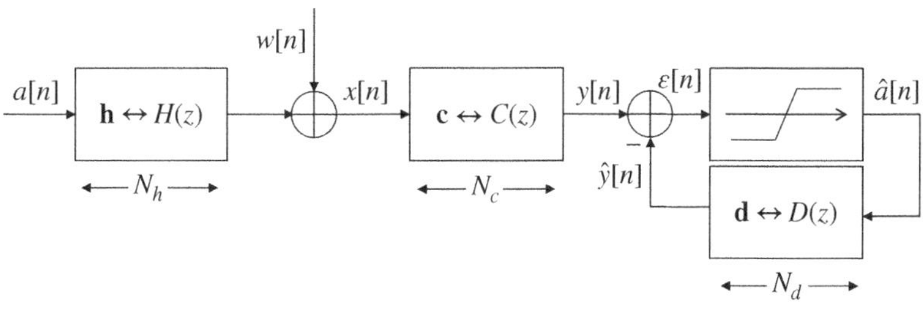

Figure 17.4 Finite length MMSE–DFE.

Rewriting explicitly the MSE (17.3)

$$MSE(\tilde{\mathbf{d}}) = \mathbf{R}_{11} + \mathbf{d}^H\mathbf{R}_{21} + \mathbf{R}_{12}\mathbf{d} + \mathbf{d}^H\mathbf{R}_{22}\mathbf{d}$$

after partitioning

$$\mathbf{R} = \begin{bmatrix} \underset{1\times 1}{\mathbf{R}_{11}} & \underset{1\times N_d}{\mathbf{R}_{12}} \\ \underset{N_d\times 1}{\mathbf{R}_{21}} & \underset{N_d\times N_d}{\mathbf{R}_{22}} \end{bmatrix}$$

it yields the solution

$$\mathbf{d}_{opt} = -\mathbf{R}_{22}^{-1}\mathbf{R}_{21}$$
$$\mathbf{c}_{opt} = \mathbf{R}_{xx}^{-1}\mathbf{R}_{x\tilde{a}} \begin{bmatrix} 1 \\ \mathbf{d}_{opt} \end{bmatrix}$$

and the MSE for this optimized solution becomes $MSE_{\min} = \mathbf{R}_{11} - \mathbf{R}_{12}\mathbf{R}_{22}^{-1}\mathbf{R}_{21}$.
In the case that the channel has length N_h samples, the empirical rules are

$$N_d \simeq N_h$$
$$N_c \simeq 2 \div 3N_h$$

with $N_1 \simeq N_2$. However, since the optimal MSE has a closed form, the filters' length can be optimized for the best (minimum) $MSE_{\min}$ within a predefined set; this is the preferred solution whenever this is practicable.

17.2.4 Asymptotic Performance for Infinite-Length Equalizers

The performance is evaluated in terms of MSE. For a linear MMSE equalizer, the MSE is

$$MSE_{MMSE} = \int_{-1/2}^{1/2} S_e(f|G_{MMSE})df$$

where $S_e(f|G_{MMSE}) = S_e(z = e^{j2\pi f}|G_{MMSE})$, and by using the solutions derived above (argument z is omitted in z-transforms)

$$\begin{aligned} S_e(z|G_{MMSE}) &= \mathbb{E}[E[G^*_{MMSE}W^* + (1 - H^*G^*_{MMSE})A^*]] \\ &= \underbrace{\mathbb{E}[G^*_{MMSE}E[W^* - H^*A^*]]}_{\text{orthogonality}} + \mathbb{E}[EA^*] = \mathbb{E}[EA^*] \\ &= \sigma_a^2\left(1 - \frac{\phi_H(z)}{\phi_H(z)+\eta}\right) = \sigma_a^2\frac{\eta}{\phi_H(z)+\eta} \end{aligned}$$

it is

$$MSE_{MMSE} = \int_{-1/2}^{1/2} \frac{\sigma_w^2}{\eta + |H(f)|^2}df.$$

The steps for the MMSE–DEF equalizer are the same as above by replacing the solutions for the $S_e(z|C_{MMSE}(z), D_{MMSE}(z))$ and after some algebra [74]:

$$MSE_{MMSE-DFE} = \exp\left\{\int_{-1/2}^{1/2} \ln\left(\frac{\sigma_w^2}{\eta + |H(f)|^2}\right) df\right\}$$

Since

$$\exp\left\{\int_{-1/2}^{1/2} \ln \phi(f) df\right\} \leq \int_{-1/2}^{1/2} \exp(\ln \phi(f)) df = \int_{-1/2}^{1/2} \phi(f) df$$

according to the Jensen inequality for convex transformation exp(.) (Section 3.2) for $\phi(f) = \frac{\sigma_w^2}{\eta+|H(f)|^2}$ this proves that

$$MSE_{MMSE-DFE} \leq MSE_{MMSE}$$

The DFE should always be preferred to the linear MMSE equalizer except when noise is too large to accumulate errors.

17.3 MIMO Linear Equalization

MIMO refers to linear systems with multiple input/output as discussed in model definition in Section 5.3, a block of N symbols $\mathbf{a}$ are filtered by an $M \times N$ matrix $\mathbf{H}$ with $M \geq N$ (i.e., more receivers that transmitted symbols) and received signals are

$$\mathbf{x} = \mathbf{Ha} + \mathbf{w} \tag{17.4}$$

with noise $\mathbf{w} \sim \mathcal{CN}(0, \mathbf{R}_{ww})$. MIMO equalization (Figure 17.5) refers to the linear or non-linear procedures to estimate $\mathbf{a}$ from $\mathbf{x}$ provided that $\mathbf{H}$ is known. In some contexts the interference reduction of the equalization is referred as *MIMO decoding* if the procedures are applied to $\mathbf{x}$, and *MIMO precoding* if applied to $\mathbf{a}$ before being transmitted as transformation $\mathbf{g}(\mathbf{a})$ with model $\mathbf{x} = \mathbf{H} \cdot \mathbf{g}(\mathbf{a}) + \mathbf{w}$, possibly linear $\mathbf{g}(\mathbf{a}) = \mathbf{Ga}$. The processing variants for MIMO systems are very broad and are challenging many researchers in seeking different degrees of optimality. The interested reader might start from [72, 73].

To complete the overview, the model (17.4) describes the block (or packet-wise) equalization when a block of N symbols $\mathbf{a}$ are arranged in a packet filtered with a channel of $L+1$ samples, and the overall range is $M = N + L$; the matrix $\mathbf{H}$ in (17.4) is the convolution matrix in block processing.

17.3.1 ZF MIMO Equalization

Block linear equalization is based on a transformation $\mathbf{G} \in \mathbb{C}^{N \times M}$ such that the ZF condition holds: $\mathbf{GH} = \mathbf{I}_N$. The estimate is

$$\hat{\mathbf{a}} = \mathbf{Gx} = \mathbf{a} + \mathbf{Gw}$$

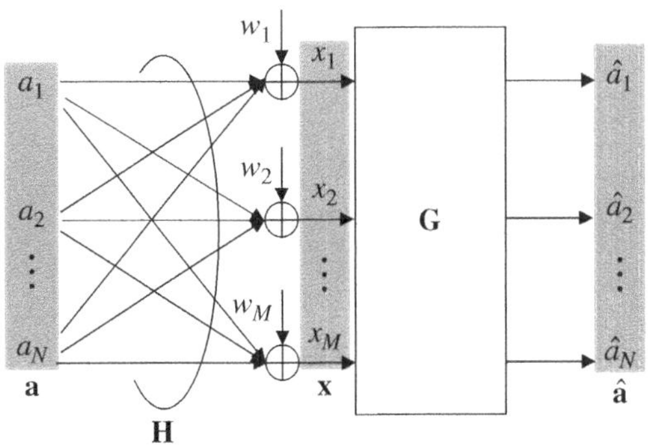

Figure 17.5 Linear MIMO equalization.

and the metric to be optimized is the norm of the filtered noise $\mathbb{E}[||\mathbf{a}-\hat{\mathbf{a}}||^2] = \mathbb{E}[||\mathbf{G}\mathbf{w}||^2]$. The optimization is constrained as

$$\mathbf{G}_{ZF} = \arg\min_{\mathbf{G}} \mathbb{E}[||\mathbf{G}\mathbf{w}||^2] \quad \text{s.t. } \mathbf{G}\mathbf{H} = \mathbf{I}_N$$

to be solved by the Lagrange multiplier method by augmenting the metric with N^2 constraints using the $N \times N$ matrix $\mathbf{\Lambda}$:

$$\mathcal{L}(\mathbf{G}, \mathbf{\Lambda}) = \text{tr}(\mathbf{G}\mathbf{R}_{ww}\mathbf{G}^H) + \text{Re}\left(\text{tr}\left(\mathbf{\Lambda}^H\left(\mathbf{G}\mathbf{H} - \mathbf{I}_N\right)\right)\right)$$

Optimization wrt $\mathbf{G}$ is (Section 1.8 and Section 1.8.2)

$$\frac{\partial \mathcal{L}(\mathbf{G}, \mathbf{\Lambda})}{\partial \mathbf{G}} = \mathbf{G}^*\mathbf{R}_{ww}^T + \frac{1}{2}\mathbf{\Lambda}^*\mathbf{H}^T = \mathbf{0} \rightarrow \mathbf{G} = -\frac{1}{2}\mathbf{\Lambda}\mathbf{H}^H\mathbf{R}_{ww}^{-1}$$

using the constraint condition

$$\mathbf{G}\mathbf{H} = \mathbf{I}_N \rightarrow -\frac{1}{2}\mathbf{\Lambda}\mathbf{H}^H\mathbf{R}_{ww}^{-1}\mathbf{H} = \mathbf{I}_N \rightarrow \mathbf{\Lambda} = -\frac{1}{2}(\mathbf{H}^H\mathbf{R}_{ww}^{-1}\mathbf{H})^{-1}$$

it yields the ZF equalization

$$\mathbf{G}_{ZF} = (\mathbf{H}^H\mathbf{R}_{ww}^{-1}\mathbf{H})^{-1}\mathbf{H}^H\mathbf{R}_{ww}^{-1}$$

Note that the ZF solution can be revised as follows. Since $\mathbf{R}_{ww} \succ \mathbf{0}$, the Cholesky factorization $\mathbf{R}_{ww} = \mathbf{R}_{ww}^{1/2}\mathbf{R}_{ww}^{H/2}$ into lower/upper triangular matrixes is unique, and similarly $\mathbf{R}_{ww}^{-1} = \mathbf{R}_{ww}^{-H/2}\mathbf{R}_{ww}^{-1/2}$. The ZF estimate is

$$\hat{\mathbf{a}}_{ZF} = (\tilde{\mathbf{H}}^H \tilde{\mathbf{H}})^{-1} \tilde{\mathbf{H}}^H \tilde{\mathbf{x}}$$

where $\tilde{\mathbf{H}} = \mathbf{R}_{ww}^{-1/2} \mathbf{H}$ and $\tilde{\mathbf{x}} = \mathbf{R}_{ww}^{-1/2} \mathbf{x}$; it can be interpreted as the pseudoinverse of $\tilde{\mathbf{H}}$ for the pre-whitened observation $\tilde{\mathbf{x}}$.

17.3.2 MMSE MIMO Equalization

The linear MMSE estimator has already been derived in many contexts. To ease another interpretation, the estimator is the minimization of the MSE

$$MSE(\mathbf{G}) = \mathbb{E}[(\mathbf{G}\mathbf{x} - \mathbf{a})^H (\mathbf{G}\mathbf{x} - \mathbf{a})$$

for the symbols with correlation $\mathbb{E}[\mathbf{a}\mathbf{a}^H] = \mathbf{R}_{aa}$. The MMSE solution follows from the orthogonality (Section 11.2.2)

$$\hat{\mathbf{a}}_{MMSE} = (\mathbf{H}^H \mathbf{R}_{ww}^{-1} \mathbf{H} + \mathbf{R}_{aa}^{-1})^{-1} \mathbf{H}^H \mathbf{R}_{ww}^{-1} \mathbf{x}$$

This degenerates into the ZF for large signals (or $\mathbf{R}_{aa}^{-1} \to 0$) as $\hat{\mathbf{a}}_{MMSE} \to (\mathbf{H}^H \mathbf{R}_w^{-1} \mathbf{H})^{-1} \mathbf{H}^H \mathbf{R}_{ww}^{-1} \mathbf{x}$; and for large noise it becomes $\hat{\mathbf{a}}_{MMSE} \to \mathbf{H}^H \mathbf{R}_{ww}^{-1} \mathbf{x}$. The MSE for $\hat{\mathbf{a}}_{MMSE}$ can be evaluated after the substitutions.

17.4 MIMO–DFE Equalization

The MIMO–DFE in Figure 17.6 is just another way to implement the cancellation of interference embedded into MIMO equalization, except that the DFE cancels the interference from decisions—possibly the correct ones. In MIMO systems this is obtained from the ordered cancellation of samples, where the ordering is a degree of freedom that does not show up in equalization as in these systems, samples are naturally ordered by time. Before considering the DFE, it is essential to gain insight into analytical tools to establish the equivalence between DFE equalization and its MIMO counterpart.

17.4.1 Cholesky Factorization and Min/Max Phase Decomposition

The key algebraic tool for MIMO–DFE is the Cholesky factorization that transforms any $\mathbf{R} \succ \mathbf{0}$ into the product of two triangular matrixes, one the Hermitian transpose of the other—sometimes referred as the (unique) square-root of $\mathbf{R}$:

$$\mathbf{R} = \mathbf{R}^{1/2} \mathbf{R}^{H/2} = \mathbf{L} \mathbf{\Sigma}^2 \mathbf{L}^H \text{ for } \mathbf{R} \succ 0$$

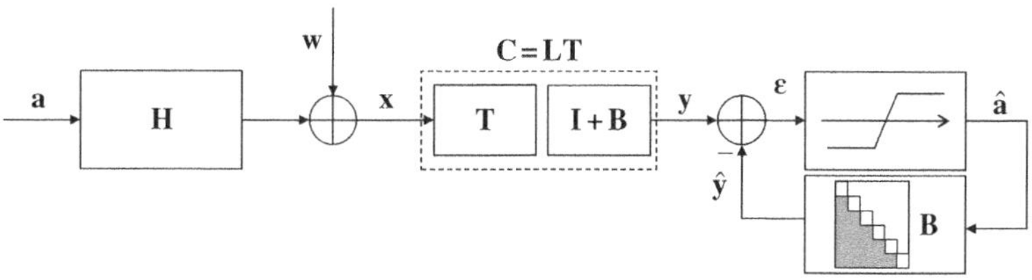

Figure 17.6 MIMO–DEF equalization.

$$[\mathbf{L}]_{ij} = \ell_{ij} = \begin{cases} 1 \text{ for } i = j \\ 0 \text{ for } \forall j > i \end{cases} \quad \text{lower triangular matrix}$$

where the entries of the diagonal matrix $\mathbf{\Sigma}^2 = \text{diag}(\sigma_1^2, \sigma_2^2, \ldots, \sigma_N^2)$ are all positive. For zero-mean rvs, Cholesky factorization of the correlation matrix is the counterpart of min/max phase decomposition of the autocorrelation sequence. To prove this, let $\mathbf{x} = [x_1, x_2, \ldots, x_N]^T \sim \mathcal{CN}(0, \mathbf{R}_{xx})$ be an $N \times 1$ vector with covariance $\mathbf{R}_{xx} = \mathbf{\Sigma}_{xx}^2 = \text{diag}(\sigma_1^2, \sigma_2^2, \ldots, \sigma_N^2)$. Transformation by a lower triangular matrix $\mathbf{L}$ with normalized terms along the diagonal is

$$y_i = x_i + \sum_{k=1}^{i-1} \ell_{ik} x_k \rightarrow \mathbf{y} = \mathbf{L}\mathbf{x}$$

and the covariance is

$$\mathbf{R}_{yy} = \mathbf{L}\mathbf{\Sigma}_{xx}^2 \mathbf{L}^H$$

If it is now assumed that a vector $\mathbf{y}$ has covariance $\mathbf{R} = \mathbf{R}_{yy} = \mathbf{L}\mathbf{\Sigma}_{xx}^2\mathbf{L}^H$, the transformation can be inverted as $\mathbf{x} = \mathbf{L}^{-1}\mathbf{y}$—still lower triangular ($[\mathbf{L}^{-1}]_{ij} = \bar{\ell}_{ij} = 0$ for $\forall j > i$). The ith entry is

$$x_i = y_i + \underbrace{\sum_{k=1}^{i-1} \bar{\ell}_{ik} y_k}_{-\hat{y}_i(\mathbf{L}^{-1})} = (\mathbf{e}_i^T \mathbf{L}^{-1})\mathbf{y}$$

where $(\mathbf{e}_i^T \mathbf{L}^{-1})$ extracts the ith row from $\mathbf{L}^{-1}$. Since $\mathbb{E}[|x_i|^2] = \sigma_i^2$ and $\mathbb{E}[\mathbf{x}\mathbf{x}^H] = \mathbf{\Sigma}_{xx}^2$ then $(\mathbf{e}_i^T \mathbf{L}^{-1})$ is a linear predictor that uses all the samples $1, 2, ..., i-1$ (similar to a causal predictor over a limited set of samples) to estimate y_i, and σ_i^2 is the variance of the prediction error. If choosing $\mathbf{L}^{-H}\mathbf{y}$ the linear predictor makes use of the complementary samples $i+1, i+2, ..., N$ to estimate y_i and the predictor would be reversed (anti-causal). For a WSS random process, the autocorrelation can be factorized into the convolution of min/max phase sequences, and these coincide with the Cholesky factorization of the Toeplitz structured correlation matrix $\mathbf{R}$ for $N \rightarrow \infty$. However when N is small, the Cholesky factorization of the correlation matrix $\mathbf{R}$ takes the boundary effects into account, and the predictors (i.e., the rows of $\mathbf{L}^{-1}$) are not just the same shifted copy for every entry.

17.4.2 MIMO–DFE

In MIMO–DFE, the received signal $\mathbf{x}$ is filtered by a compound filter to yield

$$\mathbf{y} = \mathbf{C}\mathbf{x} = \mathbf{L}\mathbf{T}\mathbf{x} = (\mathbf{I} + \mathbf{B})\mathbf{T}\mathbf{x}$$

where the normalized ($[\mathbf{L}]_{ii} = 1$ for $\forall i$) lower triangular matrix $\mathbf{L}$ is decoupled into the strictly lower triangular $\mathbf{B}$ and the identity $\mathbf{I}$ to resemble the same structure as DFE in Section 17.2. The matrix DFE acts sequentially on decisions as the product

$$\mathbf{L}\hat{\mathbf{a}} = \begin{bmatrix} \hat{a}_1 \\ \hat{a}_2 + \ell_{21}\hat{a}_1 \\ \vdots \\ \hat{a}_N + \sum_{k=1}^{N-1} \ell_{Nk}\hat{a}_k \end{bmatrix} = \hat{\mathbf{a}} + \underbrace{\begin{bmatrix} 0 \\ \ell_{21}\hat{a}_1 \\ \vdots \\ \sum_{k=1}^{N-1} \ell_{Nk}\hat{a}_k \end{bmatrix}}_{\mathbf{B}\hat{\mathbf{a}}} = (\mathbf{I}+\mathbf{B})\hat{\mathbf{a}}$$

isolates the contributions and $\mathbf{B}\hat{\mathbf{a}}$ sequentially orders the contributions arising from the upper lines toward the current one: $\sum_{k=1}^{i-1} \ell_{ik}\hat{a}_k$. The fundamental equation of DFE is the equality of the decision variable from the previous decisions carried out from the upper lines:

$$\hat{\mathbf{a}} = \mathbf{C}\mathbf{x} - \mathbf{B}\hat{\mathbf{a}}.$$

From the partition above:

$$\mathbf{L}\hat{\mathbf{a}} = \mathbf{L}\mathbf{T}\mathbf{x}$$

according to the model

$$\mathbf{L}\hat{\mathbf{a}} = \mathbf{L}\mathbf{T}\mathbf{H}\mathbf{a} + \mathbf{L}\mathbf{T}\mathbf{w}$$

There are two metrics to be optimized based on the double constraint on the structure of $\mathbf{L}$ according to the ZF or MMSE constraint:

$$\text{ZF:} \begin{cases} \hat{\mathbf{a}} = \mathbf{T}\mathbf{H}\mathbf{a} + \mathbf{T}\mathbf{w} \longrightarrow \mathbf{T}_{ZF} = \arg\min_{\mathbf{T}}\{\mathbb{E}[||\mathbf{T}\mathbf{w}||^2]\} \;\; \text{st } \mathbf{T}\mathbf{H} = \mathbf{I} \\ \mathbf{L}_{ZF} = \arg\min_{\mathbf{L}}\{\mathbb{E}[||\mathbf{L}\mathbf{T}_{ZF}\mathbf{w}||^2]\} \end{cases}$$

$$\text{MMSE:} \begin{cases} \hat{\mathbf{a}} = \mathbf{T}\mathbf{x} \longrightarrow \mathbf{T}_{MMSE} = \arg\min_{\mathbf{T}}\{\mathbb{E}[||\mathbf{T}\mathbf{x} - \mathbf{a}||^2]\} \\ \mathbf{L}_{MMSE} = \arg\min_{\mathbf{L}}\{\mathbb{E}[||\mathbf{L}(\mathbf{T}_{MMSE}\mathbf{x} - \mathbf{a})||^2]\} \end{cases}$$

For both criteria the second step, which is based on the first solution, is peculiar for DFE and it implies the optimization of a metric with the constraint on the normalized lower triangular structure of the matrix $\mathbf{L}$. Namely, in both cases it follows the minimization

$$\min_{\mathbf{L}}\{\text{tr}[\mathbf{L}\boldsymbol{\Phi}\mathbf{L}^H]\} \tag{17.5}$$

for a certain positive definite matrix $\boldsymbol{\Phi}$ that is, for the two cases and $\mathbf{R}_{aa} = \sigma_a^2\mathbf{I}$:

$$\begin{aligned} \boldsymbol{\Phi}_{ZF} &= \mathbf{T}_{ZF}\mathbf{R}_w\mathbf{T}_{ZF}^H \\ \boldsymbol{\Phi}_{MMSE} &= \sigma_a^2(\mathbf{I} - \mathbf{T}_{MMSE}\mathbf{H}) \end{aligned}$$

Since $\boldsymbol{\Phi} \succ \mathbf{0}$, its Cholesky factorization, is

$$\boldsymbol{\Phi} = \mathbf{D}\boldsymbol{\Sigma}_{\Phi}\mathbf{D}^H$$

where $\mathbf{\Sigma}_\Phi = \text{diag}(\phi_1, \phi_2, ...\phi_N)$ with $\phi_k > 0$. On choosing

$$\mathbf{L} = \mathbf{D}^{-1}$$

it follows that

$$\text{tr}[\mathbf{L}\mathbf{\Phi}\mathbf{L}^H]\Big|_{\mathbf{L}=\mathbf{D}^{-1}} = \text{tr}[\mathbf{\Sigma}_\Phi]$$

Solutions for AWGN $\mathbf{w} \sim \mathcal{CN}(0, \mathbf{I}\sigma_w^2)$ are derived below.

ZF–DFE for AWGN: $\mathbf{w} \sim \mathcal{CN}(\mathbf{0}, \mathbf{I}\sigma_w^2)$
The optimization

$$\mathbf{T}_{ZF} = \arg\min_{\mathbf{T}}\{\text{tr}[\mathbf{T}\mathbf{T}^H]\} \text{ st } \mathbf{T}\mathbf{H} = \mathbf{I}$$

yields the same solution as in Section 17.3.1:

$$\mathbf{T}_{ZF} = (\mathbf{H}^H\mathbf{H})^{-1}\mathbf{H}^H = \mathbf{H}^\dagger$$

Since the noise after filtering $\mathbf{T}_{ZF}$ is $\mathbf{w}' = \mathbf{T}_{ZF}\mathbf{w}$, the metric of (17.5) becomes $\mathbf{\Phi}_{ZF} = (\mathbf{H}^H\mathbf{H})^{-1}$ and the filter $\mathbf{L}_{ZF}$ follows from its Cholesky factorization

$$\mathbf{L}_{ZF}^H\mathbf{\Sigma}_\Phi^{-1}\mathbf{L}_{ZF} = \mathbf{H}^H\mathbf{H}$$

MMSE–DFE for AWGN: $\mathbf{w} \sim \mathcal{CN}(\mathbf{0}, \mathbf{I}\sigma_w^2)$
The solution of the first term is straightforward from orthogonality (or the MMSE–MIMO solution):

$$\mathbf{T}_{MMSE} = \arg\min_{\mathbf{T}}\{\mathbb{E}[||\mathbf{T}\mathbf{x} - \mathbf{a}||^2]\} = (\mathbf{H}^H\mathbf{H} + \gamma\mathbf{I})^{-1}\mathbf{H}^H = \mathbf{H}^H(\mathbf{H}\mathbf{H}^H + \gamma\mathbf{I})^{-1}$$

using here any of the two equivalent representations (Section 11.2.2), with $\gamma = \sigma_n^2/\sigma_a^2$. The matrix $\mathbf{\Phi}_{MMSE} = \sigma_a^2(\mathbf{I} - \mathbf{H}^H(\mathbf{H}\mathbf{H}^H + \gamma\mathbf{I})^{-1}\mathbf{H}) = (\mathbf{H}^H\mathbf{H} + \gamma\mathbf{I})^{-1}$, where the second equality follows from the matrix inversion lemma (Section 1.1.1), and $\mathbf{L}_{MMSE}$ is from the Cholesky factorization.

18

2D Signals and Physical Filters

A 2D signal is any function that depends on two variables: these could be either space-space, or space-time, or any other. 2D signals are very common in statistical signal processing and this is a situation where multidimensionality offers remarkable benefits, but specific processing tools need to be designed. Properties of 2D signals are very specific—see [77] for a comprehensive treatment. Most of the properties are just an extension of 1D signals, but there are specific properties that are a consequence of the augmented dimension. Appendix A covers the main properties useful here. Images are the most common 2D signals, and these have specific processing tools for image manipulation and enhancement that are not considered extensively here as they are too specific, but [78, 80] are excellent references on the topic. Statistical signal processing is known to be far more effective when considering parametric models for the data, and this is the subject of this chapter where 2D signals are generated and manipulated according to their physical generation models based on partial differential equations (PDEs). Furthermore, there is strong interest in revisiting PDEs as some 2D filtering in image processing can be reduced to smoothing PDE and smoothing-enhancing PDE. The reader is encouraged to read the reference by G. Aubert and P. Kornprobst [81].

The 2D Fourier transform of any 2D signal (Appendix B) is

$$s(x,y) \leftrightarrow S(\omega_x, \omega_y) = S(u,v)$$

where the angular frequency[1] ω_x, ω_y (rad/m) and frequency u, v (cycles/m) are mutually related to one another as

$$\begin{aligned} \omega_x &= 2\pi u \\ \omega_y &= 2\pi v \end{aligned}$$

When one of the variables is time, the 2D signal is

$$s(x,t) \leftrightarrow S(\omega_x, \omega_t) = S(u,f)$$

and the unit of the Fourier transform of time variable ω_t (or simply ω) is in rad/sec, or frequency f (cycles/sec or Hz). Representation of 2D signals (Figure 18.1) is either in the

1 In some contexts this is referred to as wavenumber to highlight the spatial nature, and the notation is K_x, K_y. Its inverse is the wavelength $\lambda_x = 2\pi/K_x$ and $\lambda_y = 2\pi/K_y$ rather than the spatial period.

Statistical Signal Processing in Engineering, First Edition. Umberto Spagnolini.

Companion website: www.wiley.com/go/spagnolini/signalprocessing

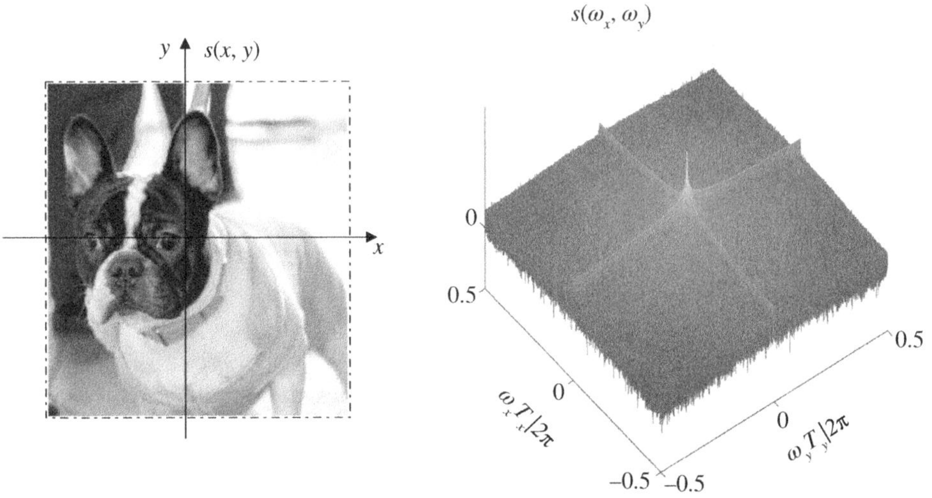

Figure 18.1 2Df signal $s(x,y)$ and its 2D Fourier transform $S(\omega_x,\omega_y)$.

form of gray images as illustrated (note that color images add the color to luminance intensity, but in any case color is another superimposed 2D signal), or a 3D graphical representation as for $|S(\omega_x,\omega_y)|$ paired in the same figure. Some of the properties of 2D signals are extensions from 1D, others are peculiar to 2D; all the most relevant properties are summarized in Appendixes A/B or [77, 83], and readers not skilled in 2D signals and transformations are encouraged to revise therein first.

18.1 2D Sinusoids

A 2D (complex) sinusoid is the signal

$$s(x,y) = e^{j(\bar{\omega}_x x + \bar{\omega}_y y)} = e^{j\bar{\omega}_x x} e^{j\bar{\omega}_y y}$$

with angular frequency (ω_x,ω_y), which is separable with periods

$$\bar{T}_x = \frac{2\pi}{\bar{\omega}_x}$$
$$\bar{T}_y = \frac{2\pi}{\bar{\omega}_y}$$

along the two directions. Transforming into polar coordinates for both space and frequency:

$$\begin{cases} x = \rho\cos\theta \\ y = \rho\sin\theta \end{cases} \leftrightarrow \begin{cases} \overline{\omega}_x = \overline{\Omega}\cos\overline{\varphi} \\ \overline{\omega}_y = \overline{\Omega}\sin\overline{\varphi} \end{cases}$$

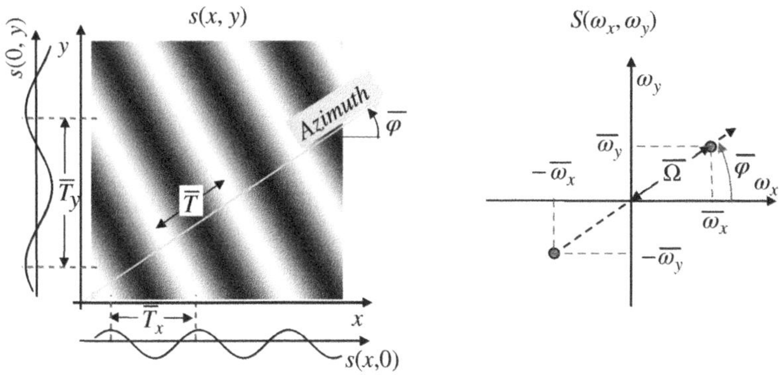

Figure 18.2 2D sinusoid $\cos(\bar{\omega}_x x + \bar{\omega}_y y)$ and the 2F Fourier transform $\frac{1}{2}\delta(\omega_x \pm \bar{\omega}_x, \omega_y \pm \bar{\omega}_y)$ (grey dots).

the sinusoid in polar coordinates is

$$s(x,y) = s_p(\rho,\theta) = e^{j\rho\bar{\Omega}\cos(\theta-\bar{\varphi})}$$

and changing the angular direction θ modifies the angular frequency:

$$\tilde{\Omega} = \bar{\Omega}\cos(\theta - \bar{\varphi})$$

Frequency depends on the angular direction $\bar{\varphi}$, which is max/min for two settings

$$\tilde{\Omega}_{\text{max}} = \bar{\Omega} = \frac{2\pi}{\bar{T}} \text{ for } \theta = \bar{\varphi}$$
$$\tilde{\Omega}_{\text{min}} = 0 \text{ for } \theta = \bar{\varphi} \pm \frac{\pi}{2}$$

Namely, a sinusoid is characterized by the angular direction of minimum period (or maximum frequency) $\theta = \bar{\varphi}$, or the *azimuth* of the sinusoid. The orthogonal direction is the equi-phase direction and the signal is cylindrical. Some of the geometrical properties of the 2D sinusoid $\cos(\bar{\omega}_x x + \bar{\omega}_y y)$ are illustrated in Figure 18.2.

The Fourier transforms for complex sinusoids (Appendix B):[2]

$$e^{j(\bar{\omega}_x x+\bar{\omega}_y y)} \leftrightarrow \delta(\omega_x - \bar{\omega}_x, \omega_y - \bar{\omega}_y)$$
$$\cos(\bar{\omega}_x x + \bar{\omega}_y y) \leftrightarrow \frac{1}{2}\delta(\omega_x \pm \bar{\omega}_x, \omega_y \pm \bar{\omega}_y)$$

are Dirac delta functions with the same orientation $\bar{\varphi}$ as the azimuth of the sinusoid(s). In general, the usage of 2D sinusoids is far more common than one expects, not only

2 Recall the shorthand notation very useful for 2D signals:

$$\delta(a \pm b) = \delta(a+b) + \delta(a-b)$$
$$\delta(a \pm b, c \pm d) = \delta(a+b, c+d) + \delta(a-b, c-d)$$

as signal decomposition by 2D Fourier transform, but to modulate signals. From the properties of the Fourier transform:

$$s(x,y)\cos(\overline{\omega}_x x+\overline{\omega}_y y) \leftrightarrow \frac{1}{2}S(\omega_x-\overline{\omega}_x,\omega_y-\overline{\omega}_y)+\frac{1}{2}S(\omega_x+\overline{\omega}_x,\omega_y+\overline{\omega}_y)$$

and this generates high/low spatial frequencies. Mutual modulation of sinusoids creates artifacts that are known as Moiré patterns, detailed below.

18.1.1 Moiré Pattern

Multiplication of two spatial sinusoids yields two other sinusoids with combined frequency and azimuths (Figure 18.3):

$$\begin{aligned}\cos(\omega_{x1}x+\omega_{y1}y)\cos(\omega_{x2}x+\omega_{y2}y) &\leftrightarrow \frac{1}{2}\delta(\omega_x\pm\omega_{x1},\omega_y\pm\omega_{y1}) ** \frac{1}{2}\delta(\omega_x\pm\omega_{x2},\\ &\qquad \omega_y\pm\omega_{y2})\\ &=\frac{1}{4}\delta(\omega_x+\omega_{x1}\pm\omega_{x2},\omega_y+\omega_{y1}\pm\omega_{y2})\\ &+\frac{1}{4}\delta(\omega_x-\omega_{x1}\pm\omega_{x2},\omega_y-\omega_{y1}\pm\omega_{y2})\end{aligned}$$

that, beyond the formula, are clearer if illustrated. Referring to Figure 18.3, the two sinusoids that follow from the product have low (with azimuth φ_b) or high (with azimuth φ_a) spatial frequencies that depend on the frequency of the original signals. The low-frequency component is usually more visible than the high-frequency when gray-scale imaging is used for displaying. The low frequency pattern, called a Moiré pattern, is more clearly visible than the high-frequency due to the low-pass sensitivity of the eyes.[3] A Moiré pattern has the ability to capture small differences in frequency and/or azimuth of two sinusoids that are not visible otherwise.

As an example, take two picket fences built parallel to each other as in Figure 18.4, but with small negligible tilting that unfortunately cannot be appreciated. The fence has a periodic blocking effect on light due to pickets with spacing T, and thus the transmitted light appears to any observer as a spatial sinusoid with values ranging from 0 (or dark as no light gets through the fence) to 1 (light fully passes through the fence). If far enough away, the pickets show as a sinusoid with angular period T/L (in rad) that is not visible if

$$\frac{\pi}{180}\frac{L}{T} > 7\ cycles/\deg$$

when the fence is appreciated as a fully shaded area due to the poor resolution of the eyes to high spatial frequency. The two picket fences now appear as the product of two sinusoids ranging between 0 and 1 with slightly different frequency due to the

3 The sensitivity of the eye is approximately low-pass with a maximum angular frequency of 6–7 cycles/deg measured with respect to the number of cycles in every degree (the center of the eye has lower spatial bandwidth than the side). Therefore, any spatial sinusoid that appears to have an angular frequency larger than 7 cycles/deg is not resolved as a sinusoid, but rather as a shade.

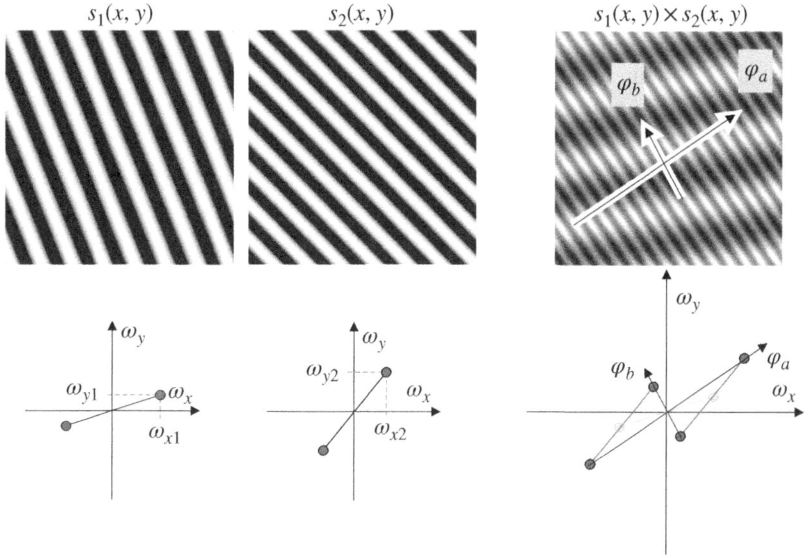

Figure 18.3 Product of 2D sinusoids.

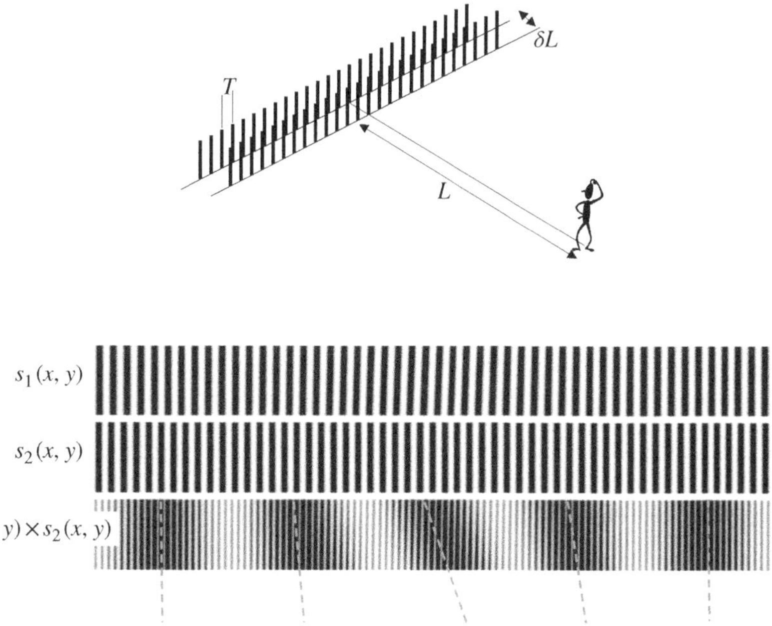

Figure 18.4 Moiré pattern from two picket fences with distance δL.

view angle. However, if one of the sinusoids has a small tilt, this becomes visible from far away as the result of the fences' alignment is two spatial sinusoids, with the low frequency one as the only one visible (dotted lines in Figure 18.4 that also shows the high-frequency). The resulting low-frequency spatial sinusoid tilts too at the center due to the pickets' tilting (here changing along the fence up to 4 deg at the center), while the spatial frequency depends on the inter-fence spacing δL as can be proved by simple geometrical reasoning. Furthermore, the Moiré pattern can spatially isolate the area where the two interfering images are changing, and this is useful to highlight the anomaly areas as here with fences.

18.2 2D Filtering

2D sinusoids are the basis for the decomposition of any 2D signal, and the 2D linear shift-invariant (LSI) system is the most common way to manipulate 2D signals. The response to a 2D signal $s(x,y)$ is from the 2D convolution

$$g(x,y) = s(x,y) ** h(x,y) \longleftrightarrow G(\omega_x,\omega_y) = S(\omega_x,\omega_y)H(\omega_x,\omega_y)$$

where $h(x,y)$ is the 2D impulse response of the LSI system. Sampled 2D signals can be represented by their samples as 2D sequences (Appendix B)

$$s[m,n] = s(mT_x, nT_y)$$

and 2D discrete convolution for a filter $h[m.n]$ is defined as

$$g[m,n] = \sum_{i,j} h[i,j]s[m-i,n-j] = h[m,n] ** s[m,n]$$

2D discrete signals defined over a limited interval can be conveniently ordered in a matrix $\mathbf{S}$ of dimensions $M \times N$, and similarly for the filter

$$\mathbf{H} = \begin{bmatrix} \times & & \times & & \times \\ & h[-1,1] & h[0,1] & h[1,1] & \\ \times & h[-1,0] & h[0,0] & h[1,0] & \times \\ & h[-1,-1] & h[0,-1] & h[1,-1] & \\ \times & & \times & & \times \end{bmatrix}$$

(only entries around $[0,0]$ are indicated here to illustrate the ordering; the others are just indicated by $\times$). Similarly to the 1D case, it is convenient to use a compact notation for 2D discrete convolution:

$$\mathbf{G} = \mathbf{H} * \mathbf{S}$$

and the extention of $\mathbf{S}$ is augmented by the range of the filter $\mathbf{H}$. In some special cases, the convolution can be written as the product of two matrixes (see Appendix C). To visualize, Figure 18.5 shows the filtering of a 2D signal represented by the dog-image (top-left) that is filtered by a low-pass filter either with Gaussian-shape $H(\omega_x,\omega_y) = \exp[-(\omega_x^2+\omega_y^2)/2\Omega_o^2]$ or a rectangular one with comparable cutoff frequency ($H(\omega_x,\omega_y)$ is in the white box). The lower part of Figure 18.5 shows the discrete derivative along x and y with impulse response $\mathbf{H} = [-1/2, 0, 1/2]$ and $\mathbf{H} = [1/2, 0, -1/2]^T$, respectively

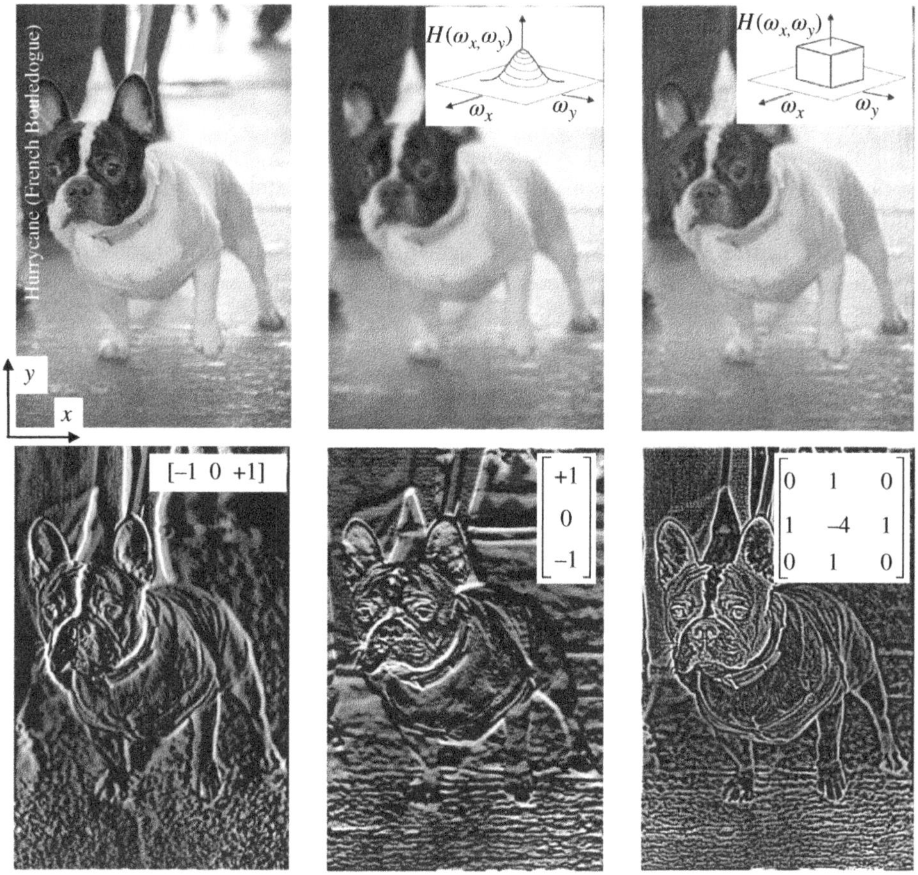

Figure 18.5 Image (top-left) and its 2D filtering.

(the scale 1/2 is to account for centering the derivative at $[0, 0]$). It also shows the discrete second order derivative (Laplace operator discussed later in chapter)

$$\mathbf{H} = \frac{1}{4}\begin{bmatrix} 0 & 1 & 0 \\ 1 & -4 & 1 \\ 0 & 1 & 0 \end{bmatrix}$$

Derivatives have the capability to highlight discontinuities in the image along each axis, and the second order derivative is a more isotropic edge enhancer.

Filters for discrete 2D signals over a rectangular grid are inherently sensitive to the grid-shape, and this creates artifacts that are oriented along the filtering direction. This can be appreciated by comparing the image after an isotropic Gaussian filter with an anisotropic rectangular one. Isotropic filtering, possibly with circular symmetry, is highly desirable in image processing, and methods have been developed to approximate a circular symmetric filter by rectangular supported ones using multiple filters (see e.g.,

[79]). Filter design is a key area in image processing, and filter responses $\mathbf{H}$ are designed depending on the objectives. The reader is referred to [78, 80] for further details.

18.2.1 2D Random Fields

2D signals are often represented by samples of a stochastic rv, and every 2D signal is just one sample of an ensemble of 2D signals with the same statistical properties. The joint pdf is difficult to handle due to the large number of variables, and thus first- and second-order central moments are used in place under the assumption of stationarity.[4]

Mean and covariance for a stationary random field are defined as

$$\mathbb{E}[s[\ell_m, \ell_n]] = \mu_s$$

$$\mathrm{cov}[s[m+\ell_m, n+\ell_n], s[m,n]] = c_{ss}[m,n]$$

These are dependent not on the absolute position (ℓ_m, ℓ_n), but rather on the displacement (m,n) and thus its covariance is shift-invariant. The covariance is central symmetric; for any (m,n) it is

$$c_{ss}[m,n] = c_{ss}[-m,-n] \text{ with } |c_{ss}[m,n]| \le c_{ss}[0,0] = \mathrm{var}[s[\ell_m, \ell_n]] = \sigma_s^2.$$

A white random field is when any two distinct elements of $s[m,n]$ are mutually uncorrelated:

$$c_{ss}[m,n] = \sigma_s^2 \delta[m]\delta[n]$$

In 2D signals the neighboring samples typically have similar values, and this defines other correlation models such as

$$\begin{array}{ll} \text{separable} & c_{ss}[m,n] = \sigma_s^2 \rho_1^{|m|} \rho_2^{|n|} \\ \text{circularly symmetric} & c_{ss}[m,n] = \sigma_s^2 \exp\{-\alpha\sqrt{m^2+n^2}\} \end{array}$$

where ρ_1 and ρ_2 define the degree of spatial correlation, with $|\rho_1| < 1$ and $|\rho_2| < 1$.

The 2D power spectral density (PSD) is the 2D FT of the covariance

$$S_{ss}(\omega_x, \omega_y) = \mathcal{F}_{2D}\{c_{ss}[m,n]\},$$

when filtering a random field $g[m,n] = s[m,n] ** h[m,n]$, the power spectral density depends on the filter

$$S_{gg}(\omega_x, \omega_y) = |H(\omega_x, \omega_y)|^2 S_{ss}(\omega_x, \omega_y)$$

4 Even though stochastic modeling of 2D signals is widely adopted, mostly in image processing, stationarity in 2D is highly questionable. The image in the example is one sample of the stochastic field—all neighboring pixels behave similarly and can be modeled with distance-dependent correlation. However, the image has areas of homogeneous gray due to the subject, and local stationarity within regions is more appropriate than global stationarity for the whole image. Methods are designed to cope with this local stationarity, by isolating the boundary of these regions (e.g., texture segmentation by edge detection) and by adapting stochastic-related processing region-by-region.

If the excitation is white, $S_{ss}(\omega_x, \omega_y) = \sigma_s^2$ and the PSD depends on the filter's characteristic $S_g(\omega_x, \omega_y) = \sigma_s^2 |H(\omega_x, \omega_y)|^2$ that shapes the PSD accordingly, and similarly for the correlation among samples as for rational PSD in 1D signals (Section 4.4). The peculiarity of 2D is that it is complex to define the notion of causality, and this limits the possibility to extend the PSD generation mechanisms that were detailed for 1D stochastic processes to 2D (e.g., the Paley–Wiener theorem, Section 4.4.3).

18.2.2 Wiener Filtering

Any estimator can be defined based on samples of a 2D signal and the central moments of the 2D signal as a random field; once defined, the notion of filtering orientation is the counterpart of causality in 1D. For instance, let $h[m,n]$ be supported over the 3rd quadrant as in Figure 18.6; the linear predictor is

$$\hat{s}[m,n] = -\sum_{i>0,j>0} a[i,j]s[m-i,n-j] = -\mathbf{a}^T\mathbf{s}[m,n]$$

with a definition of vectors $\mathbf{a}$ and $\mathbf{s}[m,n]$ that is congruent with the convolutive model, and sign is just for notation. The LMMSE predictor is just an extension of Section 12.3 that, for a zero-mean field, is

$$\hat{\mathbf{a}} = -\left(\mathbb{E}[\mathbf{s}[m,n]\mathbf{s}^T[m,n]]\right)^{-1}\mathbb{E}[\mathbf{s}[m,n]s[m,n]]$$

where correlations of the random field enter into the terms $\mathbb{E}[\mathbf{s}[m,n]\mathbf{s}^T[m,n]]$ and $\mathbb{E}[\mathbf{s}[m,n]s[m,n]]$.

The 2D Wiener filter $a[m,n]$ is not support-constrained and it is used to estimate a stationary field $g[m,n]$ from another field $s[m,n]$

$$\hat{g}[m,n] = a[m,n] ** s[m,n]$$

Following the same analytical steps of Section 12.1, the Wiener filter is defined in the spectral domain

$$A(\omega_x, \omega_y) = \frac{S_{gs}(\omega_x, \omega_y)}{S_{ss}(\omega_x, \omega_y)}$$

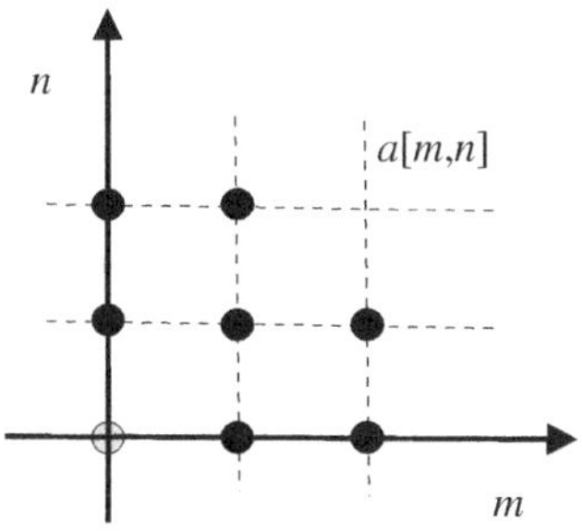

Figure 18.6 Causal 2D filter.

where

$$S_{gs}(\omega_x,\omega_y) = \mathcal{F}_{2D}\{\text{cov}[s[m+\ell_m,n+\ell_n],g[m,n]]\}$$
$$S_{ss}(\omega_x,\omega_y) = \mathcal{F}_{2D}\{\text{cov}[s[m+\ell_m,n+\ell_n],s[m,n]]\} = \mathcal{F}_{2D}\{c_{ss}[m,n]\}$$

2D deconvolution is a common application field of Wiener filtering, and it has many similarities with 1D deconvolution in Section 12.2. Let a random field $g[m,n]$ be filtered by a (known) filter $h[m,n]$ with some white Gaussian noise superimposed:

$$s[m,n] = h[m,n] ** g[m,n] + w[m,n]$$

The objective is to estimate the original field $g[m.n]$. The MMSE deconvolution is

$$A(\omega_x,\omega_y) = \frac{H^*(\omega_x,\omega_y)S_{gg}(\omega_x,\omega_y)}{|H(\omega_x,\omega_y)|^2 S_{gg}(\omega_x,\omega_y) + \sigma_w^2}$$

After some manipulation, the inverse of the filter $h[m,n]$ is

$$A(\omega_x,\omega_y) = \frac{1}{H(\omega_x,\omega_y)} \times C(\omega_x,\omega_y)$$

except for a real-valued weighting term

$$C(\omega_x,\omega_y) = \left(1 + \frac{\sigma_w^2}{|H(\omega_x,\omega_y)|^2 S_{gg}(\omega_x,\omega_y)}\right)^{-1} \leq 1$$

that depends on the power spectral density of $g[m,n]$, and filter $h[m,n]$. The term $|H(\omega_x,\omega_y)|^2 S_{gg}(\omega_x,\omega_y)$ is the PSD of the filtered signal of interest, and for spectral regions where this signal dominates, $C(\omega_x,\omega_y) \simeq 1$ and the Wiener deconvolution coincides with the inverse filtering; otherwise when the PSD of noise dominates, $C(\omega_x,\omega_y|t) \simeq 0$ and there is no noise enhancement effect.

18.2.3 Image Acquisition and Restoration

Static images are snapshots by an acquisition system (e.g., camera, smartphone, etc.) Every point in 3D space is projected onto a point in the image plane and the collection of points in space makes the image represented in luminance coding. The pinhole approximation of an imaging system in Figure 18.7 shows how every point is mapped onto a point in the image plane through a very small pinhole. Use of a lens system enables the use of the pinhole model with a higher intensity of light as this is mandatory for actual sensing devices at the image plane. Image acquisition systems have several non-idealities such as lens aberrations, defocusing, geometric distortions, and relative movements/vibrations of camera and scene, and these introduce a degradation of the images that is called blurring. *Image restoration* refers to all image manipulations to restore the quality of the image as if ideally collected before the acquisition at the object plane. Wiener filtering is a key processing tool in image restoration, but restoration algorithms are far richer than what is discussed here to illustrate a role of Wiener filtering in image deblurring.

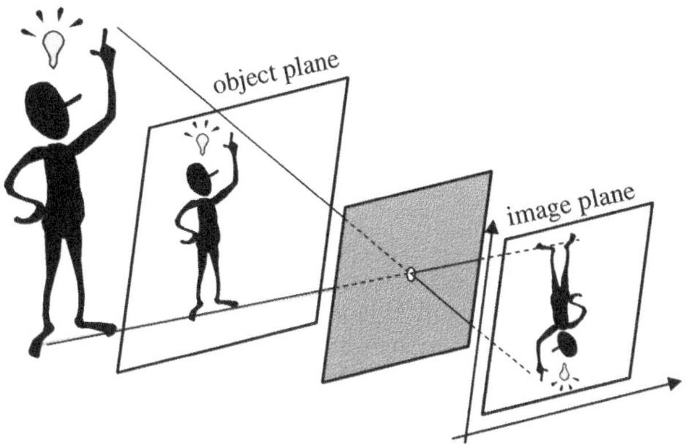

Figure 18.7 Image acquisition system.

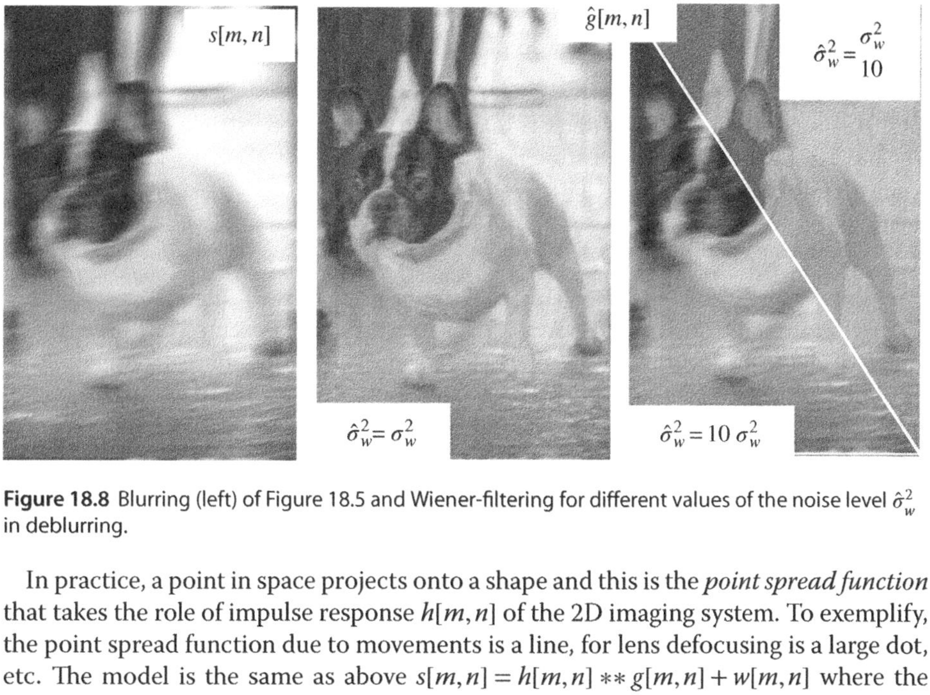

Figure 18.8 Blurring (left) of Figure 18.5 and Wiener-filtering for different values of the noise level $\hat{\sigma}_w^2$ in deblurring.

In practice, a point in space projects onto a shape and this is the *point spread function* that takes the role of impulse response $h[m,n]$ of the 2D imaging system. To exemplify, the point spread function due to movements is a line, for lens defocusing is a large dot, etc. The model is the same as above $s[m,n] = h[m,n] ** g[m,n] + w[m,n]$ where the desired (undistorted) image $g[m,n]$ is estimated by Wiener deconvolution that needs to know (or estimate separately) the filter $h[m,n]$ and the noise level σ_w^2.

Figure 18.8 of the dog illustrates the effect of blurring corresponding to an horizontal movement of approx. 5% of the whole image indicated as $s[m,n]$ at image plane, with some noise superimposed with power σ_w^2. After deblurring, the image $\hat{g}[m,n]$ is restored, and when the estimated noise power is $\hat{\sigma}_w^2 = \sigma_w^2$, the quality of the image is acceptable, and the small artifacts are due to horizontal movement. If deblurring with an estimated noise power $\hat{\sigma}_w^2$ that is over/under estimated compared to the true value, the image has greater detail (or higher frequency components) if $\hat{\sigma}_w^2 < \sigma_w^2$ as the deconvolution has larger cutoff frequency, but also higher noise. The opposite is true if $\hat{\sigma}_w^2 > \sigma_w^2$, as deconvolution shrinks the restoration bandwidth and several details are lost.

18.3 Diffusion Filtering

The PDE that describes the dynamics of diffusion in solids, or similarly for heat in 2D (e.g., within a thin plane) is

$$\frac{\partial^2 s(x,y,t)}{\partial x^2} + \frac{\partial^2 s(x,y,t)}{\partial y^2} = \frac{1}{D}\frac{\partial s(x,y,t)}{\partial t}$$

where $s(x,y,t)$ stands for density or temperature and D is a diffusion coefficient (e.g., thermal conductivity) that is considered here as constant. The use of the Fourier transform to solve this PDE is well consolidated and it was the subject of the seminal work of Joseph Fourier in 1822, *Théorie analytique de la chaleur*. Out of the many engineering problems around diffusion PDE, here we focus attention to two sets:

- $s(x,y,t=0) \longrightarrow s(x,y,t)$ evolution vs. time of density from a certain initial condition $s(x,y,t=0)$
- $s(x,y=0,t) \longrightarrow s(x,y,t)$ extrapolation of the density over a plane from the evolution measured along a line

Both these solutions have a counterpart in filtering as evolution vs. time corresponds to 2D Gaussian filters, and extrapolation is the deconvolution of 2D Gaussian filtering as in image restoration (Section 18.2.3).

18.3.1 Evolution vs. Time: Fourier Method

Evolution versus time is the problem that occurs when it is of interest to study the evolution from some initial condition $s(x,y,t=0) = s_o(x,y)$ represented by the distribution of the diffusion at the initialization (e.g., distribution of contaminant, temperature, etc.) In this case, time is the parameter, and the Fourier transform of the PDE wrt the variables x, y is

$$D(\omega_x^2 + \omega_y^2)S(\omega_x, \omega_y, t) + \frac{dS(\omega_x, \omega_y, t)}{dt} = 0$$

The differential equation has the solution

$$S(\omega_x, \omega_y, t) = S_o(\omega_x, \omega_y) \times e^{-tD(\omega_x^2+\omega_y^2)}$$

that filters the initial condition $s_o(x,y)$ according to the circular symmetric filter

$$H(\omega_x, \omega_y | t) = e^{-tD(\omega_x^2+\omega_y^2)} = e^{-\frac{\Omega^2}{2\Omega_t^2}}$$

where the cutoff-angular frequency

$$\Omega_t = (2Dt)^{-1/2}$$

decreases with time. Diffusion acts as a low-pass filter that smears the initial condition $s_o(x,y)$ isotropically in space and diffuses it over the whole space (here assumed unlimited) with pulse response:

$$h(x, y|t) = \frac{1}{4\pi Dt} exp(-(x^2 + y^2)/4Dt)$$

The time-reversed case $s(x, y, t) \longrightarrow s(x, y, t = 0)$ can be framed in the same setting, and the estimation of the initial distribution from the distribution at any arbitrary time t is a high-pass filter

$$e^{\Omega^2/2\Omega_t^2}$$

that should account for any noise by using the MMSE criteria to avoid excessive noise enhancement. Let the observation be modeled as

$$s_{obs}(x, y, t) = s(x, y, t) + w(x, y)$$

with a Gaussian spatially uncorrelated noise

$$\mathbb{E}[w(x, y)w(x', y')] = \sigma_w^2 \delta(x - x', y - y')$$

The estimate of $s(x, y, t = 0)$ is by adapting the Wiener filtering theory for deconvolution in Section 18.2.2:

$$\hat{s}_{MMSE}(x, y, t = 0) = a(x, y|t) ** s_{obs}(x, y, t)$$

The MMSE deconvolution in the frequency domain is

$$A(\omega_x, \omega_y|t) = e^{tD(\omega_x^2 + \omega_y^2)} \times C(\omega_x, \omega_y|t)$$

where the weighting term

$$C(\omega_x, \omega_y|t) = \frac{1}{1 + \sigma_w^2 / S_{ss}(\omega_x, \omega_y|t)} \leq 1$$

depends on the power spectral density of $s(x, y, t)$ that, for the special case of Gaussian and spatially uncorrelated excitation $s(x, y, t = 0)$, is

$$S_{ss}(\omega_x, \omega_y|t) = S_{ss}(\omega_x, \omega_y|t = 0) \times |H(\omega_x, \omega_y|t)|^2 = \sigma_s^2 |H(\omega_x, \omega_y|t)|^2$$

These effects are shown by the example in Figure 18.9 with the text "PdM" (which stands for *Politecnico di Milano*) that compares image restoration methods, except that the filter here is based on a physical model of the diffusion PDE that is isotropic Gaussian.

18.3.2 Extrapolation of the Density

When time evolution of the density is known along a line (e.g., a temperature is known on one side of a plate) $s(x, y = 0, t)$, it is of interest to extrapolate the density elsewhere. The density $s(x, y = 0, t)$ represents the boundary condition to extrapolate the diffusion elsewhere in $y \geq 0$, and the space coordinate y is the parameter. According to the

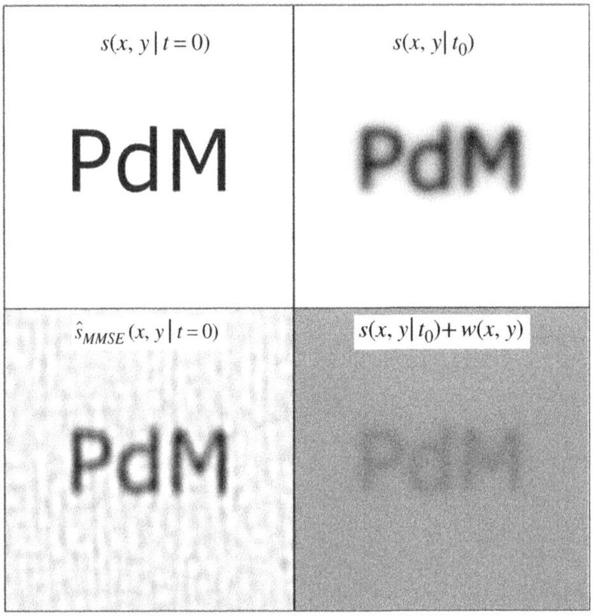

Figure 18.9 Original data (top-left), diffuse filtering (top-right), noisy diffuse filtering (bottom-right), and MMSE deconvolved data (bottom-left).

Fourier method that Fourier transforms over the variables x and t, the PDE becomes the differential equation

$$(\omega_x^2 + j\omega/D)S(\omega_x, y, \omega) = \frac{d^2S(\omega_x, y, \omega)}{dy^2}$$

that has the solution bounded by the 2D Fourier transform of the initial condition $S(\omega_x, y = 0, \omega)$:

$$S(\omega_x, y, \omega) = S(\omega_x, y = 0, \omega)e^{-y\sqrt{\omega_x^2 + j\omega/D}}$$

where the energy-diverging solutions for $y \geq 0$ are neglected. Namely, the solution is the filtered density evolution both in space and time. Again, assuming that the measurements are noisy, one can employ Wiener filtering while extrapolating the solution vs. space.

18.3.3 Effect of $\pi/4$ Phase-Shift

The spatial evolution of the diffusion has a phase-shift, which has several benefits in engineering. To simplify, let the 1D PDE be (i.e., indefinite perturbation along the y-direction):

$$D\frac{\partial^2 s(x,t)}{\partial x^2} = \frac{\partial s(x,t)}{\partial t}$$

The solution for the initial condition $s(x=0,t)$ is

$$S(x,\omega)=S(x=0,\omega)H(\omega|x)$$

where the extrapolation filter is

$$H(\omega|x)=e^{-x\sqrt{j\omega/D}}$$

Since $\sqrt{j}=e^{j\pi/4}$ is a $\pi/4$ phase-shift, the filter is the superposition

$$H(\omega|x)=e^{-x\sqrt{\frac{\omega}{2D}}}e^{-jx\sqrt{\frac{\omega}{2D}}}$$

where the first term attenuates the perturbations vs x, and the second term applies a phase-shift. One might find a combination of parameters and space x such that fluctuations with angular frequency ω_o at $x=0$ reverse their amplitudes at a certain $x=x_o$ provided that

$$x_o=\sqrt{\frac{2D}{\omega_o}(\pi+2k\pi)}$$

with decreasing deviations.

To exemplify, the daily fluctuations of temperature Δs outside of an infinite wall with thermal conductivity D_w (Figure 18.10) are given by:

$$s(x=0,t)=\bar{s}+\Delta s\cos(\omega_{day}t)$$

With $\omega_{day}=2\pi/24h$ (rad/h), this becomes

$$s(x,t)=\bar{s}+\Delta s e^{-x\sqrt{\omega_{day}/2D_w}}\cos(\omega_o t-x\sqrt{\omega_{day}/2D_w})$$

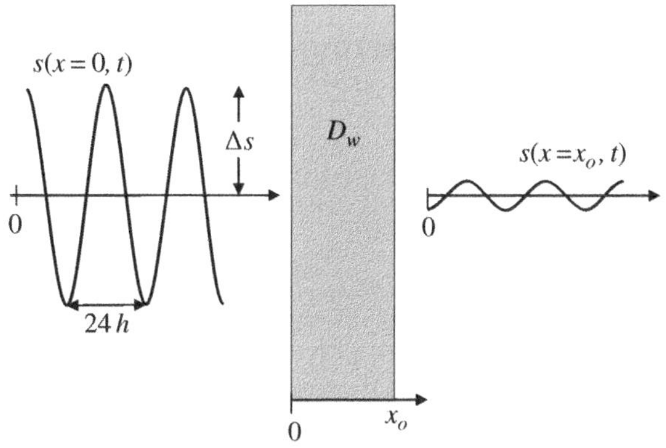

Figure 18.10 2D filtering in space and time: diffusion of the temperature.

and it changes polarity for $x_o = \pi\sqrt{2D_w/\omega_{day}}$ with 4% of the outside excursion Δs. This is what happens when in summer, the hottest outdoor temperature is temporally delayed by up to 12h (for $x_o = \pi\sqrt{2D_w/\omega_{day}}$) with very small indoor fluctuations. However, since temperature fluctuations can be appreciated when within an appreciable range, say $\Delta s/e$, the delay of the hottest indoor time is on the order of 3–4h compared to outdoor (e.g., 3–4pm compared to the hottest outside temperature being at noon), depending on the conductivity (or isolation) of the wall accounted for in D_w. In addition, when the thickness of the wall gets very large (effectively $x \to \infty$), the indoor temperature is an average $\bar{s}$ that changes mildly over a timescale of weeks. In any case, this reasoning is supportive of the good practice to keep all windows shut at noon during the hottest season!

The same example can be extended to the case of periodic spatial fluctuations with a certain period $\bar{T}_x$ in meters (and $\bar{\omega}_x = 2\pi/\bar{T}_x$) (e.g., a set of trees along a road that make the temperature fluctuate spatially because of their shade) for the 2D case with the 2D PDE solved above and

$$s(x, y = 0, t) = \bar{s} + \Delta s \cos(\omega_{day} t + \bar{\omega}_x x)$$

The diffusion filter

$$H(\bar{\omega}_x, \omega_{day}|y) = e^{-y\sqrt{\bar{\omega}_x^2 + j\omega_{day}/D}} = e^{-y\alpha} e^{-jy\beta}$$

attenuates and phase-shifts the perturbation (i.e., temperature evolution inside the asphalt):

$$s(x, y, t) = \bar{s} + \Delta s e^{-y\alpha} \cos(\omega_{day} t + \bar{\omega}_x x - y\beta)$$

in a quite complicated way. Once again, the phase-shift translates the max/min temperature pattern spatially/temporally inside the asphalt compared to the position of the shade at the surface.

18.4 Laplace Equation and Exponential Filtering

The Laplace PDE in 3D

$$\frac{\partial^2 s(x,y,z)}{\partial x^2} + \frac{\partial^2 s(x,y,z)}{\partial y^2} + \frac{\partial^2 s(x,y,z)}{\partial z^2} = 0$$

models gravity and the static electric field. There are a class of applications where some 2D data is gathered at constant height $z = z_o$ (e.g., the micro-gravity field measured from an airplane when sensing for new sources of minerals) and these need to be extrapolated to another height, say $z = 0$, to estimate the corresponding 2D pattern when close to the sources of gravity or electric field anomalies. We can consider two problems of continuation according to the available boundary condition:

$$\text{upward continuation: } s(x,y,z=0) \longrightarrow s(x,y,z_o)$$
$$\text{downward continuation: } s(x,y,z_o) \longrightarrow s(x,y,z=0)$$

Both these problems can be solved by considering the Fourier transform method to isolate the parameter of interest (here z) provided that boundary conditions are smooth and not critical.

The FT of the PDE wrt the variables x, y

$$-(\omega_x^2 + \omega_y^2)S(\omega_x, \omega_y, z) + \frac{d^2 S(\omega_x, \omega_y, z)}{dz^2} = 0$$

transforms the PDE into a differential equation that has two solutions

$$S(\omega_x, \omega_y, z) = A(\omega_x, \omega_y)e^{-z\sqrt{\omega_x^2+\omega_y^2}} + B(\omega_x, \omega_y)e^{z\sqrt{\omega_x^2+\omega_y^2}}$$

but since for $z \to \infty$ the second term diverges, the solution would be unstable and thus $B(\omega_x, \omega_y) = 0$ leaving only the first term for boundary condition:

$$S(\omega_x, \omega_y, z) = S(\omega_x, \omega_y, z=0)H(\omega_x, \omega_y|z)$$

Inspection of the solution shows that the continuation problem is a 2D filtering of the initial solution with a circular symmetric filter

$$H(\omega_x, \omega_y|z) = e^{-z\sqrt{\omega_x^2+\omega_y^2}} = e^{-z\Omega} = e^{-\Omega/\Omega_z}$$

whose spatial frequency depends on the height z.

The Laplace PDE is a basic tool for 2D symmetric filtering with exponential shape, and the filtering properties can be summarized as follows:

- In upward continuation, the initial solution is low-pass filtered with a circular symmetric exponential response filter that reduces its pass-bandwidth $\Omega_z = 1/z$ by increasing z, and this smooths the initial condition $s(x,y,z=0)$.
- The filtering compounds with z as

$$H(\omega_x, \omega_y|z_1 + z_2) = H(\omega_x, \omega_y|z_1)H(\omega_x, \omega_y|z_2) \leftrightarrow h(x,y|z_1) ** h(x,y|z_2)$$

  and this can be used for a set of filters over a uniformly discrete step $z \in \{0, \Delta z, 2\Delta z, ..., k\Delta z, ...\}$; once derived, the numerical kernel for $h(x,y|\Delta z)$ is simply applied iteratively.
- In downward continuation, the 2D signal corresponding to the initial condition $s(x,y,z_o)$ is exponentially high-pass filtered, and this must account for the power spectral density of noise that might enhance high-frequency components, where signal has negligible spectral content (or create artifacts); see Section 18.3.1 for MMSE Wiener filtering that adapts in this case with obvious extensions.

18.5 Wavefield Propagation

The propagation of the (scalar) wavefield in a 3D homogeneous medium is governed by the wave-equation PDE:

$$\frac{\partial^2 s(x,y,z,t)}{\partial x^2} + \frac{\partial^2 s(x,y,z,t)}{\partial y^2} + \frac{\partial^2 s(x,y,z,t)}{\partial z^2} = \frac{1}{v^2}\frac{\partial^2 s(x,y,z,t)}{\partial t^2}$$

where v is the propagation speed, in m/s. To simplify here, the PDE in 2D (solution is not varying along z) drops one term to

$$\frac{\partial^2 s(x,y,z,t)}{\partial z^2} = 0$$

The PDE models the scalar wavefields such as elastic waves in fluids or acoustic waves, and it provides a powerful description in optics [83]. Elastic waves in solids and electromagnetic waves both need to account for two propagating and coupled wavefields, namely compressional-shear waves, and electric-magnetic waves, respectively. Once again, a detailed discussion is left to specialized publications.

The objective here is to address two wavefield-continuation configurations that are quite common in many remote sensing applications:

- $s(x,y,t=0) \leftrightarrows s(x,y,t)$ propagation or backpropagation of the wavefield vs. time from a certain initial condition
- $s(x,y=0,t) \rightarrow s(x,y,t)$ extrapolation of the wavefield over the space from the wavefield measured along a line by an array of sensors (see Chapter 19 or Chapter 20). Crucial is the backpropagation of the wavefield $s(x,y,t)$ to $t=0$ that corresponds to the estimation of the spatial distribution of the excitations at $t=0$ as detailed later.

18.5.1 Propagation/Backpropagation

The evolution of the wavefield given the initial condition $s(x,y,t=0)$ is carried out by assuming that, once given the boundary condition $t=0$, all excitations disappear

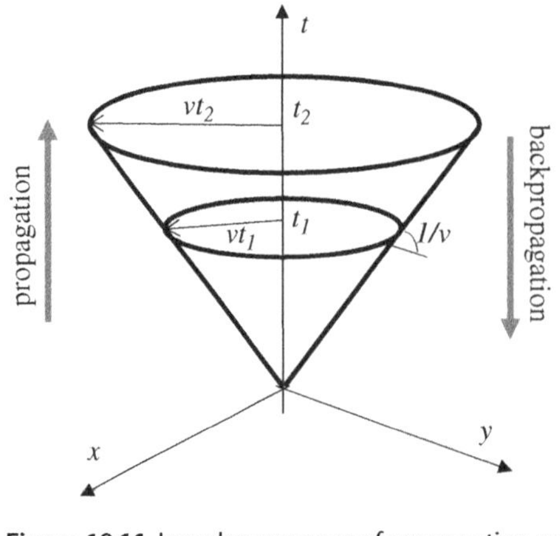

Figure 18.11 Impulse response of propagation and backpropagation.

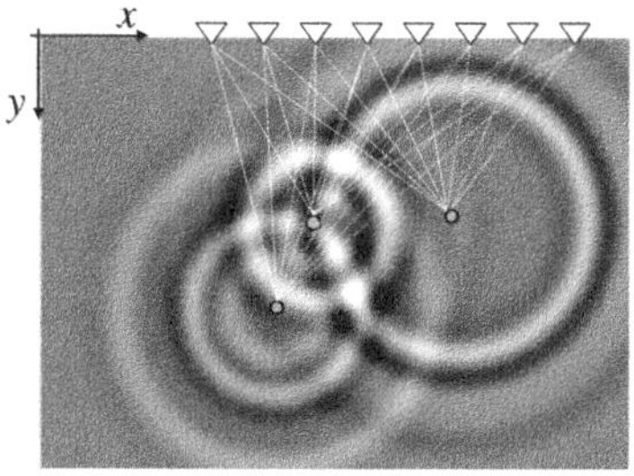

Figure 18.12 Superposition of three sources in 2D, and the measuring line of the wavefield in $y = 0$.

leaving the wavefield to propagate in the homogeneous medium. In other words, using the analogy of stones dropped into the water, at $t = 0$ stones are dropped into the water creating delta-like perturbations. Immediately after, at $t = 0^+$, all stones disappear leaving the wavefield corresponding to the superposition of spherical wavefronts that propagate. In this way, each wavefield corresponding to each delta-like source propagates independently of the others according to the linearity of the PDE, and this superposition of simple wavefields can easily create complex wavefields, and vice-versa (*Huygens principle*). In this sense, backpropagation yields the originating perturbation $s(x, y, t = 0)$ from the measurement of the 2D wavefield $s(x, y, t_o)$ at a certain time $t = t_o$ (called *snapshot*, as it is like taking a picture of the wavefield).

The Fourier transform of the PDE over the variables x and y, leaving out the time as parameter, yields the differential equation

$$(\omega_x^2 + \omega_y^2)S(\omega_x, \omega_y, t) + \frac{1}{v^2}\frac{d^2 S(\omega_x, \omega_y, t)}{dt^2} = 0$$

which has two solutions:

$$S(\omega_x, \omega_y, t) = A(\omega_x, \omega_y)e^{jvt\sqrt{\omega_x^2+\omega_y^2}} + B(\omega_x, \omega_y)e^{-jvt\sqrt{\omega_x^2+\omega_y^2}}.$$

These phase-only filters correspond to *exploding* (a wavefield that leaves the source and propagates for $t > 0$) and *imploding* (a wavefield that converges to the source and propagates for $t < 0$) waves, respectively. For a given boundary condition $s(x, y, t = 0) \leftrightarrow S(\omega_x, \omega_y, t = 0)$, the propagation

$$S(\omega_x, \omega_y, t) = S(\omega_x, \omega_y, t = 0)e^{jvt\sqrt{\omega_x^2+\omega_y^2}}$$

is a *phase-only filter* that propagates each 2D sinusoid accordingly.

For instance, for a delta-like excitation $s(x, y, t = 0) = \delta(x, y)$ (i.e., dropping a small stone into water), the solution

$$s(x, y, t) \leftrightarrow e^{jvt\sqrt{\omega_x^2+\omega_y^2}} = e^{jvt\Omega}$$

is circularly symmetric (the wavefront is circular) and increases with a slope $1/v$ as sketched in Figure 18.11, while backpropagation collapses the wavefront into the

Figure 18.13 Excitation at $t = 0$ (top-left), propagation after $t = t_o$ (top-right), noisy propagation (bottom right), and backpropagation (bottom-left).

excitation point. Multiple bandlimited sources compound similarly to stones dropped into water as in Figure 18.12, which also shows the array of sensors measuring the wavefield along one line $y = 0$.

The example in Figure 18.13 shows the propagation of excitation using a phase-only filter with the Fourier method represented by the symbols "PdM" from $t = 0$ to a certain time $t = t_o$. Notice that $s(x, y, t = t_o)$ shows the superposition of circular responses that combine to yield a complex wavefield that barely resembles the original field. After the superposition of spatially uncorrelated noise and backpropagation of the noisy observation $\hat{s}_{backprop}(x, y, t = t_o)$, the original "PdM" excitation shows up except for some artifacts due to space discretization and truncation. However, since the filter is phase-only, there is no trivial filtering that can be carried out to mitigate the noise in this case as when adopting Wiener filtering.

18.5.2 Wavefield Extrapolation and Focusing

In remote sensing experiments, one measures the wavefield along a line $s(x, y = 0, t)$ with the objective of finding the distribution of the excitations that generated the pattern in $t = 0$. This is obtained by estimating $s(x, y, t = 0)$ after extrapolating the wavefield everywhere using the transformation $s(x, y = 0, t) \rightarrow s(x, y, t)$. To better motivate the importance of this processing, let us consider the exploding reflector model first.

18.5.3 Exploding Reflector Model

In reflection seismology and remote sensing, the morphological properties of the target (e.g., layers and horizons) is inferred from the backscattered wavefield collected after

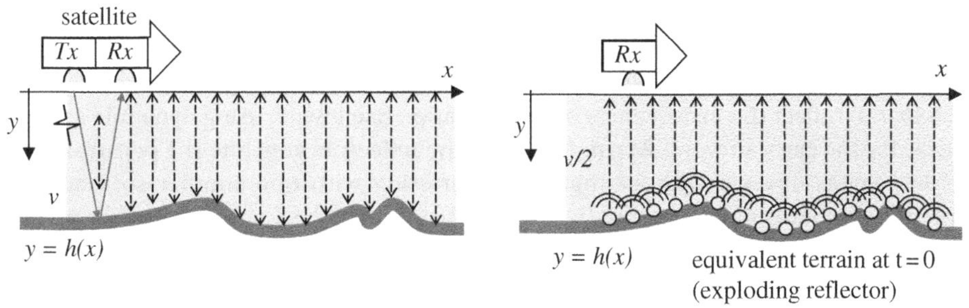

Figure 18.14 Exploding reflector model.

having illuminated the target by an appropriate source. The exploding reflector is a simple (but still rigorous) model to relate the basckscattered wavefield with the sensed medium. To exemplify, let us consider a radar system mounted as payload on a satellite (or airplane) that illuminates the ground while moving as shown in Figure 18.14. The radar is monostatic with co-located transmitting and receiving antennas (bistatic would need some additional processing steps and this complicates the exposition here) with the receiver that measures the backscattered wavefield from the terrain at distance $h(x)$ from the satellite's orbit used here as reference system in measurement. The satellite at position x_o transmits one echo (at time $t = 0$) and gets a basckscattered echo after a delay

$$\tau(x_o) = \frac{h(x_o)}{v/2}$$

from the transmitted one (see Section 5.6 for a description), the movement of the satellite tracks the delay profile of the terrain acting as a reflector. In practice, the backscattered wavefield is the superposition of multiple echoes that result from different (and complex) ground scatterers.

Considering the simple experiment above, the delay for each echo would be the same if each point along the reflector excites, or "explodes," with the same transmitted waveform and the wavefield propagates over a medium with half velocity. The exploding reflector model assumes that at time $t = 0$ all the infinite points along the reflector $h(x)$ excite simultaneously as with transmitters placed along the reflector itself, and the corresponding transmitted wavefield is measured along the line $y = 0$. The excitation wavefield is a 2D blade function conformed as the reflector

$$s(x, y, t = 0) = \delta(y - h(x))$$

and it is the superposition of infinite scatterers. Overall, it is a matter of reordering all the experimental measurements along a line $s(x, y = 0, t)$ to represent the outcome from an exploding reflector model. Estimation of the wavefield $\hat{s}(x, y, t = 0)$ yields the "image" of the reflector that originated the observations along the satellite trajectory.

Remark: The exploding reflector model is a powerful analogy that was introduced in reflection seismology where elastic propagation is over the multi-layered structure of the earth [82]. In this case, the transmitter/receiver arrangement is slightly more complex than monostatic as the backscattered signal is so weak (noise dominates) and the transmitter is so invasive (micro-explosion or vibrating plates on the earth's surface) that several receivers are deployed simultaneously over the surface for each excitation, called *shot* to reflect the invasiveness of the excitation itself when using dynamite as was the case for the early surveys. At time $t = 0$, all the reflectors generate the perturbation that propagates upward without mutually interfering with one another so that any diffraction and multiple scattering (i.e., waves that bounce back and forward between horizons) is neglected. In addition, propagation speed might change as the earth's layered structure has an increasing propagation speed vs. depth (say from 500 m/s to 3–4000 m/s), and this effect can be taken into account by some ingenious modifications of the upward propagation model, or by using finite difference methods for PDE.

18.5.4 Wavefield Extrapolation

Wavefield extrapolation is based on the Fourier transform of the PDE over the variables x and t, leaving out the space y as a parameter. The corresponding differential equation is

$$(-\omega_x^2 + \frac{\omega^2}{v^2})S(\omega_x, y, \omega) + \frac{d^2 S(\omega_x, y, \omega)}{dy^2} = 0$$

with solution (only exploding waves are considered):

$$S(\omega_x, y, \omega) = S(\omega_x, y = 0, \omega)e^{j\frac{\omega y}{v}\sqrt{1-\omega_x^2 v^2/\omega^2}}$$

The filter

$$H(\omega_x, y, \omega) = e^{j\frac{\omega y}{v}\sqrt{1-\omega_x^2 v^2/\omega^2}}$$

is a phase-only one that moves around the 2D sinusoids representing the wavefield without changing their power. However, the filter is effectively a phase-only one provided that

$$1 - \frac{\omega_x^2 v^2}{\omega^2} > 0 \rightarrow |\omega_x| < \frac{|\omega|}{v}$$

In other words, only waves with parameters (ω_x, ω) that are within the shaded region in Figure 18.15 correspond to effective waves that comply with a propagating wavefield, and all others outside of the shaded region are attenuated vs. y (*evanescent waves*), as

$$H(\omega_x, y, \omega) = e^{-\frac{\omega y}{v}\sqrt{1-\omega_x^2 v^2/\omega^2}} \text{ for } |\omega_x| > \frac{|\omega|}{v}$$

The region of propagating waves is further limited by the min/max bandwidth of the propagating waveform that is limited between $\omega_{\min}$ and $\omega_{\max}$ over ω.

To detail further, once the measurements $s(x, y = 0, t)$ have been collected, if related to a propagation experiment with velocity v, its 2D Fourier transform $S(\omega_x, y = 0, \omega)$ should be fully bounded within the area $|\omega_x| < |\omega|/v$ that corresponds to a propagating

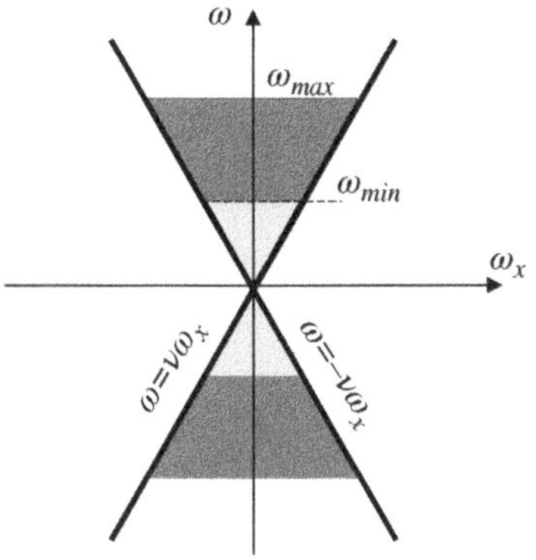

Figure 18.15 Propagating waves region.

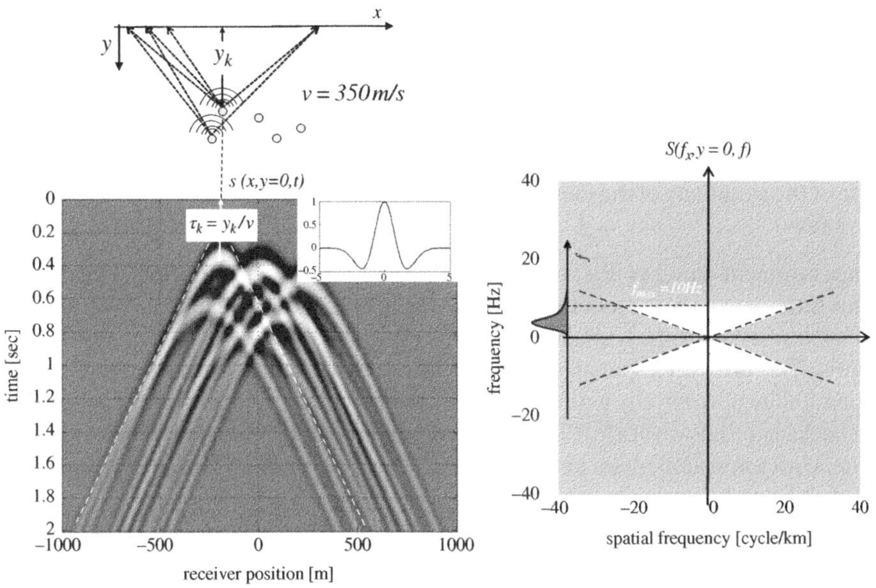

Figure 18.16 Example of $s(x, y = 0, t)$ measured along a line from multiple scattering points: experimental setup with Ricker waveform (left) and the $\mathcal{F}_{2D}\{s(x, y = 0, t)\}$ (right).

wave and zero elsewhere, as the waves do not propagate but attenuate with space. The example below shows some scattering points that represent acoustic sources with propagation velocity $v = 350\ m/s$; each source generates a Ricker waveform at $t = 0$ and the wavefield is measured by an array $y = 0$ showing the characteristic hyperbolic pattern (see Chapter 20 and Section 20.5.1). The 2D Fourier transform of the backscattered field $s(x, y = 0, t)$ (left figure) is $S(f_x, y = 0, f)$ (right figure) is fully bounded within the fan $f_x = \pm f/v$ up to the maximum frequency here of $f_{max} = 10\ Hz$ (see the PSD of the temporal signal on side of the figure).

18.5.5 Wavefield Focusing (or Migration)

The final step of any radar and geophysical imaging is to get the image of the excitation reflectors according to the exploding reflector model. This is the obvious consequence of wavefield extrapolation from FT properties

$$S(\omega_x, y, t=0) = \int S(\omega_x, y, \omega)\frac{d\omega}{2\pi}$$

This step is called *focusing* (in radar) or *migration* (in geophysics) as it collapses all the energy of the sinusoids onto the reflector that originated the wavefield. The spatial image of the reflector is, after an inverse 1D Fourier transform:

$$s(x, y, t=0) = \int S(\omega_x, y, t=0)e^{j\omega_x x}\frac{d\omega_x}{2\pi}$$

Plugging in the wavefield extrapolation filter, the full focusing equation is

$$s(x, y, t=0) = \int \left(\int S(\omega_x, y=0, \omega)e^{j\frac{\omega y}{v}\sqrt{1-\omega_x^2 v^2/\omega^2}}\frac{d\omega}{2\pi}\right) e^{j\omega_x x}\frac{d\omega_x}{2\pi}$$

to be applied only for propagating waves in $|\omega_x| < |\omega|/v$.

In spite of the simplicity of the focusing step, there are several pitfalls to be considered; here are a few:

- The measurement along y=0 is over a limited spatial interval by an array of sensors, and this implies a truncation of the visibility of the (exploding) reflector with a severe loss of resolution over the lateral dimension.
- Measurements are over a spatially sampled line by an array of sensors (or in the example, the satellite periodically gathers measurements with an implicit spatial sampling of the backscattered wavefield). The sampling implies that all Fourier transforms are discrete, with some embedded periodicities in space and time that should be properly handled during processing to avoid artifacts.
- For every choice of y, the wavefield extrapolation filter collects some samples of $S(\omega_x, y=0, \omega)$. Interpolation in the Fourier transformed domain is necessary and this could introduce some artifacts, mostly through incorrect handling of the transition between propagating and evanescent waves (as the filter experiences an abrupt transition between phase-only and attenuation regions).

Appendix A: Properties of 2D Signals

These properties can be found in many textbooks such as [77, 78, 83]. Below is a summary of the main properties:

- **Separability**: this is the closest property that extends to 2D the properties of a 1D signal; a separable signal is

$$s(x, y) = a(x)b(y)$$

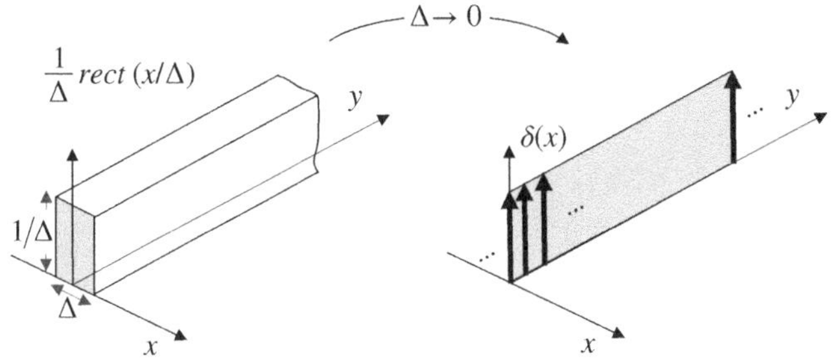

Figure 18.17 2D blade $\delta(x)$.

and by extension any sum of separable signals is given by

$$s(x,y) = \sum_k a_k(x)b_k(y)$$

- **Blades and 2D delta** (Figure 18.17): a blade is

$$\delta(x) = \lim_{\Delta\to 0} \frac{1}{\Delta} rect(x/\Delta)$$
$$\delta(y) = \lim_{\Delta\to 0} \frac{1}{\Delta} rect(y/\Delta)$$

that is a signal invariant over one dimension and singular wrt the other as sketched in the figure.

Crossing two blades yields the 2D delta

$$\delta(x,y) = \delta(x)\delta(y)$$

with the following properties:

$$\iint \delta(x,y)dxdy = 1 : \text{volume}$$
$$\iint s(x,y)\delta(x-x_o,y-y_o)dxdy = s(x_o,y_o) : \text{traslation and sampling}$$
$$s(x,y) = \iint s(\zeta,\eta)\delta(x-\zeta,y-\eta)d\zeta d\eta : \text{equivalent representation}$$
$$\text{(signal sum of 2D deltas)}$$

Crossing of two blades with angle φ still gives a 2D delta with a volume that depends on the angle (proof is by simple geometrical reasoning):

$$\delta(x)\delta(y\cos\varphi - x\sin\varphi) = \lim_{\Delta\to 0} \frac{1}{\Delta^2} rect(x/\Delta)rect((y\cos\varphi - x\sin\varphi)/\Delta) = \frac{1}{|\sin\varphi|}\delta(x,y)$$

- **Amplitude varying blade** (Figure 18.18):

$$s(x,y) = a(x)\delta(y)$$

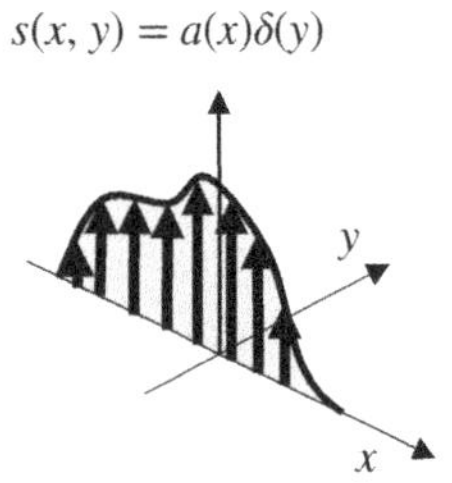

Figure 18.18 Amplitude varying 2D blade.

is separable. A further property of blades is projection:

$$\iint s(x,y)\delta(x)dxdy = \iint s(0,y)dy = p_y(x=0) : \text{projection along } y$$

$$\iint s(x,y)\delta(x-x_o)dxdy = \iint s(x_o,y)dy = p_y(x_o) : \text{arbitrary projection along } y$$

similar properties hold over any direction. Recall that $\delta(x,y) \neq \delta(x)$ and similarly $\delta(x,y) \neq \delta(y)$: this is a common mistake that always gives completely erroneous conclusions in 2D.

- **Rotation** (Section 2.2): any 2D signal can be arbitrarily rotated by an angle φ

  $$s(x,y) \to s(x\cos\varphi - y\sin\varphi, x\sin\varphi + y\cos\varphi)$$

  A simple way to see this is by representing the signal in polar coordinates (Figure 18.19) so that

  $$s(x,y) = s_p(\rho,\theta)$$

  $$\text{where: } \begin{cases} \rho = \sqrt{x^2+y^2} \\ \tan\theta = y/x \end{cases}$$

  and thus rotation is simply

  $$s_p(\rho,\theta+\varphi)$$

- **Cylindrical signal** (Figure 18.20): this is any signal that does not change along one direction (cylindrical direction). A cylindrical signal along y is

  $$s(x,y) = a(x)$$

  and it can be arbitrarily rotated to have a cylindrical signal over φ by rotating $\pi/2 - \varphi$:

  $$s(x,y) = a(y\cos\varphi - x\sin\varphi)$$

- **Convolution:** this is the leading operation for linear shift-invariant systems:

  $$g(x,y) = \iint s(\zeta,\eta)h(x-\zeta,y-\eta)d\zeta d\eta = s(x,y) ** h(x,y)$$

  with symbol "**" to denote convolution for 2D signals. Further properties are

  $$(a_1(x)b_1(y)) ** (a_2(x)b_2(y)) = (a_1(x) * a_2(x)) \cdot (b_1(y) * b_2(y)) \ : \text{convolution for separable signals}$$

  $$s(x,y) ** s(-x,-y) = \iint s(\zeta,\eta)s(\zeta+x,\eta+y)d\zeta d\eta : \text{2D autocorrelation}$$

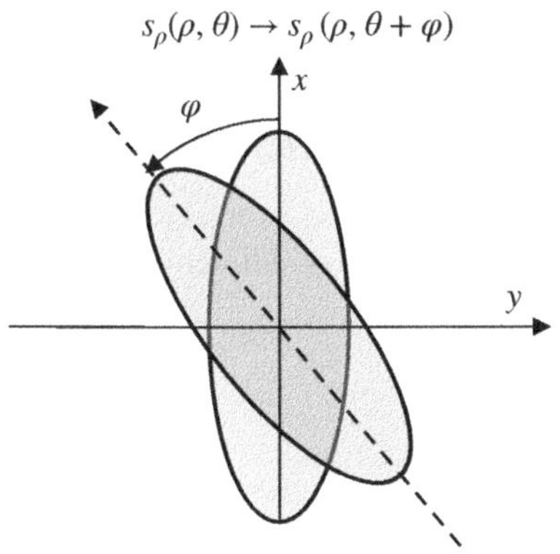

Figure 18.19 Rotation.

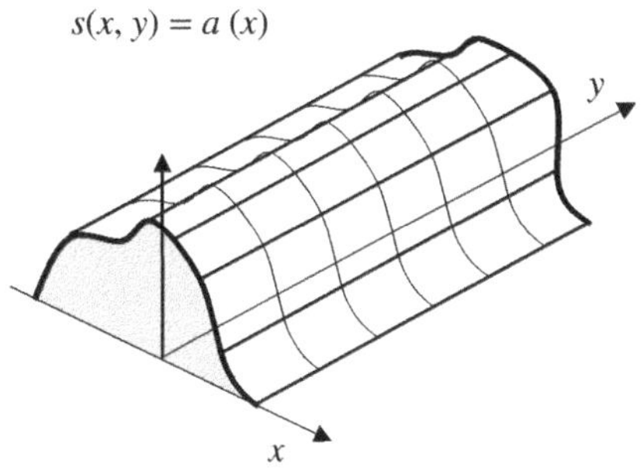

Figure 18.20 Cylindrical signal along y: $s(x, y) = a(x)$.

- **Projection:** this is a cylindrical signal as

  $$p_y(x) = s(x, y) ** \delta(x)$$

  not to be confused with slicing:

  $$s(x, y) \cdot \delta(x) = s(x = 0, y) \cdot \delta(x)$$

  which gives an amplitude varying blade.

- **Periodic 2D signal:** this is obtained by any periodic repetition along spaces according to a pre-definite lattice using the translation property:

  $$s(x, y) ** \delta(x - x_o, y - y_o) = s(x - x_o, y - y_o)$$

  To simplify, the rectangular lattice is given by a set of points

  $$\{x = mT_x, y = nT_y\}$$

  such that the 2D periodic signal $\tilde{s}(x, y)$ derived from $s(x, y)$ with period T_x and T_y along the two dimensions is

  $$\tilde{s}(x, y) = s(x, y) ** \sum_{m,n} \delta(x - mT_x, y - nT_y) = \sum_{m,n} s(x - mT_x, y - nT_y)$$

  and the property holds for any arbitrary lattice.

- **Sampling:** this is the dual of periodic, as a lattice is used to extract (sample) the values of the 2D signal as

$$s(x,y)\times\sum_{m,n}\delta(x-mT_x,y-nT_y)=\sum_{m,n}s(mT_x,nT_y)\delta(x-mT_x,y-nT_y)$$

- **Symmetries:** symmetries of a 2D signal are very peculiar and very different from 1D (consisting of odd and even symmetries only); a few examples are:

circular	$s_p(\rho,\theta)=s_p(\rho)$ for $\forall\theta$
central (even)	$s(x,y)=s(-x,-y)$
quadrantal (even)	$s(\pm x,\pm y)=s(x,y)$

Appendix B: Properties of 2D Fourier Transform

The definition of the 2D Fourier transform is

$$S(\omega_x,\omega_y)=\iint s(x,y)e^{-j(\omega_x x+\omega_y y)}dxdy$$

$$s(x,y)=\iint S(\omega_x,\omega_y)e^{j(\omega_x x+\omega_y y)}\frac{d\omega_x d\omega_y}{4\pi^2}$$

and it is common in the notation to denote the switch between the two representations

$$s(x,y)\leftrightarrow S(\omega_x,\omega_y)$$

In addition, note the complex conjugate central symmetry of the 2D FT for real-valued $s(x,y)$:

$$S(\omega_x,\omega_y)=S^*(-\omega_x,-\omega_y)$$

which is useful in many contexts. Some common properties are listed in Table 18.1; proof is by recalling the definitions (or see any textbook, such as [77, 83]).

2D Sampling

Another important aspect of 2D Fourier transforms follows from sampling. Gathering and processing of 2D signals is usually over a regular grid or lattice; these signals are defined over the sampling grid and so is the 2D Fourier transform. For rectangular grid-sampling ($T_x\times T_y$) as in Figure 18.21, the properties are

$$s_d(x,y)=s(x,y)\times\sum_{m,n}\delta(x-mT_x,y-nT_y)=\sum_{m,n}s(mT_x,nT_y)\delta(x-mT_x,y-nT_y)$$

$$S_d(\omega_x,\omega_y)=S(\omega_x,\omega_y)**\frac{1}{T_xT_y}\sum_{k,l}\delta(\omega_x-k\frac{2\pi}{T_x},\omega_y-l\frac{2\pi}{T_y})$$

$$=\frac{1}{T_xT_y}\sum_{k,l}S(\omega_x-k\frac{2\pi}{T_x},\omega_y-l\frac{2\pi}{T_y})$$

Table 18.1 2D Fourier transform properties.

$s(x,y)$	$S(\omega_x,\omega_y)$	
$a(x)b(y)$	$A(\omega_x)B(\omega_y)$	separability reduces 2D FT onto two 1D FT
$\delta(x,y)$	1	
$\delta(x)$	$\delta(\omega_y)$	blade maps onto a blade
$h(x)$	$H(\omega_x)\delta(\omega_y)$	cylindrical maps onto a blade
$s(x,y) ** h(x,y)$	$S(\omega_x,\omega_y)H(\omega_x,\omega_y)$	2D filtering
$s(x,y)\times h(x,y)$	$S(\omega_x,\omega_y) ** H(\omega_x,\omega_y)$	2D multiplication
$s(x,y)\times e^{-j(\bar{\omega}_x x+\bar{\omega}_y y)}$	$S(\omega_x-\bar{\omega}_x,\omega_y-\bar{\omega}_y)$	2D modulation (sinusoid)
$\iint \lvert s(x,y)\rvert^2 dxdy = \iint \lvert S(\omega_x,\omega_y)\rvert^2 \frac{d\omega_x d\omega_y}{4\pi^2}$		Parseval equality
$\frac{\partial}{\partial x}s(x,y)$	$j\omega_x S(\omega_x,\omega_y)$	Partial derivative wrt x
$\frac{\partial}{\partial y}s(x,y)$	$j\omega_y S(\omega_x,\omega_y)$	Partial derivative wrt y
$\frac{\partial^2}{\partial x\partial y}s(x,y)$	$-\omega_x\omega_y S(\omega_x,\omega_y)$	Partial derivatives (mixed)
$s_p(\rho,\theta+\bar{\varphi})$	$S_p(\Omega,\varphi+\bar{\varphi})$	Rotation by $\bar{\varphi}$ (polar coordinates)

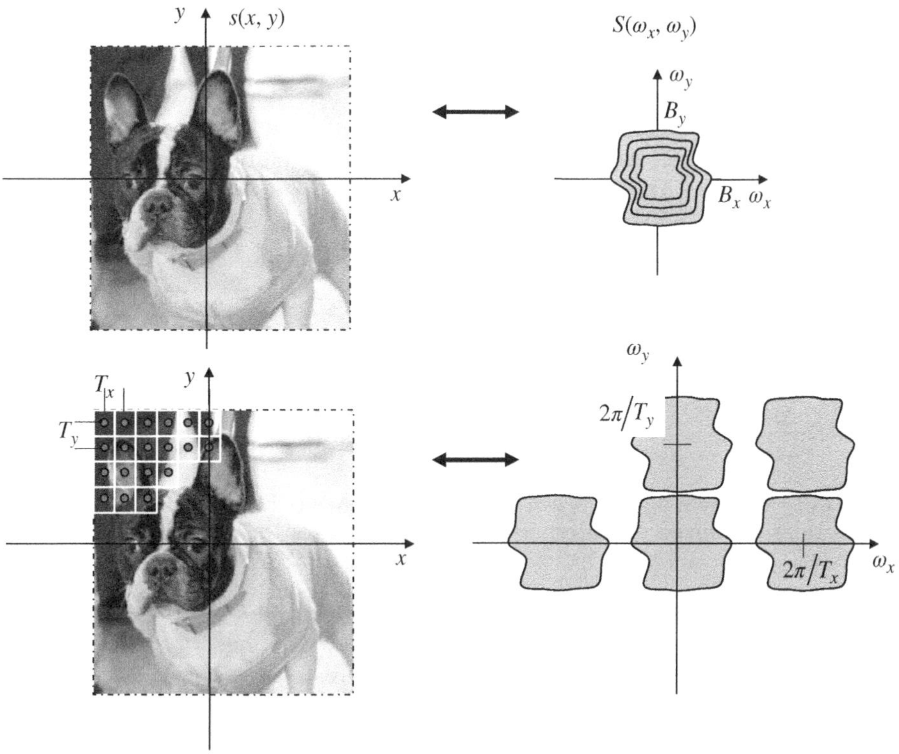

Figure 18.21 2D sampling and the corresponding Fourier transform.

and the FT becomes periodic with period $(\frac{2\pi}{T_x}, \frac{2\pi}{T_y})$ as illustrated below. The sampling theorem can be reduced to the (sufficient) conditions

$$\begin{cases} T_x \leq \frac{2\pi}{B_x} \\ T_y \leq \frac{2\pi}{B_y} \end{cases}$$

for the bandwidth $(B_x \times B_y)$, even if these conditions can be largely relaxed for some signals according to the shape of their FT. It is important to ensure that the sampling density is evaluated in terms of the number of samples per square meter, as this metric defines the cost of the sampling process and the corresponding processing; for rectangular sampling it is given by

$$density = \frac{1}{T_x T_x} \leq \frac{B_x B_y}{4\pi^2}$$

Appendix C: Finite Difference Method for PDE-Diffusion

Numerical methods to solve for PDEs are very broad and their usage is a competence itself that goes beyond the scope here. However for diffusion, the finite difference method is simple enough to outline as it will be the basis for cooperative estimation methods (Chapter 22). In PDE, the derivatives are replaced by finite difference approximations over the uniformly sampled square grid $x \in \{0, \Delta r, 2\Delta r, ...\}$ and $y \in \{0, \Delta r, 2\Delta r, ...\}$, so that the time evolution over the sampled times $t \in \{..., (k-1)\Delta t, k\Delta t, (k+1)\Delta t, ...\}$ becomes

$$s[i,j,k+1] = s[i,j,k] + \rho \big[s[i+1,j,k] + s[i-1,j,k] + s[i,j+1,k] + s[i,j-1,k] \\ -4s[i,j,k]\big] \tag{18.1}$$

where

$$\rho = D\frac{\Delta t}{\Delta r^2}$$

Method (18.1) is known as the *forward time center space* (FTCS) method due to the approximations employed. An alternative representation for the updating of temporal evolution is

$$s[i,j,k+1] = (1-4\rho)s[i,j,k] + \rho \big[s[i+1,j,k] + s[i-1,j,k] + s[i,j+1,k] + s[i,j-1,k]\big]$$

showing that the new value at time $t = (k+1)\Delta t$ is the weighted sum of the current value $s[i,j,k]$ and the average value of the surrounding point. Updating vanishes when $s[i,j,k] = const$ for k large enough (in principle $k \to \infty$), provided that boundary conditions are properly matched when handling finite spatial support.

A convenient way to implement the finite difference is by rearranging the entries into the $M \times N$ matrix

$$\mathbf{S}(k) = \begin{bmatrix} s[1,1,k] & \cdots & s[1,j-1,k] & s[1,j,k] & s[1,j+1,k] & \cdots & s[1,N,k] \\ \vdots & \ddots & \vdots & \vdots & \vdots & & \vdots \\ & & s[i-1,j-1,k] & s[i-1,j,k] & s[i-1,j+1,k] & & \\ & & s[i,j-1,k] & s[i,j,k] & s[i,j+1,k] & & \\ & & s[i+1,j-1,k] & s[i+1,j,k] & s[i+1,j+1,k] & & \\ \vdots & & \vdots & \vdots & \vdots & \ddots & \vdots \\ s[M,1,k] & \cdots & s[M,j-1,k] & s[M,j,k] & s[M,j+1,k] & \cdots & s[M,N,k] \end{bmatrix}$$

with the $M \times M$ matrix operator for the centered second order derivative:

$$\ddot{\mathbf{D}}_M = \begin{bmatrix} -2 & 1 & 0 & \cdots & 1 \\ 1 & -2 & 1 & \cdots & 0 \\ \vdots & \vdots & \ddots & & \vdots \\ 0 & 0 & \cdots & -2 & 1 \\ 1 & 0 & \cdots & 1 & -2 \end{bmatrix}$$

made circular here to wrap around the boundary condition. The finite difference approximation of the Laplace operator is simply

$$\ddot{\mathbf{D}}_M\mathbf{S}(k) + \mathbf{S}(k)\ddot{\mathbf{D}}_N = \begin{bmatrix} \times & \times & \times \\ & s[i-1,j,k] & \\ \times & -2s[i,j,k] & \times \\ & +s[i+1,j,k] & \\ \times & \times & \times \end{bmatrix} + \begin{bmatrix} \times & \times & \times \\ & s[i,j-1,k] & \\ \times & -2s[i,j,k] & \times \\ & +s[i,j+1,k] & \\ \times & \times & \times \end{bmatrix}$$

where entries other than (i,j) are indicated by $\times$. The finite difference time evolution (18.1) is

$$\mathbf{S}(k+1) = \mathbf{S}(k) + \rho\left(\ddot{\mathbf{D}}_M\mathbf{S}(k) + \mathbf{S}(k)\ddot{\mathbf{D}}_N\right) \tag{18.2}$$

which is very compact to implement in any software code.

The structure (18.2) is the basis for some investigation on estimation and convergence. The vectorization (see Section 1.1.1) of the update with $\mathbf{s}(k) = \text{vec}(\mathbf{S}(k))$ shows

$$\mathbf{s}(k+1) = \mathbf{s}(k) + \rho\left(\mathbf{I}_N \otimes \ddot{\mathbf{D}}_M + \ddot{\mathbf{D}}_N^T \otimes \mathbf{I}_M\right)\mathbf{s}(k) = \mathbf{A}_\rho \mathbf{s}(k)$$

where $\mathbf{A}_\rho = \mathbf{I}_{MN} + \rho\left(\mathbf{I}_N \otimes \ddot{\mathbf{D}}_M + \ddot{\mathbf{D}}_N^T \otimes \mathbf{I}_M\right)$ is the scaling matrix that depends on the choice of ρ.

The iterations are $\mathbf{s}(k) = \mathbf{A}_\rho^k \mathbf{s}(0)$, and for convergence the discretization of the PDE one should choose ρ such that

$$\lambda_{\max}(\mathbf{A}_\rho) < 1$$

Further analytical inspection (based on the Gershgorin circle theorem—Section 1.3) proves that convergence is guaranteed for time discretization such that $\rho < 1/4$, or

$$\Delta t < \frac{\Delta r^2}{4D}$$

and the asymptotic solution for $k \to \infty$ is

$$\mathbf{s}(k) \to \frac{\mathbf{1}^T \mathbf{s}(0)}{\mathbf{1}^T \mathbf{1}} \mathbf{1}$$

which corresponds to the average of the initial condition over all points.

19

Array Processing

In wavefield processing, array processing refers to the inference and processing methods for a set of sensors (these can be antennas for radar and wireless communication systems, acoustic microphones, hydrophones for sonar systems, geophones for elastic waves, etc.) when the set of signals are gathered synchronously and processed all together by accounting for the propagation model. The sensors of the array spatially sample the impinging wavefield over a discrete set of points. The arrangement of the sensors (*array geometry*, Figure 19.1) depends on the application and the parameters to be estimated, or the processing to be employed. There are linear, planar, or even spatial arrays depending on whether sensors are arranged along a line in space, or plane, or volume. A uniform array refers to uniform spatial sampling and it is by far the most common situation, but irregular or pseudo-irregular arrays are equally possible.

A general treatment of array processing would be broad and far too ambitious as it depends on the nature of the wavefield and the scope of the wavefield propagation. However, one common scenario is to consider the wavefield generated by some (countable set of) sources in space modeled as a superposition of L independent wavefields, one for each source, with time-varying signatures $s_1(t), \ldots, s_L(t)$. The wavefield from each source is spherical and the corresponding wavefront is circular even if in principle it could be arbitrarily curved (e.g., for multipath and/or anisotropic propagation) as in Figure 19.2. Each source can be placed far away from the array in the so called far-field and the resulting curved wavefront can be locally approximated as planar. The signatures $s_i(t)$ can be either deterministic or random, known or not, depending on the specific application, but in any case the array behavior is modeled using a simple geometrical propagation that accounts only for the signature delays due to the source-array distance and the propagation velocity v.

This chapter covers the fundamentals of array processing for narrowband signals and linear arrays. More advanced topics in narrowband or wideband array processing can be found in the books by Van Trees [101] or Haykin [102]. The most useful tools for wideband arrays are the delay estimation methods discussed in Chapter 20.

19.1 Narrowband Model

In array processing, the signals originated from sources are collected by a set of receivers arranged over a known geometry that are measuring the received signals impinging on the array. To simplify, let us consider one narrowband source $x(t) = s(t)e^{j\omega_o t}$ that

Statistical Signal Processing in Engineering, First Edition. Umberto Spagnolini.

Companion website: www.wiley.com/go/spagnolini/signalprocessing

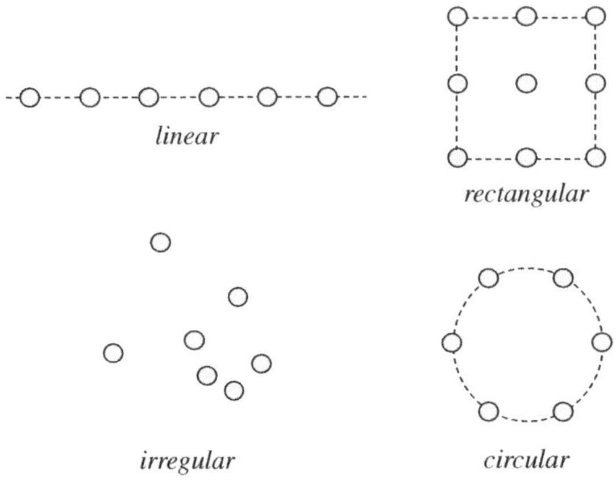

Figure 19.1 Array geometry.

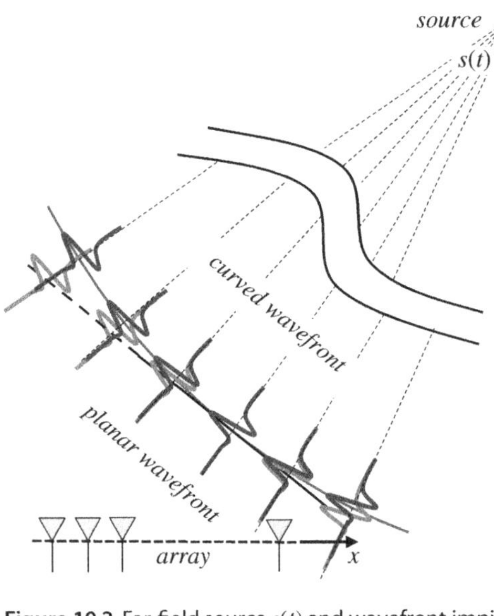

Figure 19.2 Far-field source $s(t)$ and wavefront impinging onto the uniform linear array.

is deployed far from the array to assume that the impinging wavefront arising from the wavefield propagation is linear (*far-field approximation*). The narrowband signals are represented in terms of a complex envelope where the smooth complex-valued envelope $s(t)$ (i.e., the modulating signal due to the specific phenomenon or the source of information) is modulated by the sinusoidal signal at angular frequency ω_o as $s(t)e^{j\omega_o t}$. A pragmatic definition of the narrowband approximation is that when considering a delay $\bar{\tau}$ of some periods of the sinusoids (namely, $|\bar{\tau}| \ll 100 \times 2\pi/\omega_o$), the delayed modulated waveform is

$$s(t-\bar{\tau})e^{j\omega_o(t-\bar{\tau})} \simeq s(t)e^{j\omega_o(t-\bar{\tau})}$$

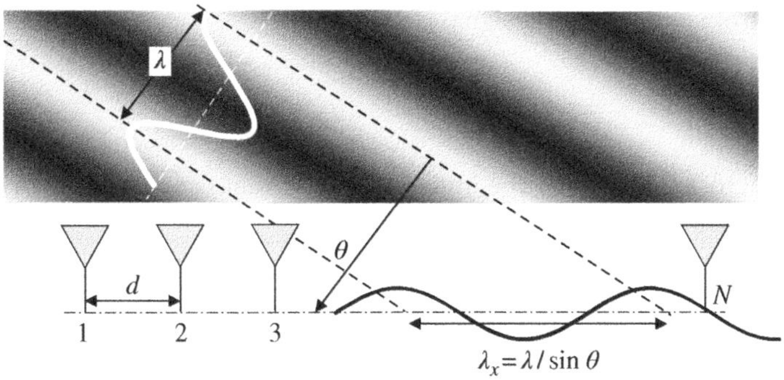

Figure 19.3 Narrowband plane wavefront of wavelength λ from DoA θ.

In other words, the envelope is so smooth that it does not change within some cycle of sinusoids: $s(t-\bar{\tau}) \simeq s(t)$.

Making reference to the Figure 19.3, we consider a uniform linear array (ULA) with N sensors deployed along a line with spatial direction x; sensors are omnidirectional (i.e., there is no preferred sensing-direction of the sensor or any different gain wrt direction) and their spacing is d. The signal received by the kth sensor of the array from a source in far-field is a delayed copy of the transmitted signal with some noise

$$y_k(t) = s(t-\tau_k)e^{j\omega_o(t-\tau_k)} + w_k(t)$$

where the delay τ_k is due to the propagation from the source and it depends on the propagation velocity v at frequency ω_o. When the max/min delays across the array are small compared to the envelope $s(t)$ so that $s(t-\max_k(\tau_k)) \simeq s(t-\min_k(\tau_k))$, the array-size is compact and the delay of the envelope $s(t)$ is usually neglected compared to the influence of the delays for the sinusoidal signals, and the received signal becomes

$$y_k(t) = s(t)e^{-j\omega_o\tau_k}e^{j\omega_o t} + w_k(t)$$

The transmitted signal (possibly delayed due to the propagation) $s(t)e^{j\omega_o t}$ is affected by a phase-delay term ($e^{-j\omega_o\tau_k}$) that depends on the position of the source and the arrangement of the sensors over the array. The phase evolution vs. sensor depends on the direction of arrival (DoA) of the impinging wavefront θ as in Figure 19.3. Since the spacing is regular (d), the delay at the kth sensor with respect to the first sensor used as reference is

$$\tau_k = \tau_1 - \frac{d(k-1)}{v}\sin\theta$$

and the received signal becomes

$$y_k(t) = s(t)e^{j\omega_o\frac{d}{v}(k-1)\sin\theta}e^{j\omega_o t} + w_k(t)$$

where the phase ($e^{-j\omega_o\tau_1}$) due to the arbitrary choice of the reference sensor has been included into the complex envelope by redefinition: $s(t)e^{-j\omega_o\tau_1} \rightarrow s(t)$.

The spatial wavelength from a narrowband signal is

$$\lambda = \frac{2\pi v}{\omega_o}$$

and thus the phase-shift across the array of the impinging wavefront can be rewritten in terms of geometrical parameters

$$y_k(t) = s(t)e^{j2\pi\frac{d}{\lambda}(k-1)\sin\theta}e^{j\omega_o t} + w_k(t)$$

The signal along the line of sensors is a spatial sinusoid, which for constant time t (say, by taking a snapshot at time t) has a spatial angular frequency

$$\omega_x(\theta) = \frac{2\pi}{\lambda_x} = \frac{2\pi}{\lambda}\sin\theta \text{ [rad/m]}$$

that depends on the DoA θ, or equivalently the (spatial) period measured along the array is

$$\lambda_x(\theta) = \frac{\lambda}{\sin\theta}$$

In other words, the array *slices* the plane wavefront with wavelength λ [meters] along the line of the array and the resulting signal is still a spatial sinusoid with wavelength λ_x (see Figure 19.3). The sensors spatially sample the sinusoid with a spatial sampling interval d, and choice of d depends on the sampling for a (complex) sinusoid with an angular frequency $\omega_x(\theta)$, that in turn depends on the DoA θ.

A common compact notation is by arranging the observations into a $N \times 1$ vector as

$$\mathbf{y}(t) = s(t)\mathbf{a}(\theta) + \mathbf{w}(t)$$

where

$$\mathbf{a}(\theta) = \begin{bmatrix} 1 \\ e^{j\omega_o\frac{d}{v}\sin\theta} \\ \vdots \\ e^{j\omega_o\frac{d}{v}(N-1)\sin\theta} \end{bmatrix}$$

is the vector[1] collecting the sampled sinusoid with spatial frequency $\omega_x(\theta)$, and noise $\mathbf{w}(t)$ is Gaussian characterized only by the spatial covariance $\mathbf{R}_w = \mathbb{E}[\mathbf{w}(t)\mathbf{w}^H(t)]$; as temporally uncorrelated, $\mathbb{E}[\mathbf{w}(t)\mathbf{w}^H(\tau)] = \mathbf{R}_w\delta(t-\tau)$:

$$\mathbf{w}(t) \sim \mathcal{CN}(0, \mathbf{R}_w)$$

1 Vector $\mathbf{a}(\theta)$ represents the array response to a source in θ, and by changing θ within the region of admissible values ($\theta \in \mathcal{I}_\theta$), the $\mathbf{a}(\theta)$ spans a hypersurface in $\mathbb{C}^N$ called the *array manifold*. In practical applications, the array manifold is very complex and needs to be experimentally evaluated as it depends on several experimental parameters: array geometry, calibration of the sensors' response, mutual coupling among sensors, etc.

Namely, by changing the time for the time snapshots $t_1, t_2, ..., t_M$, the set of observations $\mathbf{y}(t_1), ..., \mathbf{y}(t_M)$ are uncorrelated (provided that the samples are chosen so that $\mathbb{E}[s(t_i)s^*(t_j)] = 0$, for $t_i \neq t_j$) and share the same spatial frequency $\omega_x(\theta)$. The statistical model routinely adopted is one with a random source signal $s \sim \mathcal{CN}(0, \sigma_s^2)$ having power σ_s^2:

$$\mathbf{y} = s \cdot \mathbf{a}(\theta) + \mathbf{w}$$

Any randomness of phase due to the delay or the choice of the reference sensor is accounted for by the randomness of the random process accounting for the source signal s.

19.1.1 Multiple DoAs and Multiple Sources

As a generalization, multiple signals can impinge on the array with different DoAs; the compound signal with L sources

$$\mathbf{y}(t) = \sum_{l=1}^{L} \mathbf{a}(\theta_l)s_l(t) + \mathbf{w}(t) = \underbrace{[\mathbf{a}(\theta_1), ..., \mathbf{a}(\theta_L)]}_{\mathbf{A}(\theta)} \underbrace{\begin{bmatrix} s_1(t) \\ \vdots \\ s_L(t) \end{bmatrix}}_{\mathbf{s}(t)} + \mathbf{w}(t)$$

is the linear combination of columns $\{\mathbf{a}(\theta_1), ..., \mathbf{a}(\theta_L)\}$ that sets some algebraic properties of the problem at hand (see Section 16.1), and the source signals are in $\{s_l(t)\}$. Following the same steps above for $L = 1$ DoA, the general statistical model is

$$\mathbf{y} = \mathbf{A}(\theta)\mathbf{s} + \mathbf{w}$$

where the sources are modeled as

$$\mathbf{s} \sim \mathcal{CN}(\mathbf{0}, \mathbf{R}_s)$$

and the received signal is

$$\mathbf{y} \sim \mathcal{CN}(\mathbf{0}, \mathbf{R}_w + \mathbf{A}(\theta)\mathbf{R}_s\mathbf{A}^H(\theta))$$

Sources are typically associated to distinct DoAs and are usually independent with $\mathbf{R}_s = \text{diag}(\sigma_1^2, \sigma_2^2, ..., \sigma_L^2)$. In this case, $\text{rank}\{\mathbf{A}(\theta)\mathbf{R}_s\mathbf{A}^H(\theta))\} = L$ and this is the general case of multiple DoAs with independent sources, but there are exceptions as illustrated in Figure 19.4.

In DoA alignment (Figure 19.4) two distinct and uncorrelated sources (say $s_{L-1} \neq s_L$) have the same DoAs ($\theta_{L-1} = \theta_L$); the model is

$$\mathbf{y} = \mathbf{a}(\theta_1)s_1 + ... + \mathbf{a}(\theta_{L-1})(s_{L-1} + s_L) + \mathbf{w}$$

and both sources act as one equivalent stochastic source $s_{L-1} + s_L \rightarrow s_{L-1}$ at DoA $\theta_{L-1} = \theta_L$ with power $\mathbb{E}[|s_{L-1} + s_L|^2] = \sigma_{L-1}^2 + \sigma_L^2 \rightarrow \sigma_{L-1}^2$. The number of uncorrelated sources from the array is $L - 1$ as:

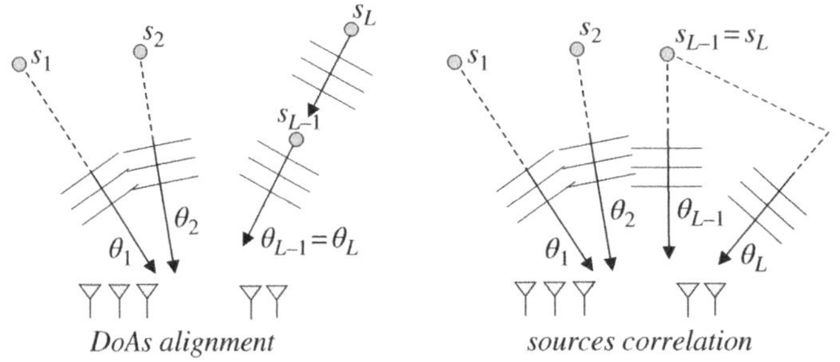

Figure 19.4 Rank-deficient configurations.

$$\mathbf{y}=\sum_{l=1}^{L-1}\mathbf{a}(\theta_l)s_l+\mathbf{w}\sim\mathcal{CN}(\mathbf{0},\mathbf{R}_w+\mathbf{A}(\boldsymbol{\theta}_L)\mathbf{R}_{s_L}\mathbf{A}^H(\boldsymbol{\theta}_L))$$

where $\mathbf{A}(\boldsymbol{\theta}_L)=[\mathbf{a}(\theta_1),...,\mathbf{a}(\theta_{L-1})]$ has one column less, and $\mathbf{R}_{s_L}=diag(\sigma_1^2,\sigma_2^2,...,\sigma_{L-1}^2+\sigma_L^2)$.

Another special case in Figure 19.4 is when sources are fully correlated as happens when the same source (say $s_{L-1}=s_L$) impinges from two distinct DoAs $\theta_{L-1}\neq\theta_L$ (e.g., in a multipath propagation environment as in the figure); then the covariance is $rank(\mathbf{R}_s)=L-1$ and the model is

$$\mathbf{y}=\mathbf{a}(\theta_1)s_1+...+(\mathbf{a}(\theta_{L-1})+\mathbf{a}(\theta_L))s_{L-1}+\mathbf{w}$$

with $L-1$ independent columns $[\mathbf{a}(\theta_1),...,\mathbf{a}(\theta_{L-2}),(\mathbf{a}(\theta_{L-1})+\mathbf{a}(\theta_L))]$.

19.1.2 Sensor Spacing Design

The set of observations $\mathbf{y}=[y_1,...,y_N]^T$ for $L=1$ DoA is a spatial sinusoid with spatial frequency

$$f_x=\frac{f_0}{c}\sin\theta\ \text{[cycle/m]}$$

(or period $\lambda_x=\lambda_x(\theta)$ [meters]) that depends on the DoA θ and all is sampled at spacing d (or sampling frequency $1/d$). Angular spatial frequency can be normalized as is customary for sampled signals (Appendix B of Chapter 4); the normalized angular frequency is one of the equivalent relationships

$$u=\omega_x(\theta)d=2\pi\frac{d}{\lambda}\sin\theta=2\pi\frac{d}{\lambda_x}=\frac{\omega_0 d}{v}\sin\theta=u_{\max}\sin\theta\quad\text{[rad/sample]}$$

where the mapping between DoA θ and normalized spatial frequency u is through the non-linear sin(.) mapping (to avoid confusion, it is advisable not to use radians as unit for DoAs, but rather degrees). According to the sampling theorem, spatial aliasing is avoided if

$$\left|\frac{\omega_0 d}{v}\sin\theta\right| \le \pi \Rightarrow d \le \frac{\lambda_x}{2} = \frac{\lambda/2}{|\sin\theta|}$$

and the sensors' spacing should be tailored depending on the DoA—but in any case $d \le \lambda/2$ is conservative for any DoA.

For multiple sources with DoAs $\Theta = \{\theta_1, \theta_2, \dots\}$, the spacing should be

$$\max_{\theta\in\Theta}\left\{\frac{\omega_0 d}{c}|\sin\theta|\right\} = \frac{\omega_0 d}{c}\max_{\theta\in\Theta}\{|\sin\theta|\} \le \pi \Rightarrow d \le \frac{\lambda}{2}\cdot\frac{1}{\max_{\theta\in\Theta}\{|\sin\theta|\}}$$

To guarantee a correct sampling for all the spatial sinusoids associated to all possible DoAs, $\max_{\theta\in\Theta}\{|\sin\theta|\} = 1$, and thus follows the condition

$$d \le \frac{\lambda}{2}$$

widely adopted in many situations. Recall that for the choice $d = \lambda/2$, $u_{\max} = \pi$ and the signal model for the spatial sinusoids is

$$y_k = \sum_{\ell} s_\ell \exp\left(j\pi\sin\theta_\ell(k-1)\right) + w_k$$

All methods for estimating the (spatial) frequency of sinusoids $\{u_1, u_2, \dots\} = \{\pi\sin\theta_1, \pi\sin\theta_2, \dots\}$ (Chapter 16), as well as the spectral analysis methods (Sections 14.1 and 14.3) can be straightforwardly applied here. For instance, the periodogram is unbiased when the (spatial) frequency is a multiple of $2\pi/N$, or equivalently

$$\frac{Nu_\ell}{2\pi} = \frac{N}{2}\sin\theta_\ell = 0, 1, 2, \dots, N-1$$

and the DoAs for unbiased condition is non-uniformly spaced angularly.

19.1.3 Spatial Resolution and Array Aperture

Resolution denotes the capability of the array to distinguish two (or more) signals with closely spaced DoAs. Let θ_1 and θ_2 be two DoAs illustrated in Figure 19.5 spaced apart by $\Delta\theta$ with a center DoA θ_o: $\theta_1 = \theta_o - \Delta\theta/2$ and $\theta_2 = \theta_o + \Delta\theta/2$. The normalized frequencies are

$$u_1 = \frac{\omega_0 d}{c}\sin\theta_1 = u_{\max}\sin\left(\theta_o - \Delta\theta/2\right)$$

$$u_2 = \frac{\omega_0 d}{c}\sin\theta_2 = u_{\max}\sin\left(\theta_o + \Delta\theta/2\right)$$

and the frequency spacing between the two line spectra (see Figure 19.5) is

$$\Delta u = u_{\max}\left[\sin\left(\theta_o + \Delta\theta/2\right) - \sin\left(\theta_o - \Delta\theta/2\right)\right] = 2u_{\max}\cos\theta_o\sin(\Delta\theta/2)$$

which depends on the bisecting angle θ_o and the DoAs' spacing $\Delta\theta$ by some non-linear trigonometric functions. For large frequency spacing Δu the two lines are separated,

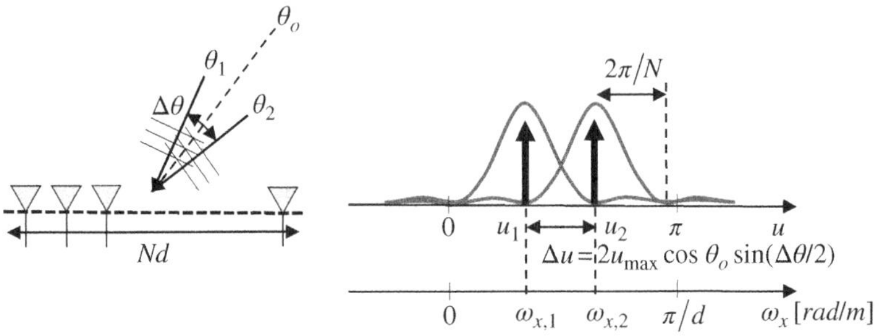

Figure 19.5 Spatial resolution of the array and spatial frequency resolution.

and it is expected that the two frequencies can be distinguished from one another. The maximum frequency spacing is for $\theta_o = 0$ (the two DoAs are symmetric in $\theta_1 = -\theta_2$ and thus $u_1 = -u_2$) with $\Delta u = 2u_{max}\sin(\Delta\theta/2)$, while for $\theta_o \rightarrow \pi/2$ the two sinusoids degenerate into one ($\Delta u \rightarrow 0$) and any possibility to distinguish the two DoAs vanishes. The capability to distinguish the two DoAs in the frequency-plane depends on the capability to have both sinusoids with a spatial frequency spacing Δu that coincides with the minimal bias due to the frequency grid spacing $2\pi/N$ as for the periodogram (Figure 19.5). Therefore:

$$2u_{max}\cos\theta_o \sin(\Delta\theta/2) = \frac{2\pi}{N}$$

and solving for $\Delta\theta$ yields

$$\Delta\theta_{min} \simeq \frac{\lambda}{Nd\cos\theta_o} \tag{19.1}$$

where the resolution depends on the inverse of the *effective array aperture* $Nd\cos\theta_o$ (i.e., the aperture of the array from the angle θ_o).

Analysis of the resolution (19.1) shows that there is a definite benefit to maximizing the array aperture (*Nd*) and having the DoAs of interest almost perpendicular to the array (say $\theta_o \simeq 0$). If possible (e.g., in a radar system with rotating antennas), the array can be rotated up to $(\theta_1 + \theta_2)/2$ to maximize the DoAs resolution and have the maximal separation $\Delta u = 2u_{max}\left|\sin\left((\theta_2 - \theta_1)/2\right)\right|$. Whenever the sensors' spacing is a free design parameter, this could be arranged to maximize the resolution as

$$d \leq \frac{\lambda}{2} \cdot \frac{1}{\left|\sin\left((\theta_2 - \theta_1)/2\right)\right|}$$

as this choice guarantees the maximal resolution for DoAs within the interval $(\theta_2 - \theta_1)/2$, and spatially aliased for larger DoAs.

19.2 Beamforming and Signal Estimation

Multiple signals impinge on the array with different DoAs

$$\mathbf{y}(t) = \mathbf{A}(\boldsymbol{\theta})\mathbf{s}(t) + \mathbf{w}(t)$$

originated from different sources. One purpose of array processing is to estimate one signal out of multiple signals from the superposition $\mathbf{y}(t)$; the linear estimator is called *beamforming*. Let θ_1 be the DoA for the signal of interest with all the others being just interfering signals, the received signal can be partitioned as

$$\mathbf{y}(t) = \mathbf{a}(\theta_1)s_1(t) + \underbrace{\sum_{l=2}^{L} \mathbf{a}(\theta_l)s_l(t)}_{\mathbf{A}(\boldsymbol{\theta}_1)\mathbf{s}_1(t)} + \mathbf{w}(t) = \mathbf{a}(\theta_1)s_1(t) + \bar{\mathbf{w}}(t) \tag{19.2}$$

where $\mathbf{A}(\boldsymbol{\theta}_1) = \left[\mathbf{a}(\theta_2), ..., \mathbf{a}(\theta_L)\right]$ and $\mathbf{s}_1(t) = [s_2(t), ..., s_L(t)]^T$ models the structured interference. Namely, the overall noise and interference is

$$\bar{\mathbf{w}}(t) = \mathbf{A}(\boldsymbol{\theta}_1)\mathbf{s}_1(t) + \mathbf{w}(t) \sim \mathcal{CN}(\mathbf{0}, \mathbf{R}_{\bar{w}})$$

where the combined covariance is

$$\mathbf{R}_{\bar{w}} = \mathbf{R}_w + \mathbf{A}(\boldsymbol{\theta}_1)\mathbf{R}_{s1}\mathbf{A}^H(\boldsymbol{\theta}_1)$$

The configuration of DoAs reflects on the algebraic properties of the covariance $\mathbf{R}_{\bar{w}}$; in detail:

- When DoAs of the interferers $\boldsymbol{\theta}_1 = [\theta_2, \theta_3, ..., \theta_L]^T$ are isotropically distributed with comparable power ($\sigma_2^2 \simeq \sigma_3^2 \simeq ... \simeq \sigma_L^2$), the term $\mathbf{A}(\boldsymbol{\theta}_1)\mathbf{R}_{s1}\mathbf{A}^H(\boldsymbol{\theta}_1) \to \sigma_{int}^2\mathbf{I}$, for a large number of interferers ($L \to \infty$). In other words, the uncorrelated noise settings describe the case when there are many sources with DoAs within the range $[-\pi/2, \pi/2)$, all from uncorrelated sources with (approximately) the same power.
- When $\mathbf{w}(t) \simeq 0$ (noise-free case) and the interference dominates, the covariance $\mathbf{R}_{\bar{w}}$ is structured and the rank depends on the specific interference configuration: $\text{rank}(\mathbf{A}(\boldsymbol{\theta}_1)\mathbf{R}_{s1}\mathbf{A}^H(\boldsymbol{\theta}_1)) = \min(\text{rank}(\mathbf{A}(\boldsymbol{\theta}_1)), \text{rank}(\mathbf{R}_{s1})) \leq L-1$. The rank-deficient configuration of Figure 19.4 applies: if the same interferer is impinging on the array with two DoAs, $\text{rank}(\mathbf{R}_{s1}) < L-1$; or if two interferers have the same DoAs, they are not distinguishable and thus $\text{rank}(\mathbf{A}(\boldsymbol{\theta}_1)) < L-1$. The analysis of the structure of the covariance $\mathbf{R}_y$, or the analysis of its eigenvalues, can help to capture the structure of the sources and/or DoAs of the signal of interest and interferers.

Beamforming is based on a set of complex-valued weights $\mathbf{b}$ used as a linear estimator of the signal of interest (here $s_1(t)$) and/or to attenuate the impairments of noise and interferers in $\bar{\mathbf{w}}(t)$ as illustrated in Figure 19.6. In this case the DoA for the signal of

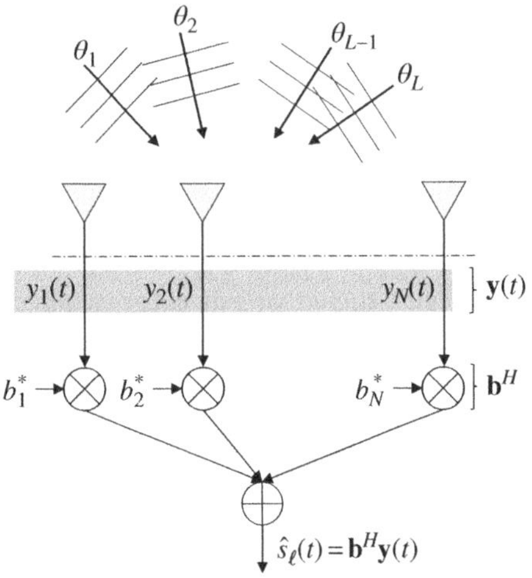

Figure 19.6 Beamforming configuration.

interest θ_1 is known (or estimated separately as detailed below), and the estimate of the signal

$$\hat{s}_1(t) = \mathbf{b}^H\mathbf{y}(t) = \underbrace{\mathbf{b}^H\mathbf{a}(\theta_1)s_1(t)}_{\text{signal}} + \underbrace{\mathbf{b}^H\bar{\mathbf{w}}(t)}_{\text{noise}}$$

is obtained by maximizing the power of the signal of interest (P_s) and/or minimizing the power of the noise (P_n) measured after the beamforming:

$$P_s = \mathbb{E}[|\mathbf{b}^H\mathbf{a}(\theta_1)s_1(t)|^2] = \sigma_s^2|\mathbf{b}^H\mathbf{a}(\theta_1)|^2$$
$$P_n = \mathbb{E}[\mathbf{b}^H\bar{\mathbf{w}}(t)\bar{\mathbf{w}}^H(t)\mathbf{b}] = \mathbf{b}^H\mathbf{R}_{\bar{w}}\mathbf{b}$$

It is of interest to evaluate the performance after beamforming as the power of the signal and noise at N sensors before beamforming are $P_{s,in} = N\sigma_s^2$ and $P_{n,in} = \text{tr}[\mathbf{R}_{\bar{w}}]$, respectively (for spatially uncorrelated noise $\mathbf{R}_{\bar{w}} = \sigma^2\mathbf{I}$, $P_{n,in} = N\sigma^2$, and the signal to noise ratio is σ_s^2/σ^2).

The design of the beamformer $\mathbf{b}$ depends on the criteria of what is considered the signal of interest and the spatial structure of the signal and interference, as specified later. However, assuming that the beamformer is designed for one specific DoA θ considered as the DoA of signal of interest, say $\mathbf{b}(\theta)$, the power of the signal for a set of M multiple snapshots (possibly independent) is

$$\hat{P}_y(\theta) = \frac{1}{M}\sum_{t=1}^{M}|\mathbf{b}^H\mathbf{y}(t)|^2 = \mathbf{b}^H\left(\frac{1}{M}\sum_{t=1}^{M}\mathbf{y}(t)\mathbf{y}(t)^H\right)\mathbf{b} = \mathbf{b}^H\hat{\mathbf{R}}_y\mathbf{b}$$

where $\hat{\mathbf{R}}_y$ is the sample correlation from the M observations. The inspection of $\hat{P}_y(\theta)$ versus the DoA θ used for beamforming design can be used to estimate the DoA (or multiple DoAs) as the value θ that maximize the received power of the signal $\hat{P}_y(\theta)$ after beamforming for the corresponding DoA. Resolution for DoA estimation depends on the beamforming method. Since any arbitrary scaling factor does not change the signal to noise ratio after beamforming, a convenient constraint is the norm of **b** being unitary: $||\mathbf{b}||^2 = 1$.

19.2.1 Conventional Beamforming

Conventional beamforming sets **b** to compensate for the phase evolution across the array corresponding to the DoA θ:

$$\mathbf{b} = \frac{1}{\sqrt{N}}\mathbf{a}(\theta)$$

with $P_s = N\sigma_s^2 = P_{s,in}$ (same as the input) for $\theta = \theta_1$ and $P_n = \mathbf{a}(\theta)^H \mathbf{R}_{\bar{w}} \mathbf{a}(\theta)/N$. The beamformer $\mathbf{b} = \mathbf{a}(\theta_1)/\sqrt{N}$ is the ML estimator of the signal of interest $s_1(t)$ when the noise is spatially uncorrelated with $\mathbf{R}_{\bar{w}} = \sigma^2\mathbf{I}$, as conventional beamforming resembles ML for frequency estimation by Fourier decomposition as discussed in Section 14.1. Furthermore, the array-gain is usually evaluated in term of gain function vs. the beamforming DoA θ for the signal impinging from θ_1 (for $d = \lambda/2$):

$$G(\theta|\theta_1) = |\mathbf{b}^H\mathbf{a}(\theta)|^2 = \frac{1}{N}\left|\sum_{k=1}^{N} \exp(j\pi(\sin\theta - \sin\theta_1)(k-1))\right|^2$$

$$= \frac{1}{N}\left|\frac{\sin\left[\pi(\sin\theta - \sin\theta_1)N/2\right]}{\sin\left[\pi(\sin\theta - \sin\theta_1)/2\right]}\right|^2$$

that is, the signal of interest is from DoA θ_1 and the beamforming steers this toward θ with $G(\theta|\theta_1) \le G(\theta_1|\theta_1) = N$, because $\theta = \theta_1$ is the largest gain, and attenuated elsewhere. Notice that this is a periodic sinc(.) that depends on the normalized spatial frequencies $\pi(\sin\theta - \sin\theta_1)$ for half-wavelength spacing $d = \lambda/2$, and thus the spacing DoAs θ for zero gain depends on the DoA θ_1 for the signal of interest. A simple case is for $\theta_1 = 0$ and the angular positions of the nulls of $G(\theta|\theta_1)$ is obtained by solving the relationship $\pi\sin\theta = \frac{2\pi}{N}\ell$ wrt θ for $\ell = 1, 2, ..., N$.

Example: The array-gain can be represented graphically in polar coordinates centered on the array where the radial distance is the gain $G(\theta|\theta_1)$. Figure 19.7 shows the array-gain $G(\theta|\theta_1)$ for $\theta_1 = 18$ deg and N = 8 (white line), and the wavefield correspondingly attenuated vs. angle in gray scale simply to visualize the wavefield attenuation vs. angle. The beamwidth $\Delta\theta_o$ between the first two zeros of $G(\theta|\theta_1)$ is computed by solving for $G(\theta_1 \pm \Delta\theta_o/2|\theta_1) = 0$:

$$\sin(\theta_1 \pm \Delta\theta_o/2) = \sin\theta_1 \pm \frac{2}{N}$$

which can be evaluated for the case here $\Delta\theta_o \simeq 30\,\text{deg}$, and the first null is for $\theta_1 \pm \Delta\theta_o/2 = \{3\text{ deg}, 33\text{ deg}\}$.

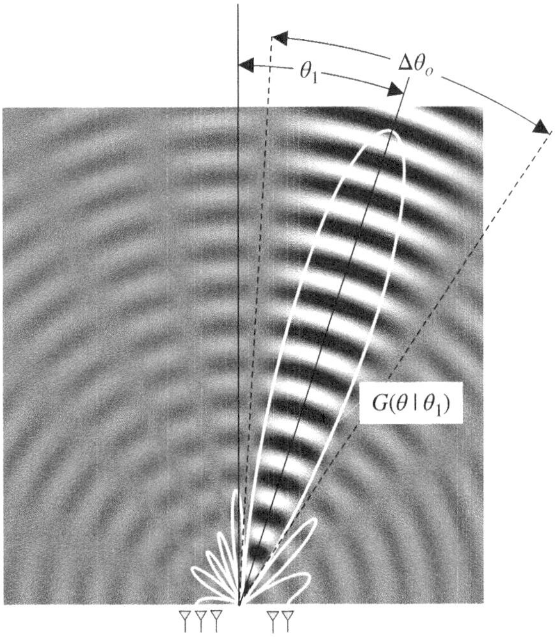

Figure 19.7 Conventional beamforming and the array-gain vs. angle pattern $G(\theta|\theta_1)$ for $N = 8$ sensors.

To gain further insight into the interference rejection capability of beamforming, let the noise be structured as corresponding to a set of interferers. To simplify, for one interferer assume that

$$\bar{\mathbf{w}}(t) = \mathbf{a}(\theta_n)s_n(t) \Rightarrow \mathbf{R}_{\bar{w}} = \sigma^2\mathbf{a}(\theta_n)\mathbf{a}(\theta_n)^H$$

then the power of the interferer after the beamforming

$$P_n = \sigma^2 \underbrace{|\mathbf{a}(\theta_1)^H\mathbf{a}(\theta_n)|^2/N}_{G(\theta_n|\theta_1)}$$

depends on the directional gain function $G(\theta_n|\theta_1)$ for the DoA of the interferers θ_n when the beamforming is pointing toward the DoA of interest θ_1. Since $G(\theta|\theta_1) \leq G(\theta_1|\theta_1)$, the interferer is attenuated if $\theta_n \neq \theta_1$, and it can be nullified if the DoA is such that $G(\theta_n|\theta_1) = 0$. Unfortunately, conventional beamforming has no control over where to angularly position the nulls of $G(\theta|\theta_1)$ to match with the DoAs of the interferers.

19.2.2 Capon Beamforming (MVDR)

Beamforming can be designed to preserve the signal from the DoA of interest, and minimize the overall interference arising from all the remaining DoAs (handled as interference). Minimum variance distortionless (MVDR) beamforming, also known as Capon beamforming, minimizes the overall mean power after spatial filtering with the constraint that for the DoA of interest θ, the beamforming response has unit gain (or equivalently, it is *distortionless* for the DoA of interest):

$$\min_{\mathbf{b}}\{\mathbb{E}[|\mathbf{b}^H\mathbf{y}(t)|^2]\} \quad \text{s.t. } \mathbf{b}^H\mathbf{a}(\theta)=1$$

Since $\mathbb{E}[|\mathbf{b}^H\mathbf{y}(t)|^2]=\mathbf{b}^H\mathbf{R}_y\mathbf{b}$, the overall problem is a constrained optimization of the quadratic $\mathbf{b}^H\mathbf{R}_y\mathbf{b}$; the solution is (see Section 1.8.1):

$$\mathbf{b}_{MVDR}=\frac{\mathbf{R}_y^{-1}\mathbf{a}(\theta)}{\mathbf{a}^H(\theta)\mathbf{R}_y^{-1}\mathbf{a}(\theta)}$$

MVDR beamforming depends on the inverse of the covariance of the received signals $\mathbf{R}_y$. In practice, the covariance is estimated from the sample mean $\hat{\mathbf{R}}_y$, and thus the MVDR solutions are obtained by the substitution $\mathbf{R}_y \to \hat{\mathbf{R}}_y$. In a non-stationary DoA environment, the number M of snapshots $\{\mathbf{y}(t)\}_{t=1}^M$ should be large enough to have a reliable estimate, but still small to capture the non-stationarity of the interference (in any case $M \geq N$ to have full-rank $\hat{\mathbf{R}}_y$). Still in a non-stationary environment, one can compute and update iteratively the inverse of the covariance using the inversion lemma as for the RLS algorithm (Section 15.10).

It is interesting to evaluate the interference rejection capability of MVDR beamforming when the DoA of interest of MVDR coincides with θ_1 and all the others are interference. According to the model (19.2), the covariance $\mathbf{R}_y=\mathbf{R}_{\tilde{w}}+\sigma_s^2\mathbf{a}(\theta_1)\mathbf{a}(\theta_1)^H$ contains both the signal of interest and the interference. The interference rejection follows from the array-gain vs. angle when the MVDR is designed for θ_1

$$G(\theta|\theta_1)=|\mathbf{b}^H\mathbf{a}(\theta)|^2=\left|\frac{\mathbf{a}^H(\theta)\mathbf{R}_y^{-1}\mathbf{a}(\theta_1)}{\mathbf{a}^H(\theta_1)\mathbf{R}_y^{-1}\mathbf{a}(\theta_1)}\right|^2$$

Unfortunately this has no simple form as the solution adapts to nullify the interference pattern as detailed below in some special cases.

Example

To numerically evaluate the effect of MVDR beamforming, let the example here be

$$\mathbf{y}=\mathbf{a}(\theta_1)s_1+\mathbf{a}(\theta_o)s_o+\mathbf{w}$$

where the DoA of interest is $\theta_1=18$ deg, the interferer is $\theta_o=12$ deg with power $\mathbb{E}[|s_o|^2]=\sigma_o^2$, the power of the signal of interest is normalized ($\sigma_1^2=1$), and the noise is spatially uncorrelated $\mathbf{w}\sim\mathcal{CN}(\mathbf{0},\sigma_w^2\mathbf{I})$. The experimental setup is worse (here $N=6$) than the example of conventional beamforming where the first null is for 3 deg, clearly not enough to effectively mitigate the interference. Figure 19.8 shows the MVDR gain $G(\theta|\theta_1)$ for varying $\sigma_o^2=\{1/10,1/2,1,2\}$ and $\sigma_w^2=1$ (left figure), and for varying $\sigma_w^2=\{1/10,1/2,1,2\}$ and $\sigma_o^2=1$ (right figure) compared to the gain $G(\theta|\theta_1)$ for conventional beamforming (gray line). When the power of the interference is $\sigma_o^2>\sigma_w^2$ in both examples the array-gain for the interferer $G(\theta_o|\theta_1)$ is close to zero (or $G(\theta_o|\theta_1)\ll G(\theta_1|\theta_1)=1$

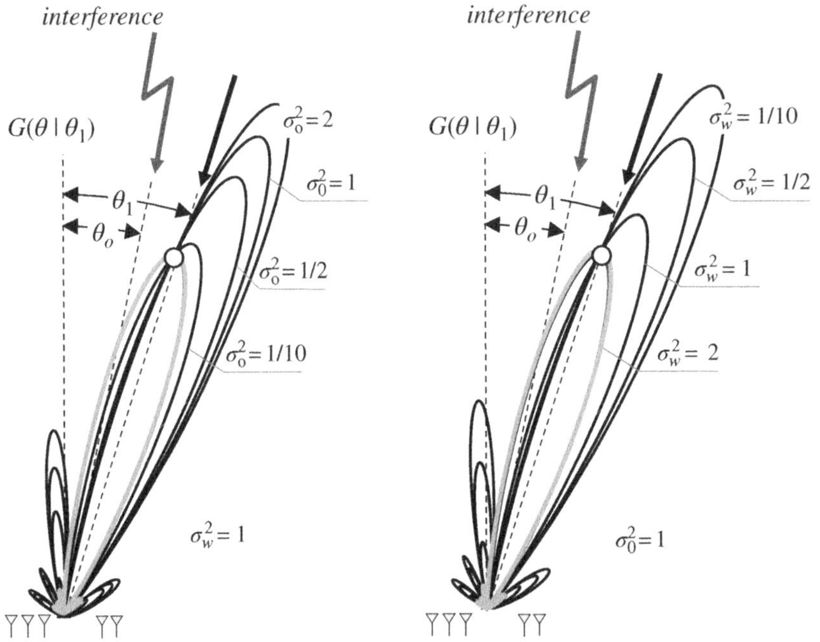

Figure 19.8 Array-gain $G(\theta|\theta_1)$ for MVDR beamforming for varying parameters.

according to the MVDR constraint, indicated by a dot in Figure 19.8). On the other hand, when the compound noise is dominated by the spatial uncorrelation and $\sigma_w^2 > \sigma_o^2$, the array-gain is similar to conventional beamforming. Notice that the array-gain can be large, say $G(\theta|\theta_1) > 1$ for some DoA where there is no interference but noise, when the noise σ_w^2 is small (or $\sigma_w^2 < \sigma_o^2$). This behavior should not be surprising as MVDR beamforming trades interference rejection for noise rejection.

The overall optimization problem of MVDR can be revisited as the design of a linear spatial filter that places a constraint on the DoA of interest to preserve the signal therein, without any control on the filter's behavior away from the constraint. Therefore, the array-gain $G(\theta|\theta_1)$ can have gain values $G(\theta|\theta_1) \gg 1$ for those DoAs where there is no directional interference embedded in the covariance $\mathbf{R}_y$, and the sum-power metric $\mathbf{b}^H\mathbf{R}_y\mathbf{b}$ is loosely affected by these values. However, if the MVDR is adopted in a non-stationary environment where there could be interferers arising or disappearing (e.g., to exemplify, if using an array for acoustic processing such as for music, the musical instruments or voices activate/deactivate asynchronously, each with its own DoA compared to the others), the overall interference pattern might quickly change and an MVDR designed for a stationary environment could have uncontrolled large gains for these temporary interferers, with disruptive effects. In this case, the beamforming design can be changed by adding at least an uncorrelated noise that accounts for the randomness of the interference and thus augmenting the covariance with uncorrelated noise (*diagonal loading*): $\mathbf{R}_y \rightarrow \mathbf{R}_y + \sigma_{load}^2\mathbf{I}$. Alternatively, the MVDR design can introduce an additional constraint on the beamforming norm $\mathbf{b}^H\mathbf{b} = \alpha$ (with arbitrary α), yielding the same solution as diagonal loading.

MVDR and Isotropic Sources

Consider MVDR for spatially uncorrelated noise ($\mathbf{R}_{\tilde{w}} = \sigma^2\mathbf{I}$). Spatially uncorrelated noise arises when DoAs are isotropically distributed, and the MVDR beamforming follows from

$$\min_{\mathbf{w}}\{\sigma^2\mathbf{b}^H\mathbf{b}+\sigma_s^2|\mathbf{b}^H\mathbf{a}(\theta_1)|^2\} \quad \text{s.t.} \quad \mathbf{b}^H\mathbf{a}(\theta_1)=1$$

The solution is $\mathbf{b} = \frac{1}{\sqrt{N}}\mathbf{a}(\theta_1)$, and the MVDR coincides with the conventional beamforming for spatially correlated noise, but it attains this solution when the interference is less structured as for the example above when $\sigma_o^2 \ll \sigma_w^2$. An intuitive explanation helps to gain insight on the MVDR criterion. When there are few interferers, beamforming attempts to minimize the overall power by exploiting the $N-1$ degrees of freedom (i.e., one is used for the constraint on the DoA of interest) of the N beamforming weights $\mathbf{b}$ by placing nulls or minima on the interfering DoAs up to the total number of $N-1$. On the contrary, when these interferers emerge as spatially uncorrelated, there are not enough degrees of freedom to null everywhere and the MVDR favors only the DoA of interest θ_1 and minimizes the array-gain wrt to the other DoAs.

19.2.3 Multiple-Constraint Beamforming

The MVDR can be generalized by placing multiple constraints on one or more DoAs of interest, and/or on one or more interferers, and/or adding some degree of uncertainty. The beamforming problem follows from the constrained optimization

$$\min_{\mathbf{b}}\{\mathbb{E}[|\mathbf{b}^H\mathbf{y}(t)|^2]\} \quad \text{s.t.} \quad \mathbf{a}(\theta_m)^H\mathbf{b} = c_m, \text{ for } m = 1,...,M$$

with $M \leq N-1$ constraints. The constrained optimization

$$\min_{\mathbf{b}}\{\mathbf{b}^H\mathbf{R}_y\mathbf{b}\} \quad \text{s.t.} \quad \mathbf{A}\mathbf{b} = \mathbf{c}$$

yields the following beamforming equation (see Section 1.8.2)

$$\mathbf{b}_{opt} = \mathbf{R}_y^{-1}\mathbf{A}^H(\mathbf{A}\mathbf{R}_y^{-1}\mathbf{A}^H)^{-1}\mathbf{c}$$

where the $M \leq N-1$ constraints $\mathbf{c}$ can be tailored to the specific problem as detailed below.

- Assuming that the beamforming is designed to null the DoAs of the interferers $\{\theta_2, \theta_3, .., \theta_M\}$ (assumed known), the set of constraints are (*null-steering*):

$$\begin{cases} \mathbf{a}(\theta_1)^H\mathbf{b} = 1 \\ \mathbf{a}(\theta_m)^H\mathbf{b} = 0 \quad \text{for } m = 2,...,M \end{cases}$$

- In a multipath environment, there could be more than one DoA of interest referred to the same source; this could yield to *multiple distortionless constraints*:

$$\mathbf{a}(\theta_m)^H\mathbf{b} = 1 \quad \text{for} \quad m = 1, ..., M$$

or any hybrid setting between the null-steering for interferers and no distortion for DoAs of interest.

- The DoA of interest θ_1 could be known with some uncertainty, and thus the distortionless condition can be tight within an interval $\pm\Delta\theta$ judiciously chosen based on the problem; the constraints become

$$\begin{cases} \mathbf{a}(\theta_1 - \Delta\theta)^H\mathbf{b} = 1 \\ \mathbf{a}(\theta_1)^H\mathbf{b} = 1 \\ \mathbf{a}(\theta_1 + \Delta\theta)^H\mathbf{b} = 1 \end{cases}$$

The constraint could be softer such as $\mathbf{a}(\theta_1 \pm \Delta\theta)^H\mathbf{b} = \gamma \leq 1$ so as to guarantee better control of the gain $G(\theta|\theta_1)$ outside of the range $[\theta_1 - \Delta\theta, \theta_1 + \Delta\theta]$. The control of the gain for the interfering DoAs is even more complex as null-steering is known to be loosely constrained around the nulling DoA, and a small mismatch between true and design DoA can vanish all the benefits of the beamforming. In this case the null-steering can be augmented by the constraints $\mathbf{a}(\theta_m \pm \Delta\theta)^H\mathbf{b} = 0$ for an appropriate interval $\pm\Delta\theta$. Needless to say, one could augment the constraints with the derivatives of gain to flatten the angular response, such as

$$\begin{aligned} \frac{d}{d\theta}\mathbf{a}(\theta)^H\mathbf{b}\Big|_{\theta=\theta_m} &= 0 \\ \frac{d^2}{d\theta^2}\mathbf{a}(\theta)^H\mathbf{b}\Big|_{\theta=\theta_m} &= 0 \\ &\vdots \end{aligned}$$

these conditions are quite relevant in null-steering.

Even if using more evolved beamforming design criteria, the angular beamwidth of the array-gain $G(\theta|\theta_1)$ around the DoA of interest (here θ_1) is not much different from (19.1):

$$\Delta\theta_{bw} \simeq \frac{\lambda}{Nd\cos\theta_1}$$

This is a convenient rule of thumb in many situations as it is dependent on the number of sensors N with a beamwidth stretching $1/\cos\theta_1$ that significantly broadens the beamwidth for $\theta_1 \neq 0$. The constrained beamforming acts by tilting the beamwidth to avoid the interferers as illustrated in the example in Section 19.2.2, and the beamwidth $\Delta\theta_{bw}$ is loosely dependent on the interfering pattern. Making reference to Figure 19.9 that approximates the array-gain (gray line) by a piecewise gain (black-line), any interferer with a DoA $\bar{\theta}$ close to the pointing beamwidth (in-beam interferer: $\bar{\theta} \in [\theta_1 - \Delta\theta_{bw}/2, \theta_1 + \Delta\theta_{bw}/2]$) cannot be rejected by the beamforming and $G(\bar{\theta}|\theta_1) \simeq 1$, while all the interferers with out-of-beam DoAs ($\bar{\theta} \notin [\theta_1 - \Delta\theta_{bw}/2, \theta_1 + \Delta\theta_{bw}/2]$) are attenuated by the array with an attenuation factor $G(\bar{\theta}|\theta_1) \ll 1$. This simple reasoning approximates fairly well the behavior of the array-gain vs. angle both for qualitative assessments and quantitative analysis [103].

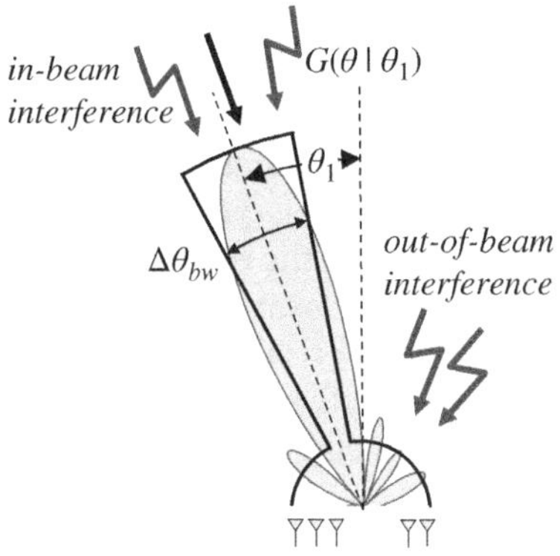

Figure 19.9 Approximation of the array-gain.

19.2.4 Max-SNR Beamforming

So far, the beamforming design has been based on knowledge of the DoAs, and in some contexts this is not practical as it involves a preliminary stage of DoA estimation and their classification into signal of interest and interference. Let the general model be

$$\mathbf{y}(t) = \underbrace{\sum_{l=1}^{L} \mathbf{a}(\theta_l)s(t)}_{\mathbf{x}(t)} + \mathbf{w}(t) = \mathbf{x}(t) + \mathbf{w}(t)$$

where the signal of interest is the superposition of one or multiple arrivals with different DoAs and same (or comparable) delays: $\mathbf{x}(t) = \sum_l \mathbf{a}(\theta_l)s(t)$. Assuming a Gaussian model of signal and noise (possibly augmented by interferers) with known covariances

$$\begin{aligned}\mathbf{x}(t) &\sim \mathcal{CN}(\mathbf{0}, \mathbf{R}_x)\\ \mathbf{w}(t) &\sim \mathcal{CN}(\mathbf{0}, \mathbf{R}_w)\end{aligned}$$

the signal to noise ratio (SNR) after the beamforming

$$SNR(\mathbf{b}) = \frac{P_x}{P_n} = \frac{\mathbf{b}^H\mathbf{R}_x\mathbf{b}}{\mathbf{b}^H\mathbf{R}_w\mathbf{b}}$$

depends on the beamforming weights $\mathbf{b}$. The SNR is the ratio of the two quadratic forms wrt $\mathbf{b}$ (Rayleigh quotient, see Section 1.7) and its maximization depends on the eigenvector decomposition of $\mathbf{R}_w^{-H/2}\mathbf{R}_x\mathbf{R}_w^{-1/2}$:

$$\mathbf{b}_{opt} = \mathbf{R}_w^{-1/2}\text{eig}_{\max}\{\mathbf{R}_w^{-H/2}\mathbf{R}_x\mathbf{R}_w^{-1/2}\}$$

and the corresponding optimal SNR is

$$SNR_{\max} = SNR(\mathbf{b}_{opt}) = \lambda_{\max}\{\mathbf{R}_w^{-H/2}\mathbf{R}_x\mathbf{R}_w^{-1/2}\}$$

Max-SNR beamforming exploits the difference in the structure of the covariance of signal and noise, and separates the two contributions provided that both are structurally different and separable. On the other hand, max-SNR has the benefit that it does not require to know the array manifold and the array itself could be uncalibrated (e.g., each sensor can have a different gain or phase-shift) or sensors could be mis-positioned.

In practice, the covariance $\mathbf{R}_x$ is not known, but the covariance of the received signal $\mathbf{R}_y$ (signal + noise) and noise $\mathbf{R}_w$ are known (or estimated as sample covariance). Making use of the difference from $\mathbf{R}_y$ as $\mathbf{R}_x = \mathbf{R}_y - \mathbf{R}_w$ could easily incur errors as there is no guarantee that the difference is positive definite. However, the objective can be redefined by considering the ratio between the power of the overall signal including the noise; this yields the new metric

$$\frac{P_y}{P_n} = 1 + \frac{P_x}{P_n} = 1 + SNR(\mathbf{b})$$

and the equivalence with max-SNR is trivial—but at least the new Rayleigh quotient is positive definite. The optimization of

$$\frac{P_y}{P_n} = \frac{\mathbf{b}^H \mathbf{R}_y \mathbf{b}}{\mathbf{b}^H \mathbf{R}_w \mathbf{b}}$$

yields the solution

$$\mathbf{b}_{opt} = \mathbf{R}_w^{-1/2} \text{eig}_{\max} \{ \mathbf{R}_w^{-H/2} \mathbf{R}_y \mathbf{R}_w^{-1/2} \}$$

that coincides with the max-SNR.

Max-SNR can be better used in practical systems where there is some control on the signals. For instance, one can periodically silence $\mathbf{x}(t)$ to estimate $\mathbf{R}_w$ with some periodicity related to the needed confidence and estimate $\mathbf{R}_y$ when $\mathbf{x}(t)$ is present. Alternatively, use a known signal $\mathbf{x}(t)$ for the sample estimate of $\mathbf{R}_x$ as in wireless systems when in training-mode (Section 15.2).

19.3 DoA Estimation

The model for DoA coincides with the superposition of complex sinusoids in line spectrum analysis in Chapter 16, and a review of Section 16.2 is recommended before reading this section. Let the sensors' spacing be half-wavelength $d = \lambda/2$, the signal at the kth sensor is

$$y_k(t) = \sum_{\ell=1}^{L} s_\ell(t) e^{j\pi \sin(\theta_\ell)(k-1)} + w_k(t)$$

and it resembles the sinusoids with angular frequency

$$u_\ell = \pi \sin \theta_\ell$$

evaluated here vs. the sensors' position. The collection of N signals from the sensors is

$$\mathbf{y}(t) = \mathbf{A}(\theta)\mathbf{s}(t) + \mathbf{w}(t)$$

where the sample $\mathbf{y}(t) \in \mathbb{C}^N$ is a snapshot that depends on the source modeled as $\mathbf{s}(t) \sim \mathcal{CN}(0, \mathbf{R}_s)$. A common assumption is to consider sampling intervals $t \in \{0, \pm\Delta t, \pm 2\Delta t, ...\}$ to have uncorrelated snapshots $\mathbf{y}(0), \mathbf{y}(\Delta t), \mathbf{y}(2\Delta t), ...$ where the uncorrelation is either for noise or signals:

$$\mathbb{E}[\mathbf{y}(n\Delta t)\mathbf{y}^H((n+k)\Delta t)] = \begin{cases} \mathbf{A}(\theta)\mathbf{R}_s\mathbf{A}^H(\theta) + \mathbf{R}_w & \text{for } k = 0 \\ 0 & \text{for } k \neq 0 \end{cases}$$

and this depends on the correlation properties of processes. The statistical model for DoA is the unconditional model (Section 16.1.2)

$$\mathbf{y} \sim \mathcal{CN}(\mathbf{0}, \mathbf{A}(\boldsymbol{\theta})\mathbf{R}_s\mathbf{A}^H(\boldsymbol{\theta}) + \sigma_w^2\mathbf{I})$$

as typically sources $\mathbf{s}(t)$ are modelled as random and, if needed, are estimated separately by beamforming methods after DoAs have been estimated.

The sample estimate of the covariance from the snapshots is

$$\hat{\mathbf{R}}_y = \frac{1}{T}\sum_{n=1}^{T} \mathbf{y}(n\Delta t)\mathbf{y}^H(n\Delta t)$$

and this converges (with probability 1) to the covariance of the array experiment

$$\hat{\mathbf{R}}_y \underset{T\to\infty}{\to} \mathbf{R}_y = \mathbf{A}(\theta)\mathbf{R}_s\mathbf{A}^H(\theta) + \sigma_w^2\mathbf{I}.$$

19.3.1 ML Estimation and CRB

The statistical model for DoA estimation is the one for stochastic (or unconditional) ML

$$\hat{\boldsymbol{\theta}} = \arg\min_{\boldsymbol{\theta}} \ln |\mathbf{A}(\boldsymbol{\theta})\hat{\mathbf{R}}_s\mathbf{A}^H(\boldsymbol{\theta}) + \hat{\sigma}^2\mathbf{I}|$$

see Section 16.2.3 for the definitions of $\hat{\mathbf{R}}_s$ and $\hat{\sigma}^2$ if both are unknown nuisance parameters. Similarly to line spectra estimation, optimization is complex and other methods are far more easily implemented. Adaptation of MUSIC (Section 16.3.3) or ESPRIT (Section 16.3.4) to DoA is straightforward provided that one maps the variables

$$\omega_\ell \Leftrightarrow u_\ell = \pi \sin\theta_\ell$$

The CRB for the unconditional model $CRB_u(\boldsymbol{\theta})$ is in Section 16.2.4 and it shows that the stochastic ML has to be preferred to a constrained ML (i.e., based on the model $\mathbf{y} \sim \mathcal{CN}(\mathbf{A}(\boldsymbol{\theta})s(t), \sigma_w^2\mathbf{I})$) since

$$\text{cov}(\hat{\boldsymbol{\theta}}_c) \succeq cov(\hat{\boldsymbol{\theta}}_u) \underset{N\to\infty}{\longrightarrow} CRB_u(\boldsymbol{\theta})$$

hence the unconditional ML is uniformly better than the conditional ML, and asymptotically attains the CRB.

The variance can be adapted from the CRB in Section 16.2.4 using the transformation $\omega_\ell = \pi \sin\theta_\ell$ and the local linearization (Section 8.3):

$$\mathrm{var}(\hat{\omega}_\ell) = \pi^2 \cos^2\theta_\ell \times \mathrm{var}(\hat{\theta}_\ell)$$

thus

$$\mathrm{var}(\hat{\theta}_\ell) = \frac{\mathrm{var}(\hat{\omega}_\ell)}{\pi^2 \cos^2\theta_\ell}$$

In other words, the variance depends on the effective aperture and scales with the inverse of $\cos\theta_\ell$; it is minimized for $\theta_\ell = 0$ and it becomes uncontrollably large for $\theta_\ell \to \pi/2$. For a single DoA in uncorrelated noise $\mathbf{w}(t) \sim \mathcal{CN}(\mathbf{0}, \sigma^2\mathbf{I})$, the CRB scales as $1/N^3$.

19.3.2 Beamforming and Root-MVDR

One alternative way to estimate the DoAs is by considering those DoAs where the power after a beamforming $\hat{P}_y(\theta) = \frac{1}{M}\sum_{t=1}^{M} |\mathbf{b}^H\mathbf{y}(t)|^2$ peaks. For MVDR, the angle for constraint θ is the DoA used for designing the beamformer, and the beamformer's power is

$$\hat{P}_{MVDR}(\theta) = \mathbf{b}^H_{MVDR}(\theta)\hat{\mathbf{R}}_y\mathbf{b}_{MVDR}(\theta) = \frac{1}{\mathbf{a}^H(\theta)\hat{\mathbf{R}}_y^{-1}\mathbf{a}(\theta)}$$

with the advantage that out-of-beam interference is rejected, or possibly mitigated, and in any case accounted in the beamforming design for every choice of θ.

Rather than searching for the L largest peaks of $\hat{P}_{MVDR}(\theta)$, one can search for the L DoAs θ that minimize the function

$$\hat{Q}(\theta) = \mathbf{a}^H(\theta)\hat{\mathbf{R}}_y^{-1}\mathbf{a}(\theta)$$

This metric can be rewritten as

$$\hat{Q}(\theta) = \sum_{m=0}^{N-1}\sum_{n=0}^{N-1} r[m,n]e^{-jm\theta}e^{jn\theta}$$

where $r[m,n] = [\hat{\mathbf{R}}_y^{-1}]_{mn}$ is the inverse sample covariance $\hat{\mathbf{R}}_y$. Let

$$z = e^{j\theta}$$

be complex exponentials, this becomes

$$\hat{Q}_z(z) = \sum_{m=0}^{N-1}\sum_{n=0}^{N-1} r[m,n]z^{n-m} = \sum_{k=-(N-1)}^{N-1} g[k]z^k$$

where

$$g[k] = \begin{cases} \sum_{n=k}^{N-1} r[n-k,n] & k = 0,1,\dots,N-1 \\ \sum_{n=0}^{k+N-1} r[n-k,n] & k = -(N-1),\dots,-1 \end{cases}$$

Since $\hat{\mathbf{R}}_y^{-1}$ is symmetric Hermitian, the polynomial $\hat{Q}_z(z)$ is complex conjugate symmetric ($g[-k] = g^*[k]$) and it has the min/max phase factorization (with $Q_{\max}(z) = Q^*_{\min}(1/z)$, see Appendix B of Chapter 4):

$$\hat{Q}_z(z) = Q_{\min}(z) \cdot Q^*_{\min}(1/z)$$

where $Q_{\min}(z)$ has N roots $\{\hat{z}_k\}$ inside the unit-radius circle (and symmetrically for $Q_{\max}(z) = Q^*_{\min}(1/z)$). The root-MVDR algorithm for DoA estimation is based on the angle of the roots $\{\hat{z}_k\}$:

$$\hat{\theta}_k = \measuredangle \hat{z}_k$$

Their number depends on that subset of L roots closer to the unit circle out of the total of N roots (note that radial errors of roots are ineffective in DoAs). The performance of Root-MVDR is better than spectral MVDR (from the peaks of $\hat{P}_{MVDR}(\theta)$) and attains the CRB even for small signal to noise ratios (e.g., below the threshold region).

20

Multichannel Time of Delay Estimation

Time of delay (ToD) estimation is a very common problem in many applications that employ wavefield propagation such as radar, remote sensing, and geophysics (Section 18.5) where the information of interest is mapped onto ToDs. Features of the wavefields call for measurement redundancy from a multichannel system where there are $M \gg 1$ signals that are collected as $s_1(t), s_2(t), ..., s_M(t)$, and this leads to massive volumes of data for estimating the ToD metric. The uniform linear array in Chapter 19 represents a special case of a multichannel system where the source(s) is narrowband and the only parameter(s) that can be estimated is the DoA(s). Here the narrowband constraint is removed while still preserving the multidimensionality of the signals in order to have the needed redundancy for estimating the ToD variation vs. measuring channels (e.g., sensors) with known geometrical positions.

The multichannel concept is illustrated in Figure 20.1 from seismic exploration (see Section 18.5 or [82]): the wavefield generated at the surface propagates in the subsurface and is backscattered at depth discontinuities, and measured with sensors at the surface. The wavefield is collected by a regularly spaced 2D array of sensors to give the measured wavefields $s(x, y, t)$ in space (x, y) and time t in the figure. Note that in Figure 20.1 the 3D volume $s(x, y, t)$ of 1.3 km×10 km is sliced along planes to illustrate by gray-scale images the complex behavior of multiple backscattered wavefronts, and the corresponding ToD discontinuities of wavefronts due to subsurface faults. Multichannel ToD estimation refers to the estimation of the ensemble of ToDs pertaining to the same wavefront represented by a 2D surface $\tau_\ell(x, y)$ that models the ℓth delayed wavefront characterized by a spatial continuity.[1] The measured wavefield can be modeled as a superposition of echoes (for any position x, y)

$$s(x, y, t) = \sum_\ell \alpha_\ell(x, y) g(t - \tau_\ell(x, y)) + w(x, y, t)$$

where the amplitudes $\alpha_\ell(x, y)$ and ToDs $\tau_\ell(x, y)$ are wavefront and sensor dependent, and $g(t)$ is the excitation waveform.

The ToD from multiple measurements is by chaining multiple distinct sensors. A radar system can be interpreted as a multichannel model when estimating the ToD variation vs. the scanning time when the target(s) move(s) from one time-scan to

1 In seismic exploration, the function $\tau_\ell(x, y)$ defines the locus of backscattered echoes due to the same in-depth surface, and these surfaces are called horizons to resemble their physical meaning.

Statistical Signal Processing in Engineering, First Edition. Umberto Spagnolini.

Companion website: www.wiley.com/go/spagnolini/signalprocessing

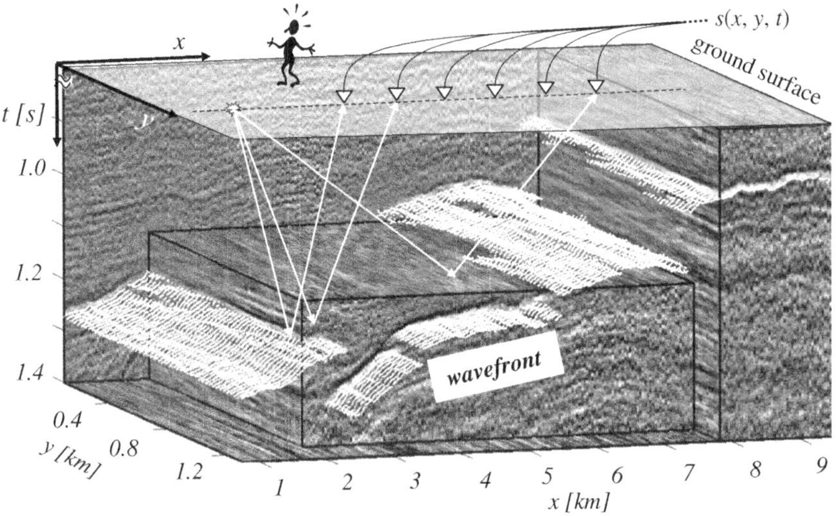

Figure 20.1 Multichannel measurements for subsurface imaging.

another, or the radar antenna moves for a still target (e.g., satellite remote sensing of the earth). Regardless of the origin of the multiplicity of measurements, the most general model for multichannel ToD for the ith measurement (Figure 20.2):

$$s_i(t) = \sum_{\ell=1}^{L} \alpha_{\ell,i} g_\ell(t - \tau_{\ell,i}) + w_i(t)$$

is the superposition of L echoes each with *waveform* $g_\ell(t)$, amplitude $\alpha_{\ell,i}$, and ToD $\tau_{\ell,i}$ as sketched below in gray-intensity plot with superimposed the trajectory of the ToD $\{\tau_{\ell,i}\}$ vs. the sensor's position (black-line). ToD estimation refers to the estimation of the delays $\{\tau_{\ell,i}\}_{\ell=1}^{L}$ for a specific measurement, and the amplitudes $\{\alpha_{\ell,i}\}_{\ell=1}^{L}$ could be nuisance parameters. However, when considering a collection of measurements $s_1(t), s_2(t), ..., s_M(t)$, multichannel ToD estimation refers to the estimation of the ToD profile for the ℓth *wavefront* (i.e., a collection of ToD vs. sensor):

$$\boldsymbol{\tau}_\ell = [\tau_{\ell,1}, \tau_{\ell,2}, ..., \tau_{\ell,M}]^T$$

Parametric wavefront: Estimation of ToDs of the wavefront(s) as a collection of the corresponding delays. The wavefront can be parameterized by a set of wavefront-dependent geometrical and kinematical *shape parameters* $\boldsymbol{\theta}_\ell$ (e.g., sensors position, sources position, and propagation velocity) as illustrated in Section 18.5, so that the ToD dependency is

$$\boldsymbol{\tau}_\ell = \boldsymbol{\tau}(\boldsymbol{\theta}_\ell)$$

To exemplify, one can use a linear model for the ToDs $\tau_{\ell,i} = \theta_{\ell,0} + i \cdot \theta_{\ell,1}$ so that the wavefront can be estimated by estimating the two parameters $\boldsymbol{\theta}_\ell = [\theta_{\ell,0}, \theta_{\ell,1}]^T$ instead. The estimation of the shape parameters from a set of redundant ($M \gg 1$) measurements

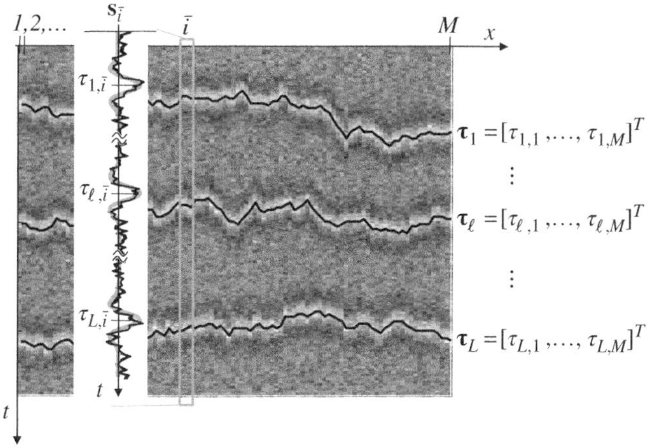

Figure 20.2 Multichannel ToD model.

Table 20.1 Taxonomy of ToD methods.

# sensors (M)	# ToD (L)	$g_\ell(t)$	*ToD estimation method*
1	1	known	MLE (Section 9.4)
2	1	not-known	Difference of ToD (DToD) (Section 20.3)
1	>1	$g_\ell(t) = g(t)$	MLE for multiple echoes (Section 20.2)
>2 (wavefront)	1	known	MLE/Shape parameters estimation (Section 20.5)
>2 (wavefront)	1	non-known	Shape parameters estimation (Section 20.6)
>2 (wavefront)	>1	known	Multi-target ToD tracking [92]

has the benefit of tailoring ToD methods to estimate the sequence of ToDs vs. a measurement index as in Figure 20.2. Applications are broad and various such as wavefield based remote sensing, radar, and ToD tracking using Bayesian estimation methods (Section 13.1). To illustrate, when two flying airplanes cross, one needs to keep the association between the radar's ToDs and the true target positions as part of the array processing (Chapter 19). Only the basics are illustrated in this chapter, leaving the full discussion to application-related specialized publications.

Active vs. passive sensing systems: The waveforms $g_1(t), g_2(t), ..., g_L(t)$ can be different to one another and dependent on the sources, which are not necessarily known. A radar/sonar system (Section 5.6) or seismic exploration (Section 18.5) are considered as *active systems* as the waveforms are generated by a controlled source and all waveforms are identical ($g_\ell(t) = g(t)$ for $\forall \ell$), possibly designed to maximize the effective bandwidth (Section 9.4) and thus the resolution for multiple echoes. In contrast to active systems, in *passive systems* the waveforms are different from one another, modeled as stochastic processes characterized in terms of PSD. An example of a passive system is the acoustic signal in an concert hall, which is the superposition of signals from many musical instruments. The scenario is complex, and Table 20.1 offers a (limited and essential) taxonomy of the methods and summarizes those discussed in this chapter.

20.1 Model Definition for ToD

The ToD problem for multiple echoes is continuous-time estimation that must be modified to accommodate discrete-time data. The time-sampled signals are modeled as in Section 5.6:

$$\mathbf{s}_i = \sum_{\ell=1}^{L} \alpha_{\ell,i}\mathbf{g}_\ell(\tau_{\ell,i}) + \mathbf{w}_i = \underbrace{[\mathbf{g}_1(\tau_{1,i}), ..., \mathbf{g}_L(\tau_{L,i})]}_{\mathbf{G}_i} \cdot \underbrace{\begin{bmatrix} \alpha_{1,i} \\ \vdots \\ \alpha_{L,i} \end{bmatrix}}_{\boldsymbol{\alpha}_i} + \mathbf{w}_i$$

where $\mathbf{g}_\ell(\tau_{\ell,i}) = [g_\ell(t_1 - \tau_{\ell,i}), g_\ell(t_2 - \tau_{\ell,i}), ...g_\ell(t_N - \tau_{\ell,i})]^T \in \mathbb{R}^N$ is the column that orders the ℓth delayed waveform sampled in $\{t_1, t_2, ..., t_N\}$, and noise $\mathbf{w}_i \sim \mathcal{N}(0, \mathbf{C}_w)$. Dependence of the ToDs is non-linear and it is useful to exploit the quadratic dependency on the amplitudes $\boldsymbol{\alpha}_i$ in the ML method (Section 7.2.2):

$$\hat{\boldsymbol{\alpha}}_i = (\mathbf{G}_i^T\mathbf{C}_w^{-1}\mathbf{G}_i)^{-1}\mathbf{G}_i^T\mathbf{C}_w^{-1}\mathbf{s}_i$$

The MLE follows from the optimization

$$(\hat{\tau}_{1,i}, ..., \hat{\tau}_{L,i}) = \arg \min_{(\tau_{1,i},...,\tau_{L,i})} \{\mathbf{s}_i^T(\mathbf{I} - \mathbf{P}_{\mathbf{G}_i})^T\mathbf{C}_w^{-1}(\mathbf{I} - \mathbf{P}_{\mathbf{G}_i})\mathbf{s}_i\}$$

where $\mathbf{P}_{\mathbf{G}_i} = \mathbf{G}_i(\mathbf{G}_i^T\mathbf{C}_w^{-1}\mathbf{G}_i)^{-1}\mathbf{G}_i^T\mathbf{C}_w^{-1}$. In the case of a white Gaussian noise process ($\mathbf{C}_w = \sigma_w^2\mathbf{I}$), the optimization for ML ToD estimation reduces to the maximization of the metric

$$(\hat{\tau}_{1,i}, ..., \hat{\tau}_{L,i}) = \arg \max_{(\tau_{1,i},...,\tau_{L,i})} \text{tr}\{\mathbf{P}_{\mathbf{G}_i}\mathbf{s}_i\mathbf{s}_i^T\}$$

where the projection matrix $\mathbf{P}_{\mathbf{G}_i} = \mathbf{G}_i(\mathbf{G}_i^T\mathbf{G}_i)^{-1}\mathbf{G}_i^T$ imbeds into the matrix $\mathbf{G}_i^T\mathbf{G}_i$ all the products $\mathbf{g}_n^T(\tau_{n,i})\mathbf{g}_m(\tau_{m,i})$ that account for the mutual interference of the different waveforms. A special case is when the amplitudes are changing, but the delays $(\tau_1, ..., \tau_L)$ are not:

$$\mathbf{s}_i = \underbrace{[\mathbf{g}_1(\tau_1), ..., \mathbf{g}_L(\tau_L)]}_{\mathbf{G}(\tau_1,...,\tau_L)} \cdot \boldsymbol{\alpha}_i + \mathbf{w}_i$$

This model reduces to the MLE in Section 7.2.3 and gives

$$(\hat{\tau}_1, ..., \hat{\tau}_L) = \arg \max_{(\tau_1,...,\tau_L)} \text{tr}\{\mathbf{P}_{\mathbf{G}} \sum_i \mathbf{s}_i\mathbf{s}_i^T\}$$

The influence of noise and amplitude fluctuations is mitigated by the averaging of the outer products $\sum_i \mathbf{s}_i\mathbf{s}_i^T$.

Recall (Section 9.4) that for $L = 1$ and $\mathbf{C}_w = \sigma_w^2 \mathbf{I}$, the MLE depends on the maximization of

$$\phi_{sg}(\tau) = \sum_n s[n]\, g(\zeta)|_{\zeta = n\Delta t - \tau} \qquad (20.1)$$

that is the cross-correlation between the samples of $\{s[n]\}$ and the continuous-time waveform $g(t)$ arbitrarily shifted (by τ) and regularly sampled by the same sampling interval Δt of $x[n] = x(n \cdot \Delta t)$. However, if the waveform is only available as sampled $g[k]$ at sampling interval Δt, the cross-correlation (20.1) can be evaluated over a coarse sample-grid (Section 10.3), or it requires that the discrete-time waveform is arbitrarily shifted by a non-integer and arbitrary delay τ, and this can be obtained by interpolation and resampling (Appendix B).

20.2 High Resolution Method for ToD (L=1)

Let the model be

$$s(t) = \sum_{\ell=1}^{L} \alpha_\ell g(t - \tau_\ell) + w(t) \qquad \text{for } t \in [0, T] \qquad (20.2)$$

with known waveform $g(t)$. The MLE of the ToDs

$$(\hat{\tau}_1, \ldots, \hat{\tau}_L) = \arg \max_{(\tau_1, \ldots, \tau_L)} \text{tr}\{\mathbf{P_G} \mathbf{s}\mathbf{s}^T\}$$

for $\mathbf{G}(\tau_1, \ldots, \tau_L) = [\mathbf{g}(\tau_1), \ldots, \mathbf{g}(\tau_L)]$ is a multidimensional search over L-dimensions (see Section 9.4 for $L = 2$) that might be affected by many local maxima, and the ToD estimate could be affected by large errors when waveforms are severely overlapped. High-resolution methods adapted from the frequency estimation in Chapter 16 have been proposed in the literature and are based on some simple similarities reviewed below.

20.2.1 ToD in the Fourier Transformed Domain

The model after the Fourier transform of the data (20.2) is

$$S(\omega) = G(\omega) \sum_{\ell=1}^{L} \alpha_\ell \exp(-j\omega\tau_\ell) + W(\omega) \qquad (20.3)$$

Since the waveform is known, its Fourier transform $G(\omega)$ is known and the observation can be judiciously deterministically deconvolved (i.e., $S(\omega)$ is divided by $G(\omega)$ except when $|G(\omega)| \simeq 0$) so that after deconvolution:

$$\widetilde{S}(\omega) = \frac{S(\omega)}{G(\omega)} = \sum_{\ell=1}^{L} \alpha_\ell \exp(-j\omega\tau_\ell) + \tilde{W}(\omega)$$

This model resembles the frequency estimation in additive Gaussian noise, where the ToD τ_ℓ induces a linear phase-variation vs. frequency, rather than a linear phase-variation vs. time as in the frequency estimation model (Chapter 16).

Based on this conceptual similarity, the discrete-time model can be re-designed using the Discrete Fourier transform (Section 5.2.2) over N samples $S[k] = [DFT_N(\mathbf{s})]_k$ to reduce the model (20.2) to (20.3):

$$S[k] = G[k]\sum_{\ell=1}^{L} \alpha_\ell \exp(j\omega_\ell k) + W[k] \quad \text{for } k = -N/2, -N/2+1, ..., N/2-1 \tag{20.4}$$

where the ℓth ToD τ_ℓ makes the phase change linearly over the sample index k

$$\omega_\ell = -\frac{2\pi\tau_\ell}{T} = -\frac{2\pi\tau_\ell}{N\Delta t}$$

behaving similarly to a frequency that depends on the sampling interval Δt. If considering multiple independent observations with randomly varying amplitudes $\{\alpha_{\ell,i}\}_{i=1}^{M}$ as in the model in Section 20.1, the ToD estimation from the estimate of $\{\omega_\ell\}_{\ell=1}^{L}$ coincides with frequency estimation, and the high-resolution methods in Chapter 16 can be used with excellent results.

In many practical applications, there is only one observation ($M = 1$) and the model (20.4) can be represented in compact notation (N is even):

$$\begin{bmatrix} S[-N/2] \\ S[-N/2+1] \\ \vdots \\ S[N/2-1] \end{bmatrix} = \mathbf{S} = \mathbf{G}\sum_{\ell=1}^{L} \alpha_\ell \mathbf{a}(\omega_\ell) + \mathbf{W}$$

where

$$\mathbf{a}(\omega_\ell) = [e^{j\omega_\ell(-N/2)}, e^{j\omega_\ell(-N/2+1)}, ..., e^{j\omega_\ell(N/2-1)}]^T$$

and

$$\mathbf{G} = \text{diag}(G[-N/2], G[-N/2+1], ..., G[N/2-1])$$

contains the DFT of the waveform $G[k] = [DFT_N(\mathbf{g}(0))]_k$. The estimation of multiple ToDs follows from LS optimization of the cost function:

$$J(\boldsymbol{\alpha}, \boldsymbol{\omega}) = \left\| \mathbf{S} - \mathbf{G}\sum_{\ell=1}^{L} \alpha_\ell \mathbf{a}(\omega_\ell) \right\|^2$$

The LS method coincides with MLE when the noise samples $W[k]$ are Gaussian and mutually uncorrelated. Since the DFT of Gaussian white process $w(n\Delta t)$ leaves the samples $W[k]$ to be uncorrelated due to the unitary transformation property of DFT, the minimization of $J(\boldsymbol{\alpha},\boldsymbol{\omega})$ coincides with the MLE.

One numerical method to optimize $J(\boldsymbol{\alpha},\boldsymbol{\omega})$ is by the weighted Fourier transform RELAXation (W-RELAX) iterative method that resembles the iterative steps of EM (Section 11.6), where at every iteration the estimated waveforms are stripped from the observations for estimation of one single parameter [95]. In detail, let the set $\{\hat{\alpha}_i,\hat{\omega}_i\}_{i=1,i\neq\ell}^{L}$ be known as estimated at previous steps, after stripping these waveforms

$$\mathbf{S}_\ell = \mathbf{S} - \mathbf{G}\sum_{i=1,i\neq\ell}^{L}\hat{\alpha}_i\mathbf{a}(\hat{\omega}_i)$$

one can use a local metric for the ℓth ToD and amplitude

$$J_\ell(\alpha_\ell,\omega_\ell) = \left\|\mathbf{S}_\ell - \alpha_\ell\mathbf{G}\mathbf{a}(\omega_\ell)\right\|^2$$

As usual, this is quadratic wrt α_ℓ and it can be easily solved for the frequency ω_ℓ (i.e., the ℓth ToD)

$$\hat{\omega}_\ell = \arg\max_{\omega_\ell}|\mathbf{a}(\omega_\ell)^H\mathbf{G}^*\mathbf{S}_\ell|$$

and the amplitude follows once $\hat{\omega}_\ell$ has been estimated as

$$\hat{\alpha}_\ell = \frac{\mathbf{a}(\hat{\omega}_\ell)^H\mathbf{G}^*\mathbf{S}_\ell}{||\mathbf{G}||^2}$$

Iterations are carried out until some degree of convergence is reached. Improved and faster convergence is achieved by adopting IQML (Section 16.3.1) in place of the iterations of W-RELAX.

The methods above can be considered extensions of statistical methods discussed so far, but some practical remarks are necessary:

- Any frequency domain method needs to pay attention to the shape of $G(\omega)$, or the equivalent DFT samples $G[k]$. In definitions, the range of $\mathbf{G}$ should be limited to those values that are meaningfully larger than zero to avoid instabilities in convergence. However, restricting the range for ω reduces ToD accuracy considerably as it is dominated by higher frequencies in the computation of the effective bandwidth (Section 9.4). Similar reasoning holds when performing deterministic deconvolution if using other high-resolution methods, as the deconvolution $1/G(\omega)$ can be stabilized by *conditioning* that is adding a constant term to avoid singularities such as $1/[G(\omega)+\eta]$, where η is a small fraction of $\max\{|G(\omega)|\}$.
- Conditioning and accuracy are closely related. In some cases in iterative estimates it is better to avoid temporary ToD estimates getting too close one another: one can set a ToD threshold, say $\Delta\omega$, that depends on the waveforms. In this case, the estimates from the iterative methods above that are too close, say $\hat{\omega}_1$ and $\hat{\omega}_2$ with $|\hat{\omega}_1-\hat{\omega}_2|<\Delta\omega$ and $\hat{\omega}_1<\hat{\omega}_2$, are replaced by $\hat{\omega}_1-\Delta\omega/2$ and $\hat{\omega}_2+\Delta\omega/2$ to improve convergence and accuracy.

20.2.2 CRB and Resolution

As usual, the CRB is a powerful analytic tool to establish the accuracy bound. In the case of ToD, the performance depends on the waveforms and their spectral content as detailed in Section 9.4 for $L = 1, 2$ ToDs. When considering multiple delays, the CRB depends on the FIM entries, which for white Gaussian noise with power spectral density $N_0/2$ can be evaluated in closed form as:

$$[\mathbf{J}]_{\alpha_i\alpha_j} = \frac{2}{N_0}\iint g(t-\tau_i)g(\alpha-\tau_j)dtd\alpha = \frac{2}{N_0}\phi_{gg}(\tau_i-\tau_j)$$
$$[\mathbf{J}]_{\alpha_i\tau_j} = -\frac{2}{N_0}\alpha_j\iint g(t-\tau_i)\dot{g}(\alpha-\tau_j)dtd\alpha = -\frac{2}{N_0}\alpha_j\phi_{g\dot{g}}(\tau_i-\tau_j)$$
$$[\mathbf{J}]_{\tau_i\tau_j} = \frac{2}{N_0}\alpha_i\alpha_j\iint \dot{g}(t-\tau_i)\dot{g}(\alpha-\tau_j)dtd\alpha = \frac{2}{N_0}\alpha_i\alpha_j\phi_{\dot{g}\dot{g}}(\tau_i-\tau_j)$$

Recall that the choice of the waveform is a powerful degree of freedom in system design to control ToD accuracy when handling multiple and interfering echoes. Choice of waveform and range can satisfy $\phi_{gg}(\tau_i-\tau_j)=0$ and $\phi_{\dot{g}\dot{g}}(\tau_i-\tau_j)=0$ so that two (or more) ToDs are decoupled and the correlation-based ML estimator can be applied iteratively on each echo. In all the other contexts when $\phi_{gg}(\tau_i-\tau_j)\neq 0$ and/or $\phi_{\dot{g}\dot{g}}(\tau_i-\tau_j)\neq 0$, the CRB for multiple echoes is larger than the CRB for one isolated echo due to the interaction of the waveforms, and it can be made dependent on their mutual interaction as illustrated in the simple example in Section 9.4 with $L = 2$.

With multiple echoes, resolution is the capability to distinguish all the echoes by their different ToDs. Let a measurement be of $L = 2$ echoes in τ_1 and τ_2 ($\tau_1 < \tau_2$) with the same waveform; it is obvious that when the two ToDs are very close, if not coincident, the two waveforms coincide and are misinterpreted as one echo, possibly with a small waveform distortion that might be interpreted as noise. Most of the tools to evaluate the performance in this situation are by numerical experiments using common sense and experience. To exemplify in Figure 20.3, if the variance from the CRB is $\text{var}(\hat{\tau}_1) =$

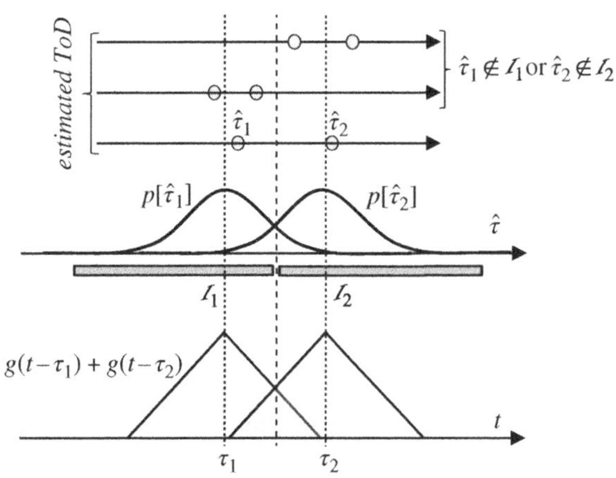

Figure 20.3 Resolution probability.

$\text{var}(\hat{\tau}_1) = \sigma_\tau^2$ and $\sigma_\tau > |\tau_2 - \tau_1|/2$ the two echoes cannot be resolved as two separated echoes and estimates interfere severely with one another. On the contrary, when noise is small enough to make $\sigma_\tau < |\tau_2 - \tau_1|/2$, one can evaluate the ***resolution probability*** for the two ToD estimates $\hat{\tau}_1$ and $\hat{\tau}_2$. A pragmatic way to do this is by choosing two disjoint time intervals $\mathcal{I}_1$ and $\mathcal{I}_2$ around τ_1 and τ_2; the resolution probability P_{res} follows from the frequency analysis that $\hat{\tau}_1$ and $\hat{\tau}_2$ are into the two intervals:

$$P_{res} = \Pr(\hat{\tau}_1 \in \mathcal{I}_1, \hat{\tau}_2 \in \mathcal{I}_2),$$

and the corresponding MSE in a Montecarlo analysis is evaluated for every simulation run conditioned to the resolution condition. Usually $\mathcal{I}_1$ and $\mathcal{I}_2$ have the same width $\Delta\tau_{12} = \tau_2 - \tau_1$ around each ToD: $\mathcal{I}_1 = \{\tau_1 - \Delta\tau_{12}/2, \tau_1 + \Delta\tau_{12}/2\}$ and $\mathcal{I}_2 = \{\tau_2 - \Delta\tau_{12}/2, \tau_2 + \Delta\tau_{12}/2\}$.

20.3 Difference of ToD (DToD) Estimation

In ToD when $L = 1$, the absolute delay is estimated wrt a reference time and this is granted by the knowledge of the waveform that is possible for an active system. $L = 2$ traces $s_1(t), s_2(t)$ collected by two sensors allow one to estimate the difference of ToDs even when the waveform $g(t)$ is not know, as one trace acts as reference waveform wrt the other. The degenerate array with two sensors is an exemplary case: the DToD for the same source signal depends on the angular position that can be estimated in this way (e.g., the human sense of hearing uses the ear to collect and transduce sound waves into electrical signals that allow the brain to perceive and spatially localize sounds). Let a pair of signals be

$$s_1(t) = g(t) + w_1(t)$$
$$s_2(t) = g(t - \tau) + w_2(t)$$

measured over a symmetric interval $[-T/2, T/2]$ where the waveform $g(t)$ can be conveniently considered a realization of a stochastic process characterized by the autocorrelation $\phi_{gg}(t) = \mathbb{E}[g(\xi)g(\xi + t)]$ (*but the sample waveform is not known!*) and its power spectral density $S_{gg}(f) = \mathcal{F}\{\phi_{gg}(t)\}$. Additive impairments $w_1(t)$ and $w_2(t)$ are Gaussian and uncorrelated with the waveform. The obvious choice is to resume the cross-correlation estimator (20.1) by using one signal as reference for the other, this is evaluated first in terms of performance. Later in the section, the elegant and rigorous MLE for DToD will be derived.

20.3.1 Correlation Method for DToD

In DToD, what matters is the relative ToD of one signal wrt the other as the two measurements are equivalent, and there is no preferred one. The cross-correlation method is

$$\hat{\tau} = \arg \max_{|\zeta| \le T/2} \left\{ \int_{-T/2}^{T/2} s_1(t) s_2(t + \zeta) dt \right\} \tag{20.5}$$

where the two signals are shifted against one another until the similarities of the two signals are proved by the cross-correlation metric. For discrete-time sequences, the method can be adapted either to evaluate the correlation for the two discrete-time signals, or to interpolate the signals while correlating (Appendix B). Usually the (computationally expensive) interpolation is carried out as second step after having localized the delay and a further ToD estimation refinement becomes necessary for better accuracy.

Performance Analysis

The cross-correlation among two noisy signals (20.5) is expected to have a degradation wrt the conventional correlation method for ToD, at least for the superposition of noise and the unknown waveform. Noise is uncorrelated over the two sensors, $\mathbb{E}[w_1(t)w_2(t+\alpha)] = 0$ for every α, but it has the same statistical properties. The MSE performance is obtained by considering the cross-correlation for all the terms that concur in $s_1(t)$ and $s_2(t)$:

$$\varphi_{s_1s_2}(\hat{\tau}) = \int_{-T/2}^{T/2} s_1(t)s_2(t+\hat{\tau})dt = \varphi_{gg}(\hat{\tau}) + \varphi_{gw_2}(\hat{\tau}) + \varphi_{w_1g}(\hat{\tau}) + \varphi_{w_1w_2}(\hat{\tau})$$

and using a sensitivity analysis around the estimated value $\hat{\tau}$ of DToD that peaks the cross-correlation, from the derivative:

$$\dot{\varphi}_{s_1s_2}(\hat{\tau}) = 0$$

Recall that $\varphi_{s_1s_2}(\tau)$ is the sample estimate of the stochastic cross-correlation for a stationary process:

$$\lim_{T\to\infty} \frac{\varphi_{s_1s_2}(\tau)}{T} = \phi_{s_1s_2}(\tau) = \mathcal{F}^{-1}\{S_{s_1s_2}(f)\}$$

To simplify the notation, let the DToD be $\tau = 0$; then the derivative of the correlation can be expanded into terms[2]

$$\begin{aligned}\dot{\varphi}_{s_1s_2}(\hat{\tau}) &= \dot{\varphi}_{gg}(\hat{\tau}) + \dot{\varphi}_{w_1g}(\hat{\tau}) - \dot{\varphi}_{w_2g}(-\hat{\tau}) + \dot{\varphi}_{w_1w_2}(\hat{\tau}) = \\ &\simeq \ddot{\varphi}_{gg}(0)\hat{\tau} + \dot{\varphi}_{w_1g}(0) - \dot{\varphi}_{w_2g}(0) + \dot{\varphi}_{w_1w_2}(0) = 0\end{aligned}$$

where the Taylor series has been applied only for the deterministic term on true DToD ($\tau = 0$): $\dot{\varphi}_{gg}(\hat{\tau}) \simeq \dot{\varphi}_{gg}(0) + \ddot{\varphi}_{gg}(0)\hat{\tau} = \ddot{\varphi}_{gg}(0)\hat{\tau}$. For performance analysis of DToD, it is better to neglect any fluctuations of the second order derivative of the correlation of the waveform, which are replaced by its mean value $\ddot{\varphi}_{gg}(0) \simeq T\ddot{\phi}_{gg}(0)$, and thus

2 Recall the property:

$$\begin{aligned}\dot{\varphi}_{gw_1}(\tau) &= \frac{d}{d\tau}\int_{-T/2}^{T/2} g(t)w_1(t+\tau)dt = \varphi_{g\dot{w}_1}(\tau) \\ &= \frac{d}{d\tau}\int_{-T/2}^{T/2} g(t-\tau)w_1(t)dt = -\varphi_{w_1\dot{g}}(-\tau) = -\dot{\varphi}_{w_1g}(-\tau)\end{aligned}$$

$$\hat{\tau} \simeq -\frac{\dot{\varphi}_{w_1g}(0) - \dot{\varphi}_{w_2g}(0) + \dot{\varphi}_{w_1w_2}(0)}{T\ddot{\phi}_{gg}(0)}$$

It is now trivial to verify that the DToD is unbiased, while its variance (see properties in Appendix A, with $\mathbb{E}[|\dot{\varphi}_{w_1g}(0)|^2] = \mathbb{E}[|\dot{\varphi}_{w_2g}(0)|^2]$)

$$\begin{aligned}\text{var}[\hat{\tau}] &= \frac{\mathbb{E}[|\dot{\varphi}_{w_1g}(0)|^2]}{T^2|\ddot{\phi}_{gg}(0)|^2} + \frac{\mathbb{E}[|\dot{\varphi}_{w_2g}(0)|^2]}{T^2|\ddot{\phi}_{gg}(0)|^2} + \frac{\mathbb{E}[|\dot{\varphi}_{w_1w_2}(0)|^2]}{T^2|\ddot{\phi}_{gg}(0)|^2} \\ &= \frac{2\int_{-\infty}^{\infty}(2\pi f)^2 S_{ww}(f)\cdot S_{gg}(f)df + \int_{-\infty}^{\infty}(2\pi f)^2\left(S_{ww}(f)\right)^2 df}{T\left[\int_{-\infty}^{\infty}(2\pi f)^2 S_{gg}(f)df\right]^2}\end{aligned} \tag{20.6}$$

depends on the power spectral densities of stochastic waveform $S_{gg}(f)$ and noise $S_{ww}(f)$. Once again, the effective bandwidth for stochastic processes for signal and noise

$$\beta_g^2 = \frac{\int_{-\infty}^{\infty}(2\pi f)^2 S_{gg}(f)df}{\int_{-\infty}^{\infty} S_{gg}(f)df}$$

$$\beta_w^2 = \frac{\int_{-\infty}^{\infty}(2\pi f)^2 S_{ww}(f)df}{\int_{-\infty}^{\infty} S_{ww}(f)df}$$

plays a key role in the DToD method. When considering bandlimited processes sampled at sampling interval Δt and bandwidth smaller than $1/2\Delta t$ (Nyquist condition), the effective bandwidth can be normalized to the sampling frequency to link the effective bandwidth for the sampled signal (say $\tilde{\beta}_g^2$ or $\tilde{\beta}_w^2$ for signal and noise)

$$\beta_g^2 = \frac{1}{(\Delta t)^2}\cdot\tilde{\beta}_g^2$$

$$\beta_w^2 = \frac{1}{(\Delta t)^2}\cdot\tilde{\beta}_w^2$$

Some practical examples can help to gain insight.

DToD for Bandlimited Waveforms

Let the noise be spectrally limited to B_w in Hz, with power spectral density $S_{ww}(f) = \bar{S}_w$ for $|f| < B_w$, and the spectral power density of the waveform upper limited to B_g (to simplify, $B_w \geq B_g$) with signal to noise ratio

$$\rho = \frac{\phi_{gg}(0)}{\phi_{ww}(0)} = \frac{\phi_{gg}(0)}{2B_w\bar{S}_w}$$

The variance of DToD from (20.6) is

$$\text{var}[\hat{\tau}] = \frac{1}{T\beta_g^2 B_w\rho}\left(1 + \frac{\beta_w^2}{2\beta_g^2\rho}\right) \tag{20.7}$$

This decreases with the observation length T and ρ, when the term in brackets is negligible as for large ρ, or when the noise is wideband ($B_w \geq B_g$) and the ratio β_w^2/β_g^2 cannot be neglected anymore. For small signal to noise ratio it decreases as $1/\rho^2$.

Another relevant example is when all processes have the same bandwidth ($B_g = B_w = B$) with the same power spectral density $S(f)$ up to a scale factor: $S_{gg}(f) = \rho S_{ww}(f) = \rho S(f)$, when

$$\mathrm{var}[\hat{\tau}] = \frac{1+2\rho}{T\rho^2} \times \frac{\int_{-\infty}^{\infty}(2\pi f)^2 \left(S(f)\right)^2 df}{\left[\int_{-\infty}^{\infty}(2\pi f)^2 S(f) df\right]^2}$$

From the Schwartz inequality $[\int_{-\infty}^{\infty}(2\pi f)^2 S(f)df]^2 \leq \int_{-\infty}^{\infty}(2\pi f)^2 \left(S(f)\right)^2 df \cdot \int_{-B}^{B}(2\pi f)^2 df$, the variance is lower bounded by

$$\mathrm{var}[\hat{\tau}] \geq \frac{1+2\rho}{T\rho^2} \cdot \frac{1}{\int_{-B}^{B}(2\pi f)^2 df} = \frac{1+2\rho}{T\rho^2} \cdot \frac{3}{8\pi^2 B^3} \tag{20.8}$$

This value is attained only for $S(f) = \bar{S} = \text{const}$.

The case of time sequences with white Gaussian noise can be derived from the results above. In this case the overall number of samples is $N = T/\Delta t$ with noise bandwidth $B_w = 1/2\Delta t$ and power $\sigma_w^2 = 2B_w \bar{S}_w = \bar{S}_w/\Delta t$. The effective bandwidth of the noise is $\beta_w^2 = \pi^2/(3\Delta t^2)$ and for the signal is $\beta_g^2 = \tilde{\beta}_g^2/(\Delta t)^2$ so that the variance of DToD follows from (20.7):

$$\mathrm{var}[\hat{\tau}] = \frac{2(\Delta t)^2}{N\tilde{\beta}_g^2 \rho}\left(1 + \frac{\pi^2}{6\tilde{\beta}_g^2 \rho}\right)$$

When considering sampled signals, it is more convenient to scale the variance to the sample interval so as to have the performance measured in sample units:

$$\mathrm{var}\left[\frac{\hat{\tau}}{\Delta t}\right] = \frac{2}{N\tilde{\beta}_g^2 \rho}\left(1 + \frac{\pi^2}{6\tilde{\beta}_g^2 \rho}\right)$$

This depends on the number of samples (N) and the signal to noise ratio ρ; again the term $1/\rho^2$ disappears for large ρ. In the case that both waveform and noise are white ($B_g = B_w = 1/2\Delta t$), $\tilde{\beta}_w^2 = \pi^2/3$ and thus

$$\mathrm{var}\left[\frac{\hat{\tau}}{\Delta t}\right] = \frac{1}{N}\frac{3}{\pi^2}\left(\frac{1+2\rho}{\rho^2}\right) \tag{20.9}$$

More details on DToD in terms of bias (namely when $\tau \neq 0$) and variance are in [96].

20.3.2 Generalized Correlation Method

The cross-correlation method for DToD is the most intuitive extension of ToD, but there is room for improvement if rigorously deriving the DToD estimator from ML theory.

Namely, the ML estimator was proposed by G.C. Carter [97] who gave order to the estimators and methods discussed within the literature of the seventies. We give the essentials here while still preserving the formal derivations.

The basic assumption is that the samples of the Fourier transform of data $s_i(t)$ (for $i = 1, 2$) evaluated at frequency binning $\Delta f = 1/T$

$$S_i[k] = S_i(k\Delta f) = \int_{-T/2}^{T/2} s_i(t)e^{-j2\pi kt/T}dt$$

are zero-mean Gaussian random variables. Provided that T is large enough to neglect any truncation effect due to the DToD τ (see Section 20.4), the correlation depends on the cross-spectrum

$$\mathcal{S}_{S_1S_2}(f) = \mathcal{F}\{\mathbb{E}[s_1(t)s_2^*(t+\tau)]\}$$

and it follows that

$$\mathbb{E}[S_1[k]S_2[m]^*] \simeq T\mathcal{S}_{S_1S_2}(k \cdot \Delta f)\delta[m-k]$$

is uncorrelated vs. frequency bins. The rv at the kth frequency bin is

$$\mathbf{S}[k] = \begin{bmatrix} S_1[k] \\ S_2[k] \end{bmatrix} \sim \mathcal{CN}(0, T\mathbf{C}_S(k))$$

where

$$\mathbf{C}_S[k] = \begin{bmatrix} \mathcal{S}_{S_1S_1}(k \cdot \Delta f) & \mathcal{S}_{S_1S_2}(k \cdot \Delta f) \\ \mathcal{S}^*_{S_1S_2}(k \cdot \Delta f) & \mathcal{S}_{S_2S_2}(k \cdot \Delta f) \end{bmatrix}$$

The set of rvs for all the N frequency bins $\mathbf{S} = [\mathbf{S}[-N/2]^T, \mathbf{S}[-N/2+1]^T, \dots, \mathbf{S}[N-1]^T]^T \in \mathbb{C}^{2N}$ are Gaussian and independent, and the log-likelihood is (apart from constant terms)

$$\ln p(\mathbf{S}|\tau) = -\frac{1}{2T}\sum_{k=-N/2}^{N/2-1} \mathbf{S}^H[k]\mathbf{C}_S^{-1}[k]\mathbf{S}[k]$$

Still considering a large interval T, the ML metric for $T \to \infty$ becomes

$$\mathcal{L}(\tau) = -\frac{1}{2T}\sum_{k=-N/2}^{N/2-1} \mathbf{S}^H[k]\mathbf{C}_S^{-1}[k]\mathbf{S}[k] \to -\frac{1}{2}\int \mathbf{S}^H(f)\mathbf{C}_S^{-1}(f)\mathbf{S}(f)df$$

where the dependency on the DToD τ is in the phase of $\mathbf{S}(f)$ as

$$\mathcal{S}_{S_1S_2}(f) = \mathcal{S}_{gg}(f)e^{j2\pi f\tau}$$

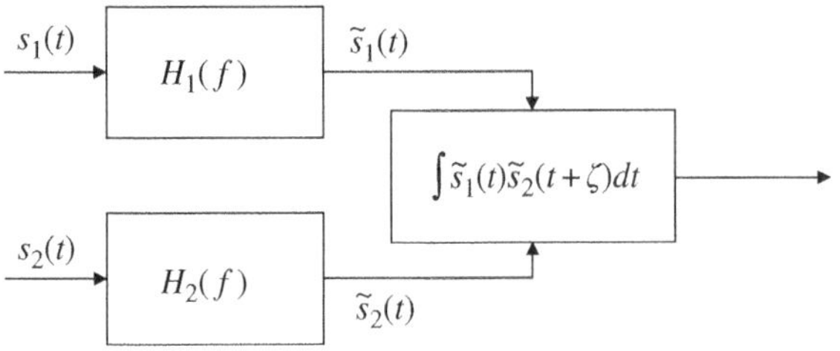

Figure 20.4 Generalized correlation method.

The inverse of the covariance

$$\mathbf{C}_S^{-1}(f) = \frac{1}{1-|\Gamma_{S_1S_2}(f)|^2}\begin{bmatrix} \frac{1}{S_{S_1S_1}(f)} & \frac{-S_{S_1S_2}(f)}{S_{S_1S_1}(f)S_{S_2S_2}(f)} \\ \frac{-S^*_{S_1S_2}(f)}{S_{S_1S_1}(f)S_{S_2S_2}(f)} & \frac{1}{S_{S_2S_2}(f)} \end{bmatrix}$$

depends on the *coherence* of the two rvs defined as:

$$\Gamma_{S_1S_2}(f) = \frac{S_{S_1S_2}(f)}{[S_{S_1S_1}(f)S_{S_2S_2}(f)]^{1/2}}$$

The ML metric $\mathcal{L}(\tau)$ can be expanded (recall the property: $S(f)^* = S(-f)$) to isolate only the delay-dependent term:

$$\mathcal{L}(\tau) = \int S_1(f)S_2(f)^* \underbrace{\left(\frac{1}{|S_{S_1S_2}(f)|}\cdot\frac{|\Gamma_{S_1S_2}(f)|^2}{1-|\Gamma_{S_1S_2}(f)|^2}\right)}_{\Psi(f)} e^{j2\pi f\tau}df + const \tag{20.10}$$

The MLE is based on is the search for the delayed cross-correlation of the two signals (term $S_1(f)S_2(f)^*$) after being filtered by $\Psi(f)$. In practice, the DToD is by the correlation method provided that the two signals are preliminarily filtered by two filters $H_1(f)$ and $H_2(f)$ such that $H_1(f)\times H_2^*(f) = \Psi(f)$ as sketched in Figure 20.4.

Filter Design

Inspection of the coherence term based on the problem at hand

$$|\Gamma_{S_1S_2}(f)| = \frac{\rho(f)}{1+\rho(f)} \leq 1$$

shows that it depends on the signal to noise ratio for every frequency bin

$$\rho(f) = \frac{S_{gg}(f)}{S_{ww}(f)}$$

and it can be approximated as

$$|\Gamma_{S_1S_2}(f)| \simeq \begin{cases} 1 \text{ for } \rho(f) \gg 1 \\ 0 \text{ for } \rho(f) \ll 1 \end{cases}$$

it downweights the terms when noise dominate the signal. The filter $\Psi(f)$ is

$$\Psi(f) = \frac{1}{S_{gg}(f)} \frac{\rho(f)^2}{1+2\rho(f)}$$

and a convenient way to decouple $\Psi(f)$ into two filters is

$$H_1(f) = H_2(f) = H(f) = \frac{1}{\sqrt{S_{gg}(f)}} \cdot \frac{\rho(f)}{\sqrt{1+2\rho(f)}} \tag{20.11}$$

where the first term decorrelates the signal for the waveform, and the second term accounts for the signal to noise ratio. If $\rho(f)$ is independent of frequency, the second term is irrelevant for the MLE of DToD.

CRB for DToD

CRB can be derived from the log-likelihood term that depends on the delay (20.10) and it follows from the definitions in Section 8.1:

$$\begin{aligned} \text{var}[\hat{\tau}] &\geq -\left(\mathbb{E}[d^2 \ln(p(\mathbf{S}|\tau)/d\tau^2])\right)^{-1} = -\left(\mathbb{E}[d^2\mathcal{L}(\tau)/d\tau^2])\right)^{-1} \\ &= \left[T\int (2\pi f)^2 \frac{|\Gamma_{S_1S_2}(f)|^2}{1-|\Gamma_{S_1S_2}(f)|^2} df\right]^{-1} \end{aligned}$$

or compactly

$$\text{var}[\hat{\tau}] \geq \left[T\int (2\pi f)^2 \frac{\rho(f)^2}{1+2\rho(f)} df\right]^{-1} \tag{20.12}$$

To exemplify, consider the case where signal and noise have the same bandwidth (e.g., this might be the case that lacks of any assumption on what is signal and what is noise, and measurements are prefiltered within the same bandwidth to avoid "artifacts") $B_g = B_w = B$ with $\rho(f) = \rho$ (i.e., this does not imply that $S_{gg}(f) = const$), then

$$\text{var}[\hat{\tau}] \geq \left[T\frac{\rho^2}{1+2\rho}\int_{-B}^{B}(2\pi f)^2 df\right]^{-1} = \frac{1+2\rho}{\rho^2}\frac{3}{8\pi^2 TB^3}$$

coincides with the variance of correlation estimator (20.8) derived from a sensitivity analysis, except that any knowledge of the power spectral density of the waveform $S_{gg}(f)$ could use now the prefiltering (20.11) to attain an improved performance for the DToD correlator. If sampling the signals at sampling interval $\Delta t = 1/2B$ (i.e., maximal sampling

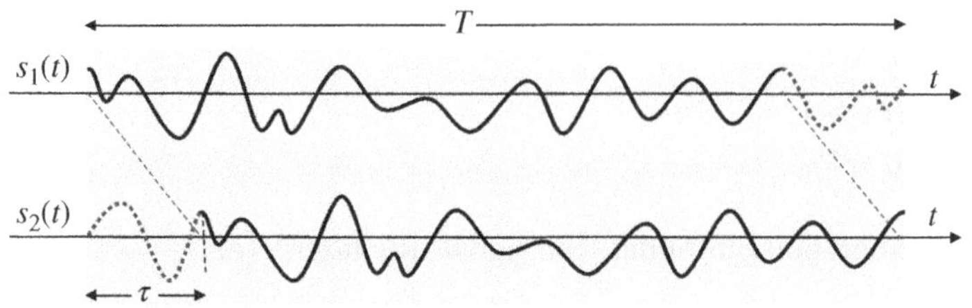

Figure 20.5 Truncation effects in DToD.

rate compatible with the processes here), the sequences have $N = T/\Delta t$ samples in total and the CRB coincides with (20.9).

Even if the generalized correlation method is clearly the MLE for DToD, there are some aspects to consider when planning its use. The filtering (20.11) needs to have a good control of the model, and this is not always the case as the difference ToD is typically adopted in passive sensing systems where there is poor control of the waveform from the source, and the use of one measurement as reference for the others is a convenient and simple trick, namely when measurements are very large. In addition, the filtering introduces edge effects due to the filters' transitions $h(t)$ that limits the valid range T of the measurements. When filters are too selective, the filter response $h(t)$ can be too long for the interval T and the simple correlation method (20.5) is preferred.

20.4 Numerical Performance Analysis of DToD

Performance of DToD depends on the observation interval T as this establishes the interval where the two waveforms maximally correlate. However, in the case that the true DToD τ is significantly different compared to T (Figure 20.5), the overlapping region within the area showing the same delayed waveform reduces by $T - |\tau|$ and thus the variance (20.8) scales accordingly as the non-overlapped region acts as self-noise:

$$\text{var}[\hat{\tau}] \geq \frac{1+2\rho}{T(1-|\tau|/T)\rho^2} \cdot \frac{3}{8\pi^2 B^3}$$

The same holds for any variance computation. This loss motivates the choice of "slant windowing" whenever one has the freedom to choose the two portions of observations $s_1(t)$ and $s_2(t)$ to correlate.

Measurements are usually gathered as sampled and all methods above need to be implemented by taking the sampling into account. Correlation for sampled signals needs to include the interpolation and resampling as part of the DToD estimation step as for the ToD method. However, the filtering might introduce artifacts that in turn degrade the overall performance except for the case when the DToD is close to zero. This is illustrated in the example below by generating a waveform from a random process with PSD within $B_g = 1/20$ and white noise ($B_w = 1/2$). The signal is generated by filtering a white stochastic process with a Butterworth filter of order 8, and the second signal has been delayed using the DFT for fractional delay (see Appendix B). The fractional delay is estimated by using the parabolic regression discussed in Section 9.4.

Below is the Matlab code for the Montecarlo test to evaluate the performance wrt the CRB.

```
N=64; % # samples observation window
M=5000; % # intervals
Ntot=M*N;
SNR_dB=[-10:2:40];
tau=.2;

% Filter design for waveform generation
Bg=.05; %Bandwidth for random waveform g[n]
[b,a]=butter(8,2*Bg);
g=filter(b,a,randn(Ntot,1));
g1=g;
% Delay (periodic) over DFT domain w/interpolation for g2(t)
G=fft(g);
k=[0:Ntot/2,-Ntot/2+1:-1]'; G2=G.*exp(1i*2*pi*tau*k/Ntot);
g2=real(ifft(G2));
w1=randn(Ntot,1); w2=randn(Ntot,1); % noise

for isnr=1:length(SNR_dB),
    x1=10^(SNR_dB(isnr)/20)*g1+w1;
    x2=10^(SNR_dB(isnr)/20)*g2+w2;
    for m=1:M,
        c=xcorr(x1((m-1)*N+1:m*N),x2((m-1)*N+1:m*N));
        i=find(c==max(c));
        dt=.5 * (c(i-1)-c(i+1)) / (c(i-1)+c(i+1)-2*c(i)); %
parabolic regression
        tau_est(m)=i+dt-N;
    end
    MSE_DToD(isnr)=mean((tau_est-tau).^2);
end

% CRB computation (see text)
snr=10.^(SNR_dB/10);
Bg_Butter=1.1*Bg; % correcting term for Butterworth PSD
CRBt=(3/(8*N*(pi^2)*((Bg_Butter)^3))) *
((1+2*snr)./(snr.^2));

semilogy(SNR_dB,sqrt(MSE_DToD),'-',SNR_dB,sqrt(CRBt),'--')
xlabel('SNR [dB]'); ylabel('RMSE [samples]')
```

Figure 20.6 shows the performance in terms of RMSE (measured in samples) compared to the CRB (dashed lines) for varying choices of the delay $\tau = 0 : .1 : 1$ and $N = 256$ samples. The performance attains the CRB for large SNR and $\tau = 0$, while there is a small bias that justifies the floor for large SNR due to the parabolic regression on correlation evaluated for discrete-time signals.

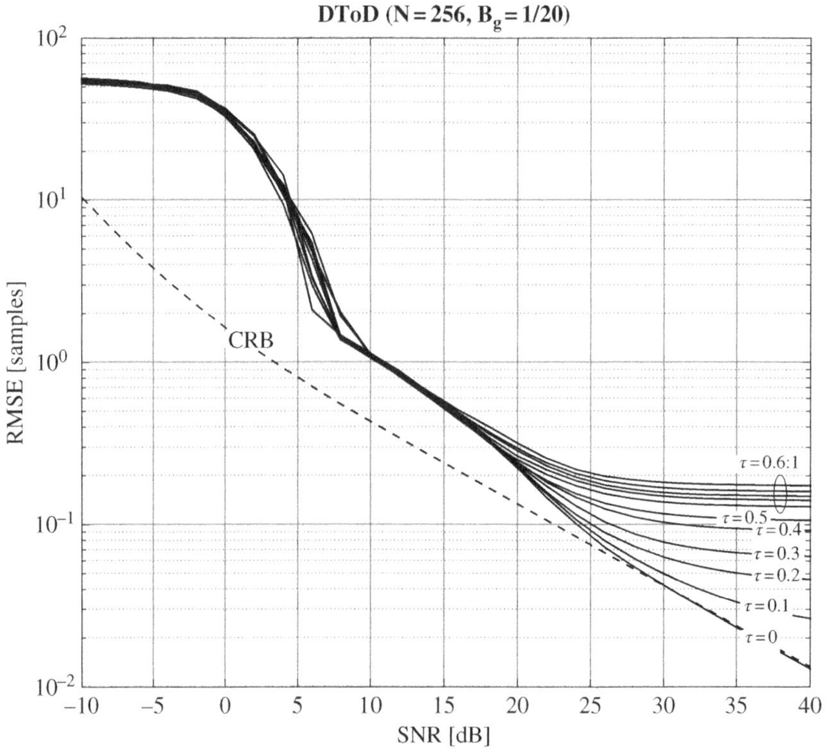

Figure 20.6 RMSE vs. SNR for DToD (solid lines) and CRB (dashed line).

20.5 Wavefront Estimation: Non-Parametric Method (L=1)

When the waveform is not known and the measurements are highly redundant ($M \gg 1$), the wavefront can be estimated by iterated DToD among all measurements [98] even if, because of small waveform distortion vs. distance (see Figure 20.8), it would be preferable to use the DToD of neighboring signals. The set of delayed waveforms over a wavefront

$$\tau(x, y)$$

depends on two geometrical parameters x, y according to the planar coordinates of the sensors (see Figure 20.7). For every pair of sensors out of M, say i and j at geometrical positions $\mathbf{p}_i$ and $\mathbf{p}_j$, the correlation estimator on $s(t; \mathbf{p}_i)$ and $s(t; \mathbf{p}_j)$ yields the estimate of the DToD $\Delta\tau_{i,j}$ that is dependent on the DToD of the true delays

$$\tau_i - \tau_j = \Delta\tau_{i,j} + \alpha_{i,j}$$

apart from an error $\alpha_{i,j}$ that depends on the experimental settings (waveform, noise, any possible distortion, and even delays) as proved in (20.12).

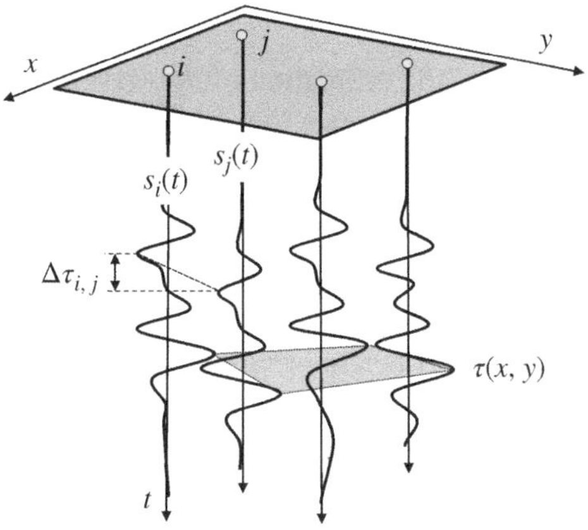

Figure 20.7 Wavefront estimation from multiple DToD.

All the pairs that are mutually correlated can be used to estimate the DToDs up to a total of $M(M-1)/2$ estimates (if using all-to-all pairs) to build a linear system

$$\underbrace{\begin{bmatrix} -1 & 1 & 0 & \dots & 0 \\ -1 & 0 & 1 & \dots & 0 \\ \vdots & \vdots & \vdots & \ddots & \vdots \\ -1 & 0 & 0 & \dots & 1 \\ 0 & -1 & 1 & \dots & 0 \\ \vdots & \vdots & \vdots & \ddots & \vdots \\ 0 & 0 & 0 & \dots & 1 \end{bmatrix}}_{\mathbf{A}} \cdot \underbrace{\begin{bmatrix} \tau_1 \\ \tau_2 \\ \tau_3 \\ \vdots \\ \tau_M \end{bmatrix}}_{\tau} = \underbrace{\begin{bmatrix} \Delta\tau_{1,2} \\ \Delta\tau_{1,3} \\ \vdots \\ \Delta\tau_{1,M} \\ \Delta\tau_{2,3} \\ \vdots \\ \Delta\tau_{M-1,M} \end{bmatrix}}_{\Delta\hat{\tau}} + \underbrace{\begin{bmatrix} \alpha_{1,2} \\ \alpha_{1,3} \\ \vdots \\ \alpha_{1,M} \\ \alpha_{2,3} \\ \vdots \\ \alpha_{M-1,M} \end{bmatrix}}_{\alpha} \tag{20.13}$$

where the matrix $\mathbf{A} \in \mathbb{R}^{M(M-1)/2\times M}$ has values ± 1 along every row to account for the pair of sensors used in DToD, and all the other terms have obvious meanings. Assuming that the DToD estimates are

$$\boldsymbol{\alpha} \sim \mathcal{N}(0, \mathbf{C}_\alpha)$$

then the LS estimate of the ensemble of ToDs is

$$\hat{\boldsymbol{\tau}} = (\mathbf{A}^T\mathbf{C}_\alpha^{-1}\mathbf{A})^{-1}\mathbf{A}^T\mathbf{C}_\alpha^{-1}\boldsymbol{\Delta}\hat{\boldsymbol{\tau}}$$

Notice that the wavefront $\hat{\boldsymbol{\tau}}$ is estimated up to an arbitrary translation as DToDs are not sensitive to a constant, and this reflects to a null eigenvalue of $\mathbf{A}^T\mathbf{A}$ that calls for the LS estimate using the pseudoinverse (Section 2.7). Any iterative estimation method for the solution of the LS estimate (e.g., iterative methods in Section 2.9) are not sensitive to null eigenvalues as that part of the solution in $\mathcal{N}(\mathbf{A}^T)$ corresponding to the eigenvector of the constant value of $\boldsymbol{\tau}$ is not updated.

20.5.1 Wavefront Estimation in Remote Sensing and Geophysics

In remote sensing and geophysical signal processing, the estimation of the wavefront of delays (in seismic exploration these are called horizons since they define the ToD originated by the boundary between different media, as illustrated at the beginning of this Chapter) is very common and of significant importance. Even if these remote sensing systems are usually active with known waveform, the propagation over the dispersive medium and the superposition of many backscattered echoes severely distort the waveform, and make it almost useless for ToD estimation. However, in DToD estimation the waveform is similarly distorted for two neighboring sensors. This is exemplified in Figure 20.8 where the waveforms are not exactly the same shifted copy, but there is some distortion both in wavefront and in waveform (see the shaded area that visually compares the waveforms at the center and far ends of the array).

One common way to handle these issues is by preliminarily selecting those pair of sensors where the waveforms are not excessively distorted between one another, and this reduces the number of equations associated with the selected pairs of the overdetermined problem (20.13) to be smaller than $M(M-1)/2$ (*all-to-all configuration*). This has a practical implication as the number of scans can be $M \simeq 10^5 \div 10^6$ as in the seismic image at the beginning in Figure 20.1, and the use of $M^2/2$ would be intolerably complex. Selecting the K neighbors (say $K = 4 \div 10$) can reduce the complexity to manageable levels $KM/2$. If M is not excessively large, one could still use all the $M(M-1)/2$ pairs and account separately for the waveform distortion by increasing the DToD variance (20.12). As a rule of thumb, the distortion increases with the sensors' distance $|\mathbf{p}_i - \mathbf{p}_i|$, and this degrades the variance proportionally

$$\text{var}(\Delta\hat{\tau}_{i,j}) \simeq \frac{1}{T(2\pi)^2 \int f^2 \frac{\rho_{i,j}(f)^2}{1+2\rho_{i,j}(f)} df} \cdot \frac{1}{1-|\Delta\tau_{i,j}|/T} + \eta|\mathbf{p}_i - \mathbf{p}_i|^\gamma$$

for an empirical choice of the scaling term η and exponent γ.

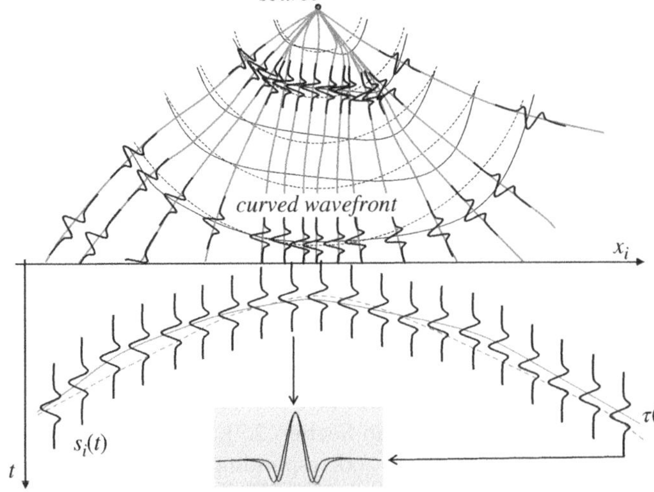

Figure 20.8 Curved and distorted wavefront (upper part), and measured data (lower part).

It is quite common to have outliers in DToD and this calls for some preliminary data classification (Chapter 23). In addition to the robust estimation methods (Section 7.9), another alternative is to solve for (20.13) by using an L1 norm. These methods are not covered here as they are based on the specific application, but usually they are not in closed form even if the optimization is convex, and are typically solved by iterative methods.

20.5.2 Narrowband Waveforms and 2D Phase Unwrapping

In narrowband signals, the waveform is the product of a slowly varying envelope $a(t)$ with a complex sinusoid:

$$g(t) = a(t)e^{j\omega_o t}$$

where ω_o is the angular frequency. The delays are so small compared to the envelope variations that the signals at the sensors (i,j) are

$$s_i(t) = g(t + \tau_{ij}/2) + w_i(t) \simeq a(t)e^{j\omega_o(t+\tau_{ij}/2)} + w_i(t)$$
$$s_j(t) = g(t - \tau_{ij}/2) + w_j(t) \simeq a(t)e^{j\omega_o(t-\tau_{ij}/2)} + w_j(t)$$

and the correlator-based DToD estimator (neglecting the noise only for the sake of reasoning)

$$s_i(t)s_i^*(t) \simeq |a(t)|^2 e^{j\omega_o \tau_{ij}}$$

yields the estimate of the phase difference

$$[\Delta\vartheta_{i,j}]_{2\pi} = \arg[s_i(t)s_i^*(t)] = \omega_o \tau_{ij} \bmod 2\pi \tag{20.14}$$

that is directly related to the DToD. Wavefront estimation is the same as above with some adaptations [99]: phase difference (and phase in general[3]) is defined modulo-2π from the *arg*[.] of the correlation of signals. Since $\Delta\vartheta_{i,j} \in [-\pi, \pi)$ phase is wrapped within $[-\pi, \pi)$ (herein indicated as $[.]_{2\pi}$) so that the true measured phase difference differs from the true one by multiple of 2π, unless $|\omega_o \tau_{ij}| < \pi$. 2D phase unwrapping refers to the estimation of the continuous phase-function

$$\vartheta(\mathbf{p}_i)$$

(or the wavefront $\tau(\mathbf{p}_i) = \vartheta(\mathbf{p}_i)/\omega_o$) from the phase-differences (20.14) evaluated modulo-2π (also called *wrapped phase*). The LS unwrapping is similar to problem (20.13) by imposing for any pair of measurements the equality

---

3 Phase is measured in the interval $[-\pi, \pi)$ from arg[.] of complex-valued signals and any measured (wrapped) phase $[\vartheta]_{2\pi}$ value differs from the *true* phase ϑ by a multiple of 2π such that $-\pi \leq [\vartheta]_{2\pi} < \pi$:

$$[\vartheta]_{2\pi} = \vartheta + 2k\pi$$

Modulo-2π acts as a non-linear wrapping operator on phase values, it leads to a discontinuous phase behavior that usually hides the true meaning of the phase information.

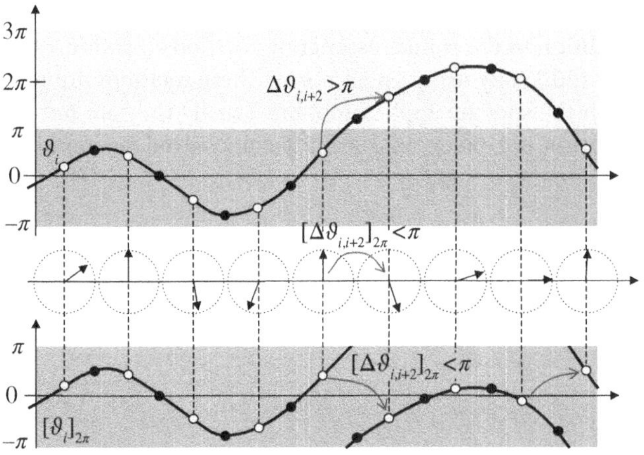

Figure 20.9 Phase-function and wrapped phase in $[-\pi, \pi)$ from modulo-2π.

$$\vartheta(\mathbf{p}_i) - \vartheta(\mathbf{p}_j) = [\Delta\vartheta_{\mathbf{i},\mathbf{j}}]_{2\pi} \tag{20.15}$$

in the LS sense. The system becomes

$$\mathbf{A} \cdot \boldsymbol{\vartheta} = [\boldsymbol{\Delta\vartheta}]_{2\pi} \tag{20.16}$$

that yields the LS solution

$$\hat{\boldsymbol{\vartheta}} = (\mathbf{A}^T\mathbf{A})^{-1}\mathbf{A}^T[\boldsymbol{\Delta\vartheta}]_{2\pi} \tag{20.17}$$

Since phase is measured modulo-2π, the choice of the sensor's pairs is not extended to all the sensors, but rather to the neighborhood of each sensor as the phase-variation among neighbor sensors is expected to be *mostly* smaller than π.

The concept of *phase aliasing* is illustrated by the example in Figure 20.9 where phase ϑ_i increases along the sensors' position and the wrapped phase $[\vartheta_i]_{2\pi}$ has 2π jumps when close to $\pm\pi + 2k\pi$. If considering the angular difference of the phase between two points coarsely sampled 1:2 (gray points with double-spacing $i, i+2, i+4, \ldots$), the modulo-2π value coincides with the true value except where phase difference $|\Delta\vartheta_{i,i+2}| > \pi$ (gray lines) while the wrapped phase difference $|[\Delta\vartheta_{i,j}]_{2\pi}| < \pi$ (i.e., the wrapped phase difference can be considered as the smallest angular difference between two consecutive vectors). Since the phase difference estimated from the wrapped phase differs from the true one by (a multiple of) 2π, there is no way to retrieve these $\pm 2\pi$ jumps that impair the overall estimated phase-function by the unwrapping algorithm. This is called *phase aliasing* condition. If sampling is dense enough (gray and black points in Figure 20.9), the phase-differences are likely to be coincident with the true (unwrapped) phase without any artifacts.

20.5.3 2D Phase Unwrapping in Regular Grid Spacing

In addition to wavefront estimation, 2D phase unwrapping has several notable applications in magnetic resonance imaging, optical interferometry, and satellite synthetic aperture radar (SAR) imaging. All these applications collect measurements over a regular grid spacing and the phase-differences (20.15) are computed between neighboring points along two orthogonal directions. Since the transformation $\mathbf{A}$ is the derivative along the two dimensions, the phase differences can be partitioned along these two directions as

$$\begin{bmatrix} \mathbf{A}_x \\ \mathbf{A}_y \end{bmatrix} \cdot \vartheta = \begin{bmatrix} [\mathbf{\Delta}\vartheta_x]_{2\pi} \\ [\mathbf{\Delta}\vartheta_y]_{2\pi} \end{bmatrix}$$

so that the LS estimate becomes

$$\left(\mathbf{A}_x^T\mathbf{A}_x + \mathbf{A}_y^T\mathbf{A}_y\right)\hat{\vartheta} = \mathbf{A}_x^T[\mathbf{\Delta}\vartheta_x]_{2\pi} + \mathbf{A}_y^T[\mathbf{\Delta}\vartheta_y]_{2\pi} \tag{20.18}$$

which can be solved efficiently by exploiting the structure of the linear operators and their sparsity—see Appendix C.

Inspection of the term $\mathbf{A}_x^T\mathbf{A}_x + \mathbf{A}_y^T\mathbf{A}_y$ shows that it contains the second order derivatives (i.e., discrete Laplace operator) of the unwrapped phase field $\hat{\vartheta}$, while the terms on the right of (20.18) are the first-order derivatives of the phase-differences. This is the basis for several detailed lines of thought on the impact of phase-difference artifacts such as the 2π discontinuities on the final solutions. The reader is referred to the copious literature on this fascinating topic that deals with the phase field as a continuous field that is sampled, and phase unwrapping artifacts are the results of coarse sampling of these phase fields.

The example in Figure 20.10 shows the consequences of phase aliasing from the wrapped phase image $[\vartheta(x,y)]_{2\pi}$ using (20.18) when the phase image is down-sampled 1:n (i.e., take one sample out of n in both directions) to simulate phase aliasing when evaluating phase-differences. The estimated phase $\hat{\vartheta}(x,y)$ is (visually) not affected by 1:2 decimation even if some residual is still visible on residual $[\hat{\vartheta}(x,y) - \vartheta(x,y)]_{2\pi}$, while it is more severely affected when decimation is larger, as for 1:5, most of the artifacts are in those areas where there is a sudden change of the phase-function (e.g., in the face, where luminance is abruptly changing from black to white). Residuals for 1:2 or 1:5 show a pattern that resembles the electric field with dipoles, there is a close analytical relationship, but the explanation is beyond the scope of the introductory discussion.

One might argue that the solution to these artifacts could be to oversample the wavefields, but this is not always possible as the sensors' density is given by the application. The interpolation of phase images cannot provide the solution as unfortunately phase information is related to the dynamics of the phase and not (easily) to the spectrum of the signals. Since phase aliasing introduces missing 2π jumps that are smeared by the LS method as diffused error (see 1:5 decimation around the face with abrupt black-white discontinuities), these can be interpreted as outliers that are surely not-Gaussian and the L2-norm can be replaced with L1 metrics, possibly iterated. There are methods that solve for the LS phase estimate by cosine-transforms, but this method is questionable for small images due to the poor control of artifacts at the boundary as consequence of the periodicity of the cosine-transformation.

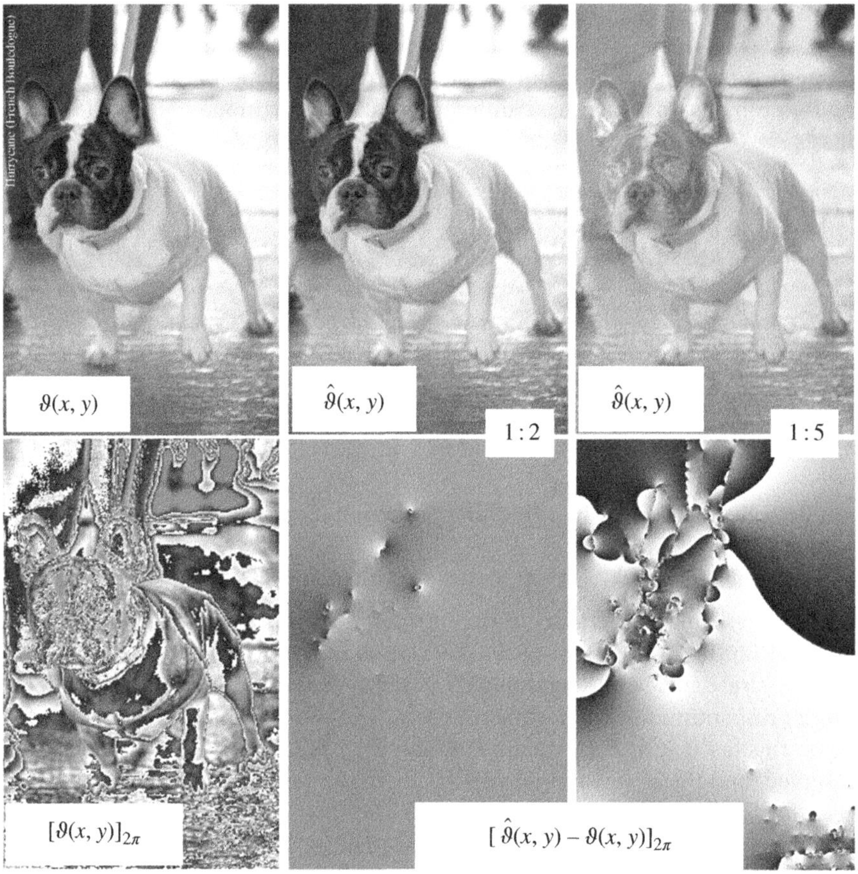

Figure 20.10 Example of 2D phase unwrapping.

20.6 Parametric ToD Estimation and Wideband Beamforming

The ToDs $\boldsymbol{\tau}_\ell = \boldsymbol{\tau}(\boldsymbol{\theta}_\ell)$ depend on a set of shape parameters that fully characterize each waveform. A common and intuitive, but not unique, example is the wavefield originated by point sources at distances h_ℓ from a linear array of sensors located at $y = 0$ (see Figure 20.11). The 2D ray-geometrical model of propagation in a homogeneous medium with propagating velocity v (in m/s) gives the hyperbolic ToD vs. array space x (here assumed as continuous just to ease the notation) in the two equivalent formulations:

$$\tau_\ell(x) = \tau(x|\boldsymbol{\theta}_\ell) = \sqrt{\frac{h_\ell^2 + (x - x_{\ell,0})^2}{v^2}} = \sqrt{\tau_{\ell,0}^2 + \frac{(x - x_{\ell,0})^2}{v^2}}$$

where the two shape parameters

$$\boldsymbol{\theta}_\ell = [x_{\ell,0}, h_\ell]^T$$

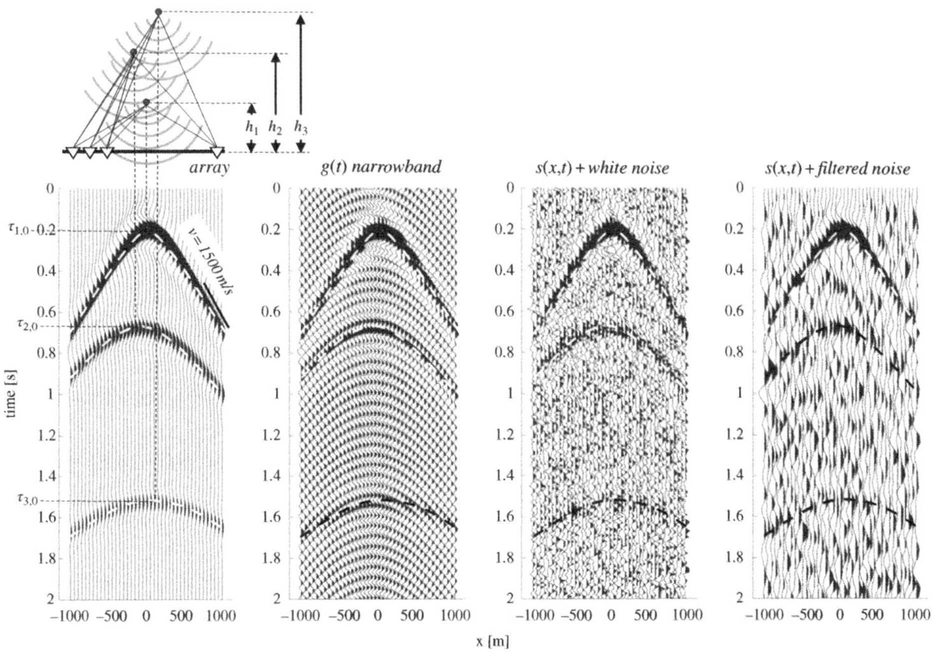

Figure 20.11 Example of delayed waveforms (dashed lines) from three sources impinging on a uniform linear array.

are the geometrical position of the source, or alternatively when v is not known, the ToD at the vertex of the hyperbola $\tau_{\ell,0} = h_\ell / v$: $\boldsymbol{\theta}_\ell = [x_{\ell,0}, \tau_{\ell,0}]^T$. Note that in general, the wavefront is curved and, for large distances (*far-field approximation* for the wavefront) $|x - x_{\ell,0}| \gg h_\ell$, it is linear with a slope

$$\left.\frac{d\tau_\ell(x)}{dx}\right|_{x=x_i} = 1/v$$

that depends on the *slowness* $1/v$ of the propagating medium. The example in Figure 20.11 shows the wavefields from 3 sources for wideband $g(t)$ (left) and narrowband waveforms, and wideband $s(x, t)$ with white noise or the same after filtering the noisy wavefield within the bandwidth of the signal. All slopes of the wavefronts are below the value defined by the propagation velocity v =1500 m/s that mimics the acoustic waves in water as for a sonar sensing system, and hyperbolic ToDs $\tau(x|[x_{\ell,0}, \tau_{\ell,0}])$ are indicated by the dashed lines superimposed on the wavefield $s(x, y = 0, t)$ in various settings.

Furthermore, there are applications such as in subsurface remote sensing where backscattered wavefields are used for depth imaging and the propagation velocity v is unknown and this becomes an additional parameter to be estimated—or even worse (but unfortunately very common in many geophysical seismic exploration applications), the propagating medium is not homogeneous, the wavefield is not spherical, and the wavefront deviates from the hyperbolic shapes of the Figure 20.11. In these contexts, the number of parameters of a parametric ToD model become very complex and this is not covered herein. However, the approach is still based on estimation of the deviations from a nominal hyperbolic ToD shape based on some initial assumptions of the propagation velocity.

In complex contexts, there are many wavefronts impinging on an array of sensors and the signal at position $(x, y = 0)$

$$s(t,x) = \sum_{\ell=1}^{L} \alpha_\ell(x) g_\ell(t - \tau(x|\boldsymbol{\theta}_\ell)) + w(t,x) \tag{20.19}$$

is characterized by the superposition of these delayed waveforms. The waveforms are not known but redundantly measured by the array to guarantee the possibility of estimating one waveform and mitigating (or attenuating) the others. Wideband beamforming refers to the method to delay and combine the multiple signals (20.19) to estimate the waveform of interest (say $g_{\bar{\ell}}(t)$ for the $\bar{\ell}$-th wavefront), and possibly mitigate the others considered as interference.

20.6.1 Delay and Sum Beamforming

Delay and sum beamforming extends the beamforming methods in Chapter 19 to wideband waveforms. Making reference to the model (20.19), the wideband beamforming method delays each signal by a certain shape parameters $\hat{\boldsymbol{\theta}} = [\hat{x}_0, \hat{\tau}_0]^T$ and sums all the N delayed signals from all sensors:

$$s_{out}(t|\hat{\boldsymbol{\theta}}) = \frac{1}{N} \sum_{x} s(t + \tau(x|\hat{\boldsymbol{\theta}}), x) \tag{20.20}$$

When the shape parameters match the $\bar{\ell}$th wavefield $\hat{\boldsymbol{\theta}} = \boldsymbol{\theta}_{\bar{\ell}}$, this value flattens $s(t,x)$ in (20.20) and the sum enhances it; the resulting signal becomes

$$s_{out}(t|\hat{\boldsymbol{\theta}}) = \left(\frac{\sum_x \alpha_{\bar{\ell}}(x)}{N}\right) g_{\bar{\ell}}(t) + \sum_{\ell=1,\ell\neq\bar{\ell}}^{L} \frac{\sum_x \alpha_\ell(x) g_\ell(t + \tau(x|\boldsymbol{\theta}_{\bar{\ell}}) - \tau(x|\boldsymbol{\theta}_\ell))}{N} + \frac{\sum_x w(t + \tau(x|\boldsymbol{\theta}_{\bar{\ell}}), x)}{N} \tag{20.21}$$

where the three terms are the signal of interest, interference from the other wavefronts $(\ell \neq \bar{\ell})$ that are reduced by the summing, and the noise that is reduced by $1/N$. When $\hat{\boldsymbol{\theta}} \neq \boldsymbol{\theta}_{\bar{\ell}}$, the signal is somewhat attenuated by the sum.

Wideband beamforming is shown in the example in Figure 20.12 for a point source at $x_0 = -100\ m$ and $h = 600\ m$ (or $\tau_0 = 400\ ms$ as from the vertex of the hyperbolic wavefield) from the linear array, and propagation velocity is $v = 1500\ m/s$ as for ultrasound systems with Ricker waveform (Section 9.4). The delay and sum beamforming $s_{out}(t|[\hat{x}_0, \hat{\tau}_0 = \tau_0])$ is for varying source point $\hat{x}_0$ and the clearest waveform is when $\hat{x}_0 = -100\ m$ as expected, with small artifacts when $\hat{x}_0 \neq -100\ m$ that decrease with N. The interference mitigation capability of delay and sum beamforming is shown for the same example when considering the presence on another interfering wavefield at

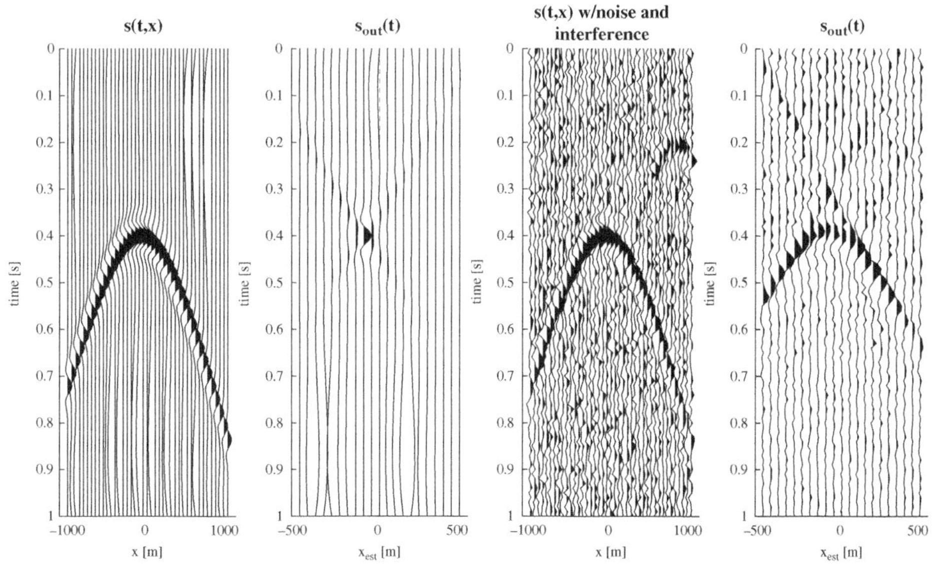

Figure 20.12 Delay and sum beamforming for wideband signals.

$x_1 = +800$ m and $\tau_1 = 200$ ms, and noise $s_{out}(t|[\hat{x}_0, \hat{\tau}_0 = \tau_0])$ is affected by the impairment, but surely much less than the original observation $s(t, x)$.

The corresponding Matlab code for delay and sum beamforming described here follows, together with the generation of the signal (a gray-scale image is used in place of wiggle plot in the figure, which enhances any waveform distortion after the sum).

```
x=-1000:50:1000;
v=1500; %m/s (acoustic wavefield in water) % v=350; %m/s
(acoustic wavefield in air)
dT=10E-3; % sampling interval (10ms)
t=[0:dT:1]; % recording time T=1sec
Tw=20E-3; % Tw=20ms Ricker waveform
tau_o=.400; x_o=-100; A=1; % source parameters

D=zeros(length(t),length(x));
for ix=1:length(x);
    tau=sqrt(tau_o^2+((x(ix)-x_o)/v)^2);
    D(:,ix)=D(:,ix)+(A/tau)*Ricker((t-tau)/Tw)'; % 1/r^2
energy attenuation
end

% Delay and sum BF
D_BF=zeros(length(t),length(x));
xBF=[-500:50:500];
BF_out=zeros(length(t),length(xBF));
for iBF=1:length(xBF),
    tBF=sqrt(tau_o^2+((x-xBF(iBF))./v).^2);
```

```
    for ix=1:length(x)
        D_BF(:,ix)=delayseq(D(:,ix),(tau_o-tBF(ix))/dT); %
delay-part
    end
    BF_out(:,iBF)=sum(D_BF'); % sum-part
end

subplot(121)
% imagesc(x,t,D); colormap('gray') % optional plot in gray
scale
wiggle(t,x,D,'2') % wiggle plot with black-filled positive
waveshape
xlabel('x [m]'); ylabel('time [s]'); title('s(t,x)
[v=1500m/s]')
subplot(122)
% imagesc(xBF,t,BF_out); colormap('gray') % optional plot in
gray scale
wiggle(t,xBF,BF_out,'1')
xlabel('x_e_s_t [m]'); ylabel('time [s]')
```

Remark 1: The signal $s_{out}(t|\hat{\theta})$ after the delay and sum beamforming is maximized for those choices of the shape parameters $\hat{\theta}$ that flattens the wavefields $s(t+\tau(x|\hat{\theta}),x)$ before summing up. The energy of $s_{out}(t|\hat{\theta})$ within a predefined window can be used to estimate the presence or not of one source for a certain combination of shape parameters. The search is over the bidimensional pairs $[\hat{x}_0,\hat{\tau}_0]$ for the value(s) that peak(s) the energy metric, and time resolution is limited by the time window adopted in the energy metric.

Remark 2: After delaying, the sum in (20.21) can take into account the different degree of noise and interference occurring in the flattened wavefield. This is obtained by weighting the terms before summing, and the weighting can be designed according to the degree of noise and interference that is experienced in the sum. The optimal weighting is the same as for BLUE (Section 6.5) and depends on the inverse of the power of the noise and interference $\sigma^2(x,\hat{\theta})$ for the specific shape parameters $\hat{\theta}$:

$$s_{out}(t|\hat{\theta}) = \gamma \sum_x \frac{s(t+\tau(x|\hat{\theta}),x)}{\sigma^2(x,\hat{\theta})}$$

with normalization $\gamma = \sum_x \sigma^{-2}(x,\hat{\theta})$

20.6.2 Wideband Beamforming After Fourier Transform

Implementation of delay and sum beamforming for discrete-time signals needs to interpolate the signals when the delay is fractional. This is carried out by appropriate filtering as part of the delay for discrete-time signals as detailed in Appendix B.

Alternatively, one can use the delay after the Fourier transform of the wavefields vs. time, which has several benefits. The model (20.19) after FT is

$$S(f,x) = \sum_{\ell=1}^{L} G_\ell(f)\alpha_\ell(x)\exp(-j2\pi f\tau(x|\theta_\ell)) + W(f,x)$$

and, for every frequency, it is the superposition of spatial sinusoids each with appropriate spatial varying amplitudes $\alpha_\ell(x)$. A wideband beamforming becomes:

$$s_{out}(t|\hat{\theta}) = \frac{1}{N}\mathcal{F}^{-1}\{\sum_x S(f,x)\exp(j2\pi f\tau(x|\hat{\theta}))\}$$

and avoids any explicit interpolation step.

Another simplification of wideband modeling is for far-field when the wavefront is planar, or around a certain position $x = \bar{x}$.

$$\tau(x|\theta_\ell) \simeq \tau(\bar{x}|\theta_\ell) + \frac{(x-\bar{x})}{v}$$

When the amplitude variations across space is negligible ($\alpha_\ell(x) \simeq \alpha_\ell(\bar{x})$) the model is

$$S(f,x) = \sum_{\ell=1}^{L} \Upsilon_\ell(f|\theta_\ell)\exp(-jK_\ell(f)x) + W(f,x)$$

where

$$\Upsilon_\ell(f|\theta_\ell) = G_\ell(f)\alpha_\ell(\bar{x})\exp[-j2\pi f(\tau(\bar{x}|\theta_\ell) - \dot{\tau}(\bar{x}|\theta_\ell)\bar{x})]$$
$$K_\ell(f) = 2\pi f/v$$

These are the sum of spatial sinusoids that can be handled using the theory of narrowband array processing as illustrated in Chapter 19.

Appendix A: Properties of the Sample Correlations

Preliminarily, we can notice that for any WSS process $x(t)$ the mean values can be evaluated as

$$\mathbb{E}[(\int_{-T/2}^{T/2} x(t)dt)^2] = \mathbb{E}[\iint_{-T/2}^{T/2} x(t)x(\xi)d\xi dt] = T\int_{-T}^{T}\left(1-\frac{|p|}{T}\right)\phi_{xx}(p)dp$$

where the last equality is due to the limited range of the autocorrelation within $\pm T/2$. If the interval T is much larger than the decorrelation of the process:

$$T\int_{-T}^{T}\left(1-\frac{|p|}{T}\right)\phi_{xx}(p)dp \simeq T\int_{-T}^{T}\phi_{xx}(p)dp \simeq T\int_{-\infty}^{\infty}\phi_{xx}(p)dp$$

These approximations ease the computations of the terms:

$$\mathbb{E}[|\dot{\varphi}_{w_1g}(0)|^2] = \mathbb{E}[(\int_{-T/2}^{T/2} w_1(t)\cdot\dot{g}(t)dt)^2] \simeq T\int_{-\infty}^{\infty}\phi_{w_1w_1}(p)\cdot\phi_{\dot{g}\dot{g}}(p)dp$$
$$= T\int_{-\infty}^{\infty}(2\pi f)^2 S_{w_1w_1}(f)\cdot S_{gg}(f)df$$

$$\mathbb{E}[|\dot{\varphi}_{w_1w_2}(0)|^2] = \mathbb{E}[(\int_{-T/2}^{T/2} w_1(t)\cdot \dot{w}_2(t)dt)^2] \simeq T\int_{-\infty}^{\infty} \phi_{w_1w_1}(p)\cdot \phi_{\dot{w}_2\dot{w}_2}(p)dp$$
$$= T\int_{-\infty}^{\infty} (2\pi f)^2 S_{w_1w_1}(f)\cdot S_{w_2w_2}(f)df = T\int_{-\infty}^{\infty} (2\pi f)^2 \left(S_{ww}(f)\right)^2 df$$

that reduce to those in the main text.

Appendix B: How to Delay a Discrete-Time Signal?

Fractional delay (not sample-spaced) of a discrete-time signal is very frequent in statistical signal processing. Let $x_1, x_2, ..., x_N$ be a sequence to be delayed by τ samples, with τ non-integer (otherwise, this would be simply a time-shift of the samples) and sample index indicated by subscript. The solution is to interpolate the sequence with an appropriate continuous-time interpolating function $h(t)$ that estimates *at best* the continuous-time signal:

$$\hat{x}(t) = \sum_{k=1}^{N} x_k h(t-k)$$

assuming that the sampling interval is unitary. The interpolated signal $\hat{x}(t)$ can now be arbitrarily delayed

$$y(t) = \hat{x}(t-\tau) = \sum_{k=1}^{N} x_k h(t-\tau-k)$$

and the delayed signal is uniformly resampled to have the delayed sequence:

$$y_n = y(t=k) = \sum_{k=1}^{N} x_k h(n-k-\tau) = x_n * h_n(\tau)$$

The last equality shows that the delayed sequence is the convolution of the original sequence and the interpolating function delayed by τ and resampled.

Accuracy of fractional delay depends on the choice of the interpolating function $h(t)$ that must honor the original samples (i.e., $\hat{x}(t=n) = x_n$) as discussed in Section 5.5. The ideal interpolating function

$$h(t) = \frac{\sin(\pi t)}{\pi t}$$

is not commonly used as the tails decay too slowly and the sequence needs to be very long to avoid edge effects. One of the most common ways to interpolate is by using the spline (see the book by de Boor [100] for an excellent overview).

A very pragmatic way to delay a sequence is by the use of the discrete Fourier transform (DFT) under the assumption of periodicity of the sequence. Namely, from the sequence $x_1, x_2, \ldots, x_N$, the DFT is computed and the result in turn is delayed by a linear phase:

$$Y(k) = X(k)e^{-j2\pi k\tau/N} = \left(\sum_{n=1}^{N} x_n e^{-j2\pi kn/N}\right) e^{-j2\pi k\tau/N}$$

In spite of its simplicity, there are pitfalls to be considered to avoid "surprises" after fractional delays. Namely, the use of the DFT still uses an interpolation, which can be easily seen from the definition of the inverse DFT of $Y(k)$:

$$h(t) = \frac{\sin(\pi t)}{N\sin(\pi t/N)}$$

which is periodic over N samples. Namely, when using the DFT all sequences are periodic, even if not originally. If the delay τ is small and the original sequence tapered to zero, these wrap-around effects are negligible. Alternatively, the original sequence can be padded by zeros up to N' (with $N' > N$) to make the implicit periodic sequence long enough to reduce these artifacts (but still not zero after non-fractional delays). Needless to say, the DFT must be computed with basis N'.

Appendix C: Wavefront Estimation for 2D Arrays

Let $\hat{\boldsymbol{\tau}} = \left[\hat{\tau}(i,j)\right] \in \mathbb{R}^{M\times N}$ denote the matrix of (unknown) delay function (bold font indicates matrix), $\boldsymbol{\Delta\tau}_x = \left[T_x(i,j)\right] \in \mathbb{R}^{(M-1)\times N}$ and $\boldsymbol{\Delta\tau}_y = \left[T_y(i,j)\right] \in \mathbb{R}^{M\times(N-1)}$ the matrixes of gradients, and $\mathbf{J}_N \in \mathbb{R}^{N\times N}$ the shift matrix (i.e., $\left[\mathbf{J}_N\right]_{i,j} = \delta_{i-j+1}$). The ToD of the wavefront can be related to the DToD along the two directions for the neighboring points (gradients of the wavefront):

$$\begin{cases} \hat{\boldsymbol{\tau}}^T(\mathbf{I}_M - \mathbf{J}_M)\mathbf{J}_M^T = \left[\hat{\boldsymbol{\Delta\tau}}_x^T, \mathbf{0}_N\right] = \tilde{\mathbf{T}}_x^T \\ \hat{\boldsymbol{\tau}}(\mathbf{I}_N - \mathbf{J}_N)\mathbf{J}_N^T = \left[\hat{\boldsymbol{\Delta\tau}}_y, \mathbf{0}_M\right] = \tilde{\mathbf{T}}_y \end{cases} \tag{20.22}$$

or equivalently

$$\begin{cases} \mathbf{D}_x\hat{\boldsymbol{\tau}} = \tilde{\mathbf{T}}_x \\ \hat{\boldsymbol{\tau}}\mathbf{D}_y = \tilde{\mathbf{T}}_y \end{cases} \tag{20.23}$$

where $\mathbf{D}_x = \mathbf{J}_M(\mathbf{I}_M - \mathbf{J}_M^T)$ and $\mathbf{D}_y = (\mathbf{I}_N - \mathbf{J}_N)\mathbf{J}_N^T$ are operators for the evaluation of the gradients from the wavefront $\hat{\boldsymbol{\tau}}$. The equation (20.23) represents an overdetermined set of $2(M-1)(N-1)$ linear equations and MN unknowns that can be rearranged using the properties of the Kronecker products (Section 1.1.1)

$$\begin{bmatrix} \mathbf{A}_x \\ \mathbf{A}_y \end{bmatrix} \mathrm{vec}(\hat{\boldsymbol{\tau}}) = \begin{bmatrix} \mathrm{vec}(\tilde{\mathbf{T}}_x) \\ \mathrm{vec}(\tilde{\mathbf{T}}_y) \end{bmatrix} \tag{20.24}$$

here $\mathbf{A}_x = \mathbf{I}_N \otimes \mathbf{D}_x$ and $\mathbf{A}_y = \mathbf{D}_y^T \otimes \mathbf{I}_M$. The LS wavefront estimator is

$$(\mathbf{A}_x^T \mathbf{A}_x + \mathbf{A}_y^T \mathbf{A}_y)\text{vec}(\hat{\boldsymbol{\tau}}) = \mathbf{A}_x^T \text{vec}(\tilde{\mathbf{T}}_x) + \mathbf{A}_y^T \text{vec}(\tilde{\mathbf{T}}_y) \tag{20.25}$$

or equivalently

$$\mathbf{D}_x^T \mathbf{D}_x \hat{\boldsymbol{\tau}} + \hat{\boldsymbol{\tau}} \mathbf{D}_y \mathbf{D}_y^T = \mathbf{D}_x^T \tilde{\mathbf{T}}_x + \tilde{\mathbf{T}}_y \mathbf{D}_y^T \tag{20.26}$$

which is the discrete version of the partial differential equation with boundary conditions [99], where $-\mathbf{D}_x^T\mathbf{D}_x$ (or $-\mathbf{D}_y\mathbf{D}_y^T$) is the discrete version of the second order derivative along x (or y). The matrixes are large but sparse, and this can be taken into account when employing iterative methods for the solution (Section 2.9) that converge even if $\mathbf{A}$ is ill conditioned (i.e., the update rule is independent of any constant additive value).

The Matlab code below exemplifies how to build the numerical method to solve for the wavefront from the set of DToD, or for 2D phase unwrapping when the phase-differences $\widehat{\Delta\boldsymbol{\tau}}$ are evaluated modulo-2π. Since all matrixes $\mathbf{D}_x$ and $\mathbf{D}_y$ are sparse, its is (highly) recommended to build these operators using the `sparse` toolbox (or any similar toolbox).

```
% Preparation of algebraic operators
I_N=sparse(eye(N,N));
J_N=zeros(N,N);
I_M=sparse(eye(M,M));
J_M=zeros(M,M);
for i=1:N, J_N(i,i+1)=1; end
for i=1:M, J_M(i,i+1)=1; end
J_N=sparse(J_N(1:N,1:N));
J_M=sparse(J_M(1:M,1:M));
Dx=sparse(J_M*(I_M-J_M'));
Dy=sparse((I_N-J_N)*J_N');
DDx=Dx'*Dx;
DDy=Dy*Dy';

% Tx and Ty are the DToD or the phase-differences
A=kron(sparse(I_N),DDx)+kron(DDy,sparse(I_M)); %Laplace
operator
B=kron(sparse(I_N),Dx')*Tx(:)+kron(Dy,sparse(I_M))*Ty(:);
T=A \ B;
T=reshape(T, [M,N]);
```

21

Tomography

Tomography is the method of getting an inner image of an object by external observations called projections. The word comes from the Greek "tomos," which means *slice*, so tomography slices an object from external measurements and gets an image by reconstructing the image from the projections. In radiography diagnostics it is known as computer tomography (CT), and it was investigated by Allan M. Cormack (1963) and later by Godfrey N. Hounsfield (1972), and their research was awarded by the Nobel prize in 1979. CT is a leading diagnostic tool together with magnetic resonance imaging (MRI), which can be considered more complex in terms of imaging method. A lot of progress has been made since the early tomography; the early 2D slices have been enriched by the third dimension to get volumetric (or 3D) CT. The importance of tomographic methods goes beyond diagnostic medical imaging: one can find engineering applications in non-destructive testing, geoscience imaging, ultrasound imaging, and recently radio imaging. The wide usage of tomographic methods in engineering justifies the discussion here of the basic principles of tomographic imaging methods. An excellent methodological reference is the book by Kak and Slaney [84], while the basics of the physics of the interaction of radiation with matter is in [85], or others.

Tomography is based on the measurement of the projections inferred from excitations external to the body. An exemplary CT experiment of X-ray radiation absorption is illustrated in Figure 21.1. Let $f(x,y)$ be the specific absorption of the body in position (x,y) so that an X-ray transmitter at position Tx emits radiation with intensity I_o (photon energy). The received intensity at Rx is reduced (attenuated) by the specific absorption $f(x,y)$ along the path Γ according to the Beer–Lambert (J.H. Lambert, 1760; A. Beer, 1852) law:

$$I(\Gamma) = I_o \times \exp[-\int_{\Gamma} f(x,y)d\ell]$$

the attenuation in a logarithmic scale

$$-\ln\frac{I(\Gamma)}{I_o} = \int_{\Gamma} f(x,y)d\ell$$

is the integral along the "absorbing tube" Γ of the 2D specific absorption. In other words, any measurement of the attenuation depends on the specific absorption function $f(x,y)$, that in turn depends on some physical parameters of the medium such as

Statistical Signal Processing in Engineering, First Edition. Umberto Spagnolini.

Companion website: www.wiley.com/go/spagnolini/signalprocessing

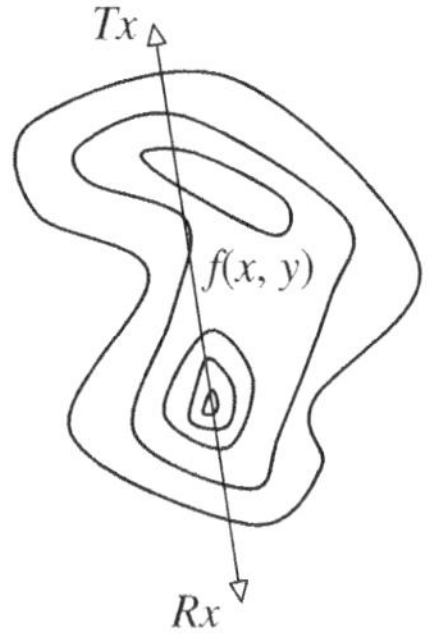

Figure 21.1 X-ray transmitter (Tx) and sensor (Rx), the X-ray is attenuated along the line according to the Beer–Lambert law.

X-ray intensity, and the combination of the absorption characteristics of the chemical elements (e.g., water content and fat, if tissues) weighted by the corresponding fractional density [86]. The intensity of excitation I_o and measurement $I(\Gamma)$ is subject to exponential decay due to the physics of interaction with matter, and this makes the intensities to be modeled random variables with Poisson distributions.[1]

In nuclear medicine, the emission imaging is from a radioactive isotope introduced into the body that forms an unknown emitter density that depends on the body metabolism. High metabolic activity is imaged by an anomalous concentration of the isotope, which generates local radiation that escapes the body almost without any attenuation or scatter. The decaying is stochastic by nature following a Poisson distribution for every unit volume (voxel), with a half-life decay that is matched to the duration of the patient examination. Single-photon emission computed tomography (SPECT) and photon emission tomography (PET) use two different criteria: SPECT images the gamma-rays from a gamma-emitting radioisotope, while PET images the two gamma-rays emitted back-to-back when a positron annihilates. In both cases the image is obtained as emission tomography.

Radiotheraphy (RT) is one therapeutic weapon against cancer. The task is to deposit the energy within the volume of the tumor (target region) and minimize the radiation to the peripheral tissues. RT takes advantage of a non-linear behavior of cell survival probability vs. radiation dose: if healthy tissues are radiated with a half-dose compared to the tumor, the death-rate of these tissues can be as low as 10% vs. 90%, respectively. The energy source in RT needs to be collimated to the target region, and the radiation sources are gamma-rays or an electron beam, or even a radiactive source, either external or internal with in-situ placement (brachytheraphy). Energy collimation is by designing the beam conformation to the target zone to radiate the target region using energy

1 The Poisson distribution with parameter λ for countable infinite sample is

$$Pr(X = x) = \frac{\lambda^x e^{-\lambda}}{x!} \text{ for } x = 0,1,2,3,..$$

with the properties $\mathbb{E}[X] = \text{var}[X] = \lambda$ and $\sum_{x=0}^{\infty} Pr(X = x) = 1$. In short, $X \sim Poiss(\lambda)$ denotes that the rv is Poisson distributed with parameter (or mean) λ. The sum of independent Poisson rvs is Poisson: if $X_i \sim Poiss(\lambda_i)$ then $Y = \sum_{i=1}^{N} X_i \sim Poiss(\lambda)$ with $\lambda = \sum_{i=1}^{N} \lambda_i$.

uniformity as the optimization metric. Beam conformation in RT is an equivalent problem to X-ray imaging [87].

However, tomography in engineering has a much broader meaning than just medical imaging. The tomographic method refers to any application where the inner parameters are inferred from external measurements with experiments purposely tailored to sense a subset of the parameters. These estimation problems are more common that one might imagine, and the estimation of inner parameters requires designing a model and experiments accordingly. Some examples from networking tomography, geoscience, and ultrasound are discussed below.

21.1 X-ray Tomography

In X-ray systems, the generation and absorption contains a stochastic component characterized by the exponential decaying of radiation, which is modeled by a Poisson distribution. The intensity $I(\Gamma)$ is Poisson distributed depending on the (unknown) absorption along the (known) direction Γ, this sets the basis for MLE once the corresponding model has been developed.

21.1.1 Discrete Model

A straightforward method for estimating the absorption is by building an experiment such as the one illustrated in Figure 21.2. One radiation transmitter at position $Tx(m)$ and a set of receiving X-ray detectors at position $Rx(n)$ are arranged in a circle on the tomographer; transmitter and receivers possibly rotate rigidly around the body. The transmitter radiates and the receivers collect the corresponding radiation flux. The intensity at the transmitter is known, as well as any difference due to the emission angle

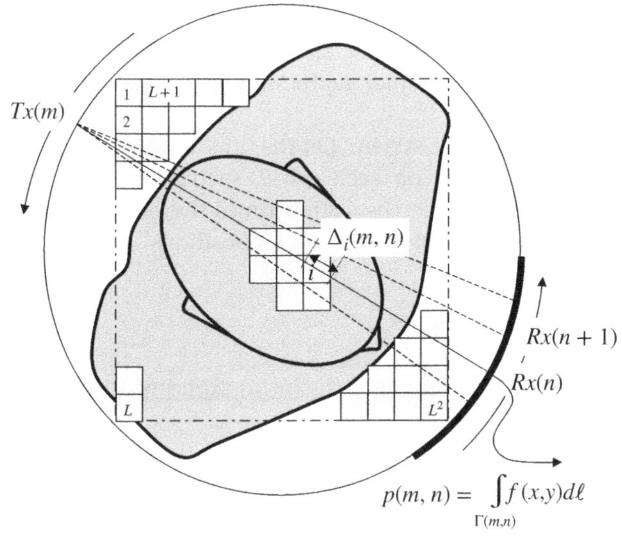

Figure 21.2 X-ray tomographic experiment.

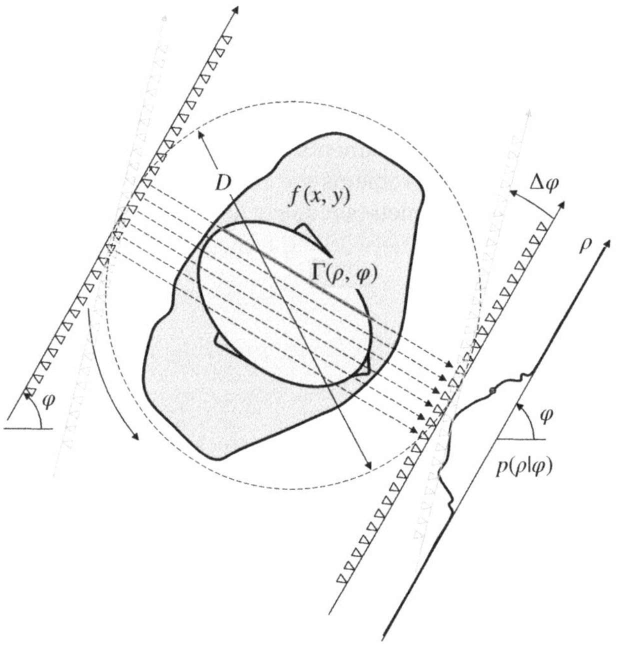

Figure 21.3 Parallel plane acquisition system and projection.

or distance due to the acquisition geometry, so that (after some compensation of the geometrical setting), the measured attenuation

$$p(m,n) = \int_{\Gamma(m,n)} f(x,y)d\ell$$

depends on the integral along the geometrical line $\Gamma(m,n)$ connecting $Tx(m)$ and $Rx(n)$. These measurements are called *projections* as will be clarified later. An alternative agreement is in Figure 21.3 for parallel beam scanner geometry with a uniform linear array of transmitters and receivers.

These models can be framed into an algebraic system. Let the (unknown) absorption be divided into cells with specific absorption on each cell θ_i such that $\theta_i \geq 0$; the measured attenuation is the sum of the specific absorption times the (known) line-length within the cell $\Delta_i(m,n)$, which depends on the Tx and Rx positions

$$p(m,n) = \sum_i \Delta_i(m,n)\theta_i = \mathbf{A}(m,n)\boldsymbol{\theta}$$

where $\mathbf{A}(m,n)$ is the row of the matrix corresponding to the (m,n) experiment pair, and all the absorptions are in $\boldsymbol{\theta}$. The matrix

$$\mathbf{A} = [\Delta_i(m,n)]$$

sets all the geometrical parameters that depend on the geometry of the experiment. Cells are square and line-length $\Delta_i(m,n)$ depends purely on geometrical parameters.

However, other choices of cells could be feasible to account for some collimation errors that makes each tube a "fat tube" that is slightly broader than in the figure, covering multiple cells This complicates the setting of model matrix $\mathbf{A}$. Note that the cells discretize the full area of diameter D, and the boundary between the body ($\theta_i \geq 0$) and the air ($\theta_i = 0$) is not known a-priori and implicitly enters into the unknowns.

21.1.2 Maximum Likelihood

The intensity for any (m,n) pair depends on the photon-counting mechanism of the X-ray detectors, and it is subject to some photon granularity that is modeled as Poisson

$$I(m,n) \sim Poiss(\lambda(m,n))$$

with parameter

$$\lambda(m,n) = \mathbb{E}[I(m,n)] = \bar{I}_o \times \exp[-\mathbf{A}(m,n)\boldsymbol{\theta}]$$

where the mean intensity $\bar{I}_o = \mathbb{E}[I_o]$ is a positive constant and known. The likelihood of each intensity measurement is

$$p(I(m,n)|\boldsymbol{\theta}) = \frac{\lambda(m,n)^{I(m,n)} e^{-\lambda(m,n)}}{I(m,n)!}$$

and, as a consequence of the statistical independence of the intensities $\{I(m,n)\}$, the likelihood function of the ensemble of the measurements is

$$p[\{I(m,n)\}|\boldsymbol{\theta}] = \prod_{m,n} \frac{\lambda(m,n)^{I(m,n)} e^{-\lambda(m,n)}}{I(m,n)!}$$

The log-likelihood becomes

$$\begin{aligned}\mathcal{L}[\{I(m,n)\}|\boldsymbol{\theta}] &= \textstyle\sum_{m,n} I(m,n) \ln\left(\bar{I}_o \times \exp[-\mathbf{A}(m,n)\boldsymbol{\theta}]\right) - \left(\bar{I}_o \times \exp[-\mathbf{A}(m,n)\boldsymbol{\theta}]\right) \\ &= -\textstyle\sum_{m,n} I(m,n)\mathbf{A}(m,n)\boldsymbol{\theta} + \left(\bar{I}_o \times \exp[-\mathbf{A}(m,n)\boldsymbol{\theta}]\right)\end{aligned}$$

which should be maximized wrt the absorptions $\boldsymbol{\theta}$. Due to the cumbersome structure of the optimization problem, a widely investigated method in the literature is the EM method discussed in Section 11.6. The EM approach [88] defines the complete observation by augmenting the observed radiation intensity $\{I(m,n)\}$ with the unobserved intensities entering and leaving every single cell along the line $\Delta_i(m,n)$ to form the complete set. The E-step of the EM method estimates the radiation intensities entering and leaving every cell, conditioned to the measured data. The EM iterations offer an excellent tool to control the positivity of the absorption at every iteration.

21.1.3 Emission Tomography

Emission tomography (SPECT and PET) is based on local radiation emission and the corresponding photon-counting mechanism. The analytical infrastructure is similar to

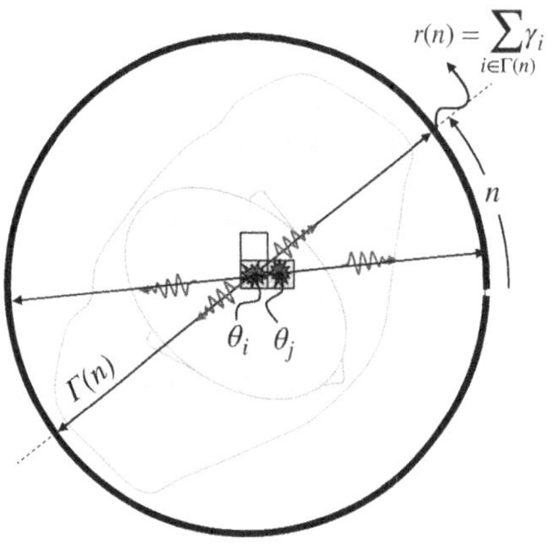

Figure 21.4 Emission tomography.

transmission tomography discussed so far, and shares several commonalities that can be found in [88]; ML is discussed in [90]. The objective is to evaluate the emission density in every voxel that depends on local density of the radioisotope. To exemplify here, in PET the gamma-rays are emitted symmetrically back-to-back as in Figure 21.4 (apart from a small error-angle [85]) and radiation is captured by the photon-counting devices around the body.

Due to the local decaying, the source intensity in the ith voxel is Poisson distributed $\gamma_i \sim Poiss(\theta_i)$ with average emission θ_i to be estimated. Each external measurement at the nth position (Figure 21.4)

$$r(n) = \sum_{i \in \Gamma(n)} \gamma_i$$

is the sum of all emissions along the line-direction $\Gamma(n)$, which now depends only on the sensing position according to the geometrical arrangement chosen here. Emission levels in every voxel are independent, and from the properties of Poisson rvs, the sum is

$$r(n) \sim Poiss(\sum_{i \in \Gamma(n)} \theta_i)$$

which sets the statistical model for MLE. The log-likelihood is

$$\mathcal{L}[\{r(n)\}|\boldsymbol{\theta}] = \sum_n \left[r(n) \ln\left(\sum_{i \in \Gamma(n)} \theta_i \right) - \sum_{i \in \Gamma(n)} \theta_i \right]$$

apart from scaling factors irrelevant for MLE. Maximization of $\mathcal{L}[\{r(n)\}|\boldsymbol{\theta}]$ is complicated, and is again by the EM method (the reader is referred to the literature, e.g., [88]).

21.2 Algebraic Reconstruction Tomography (ART)

The ML for transmission tomography is compliant to the stochastic model of the physics of radiation, but it can be complex, namely if augmenting the model with noise. Alternatively, the log-intensity ratio $p(m,n) = \mathbf{A}(m,n)\theta$ can be considered as deterministic and this can be arranged in a compact notation

$$\mathbf{p} = \mathbf{A}\boldsymbol{\theta} + \mathbf{w} \tag{21.1}$$

by collecting in vectors all measurements of attenuations $\mathbf{p}$, possibly affected by measurement noise $\mathbf{w}$. The statistical properties of noise are not trivial, namely because intensity measurements are in logarithmic scale and a Gaussian noise superimposed on power does not preserve the statistical properties in projection. The LS method is a reasonable strategy to estimate the absorption $\boldsymbol{\theta}$ in this case.

The estimate is purely algebraic with a tall matrix $\mathbf{A}$, and the problem is solved by the LS method (Section 2.7). There are many tomographic projections from different views to better constrain the solution, but this makes the linear system very large, hence iterative methods to solve for the LS solution are more appropriate (Section 2.9). Estimation of the absorption $f(x,y)$ over a grid by multiple views all around the target, by solving the linear system (21.1), is referred to as algebraic reconstruction tomography (ART). Most of the effort of the scientific community is to solve the linear system (21.1) iteratively with the image of $f(x,y)$ that slowly appears from projections, or adapting the choice of the projections to enable a uniform coverage of the image $f(x,y)$, or sampling the image $f(x,y)$ according to the coverage (e.g., to avoid rank-deficiency of $\mathbf{A}$). This is all to guarantee that the entries of the estimated $\hat{\boldsymbol{\theta}}$ are positive semidefinite. In any case, the ART method is unavoidably compute-intensive.

A this stage, there are many remarks that could be made. In brief: (1) cell-binning is a spatial sampling that should fulfill the sampling theorem, and should be comparable to the thickness of the integrating tube; (2) raypaths of tubes have an unequal distribution of coverage for cells—some cells are covered densely when close to the source and coarser when closer to the receiving devices; (3) the accuracy of the estimate on cells depends on their coverage (e.g., the number of raypaths that hit a cell on the overall survey with multiple view angles); low coverage might result in poor conditioning of $\mathbf{A}$ and higher noise or artifacts that can be avoided by unequal binning; (4) the non-negativity constraints of the entries of absorption $\boldsymbol{\theta}$ (or null-absorption as external-body constraints) introduce significant complexity into numerical methods.

21.3 Reconstruction From Projections: Fourier Method

Let us consider again the parallel beam scanner geometry in Figure 23.1. The radiation is emitted to be collimated so that multiple transmitters and receivers collect the corresponding radiations, (ideally) without angular dispersions. When the tomographer is oriented with an angle φ, the receiver at position ρ along the receiving line collects the line integral of the corresponding absorption

$$p(\rho|\varphi) = \int_{\Gamma(\rho,\varphi)} f(x,y)d\ell$$

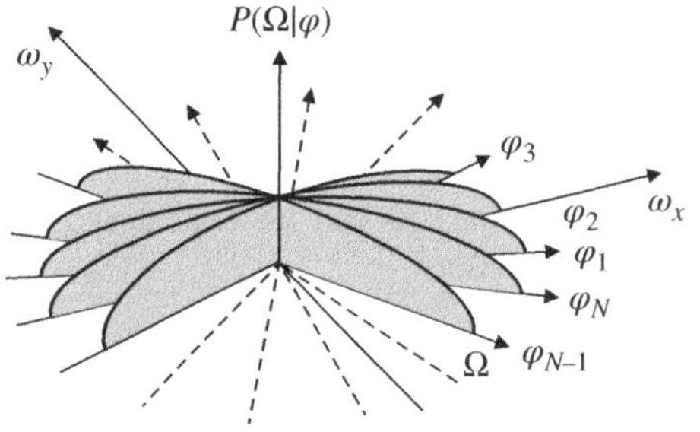

Figure 21.5 Reconstruction from Fourier transform of projections.

along $\Gamma(\rho,\varphi)$. The projection $p(\rho|\varphi)$ can be measured for varying angles and it depends on the absorption $f(x,y)$, as illustrated below.

The Fourier slice theorem (or projection theorem) [84] states that the 1D Fourier transform of the projection $p(\rho|\varphi)$ is a slice of the 2D Fourier transform of $f(x,y)$ along the same angle:

$$\mathcal{F}\{p(\rho|\varphi)\} = P(\Omega|\varphi) = F(\omega_x,\omega_y)\Big|_{\substack{\omega_x = \Omega\cos\varphi \\ \omega_y = \Omega\sin\varphi}}$$

The way to prove this is by considering the case $\varphi = 0$ and thus $p(\rho|\varphi = 0) = f(x,y) ** \delta(x)|_{x=\rho}$; hence the 2D Fourier transforms is

$$\mathcal{F}_{2D}\{f(x,y) ** \delta(x)\} = F(\omega_x,\omega_y)\times\delta(\omega_y) \rightarrow F(\omega_x,0) = P(\omega_x|0)$$

The result is independent of orientation and thus the theorem follows.

The Fourier slice theorem in words states that by measuring multiple projections around the object, one samples angularly the 2D FT of the object itself taken from the set of 1D FT, each projection for each angle. Every projection gathers an independent "view" of the object; putting together all the views one gets an estimate of the object itself that is angularly sampled at step $\Delta\varphi$. This is sketched by Figure 21.5.

The slice theorem suggests a simple way to estimate the image by interpolating from uniformly polar 2D FT onto a regular grid 2D FT. This is a 2D interpolation of non-uniform samples that is densely sampled at low frequency (or the smooth part of the image) and coarsely sampled at high-frequency. Accuracy depends on the interpolation method.

21.3.1 Backprojection Algorithm

The way to estimate the function from projections is by the backprojection algorithm. From the definition of the inverse FT:

$$f(x,y) = \iint F(u,v)e^{j2\pi(ux+vy)}du\,dv$$

in polar coordinates

$$u = w\cos\varphi$$
$$v = w\sin\varphi$$

the Jacobian of the integral becomes

$$du\,dv = w\,dw\,d\varphi$$

so rewriting the inverse Fourier transform in polar coordinates

$$F_p(w,\varphi) = F(w\cos\varphi, w\sin\varphi)$$

gives

$$f(x,y) = \int_0^{\pi}\left[\int_{-\infty}^{\infty} F_p(w,\varphi)e^{j2\pi w(x\cos\varphi + y\sin\varphi)}|w|dw\right]d\varphi$$

To derive the relationship above, the following property on the even symmetry of the sign holds specifically for the polar coordinates:

$$F_p(w,\varphi+\pi) = F_p(-w,\varphi)$$

But $P(\Omega|\varphi) = F_p(w,\varphi)$, and thus the backprojection relationship is

$$f(x,y) = \int_0^{\pi}\left[\int_{-\infty}^{\infty} P(w|\varphi)e^{j2\pi w(x\cos\varphi + y\sin\varphi)}|w|dw\right]d\varphi$$

for the FT of the projections.

The integral is more meaningful if separated into

$$f(x,y) = \int_0^{\pi} Q(x\cos\varphi + y\sin\varphi|\varphi)\,d\varphi$$

where

$$Q(\rho|\varphi) = \int_{-\infty}^{\infty} P(w|\varphi)e^{j2\pi w\rho}|w|dw$$

is called the filtered projection at radial position ρ. Every projection for the specific angle φ is filtered by a filter with frequency response $|w|$ to get $Q(\rho|\varphi)$. Once filtered, all the projections are summed along a trajectory that depends on the point $(\bar{x},\bar{y})$: that is, for every angle φ, the value that contributed to the (filtered) projection in $\bar{\rho} = \bar{x}\cos\varphi + \bar{y}\sin\varphi$. This step is called *backprojection*, and the overall result is the filtered backprojection. Figure 21.6 can help to illustrate which values are contributing to what, mostly if using the polar coordinate for the point itself

$$\bar{x} = r\cos\theta \qquad \bar{y} = r\sin\theta$$

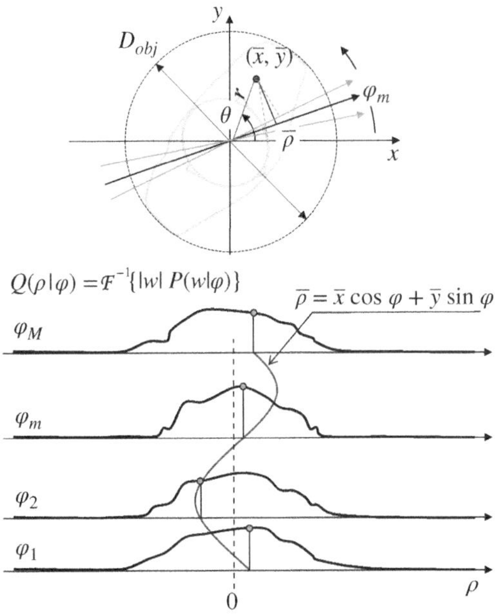

Figure 21.6 Filtered backprojection method.

so that

$$\bar{\rho} = \bar{x}\cos\varphi + \bar{y}\sin\varphi = r\cos(\theta - \varphi)$$

is the locus of the projection of this point by varying the projection angle φ.

Rho-Filter

The filter with frequency response $|w|$ resembles a derivative except for the missing 90 deg phase-shift. This is called the *rho-filter* and it accounts for the different densities of spectral measurements when changing the coordinate from polar to rectangular. If

$$h(t) \leftrightarrow H(w) = |w|$$

is the pulse response of the rho-filter, the filtered projections are obtained as

$$Q(\rho|\varphi) = h(\rho) * p(\rho|\varphi).$$

The inverse FT of $H(w) = |w|$ does not exist as it is not energy-limited, however one can notice that

$$|w| = \left(\frac{-j}{2\pi}sign(w)\right) j2\pi w \leftrightarrow \frac{1}{2\pi^2\rho} * \frac{\partial}{\partial\rho}$$

so that the filtered projection is

$$Q(\rho|\varphi) = \frac{1}{2\pi^2 \rho} * \frac{\partial}{\partial \rho} p(\rho|\varphi)$$

which is the Hilbert transform of the derivative of the projection.

The pulse response of the rho-filter can be derived for bandlimited projections up to $\Omega_{max} = 2\pi/D_{obj}$, where D_{obj} is the diameter of the object as detailed below. The rho-filter has a sinc-like behavior with some fluctuating values. A simple interpretation of the need for the rho-filter, and these fluctuating terms, is by considering the tomography of a $f(x,y) = \delta(x - \overline{x}, y - \overline{y})$ that gives a blade as projection $p(\rho|\varphi) = \delta(\rho - \overline{x}\cos\varphi - \overline{y}\sin\varphi)$. Any backprojection of $p(\rho|\varphi) = \delta(\rho - \overline{x}\cos\varphi - \overline{y}\sin\varphi)$ without the rho-filter would smear the estimated value around $(\overline{x}, \overline{y})$, while the rho-filter introduces some zero-mean fluctuating terms that mutually cancel except in $(\overline{x}, \overline{y})$, as expected. Observe that the rho-filter cancels the zero-frequency, and the lack of a DC component distorts the estimation and this needs to be resolved ad-hoc as detailed below.

21.3.2 How Many Projections to Use?

Projections sample the 2D FT of the body in polar coordinates, and their number M is important to define the complexity and the existence of artifacts and aliasing. A rigorous analysis to define M is quite complex, but [89] offers a simple and pragmatic model.

Angles of projection are uniformly distributed with angular spacing

$$\Delta\varphi = \frac{\pi}{M}$$

The object is within the sensing area of diameter D, but for imaging we need to consider the frequency sampling and thus any periodic repetition of the object beyond D would overlap with the original image and create artifacts. This means that the FT of the projection can be sampled with angular frequency sampling

$$\Delta\Omega \leq \frac{2\pi}{D}$$

as the periodic repetition resulting from frequency sampling with period D would not introduce any (spatial) aliasing. Now we need to compare the physical size of the object with the maximum frequency. Let the diameter of the object be D_{obj} (obviously $D \geq D_{obj}$, to fit the sensing area); the spectrum allocation is approx. (think about the FT of $rect(r/D_{obj})$)

$$|\Omega| \leq \Omega_{max} = \frac{2\pi}{D_{obj}}$$

The angular sampling with spacing $\Delta\varphi$ slices the 2D FT along radial lines as in Figure 21.7, and the largest spacing along the tangential directions are

$$\Delta\varphi \times \Omega_{max}$$

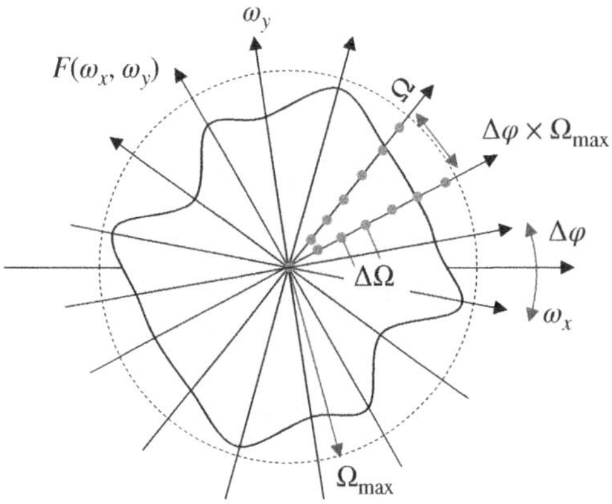

Figure 21.7 Angular sampling in tomography.

This frequency spacing corresponds to a spatial repetition whose limit is set by the frequency sampling $\Delta\Omega$; this yields the condition

$$\Delta\varphi \times \Omega_{max} \leq \Delta\Omega \rightarrow M \geq \pi \frac{D}{D_{obj}}$$

that defines the minimum number of projections to use to avoid aliasing. Needless to say, an increase in the number of projections M improves the definition of the body and improves the image quite considerably.

There is another reason to make M large. The object is estimated within the support D_{obj}, but since the rho-filter removes the DC component, the estimated image has no DC component, and the average estimate of the object $f(x,y)$ is zero. To avoid the corresponding artifacts, the easiest way is to estimate the object within the area of double the diameter; in this way, the ring surrounding D_{obj} (which holds no interest) contains the negative values that counterbalance the positive values (that are of interest) in the center.

21.4 Traveltime Tomography

When wavefields propagate over a non-homogeneous medium with varying propagation velocity $v(x,y)$, the wavefronts are distorted (Section 20.5.1). This distortion appears as a roughness in ToDs, which can be used to infer the properties of the medium when the measurements are made redundant from several viewpoints. Making reference to Figure 21.8 illustrating a box (e.g., a room, a steel-reinforced concrete block, a pillar, or even a sketch of a body), with some inclusions, the propagation velocity can be modeled as

$$v(x,y) = v_o + \delta v(x,y)$$

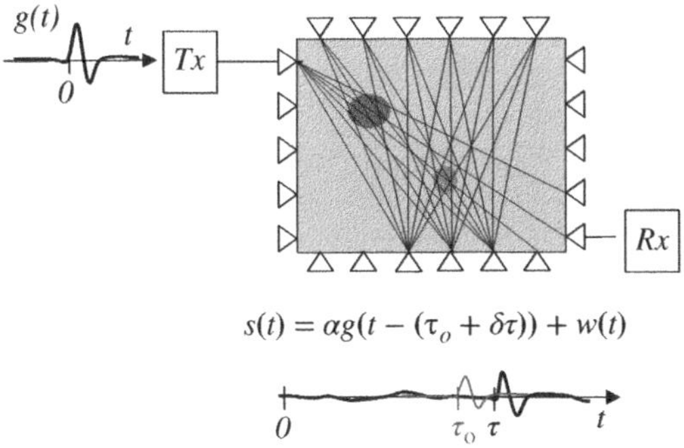

Figure 21.8 Traveltime tomography.

where v_o is the overburden homogeneous velocity (or the average), and $\delta v(x,y)$ models the in-homogeneities. The wavefield in this case can be very complex to be modeled as it includes diffractions and multiple reflections bouncing among the inclusions. A useful simplification is the Born approximation [84] that considers the inclusions sufficiently small in size and (mostly) in velocity contrast (say $|\delta v(x,y)|/v_o < 1/10$) to separate the contribution of these two terms and account for each one individually by adopting the principle of linearity. The Born approximation for diffraction is far more complex than simplified herein, but by reasoning on the wavefronts we can capture the complexity of these tomographic methods, which are sometimes called linearized Born models.

Let Γ be the raypath between transmitter (Tx) and receiver (Rx) in the figure; the time of delay (or traveltime)

$$\tau = \int_{\Gamma} \frac{d\ell}{v(x,y)} \simeq \int_{\Gamma_o} \frac{d\ell}{v_o} + \int_{\Gamma_o} \frac{d\ell}{\delta v(x,y)} = \tau_o + \delta\tau$$

can be decoupled onto the two contributions over the same raypath Γ_o for homogeneous medium ($\Gamma_o \simeq \Gamma$) that is just a geometrical line (Figure 21.8). The traveltime anomalies wrt the homogeneous medium contain the line integral of the velocity anomalies. Multiple views add the necessary redundancy to build an overdetermined linear system to set the LS estimate, and uses the methods described so far except that the measurements are traveltime, obtained by any of the ToD methods discussed in Chapter 20 from a received signal-type

$$s(t) = \alpha g(t - \tau) + w(t)$$

for excitation $g(t)$. To reduce the traveltime model to be superimposed to transmission tomographic models, a common approach is to rewrite the problem in terms of slowness $1/v(x,y)$.

In spite of the conceptual simplicity, (transmission) traveltime tomography is far from being easy to engineer, despite there being many applications such as ultrasound tomography (e.g., non-destructive testing in civil engineering), and microwave tomography (e.g., breast cancer diagnosis), just to mention a couple. The wavefield can be acoustic or electromagnetic depending on the object and the physical parameters to investigated,

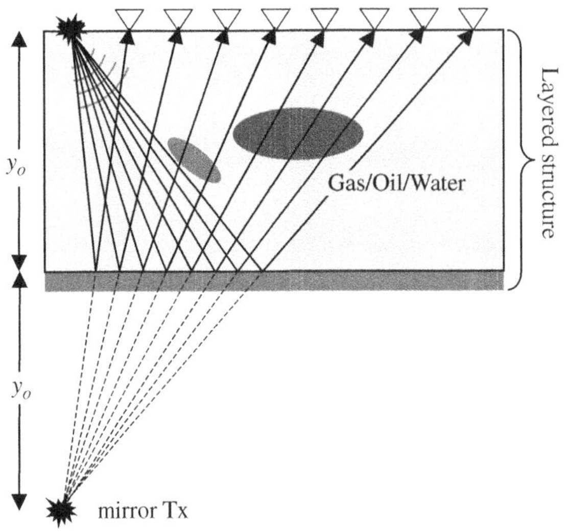

Figure 21.9 Reflection tomography.

but some of these transducers do not comply simply with the geometrical model in Figure 21.8, and the wavefields are mutually coupled between each other as if using multiple transmitters activated simultaneously. In addition, the transducers are placed on the surface, and waves are not simply penetrating the medium, but rather propagate everywhere (e.g., along surfaces). Therefore, there are multiple arrivals due to several scatterings and the tomographic model above holds only for those arrivals that travel through the inhomogeneity. The ToD method should separate the (many) arrivals to avoid irrelevant traveltimes entering into the estimation loop. The iterative method of refinement can help here. Needless to say, accuracy in ToD estimation directly impairs the quality of the estimate $\hat{\delta v}(x,y)$ (or the slowness), and the bandwidth of the excitation $g(t)$ matters as variance is largely reduced for wideband waveforms $g(t)$. On the contrary, wideband waveforms can be complex or expensive to generate, or quickly attenuating in propagation over a dispersive medium.

Another topic is reflection traveltime tomography illustrated in Figure 21.9. In applications where there is no access on the other side of the body such as when sensing the subsurface, the reflections from interfaces among media with different propagation velocities are used as wavefield "mirrors" (with some attenuation). Figure 21.9 shows an example of a layered structure that contains inclusions (e.g., oil, gas, water, or even unexploded ordnances). The excitation on the surface generates a wavefield that propagates spherically down to the planar discontinuity at depth y_o; part of the waves impinging on the interface at y_o are reflected back to the surface where receivers are collecting the backscattered signals. The mirror-like reflections can be interpreted by modeling the wavefield as generated by a transmitter positioned at $2y_o$ from the surface as in Figure 21.9. The inclusions distort the traveltimes as for transmission traveltime tomography and the model is the same.

The uncertainty in the depth position of the reflector, and its shape, pose more complexity for reflection tomography, which can be solved either by increasing the redundancy by using several measurements (that unfortunately are all on one side), or/and by placing some a-priori constraints on the shape/smoothness and the slowness $1/\delta v(x, y)$. Linear models in reflection tomography are often under-determined, and need to be solved by judicious choice of dumping parameters that should match with the experiment coverage.

21.5 Internet (Network) Tomography

The Internet is a massive and distributed communication network that forwards information in the form of packets from source to destination over a mesh of connections. Each link is characterized by its own properties in terms of reliability, error probability, latency etc. Network engineers have developed tools for active and passive measurement of the network. There are probing network tools such as `ping` and `traceroute` that measure end-to-end packet transport attributes. Network tomography estimates the properties of every link from these aggregated measurements—possibly from a very large set of measurements to compensate for inaccuracies due to redundancy, see [91].

A network is composed of nodes (computer and/or servers and/or routers) that are connected over paths; when considering M nodes there are at most $M(M-1)$ links if connecting all-to-all. However, the number of links N are much less as the connectivity is limited to some neighbors in a geographically deployed network. Each path consists of more links that chain a set of physical connections over multiple nodes (Figure 21.10). Packets intended for a specific destination can be routed over these paths and every packet is routed (almost) independently of the others. A statistical model to characterize link properties is the most appropriate model as the nature of packet communication is inherently stochastic. Typical network parameters of each link are the link-by-link delay (or latency), loss-rates by counting lost packets, and traffic in terms of successful packets/sec. Measurement of these end-to-end statistical parameters $x(\ell)$ are the sum of the link-parameters θ_k along the path. The linear model becomes

$$\mathbf{x} = \mathbf{A}\boldsymbol{\theta} + \mathbf{w}$$

where matrix $\mathbf{A}$ is the routing matrix that captures the topology of the network for every measurement. Matrix entries are binary-valued with N columns, and the entries in each row are 0 or 1 depending on which link is active along each path. The additive term $\mathbf{w}$ arises from the effective noise, or the random perturbations around their mean value as

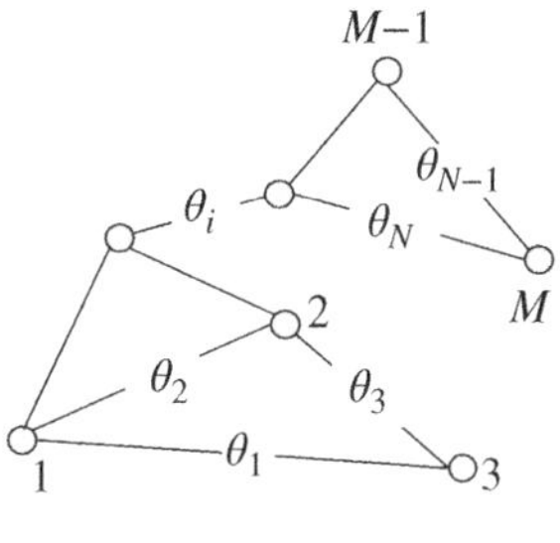

Figure 21.10 Internet tomography.

detailed below. Once again, the size of the problem can range from few tens of edges in a small network, to several thousands in a large Internet network.

Remark 1. Note that the network (routing) topology is expressed by the matrix $\mathbf{A}$. The topology is extracted from network tools like `traceroute` that requests to all routers and servers to make available the history of packets routes. In some cases routers are not reliable or non-cooperative in releasing this information, and the network topology is uncertain. In these cases, the network topology could be one additional parameter to estimate and the estimation problem gets definitely more complex.

21.5.1 Latency Tomography

The statistical properties of each parameter over a link is random, and one is usually interested in their mean. For instance, the latency on every link τ_ℓ is an rv, say $\tau_\ell \sim \mathcal{N}(\theta_\ell, \sigma_\ell^2)$, and the overall latency along a route Γ_k is

$$d_k = \sum_{\ell \in \Gamma_k} \tau_\ell \sim \mathcal{N}(\sum_{\ell \in \Gamma_k} \theta_\ell, \sum_{\ell \in \Gamma_k} \sigma_\ell^2)$$

The tomographic model obtained by averaging L latency measurements $\{d_k(l)\}_{l=1}^L$ that refer to the same route Γ_k in $x_k = \sum_{l=1}^L d_k(l)/L$, so that the model becomes

$$x_k = \sum_{\ell \in \Gamma_k} \theta_\ell + w_k$$

where

$$w_k \sim \mathcal{N}(0, \sum_{\ell \in \Gamma_k} \sigma_\ell^2 / L)$$

Therefore, each of the measurements in $\{x_k\}$ has a different variance depending on the path and the number of measurements used for the same path, and the ML method can be used for this inference problem.

21.5.2 Packet-Loss Tomography

Delivery of packets can be affected by errors, malfunctioning, etc. that is referred as packet-loss and it is governed by probabilistic models. One could be interested in the estimation of success probability (or packet-loss probability) p_ℓ for every link in a network from the compound success (or packet-loss) probability over different paths. The property used here is that the success probability compounds linearly along every path when on a log-scale.[2]

---

2 Consider a cascade of N links. The packet success probability of every link is p_ℓ and it is modeled as a Bernoulli pmf; the success probability of the chain of the N statistical independent links is

$$p_{tot} = \prod_{\ell=1}^{N} p_\ell$$

that in log-scale becomes $\ln p_{tot} = \sum_{\ell=1}^N \ln p_\ell$.

Once again, along any of the path, say Γ_k, the number of K successful packets out of N packets sent is binomial

$$Pr(K|N) = \binom{N}{K} p_{\Gamma_k}^K (1 - p_{\Gamma_k})^{N-K}$$

It depends on the compound success probability along the Γ_k path

$$p_{\Gamma_k} = \prod_{\ell \in \Gamma_k} p_\ell$$

The estimate of the success probability $\hat{p}_{\Gamma_k}$ is the ratio of successful packets $\overline{K}$ out of the N total packets sent (Section 9.6). The variance is bounded by $\text{var}(\hat{p}_{\Gamma_k}) \geq p_{\Gamma_k}(1 - p_{\Gamma_k})/N$ and the distribution is Gaussian as a binomial distribution is approximated by a Gaussian pdf for large N. The model is linear in log-probability

$$\ln \hat{p}_{\Gamma_k} = \sum_{\ell \in \Gamma_k} \ln p_\ell + w_k$$

and it reduces to the linear model $\mathbf{x} = \mathbf{A}\boldsymbol{\theta} + \mathbf{w}$ for log-success probability of every link. An LS solution is usually employed as noise $\mathbf{w}$ is non-Gaussian when the problem is solved in log-probability scale. However, since $\hat{p}_{\Gamma_k} = p_{\Gamma_k} + \delta p_{\Gamma_k}$ with $\text{var}(\delta p_{\Gamma_k}) \simeq p_{\Gamma_k}(1 - p_{\Gamma_k})/N$ for large N, the log-probability is

$$\ln \hat{p}_{\Gamma_k} = \ln p_{\Gamma_k} + \ln(1 + \delta p_{\Gamma_k}/p_{\Gamma_k}) \simeq \ln p_{\Gamma_k} + \delta p_{\Gamma_k}/p_{\Gamma_k}$$

where the last equality holds when

$$\text{var}(\delta p_{\Gamma_k}/p_{\Gamma_k}) \simeq \frac{1 - p_{\Gamma_k}}{N p_{\Gamma_k}} \ll 1$$

This sets the minimum number of packets N to be used to estimate the log-probability on each link, and the corresponding variance of the additive noise: $\text{var}(w_k) \simeq (1 - p_{\Gamma_k})/N p_{\Gamma_k}$.

22

Cooperative Estimation

Cooperative estimation refers to system where there is a set of nodes accessing local measurements related to a common phenomenon. Each node can estimate locally the set of parameters in the conventional way, but cooperation among the nodes gives a benefit to all as the accuracy of the estimates refined from cooperation is largely improved wrt the individuals. Nodes are complex entities as they have a sensing system for measuring, processing for estimating, and communication capability to enable the cooperation with other nodes. In cooperative estimation, each node cooperates with others to gain a benefit without exchanging the local measurements as this would infringe some confidentiality issues related to the measurements, or the availability of a local model. One example is when every node has a partial coverage of the parameters and cannot estimate them all as the number of measurements is insufficient for a single node, but cooperation with others lets every node act as if all measurements were put in common, thus gaining measurements and improving the accuracy.

The reference model is illustrated by Figure 22.1, where there are a set of K measuring points for the same set of parameters $\boldsymbol{\theta}$. In the additive noise model, the data at the ith node is

$$\mathbf{x}_i = \mathbf{h}_i(\boldsymbol{\theta}) + \mathbf{w}_i$$

and the relationship is in general non-linear. Each node can carry out the local estimate by using any known method such as MLE in Chapter 7, to arrive at the estimate

$$\hat{\boldsymbol{\theta}}_i = \phi_i(\mathbf{x}_i)$$

characterized by the local covariance

$$\text{cov}(\hat{\boldsymbol{\theta}}_i) = \mathbf{C}_i$$

Cooperation among nodes starts from the local estimate in an iterative manner. The local estimate becomes the initialization of an iterative method that refines the estimate of $\hat{\boldsymbol{\theta}}_i$ by exchanging the estimate within a predefined set of communication links (dashed lines) in Figure 21.1. More specifically, let $\hat{\boldsymbol{\theta}}_i(0) = \hat{\boldsymbol{\theta}}_i$ the estimate at initialization (iteration=0); the estimation is refined from the estimate of a set of nodes $\mathcal{K}_i$ that are typically within the communication range of the ith node (dashed line arrows in Figure 21.1):

Statistical Signal Processing in Engineering, First Edition. Umberto Spagnolini.

Companion website: www.wiley.com/go/spagnolini/signalprocessing

$$\hat{\boldsymbol{\theta}}_i(\ell+1) = f_i[\{\hat{\boldsymbol{\theta}}_i(\ell)\}_{i\in\mathcal{K}_i}]$$

where the manipulation $f_i[.]$ is appropriately chosen. Cooperation is beneficial if

$$\text{cov}(\hat{\boldsymbol{\theta}}_i(\ell)) \prec \mathbf{C}_i$$

for large enough ℓ. In addition, all nodes attain similar estimates and similar accuracy for $\ell \to \infty$ so that the cooperative method can be interpreted as a way to distribute the estimation among distinct entities as for *distributed estimation*. Even if one could distinguish between distributed and cooperative depending on what every node is supposed to do and for what purpose, here both are used interchangeably with the same meaning.

Cooperative estimation should be compared with a performance bound from a centralized estimation where all the measurements are forwarded to one fusion center that can estimate the set of parameters in a centralized fashion exploiting all the measurements/information available, so that

$$\hat{\boldsymbol{\theta}}_C = \phi_C(\mathbf{x}_1, \mathbf{x}_2, ..., \mathbf{x}_K)$$

The covariance of the centralized estimate is

$$\text{cov}(\hat{\boldsymbol{\theta}}_C) = \mathbf{C}_C$$

The cooperative estimation is lower bounded by the centralized method so that

$$\text{cov}(\hat{\boldsymbol{\theta}}_i(\ell \to \infty)) \succeq \mathbf{C}_C$$

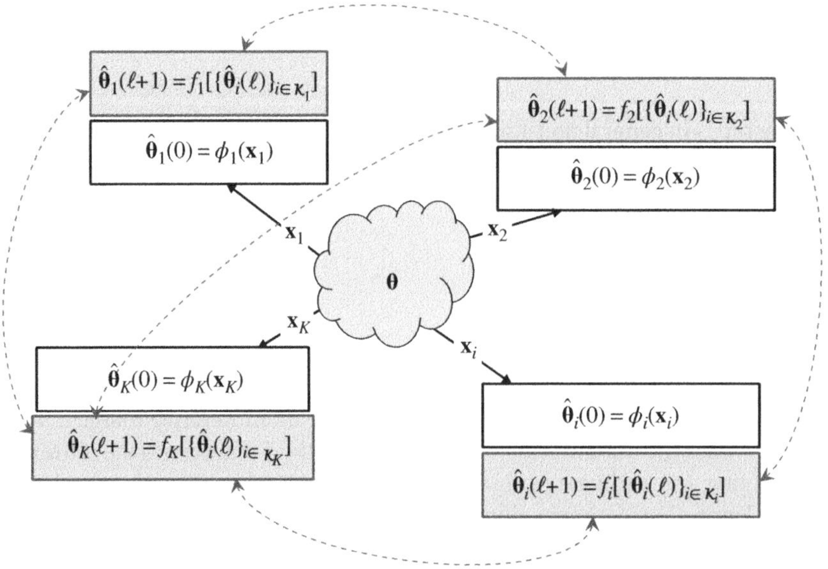

Figure 22.1 Cooperative estimation among interconnected agents.

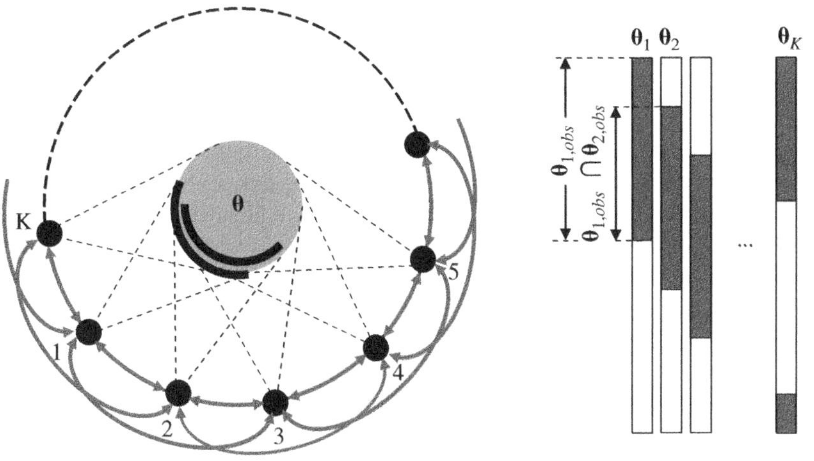

Figure 22.2 Multiple cameras cooperate with neighbors to image the complete object.

where the equality denotes the efficiency of the cooperative estimation method for a large number of iterations. Of course, the centralized method is in turn bounded from estimation theory (e.g., Chapter 8).

There are many applications for cooperative estimation, and others are expected to be promoted by the availability of smart phones with processing and connectivity capability. The distributed model is representative of some engineering problems that involve data-sharing. The example in Figure 22.2 is the case of multiple cameras surrounding an object (here a cylinder) taking pictures (here the measurements $\{\mathbf{x}_i\}$) from different views [107]. The set of p parameters $\boldsymbol{\theta}$ is the full texture of the object, but each camera has only a partial view of the whole object and thus from the snapshot in the ith position only a subset $\boldsymbol{\theta}_{i,obs}$ can be estimated as visible, and not the others as shaded. However, since all the cameras are deployed around the objects, the ensemble of all have full visibility of the object and sharing their estimates could allow each node to gain estimates of those shaded portions in exchange for its own estimates. At the same time, none of the nodes are inclined to exchange the snapshot $\mathbf{x}_i$ as it may contain some personal information (e.g., each snapshot might be taken by different individuals and contain parents or relatives superimposed on the object) but gets a benefit from exchanging the estimate of the cylinder as it gains full visibility. Figure 22.2 is an example of *ring-topology* where the exchange could be among $d = 4$ neighbors as indicated by the gray arrows up to a certain degree of accuracy. Looking at the views from nodes 1 and 2, there is one subset of visible parameters is common to both views, that is $\boldsymbol{\theta}_{1,obs} \bigcap \boldsymbol{\theta}_{2,obs}$, and cooperation can only improve the accuracy for this subset, while for disjoint parameters it is only possible to forward a portion of the estimates from one node to the other. This example is simple enough to provide a guideline to gain insight into cooperative estimation methods. Another example is from cooperative localization, where a large network of nodes can estimate the position of all the nodes by simply exchanging locally their mutual distance based on their limited visibility/accessibility [105, 106]. Further developments on distributed estimation are based on the exchange of the conditional pdfs following a Bayesian network approach, for example, see [108, 109].

22.1 Consensus and Cooperation

Consensus refers to the aggregation of information among nodes (also called *agents* as each one acts according to the information exchanged) each with uncertainties to achieve an agreement regarding a quantity of interest, according to the state of all agents. Consensus theory involves the mechanism that controls the dynamic behavior of the ensemble from the degree of information exchanged, and the local correction applied to reach alignment of opinions. Consensus problems have been widely investigated in different areas involving the dynamic behavior of a network of agents [110] such as in management science, distributed computing, collaborative information processing in networks, wireless sensor networks, biology, social sciences, medical decisions, and politics. The driving principle for cooperation is that there is a benefit for a community to find the consensus of information after some trading and adaptations based on a global metric that is individually optimized. There is the notion of cost/benefit for the ensemble of agents, the dynamic of their update, and finally the degree of mutual interaction. Information consensus is a fairly simple but still general method to introduce the principle of cooperation in estimation where the consensus is reached on the estimate. As usual, there are many references as this is a topic investigated in several disciplines with context-related specificities; a fairly general reference is [111]. The consensus property depends on the mutual connectivity among the nodes, and this is described by the *network graph*. Basics of graph theory and related notation strictly necessary for the chapter are in Appendix A.

22.1.1 *Vox Populi*: The Wisdom of Crowds

In 1906, Dr. Francis Galton published a colorful experiment in Nature (vol.75, pp.450–451). In a weight-judging competition at Plymouth in England among butchers and farmers, people were asked to guess the weight of a fat ox by card—each could make one vote only. The guesses of the 787 different persons were arranged into a distribution, and the average of the ensemble of voters (*vox populi*) was 1207 lb, while the weight of the ox proved to be 1198 lb, so the vox populi was 9lb (or 0.8%) too high of the actual weight. The dispersion of values was a skewed Gaussian and approx. 37 lb around the mean.

The experiment by Galton was based only on the sight of the ox. If persons can iterate their votes by interacting with one another they can refine their estimate and come to a new set of values that are significantly less dispersed and closer to the vox populi. This experiment was used by later researchers to prove that the aggregation of people cancels out idiosyncratic noise associated with each individual judgment. This gave raise to later investigations in different fields related to social sciences. In statistical signal processing, consensus algorithms mimic this crowd behavior using distributed nodes that exchange their information to improve their local estimate of a common parameter (or a subset) by exchanging with others.

22.1.2 Cooperative Estimation as Simple Information Consensus

Let $\theta_1(0), \theta_2(0), ..., \theta_K(0)$ be a set of values from different agents that are mutually connected over a network with a topology defined by the graph on the node set $\{1, 2, ..., K\}$. Mutual exchange of values among neighboring agents enables an iterative updating procedure that modifies and refines these values until reaching the consensus for a large number of iteration exchanges ℓ:

$$\theta_1(\ell \to \infty) = \theta_2(\ell \to \infty) = \dots = \theta_K(\ell \to \infty) = \theta_\infty$$

In consensus algorithms, the ith node exchanges with its neighbors in the set $\mathcal{K}_i$ the values that are updated by all nodes according to the iterative discrete-time relationship

$$\theta_i(\ell+1) = \theta_i(\ell) + \varepsilon \sum_{j \in \mathcal{K}_i} (\theta_j(\ell) - \theta_i(\ell))$$

for an appropriate scaling term ε. Each node corrects the local value based on the average of the errors $\theta_j(\ell) - \theta_i(\ell)$ wrt the neighbors ($j \in \mathcal{K}_i$) so that, at convergence, there is no more need to correct. Rearranging this relationship, it is a recursive update:

$$\theta_i(\ell+1) = (1-\varepsilon_i)\theta_i(\ell) + \varepsilon_i \frac{\sum_{j \in \mathcal{K}_i} \theta_j(\ell)}{|\mathcal{K}_i|}$$

where the current value $\theta_i(\ell)$ is updated with the average of the neighboring nodes with scaling

$$\varepsilon_i = |\mathcal{K}_i|\varepsilon$$

All nodes update their values at the same time, and calculate the updated value $\theta_i(\ell+1)$ once they have received all the estimates from the neighboring nodes $\{\theta_j(\ell)\}_{j \in \mathcal{K}_i}$ and their own. Even if simultaneous update is not strictly necessary for convergence, the update $\ell \to \ell+1$ is carried out independently by all the nodes, and this implies that all nodes have the same time-basis and are synchronized using any of the methods such as the one in Section 22.3. To avoid instability, the update with the average of the neighbors should be

$$\varepsilon_i \le 1 \to \varepsilon \le \frac{1}{\max_i |\mathcal{K}_i|}$$

which is the general condition based on the maximum degree of the graph (see Appendix A).

In order to capture the collective dynamics of the network, the ensemble of values of all the K nodes can be collected in a vector with

$$\boldsymbol{\theta}(\ell) = \begin{bmatrix} \theta_1(\ell) \\ \theta_2(\ell) \\ \vdots \\ \theta_K(\ell) \end{bmatrix}$$

so that the consensus iterations are

$$\boldsymbol{\theta}(\ell+1) = (\mathbf{I} - \varepsilon\mathbf{L})\boldsymbol{\theta}(\ell)$$

where $\mathbf{L}$ is the $K \times K$ Laplace matrix of the graph that captures some properties of consensus iterations in Appendix A. Consensus iterations are the same as the optimization of the disagreement function, with the following properties:

$$\boldsymbol{\theta}_\infty = \frac{\mathbf{1}^T \boldsymbol{\theta}(0)}{\mathbf{1}^T \mathbf{1}} \mathbf{1} = \frac{\mathbf{1}}{K} \sum_{i=1}^{K} \theta_i(0) \tag{22.1}$$

The speed of convergence depends on the connectivity of the graph defined by the connectivity eigenvalue $\lambda_2(\mathbf{L})$, and the number of iterations for convergence is approx.

$$T_{conv} \simeq \frac{1}{\varepsilon \cdot \lambda_2(\mathbf{L})}$$

After this review of the consensus algorithm, its use for estimation should take into account the capability of the consensus to average a set of independent values $\boldsymbol{\theta}(0)$ as in (22.1). Assuming that the initialization represents the local estimate, the consensus algorithm has the capability to converge to the consensus value given by the sample mean of the local estimates obtained by exchanging and refining the local estimates with initialization on

$$\theta_i(0) = \hat{\theta}_i$$

so that the consensus is the value

$$\hat{\theta}_\infty = \frac{1}{K} \sum_{i=1}^{K} \hat{\theta}_i \tag{22.2}$$

for a large enough number of exchanges. In other words, the consensus algorithm computes iteratively the sample mean of initialization values, and thus it is the basic tool to compute the sample mean by exchanging values with neighboring nodes only. From standard estimation theory:

$$\text{var}(\hat{\theta}_\infty) = \frac{1}{K^2} \sum_{i=1}^{K} \text{var}(\hat{\theta}_i)$$

and the sample mean cannot, in general, be considered the optimal choice (see the examples in Section 6.7).

Example of Distributed Estimate of the Number of Nodes

Consensus is quite common to model population behavior, and the dynamic depends on the connectivity. The example herein is on the estimation of the number of nodes in a network when each node has a limited visibility of the overall network. In networks, nodes are labeled by a number that identifies the address, and numbers are assigned to be distinct and non-duplicated. Let every node be labeled by its number $i = 1, 2, \ldots$ but none of the nodes know how many nodes are part of the network and the overall number K is not known to anyone.

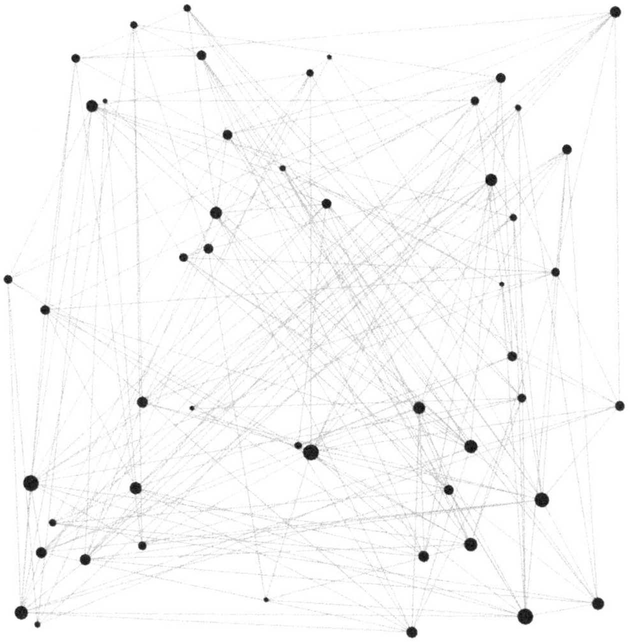

Figure 22.3 Example of random nodes mutually connected with others (radius of each node is proportional to the degree of each node).

To estimate K in a distributed fashion by consensus, each node initializes the estimate with the node-numbering $\theta_i(0) = i$ and starts exchanging these values as initial values. The asymptotic value is the consensus

$$\theta_\infty = \frac{1}{K}\sum_{i=1}^{K}\theta_i(0) = \frac{1}{K}\sum_{i=1}^{K} i = \frac{K+1}{2}$$

which provides an indirect estimate of the total number of active nodes as $\hat{K} = 2\theta_\infty - 1$.

An illustrative example is provided in Figure 22.3 for $K = 50$ nodes randomly deployed over an area and randomly connected to one another. Each node is denoted by a dot, and the radius of each dot is proportional to the degree of the node (the links are indicated by gray shading). The nodes are numbered increasingly $i = 1, 2, ..., K = 50$ but none of the nodes know the overall number (i.e., $K = 50$) and each exchanges the node-numbering with the others.

The consensus iterations are shown in Figure 22.4 with the initialization $\theta_i(0) = i$ that attains the convergence value of $\theta_\infty = (K+1)/2 = 51/2$ indicated by the dashed gray line. The time of convergence T_{conv} depends on the Laplacian matrix for the example at hand; the convergence behavior is approx. $\exp(-\ell\lambda_{\max}/2\lambda_2)$ and the complete convergence is reached after 9–10 iterations as shown in Figure 22.4.

22.1.3 Weighted Cooperative Estimation ($p = 1$)

The simple consensus converges to the mean of the initialization values, which are the local estimates (22.2). However, the sample mean can be impaired by the least accurate

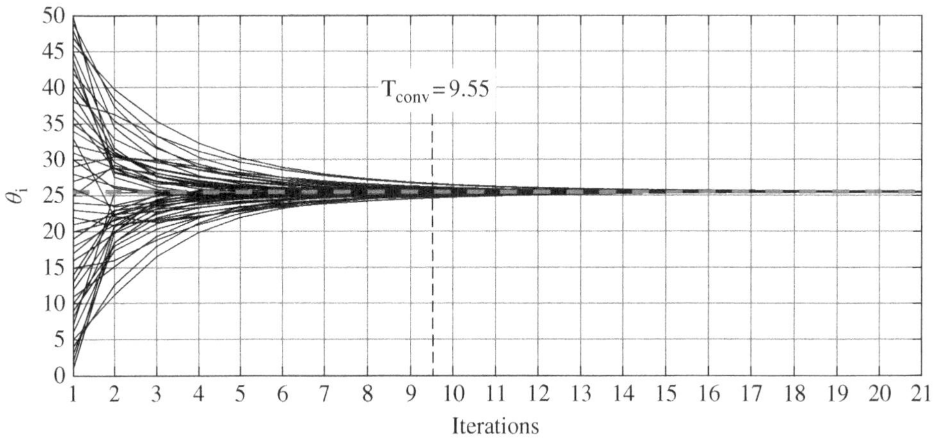

Figure 22.4 Consensus iterations for the graph in Figure 22.3.

estimates and the weighted mean is more appropriate (Section 6.7). The weighted consensus algorithm is

$$\theta_i(\ell+1) = \theta_i(\ell) + \varepsilon \sum_{j\in\mathcal{K}_i} q_j(\theta_j(\ell) - \theta_i(\ell)) = (1 - \varepsilon \sum_{k\in\mathcal{K}_i} q_k) \cdot \theta_i(\ell) + \varepsilon \sum_{j\in\mathcal{K}_i} q_j \cdot \theta_j(\ell)$$

where q_j is a positive weight that depends on the jth node that is chosen below to guarantee some degree of optimality. Recast in terms of consensus iterations $\boldsymbol{\theta}(\ell+1) = (\mathbf{I} - \varepsilon\mathbf{L})\boldsymbol{\theta}(\ell)$, the ith row of the Laplace matrix is

$$[\mathbf{L}]_{i,j} = \begin{cases} -q_j & j \in \mathcal{K}_i \\ \sum_{k\in\mathcal{K}_i} q_k & i = j \\ 0 & j \notin \mathcal{K}_i \end{cases} \tag{22.3}$$

and it sums to zero as $\mathbf{L}\mathbf{1} = \mathbf{0}$. A special case is for a fully connected graph ($|\mathcal{K}_i| = K-1$), when the $K \times K$ Laplacian is

$$\mathbf{L} = \begin{bmatrix} \sum_{j=1,j\neq 1}^{K} q_j & -q_2 & -q_3 & \cdots & -q_K \\ -q_1 & \sum_{j=1,j\neq 2}^{K} q_j & -q_3 & & -q_K \\ -q_1 & -q_2 & \sum_{j=1,j\neq 3}^{K} q_j & \cdots & -q_K \\ \vdots & & \vdots & \ddots & \vdots \\ -q_1 & -q_2 & -q_3 & & \sum_{j=1,j\neq K}^{K} q_j \end{bmatrix}$$

which is non-symmetric but still $\mathbf{L}\mathbf{1} = \mathbf{0}$. Observe that a less connected graph has missing entries along the columns corresponding to the missing links and diagonal entries are adapted accordingly.

Based on the structure of the Laplace matrix, the vector

$$\mathbf{q} = \frac{1}{\sum_{j=1}^{K} q_j} \begin{bmatrix} q_1 \\ q_2 \\ \vdots \\ q_K \end{bmatrix}$$

here normalized to have $\mathbf{q}^T \mathbf{1} = 1$ (Appendix A) is the left eigenvector of the Laplace matrix, as

$$[\mathbf{q}^T \mathbf{L}]_i = q_i \sum_{k \in \mathcal{K}_i} q_k - q_i \sum_{j \in \mathcal{K}_i} q_j = 0 \rightarrow \mathbf{q}^T \mathbf{L} = [0, 0, \ldots, 0]$$

Similarly, $\mathbf{1}$ is the right eigenvector as

$$\mathbf{L1} = \mathbf{0}$$

According to the Perron–Frobenius theorem (Appendix A), the consensus is

$$\boldsymbol{\theta}_\infty = [\mathbf{q}^T \boldsymbol{\theta}(0)]\mathbf{1}$$

where all nodes converge to the same consensus value

$$\hat{\theta}_\infty = \frac{\sum_{i=1}^{K} q_i \hat{\theta}_i}{\sum_{i=1}^{K} q_i}$$

which is the weighted average of the local estimates by every node that, at consensus, is the steady-state variable at every node. This is the basis for the distributed inference method.

22.1.4 Distributed MLE ($p = 1$)

The weights of weighted consensus (Section 22.1.3) are free parameters that can be chosen to enable the optimum estimate. From MLE or BLUE examples in Section 6.7, the optimal average of uncorrelated Gaussian rvs is by weighting for the inverse of the variance. In cooperative estimation with

$$\hat{\theta}_i \sim \mathcal{N}(\theta, \sigma_i^2)$$

this is equivalent in choosing the weights

$$q_i = \frac{1}{\sigma_i^2}$$

since this converges to the MLE as for the centralized method without exchanging measurements, but only as a weighted consensus method.

A careful inspection of the consensus iteration for any node shows that

$$\theta_i(\ell+1) = \theta_i(\ell) + \varepsilon \sum_{j \in \mathcal{K}_i} \frac{1}{\sigma_j^2}(\theta_j(\ell) - \theta_i(\ell))$$

so every node in $\mathcal{K}_i$ should preliminarily signal to the ith node the degree of accuracy of its estimate in the form of its variance in order to compute the weights before the consensus iterations are carried out. This is a preliminary step to the consensus iterations (and thus an additional signaling step) with significant benefits. More specifically, the consensus estimate from weighted averaging is

$$\hat{\theta}_\infty = \frac{\sum_{i=1}^K \hat{\theta}_i/\sigma_i^2}{\sum_{i=1}^K 1/\sigma_i^2}$$

which coincides with centralized MLE when all nodes forward all their measurements to a centralized estimator. In addition,

$$\text{var}(\hat{\theta}_\infty) = \left(\sum_{j=1}^{K} \frac{1}{\sigma_j^2}\right)^{-1} = CRB(\hat{\theta})$$

attains the CRB and this sets the optimality of the distributed method. This principle can be generalized to any number of parameters for a linear model as detailed below.

22.2 Distributed Estimation for Arbitrary Linear Models (p>1)

Let the real-valued measurement at the ith node be modeled as a linear combination of the set $\boldsymbol{\theta} = [\theta_1, ..., \theta_p]^T$ of p real-valued parameters common to all observations

$$\mathbf{x}_i = \mathbf{H}_i \boldsymbol{\theta} + \mathbf{w}_i$$

for the N measurements $\mathbf{x}_i = [x_i(1), x_i(2), ..., x_i(N)]^T$, and the $N \times p$ regressor

$$\mathbf{H}_i = \begin{bmatrix} \mathbf{h}_i^T(1) \\ \vdots \\ \mathbf{h}_i^T(N) \end{bmatrix}$$

that aggregates N regressors $\mathbf{h}_i(1), ..., \mathbf{h}_i(N)$ (e.g., these N instances could be obtained at N different times). Noise $\mathbf{w}_i \sim \mathcal{N}(0, \mathbf{C}_w)$ is arbitrarily correlated with the same covariance $\mathbf{C}_w$ independent of nodes (extension to arbitrary node-dependent covariance is straightforward, just with more complex notation). Based on this linear model, standard estimation theory proves that the *local MLE* on each node based on the local set of N measurements is (Section 7.2)

$$\hat{\theta}_i = (\mathbf{H}_i^T \mathbf{C}_w^{-1} \mathbf{H}_i)^{-1} \mathbf{H}_i^T \mathbf{C}_w^{-1} \mathbf{x}_i$$

In this linear setting, the covariance of the estimate $\text{cov}(\hat{\boldsymbol{\theta}}_i)$ reaches the CRB (Section 7.2)

$$\text{cov}(\hat{\boldsymbol{\theta}}_i) = \mathbf{C}_i = \left(\mathbf{H}_i^T \mathbf{C}_w^{-1} \mathbf{H}_i\right)^{-1} \tag{22.4}$$

The local MLE can be conveniently written for cooperative settings as

$$\hat{\boldsymbol{\theta}}_i = \mathbf{C}_i \times \mathbf{H}_i^T \mathbf{C}_w^{-1} \mathbf{x}_i \tag{22.5}$$

which highlights the relationships between the local estimate and the CRB for the same estimate.

To simplify the reasoning and without any loss of generality, the noise can be assumed to be uncorrelated with $\mathbf{C}_w = \mathbf{I}$, so that any of the properties derived herein can be made dependent solely on the property of the regressors $\{\mathbf{h}_i(1), \ldots, \mathbf{h}_i(N)\}$. This assumption will be adopted herein, where specified.

22.2.1 Centralized MLE

Assuming that a fusion center collects all the N measurements from every node to estimate the p parameters $\boldsymbol{\theta}$ common to all K nodes, the augmented model of all KN measurements is

$$\underbrace{\begin{bmatrix} \mathbf{x}_1 \\ \mathbf{x}_2 \\ \vdots \\ \mathbf{x}_K \end{bmatrix}}_{\mathbf{X}} = \underbrace{\begin{bmatrix} \mathbf{H}_1 \\ \mathbf{H}_2 \\ \vdots \\ \mathbf{H}_K \end{bmatrix}}_{H} \boldsymbol{\theta} + \underbrace{\begin{bmatrix} \mathbf{w}_1 \\ \mathbf{w}_2 \\ \vdots \\ \mathbf{w}_K \end{bmatrix}}_{\mathbf{W}} \tag{22.6}$$

where $\mathbf{W} \sim \mathcal{N}(0, \mathbf{I}_K \otimes \mathbf{C}_w)$ from the node-invariance of covariance. The MLE at the fusion center (also referred to as *global MLE*) follows from the augmented linear model

$$\hat{\theta}_C = \left(\sum_{i=1}^{K} \mathbf{H}_i^T \mathbf{C}_w^{-1} \mathbf{H}_i\right)^{-1} \sum_{i=1}^{K} \mathbf{H}_i^T \mathbf{C}_w^{-1} \mathbf{x}_i$$

After some manipulations, the global MLE can be rewritten in terms of the weighted sum of local MLEs:

$$\hat{\theta}_C = \left(\sum_{i=1}^{K} \mathbf{C}_i^{-1}\right)^{-1} \sum_{i=1}^{K} \mathbf{C}_i^{-1} \times \hat{\boldsymbol{\theta}}_i \tag{22.7}$$

This structure proves that the weighted average of the local estimates $\hat{\boldsymbol{\theta}}_i$ yields the same performance of the global MLE provided that the fusion center is made available of both the local estimates $\{\hat{\boldsymbol{\theta}}_i\}$ and their covariances $\{\mathbf{C}_i\}$, the latter acting as weightings. For linearity, the covariance of MLE coincides with the CRB:

$$\text{cov}(\boldsymbol{\theta}_C) = \left(\sum_{i=1}^{K} \mathbf{C}_i^{-1}\right)^{-1} \tag{22.8}$$

The covariance (22.8) sets the lower bound of any consensus-based estimator for the problem at hand, as clearly a centralized estimator that has access to all information from all nodes can perform the global MLE.

Remark: In the case of ideal regressors $\mathbf{H}_i^T\mathbf{H}_i = N\mathbf{I}$ and uncorrelated noise $\mathbf{C}_w = \sigma_w^2\mathbf{I}$, the covariance for local estimates $\mathbf{C}_i = \sigma_w^2\mathbf{I}/N$ scales with the number of samples as 1/N. The centralized covariance when $\mathbf{C}_i = \mathbf{C}$ scales with the number of nodes $\text{cov}(\boldsymbol{\theta}_C) = \mathbf{C}/K$ and for the ideal regressors and white noise is $\mathbf{C} = \sigma_w^2\mathbf{I}/N$, and thus the covariance $\text{cov}(\boldsymbol{\theta}_C) = \sigma_w^2\mathbf{I}/KN$ scales with respect to the overall number of measurements KN collected by all the K nodes.

22.2.2 Distributed Weighted LS

The aim of cooperative estimation is the iterative exchange and update of the local estimates by distinct agents to yield asymptotically to the same value $\boldsymbol{\theta}_\infty$ on every node of the network, such that this value is better (i.e., with lower covariance) than any local estimate carried out independently by each agent without any cooperation with the others. Again, every node first carries out a local MLE $\hat{\boldsymbol{\theta}}_i$ from N samples achieving a covariance $\mathbf{C}_i$ that coincides with the CRB from the linearity of the model. Similarly to the case of $p = 1$ parameter (Section 22.1.3), a preliminary stage where nodes exchange their accuracy is beneficial to attain optimal performance. However, for $p > 1$ each node exchanges the covariance $\mathbf{C}_i$ with the set $\mathcal{K}_i$ of neighboring nodes so that after this step the ith node has the knowledge of the set $\{\mathbf{C}_j\}_{j=1}^{\mathcal{K}_i}$ in light of the reciprocity of links $i \longrightarrow j \in \mathcal{K}_i$. These covariances are collected to weight the error updating in vector-like consensus iterations as detailed below.

Let the consensus at the ℓth iteration weight each error as follows:

$$\boldsymbol{\theta}_i(\ell+1) = \boldsymbol{\theta}_i(\ell) + \varepsilon \sum_{j\in\mathcal{K}_i} \mathbf{Q}_j(\boldsymbol{\theta}_j(\ell) - \boldsymbol{\theta}_i(\ell)) \tag{22.9}$$

The weight matrixes $\{\mathbf{Q}_j\}_{j=1}^{\mathcal{K}_i}$ are node-dependent and symmetric ($\mathbf{Q}_j = \mathbf{Q}_j^T$), and the consensus is initialized from the local estimates $\boldsymbol{\theta}_i(0) = \hat{\boldsymbol{\theta}}_i$ so that, at convergence:

$$\boldsymbol{\theta}_1(\ell \to \infty) = \boldsymbol{\theta}_2(\ell \to \infty) = \ldots = \boldsymbol{\theta}_K(\ell \to \infty) = \boldsymbol{\theta}_\infty$$

For the convergence analysis to derive $\boldsymbol{\theta}_\infty$, the consensus iterations needs to be rewritten after vectorizing the parameters to isolate the Perron matrix (Appendix A):

$$\boldsymbol{\phi}(\ell+1) = (\mathbf{I} - \varepsilon\mathbf{L})\boldsymbol{\phi}(\ell) \tag{22.10}$$

where the ensemble of estimates are collected in a block-vector where parameters are ordered node-wise

$$\boldsymbol{\phi}(\ell) = \begin{bmatrix} \theta_1(\ell) \\ \theta_2(\ell) \\ \vdots \\ \theta_K(\ell) \end{bmatrix} \in \mathbb{R}^{Kp\times 1}$$

and the Laplace matrix $\mathbf{L}$ accounts for the connectivity pattern of the network. For the specific problem at hand, $\mathbf{L}$ is block-partitioned into $K \times K$ blocks; the single $p \times p$ block is:

$$\mathbf{L}_{i,j} = \begin{cases} -\mathbf{Q}_j & j \in \mathcal{K}_i \\ \sum_{k\in\mathcal{K}_i} \mathbf{Q}_k & i = j \\ 0 & j \notin \mathcal{K}_i \end{cases} \tag{22.11}$$

which resembles the block-partitioned Laplace (22.3).

Similarly to the case of $p = 1$, the Laplace matrix for the special case of a fully connected graph is

$$\mathbf{L} = \mathbf{I}_N \otimes (\mathbf{Q}^T\mathbf{U}) - \mathbf{U}\mathbf{Q}^T \tag{22.12}$$

where

$$\mathbf{U} = (\mathbf{1}_K \otimes \mathbf{I}_p) \in \mathbb{R}^{Kp\times p} \tag{22.13}$$

$$\mathbf{Q} = \begin{pmatrix} \mathbf{Q}_1 \\ \vdots \\ \mathbf{Q}_K \end{pmatrix} \in \mathbb{R}^{Kp\times p} \tag{22.14}$$

Convergence is guaranteed for $0 < \varepsilon < 1/\lambda_{\max}(\mathbf{L})$. The convergence property follows from inspection of the right and left eigenvectors of $\mathbf{L}$ as $\mathbf{L}\mathbf{U} = \mathbf{0}$ and $\mathbf{V}^T\mathbf{L} = \mathbf{0}$. More specifically, the consensus value depends on the initial estimate as $\boldsymbol{\phi}(\ell \to \infty) = \mathbf{U}(\mathbf{V}^T\mathbf{U})^{-1}\mathbf{V}^T\boldsymbol{\phi}(0)$ (see [111]). Interestingly for the estimation problem, the weights $\mathbf{Q}$

$$\mathbf{Q}^T\mathbf{L} = \left[\ldots, \mathbf{Q}_j \cdot \sum_{k\in\mathcal{K}_i} \mathbf{Q}_k - \sum_{k\in\mathcal{K}_i} \mathbf{Q}_k \cdot \mathbf{Q}_j, \ldots \right] \tag{22.15}$$

coincide with the left eigenvectors

$$\mathbf{Q}^T\mathbf{L} = \mathbf{0} \tag{22.16}$$

when the matrixes commute[1]

$$\mathbf{Q}_j \cdot \sum_{k\in\mathcal{K}_i} \mathbf{Q}_k = \sum_{k\in\mathcal{K}_i} \mathbf{Q}_k \cdot \mathbf{Q}_j$$

and this happen in these two special (but common in applications) cases:

$$\text{case 1: } \mathbf{Q}_i = \text{diag}(q_{i,1}, q_{i,2}, \ldots, q_{i,p}) \tag{22.17}$$

$$\text{case 2: } \mathbf{Q}_i = q_i \mathbf{Q}_o \tag{22.18}$$

namely weights are diagonal (22.17), or for the same weight matrix (22.18) apart from an arbitrary scaling factor. In these two scenarios, the consensus is a weighted average of the local estimates

$$\boldsymbol{\theta}_k(\ell \to \infty) = \boldsymbol{\theta}_\infty = \boldsymbol{\Gamma} \sum_{i=1}^{K} \mathbf{Q}_i \hat{\boldsymbol{\theta}}_i \tag{22.19}$$

with scaling term $\boldsymbol{\Gamma} = \left(\sum_{i=1}^{K} \mathbf{Q}_i\right)^{-1}$ and covariance

$$\text{cov}(\boldsymbol{\theta}_\infty) = \boldsymbol{\Gamma}(\sum_{i=1}^{K} \mathbf{Q}_i \mathbf{C}_i \mathbf{Q}_i)\boldsymbol{\Gamma} \tag{22.20}$$

The weighting matrix is a free parameter detailed below.

22.2.3 Distributed MLE

Considering now the choice for the weights

$$\mathbf{Q}_i = \mathbf{C}_i^{-1}$$

that resembles the accuracy-dependent weight for $p = 1$, the weighted estimate guarantees the convergence to the unbiased estimate that attains the global MLE (22.7) for the cases (22.17–22.18)

$$\begin{aligned} \boldsymbol{\theta}_k(\ell \to \infty) &= \boldsymbol{\theta}_\infty = \boldsymbol{\theta}_C \\ \text{cov}(\boldsymbol{\theta}_\infty) &= \text{cov}(\boldsymbol{\theta}_C) \end{aligned}$$

This can be easily proved by substitutions. Needless to say, in case that the matrixes do not commute, the weighted LS does not attain the CRB but the convergence is fast and the loss is negligible for practical purposes [112].

1 Two Hermitian matrixes $\mathbf{A}$ and $\mathbf{B}$ commute if they share the same eigenspace. Since from the factorization $\mathbf{A} = \mathbf{U}\boldsymbol{\Lambda}_A\mathbf{U}^H$ and $\mathbf{B} = \mathbf{U}\boldsymbol{\Lambda}_B\mathbf{U}^H$:

$$\mathbf{AB} = \mathbf{U}\boldsymbol{\Lambda}_A\mathbf{U}^H\mathbf{U}\boldsymbol{\Lambda}_B\mathbf{U}^H = \mathbf{U}\boldsymbol{\Lambda}_B\mathbf{U}^H\mathbf{U}\boldsymbol{\Lambda}_A\mathbf{U}^H = \mathbf{BA}$$

and simple examples are for diagonal matrixes, or identical matrixes apart from a scale factor.

Simple (or unweighted) consensus is the update (22.9) without any weighting, or $\mathbf{Q}_i = \mathbf{I}$; the estimate on each node is the sample mean of the local estimates

$$\theta_k(\ell \to \infty) = \theta_{sc} = \frac{\sum_{i=1}^{K} \hat{\theta}_i}{K}$$

and the covariance is

$$\text{cov}(\theta_{sc}) = \frac{\sum_{i=1}^{K} \mathbf{C}_i}{K^2} \geq \left(\sum_{i=1}^{K} \mathbf{C}_i^{-1} \right)^{-1} = \text{cov}(\theta_C)$$

where $\text{cov}(\theta_C)$ is the CRB for the centralized MLE that is attained when choosing $\mathbf{Q}_i = \mathbf{C}_i^{-1}$ as shown above. When all the nodes have the same covariance $\mathbf{C}_i = \mathbf{C}$, $\text{cov}(\theta_{sc}) = \text{cov}(\theta_C) = \mathbf{C}/K$ and simple consensus is sufficient. The price to be paid in exchange of a remarkable improvement between the simple (choosing $\mathbf{Q}_i = \mathbf{I}$) and optimally weighted (choosing $\mathbf{Q}_i = \mathbf{C}_i^{-1}$) consensus algorithm in distributed estimation is the exchange between a pair of nodes of their covariances $\mathbf{C}_i$ that reflect the individual confidence of local estimate.

Remark: In practical systems, the nodes might exchange values that are affected by stochastic perturbation (e.g., additive noise in communication among nodes), which can be modeled link-dependent additive Gaussian noise $\mathbf{v}_{j|i}(\ell) \sim \mathcal{N}(0, \mathbf{C}_{j|i})$. In each consensus step (22.9), the update variable at the ith node is $\theta_{j|i}(\ell) = \theta_j(\ell) + \mathbf{v}_{j|i}(\ell)$, and the consensus steps become node and noise dependent:

$$\theta_i(\ell + 1) = \theta_i(\ell) + \varepsilon \sum_{j \in \mathcal{K}_i} \mathbf{Q}_{j|i}(\theta_{j|i}(\ell) - \theta_i(\ell))$$

where the new weight should account for the noise $\mathbf{Q}_{j|i} = \left(\mathbf{C}_i(\ell) + \mathbf{C}_{j|i}\right)^{-1}$.

22.2.4 Distributed Estimation for Under-Determined Systems

Let us consider the case where the ensemble of K sensors acquire enough measurements to estimate the set of p-parameters θ as an overdetermined set ($p < KN$), but the set of N measurements acquired by each sensor makes the estimation problem on each node under-determined:

$$N < p < KN$$

This case follows the same steps as above [106, 113]: each node solves locally an under-determined least-norm, and the set of these estimates are exchanged with the neighbors jointly with the subspace corresponding to the eigenvectors of the regressors. The weighted consensus iterations tailored for these settings refine these estimates up to the convergence and, for a network of connected nodes, the method attains the Cramér–Rao bounds as for a centralized estimate within a small set of iterations.

In an under-determined linear problem, the estimation of the parameter at every node needs to solve for the corresponding min-norm optimization

$$\hat{\theta}_i = \arg\min_{\theta} ||\theta||^2 \text{ s.t. } \mathbf{x}_i = \mathbf{H}_i\theta \tag{22.21}$$

that yields the solution based on the right pseudoinverse of $\mathbf{H}_i$

$$\hat{\theta}_i = \mathbf{H}_i^T(\mathbf{H}_i\mathbf{H}_i^T)^{-1}\mathbf{x}_i \tag{22.22}$$

The $N \times N$ matrix $\mathbf{H}_i\mathbf{H}_i^T$ is full-rank except for some special cases.

At consensus setup, each node broadcasts toward the set of neighboring nodes $\mathcal{K}_i$ the $p \times p$ matrix $\mathbf{H}_i^T\mathbf{H}_i$ so that the ith node has the knowledge of the set $\{\mathbf{H}_j^T\mathbf{H}_j\}_{j\in\mathcal{K}_i}$ in this case. The consensus iterations update the local estimate from the ensemble of estimates from neighbors at iteration ℓ as:

$$\theta_i(\ell+1) = (1-\varepsilon)\theta_i(\ell) + \varepsilon\boldsymbol{\Gamma}_i \sum_{j\in\mathcal{K}_i} \mathbf{Q}_{ij}\theta_j(\ell) \tag{22.23}$$

where $\mathbf{Q}_{ij}= \mathbf{H}_j^T\mathbf{H}_j$ denotes the weighting factors in consensus and the scaling factor is

$$\boldsymbol{\Gamma}_i = \left(\sum_{k\in\mathcal{K}_i} \mathbf{H}_k^T\mathbf{H}_k\right)^{-1} \tag{22.24}$$

The updating term ε is chosen to guarantee the convergence to the consensus $\theta_i(\ell) \to \theta_\infty$ (Appendix A).

Choice of weighting $\mathbf{Q}_{ij}= \mathbf{H}_j^T\mathbf{H}_j$ can be justified by simple reasoning. Let $\mathbf{H}_j = \mathbf{U}_j\boldsymbol{\Sigma}_j\mathbf{V}_{j1}^T$ be the singular value decomposition for $N < p$ (Section 2.4); here the left ($\mathbf{U}_j$) and right ($\mathbf{V}_{j1}$) eigenvectors are from the eigenvector decomposition of $\mathbf{H}_j\mathbf{H}_j^T = \mathbf{U}_j\boldsymbol{\Sigma}_j^2\mathbf{U}_j^T$ and $\mathbf{H}_j^T\mathbf{H}_j = \mathbf{V}_{j1}\boldsymbol{\Sigma}_j^2\mathbf{V}_{j1}^T$, and $\boldsymbol{\Sigma}_j = \text{diag}(\sigma_{j,1}, \sigma_{j,2}, ..., \sigma_{j,N})$ collects the N singular values. The min-norm estimate for the local problem (22.22) is revisited as

$$\hat{\theta}_j = \mathbf{V}_{j1}\boldsymbol{\Sigma}_j^{-1}\mathbf{U}_j^T\mathbf{x}_i = \mathbf{V}_{j1}\mathbf{V}_{j1}^T\theta + \mathbf{n}_j \tag{22.25}$$

The estimate is unbiased provided that bias is evaluated within the row-space of $\mathbf{H}_j$ as $\mathbf{V}_{j1}\mathbf{V}_{j1}^T$ is the corresponding projection matrix, and the stochastic term $\mathbf{n}_j = \mathbf{V}_{j1}\boldsymbol{\Sigma}_j^{-1}\mathbf{U}_j^T\mathbf{w}_j$ for $\mathbf{w}_j \sim \mathcal{N}(0, \mathbf{I})$ is $\text{cov}(\mathbf{n}_j) = \mathbf{V}_{j1}\boldsymbol{\Sigma}_j^{-2}\mathbf{V}_{j1}^T$.

Since weighting $\mathbf{Q}_{ij} = \mathbf{H}_j^T\mathbf{H}_j = \mathbf{V}_{j1}\boldsymbol{\Sigma}_j^2\mathbf{V}_{j1}^T$ spans the same subspace of $\hat{\theta}_j$, it does not modify the information gathered in the consensus as from (22.25) $\mathbf{H}_j^T\mathbf{H}_j\hat{\theta}_j = \mathbf{V}_{j1}\boldsymbol{\Sigma}_j^2\mathbf{V}_{j1}^T\theta + \tilde{\mathbf{n}}_j$ with $\text{cov}(\tilde{\mathbf{n}}_j) = \mathbf{V}_{j1}\boldsymbol{\Sigma}_j^2\mathbf{V}_{j1}^T$, but rather it projects each contribution $\theta_j(\ell)$ onto the row-space of $\mathbf{H}_j$ when updating $\theta_i(\ell)$ in (22.23), even if $\theta_j(\ell)$ is spanning complementary subspaces during iterations. The drawback of narrowing the updating to the corresponding row-spaces is a larger MSE for the degree of the graph $|\mathcal{K}_i| \simeq p/T$, or equivalently a larger connectivity (or an augmented data exchange) is necessary to have MSE reaching the CRB.

Note that distributed methods for an under-determined model need a minimum connectivity to guarantee the convergence with some excess wrt the CRB (22.8) that sets the maximum degree of the graph $d_{max} \geq p/N$, and in some situations the CRB can be attained even for under-determined systems in ring-topology [113]. In any case, the under-determined context opens up many applications, but the convergence is quite delicate.

22.2.5 Stochastic Regressor Model

Even if the regressor $\mathbf{h}_i(t)$ is deterministic (and known) at every node, it could be convenient to detail the problem when it is (or can be considered as) stochastic

$$\mathbf{h}_i(t) \sim \mathcal{N}(0, \alpha_i \mathbf{R}_i) \tag{22.26}$$

where $\mathbf{R}_i$ accounts for the correlation properties, and the scaling term α_i sets the node-dependent signal to noise ratio. In addition, the regressor is statistically independent over time $\mathbb{E}[\mathbf{h}_i(t)\mathbf{h}_i^T(\tau)] = \alpha_i \mathbf{R}_i \delta(t-\tau)$. In this situation, the covariance of the local estimate (22.4) can be approximated by its asymptotic bound from the Jensen inequality (Section 3.2) that for $\mathbf{C}_w = \mathbf{I}$ becomes

$$\mathbf{C}_i \simeq (N\alpha_i \mathbf{R}_i)^{-1} \tag{22.27}$$

Recall that this approximation follows from the asymptotic properties (for $N \to \infty$) of sample correlation $\mathbf{H}_i^T \mathbf{H}_i / N \to \alpha_i \mathbf{R}_i$, and it sets the asymptotic CRB that is typical for these configurations as in (Section 10.1).

The global covariance can be similarly approximated (valid for $N \to \infty$) as $\text{cov}(\boldsymbol{\theta}_C) \simeq \mathbf{C}_\infty$ where

$$\mathbf{C}_\infty = \frac{1}{NK}\left(\sum_{i=1}^{K} \alpha_i \mathbf{R}_i\right)^{-1} \tag{22.28}$$

For the special case of $\mathbf{R}_i = \mathbf{R}$, independent of the node, the weighted LS attains the centralized MLE and the covariance simplifies to $\mathbf{C}_\infty = (KN\bar{\alpha})^{-1}\mathbf{R}^{-1}$ thus scaling with respect to the total number of measurements N and nodes K, where $\bar{\alpha} = \sum_{i=1}^{K} \alpha_i / K$ is the average gain factor. One simple and quite illustrative model is the AR(1) model with correlation ρ; the matrix is Toeplitz structured with entry $[\mathbf{R}]_{ij} = \rho^{|i-j|}$. The ideal regressor with $\rho = 0$ and $\mathbf{H}_i^T \mathbf{H}_i = N\mathbf{I}$ provides the best estimate with $\mathbf{C}_i = \mathbf{I}/(\mathrm{N}\alpha_i)$ and global covariance $\text{cov}(\boldsymbol{\theta}_C) = (NK\bar{\alpha})^{-1}\mathbf{I}$.

22.2.6 Cooperative Estimation in the Internet of Things (IoT)

Devices measuring environmental parameters (e.g., temperature, wind, pollution, rainfall, etc.) are more effective if sensing capability is paired with local processing and mutual connectivity. The wireless sensor paradigm is based on the mutual interconnection of sensing nodes that might be not very accurate if taken individually, but the ensemble of many inaccurate nodes compensates for the limited capability of each. Nodes are connected in a network and communicate with each other using

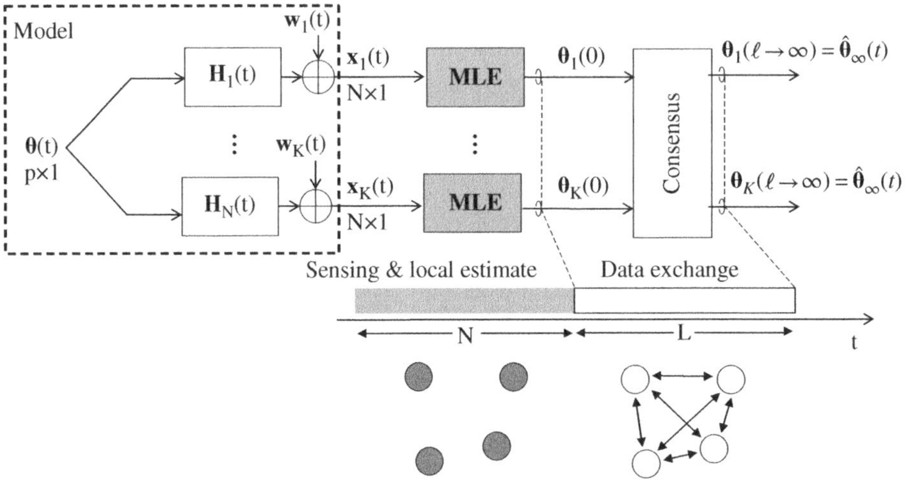

Figure 22.5 Cooperative estimation in distributed wireless system.

Internet connectivity, and this is known as the Internet of Things (IoT), as "things" embraces all possible devices and applications. IoT devices are very popular due to the availability of simple software controlled hardware that is flexible to enable complex functions. Applications of IoT with multiple sensing capability range from measurement of environmental parameters to some advanced industrial situations as for the oil industry and factories.

Regardless of the context, the functionality of each node can be classified as alternating the sensing stage when measuring the parameter(s) of interest, and the interaction with other nodes. Since nodes are battery-powered with a limited lifetime, communication is energy consuming if transmission ranges are too high, and periodic battery-maintenance can be minimized if the radio sections of these nodes are turned off for most of the operating time, and use low power (i.e., short-range communications). Cooperative processing is a natural application in IoT, and nodes cooperate with neighbors only in order to minimize battery usage.

Figure 22.5 illustrates the scenario where each node deployed in space senses a linear combination of the parameters collecting a block of N samples that can be referred to time t:

$$\mathbf{x}_i(t) = \mathbf{H}_i(t)\boldsymbol{\theta}(t) + \mathbf{w}_i(t)$$

Each sensor elaborates $\mathbf{x}_i(t)$ after collecting all samples (or while collecting, using adaptive estimation methods, Chapter 15) and provides a very noisy MLE $\hat{\boldsymbol{\theta}}_i(t)$ based only on local measurements. The nodes exchange their estimates with the purpose of improving them, and this corresponds to the activation of a distributed estimation step initialized by the estimate $\boldsymbol{\theta}_i(0|t) = \hat{\boldsymbol{\theta}}_i(t)$; each node performs consensus iterations:

$$\boldsymbol{\theta}_i(\ell+1|t) = \boldsymbol{\theta}_i(\ell|t) + \varepsilon \sum_{j\in\mathcal{K}_i} \mathbf{Q}_j(\boldsymbol{\theta}_j(\ell|t) - \boldsymbol{\theta}_i(\ell|t))$$

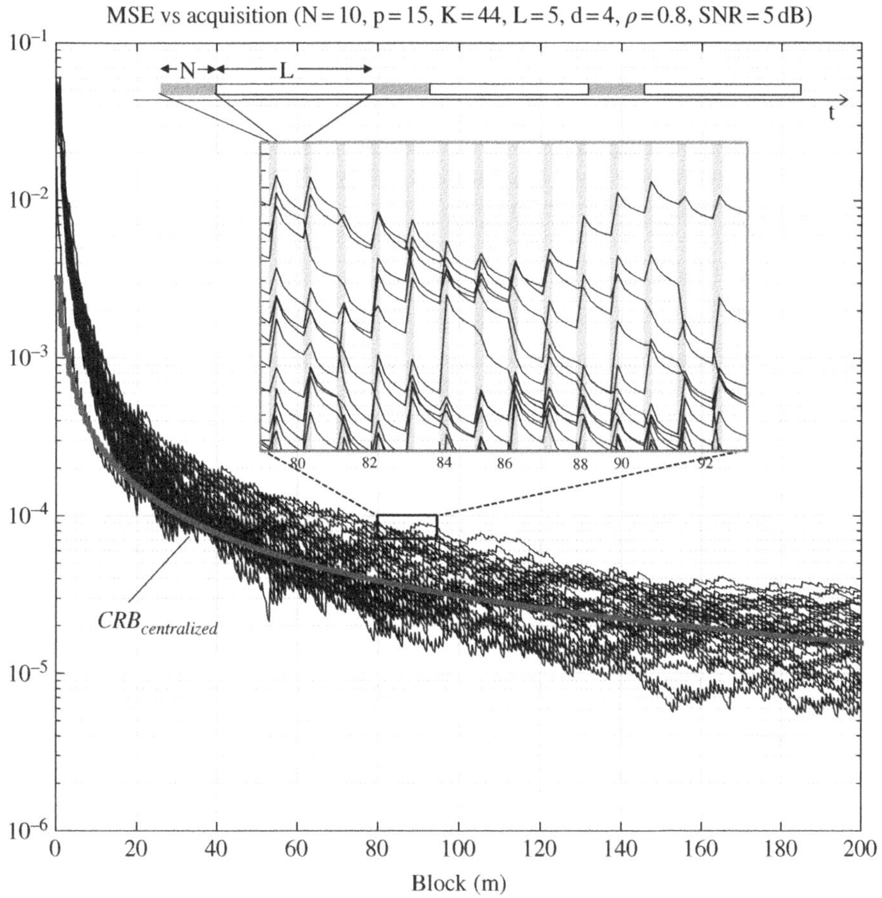

Figure 22.6 MSE vs. blocks in cooperative estimation (100 Montecarlo runs) and CRB (thick gray line). The insert details the behavior between sensing new measurements (i.e., collecting $N = 10$ samples at time) and $L = 5$ exchanging the local estimates.

using an appropriate weighting $\mathbf{Q}_j$, possibly optimal (Sections 22.2.3, 22.2.4). After L iterations, possibly large enough to have convergence of the network to $\boldsymbol{\theta}_i(L|t) = \boldsymbol{\theta}_\infty(t)$, this convergence value is used as the estimate of the ensemble of nodes, after mutual cooperation. Nodes alternate sensing and communication—these two stages can be temporally coincident or interleaved, and this depends on the application. In the case of time-varying phenomena, the measurement and consensus intervals N and L can be optimized to capture the time-varying nature of the specific phenomena $\boldsymbol{\theta}(t)$ by trading accuracy of individual estimates with convergence speed (which in turn depends on the connectivity of the nodes).

22.2.7 Example: Iterative Distributed Estimation

A comprehensive example based on wireless sensing IoT devices can be used to clarify. Let $\boldsymbol{\theta}$ be a set of p parameters to be estimated by each node. The data collection and estimation is arranged into blocks of N data measurements and L consensus iterations in each block so that sensing and refinement are iterated over the time-blocks. The convenience of this iterative block-estimation is that the estimate on each node can be overdetermined only after p/N time-blocks, and in this way each node can have reliable estimation outcomes earlier than waiting for a long acquisition (say $p \gg N$), by refining the estimates with neighboring nodes even if not highly accurate, as only the ensemble has enough measurements ($p < KN$) but not individually. The data at the mth temporal block is a linear combination of the parameters

$$x_i(n|m) = \mathbf{h}_i^T(n|m)\boldsymbol{\theta} + w_i(n|m)$$

where the double index n, m stands for sample ($n = 1, 2..., N$) and block ($m = 1, 2, ...$) indexes at the ith node, and the regressor is stochastic (but known) $\mathbf{h}_i(n|m) \sim \mathcal{N}(0, \mathbf{R})$ with AR(1) correlation $[\mathbf{R}]_{ij} = \rho^{|i-j|}$ (Section 22.2.5). The goal is to evaluate the convergence versus block m for a fixed size N and iterations L compared to the covariance of the centralized MLE (22.28) that sets the bound (one block length is N acquisitions + L iterations).

Figure 22.6 shows 100 Montecarlo runs of the MSE vs. processing block for $K = 44$ nodes arranged in a ring-topology with degree $d = 4$ for each node, and stochastic regressor with $\rho = 0.8$. Comparison with the CRB for a centralized MLE that in this case is decreasing vs. the block index m (thick gray line) highlights that for a small number of processing blocks (say below 20), the number of consensus iterations $L = 5$ is largely not enough to reach the convergence to the centralized estimate with $K = 44$ nodes and degree of the graph only $d = 4$. Therefore, for small processing blocks, the MSE in Figure 22.6 is close to the local MLE (22.4) and improves due to the exchange with other nodes. After 20–30 processing blocks, the MSE attains the CRB for all the nodes with some dispersions, CRB decreases vs. blocks as the system is collecting N data samples per block. A detailed view of the MSE for each estimate has a sawtooth behavior due to the alternating phases of sensing and data exchanging that further refines the estimates vs. blocks.

22.3 Distributed Synchronization

Synchronization refers to the process of achieving and maintaining time-coordination among independent local clocks via the exchange of individual time information. Synchronization schemes differ in the way such information is encoded, exchanged, and processed by each individual entity to align all the clocks. Wireless communications provide the natural platform for the exchange of local time information between synchronizing clocks. Distributed synchronization refers to the method for cooperative estimation of the common timescale as a consensus problem based on the exchange of local time information among neighboring nodes via transmission of a set of common waveforms that follows the local timings.

Synchronization of a distant clock to a reference time is a standard engineering procedure. The idea of synchronized time spurred an intense debate in physics and philosophy that eventually produced Einstein's theory of relativity [114].[2] In the years following

---

2 In this regard, it is interesting to quote H. Poincaré: "Simultaneity is a convention, nothing more than the coordination of clocks by a cross exchange of electromagnetic signals taking into account the transit time of the signal."

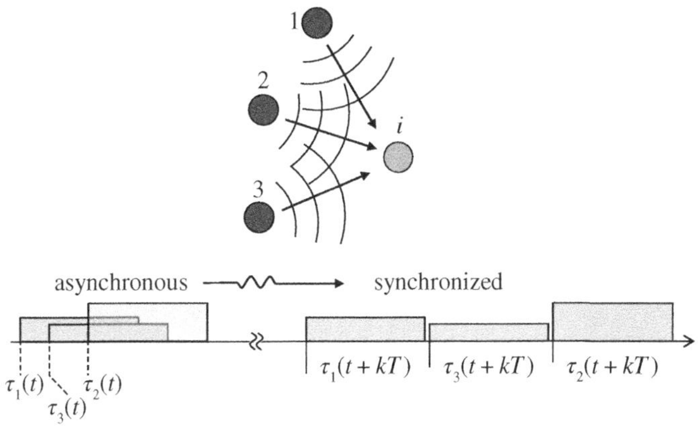

Figure 22.7 Synchronization in communication engineering.

these efforts, scientists wondered at the evidence of synchronization among *multiple* distributed periodic events in a number of natural phenomena such as the synchronous flashing of fireflies, or the activity of individual fibers in heart muscles to produce the heartbeat, or even spontaneous synchronous hand clapping in a concert hall (see e.g., [115]). Analytical modeling of the dynamic establishment of global synchrony involving multiple individual entities dates back to the landmark work by Winfree in 1967 [116] and later by Kuramoto [117].

In communication systems, synchronization is the processing step that enables receivers to align to the transmitter(s) in frequency and time to correctly sample the transmitted symbols for estimation of the transmitted information (see Section 15.2). A signal in digital communications is a cyclostationary stochastic process and this property is commonly used for synchronization by tailoring the estimators accordingly (for this specialized area the reader is referred to the comprehensive books on this topic [93, 94]). Distributed wireless networks are the communication engineering counterpart of spontaneous synchronization of independent and correlated agents, and the IoT can be an exemplary case study of distributed estimation to reach a common synchronization. Nodes deployed over a limited geographical area as in Figure 22.7 should mutually-coordinate to synchronize their communications and avoid deleterious superpositions of signals (this is commonly referred as packet collision when the signals that encode the transmitting bits are arranged in blocks as in Figure 22.7). Coordination is achieved by a consensus among the nodes on the common timescale as a necessary condition for any inter-node communication, and this is an iterative process that starts from the state of asynchrony and moves the clock of individual nodes until all have the same common clock. Distributed synchronization methods in communication date back to the early work on synchronization of satellites in the '70s [34, 118], and were revised for the synchronization of nodes in wireless networks [36, 119, 120] using the consensus method. This section follows these references, and further reading is in the cited literature therein.

22.3.1 Synchrony-States for Analog and Discrete-Time Clocks

The model for uncoupled clocks defines the behavior of clocks when each node runs its own local clock without exchanging timing information with the others. Synchrony refers to the state of clocks when these are compared and possibly mutually coupled. From the illustration in Figure 22.8 with two signals $x_1(t)$ and $x_2(t)$, one can distinguish the cases of asynchronous clocks (a), frequency synchronous (b), and fully synchronous (c), all in terms of signals clicks or sinusoidal signals.

Analog (Sinusoidal) Clocks

An analog clock, say the ith, is commonly characterized by a sinusoidal signal generated by oscillators

$$x_i(t) = \cos \phi_i(t) \tag{22.29}$$

where $\phi_i(t)$ is the instantaneous phase that, in the case of uncoupled nodes, evolves linearly and independently as

$$\phi_i(t) = \phi_i(0) + \frac{2\pi}{T_i}t + \zeta(t) \tag{22.30}$$

The period $T_i = T_o + \Delta T_i$ deviates by ΔT_i from the nominal period T_o, or equivalently, the instantaneous frequency $\omega_i = 2\pi/T_i$ is affected by a frequency skew $\Delta\omega_i$ wrt the nominal frequency $\omega_o = 2\pi/T_o$. In general, frequency error $\Delta\omega_i$ is time-varying but this is not considered here as the observation timescale is smaller than its deviation, and it can be accounted for by the random process $\zeta(t)$ modeling the phase noise [118]. Moreover, the initial time $t = 0$ is arbitrarily chosen for all the oscillators as is the initial phase $\phi_i(0)$.

For analog clocks, the two synchronization conditions that can be obtained by coupling the oscillators are:

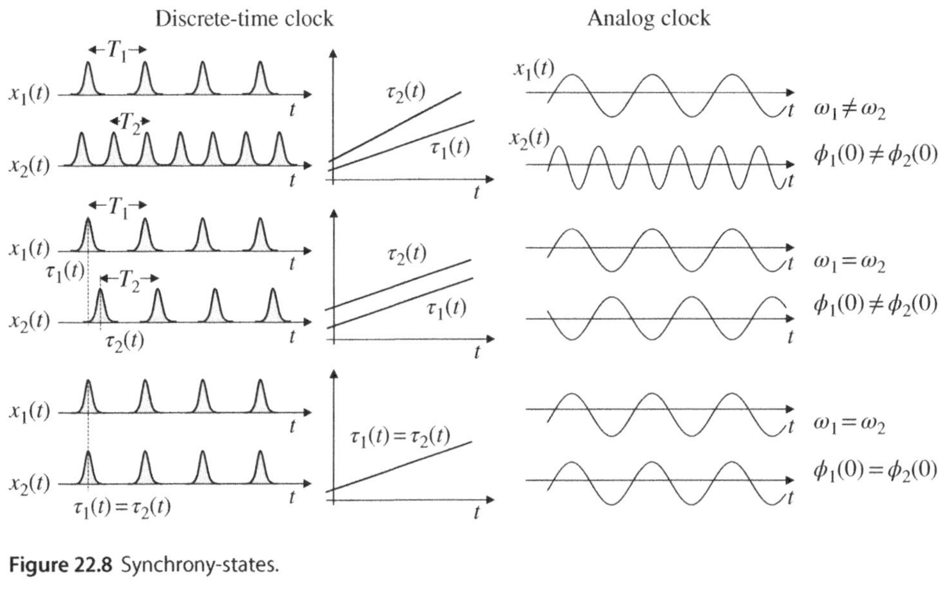

Figure 22.8 Synchrony-states.

- *Frequency synchronicity*: for t sufficiently large, there exists a common period of oscillation T for all the nodes so that

$$\phi_i(t+T)-\phi_i(t)=2\pi,\ \text{for}\ \forall i=1,2,\ldots,K \tag{22.31}$$

- *Full (frequency and phase) synchronicity:* for t sufficiently large, we have

$$\phi_i(t)=\phi_j(t)\ \text{for each}\ i\neq j \tag{22.32}$$

Notice that full synchronicity implies the existence of a common timescale at *all* times wrt an absolute time reference t.

Discrete-Time Clocks

A discrete-time clock can be seen as a sequence $\tau_i(\ell)$ of time instants of an analog clock, where index $\ell=0,1,2,\ldots$ runs over the periods of the oscillator where one can observe any phenomenon such as a waveform (possibly delta-like) that should be periodic with period T referred to an absolute time reference. In general these time instant $\tau(t)$ can be considered as time-dependent with respect to an absolute time t that ideally should be linearly increasing but it could be affected by some deviations (Figure 22.9). Locally at each node the time can be considered approximated as linear with

$$\tau_i(t)=(1+\alpha_i)t+\beta_i$$

Any node that is supposed to transmit waveforms uniformly spaced with an absolute time interval T transmits the waveforms with intervals

$$T_i(\ell)=(1+\alpha_i)T=\tau_i((\ell+1)T)-\tau_i(\ell T)$$

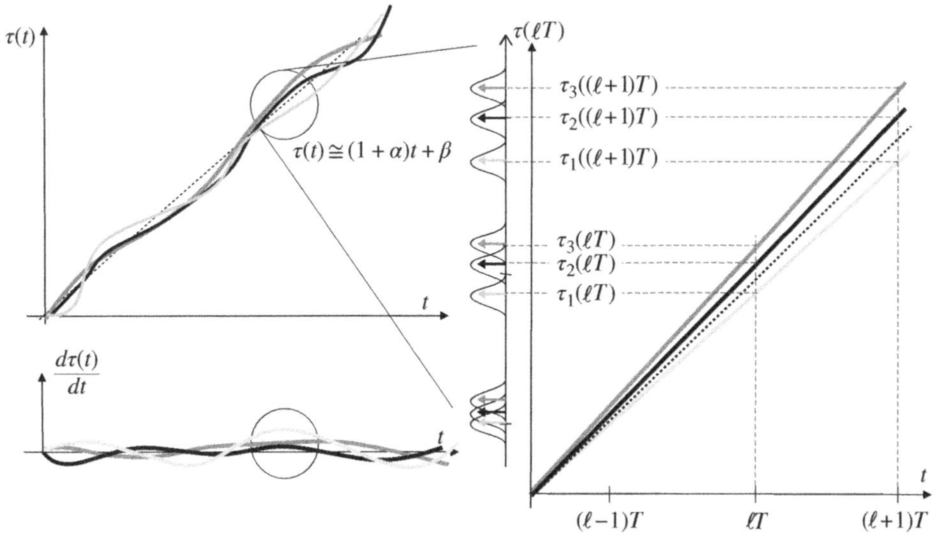

Figure 22.9 Temporal drift of time-references.

that are time-varying according to the time indexed by ℓ. When considering a set of nodes each with a local time-reference, the different time-references makes the waveforms diverge (see Figure 22.9) and/or fluctuate, and there is a need to align these time-references by exchanging these waveforms and correcting to make them mutually aligned. An uncoupled discrete-time clock evolves as

$$\tau_i(\ell) = \tau_i(0) + \ell T_i + \upsilon(\ell) \tag{22.33}$$

where $\upsilon(\ell)$ is the additive noise term that accounts for phase noise.

Nodes are said to be synchronous if they agree on the time instants $\tau_i(\ell)$ corresponding to the ticks of the local clocks, which entails that a common notion of time does not exist for the period elapsed between two ticks. More specifically, for discrete-time clocks, we have the following two conditions (Figure 22.8):

- *Frequency synchronicity*: for n sufficiently large, there exists a common period of oscillation T for all the nodes so that

  $$\tau_i(\ell+1) - \tau_i(\ell) = T \tag{22.34}$$

  where the period T can be arbitrary and not necessarily the absolute one.
- *Full (frequency and phase) synchronicity*: for n sufficiently large, we have

  $$\tau_i(\ell) = \tau_j(\ell) \text{ for } \forall i \neq j \tag{22.35}$$

22.3.2 Coupled Clocks

The goal of each receiver in distributed synchronization, say the ith, is to measure the phase or time differences between the local clock and the clocks of neighboring transmitters ($\phi_j(t) - \phi_i(t)$ or $\tau_j(\ell) - \tau_i(\ell)$, respectively), and to correct the local clock accordingly, despite the nuisance term due to propagation delays that is neglected here (see Section 22.3.3).

Analog Coupled Clocks

With coupled analog clocks, each node transmits a signal proportional to its local oscillator $s_i(t)$ in (22.29) and updates the instantaneous phase $\phi_i(t)$ based on the signal received from other nodes. Notice that this procedure assumes that each node is able to transmit and receive continuously and at the same time (full duplex, see Remark 2). The basic mechanism of continuously coupled clocks is phase-locking (see Section 7.7.2). Each node, say the ith, measures through its phase-error detector the convex combination of phase-differences

$$\Delta\phi_i(t) = \sum_{j\in\mathcal{K}_i} \gamma_{ij} \cdot f(\phi_j(t) - \phi_i(t)) \tag{22.36}$$

where $\phi_j(t) - \phi_i(t)$ is the phase difference with respect to node j, $f(\cdot)$ is a feature that depends on the phase-detector, namely a non-linear function, and γ_{ij} is a link-dependent weight normalized on each node (i.e., $\sum_{j\in\mathcal{K}_i} \gamma_{ij} = 1$ and $\gamma_{ij} \geq 0$) that recalls the weighted consensus method (Section 22.1.3). Notice that the choice of a convex combination in

(22.36) ensures that the output of the phase detector $\Delta\phi_i(t)$ takes values in the range between the minimum and the maximum of phase-differences $f(\phi_j(t)-\phi_i(t))$. Finally, the difference $\Delta\phi_i(t)$ (22.36) is fed to a loop filter ε, whose output updates the local phase as

$$\dot{\phi}_i(t)=\frac{2\pi}{T_i}+\varepsilon\sum_{j\in\mathcal{K}_i}\gamma_{ij}\cdot f(\phi_j(t)-\phi_i(t)) \tag{22.37}$$

A common model for the phase-update (22.37) is the sinusoidal phase-detector $f(x)=\sin(x)$ that corresponds to the basic Kuramoto model [117].

For the convergence analysis, the non-linear update (22.37) can be approximated by the linear one for linear phase detectors $f(x)=x$. This holds true when close to synchronization and $\sin(x)\simeq x$ for $|x|\ll 1$, and the ensemble of corrections can be cast in the linear time-invariant differential equation

$$\dot{\boldsymbol{\phi}}(t)=\boldsymbol{\omega}-\varepsilon_0\cdot\mathbf{L}\boldsymbol{\phi}(t) \tag{22.38}$$

for the vectors $\boldsymbol{\phi}(t)=[\phi_1(t)\cdots\phi_K(t)]^T$, $\boldsymbol{\omega}=[2\pi/T_1\cdots 2\pi/T_K]^T$ and $\mathbf{L}$ is the Laplacian matrix of the connectivity graph that includes the link-dependent weights γ_{ij}. The steady-state (or convergence) solution of (22.38) is characterized by frequency synchronization (22.31) with the common frequency given by the weighted combination

$$\frac{1}{T}=\sum_{i=1}^{K}v_i\frac{1}{T_i} \tag{22.39}$$

where the vector $\mathbf{v}=[v_1\cdots v_K]^T$ is the left eigenvector of $\mathbf{L}$ corresponding to the zero eigenvalue: $\mathbf{L}^T\mathbf{v}=\mathbf{0}$ (Appendix A). The common frequency $1/T$ (22.39) is a convex combination of all the local frequencies $\{1/T_i\}_{i=1}^K$. While frequency synchronization is attained, full frequency and phase synchronization (22.32) is generally not achieved, and the steady-state phases are mismatched by an amount related to the deviations between local and common frequency $\boldsymbol{\Delta\omega}=\boldsymbol{\omega}-2\pi/T\cdot\mathbf{1}$ [36]:

$$\boldsymbol{\phi}(t)\rightarrow\mathbf{1}\cdot\frac{2\pi}{T}t+\mathbf{1}\cdot\mathbf{v}^T\left(\boldsymbol{\phi}(0)-\frac{\mathbf{L}^{\dagger}\boldsymbol{\Delta\omega}}{\varepsilon_0}\right)+\frac{\mathbf{L}^{\dagger}\boldsymbol{\Delta\omega}}{\varepsilon_0} \tag{22.40}$$

Note that the second term in the right hand side of (22.40) is the phase common to all clocks and the third represents the phase mismatch. As a special case of these results, if no deviation among local frequency exists ($T_i=T_{nom}$), then from (22.39) the common frequency is $1/T=1/T_{nom}$, and, from (22.40), full frequency and phase synchronization is achieved with (recall that $\boldsymbol{\Delta\omega}=\mathbf{0}$)

$$\phi_i(t)\rightarrow\frac{2\pi}{T}t+\sum_{j=1}^{K}v_j\phi_j(0) \tag{22.41}$$

The final phase is a combination of the initial phases $\mathbf{v}^T\boldsymbol{\phi}(0)$.

Time Coupled Clocks

Similarly to the analog coupled clocks, each receiving node calculates a convex combination of the time differences (time difference detector)

$$\Delta\tau_i(n) = \sum_{j\in\mathcal{K}_i} \gamma_{ij} \cdot (\tau_j(\ell) - \tau_i(\ell)) \tag{22.42}$$

that is fed to a loop filter $\varepsilon(z)$. Considering for simplicity loop filters $\varepsilon(z) = \varepsilon_0$ (first-order PLLs), we have

$$\tau_i(\ell+1) = \tau_i(\ell) + T_i + \varepsilon_0 \cdot \sum_{j\in\mathcal{K}_i} \gamma_{ij} \cdot (\tau_j(\ell) - \tau_i(\ell)) \tag{22.43}$$

To further analyze this system, the update (22.43) is cast as a vector time-invariant difference equation:

$$\boldsymbol{\tau}(\ell+1) - \boldsymbol{\tau}(\ell) = \mathbf{T} - \varepsilon_0 \mathbf{L} \cdot \boldsymbol{\tau}(\ell) \tag{22.44}$$

for the vectors $\boldsymbol{\tau}(\ell) = [\tau_1(\ell) \cdots \tau_K(\ell)]$ and $\mathbf{T} = [T_1 \cdots T_K]^T$. From a comparison of (22.44) and (22.38), it is apparent that the same tools and results derived in the continuous case can be applied to the pulse-coupled discrete-time case. In particular, convergence is guaranteed under the same conditions (see Appendix A), and the steady-state solutions are characterized by frequency synchronization (22.34) with common frequency given by the weighted combination of local frequencies (22.39), but generally mismatched phases with

$$\boldsymbol{\tau}(\ell) \to \ell T \cdot \mathbf{1} + \mathbf{1} \cdot \mathbf{v}^T \left(\boldsymbol{\tau}(0) - \frac{\mathbf{L}^\dagger \boldsymbol{\Delta}\mathbf{T}}{\varepsilon_0} \right) + \frac{\mathbf{L}^\dagger \boldsymbol{\Delta}\mathbf{T}}{\varepsilon_0} \tag{22.45}$$

where $\boldsymbol{\Delta}\mathbf{T} = \mathbf{T} - T \cdot \mathbf{1}$ (similarly to (22.40)). As previously discussed, an important special case of these results occurs when there is no frequency mismatch between the clocks ($T_i = T_{nom}$), in this case the common frequency equals the nominal local frequency $1/T = 1/T_{nom}$, and, from (22.45), full frequency and phase synchronization is achieved (similarly to (22.41)):

$$\tau_i(\ell) \to \ell T + \sum_{j=1}^{K} v_j \tau_j(0) \tag{22.46}$$

The final phase is the weighted average of the initial time skew of each node.

22.3.3 Internet Synchronization and the Network Time Protocol (NTP)

Devices can communicate with each other by using Internet Protocol (IP), which exchanges information in packets of bits. These packets should be transmitted by multiple devices at the right time to avoid packets colliding and information being unavoidably lost. Furthermore, propagation adds a further delay to each transmission that needs to be estimated to enable the synchronization of the ensemble of devices.

Network synchronization when there is a huge number of servers connected as for the Internet (approx. 10^9 Web servers in 2014) becomes mandatory to provide the network functionality. The concept of distributing the clock among geographically distributed servers to guarantee global time alignment was introduced by the pioneering work of D. Mills [121] who proposed the Network Time Protocol (NTP)—still in place, with minor adaptations. The synchronization algorithm is basically an exchange of time-stamps among nodes with local time information and corresponding corrections. To exemplify, each person walking in the street gets the relative time from the watches of the people they meet and align their personal watch to the majority, but none knows the true time and they mutually align to an arbitrary relative time.

In packet communication, each node transmits a beacon that contains a time-stamp with the local perception of the absolute time $\bar{t}$ that represents the time used by the nodes to align their clocks (e.g., one specific interval of the frame-period). Let

$$\tau_j(\bar{t}) = (1+\alpha_j)\bar{t} + \beta_j$$

be the time a packet with a time-stamp that is transmitted to a receiving node i at absolute time $\bar{t}$ (unknown to the node). The time-stamp is received after a delay d_{ij} measured wrt the absolute time reference ($\bar{t}$) due to effective radiopropagation (notice that $d_{ij} = d_{ji}$). The absolute time for the jth node $\bar{t}_j$ is obtained by reversing the relationship:

$$\bar{t}_j = \frac{\tau_j(\bar{t}) - \beta_j}{1+\alpha_j}$$

The time-stamp of the jth node on the time-basis of the ith node with propagation delay d_{ij} is

$$\tau_{j|i}(\bar{t}) = \tau_i(\bar{t}_j + d_{ij}) = (1+\alpha_i)(\bar{t}_j + d_{ij}) + \beta_i = \frac{1+\alpha_i}{1+\alpha_j}\tau_j(\bar{t}) + \left(\beta_i - \frac{1+\alpha_i}{1+\alpha_j}\beta_j\right) + (1+\alpha_i)d_{ij}$$

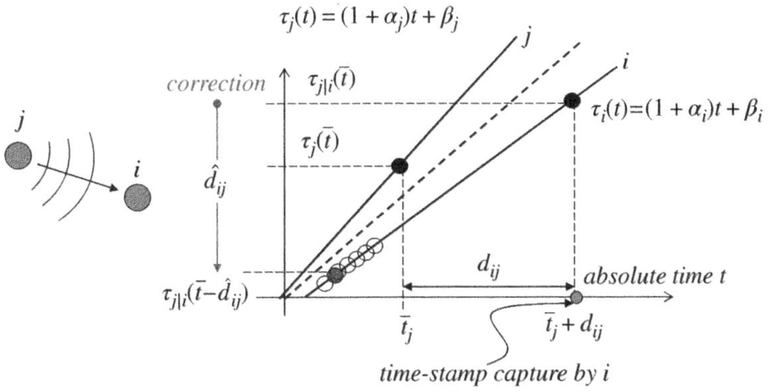

Figure 22.10 Exchange of time-stamps between two nodes with propagation delay d_{ij}.

This relationship is fundamental in clock synchronization and it decouples frequency error, phase error, and the impairments due to propagation delays, in order. Since frequency errors are small with $|\alpha_k| \ll 1$ for $\forall k$ it becomes

$$\tau_{j|i}(\bar{t}) = (1+\alpha_i-\alpha_j)\tau_j(\bar{t}) + \left(\beta_i-(1+\alpha_i-\alpha_j)\beta_j\right) + (1+\alpha_i)d_{ij}$$

Let $\hat{d}_{ij}$ be the propagation delay estimated by proper signaling as part of NTP (see below), the clock error corrected by the propagation delay for an ensemble of $\mathcal{K}_i$ nodes becomes

$$\Delta\tau_i(\bar{t}) = \frac{1}{|\mathcal{K}_i|}\sum_{j\neq i}\tau_{j|i}(\bar{t}-\hat{d}_{ij}) - \tau_i(\bar{t}).$$

This is used locally as a correcting term for consensus-based local clock update, possibly with some filtering over time (not considered here), to reach a distributed synchronization.

NTP

When considering the Internet network, simplicity is mandatory. For a pair of nodes, the propagation delayed signaling of the relative time-stamps referred to one node or two consecutive time-stamps, can let one node estimate the propagation delay and frequency drift. Consider two nodes in Figure 22.11, labeled as A and B; node A transmits to node B a first time-stamp at absolute time $\bar{t}$:

$$\tau_A(\bar{t}) = (1+\alpha_A)\bar{t} + \beta_A$$

Node B receives the delayed time-stamp that, according to its local time, corresponds to

$$\tau_B(\bar{t}) = (1+\alpha_B)(\bar{t}+d_{BA}) + \beta_B$$

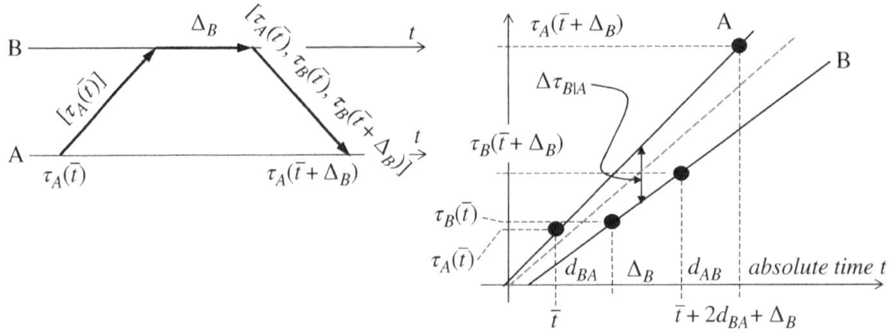

Figure 22.11 Network Time Protocol to estimate the propagation delay.

after an arbitrary time Δ_B that is based on the internal circuitry of node B, the workload of the server, etc. at local time[3]

$$\tau_B(\bar{t}+\Delta_B) = (1+\alpha_B)(\bar{t}+d_{BA}+\Delta_B)+\beta_B$$

Node B itself transmits to node A the three values of the time stamps:

$$\left[\tau_A(\bar{t}), \tau_B(\bar{t}), \tau_B(\bar{t}+\Delta_B)\right]$$

and these values are received after the propagation delay

$$\tau_A(\bar{t}+\Delta_B) = (1+\alpha_A)(\bar{t}+2d_{BA}+\Delta_B)+\beta_A$$

Node A now has a collection of four values

$$\left[\tau_A(\bar{t}), \tau_B(\bar{t}), \tau_B(\bar{t}+\Delta_B), \tau_A(\bar{t}+\Delta_B)\right]$$

As a result of the handshaking with node B and it can estimate delay

$$\hat{d}_{BA} = \frac{(\tau_A(\bar{t}+\Delta_B)-\tau_A(\bar{t}))+(\tau_B(\bar{t})-\tau_B(\bar{t}+\Delta_B))}{2}$$

and clock-offset evaluated wrt A at time $\bar{t}+d_{BA}+\Delta_B/2$

$$\begin{aligned}\Delta\tau_{B|A} &= \frac{(\tau_B(\bar{t})-\tau_A(\bar{t}))+(\tau_B(\bar{t}+\Delta_B)-\tau_A(\bar{t}+\Delta_B))}{2}\\ &= (\alpha_B-\alpha_A)(\bar{t}+d_{BA}+\frac{\Delta_B}{2})+(\beta_B-\beta_A)\end{aligned}$$

The iterative exchange of signaling lets every node estimate the necessary parameters for the relative time and drift to converge to the synchronization state.

Appendix A: Basics of Undirected Graphs

Network topologies are formally described by algebraic structures called graphs, and the corresponding properties are widely investigated in graph theory. The basics date back to 1736 by Euler and have largely evolved in the last decades to find the analytical framework for a set of network-related problems. The basic properties of an *undirected* graph are summarized here; detailed overviews are in [122, 123], among others.

A graph is a collection of nodes (or vertexes) labeled in $\mathcal{V} = \{1, \ldots, K\}$ with a set of links (or edges) in $\mathcal{E} \subseteq \mathcal{V} \times \mathcal{V}$ connecting some of the nodes. A node v_i is adjacent to v_j if (v_i, v_j) is an edge and there is a link connecting v_i and v_j. The set of neighbors of v_i is:

$$\mathcal{K}_i = \{v_j | (v_i, v_j) \in \mathcal{E}\}$$

3 The internal delay of node B is based on the local clock of node B, but this time-error is neglected.

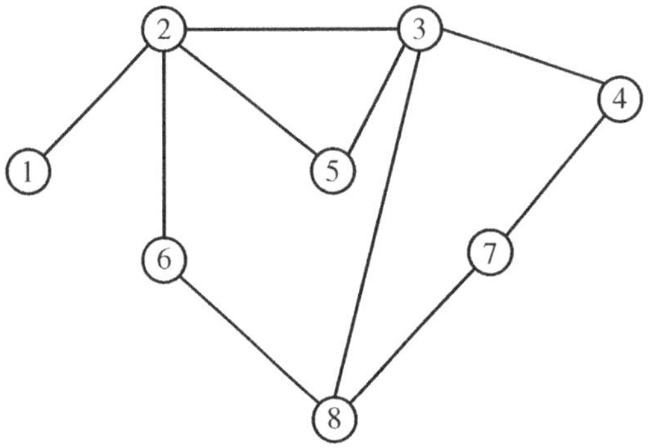

Figure 22.12 Example of undirected graph.

and the cardinality of $\mathcal{K}_i$ counts the number of neighbors and is called the degree of the vertex:

$$d_i = |\mathcal{K}_i|$$

An isolated node has degree $d_i = 0$. The maximum degree of a graph is

$$d_{max} = max\{d_i\}$$

A path is a sequence of links, and a graph is connected if there is a path connecting any two nodes. Making reference to the example in Figure 22.12, the graph on the vertex set

$$\mathcal{V} = \{1, \ldots, 8\}$$

has edge set

$$\mathcal{E} = \{(1,2),(2,3),(2,5),(2,6),(3,4),(3,8),(3,5),(4,7),(6,8),(7,8)\}$$

so any pair of vertexes are connected by a path and the graph is connected.

The adjacency matrix $\mathbf{A} \in \mathbb{R}^{N\times N}$ describes how a graph is connected and for any pair v_i, v_j it is $a_{ij} = 1$ if linked (or $(v_i, v_j) \in \mathcal{E}$) and $a_{ij} = 0$ elsewhere, included $i = j$. From the example:

$$\mathbf{A} = \begin{bmatrix} 0 & 1 & 0 & 0 & 0 & 0 & 0 & 0 \\ 1 & 0 & 1 & 0 & 1 & 1 & 0 & 0 \\ 0 & 1 & 0 & 1 & 1 & 0 & 0 & 1 \\ 0 & 0 & 1 & 0 & 0 & 0 & 1 & 0 \\ 0 & 1 & 1 & 0 & 0 & 0 & 0 & 0 \\ 0 & 1 & 0 & 0 & 0 & 0 & 0 & 1 \\ 0 & 0 & 0 & 1 & 0 & 0 & 0 & 1 \\ 0 & 0 & 1 & 0 & 0 & 1 & 1 & 0 \end{bmatrix}$$

and a generalization is by placing a positive weight for each entry. The adjacency matrix is symmetric ($\mathbf{A} = \mathbf{A}^T$), but this property is not general (for an undirected graph, $\mathbf{A} \neq \mathbf{A}^T$, see [122, 123]). The degree matrix of a graph is

$$\mathbf{D} = \text{diag}(d_1, \ldots, d_K)$$

and collects the degree of each node.

The Laplacian matrix

$$\mathbf{L} = \mathbf{D} - \mathbf{A}$$

is symmetric and positive semidefinite as

$$\mathbf{L1} = \mathbf{0} \rightarrow \lambda_1(\mathbf{L}) = 0$$

and **1** is the corresponding eigenvector (the sum along rows is zero, and from symmetry, the sum along columns is zero also). The eigenvalues of **L** are real and non-negative, upper bounded by the maximum degree of the graph (the proof is by the Gershgorin circle theorem—Section 1.3)

$$0 = \lambda_1(\mathbf{L}) \leq \lambda_2(\mathbf{L}) \leq \ldots \leq \lambda_K(\mathbf{L}) \leq 2d_{max}$$

From analysis of the eigenvalues, a graph is connected iff $\lambda_2(\mathbf{L}) > 0$: this is called the connectivity eigenvalue. For the example at hand:

$$\mathbf{L} = \begin{bmatrix} 1 & -1 & 0 & 0 & 0 & 0 & 0 & 0 \\ -1 & 4 & -1 & 0 & -1 & -1 & 0 & 0 \\ 0 & -1 & 4 & -1 & -1 & 0 & 0 & -1 \\ 0 & 0 & -1 & 2 & 0 & 0 & -1 & 0 \\ 0 & -1 & -1 & 0 & 2 & 0 & 0 & 0 \\ 0 & -1 & 0 & 0 & 0 & 2 & 0 & -1 \\ 0 & 0 & 0 & -1 & 0 & 0 & 2 & -1 \\ 0 & 0 & -1 & 0 & 0 & -1 & -1 & 3 \end{bmatrix}$$

Note that the Laplacian generalizes the second order derivative of the finite difference Laplace operator over a regular grid detailed in Appendix C of Chapter 18 to the concept of neighbors of a graph.

Optimization Posed on Graph

Consider a problem where every node has the objective to minimize the difference with the variables of adjacent nodes. The optimization can be formulated on the underlying graph according to a problem-dependent cost function. One example is optimization based on the quadratic disagreement function

$$\varphi(\mathbf{x}) = \frac{1}{2} \sum_{(i,j) \in \mathcal{E}} a_{ij}(x_i - x_j)^2 = \frac{\mathbf{x}^T \mathbf{L} \mathbf{x}}{2}$$

The minimization is for the identity $x_1 = x_2 = \ldots = x_K$ and this defines the solution; formally it is

$$\min_x \varphi(\mathbf{x}) \qquad \text{st } x_1 = x_2 = \ldots = x_K \neq 0$$

and this can be solved iteratively by the gradient search method as

$$\mathbf{x}(\ell+1) = \mathbf{x}(\ell) - \varepsilon \mathbf{L}\mathbf{x}(\ell)$$

The iterative method resembles steepest descend methods (Section 15.3) and convergence analysis follows the same steps. The solution of the minimization problem is the average of the initialization $x_1 = x_2 = \dots = x_K = x_\infty$ for $\ell \to \infty$:

$$\mathbf{x}_\infty = \frac{\mathbf{1}^T\mathbf{x}(0)}{\mathbf{1}^T\mathbf{1}}\mathbf{1}$$

By the property of the eigenvectors of the Laplacian for $\lambda_1(\mathbf{L}) = 0$, the Laplacian can be decomposed into the eigenvectors/eigenvalues

$$\mathbf{L} = \mathbf{U}_1\mathbf{L}_1\mathbf{U}_1^T = \sum_{i=2}^{K} \lambda_i \mathbf{u}_i \mathbf{u}_i^T$$

so that the iterative update can be referred wrt the principal axes of the quadratic form. The convergence depends on the largest eigenvalue, so that

$$0 < \varepsilon < \frac{2}{\lambda_{max}}$$

but according to the property above, $\lambda_{max} \leq 2d_{max}$ and thus

$$\varepsilon < \frac{1}{d_{max}}$$

The convergence speed depends on the rate for all these parallel optimizations that are converging to the solution. The time to convergence (in iterations) on the ith direction is

$$\tau_i \simeq \frac{1}{\varepsilon\lambda_i} \geq \frac{\lambda_{max}}{2\lambda_i}$$

and thus the timescale of the exponential decaying that number of iterations to converge is

$$T_{conv} \geq \frac{\lambda_{max}}{2\lambda_2}$$

Sometimes the choice is $\varepsilon = \varepsilon_o/d_{max}$ with $\varepsilon_o \leq 1$ and the timescale can be made dependent on the normalized gain ε_o:

$$T_{conv} \simeq \frac{\lambda_{max}}{2\varepsilon_o\lambda_2}$$

The convergence depends on the degree of connectivity of the graph and is faster for largest λ_2 (strongly connected graph). This clarifies the role of the connectivity eigenvalue on any processing on graphs.

Perron–Frobenius Theorem [1] for Convergence

The iterations of the search method are conveniently rewritten as

$$\mathbf{x}(\ell+1)=\mathbf{P}\mathbf{x}(\ell)$$

where

$$\mathbf{P}=\mathbf{I}-\varepsilon\mathbf{L}$$

is the Perron matrix with

$$\lambda_i(\mathbf{P})=1-\varepsilon\lambda_i(\mathbf{L})\leq 1-2\frac{\lambda_i(\mathbf{L})}{\lambda_{max}(\mathbf{L})}\leq 1$$

where the largest eigenvalue $\lambda_1(\mathbf{P})=1$ is for $\lambda_1(\mathbf{L})=0$. The eigenvectors of $\mathbf{L}$ and $\mathbf{P}$ coincide, but for the general case of a *non-symmetric Laplacian matrix,* let $\mathbf{u}_1$ and $\mathbf{v}_1$ the left and right eigenvector for $\lambda_1(\mathbf{P})=1$ so that

$$\mathbf{P}\mathbf{v}_1=\mathbf{v}_1 \qquad \mathbf{u}_1^T\mathbf{P}=\mathbf{u}_1^T$$

and provided that the eigenvectors are normalized $\mathbf{v}_1^T\mathbf{u}_1=1$:

$$\lim_{k\to\infty}\mathbf{P}^k=\mathbf{u}_1\mathbf{v}_1^T$$

and thus

$$\mathbf{x}_\infty=\mathbf{u}_1\mathbf{v}_1^T\mathbf{x}(0).$$

23

Classification and Clustering

There are many applications where a set of data should be classified as belonging to a predefined subset, or to discover the structure of the data when those subsets are not known. In signal processing, a signal can contain or not a specific waveform, or an image can be contain or not a feature of interest (e.g., a search for faces in images), or feature extraction can be used to reduce the complexity of the image itself. This chapter covers methods for classification and clustering as part of routine statistical signal processing for signal detection or feature extraction. The topic is broad, and the focus is to discuss the essential concepts collected in Table 23.1.

Classification involves the identification of group membership and it can be quantitatively stated using an example from handwriting recognition. Consider the handwritten images that correspond to each of the digits 0,1,...,9 indicated as classes; each image is digitized into a set of N pixels that form the observation vector $\mathbf{x} \in \mathbb{R}^N$. The final goal is to have a classifier that identifies which digit corresponds to a certain $\mathbf{x}$ knowing the features of all the classes. In a classification problem, the parametric pdfs $p(\mathbf{x}|\mathcal{H}_1),...,p(\mathbf{x}|\mathcal{H}_K)$ for the K classes $\mathcal{H}_1, \mathcal{H}_2, ..., \mathcal{H}_K$ are known from the problem settings and are used for the decision of $\mathbf{x}$ in favor of the class that maximizes the probability $p(\mathbf{x}|\mathcal{H}_k)$, as for maximum likelihood. Whenever the pdfs $p(\mathbf{x}|\mathcal{H}_k)$ are not known either in closed form or analytically, one can use a set of training data to infer these properties from experiments, and let the classifier learn from samples (e.g., the classifier learns the main features from a training sample of handwritten digits, and any new $\mathbf{x}$ is classified based on the features extracted from the training set); this is called *supervised learning.*

Classification when $K = 2$ is binary between two alternative hypotheses of the data. This problem has been widely investigated in the statistical signal processing literature and it is the basis of understanding more evolved classification methods with $K > 2$ (also called multiple hypothesis testing).

In Bayesian classification, each of the classes is characterized by the corresponding a-priori probability $p(\mathcal{H}_1),p(\mathcal{H}_2),...,p(\mathcal{H}_K)$, and the joint pdf $p(\mathbf{x}, \mathcal{H}_k) = p(\mathbf{x}|\mathcal{H}_k)p(\mathcal{H}_k)$ gives the complete probabilistic description of the problem at hand. Bayes' theorem for classification

$$p(\mathcal{H}_k|\mathbf{x}) = \frac{p(\mathbf{x}|\mathcal{H}_k)p(\mathcal{H}_k)}{p(\mathbf{x})} = \frac{p(\mathbf{x}|\mathcal{H}_k)p(\mathcal{H}_k)}{\sum_{n=1}^{K} p(\mathbf{x}, \mathcal{H}_n)}$$

Statistical Signal Processing in Engineering, First Edition. Umberto Spagnolini.

Companion website: www.wiley.com/go/spagnolini/signalprocessing

Table 23.1 Taxonomy of principles and methods for classification and clustering.

	Classification (Section 23.2)	Clustering (Section 23.6)
Prior knowledge	$p(\mathbf{x}\|\mathcal{H}_k)$ is known/estimated $p(\mathcal{H}_k)$ is known (Bayesian)	No a-priori classes and numbers
Application	Classify new sample $\mathbf{x}$ into $\mathcal{H}_k$	Grouping of homogeneous data (structure discovery from data)
Methods	Detection theory (Section 23.2.1) Bayesian classifiers (Section 23.4)	K-means and EM clustering
Metric	Classification probability	Minimum clusters/clustering error

is instrumental in getting the a-posteriori probability for each class, and this is useful to decide to use a-posteriori probability criteria that assign $\mathbf{x}$ to the class with the largest a-posteriori probability $p(\mathcal{H}_k|\mathbf{x})$. The classification metric is the correct or incorrect classification probability:

$$\begin{aligned}\Pr\{\mathbf{x} \text{ is } \mathcal{H}_k|\mathcal{H}_k\} &: \text{ correct classification}\\ \Pr\{\mathbf{x} \text{ is } \mathcal{H}_i|\mathcal{H}_k\} &: \text{ incorrect classification } (i \neq k)\end{aligned}$$

or any other error probability that accounts for all the wrong (Section 23.2.1).

When classification involves signals, it is called *detection theory* and good references are [37, 128, 132]. Classification is the basis of pattern recognition, and overviews are provided by Fukunaga [129] and many others.

Clustering is the method of discovering any structure from a (large) set of samples $\{\mathbf{x}_\ell\}$ when there is no a-priori information on classes and no training is available (*unsupervised learning*). Cluster analysis divides data into groups (clusters) that are meaningful and/or useful, and the clusters should capture the natural structure of the data useful for a higher level of understanding. In signal processing, clustering is used to reduce the amount of information in signals, or to discover some properties of signals. Cluster analysis plays a key role when manipulating a large amount of data such as in pattern recognition, machine learning, and data mining. Application examples in the area of clustering aim to analyze and classify a large amount of data such as genetic information (e.g., genome analysis), multimedia information for information retrieval in the World Wide Web (which consists of billions of items of multimedia content), bank transactions to detect fraud, and business to segment customers into a small number of groups for additional analysis and marketing activities. An excellent review is by Jain [131], and a discussion of the application to learning processes in machine learning and mining of large data sets can be found in [142].

23.1 Historical Notes

Detection and classification in signal processing are two sides of the same coin. Signal detection algorithms decide whether a signal consists of "signal masked by noise" or "noise only," and detection theory helps to design the algorithms that minimize the average number of decision errors. Detection and classification have their roots in statistical decision theory where they are called binary ($K = 2$) and K-ary ($K > 2$) hypothesis testing [124]. The first systematic developments in signal detection theory were around 1950 by communication and radar engineering [125, 126], but some statistical detection algorithms were implemented even before that. From 1968, H. Van Trees [37] published a set of four volumes that are widely used in engineering, and vol. 3 is on detection theory [128], where he extensively investigated all the possible cases of signal detection with/without known parameters/noise/signals, as briefly revised herein.

In general, classification in statistical signal processing problems is: (1) classifying an individual waveform out of a set of waveforms that could be linearly independent to form a basis, or (2) classifying the presence of a particular set of signals as opposed to other sets of signals, and classifying the number of signals present. Each of these problems are based on the definition of specific classes that depend on the specific signals at hand.

In pattern recognition, classification is over a set of classes. Classification for multivariate Normal distributions was proposed by Fisher in 1936 [133] with a frequentist vision, and the target is to use a linear transformation that set the basis for linear discriminant analysis (LDA). The characteristics of the classes are known, and the aim is to establish a rule whereby we can classify a new observation into one of the existing classes. The classification rule can be defined by well-defined statistical properties of the data or, in complex contexts, the classifier can be trained by using some known data–class pairs. Pattern recognition dates back to around 1967 when it was established as a dedicated Technical Committee by the IEEE Computer Society; a comprehensive book by K. Fukunaga appeared in 1972 [127], while the first dedicated IEEE journal appeared in 1979. There are many applications based on the recognition of patterns such as handwriting and optical character recognition, speech recognition, image recognition, and information retrieval, just to mention a few. Classification rules in all these applications are based on a preliminary training—supervised training is part of the *machine learning* process. Machine Learning is the field of study that gives computers the ability to learn without being explicitly programmed (this definition is attributed to A. Samuel, 1959) [130], and nowadays it is synonymous with pattern recognition. In this framework, neural networks are a set of supervised algorithms that achieve classification using the connection of a large number of simple threshold-based processors.

Clustering methods to infer the structure of data from the data itself (unsupervised learning) are of broad interest since the early paper by J. MacQueen in 1967 [141], and the use of mixture models for clustering [134][143] strengthen their performance.

23.2 Classification

23.2.1 Binary Detection Theory

The two-group (or *binary*) classification problem between two classes indexed as $\mathcal{H}_0$ and $\mathcal{H}_1$ was largely investigated in the past as being the minimal classification. Detection theory in signal processing is used to decide if $\mathbf{x}$ contains a signal+noise or just noise,

Table 23.2 Classification metrics.

Decision on x	True hypothesis $\mathcal{H}_1$	True hypothesis $\mathcal{H}_0$
$\mathbf{x} \in \mathcal{R}_1$ decide $\mathcal{H}_1$	$\Pr(\mathcal{H}_1\|\mathcal{H}_1) = \Pr(\mathbf{x} \in \mathcal{R}_1\|\mathcal{H}_1)$ true positive (detection)	$\Pr(\mathcal{H}_1\|\mathcal{H}_0) = \Pr(\mathbf{x} \in \mathcal{R}_1\|\mathcal{H}_0)$ false positive (false alarm)
$\mathbf{x} \in \mathcal{R}_0$ decide $\mathcal{H}_0$	$\Pr(\mathcal{H}_0\|\mathcal{H}_1) = \Pr(\mathbf{x} \in \mathcal{R}_0\|\mathcal{H}_1)$ false negative	$\Pr(\mathcal{H}_0\|\mathcal{H}_0) = \Pr(\mathbf{x} \in \mathcal{R}_0\|\mathcal{H}_0)$ true negative

and this is a typical problem in conventional radar systems (Section 5.6) where it is crucial to detect or not the presence of a target. In communication engineering, binary classification is used to decide if a transmitted bit is 0 or 1 (Section 17.1), or if the frequency of a sinusoid is ω_0 or ω_1 provided that the class of frequencies are $\{\omega_0, \omega_1\}$, just to mention few. Furthermore, in medicine, binary classification is used to decide if a patient is healthy or not, in bank or electronic payments if transactions are secure or not, in civil engineering if a bridge is faulty or not, etc.

The binary decision is based on the choice between two hypotheses referred as the *null hypothesis* $\mathcal{H}_0$ and the *alternative hypothesis* $\mathcal{H}_1$, and for the case of known pdfs the associated pdfs are $p(\mathbf{x}|\mathcal{H}_0)$ and $p(\mathbf{x}|\mathcal{H}_1)$. The decision criterion on $\mathbf{x}$ is based on certain disjoint decision regions $\mathcal{R}_0$ and $\mathcal{R}_1$, such that $\mathcal{R}_0 \cup \mathcal{R}_1 = \mathbb{R}^N$. The decision regions are tailored according to the decision metric, and an intuitive one is to decide in favor of $\mathcal{H}_1$ if $p(\mathbf{x}|\mathcal{H}_1) > p(\mathbf{x}|\mathcal{H}_0)$, and for $\mathcal{H}_0$ if $p(\mathbf{x}|\mathcal{H}_1) < p(\mathbf{x}|\mathcal{H}_0)$ as for maximum likelihood criteria. Since decision regions do not involve any statistical comparison on actual data, but rather the design of the regions to be used according to the decision metric, regions are more practical in decision between $\mathcal{H}_0$ vs. $\mathcal{H}_1$ yet recognizing the equivalence of comparing $p(\mathbf{x}|\mathcal{H}_0)$ vs. $p(\mathbf{x}|\mathcal{H}_1)$.

Any value $\mathbf{x}$ is classified as $\mathcal{H}_0$ if it belongs to $\mathcal{R}_0$ ($\mathbf{x} \in \mathcal{R}_0$, and $\mathbf{x} \notin \mathcal{R}_1$), and alternatively is classified as $\mathcal{H}_1$ if it belongs to $\mathcal{R}_1$ ($\mathbf{x} \in \mathcal{R}_1$, and $\mathbf{x} \notin \mathcal{R}_0$). Correct classification is when $\mathbf{x} \in \mathcal{R}_1$ and the true class is $\mathcal{H}_1$, and this is indicated as a true positive (TP). Conversely, a true negative (TN) is for $\mathbf{x} \in \mathcal{R}_0$ and the true class is $\mathcal{H}_0$. One can define the probability for TP and TN as indicated in Table 23.2. Misclassification is when $\mathbf{x} \in \mathcal{R}_1$ and the true class is $\mathcal{H}_0$, this is called a false positive (FP), and when $\mathbf{x} \in \mathcal{R}_0$ and the true class is $\mathcal{H}_1$, a false negative (FN). An alternative notation used here comes from radar contexts (but applies also to tumor detection) where $\mathcal{H}_1$ refers to a target (that is important to detect with minimal error), and $\mathcal{H}_0$ is the lack of any target: TP is true target detection, and FP is a false alarm when there is no target. Table 23.2 summarizes the cases, however statisticians uses a different notation in binary hypothesis testing: *sensitivity* is TP probability $p(\mathcal{H}_1|\mathcal{H}_1)$, *specificity* is TN probability $p(\mathcal{H}_0|\mathcal{H}_0)$.

The shape of the decision regions $\mathcal{R}_0$ and $\mathcal{R}_1$ for test hypothesis $\mathcal{H}_0$ versus $\mathcal{H}_1$ depends on the classification metrics that are based on the probabilities

$$\Pr(\mathcal{H}_n|\mathcal{H}_m) = \Pr(\mathbf{x} \in \mathcal{R}_n|\mathcal{H}_m) \text{ for } n, m \in \{0,1\}$$

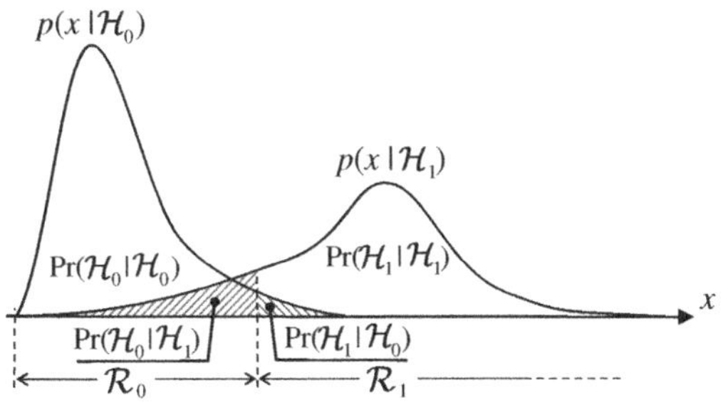

Figure 23.1 Binary hypothesis testing.

The probability of a false alarm (also called *size of test* in statistics) is (Figure 23.1)

$$\Pr(\mathcal{H}_1|\mathcal{H}_0) = \int_{\mathbf{x}\in\mathcal{R}_1} p(\mathbf{x}|\mathcal{H}_0)d\mathbf{x} = P_{FA}$$

and measures the worst-case probability of choosing $\mathcal{H}_1$ when $\mathcal{H}_0$ is in place. Detection probability (Figure 23.1)

$$\Pr(\mathcal{H}_1|\mathcal{H}_1) = \int_{\mathbf{x}\in\mathcal{R}_1} p(\mathbf{x}|\mathcal{H}_1)d\mathbf{x} = P_D$$

measures the probability of correctly accepting $\mathcal{H}_1$ when $\mathcal{H}_1$ is in force. Since $\Pr(\mathcal{H}_0|\mathcal{H}_1) + \Pr(\mathcal{H}_1|\mathcal{H}_1) = 1$ and $\Pr(\mathcal{H}_0|\mathcal{H}_0) + \Pr(\mathcal{H}_1|\mathcal{H}_0) = 1$, any complementary probability is fully defined from the pair P_D and P_{FA}. The goal is to choose the test

$$\mathbf{x} \in \mathcal{R}_0 \Longrightarrow \text{decide for } \mathcal{H}_0$$
$$\mathbf{x} \in \mathcal{R}_1 \Longrightarrow \text{decide for } \mathcal{H}_1$$

or equivalently the decision region boundary, to have the largest P_D with minimum P_{FA}. These probabilities cannot be optimized together, and conventionally one constrains $P_{FA} = \bar{P}_{FA}$ and chooses the largest P_D. This is the celebrated Neyman–Pearson (NP) test that is obtained by solving the optimization constraining $\int_{\mathbf{x}\in\mathcal{R}_1} p(\mathbf{x}|\mathcal{H}_0)d\mathbf{x} = \bar{P}_{FA}$ from the Lagrangian:

$$L(\mathcal{R}_1, \lambda) = \int_{\mathbf{x}\in\mathcal{R}_1} \left(p(\mathbf{x}|\mathcal{H}_1) + \lambda p(\mathbf{x}|\mathcal{H}_0)\right) d\mathbf{x} - \lambda\bar{P}_{FA}$$

To maximize $L(\mathcal{R}_1, \lambda)$, the argument should be $p(\mathbf{x}|\mathcal{H}_1) + \lambda p(\mathbf{x}|\mathcal{H}_0) > 0$ within the decision bound $\mathcal{R}_1$, and this is the basis for the proof that shows that $\lambda < 0$ (see e.g., [132] for details). The decision is based on the likelihood ratio $p(\mathbf{x}|\mathcal{H}_1)/p(\mathbf{x}|\mathcal{H}_0)$ and the decision in favor of $\mathcal{H}_1$ is called the likelihood ratio test (LRT)

$$\mathcal{L}(\mathbf{x}) = \frac{p(\mathbf{x}|\mathcal{H}_1)}{p(\mathbf{x}|\mathcal{H}_0)} > \gamma$$

(or alternatively, decide $\mathcal{H}_0$ if $\mathcal{L}(\mathbf{x}) < \gamma$), where the threshold $\gamma = \gamma(\bar{P}_{FA})$ is obtained by solving for the constraint

$$\int_{\mathbf{x}\in\mathcal{R}_1} p(\mathbf{x}|\mathcal{H}_0)d\mathbf{x} = \bar{P}_{FA}$$

The LRT is often indicated by highlighting the two alternative selections wrt the threshold:

$$\mathcal{L}(\mathbf{x}) = \frac{p(\mathbf{x}|\mathcal{H}_1)}{p(\mathbf{x}|\mathcal{H}_0)} \underset{\mathcal{H}_0}{\overset{\mathcal{H}_1}{\gtrless}} \gamma$$

The notation compactly specifies that for $\mathcal{L}(\mathbf{x}) > \gamma$, the decision is in favor of $\mathcal{H}_1$ and for $\mathcal{L}(\mathbf{x}) < \gamma$ it is for $\mathcal{H}_0$. The NP-test solves for the maximum P_D constrained to a certain P_{FA}, and the corresponding value of optimized P_D vs. P_{FA} can be evaluated analytically for the optimum threshold values. The plot of (optimized) P_D vs. P_{FA} (or equivalently *sensitivity* vs. *1-specificity*) is called the *receiver operating characteristic* (ROC), and it represents a common performance indicator of test statistics that removes the dependency on the threshold. The analysis of the ROC curve reveals many properties of the specific decision rule of $\mathcal{H}_1$ versus $\mathcal{H}_0$ based on the statistical properties $p(\mathbf{x}|\mathcal{H}_1)$ versus $p(\mathbf{x}|\mathcal{H}_0)$ that are better illustrated by an example from [128].

Example: Detection of a Constant in Uncorrelated Noise

Assume that the signal is represented by a constant amplitude $a > 0$ in Gaussian noise (value of a is known but not its presence), and one should decide if there is a signal or not by detecting the presence of a constant amplitude in a set of N samples $\mathbf{x} = [x_1, x_2, ..., x_N]^T$. This introductory example is to illustrate the steps to produce the corresponding ROC curves. The two alternative hypotheses for the signal $\mathbf{x}$ are

$$\begin{aligned}
\mathcal{H}_1 &: x_n = a + w_n \Longrightarrow p(\mathbf{x}|\mathcal{H}_1) = G(\mathbf{x}; a\mathbf{1}, \sigma_w^2\mathbf{I}) \\
\mathcal{H}_0 &: x_n = w_n \Longrightarrow p(\mathbf{x}|\mathcal{H}_0) = G(\mathbf{x}; \mathbf{0}, \sigma_w^2\mathbf{I})
\end{aligned}$$

The LRT is

$$\frac{G(\mathbf{x}; a\mathbf{1}, \sigma_w^2\mathbf{I}))}{G(\mathbf{x}; 0\mathbf{1}, \sigma_w^2\mathbf{I})} \underset{\mathcal{H}_0}{\overset{\mathcal{H}_1}{\gtrless}} \gamma$$

which in this case it simplifies as

$$\bar{x} = \frac{1}{N}\sum_{n=1}^{N} x_n \underset{\mathcal{H}_0}{\overset{\mathcal{H}_1}{\gtrless}} \frac{a}{2} + \frac{\sigma_w^2}{NA}\ln\gamma = \bar{\gamma}$$

The test is based on the sample mean of the signal, which can be considered as an estimate of the amplitude (Section 7.2). Since the P_D vs. P_{FA} analysis depends on the pdf of the sample mean $\bar{x}$ under the two hypotheses, these should be detailed based on the properties of the LRT manipulations. In this case, since decision statistic $\bar{x}$ is the sum of Gaussians, the pdfs for the two conditions are:

$$\mathcal{H}_1 : \bar{x} \sim \mathcal{N}(a, \sigma_w^2/N)$$
$$\mathcal{H}_0 : \bar{x} \sim \mathcal{N}(0, \sigma_w^2/N)$$

and the probabilities can be evaluated analytically vs $\bar{\gamma}$

$$P_D(\bar{\gamma}) = \Pr(\bar{x} > \bar{\gamma} | \mathcal{H}_1) = Q\left(\frac{\bar{\gamma} - a}{\sqrt{\sigma_w^2/N}}\right) \tag{23.1}$$

$$P_{FA}(\bar{\gamma}) = \Pr(\bar{x} > \bar{\gamma} | \mathcal{H}_0) = Q\left(\frac{\bar{\gamma}}{\sqrt{\sigma_w^2/N}}\right) \tag{23.2}$$

where the Q-function

$$Q(\varsigma) = \frac{1}{\sqrt{2\pi}} \int_{\varsigma}^{\infty} \exp\left(-\frac{u^2}{2}\right) du$$

is monotonic decreasing with increasing ς. The probabilities are monotonic decreasing with the threshold $\bar{\gamma}$ ranging from $P_D = P_{FA} = 1$ for $\bar{\gamma} \to -\infty$ (i.e., accept any condition as being $\mathbf{x} \in \mathcal{H}_1$, even if wrong) down to $P_D = P_{FA} = 0$ for $\bar{\gamma} \to \infty$ (i.e., reject any condition as being $\mathbf{x} \in \mathcal{H}_1$, even if wrong), but in any case the probabilities $P_D = P_{FA}$ are useless as this implies that decision $\mathcal{H}_1$ vs $\mathcal{H}_0$ is random, or equivalently the two hypotheses are indistinguishable (i.e., $a \to 0$).

To derive the ROC, consider a given P_{FA}; one can optimally evaluate the threshold γ that depends on the signal to noise ratio

$$SNR = \frac{Na^2}{\sigma_w^2}$$

and the corresponding P_D is the largest one subject to the P_{FA}. The ROC is independent of the threshold $\bar{\gamma}$, which should be removed from the dependency of P_{FA} in (23.2) using the inverse Q-function: $\bar{\gamma} = \sqrt{\sigma_w^2/N} Q^{-1}(P_{FA})$. Plugging this dependency into the P_D:

$$P_D = Q\left(Q^{-1}(P_{FA}) - \sqrt{SNR}\right) \tag{23.3}$$

yields the locus of the optimal P_D for a given P_{FA} assuming that the test hypothesis is optimized. Figure 23.2 illustrates the ROC curve for different values of SNR showing that when the SNR value is small, the ROC curve is close to the random decision (dashed line), while for given P_{FA}, the P_D increases with SNR. The ROC curve is also the performance for the optimized threshold, and for the optimum decision regions that are here transformations of the observations.

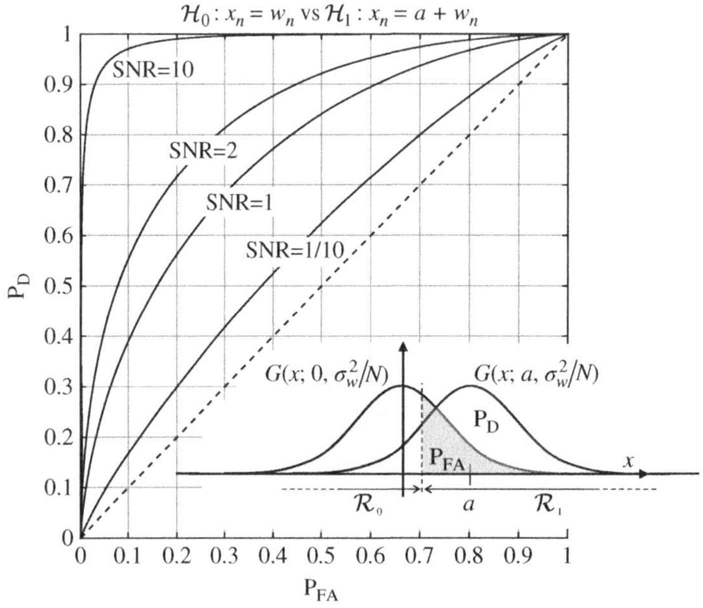

Figure 23.2 Receiver operating characteristic (ROC) curves for varying SNR.

23.2.2 Binary Classification of Gaussian Distributions

The most general problem of binary classification between two Gaussian populations with known parameters can be stated as follows. Given the two classes

$$\begin{aligned}\mathcal{H}_1 &: \mathbf{x} \sim \mathcal{N}(\boldsymbol{\mu}_1, \mathbf{C}_1)\\ \mathcal{H}_0 &: \mathbf{x} \sim \mathcal{N}(\boldsymbol{\mu}_0, \mathbf{C}_0)\end{aligned}$$

the decision rule becomes

$$\mathcal{L}'(\mathbf{x}) = \ln \frac{G(\mathbf{x}; \boldsymbol{\mu}_1, \mathbf{C}_1)}{G(\mathbf{x}; \boldsymbol{\mu}_0, \mathbf{C}_0)} \underset{\mathcal{H}_0}{\overset{\mathcal{H}_1}{\gtrless}} \ln \gamma$$

Replacing the terms it follows that (apart from a constant)

$$\mathcal{L}''(\mathbf{x}) = -\frac{1}{2}(\mathbf{x} - \boldsymbol{\mu}_1)^T \mathbf{C}_1^{-1}(\mathbf{x} - \boldsymbol{\mu}_1) + \frac{1}{2}(\mathbf{x} - \boldsymbol{\mu}_0)^T \mathbf{C}_0^{-1}(\mathbf{x} - \boldsymbol{\mu}_0) + \ln \frac{|\mathbf{C}_1|}{|\mathbf{C}_0|}$$

and the *decision boundary* that divides the decision regions can be found for $\mathcal{L}''(\mathbf{x}) =$ *const*: it is quadratic in $\mathbf{x}$ and typically its shape is quite complex.

In the case where $\mathbf{C}_1 = \mathbf{C}_0 = \mathbf{C}$, the quadratic decision function degenerates into a linear one:

$$\mathcal{L}''(\mathbf{x}) = \underbrace{(\boldsymbol{\mu}_1 - \boldsymbol{\mu}_0)^T \mathbf{C}^{-1}}_{\mathbf{a}^T}\mathbf{x} + \underbrace{\frac{1}{2}\left(\boldsymbol{\mu}_0^T \mathbf{C}^{-1}\boldsymbol{\mu}_0 - \boldsymbol{\mu}_1^T \mathbf{C}^{-1}\boldsymbol{\mu}_1\right)}_{b},$$

and the decision boundary is a hyperplane. Classification can be stated as finding the decision boundaries, and the number of parameters to define the quadratic decision function is $\mathcal{O}(N^2 + 3N)$ while for the linear one it is $\mathcal{O}(N+1)$; this is a convenience exploited in pattern analysis problems later in Section 23.5. The linear discriminant analysis (LDA) is based on the decision boundary

$$\mathbf{a}^T\mathbf{x} + b \underset{\mathcal{H}_0}{\overset{\mathcal{H}_1}{\gtrless}} \gamma' \tag{23.4}$$

that is adopted in pattern recognition, even when $\mathbf{C}_1 \neq \mathbf{C}_0$ by using

$$\mathbf{C} = \alpha\mathbf{C}_1 + (1-\alpha)\mathbf{C}_0$$

with weighting term α appropriately chosen [135, 137].

The classification of Gaussian distributions is illustrated in the example in Figure 23.3 for $N = 2$ with $\mathbf{C}_1 = \mathbf{C}_0$ for $\rho_1 = \rho_0$. The decision boundary $\mathbf{a}^T\mathbf{x} + b = \gamma'$ (gray line) between the two decision regions $\mathcal{R}_0$ and $\mathcal{R}_1$ is linear. However, for the case $\mathbf{C}_1 \neq \mathbf{C}_0$ that is here represented by the correlation $\rho_1 = -\rho_0$, the decision boundary for $\mathcal{L}''(\mathbf{x}) = 0$ is represented as a quadratic boundary, and the decision regions for $\mathbf{C}_1 \neq \mathbf{C}_0$ are multiple and their pattern is very complex for multidimensional ($N > 2$) data.

To gain insight in the decision regions as a consequence of the tails of the Gaussians, Figure 23.4 shows the decision regions $\mathcal{R}_0$ and $\mathcal{R}_1$ for the case $N = 1$ (line interval decision regions) and variances $\sigma_0^2 \neq \sigma_1^2$.

23.3 Classification of Signals in Additive Gaussian Noise

The general problem of signal classification in additive Gaussian noise can be stated for the two hypotheses:

$$\begin{aligned}
\mathcal{H}_1 &: \mathbf{x} = \mathbf{s}_1 + \mathbf{w} \Longrightarrow p(\mathbf{x}|\mathcal{H}_1) = G(\mathbf{x}; \mathbf{s}_1, \mathbf{C}_w) \\
\mathcal{H}_0 &: \mathbf{x} = \mathbf{s}_0 + \mathbf{w} \Longrightarrow p(\mathbf{x}|\mathcal{H}_0) = G(\mathbf{x}; \mathbf{s}_0, \mathbf{C}_w)
\end{aligned}$$

with known (Section 23.3.1–23.3.3) or random (Section 23.3.4) signals $\mathbf{s}_0$, $\mathbf{s}_1$ and known $\mathbf{C}_w$. Depending on the choice of the signals, this problem is representative of several common applications. The decision regions are linear and the decision criterion in closed form is (Section 23.2.2)

$$(\mathbf{s}_1 - \mathbf{s}_0)^T \mathbf{C}_w^{-1}\mathbf{x} \underset{\mathcal{H}_0}{\overset{\mathcal{H}_1}{\gtrless}} \gamma$$

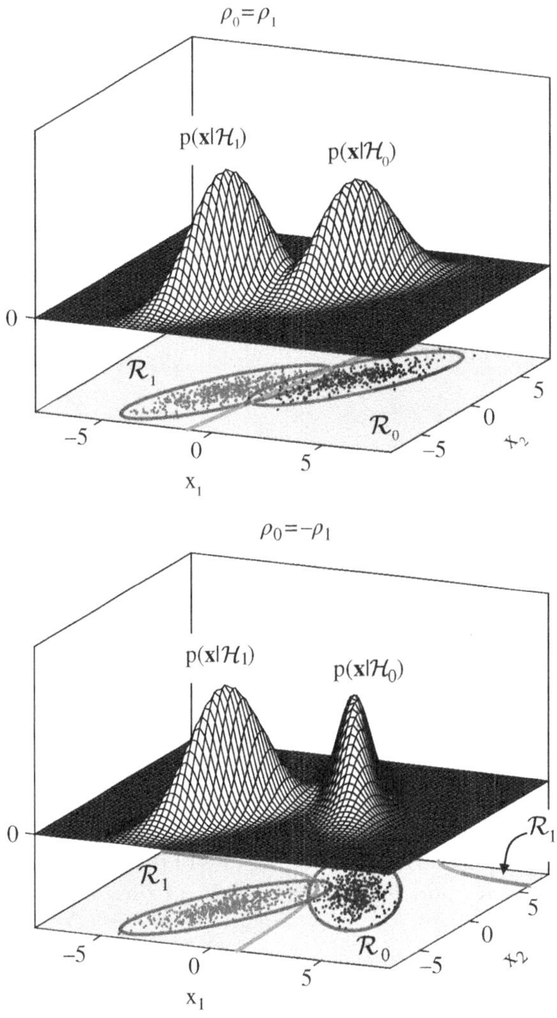

Figure 23.3 Classification for Gaussian distributions: linear decision boundary for $\mathbf{C}_1 = \mathbf{C}_0$ (upper figure) and quadratic decision boundary for $\mathbf{C}_1 \neq \mathbf{C}_0$ (lower figure).

where the threshold γ collects all the problem-specific terms for P_D vs. P_{FA}. The classification is part of the operations carried out by a receiving system that receives one of two alternative and known waveforms $\mathbf{s}_1$ and $\mathbf{s}_0$, or if the receiver should detect the presence of a backscattered waveform $\mathbf{s}_1$ in noise (when $\mathbf{s}_0 = 0$). The performance of the decision is in terms of P_D and P_{FA} for a given threshold $\bar{\gamma}$, and one can evaluate the statistical property of $(\mathbf{s}_1 - \mathbf{s}_0)^T \mathbf{C}_w^{-1} \mathbf{x} = \mathbf{a}^T \mathbf{x}$ that is Gaussian being the sum of Gaussian rvs under both hypotheses. These steps are the same as in Section 23.2.1, except analytically more complex, and one can evaluate the error probabilities P_D and P_{FA} in closed form, as well as the ROC.

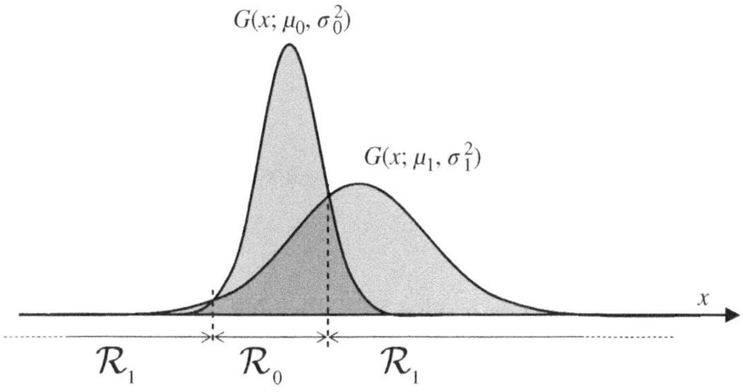

Figure 23.4 Decision regions for $N=1$ and $\sigma_0^2 \neq \sigma_1^2$.

23.3.1 Detection of Known Signal

In many applications such as radar/sonar or remote sensing systems, the interest is in detecting the presence ($\mathcal{H}_1$) of a target that backscatters the transmitted (and known) waveform $\mathbf{s}$ or not ($\mathcal{H}_0$) in noisy data with known covariance $\mathbf{C}_w$. Another context is in array processing with one active source (Section 19.1). The two hypotheses for the detection problem are

$$\begin{aligned}\mathcal{H}_1 &: \mathbf{x} = \mathbf{s}+\mathbf{w} \Longrightarrow p(\mathbf{x}|\mathcal{H}_1) = G(\mathbf{x};\mathbf{s},\mathbf{C}_w)\\ \mathcal{H}_0 &: \mathbf{x} = \mathbf{w} \Longrightarrow p(\mathbf{x}|\mathcal{H}_0) = G(\mathbf{x};\mathbf{0},\mathbf{C}_w)\end{aligned}$$

which is a special case with $\mathbf{s}_0 = \mathbf{0}$ for $\mathcal{H}_0$. Incidentally, the example in Section 23.2.1 is a special case for the choice $\mathbf{s} = a\mathbf{1}$ and $\mathbf{C}_w = \sigma_w^2\mathbf{I}$.

The detection rule is

$$h(\mathbf{x}) = \mathbf{s}^T\mathbf{C}_w^{-1}\mathbf{x} \underset{\mathcal{H}_0}{\overset{\mathcal{H}_1}{\gtrless}} \gamma \tag{23.5}$$

where the detector $h(\mathbf{x})$ cross-correlates the signal $\mathbf{x}$ with $\mathbf{s}$, with a noise pre-whitening by the $\mathbf{C}_w^{-1}$. This is a generalized correlator–detector [128]. The pdf of $h(\mathbf{x})$ under the two hypotheses is necessary to evaluate the ROC curve. The transformation $h(\mathbf{x})$ is linear, and this simplifies the analysis as the rvs are Gaussian. In detail:

$$\begin{aligned}\mathcal{H}_1 &: h(\mathbf{x}) = \mathbf{s}^T\mathbf{C}_w^{-1}(\mathbf{s}+\mathbf{w}) \sim \mathcal{N}(\mathbf{s}^T\mathbf{C}_w^{-1}\mathbf{s}, \mathbf{s}^T\mathbf{C}_w^{-1}\mathbf{s})\\ \mathcal{H}_0 &: h(\mathbf{x}) = \mathbf{s}^T\mathbf{C}_w^{-1}\mathbf{w} \sim \mathcal{N}(0, \mathbf{s}^T\mathbf{C}_w^{-1}\mathbf{s})\end{aligned}$$

and thus it is similar to (23.1, 23.2)

$$P_D = \Pr\{h(\mathbf{x}) > \gamma|\mathcal{H}_1\} = Q\left(\frac{\gamma - \mathbf{s}^T\mathbf{C}_w^{-1}\mathbf{s}}{\sqrt{\mathbf{s}^T\mathbf{C}_w^{-1}\mathbf{s}}}\right) \tag{23.6}$$

$$P_{FA} = \Pr\{h(\mathbf{x}) > \gamma | \mathcal{H}_0\} = Q\left(\frac{\gamma}{\sqrt{\mathbf{s}^T\mathbf{C}_w^{-1}\mathbf{s}}}\right) \tag{23.7}$$

that yields the same P_D vs. P_{FA} in (23.3) provided that signal to nosie ratio is redefined as

$$SNR = \mathbf{s}^T\mathbf{C}_w^{-1}\mathbf{s}$$

Since the ROC curve moves upward when increasing SNR, the SNR can be maximized by a judicious choice $\mathbf{s}$ to conform to the covariance $\mathbf{C}_w$. In detail, the choice $\mathbf{s} = \sqrt{E_s}\mathbf{q}_{min}$ where $(\lambda_{min}, \mathbf{q}_{min})$ is the minimum eigenvalue/eigenvector pair of $\mathbf{C}_w$ is optimal, $E_s = \mathbf{s}^T\mathbf{s}$ and this maximizes the SNR that in turn is $SNR = E_s/\lambda_{min}$.

This model can be specialized for detection of linear models (e.g., convolutive filtering, see Section 5.2) $\mathbf{s} = \mathbf{H}\boldsymbol{\theta}$ under $\mathcal{H}_1$ by simply re-shaping the problem accordingly. The detector is $h(\mathbf{x}) = \boldsymbol{\theta}^T\mathbf{H}^T\mathbf{C}_w^{-1}\mathbf{x}$, but since the MLE of $\boldsymbol{\theta}$ is $\hat{\boldsymbol{\theta}} = (\mathbf{H}^T\mathbf{C}_w^{-1}\mathbf{H})^{-1}\mathbf{H}^T\mathbf{C}_w^{-1}\mathbf{x}$ (Section 7.2.1), the detector (23.5) becomes

$$h(\mathbf{x}) = \hat{\boldsymbol{\theta}}^T(\mathbf{H}^T\mathbf{C}_w^{-1}\mathbf{H})\boldsymbol{\theta}$$

and the test hypothesis can be equivalently carried out on the estimates used as a sufficient statistic for $\boldsymbol{\theta}$ without any performance loss.

23.3.2 Classification of Multiple Signals

In the case of multiple distinct signals in Gaussian noise, the model is

$$\begin{aligned}
\mathcal{H}_1 &: \mathbf{x} = \mathbf{s}_1 + \mathbf{w} \Longrightarrow p(\mathbf{x}|\mathcal{H}_1) = G(\mathbf{x};\mathbf{s}_1, \mathbf{C}_w) \\
\mathcal{H}_2 &: \mathbf{x} = \mathbf{s}_2 + \mathbf{w} \Longrightarrow p(\mathbf{x}|\mathcal{H}_2) = G(\mathbf{x};\mathbf{s}_2, \mathbf{C}_w) \\
&\vdots \\
\mathcal{H}_K &: \mathbf{x} = \mathbf{s}_K + \mathbf{w} \Longrightarrow p(\mathbf{x}|\mathcal{H}_K) = G(\mathbf{x};\mathbf{s}_K, \mathbf{C}_w)
\end{aligned}$$

and classification is by identifying which hypothesis is more likely from the data. Assuming that all classes are equally likely with $p(\mathcal{H}_i) = p(\mathcal{H}_k) = 1/K$ (see Section 23.4 how to generalize in any context), the decision in favor of $\mathcal{H}_k$ is when

$$p(\mathbf{x}|\mathcal{H}_k) > p(\mathbf{x}|\mathcal{H}_i) \text{ for } \forall i \neq k$$

Since inequalities hold for log-transformations, it is straightforward to prove that the condition is

$$||\mathbf{x}-\mathbf{s}_k||^2_{\mathbf{C}_w^{-1}} < ||\mathbf{x}-\mathbf{s}_i||^2_{\mathbf{C}_w^{-1}} \text{ for } \forall i \neq k$$

which is the hypothesis for the signal $\mathbf{s}_k$ that is closest to $\mathbf{x}$, where the distance is the weighted norm.

The case $\mathbf{C}_w = \sigma_w^2\mathbf{I}$ is common in communication engineering where the transmitter maps the information of a set of bits into one signal $\mathbf{s}_k$ (modulated waveform) out of K,

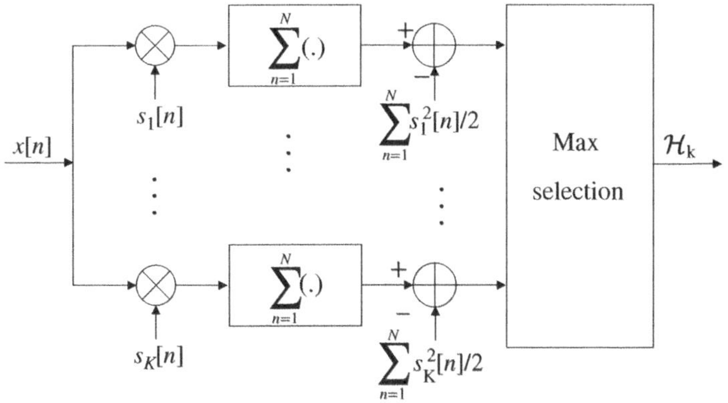

Figure 23.5 Correlation-based classifier (or decoder).

and the receiver must decide from a noisy received signal $\mathbf{x}$ which one was transmitted out of the set $\mathbf{s}_1, \mathbf{s}_2, ..., \mathbf{s}_K$. The test hypothesis for $\mathcal{H}_k$ is

$$\mathbf{s}_k^T\mathbf{x} - \frac{1}{2}||\mathbf{s}_k||^2 > \mathbf{s}_i^T\mathbf{x} - \frac{1}{2}||\mathbf{s}_i||^2 \text{ for } \forall i \neq k$$

where $\mathbf{s}_k^T\mathbf{x} = \sum_{n=1}^{N} s_k[n]x[n]$ is the correlation of $x[n]$ with $s_k[n]$ and the choice is for the hypothesis with the largest correlation, apart from a correction for the energy of each signal. The multiple test hypothesis of $\mathbf{x}$ is called *decoding* as it decodes the information contained in the bit-to-signature mapping, and the correlation-based decoder in Figure 23.5 is optimal in the sense that it minimizes wrong classifications.

The performance is evaluated in terms of misclassification probability, or error probability. Error probability analysis is very common in communication engineering as the performance of a communication system largely depends on the choice of the transmitted set $\mathbf{s}_1, \mathbf{s}_2, ..., \mathbf{s}_K$. There are several technicalities to evaluating the error probability or approximations and bounds for some specific modulations; these are out of scope here but readers are referred to the specialized texts texts [35, 37, 57, 58], mostly for continuous-time signals.

23.3.3 Generalized Likelihood Ratio Test (GLRT)

In detection theory when the signal depends on a parameter $\boldsymbol{\theta}$ that can take some values, the NP lemma no longer applies. An alternative route to the LRT is the generalized likelihood ratio test (GLRT) that uses for each data the best parameter value in any class. Let the probability be rewritten as

$$p(\mathbf{x}|\mathcal{H}_k) = p(\mathbf{x}|\boldsymbol{\theta} \in \boldsymbol{\Theta}_k)$$

where $\boldsymbol{\Theta}_k$ is a partition of the parameter space corresponding to the class $\mathcal{H}_k$. The sets are disjoint $\boldsymbol{\Theta}_k \cap \boldsymbol{\Theta}_i = \emptyset$ (to avoid any ambiguity in classification) and exhaustive $\boldsymbol{\Theta}_1 \cup \boldsymbol{\Theta}_2 \cup ... \cup \boldsymbol{\Theta}_K = \boldsymbol{\Theta}$ to account for all states. LRT classification can be equivalently stated as deciding on which class of parameters is more appropriate to describe $\mathbf{x}$ assuming that **x** is drawn from one value. However, when the choice is over many values, one can redefine the LRT using the most likely values of the unknown parameters in each class.

For binary classification with the set $\boldsymbol{\Theta}_0$ and $\boldsymbol{\Theta}_1$ for the hypotheses $\mathcal{H}_0$ and $\mathcal{H}_1$, respectively, the GLRT is

$$\mathcal{L}_g(\mathbf{x}) = \frac{\max_{\theta\in\boldsymbol{\Theta}_1} p(\mathbf{x}|\theta)}{\max_{\theta\in\boldsymbol{\Theta}_0} p(\mathbf{x}|\theta)} = \frac{p(\mathbf{x}|\hat{\theta}_1)}{p(\mathbf{x}|\hat{\theta}_0)} \underset{\mathcal{H}_0}{\overset{\mathcal{H}_1}{\gtrless}} \gamma$$

and the threshold is set to attain a predefined level of P_{FA}, provided that it is now feasible. Since each term in the ratio is the MLE, the first step of the GLRT is to perform MLE for the parameters under each of the hypotheses, and then compute the ratio of the corresponding likelihoods. The GLRT has a broad set of applications in signal processing and an excellent taxonomic review can be found in the book by S. Kay [132].

Example 1: Sinusoid Detection

The sinusoidal detection problem on N samples is

$$\begin{aligned}
\mathcal{H}_1 &: x[n] = a_o \cdot \cos(\omega_o n + \phi_o) + w[n] \\
\mathcal{H}_0 &: x[n] = w[n]
\end{aligned}$$

where $\boldsymbol{\theta} = [a, \omega, \phi]^T$ are the characterizing parameters that can be all/partially unknown. The likelihood of the two settings is (see Section 5.4 for the definition of parameters)

$$\begin{aligned}
\mathcal{H}_1 &: \mathbf{x} = \mathbf{H}(\omega_o)\boldsymbol{\alpha}_o + \mathbf{w} \Longrightarrow p(\mathbf{x}|\mathcal{H}_1) = G(\mathbf{x};\mathbf{H}(\omega_o)\boldsymbol{\alpha}_o, \mathbf{C}_w) \\
\mathcal{H}_0 &: \mathbf{x} = \mathbf{w} \Longrightarrow p(\mathbf{x}|\mathcal{H}_0) = G(\mathbf{x};\mathbf{0}, \mathbf{C}_w)
\end{aligned}$$

and the GLRT is

$$\mathcal{L}_g(\mathbf{x}) = \max_{\theta} \frac{G(\mathbf{x};\mathbf{H}(\omega)\boldsymbol{\alpha}, \mathbf{C}_w)}{G(\mathbf{x};\mathbf{0}, \mathbf{C}_w)} = \frac{G(\mathbf{x};\mathbf{H}(\hat{\omega})\hat{\boldsymbol{\alpha}}, \mathbf{C}_w)}{G(\mathbf{x};\mathbf{0}, \mathbf{C}_w)} \underset{\mathcal{H}_0}{\overset{\mathcal{H}_1}{\gtrless}} \gamma$$

as the pdf $p(\mathbf{x}|\mathcal{H}_0)$ is independent of any parameter. Recall that for uncorrelated noise $\mathbf{C}_w = \sigma_w^2\mathbf{I}$, the frequency estimation is based on the peak of the periodogram (Section 14.1, 10.2) $\widehat{S}_x(\omega)$, and the GLRT becomes

$$\max_{\omega\in(-\pi,\pi]} \frac{\widehat{S}_x(\omega)}{\sigma_w^2} \underset{\mathcal{H}_0}{\overset{\mathcal{H}_1}{\gtrless}} \ln\gamma$$

The detector compares the maximum value of the periodogram wrt a threshold that depends on the pdf of the periodogram itself that is a chi-square pdf (Section 14.1.4).

Example 2: Target Detection in Radar

A radar/sonar system can estimate the target distance using the estimated ToD Chapter 20. This example related to delay estimation is instrumental to map the properties from a discrete set of samples onto continuous-time signals. The two hypotheses within the observation time window T are

$$\mathcal{H}_1 : x(t) = \alpha_o g(t - \tau_o) + w(t)$$
$$\mathcal{H}_0 : x(t) = w(t)$$

where $w(t)$ is white as detailed in Section 9.4, and the parameters are $\boldsymbol{\theta} = [\alpha, \tau]^T$. The likelihood ratio for the two conditions is

$$\mathcal{L}(x(t)|\boldsymbol{\theta}) = \frac{p(x(t)|\mathcal{H}_1)}{p(x(t)|\mathcal{H}_0)} = \frac{\exp\left(-\frac{1}{N_0}\int_0^T (x(t) - \alpha g(t-\tau))^2 dt\right)}{\exp\left(-\frac{1}{N_0}\int_0^T x^2(t)dt\right)}$$

and simplifications yield (using the notation in Section 9.4):

$$\ln \mathcal{L}(x(t)|\boldsymbol{\theta}) = \frac{2\alpha}{N_0}\int_0^T x(t)g(t-\tau)dt - \frac{\alpha^2 E_g}{N_0}$$

Maximization wrt α yields:

$$\ln \mathcal{L}(x(t)|\hat{\alpha}, \tau) = \frac{1}{N_0 E_g}\left(\int_0^T x(t)g(t-\tau)dt\right)^2$$

and the GLRT is

$$\max_\tau \left(\int_0^T x(t)g(t-\tau)dt\right)^2 \underset{\mathcal{H}_0}{\overset{\mathcal{H}_1}{\gtrless}} N_0 E_g \ln \gamma$$

which basically implies that one estimates the ToD assuming that there is one waveform present, and then compares the value of the maximum of the cross-correlation with a threshold that depends on the pdf of the noise after the cross-correlation.

23.3.4 Detection of Random Signals

The classification here is between signal+noise and noise only, when the signal is random characterized by a known covariance $\mathbf{C}_s$ that is superimposed on the noise:

$$\mathcal{H}_1 : \mathbf{x} = \mathbf{s} + \mathbf{w} \Longrightarrow p(\mathbf{x}|\mathcal{H}_1) = G(\mathbf{x};\mathbf{0}, \mathbf{C}_s + \mathbf{C}_w)$$
$$\mathcal{H}_0 : \mathbf{x} = \mathbf{w} \Longrightarrow p(\mathbf{x}|\mathcal{H}_0) = G(\mathbf{x};\mathbf{0}, \mathbf{C}_w)$$

To generalize, the two hypotheses are:

$$\mathcal{H}_1 : \mathbf{x} \sim \mathcal{N}(\mathbf{0}, \mathbf{C}_1)$$
$$\mathcal{H}_0 : \mathbf{x} \sim \mathcal{N}(\mathbf{0}, \mathbf{C}_0)$$

that referred back to the motivating example has known covariance $\mathbf{C}_1 = \mathbf{C}_s + \mathbf{C}_w$ and $\mathbf{C}_0 = \mathbf{C}_w$. It can be shown that the NP lemma applies as covariances are known, and the log-LRT (apart from additive terms that are accounted for in the threshold γ)

$$\mathcal{L}(\mathbf{x}) = \mathbf{x}^T\mathbf{Q}\mathbf{x}$$

is a quadratic form with

$$\mathbf{Q} = \mathbf{C}_0^{-1} - \mathbf{C}_1^{-1}$$

The factorization $\mathbf{Q} = \mathbf{C}_0^{-T/2}(\mathbf{I} - \mathbf{C}_0^{T/2}\mathbf{C}_1^{-1}\mathbf{C}_0^{1/2})\mathbf{C}_0^{-1/2}$ from $\mathbf{C}_0 = \mathbf{C}_0^{1/2}\mathbf{C}_0^{T/2}$ (and $\mathbf{C}_0^{-1} = \mathbf{C}_0^{-T/2}\mathbf{C}_0^{-1/2}$) allows the log-LRT to be rewritten as

$$\mathcal{L}(\mathbf{x}) = \mathbf{x}^T\mathbf{C}_0^{-T/2}(\mathbf{I} - \mathbf{C}_0^{T/2}\mathbf{C}_1^{-1}\mathbf{C}_0^{1/2})\mathbf{C}_0^{-1/2}\mathbf{x} = \mathbf{y}^T(\mathbf{I} - \mathbf{S}^{-1})\mathbf{y} = \mathcal{L}(\mathbf{y})$$

with $\mathbf{y} = \mathbf{C}_0^{-1/2}\mathbf{x}$ (pre-whiten of $\mathbf{x}$) and $\mathbf{S}^{-1} = \mathbf{C}_0^{T/2}\mathbf{C}_1^{-1}\mathbf{C}_0^{1/2}$. The distribution of the log-LRT is a chi-square pdf, and the NP lemma for a given γ needs to detail the pdf of $\mathcal{L}(\mathbf{x})$ under $\mathcal{H}_0$ and $\mathcal{H}_1$ (see Section 4.14 of [7]).

23.4 Bayesian Classification

The case where the probabilistic structure of the underlying classes is known perfectly is an excellent theoretical reference to move on to the Bayesian classifiers characterized by the a-priori probabilities on each class. Let $K = 2$ classes, the a-priori information on classes is in form of probabilities $p(\mathcal{H}_1)$ and $p(\mathcal{H}_0)$, with known $p(\mathbf{x}|\mathcal{H}_1)$ and $p(\mathbf{x}|\mathcal{H}_0)$. The misclassification error is when $\mathbf{x} \in \mathcal{R}_0$ for class $\mathcal{H}_1$ and vice-versa, and the probability is

$$\Pr(\text{error}) = p(\mathcal{H}_0|\mathcal{H}_1)p(\mathcal{H}_1) + p(\mathcal{H}_1|\mathcal{H}_0)p(\mathcal{H}_0) \tag{23.8}$$

Minimization of Pr(error) defines the optimal decision region with minimum misclassification probability, which is minimized when assigning each value $\mathbf{x}$ to the class $\mathcal{H}_k$ that has the largest a-posteriori probability $p(\mathcal{H}_k|\mathbf{x})$. Class $\mathcal{H}_1$ is assigned when

$$p(\mathcal{H}_1|\mathbf{x}) > p(\mathcal{H}_0|\mathbf{x})$$

but according to Bayes' rule it can be rewritten in terms of likelihood ratio for the two alternative classes

$$\mathcal{L}(\mathbf{x}) = \frac{p(\mathbf{x}|\mathcal{H}_1)}{p(\mathbf{x}|\mathcal{H}_0)} \underset{\mathcal{H}_0}{\overset{\mathcal{H}_1}{\gtrless}} \frac{p(\mathcal{H}_0)}{p(\mathcal{H}1)} \tag{23.9}$$

where the threshold depends on the a-priori probabilities. In the case where $p(\mathcal{H}_1) = p(\mathcal{H}_0) = 1/2$, the decision in favor of class $\mathcal{H}_1$ is when $p(\mathbf{x}|\mathcal{H}_1) > p(\mathbf{x}|\mathcal{H}_0)$ as in Section 23.3.2. For example, the binary classification for Gaussian distributions (Section 23.2.2) that minimizes the Pr(error) yields the decision rule

$$-\frac{1}{2}(\mathbf{x} - \boldsymbol{\mu}_1)^T\mathbf{C}_1^{-1}(\mathbf{x} - \boldsymbol{\mu}_1) + \frac{1}{2}(\mathbf{x} - \boldsymbol{\mu}_0)^T\mathbf{C}_0^{-1}(\mathbf{x} - \boldsymbol{\mu}_0) + \ln\frac{|\mathbf{C}_1|}{|\mathbf{C}_0|} \underset{\mathcal{H}_0}{\overset{\mathcal{H}_1}{\gtrless}} \ln\frac{p(\mathcal{H}_0)}{p(\mathcal{H}1)}$$

Again, the decision boundary characteristics do not change between linear and quadratic.

For K classes, the misclassification error is

$$\Pr(\text{error}) = \sum_{k=1}^{K} \Pr(\mathbf{x} \notin \mathcal{R}_k | \mathcal{H}_k)$$

and it needs to compare $\mathbf{x}$ with $K-1$ decision regions (e.g., it could be too complex). Alternatively, one evaluates the probability of correct classification given by

$$\Pr(\text{correct}) = \sum_{k=1}^{K} \Pr\{\mathbf{x} \in \mathcal{R}_k | \mathcal{H}_k\}$$

which is maximized when choosing for $\mathbf{x}$ the class $\mathcal{H}_k$ that has the largest a-posteriori probability $p(\mathcal{H}_k|\mathbf{x})$ compared to all the others.

23.4.1 To Classify or Not to Classify?

Decisions should be avoided when too risky. One would feel more confident to decide in favor of $\mathcal{H}_k$ if $p(\mathcal{H}_k|\mathbf{x}) \simeq 1$ and all the others are $p(\mathcal{H}_i|\mathbf{x}) \simeq 0$ for $i \neq k$. When all the a-posteriori probabilities are comparable (i.e., for K classes, $p(\mathcal{H}_k|\mathbf{x}) \simeq p(\mathcal{H}_i|\mathbf{x}) \simeq 1/K$, still $p(\mathcal{H}_k|\mathbf{x}) > p(\mathcal{H}_i|\mathbf{x})$), any decision is too risky and the decision could be affected by errors (e.g., due to the inappropriate a-priori probabilities), and misclassification error occurs. The way out of this is to introduce the *reject* class, which express doubt on a decision. A simple approach for a binary decision is to add an interval $\Delta \geq 0$ to the decision such that the a-posteriori probabilities for the decision become

$$\frac{p(\mathcal{H}_1|\mathbf{x})}{p(\mathcal{H}_0|\mathbf{x})} = \begin{cases} > 1+\Delta & \text{decide } \mathcal{H}_1 \\ < 1+\Delta \ \& \ > 1-\Delta & \text{reject option} \\ < 1-\Delta & \text{decide } \mathcal{H}_0 \end{cases}$$

Alternatively, the reject option can be introduced by setting a minimum threshold on the a-posteriori probabilities for classifying, say p_t, such that even if one would decide for $\mathcal{H}_k$ as $p(\mathcal{H}_k|\mathbf{x}) > p(\mathcal{H}_i|\mathbf{x})$, this choice is rejected if $p(\mathcal{H}_k|\mathbf{x}) < p_t$. The threshold p_t controls the fraction to be rejected: all are rejected if $p_t = 1$, all are classified if $p_t = 0$, and for K classes the threshold is approx. $p_t > 1/K$. When the fraction of rejections is too large, further experiments or training are likely to be necessary.

23.4.2 Bayes Risk

The error probability (23.8) could be inappropriate in some contexts where errors $p(\mathcal{H}_0|\mathcal{H}_1)$ and $p(\mathcal{H}_1|\mathcal{H}_0)$ are perceived differently. To exemplify, consider medical screening for early cancer diagnosis to compare the presence of a cancer (class $\mathcal{H}_1$) compared to being healthy (class $\mathcal{H}_0$). The incorrect diagnosis of cancer when a patient is healthy ($\Pr(\mathbf{x} \in \mathcal{R}_1|\mathcal{H}_0)$) induces some further investigations and psychological discomfort; the wrong diagnosis of lack of any pathology when the cancer is already in an advanced state ($\Pr(\mathbf{x} \in \mathcal{R}_0|\mathcal{H}_1)$) delays the cure and increases secondary risks. The two errors have different impact that can be accounted for with a different *cost function* (sometime referred as loss function) C_{ki} that is the cost of choosing $\mathcal{H}_k$ when $\mathcal{H}_i$ is the true class. For cancer diagnosis, $C_{01} > C_{10}$. The optimum classification is the one that minimizes the average cost $C = \sum_{k,i} C_{ki}$, defined from the expectation and called *Bayes risk*:

$$\mathbb{E}[C] = \sum_{k,i} C_{ki} p(\mathcal{H}_k|\mathcal{H}_i) p(\mathcal{H}_i)$$

Usually $C_{ki} > C_{ii}$, but in any case it is a common choice not to apply any cost to correct decisions hence $C_{kk} = 0$. The Bayes risk for $K = 2$ is

$$\mathbb{E}[C] = C_{11}p(\mathcal{H}_1) + C_{00}p(\mathcal{H}_0) + (C_{10} - C_{00})p(\mathcal{H}_1|\mathcal{H}_0)p(\mathcal{H}_0) + (C_{01} - C_{11})p(\mathcal{H}_0|\mathcal{H}_1)p(\mathcal{H}_1) \tag{23.10}$$

(recalling that $p(\mathcal{H}_0|\mathcal{H}_i) + p(\mathcal{H}_1|\mathcal{H}_i) = 1$). The choice of the decision region to minimize (23.10) does not depend on its first term, but rather on the remaining two terms that can be rearranged as

$$p(\mathcal{H}_1|\mathcal{H}_0) \times (C_{10} - C_{00})p(\mathcal{H}_0) + p(\mathcal{H}_0|\mathcal{H}_1) \times (C_{01} - C_{11})p(\mathcal{H}_1)$$

These resemble the minimization of the error probability (23.8) after correcting the a-priori probabilities as $p(\mathcal{H}_0) \to \alpha(C_{10} - C_{00})p(\mathcal{H}_0)$ and $p(\mathcal{H}_1) \to \alpha(C_{01} - C_{11})p(\mathcal{H}_1)$, for a normalization factor α of probabilities. The LRT is

$$\mathcal{L}(\mathbf{x}) = \frac{p(\mathbf{x}|\mathcal{H}_1)}{p(\mathbf{x}|\mathcal{H}_0)} \underset{\mathcal{H}_0}{\overset{\mathcal{H}_1}{\gtrless}} \frac{(C_{10} - C_{00})p(\mathcal{H}_0)}{(C_{01} - C_{11})p(\mathcal{H}1)}$$

and the threshold is modified to take into account the cost functions applied to the decisions (for $C_{10} = C_{01}$ and $C_{00} = C_{11} = 0$ it coincides with the threshold (23.9)).

23.5 Pattern Recognition and Machine Learning

Classification when the problem is complex (K large) and data is large (N large) should use a preliminary training stage to define by samples the decision regions (or their boundaries), and classification is by recognizing the "similarity" with one of the patterns used during the training. The simplest classifier for binary classes is a linear decision boundary, which is shown in Section 23.2.2 to be optimum when Gaussian distributions of the two classes have the same covariances. The simplest pattern recognition method is the *perceptron algorithm*, which is a supervised learning algorithm for the linear binary classifier developed in the early 60s, and is celebrated as the first artificial neural network classifier. Machine learning methods have evolved a great deal since then, and the *support vector machine* (SVM) is one of the most popular ones today. An excellent overview of the methods is by Bishop [136].

23.5.1 Linear Discriminant

The linear classifier function between two classes can be stated as follows:

$$y(\mathbf{x}) = \text{sign}(h(\mathbf{x})) = \begin{cases} 1 & \text{if } \mathbf{x} \in \mathcal{R}_1 \\ -1 & \text{if } \mathbf{x} \in \mathcal{R}_0 \end{cases}$$

where

$$h(\mathbf{x}) = \mathbf{a}^T\mathbf{x} + b$$

is the linear discriminant function that depends on $(\mathbf{a}, b)$; the decision boundary $\mathbf{a}^T\mathbf{x} + b = 0$ describes a hyperplane in N-dimensional space. To illustrate the geometrical meanings, Figure 23.6 shows the case of $N = 2$. The orthogonality condition $\mathbf{a}^T\mathbf{x} + b = 0$ states that any value of $\mathbf{x}$ along the boundary is orthogonal to the unit-norm vector $\mathbf{a}/||\mathbf{a}||$. The distance of the (hyper)plane from the origin is defined by b by taking the point $\mathbf{x}_o$ perpendicular to the origin, hence

$$\frac{\mathbf{a}^T}{||\mathbf{a}||}\mathbf{x}_o = -\frac{b}{||\mathbf{a}||}$$

The signed distance of any point from the boundary is

$$\rho(\mathbf{x}) = \frac{\mathbf{a}^T}{||\mathbf{a}||}\mathbf{x} + \frac{b}{||\mathbf{a}||} = \frac{h(\mathbf{x})}{||\mathbf{a}||}$$

and this can be easily shown once decomposed $\mathbf{x} = \mathbf{x}_p + \rho(\mathbf{x})\mathbf{a}/||\mathbf{a}||$ into its value along the boundary $\mathbf{x}_p$ (projection of $\mathbf{x}$ onto the hyperplane) and its distance $\rho(\mathbf{x})\mathbf{a}/||\mathbf{a}||$.

Extending to K classes, the linear discriminant can follow the same rules as in Section 23.3.2 by defining K linear discriminants such as for the kth class

$$h_k(\mathbf{x}) = \mathbf{a}_k^T\mathbf{x} + b_k$$

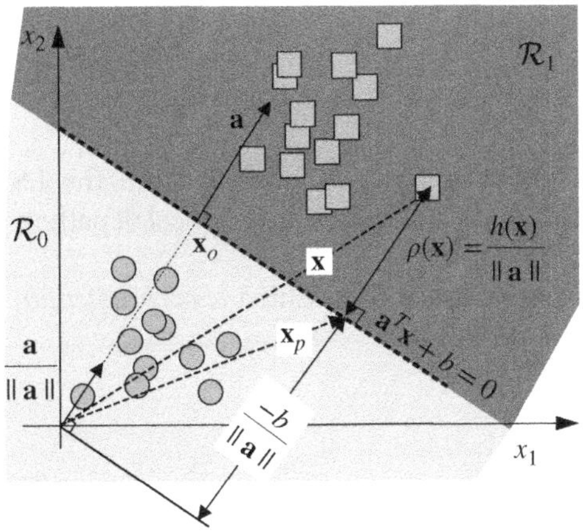

Figure 23.6 Linear discriminant for $N = 2$.

and one decides in favor of $\mathcal{H}_k$ when

$$h_k(\mathbf{x}) > h_i(\mathbf{x}) \text{ for } \forall i \neq k$$

This is called a one-versus-one classifier. This decision rule reflects the notion of distance from each hyperplane, and one chooses for the largest. The decision boundary $k - i$ can be found by solving

$$(\mathbf{a}_k - \mathbf{a}_i)^T \mathbf{x} + (b_k - b_i) = 0$$

Notice that the decision regions defined from the linear discriminants are simply connected and convex (i.e., any two points in $\mathcal{R}_k$ are connected by a line in $\mathcal{R}_k$). A one-versus-one classifier involves $K(K-1)/2$ comparisons, and one could find that there is not only one class that prevails. This creates an ambiguity that is solved by majority vote: the class that has the greatest consensus wins.

23.5.2 Least Squares Classification

Learning the parameter vectors $(\mathbf{a}_k, b_k)$ of linear classifiers is the first step in supervised pattern recognition. The LS method fits hyperplanes to the set of training vectors. Let $(\mathbf{x}_\ell, \mathbf{t}_\ell)$ be the ℓth training vector and the coding vector pair for a class $\mathcal{H}_k$; this means that the coding vector encodes the information on the class by assigning the kth entry $\mathbf{t}_\ell(k) = 1$ and $\mathbf{t}_\ell(i) = 0$ for all the others $i \neq k$. The K linear discriminants are grouped as

$$\underbrace{\begin{bmatrix} h_1(\mathbf{x}_\ell) \\ h_2(\mathbf{x}_\ell) \\ \vdots \\ h_K(\mathbf{x}_\ell) \end{bmatrix}}_{\mathbf{h}(\mathbf{x}_\ell)} = \underbrace{\begin{bmatrix} \mathbf{a}_1^T \\ \mathbf{a}_2^T \\ \vdots \\ \mathbf{a}_K^T \end{bmatrix}}_{\mathbf{A}} \mathbf{x}_\ell + \underbrace{\begin{bmatrix} b_1 \\ b_2 \\ \vdots \\ b_K \end{bmatrix}}_{\mathbf{b}}$$

The classifiers can be found by minimizing the LS from all coding vectors $\mathbf{t}_\ell$:

$$(\hat{\mathbf{A}}, \hat{\mathbf{b}}) = \arg\min_{\mathbf{A},\mathbf{b}} \left\{ \sum_{\ell \in \mathcal{T}} ||\mathbf{t}_\ell - (\mathbf{A}\mathbf{x}_\ell + \mathbf{b})||^2 \right\}$$

over all the training set $\mathcal{T}$. The decision functions $\hat{\mathbf{a}}_k^T \mathbf{x} + \hat{b}_k$ derived from the LS minimization are a combination of trainings $\{\mathbf{x}_\ell\}_{\ell \in \mathcal{T}}$, and this is quite typical of pattern recognition methods.

After the learning stage, one can recognize the pattern (also called *class prediction*) by evaluating the metric for the current $\mathbf{x}$

$$\hat{\mathbf{h}}(\mathbf{x}) = \hat{\mathbf{A}}\mathbf{x} + \hat{\mathbf{b}}$$

and choosing the largest one. Even if the entries of training $\mathbf{t}_\ell$ take values $\{0,1\}$ and resemble the classification probabilities, the discriminant $\hat{\mathbf{h}}(\mathbf{x})$ for any $\mathbf{x}$ is not guaranteed to be in the range $[0,1]$. Furthermore, LS implicitly assumes that the distribution of each class is Gaussian, and the LS coincides with the MLE in this case and is sub-optimal otherwise. The target coding scheme can be tailored differently from the choice adopted here—see [136].

23.5.3 Support Vectors Principle

Linear classifiers can be designed based on the concept of *margin*, and this is illustrated here for the binary case. Considering a training set

$$(t_1, \mathbf{x}_1), \ldots, (t_L, \mathbf{x}_L)$$

where binary valued

$$t_\ell \in \{-1, +1\}$$

encodes the two classes; the linear classifier is $(\mathbf{a}, b)$ such that the training classes are linearly separable:

$$\begin{aligned} \mathbf{a}^T\mathbf{x}_\ell + b \geq 1 \quad &\text{if} \quad t_\ell = 1 \\ \mathbf{a}^T\mathbf{x}_\ell + b \leq -1 \quad &\text{if} \quad t_\ell = -1 \end{aligned}$$

This is equivalent to rewriting the trainings into

$$t_\ell \left(\mathbf{a}^T\mathbf{x}_\ell + b\right) \geq 1 \qquad (23.11)$$

where the equality is for the margins illustrated in Figure 23.7. The optimal hyperplane is from signed $\rho(\mathbf{x})$ for the optimal margins such that the relative distance

$$\Delta\rho = \min_{\{\mathbf{x}:t=+1\}} \rho(\mathbf{x}) - \max_{\{\mathbf{x}:t=-1\}} \rho(\mathbf{x})$$

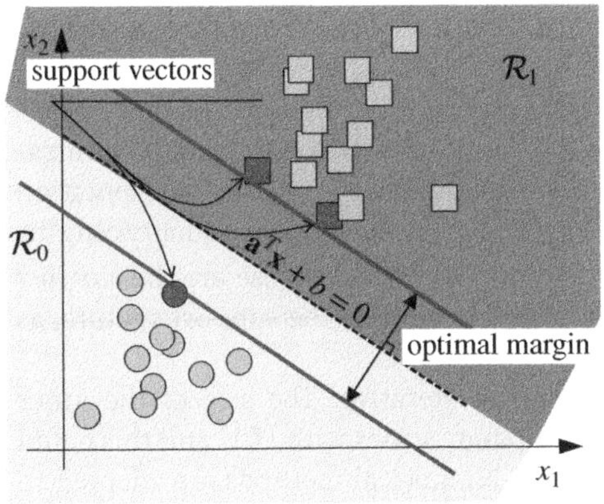

Figure 23.7 Support vectors and margins from training data.

is maximized. The maximum value is for the margins corresponding to the *support vectors* $\mathbf{x}_\ell$ such that $t_\ell\left(\mathbf{a}^T\mathbf{x}_\ell + b\right) = 1$, and the optimal margin is

$$\Delta\rho_{max} = \frac{2}{||\mathbf{a}||}$$

The optimal hyperplane is the one that minimizes the norm $||\mathbf{a}||^2 = \mathbf{a}^T\mathbf{a}$ (or equivalently, maximizes the margin) subject to the constraint (23.11), and this yields the maximum margin linear classifier.

This statement can be proved analytically, and the optimization is written using the Lagrange multiplier [137]:

$$L(\mathbf{a}, b, \lambda) = \frac{1}{2}\mathbf{a}^T\mathbf{a} - \sum_{\ell\in\mathcal{T}} \lambda_\ell t_\ell\left(\mathbf{a}^T\mathbf{x}_\ell + b\right) + \sum_{\ell\in\mathcal{T}} \lambda_\ell \tag{23.12}$$

with $\lambda_\ell \geq 0$. Taking the gradients wrt $\mathbf{a}$ yields:

$$\mathbf{a} = \textstyle\sum_{\ell\in\mathcal{T}} \lambda_\ell t_\ell \mathbf{x}_\ell$$
$$\textstyle\sum_{\ell\in\mathcal{T}} \lambda_\ell t_\ell = 0$$

which implies that the optimal hyperplane is a linear combination of the training vectors, and the classifier is

$$h(\mathbf{x}) = \sum_{\ell\in\mathcal{T}} \lambda_\ell t_\ell(\mathbf{x}_\ell^T\mathbf{x}) + b \tag{23.13}$$

in terms of weighted linear combination of the inner products with all the trainings. The SVM solves the optimization by reformulating the Lagrangian (23.12) in terms of the equivalent representation:

$$\max_{\{\lambda_\ell\}}\left\{\sum_{\ell\in\mathcal{T}} \lambda_\ell - \frac{1}{2}\sum_{i,j\in\mathcal{T}} \lambda_i\lambda_j t_i t_j \mathbf{x}_i^T\mathbf{x}_j\right\} = \max_{\lambda}\left\{\lambda^T\mathbf{1} - \frac{1}{2}\lambda^T\mathbf{D}\lambda\right\} \quad \text{st } \lambda^T\mathbf{t} = 0 \tag{23.14}$$

where $\lambda = [\lambda_1, ..., \lambda_{|\mathcal{T}|}]^T$, $\boldsymbol{t} = [t_1, ..., t_{|\mathcal{T}|}]^T$, and $[\mathbf{D}]_{ij} = t_i t_j \mathbf{x}_i^T\mathbf{x}_j$ contains all the inner products of the training vectors arranged in a matrix. This reduces to a constrained optimization for a quadratic form in λ, with the additional constraint $\lambda_\ell \geq 0$. This representation proves that the constraint $\lambda_\ell[t_\ell\left(\mathbf{a}^T\mathbf{x}_\ell + b\right) - 1] \geq 0$ is guaranteed by an equality only for $\lambda_\ell \neq 0$ where $t_\ell\left(\mathbf{a}^T\mathbf{x}_\ell + b\right) - 1 \geq 0$ is the constraint (23.11), and these are the *support vectors*. This means that the classifier (23.13) depends only on the support vectors, as for all others, $\lambda_\ell = 0$.

The optimization (23.14) is quadratic and can be iterative. The size of the inner products matrix $\mathbf{D}$ is $|\mathcal{T}| \times |\mathcal{T}|$; the memory occupancy scales with $|\mathcal{T}|^2$ and this could be very challenging when the training set is large. Numerical methods are adapted to solve this problem—such as those based on the decomposition of the optimization problem on set λ into two disjoint sets called active λ_A and inactive λ_N part [138]. The optimality condition holds for λ_A when setting λ_N=0, and the active part should include the support vector (with $\lambda_\ell \neq 0$). The active set is not known but it should be chosen arbitrarily and refined iteratively by solving at every iteration for the sub-problem that is related to the active part only.

The SVM plane (or surface) is based on a distinguishable, reliable, and noise-free training set. However, in practice the training could be unreliable, or even maliciously manipulated to affect the classification error. This calls for a robust approach to SVM that modifies the constraint on trainings (23.11) by introducing some regularization terms (see e.g., [139] [140]).

Remark: Classification surfaces are not only planar, but in general can be arbitrary to minimize the classification error and their shapes depend on the problem at hand, and in turn their (unknown) pdfs. Non-linear SVM is application-specific and it needs some skill for its practical usage for pattern-recognition. The key conceptual point is that the arrangement $[\mathbf{D}]_{ij} = t_i t_j \mathbf{x}_i^T \mathbf{x}_j$ contains the training data only as an inner product. One can apply a non-linear transformation $\phi(.)$ that maps $\mathbf{x}_\ell$ onto another space $\phi(\mathbf{x}_\ell)$ where training vectors can be better separated by hyperplanes, and one can define the corresponding inner products $\phi^T(\mathbf{x}_i)\phi(\mathbf{x}_j)$, called *kernels*, to produce a non-linear SVM. The topic is broad and problem-dependent, and the design of $\phi(.)$ needs experience. A discussion is beyond the scope here, and the reader is referred to [136][140].

23.6 Clustering

Originally clustering was considered as a signal processing method for vector quantization (or *data compression*) where it is relevant to encode each N-dimensional vector $\mathbf{x}$ by using K fixed vectors (*codevectors*) $\boldsymbol{\mu}_1, \boldsymbol{\mu}_2, ..., \boldsymbol{\mu}_K$ each of dimension N. This is based on a set of encoding regions $\mathcal{R}_1, \mathcal{R}_2, ..., \mathcal{R}_K$, possibly disjoint ($\mathcal{R}_m \cap \mathcal{R}_n = \emptyset$), associated to each codevector ($\boldsymbol{\mu}_k$ is uniquely associated to $\mathcal{R}_k$) such that when any vector $\mathbf{x}$ is in the encoding region $\mathcal{R}_k$, then its approximation is $\boldsymbol{\mu}_k$ and this is denoted as

$$Q(\mathbf{x}) = \boldsymbol{\mu}_k, \quad \text{if } \mathbf{x} \in \mathcal{R}_k$$

Representation of $\mathbf{x}$ by the K codevectors is affected by a *distortion* that is $d(\mathbf{x}, Q(\mathbf{x})) = \mathbf{x} - Q(\mathbf{x})$, and the *mean square distortion* is

$$D = \mathbb{E}_x[||\mathbf{x} - Q(\mathbf{x})||^2]$$

considering the pdf of $\mathbf{x}$ (or its sample value if unavailable, or the training set). Design of the encoding regions (Figure 23.8) is to minimize the mean square distortion D, and this in turn provides a reduced-complexity representation of any $\mathbf{x}$ by indexing the corresponding codevector $Q(\mathbf{x})$ that, being limited to K values, can be represented at most by a limited set of $\log_2 K$ binary values. Distortion D is monotonic decreasing in the number of codevectors K, but the codevectors should be properly chosen according to the pdf $p(\mathbf{x})$ as it is essentially a parsimonious (but distorted) representation of $\mathbf{x}$.

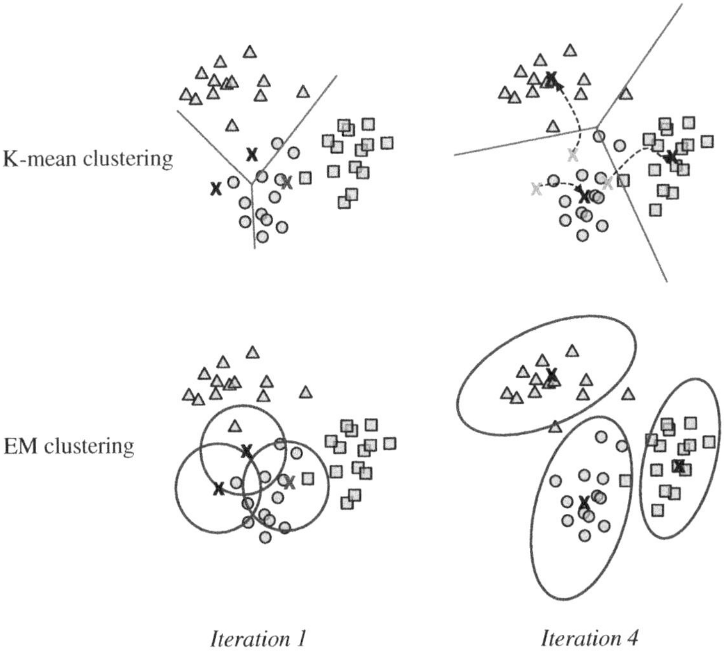

Figure 23.8 Clustering methods: K-means (left) and EM (right) vs. iterations.

23.6.1 K-Means Clustering

K-means is the simplest clustering method, which makes an iterative assignment of the L samples $\mathbf{x}_1, \mathbf{x}_2, ..., \mathbf{x}_L$ to a predefined number of clusters K. One starts from a initial set of values $\boldsymbol{\mu}_1^{(0)}, \boldsymbol{\mu}_2^{(0)}, ..., \boldsymbol{\mu}_K^{(0)}$ that represent the cluster centroids. Each sample $\mathbf{x}_\ell$ is assigned to the closest centroid, and after all samples are assigned, the new centroids are recomputed. It is convenient to use the binary variables $z_{k\ell} \in \{0,1\}$ to indicate which cluster the ℓth sample $\mathbf{x}_\ell$ belongs to: $z_{k\ell} = 1$ if $\mathbf{x}_\ell$ is assigned to the kth cluster, and $z_{i\ell} = 0$ for $j \neq k$. The mean square distortion

$$D = \sum_{\ell=1}^{L} \sum_{k=1}^{K} z_{k\ell} ||\mathbf{x}_\ell - \boldsymbol{\mu}_k||^2$$

depends on the choice of cluster centroids, which are obtained iteratively. K-means clustering in pseudo-code is (see also Figure 23.8):

Initialize by $\{\boldsymbol{\mu}_k^{(0)}\}$

1) for $\ell = 1 : L$, assign $\mathbf{x}_\ell \in \mathcal{R}_k$ if $||\mathbf{x}_\ell - \boldsymbol{\mu}_k^{(i)}||^2 < ||\mathbf{x}_\ell - \boldsymbol{\mu}_j^{(i)}||^2$ for $\forall j \neq k$: $z_{k\ell}^{(i)} = 1$; end
2) recalculate the position of the centroids:

$$\boldsymbol{\mu}_k^{(i+1)} = \frac{\sum_\ell z_{k\ell}^{(i)} \mathbf{x}_\ell}{\sum_\ell z_{k\ell}^{(i)}}$$

3) iterate to 1) until the centroids do not change vs. iterations ($\boldsymbol{\mu}_k^{(i)} = \boldsymbol{\mu}_k^{(i+1)}$ for $\forall k$).

Even if the K-means algorithm converges, the convergence clustering depends on the initial choice of centroids $\boldsymbol{\mu}_1^{(0)}, \boldsymbol{\mu}_2^{(0)}, \ldots, \boldsymbol{\mu}_K^{(0)}$. There is no guarantee that the convergence really minimizes the mean square distortion D, and to avoid convergence to bad conditions one can initialize over a random set $\{\boldsymbol{\mu}_1^{(0)}, \boldsymbol{\mu}_2^{(0)}, \ldots, \boldsymbol{\mu}_K^{(0)}\}$ and choose the best clustering based on its smallest distortion D. Notice that for a random initialization, there could be cluster centroids that have no samples assigned, and do not update. Further, the number of clusters K can be a free parameter, and clusters can be split or merge to solve for some critical settings. A good reference is Chapter 8 of [142].

23.6.2 EM Clustering

Let the data set $\mathbf{x}_1, \mathbf{x}_2, \ldots, \mathbf{x}_L$ be a sample set of the pdf $p(\mathbf{x})$ that is from the Gaussian mixture model (Section 11.6.3):

$$p(\mathbf{x}|\boldsymbol{\alpha}, \boldsymbol{\theta}) = \sum_{k=1}^{K} \alpha_k G(\mathbf{x}; \boldsymbol{\mu}_k, \mathbf{C}_k)$$

where $\boldsymbol{\theta}_k = \{\boldsymbol{\mu}_k, \mathbf{C}_k\})$ are the moments characterizing the N-dimensional Gaussian pdf of the kth class with parameters $\boldsymbol{\theta}_k$, and $\alpha_k \geq 0$ is the mixing proportions (or coefficients) that are constrained to

$$\sum_{k=1}^{K} \alpha_k = 1 \tag{23.15}$$

In expectation maximization (EM) clustering, the complete set is obtained by augmenting each sample value $\mathbf{x}_\ell$ by the corresponding missing data that describe the class of each sample to complete the data set. In detail, to complete the data there is a new set of *latent variables* $\mathbf{z}_1, \mathbf{z}_2, \ldots, \mathbf{z}_L$ that describe which class each sample $\mathbf{x}_1, \mathbf{x}_2, \ldots, \mathbf{x}_L$ belongs to by a binary class-indexing variable $z_{k\ell} \in \{0,1\}$, where $z_{k\ell} = 1$ means that $\mathbf{x}_\ell$ belongs to kth class. The marginal probability of $\mathbf{z}_\ell$ is dependent on the mixing proportions as

$$\Pr(z_{k\ell} = 1) = \alpha_k$$

and the conditional pdf is

$$p(\mathbf{x}_\ell | z_{k\ell} = 1, \boldsymbol{\alpha}, \boldsymbol{\theta}) = G(\mathbf{x}_\ell; \boldsymbol{\mu}_k, \mathbf{C}_k)$$

The complete data is $\mathcal{Y} = \{\mathbf{x}_1, \mathbf{x}_2, \ldots, \mathbf{x}_L, \mathbf{z}_1, \mathbf{z}_2, \ldots, \mathbf{z}_L\}$ and the EM method contains the E-step to estimate iteratively the latent variables $\{\mathbf{z}_1, \mathbf{z}_2, \ldots, \mathbf{z}_L\}$ and the M-step for the mixture parameters $\boldsymbol{\alpha}, \boldsymbol{\theta}$. Since the joint probability is

$$p(\mathbf{z}_\ell) = \prod_{k=1}^{K} \alpha_k^{z_{k\ell}}$$

(recall that $\alpha_k^0 = 1$ and $\alpha_k^1 = \alpha_k$), the conditional probability

$$p(\mathbf{x}_\ell | \mathbf{z}_\ell, \boldsymbol{\alpha}, \boldsymbol{\theta}) = \prod_{k=1}^{K} \left[G(\mathbf{x}_\ell; \boldsymbol{\mu}_k, \mathbf{C}_k) \right]^{z_{k\ell}}$$

is the product of all the pdfs depending on which rv in $\mathbf{z}_\ell$ is active, and the joint pdf is

$$p(\mathbf{x}_\ell, \mathbf{z}_\ell | \alpha, \boldsymbol{\theta}) = \prod_{k=1}^{K} \left[\alpha_k G(\mathbf{x}_\ell; \boldsymbol{\mu}_k, \mathbf{C}_k) \right]^{z_{k\ell}}$$

The likelihood of the complete data is:

$$p(\mathbf{x}_1, \mathbf{x}_2, ..., \mathbf{x}_L, \mathbf{z}_1, \mathbf{z}_2, ..., \mathbf{z}_L | \boldsymbol{\alpha}, \boldsymbol{\theta}) = \prod_{\ell=1}^{L} \prod_{k=1}^{K} \left[\alpha_k G(\mathbf{x}_\ell; \boldsymbol{\mu}_k, \mathbf{C}_k) \right]^{z_{k\ell}}$$

and the log-likelihood is

$$\mathcal{L}(\{\mathbf{x}_\ell\}, \{\mathbf{z}_\ell\} | \boldsymbol{\alpha}, \boldsymbol{\theta}) = \sum_{\ell=1}^{L} \sum_{k=1}^{K} z_{k\ell} \ln \left[\alpha_k G(\mathbf{x}_\ell; \boldsymbol{\mu}_k, \mathbf{C}_k) \right] \tag{23.16}$$

The **E-step** is to estimate the latent variables $z_{k\ell}$ as $\hat{z}_{k\ell} = \mathbb{E}[z_{k\ell} | \mathbf{x}_\ell; \boldsymbol{\alpha}, \boldsymbol{\theta}]$ (iteration number is omitted to avoid a cluttered notation). From Bayes, one gets the probability of $\mathbf{x}_\ell$ belonging to the kth class from the property of the mixture [45, 143]:

$$p(z_{k\ell} = 1 | \mathbf{x}_\ell, \boldsymbol{\alpha}, \boldsymbol{\theta}) = \frac{p(\mathbf{x}_\ell | z_{k\ell} = 1, \boldsymbol{\alpha}, \boldsymbol{\theta}) \Pr(z_{k\ell} = 1)}{\sum_{j=1}^{K} p(\mathbf{x}_\ell | z_{j\ell} = 1, \boldsymbol{\alpha}, \boldsymbol{\theta}) \Pr(z_{j\ell} = 1)} = \frac{\alpha_k G(\mathbf{x}_\ell; \boldsymbol{\mu}_k, \mathbf{C}_k)}{\sum_{s=1}^{K} \alpha_s G(\mathbf{x}_\ell; \boldsymbol{\mu}_s, \mathbf{C}_s)} = \hat{z}_{k\ell}$$

which yields the estimate of the latent variables based on the current mixture parameters $\boldsymbol{\alpha}, \boldsymbol{\theta}$.

The **M-step** estimates the means and covariances of the new Gaussian mixture based on the latent variables $\{\hat{z}_{k\ell}\}$ by maximizing the log-likelihood (23.16). The MLE of $\boldsymbol{\theta}$ can be obtained from the log-likelihood function

$$\mathcal{L}(\{\mathbf{x}_\ell\} | \boldsymbol{\alpha}, \boldsymbol{\theta}) = \ln p(\mathbf{x}_1, \mathbf{x}_2, ..., \mathbf{x}_L | \boldsymbol{\alpha}, \boldsymbol{\theta}) = \sum_{\ell=1}^{L} \ln \left[\sum_{k=1}^{K} \alpha_k G(\mathbf{x}_\ell; \boldsymbol{\mu}_k, \mathbf{C}_k) \right] \tag{23.17}$$

Setting the gradients to zero (see Section 1.4 for gradients)

$$\frac{\partial \mathcal{L}(\{\mathbf{x}_\ell\} | \boldsymbol{\alpha}, \boldsymbol{\theta})}{\partial \boldsymbol{\mu}_k} = -\sum_{\ell=1}^{L} \hat{z}_{k\ell} \mathbf{C}_k^{-1} (\mathbf{x}_\ell - \boldsymbol{\mu}_k) = 0$$

$$\frac{\partial \mathcal{L}(\{\mathbf{x}_\ell\} | \boldsymbol{\alpha}, \boldsymbol{\theta})}{\partial \mathbf{C}_k} = \frac{1}{2} \sum_{\ell=1}^{L} \hat{z}_{k\ell} \left[\mathbf{C}_k^{-1} (\mathbf{x}_\ell - \boldsymbol{\mu}_k)(\mathbf{x}_\ell - \boldsymbol{\mu}_k)^T \mathbf{C}_k^{-1} - \mathbf{C}_k^{-1} \right] = 0$$

one obtains the update equations

$$\boldsymbol{\mu}_k = \frac{\sum_{\ell=1}^{L} \hat{z}_{k\ell}\mathbf{x}_\ell}{\sum_{\ell=1}^{L} \hat{z}_{k\ell}} \tag{23.18}$$

$$\mathbf{C}_k = \frac{\sum_{\ell=1}^{L} \hat{z}_{k\ell}(\mathbf{x}_\ell - \boldsymbol{\mu}_k)(\mathbf{x}_\ell - \boldsymbol{\mu}_k)^T}{\sum_{\ell=1}^{L} \hat{z}_{k\ell}} \tag{23.19}$$

which are a weighted average by the a-posteriori probability $\hat{z}_{k\ell} = p(z_{k\ell} = 1|\mathbf{x}_\ell, \alpha, \boldsymbol{\theta})$ for each sample $\mathbf{x}_\ell$ and the kth class.

The mixing set $\boldsymbol{\alpha}$ can be obtained by optimizing the log-likelihood (23.17) wrt $\boldsymbol{\alpha}$, constrained by sum (23.15) using the Lagrange multiplier (Section 1.8). The mixing coefficients

$$\alpha_k = \frac{\sum_{\ell=1}^{L} \hat{z}_{k\ell}}{L} \tag{23.20}$$

are be used for the next iterations.

The EM can be interpreted as follows. At each iteration, the a-posteriori probabilities $\hat{z}_{1\ell}, \hat{z}_{2\ell}, ..., \hat{z}_{K\ell}$ on classes of each sample $\mathbf{x}_\ell$ are evaluated based on the mixture parameters previously computed. Using these probabilities, the means and covariances for each class are re-estimated (23.18, 23.19) depending on the class-assignment probabilities $\hat{z}_{1\ell}, \hat{z}_{2\ell}, ..., \hat{z}_{K\ell}$. The corresponding mixing probabilities are re-calculated accordingly as in (23.20). Notice that re-estimation of moments (23.18, 23.19) is weighted by the a-posteriori probabilities $\hat{z}_{k1}, \hat{z}_{k2}, ..., \hat{z}_{kL}$ and if the class is empty (i.e., $\hat{z}_{k\ell}$ is mostly small for all samples $\mathbf{x}_1, \mathbf{x}_2, ..., \mathbf{x}_L$, or equivalently α_k from (23.20) is very small), the EM update can be unstable.

At convergence, the estimate of the latent variables $\hat{z}_{k\ell}$ gives the a-posteriori probability for each data $\mathbf{x}_\ell$ and the kth cluster. EM clustering can be interpreted as an assignment to each cluster based on the a-posteriori probability. For comparison, K-means clustering makes a sharp assignment of each sample to the cluster, while the EM method makes a soft assignment based on the a-posteriori probabilities. EM clustering degenerates to K-means for the choice of symmetric Gaussians with $\mathbf{C}_k = \sigma^2\mathbf{I}$, and σ^2 small and fixed, the a-posteriori clustering becomes

$$\hat{z}_{k\ell} = \frac{\alpha_k \exp(-||\mathbf{x}_\ell - \boldsymbol{\mu}_k||^2/2\sigma^2)}{\sum_{s=1}^{K} \alpha_s \exp(-||\mathbf{x}_\ell - \boldsymbol{\mu}_s||^2/2\sigma^2)}$$

which for small enough σ^2 has $\hat{z}_{k\ell} \to 1$ as $||\mathbf{x}_\ell - \boldsymbol{\mu}_k||^2 \ll ||\mathbf{x}_\ell - \boldsymbol{\mu}_s||^2$ for $\forall s \neq k$.

Note that EM clustering might converge to a local optimum that depends on the initial conditions and the choice of the number of clusters K. Initialization could be non-informative with a random starting point: $\hat{z}_{k\ell} = 1/K$. The number of clusters K should be set and kept fixed over the iterations. Clusters can be annihilated during iterations when not supported by data, one component of the mixture is too weak, and one value α_k is excessively small (i.e., since α_k is the probability that one sample belongs to the kth class, it is clear that when $\alpha_k < 1/L$ there are not enough samples to update the classes). This can be achieved by setting their mixing proportion to zero [143].

References

1 Golub, G.H. and Van Loan, C.F (2013) *Matrix Computations (4th Ed.)*, Johns Hopkins University Press, Baltimore.
2 Horn, R.A. and Johnson, C.R. (2013) *Matrix Analysis (2nd Ed.)*, Cambridge Univ. Press, New York.
3 Hannah, J. (1996) A geometric approach to determinants, *The American Mathematical Monthly*, vol. 103, n. 5, pp. 401–409.
4 Brandwood, D.H. (1983) A complex gradient operator and its application in adaptive array theory, *IEE Proceedings* vol. 130, N. 1, Feb. 1983, pp. 11–16.
5 Kung, S.Y. (1998) *VLSI array processors*. Englewood Cliffs, NJ, Prentice Hall.
6 Kay, S.M. (1993) *Fundamentals of statistical signal processing: Estimation Theory (vol. 1)*, Prentice Hall Ed.
7 Scharf, L.L. (1991) *Statistical signal processing: detection, estimation, and time series analysis*, Addison-Wesley Pub.
8 Rahaman, M. and Ahsanullah, M. (1973) A note on the expected values of power of a matrix, *The Canadian Journal of Statistics*, vol. 1, n. 1, pp. 123–125.
9 Graham, A. (1981) *Kronecker product and matrix calculus*, John Wiley & Sons, Ltd.
10 David, H.A. and Nagaraja, H.N. (2003) *Order Statistics*, John Wiley & Sons, Ltd.
11 Papoulis, A. and Pillai, S.U. (2002) *Probability, random variables, and stochastic processes*, McGraw-Hill Education.
12 Oppenheim, A.V., Willsky, A.S. and Nawab, S.H. (1983) *Signals and systems* (vol.2). Englewood Cliffs, NJ: Prentice Hall.
13 Proakis, J.G. and Manolakis, D.G. (1988) *Introduction to digital signal processing*, Prentice Hall Ed.
14 Oppenheim, A.V. and Schafer, R.W. (2010) *Discrete-time signal processing*, Pearson Higher Education.
15 Schreier, P.J. and Scharf, L.L. (2010) *Statistical signal processing of complex-valued data*, Cambridge Univ.Press.
16 Boyd, S. and Vandenberghe, L. (2004) *Convex Optimization*, Cambridge University Press Ed. (http://stanford.edu/~boyd/cvxbook).
17 Porat, B. and Friedlander, B. (1986) Computation of the exact Information Matrix of Gaussian time series with stationary random components, *IEEE Trans. Acoustic Speech and Signal Processing*, vol.ASSP-34, pp. 118–130, Feb. 1986.
18 Westwater, E.R. (1978) The accuracy of water vapor and cloud liquid determination by dual-frequency ground-based microwave radiometry. *Radio Science*, 13(4): pp. 677–685.

Statistical Signal Processing in Engineering, First Edition. Umberto Spagnolini.

Companion website: www.wiley.com/go/spagnolini/signalprocessing

19 DeVore, R.A. and Temlyakov, V.N. (1996) Some remarks on greedy algorithms, *Advances in Computational Mathematics*, vol. 5, pp. 173–187.

20 Eldar, Y.C. and Kutyniok, G. (2012) *Compressed Sensing: Theory and Applications*, Cambridge Univ.Press.

21 Pillai, S.U., Suel, T. and Cha, S. (2005) *The Perron-Frobenius theorem and some of its applications*, IEEE Signal Processing Magazine, vol. 62, pp. 62–75.

22 Haylock, M.R., Hofstra, N., Klein Tank, A.M.G., Klok, E.J., Jones, P.D. and New, M. (2008) European daily high-resolution gridded data set of surface temperature and precipitation for 1950–2006, *Journal of Geophysical Research: Atmospheres*, vol. 113 n. D20, pp. 1–12.

23 Argo Project: *www.argo.ucsd.edu*

24 Marvasti, F. (Ed.) (2001) *Nonuniform sampling: theory and practice*, Springer Science, New York.

25 Fienberg, S.E. (2006) *When did Bayesian inference become "Bayesian"?*, Bayesian Analysis. vol. 1,n. 1, pp. 1–40.

26 Edwards, A.W. (1974) *The history of likelihood*, Int. Statistical Review, vol. 42, no. 1, pp. 9–15

27 Stigler, S.M. (2007) The epic story of maximum likelihood, *Statistical Science* vol. 22, n. 4, pp. 598–620.

28 Robert, C.R. and Casella, G. (2013) *Monte Carlo statistical methods*, Springer Ed., 1999 (2nd Ed., 2013).

29 Hyvärinen, A., Juha, K. and Erkki, O. (2004) *Independent component analysis*. vol.46, John Wiley & Sons, Ltd.

30 Strobach, P. (1986) Pure order recursive least squares ladder algorithms, *IEEE Transactions on Acoustics, Speech, and Signal Processing*, vol. 34, no. 4, pp. 880–897.

31 Huber, P.J. (2011) *Robust statistics*, Springer Berlin Heidelberg Ed.

32 Zoubir, A.M., Koivunen, V., Chakhchoukh, Y. and Muma, M. (2012) Robust estimation in signal processing: a tutorial-style treatment of fundamental concepts, *IEEE Signal Processing Magazine*, vol. 29, no. 4, pp. 61–80.

33 Lehmann, E.L. (1986) *Testing statistical hypothesis* (2nd ed.), John Wiley & Sons, Ltd.

34 Lindsey, W.C. and Simon, M.K. (1978) *Phase-locked loops & their application*, IEEE Communications Society, IEEE Press.

35 Meyr, H., Moeneclaey, M. and Fechtel, S.A. (1997) *Digital Communication Receivers, Synchronization, Channel Estimation, and Signal Processing*, John Wiley & Sons, Ltd.

36 Simeone, O., Spagnolini, U., Bar-Ness, Y. and Strogatz, S.H. (2008) Distributed synchronization in wireless networks, *IEEE Signal Processing Magazine*, vol. 25, n. 5, pp. 81–97.

37 Van Trees, H.L. (2001) *Detection Estimation and Modulation Theory*, part 1, John Wiley & Sons, Ltd.

38 Cramér, H. (1946) *A contribution to the theory of statistical estimation*, Aktuariestidskrift, pp. 458–463, 1946.

39 Rao, C.R. (1945) *Information and accuracy attainable in the estimation of statistical parameters, Bulletin of the Calcutta Mathematical Society*, vol. 37, no. 3, pp. 81–91.

40 Stein, M., Mezghani, A. and Nossek, J.A. (2014) A Lower Bound for the Fisher Information Measure, *IEEE Signal Processing Letters*, vol. 21, no. 7, pp. 796–799.

41 Scott, D.W. (1992) *Multivariate density estimation: theory, practice and visualization*, John Wiley & Sons, Ltd.

42 Weinstein, E. and Weiss, A.J. (1988) A general class of lower bounds in parameter estimation, *IEEE Trans. Inform.Th.*, vol. 34, n. 2, pp. 338–342.

43 Noam, Y. and Messer, H. (2009) Notes on the tightness of the Hybrid Cramér-Rao lower bound, *IEEE Trans. Signal Proc.* vol. 57, n. 6, pp. 2074–2085.

44 Rife, D. and Boorstyn, R. (1974) Single tone parameter estimation from discrete-time observations, *IEEE Transactions on Information Theory*, vol. 20, no. 5, pp. 591–598.

45 MacLachlan, G. and Krishnan, T. (1997) *The EM algorithm and extensions*, John Wiley & Sons, Inc., New York.

46 Feder, M. and Weinstein, E. (1988) Parameter estimation of superimposed signals using the EM algorithm, *IEEE Transactions on Acoustic Speech and Signal Processing*, vol. ASSP-34, pp. 477–489.

47 Fessler, J.A. and Hero, A.O. (1994) Space-alternating generalized EM algorithm, *IEEE Transactions on Signal Processing*, vol. SP-42, pp. 4664–4677.

48 Friedlander, B. (1982) Lattice filters for adaptive processing, *Proceedings of the IEEE*, vol. 70, no. 8, pp. 829–867.

49 Lev-Ari, H. and Kailath, T. (1984) Lattice filter parameterization and modeling of non-stationary processes, *IEEE Trans. Information Th.*, vol. 30, no. 1, pp. 2–16.

50 Kalman, R.E. and Bucy, R.S. (1961) New results in linear filtering and prediction theory. *ASME. J. Basic Eng.* vol. 83, n. 1, pp. 95–108.

51 Anderson, B.D.O. and Moore, J.B. (1979) *Optimal filtering*, Prentice Hall Ed.

52 P. Stoica and R. L. Moses, Spectral Analysis of Signals, Prentice Hall, 2005.

53 Tsatsanis, M.K., Giannakis, G.B. and Zhou, G. (1996) *Estimation and equalization of fading channels with random coefficients*, Proceedings of the IEEE International Conference on Acoustics, Speech, and Signal Processing, vol. 2, pp. 1093–1096, Atlanta (USA).

54 Lindbom, L. (1993) *Simplified Kalman estimation of fading mobile radio channels: high performance at LMS computational load*, Proceedings of the IEEE International Conference on Acoustics, Speech, and Signal Processing, pp. 352–355 vol. 3, Minneapolis (USA).

55 Fliege, N.J. (1994) *Multirate digital signal processing*, John Wiley & Sons, Ltd.

56 Whittle, P. (1953) Analysis of multiple stationary time series, *J. Roy. Stat. Soc.*, vol. 15, pp. 125–139.

57 Proakis, J.G., Salehi, M., Zhou, N. and Li, X. (1994) *Communication systems engineering.* Prentice Hall Ed.

58 Stüber, G.L. (2011) *Principles of mobile communication.* Springer Ed.

59 Widrow, B. and Stearns, S.D. (1985) *Adaptive signal processing*, Prentice Hall Ed., New Jersey.

60 Sayed, A.H. (2003) *Fundamentals of Adaptive Filtering.* Wiley-IEEE Press.

61 Jaffer, A.G. (1998) *Maximum likelihood direction finding of stochastic sources: A separable solution*, Proceedings of the IEEE International Conference on Acoustics, Speech, and Signal Processing (New York. NY), pp. 2893–2896.

62 Stoica, P. and Nehorai, A. (1989) MUSIC, maximum likelihood, and Cramér-Rao bound, *IEEE Transactions on Acoustics, Speech, Signal Processing* vol. 37, pp. 720–741.

63 Stoica, P. and Nehorai, A. (1990) Performance Study of Conditional and Unconditional Direction-of-Arrival Estimation, *IEEE Transactions on Acoustics, Speech and Signal Processing*, vol. 38, n. 10.

64 Besler, Y. and Makovski, A. (1986) Exact maximum likelihood parameter estimation of superimposed exponential signals in noise, *IEEE Transactions on Acoustics, Speech, and Signal Processing*, vol. ASSP-34, pp. 1081–1089.

65 Schmidt, R.O. (1979) *Multiple emitter location and signal parameter estimation*, in Proceedings of the RADC Spectral Estimation Workshop, Rome, NY, pp. 243–258.

66 Pisarenko, V.F. (1973) The retrieval of harmonics from a covariance function, *Geophysical Journal International* vol.33, n.3, pp. 347–366.

67 Stoica P. and Sharman, K.C. (1990) Novel eigenanalysis method for direction estimation *Proceedings of the IEE-F*, vol. 137, pp. 19–26.

68 Roy, R. and Kailath, T. (1989) ESPRIT – Estimation of signal parameters via rotation invariance techniques, *IEEE Transactions on Acoustics, Speech, and Signal Processing* , vol. 17, no. 7.

69 Anderson, T.W. (1963) Asymptotic theory for principal component analysis, *Ann. Math. Stat*, vol. 34, pp. 122–148.

70 Akaike, H. (1974) A new look at the statistical model identification, *IEEE Trans. Autom. Control*, vol. AC-19, pp. 716–723.

71 Rissanen, J. (1978) Modeling by shortest data description, *Automatica*, vol. 14, pp. 465–471.

72 Joham, M., Utschick, W. and Nossek, J.A. (2005) Linear transmit processing in MIMO communications systems, *IEEE Trans. Signal Processing*, vol. 53, n. 8, pp. 2700–2712.

73 Fischer, R.F.H. (2002) *Precoding and signal shaping for digital transmission*, John Wiley & Sons, Ltd.

74 Salz, J. (1973) Optimum mean square decision feedback equalization *Bell Lab Syst. Tech. Journal*, vol. 52, n. 8, pp. 1341–1373.

75 Al-Dhahir, H. and Cioffi, J.M. (1995) MMSE decision feedback equalizers: finite-length results, *IEEE Trans. Information Theory*, vol. 41, n. 4, pp. 961–975.

76 Kaleh, G.K. (1995) Channel equalization for block transmission systems, *IEEE Journal on Select. Areas on Comm.*, vol. 13, n. 1, pp. 110–121.

77 Bracewell, R.N. (2000) *The Fourier transform and its applications* (3rd ed.), McGraw-Hill.

78 Gonzales, R.C. and Woods, R.E. (2007) *Digital image processing* (3rd ed.), Prentice Hall, NJ.

79 McClellan, J.H. (1973) *The design of two-dimensional digital filters by transformations*, Proc. 7th Annual Princeton Conf. Information Science and System, pp. 247–251.

80 Jain, A.K. (1989) *Fundamentals of Digital Image Processing*, Prentice Hall Ed.

81 Aubert, G. and Kornprobst, P. (2006) *Mathematical problems in image processing: partial differential equations and the calculus of variations* (2nd ed.), Springer Ed., 2006.

82 Claerbout, J.F. (1985) *Imaging the earth's interior*, Blackwell Sci. (http://sepwww.stanford.edu).

83 Goodman, J.W. (1996) *Introduction to Fourier Optics*, McGraw-Hill Ed.

84 Kak, A.C. and Slaney, M. (2001) *Principles of Computerized Tomographic Imaging* (Classics in Applied Mathematics).

85 Allison, W. (2006) *Fundamental of Physics for Probing and Imaging*, Oxford University Press Inc., New York.
86 Alexander, L. and Klug, H.A. (1948) Basic aspects of X-ray absorption, *Analyt. Chem.*, vol. 20, pp. 886–889.
87 Borfeld, T., Bürkelbach, J., Boesecke, R. and Schlegel, W. (1990) Methods of image reconstruction from projections applied to conformation radiotherapy, *Physics in Medicine and Biology*, vol. 35, n. 10, pp. 1423–1434.
88 Kenneth, L. and Carson, R. (1984) EM reconstruction algorithms for emission and transmission tomography, *J. Comput. Assist. Tomogr.* vol. 8, n. 2, pp. 306–16.
89 Bracewell, R. (2003) *Fourier analysis and imaging*, Prentice Hall.
90 Shepp, L.A. and Vardi, Y. (1982) Maximum Likelihood Reconstruction for Emission Tomography, *IEEE Trans. on Medical Imaging*, vol. 1, no. 2, pp. 113–122.
91 Coates, M., Hero III, A.O., Nowak, R. and Yu, B. (2002) *Internet tomography, IEEE Signal Processing Magazine*, vol. 19, n. 3, pp. 47–65.
92 Nicoli, M., Rampa, V. and Spagnolini, U. (2002) Hidden Markov Model for multidimensional wavefront tracking, *IEEE Trans on Geoscience and Remote Sens.*, vol. 40, n. 3, pp. 561–662.
93 Mengali, U. and D'Andrea, A.N. (1997) *Synchronization Techniques for Digital Receivers*, Plenum Press Ed.
94 Meyr, H. and Ascheid, G. (1990) *Synchronization in Digital Communications: Phase-, frequency-locked loops and amplitude control (vol.1)*, John Wiley & Sons, Ltd.
95 Li, J. and Wu, R. (1998) An efficient algorithm for time delay estimation, *IEEE Trans. Signal Processing*, vol. 46, n. 8, pp. 2231–2235.
96 Jacovitti, G. and Scarano, G. (1993) Discrete-time techniques for time delay estimation, *IEEE Trans. Signal Processing*, vol. 41, pp. 525–533.
97 Knapp, C.H. and Carter, G.C. (1976) The generalized correlation method for estimation of time of delay, *IEEE Trans. Acoustics, Speech, and Signal Processing*, vol. 24, pp. 320–327.
98 Bienati, N. and Spagnolini, U. (2001) Multidimensional wavefront estimation from differential delays, *IEEE Trans. Geoscience and Remote Sensing*, vol. 39, n. 3, pp. 655–664.
99 Spagnolini, U. (1999) Nonparametric narrowband wavefront estimation from wavefront gradients, *IEEE Trans Signal Processing*, vol. 47, n. 11, pp. 3116–3121.
100 de Boor, C. (1978) *A practical guide to splines*, Springer-Verlag, New York.
101 Van Trees, H.L. (2004) *Detection, Estimation, and Modulation Theory, Optimum Array Processing*, John Wiley & Sons, Ltd.
102 Haykin, S. (1985) *Array signal processing*, Prentice Hall Ed.
103 Spagnolini, U. (2004) A simplified model to evaluate the probability of error in DS-CDMA systems with adaptive antenna arrays, *IEEE Trans. Wireless Comm.*, vol. 3, n. 2, pp. 578–587.
104 Weiss, A.J. and Friedlander, B. (1993) On the Cramér-Rao bound for direction finding of correlated signals, *IEEE Trans. Signal Process.*, vol. SP-41, pp. 495–499.
105 Langendoen, K. and Niels, R. (2003) Distributed localization in wireless sensor networks: a quantitative comparison, *Computer Networks*, vol. 43, pp. 499–518.

106 Soatti, G., Nicoli, M., Savazzi, S. and Spagnolini, U. (2017) Consensus-based Algorithms for Distributed Network-State Estimation and Localization, *IEEE Trans. on Signal and Information Proc. over Networks*, vol. PP, no. 99.
107 Funiak, S., Guestrin, C., Paskin, M. and Sukthankar, R. (2006) *Distributed localization of networked cameras*. In Proceedings 5th Int. Conf. on Information Processing in Sensor Networks, pp. 34–42. ACM.
108 Ihler, A.T., Fisher, J.W., Moses, R.L. and Willsky, A.S. (2005) Nonparametric belief propagation for self-localization of sensor networks, *IEEE Journal on Selected Areas in Communications*, vol. 23, n. 4, pp. 809–819.
109 Wymeersch, H., Lien, J. and Win, M.Z. (2009) Cooperative localization in wireless networks, *Proceedings of the IEEE*, vol. 97, n. 2, pp. 427–450.
110 Wei, R., Beard, R.W. and Atkins, E.M. (2005) *A survey of consensus problems in multi-agent coordination, Proceedings of the 2005 American Control Conference*, vol. 3, pp. 1859–1864.
111 Olfati-Saber,R., Fax, J.A. and Murray, R.M. (2007) Consensus and Cooperation in Networked Multi-Agent Systems, *Proceedings of the IEEE*, vol. 95, n. 1, pp. 215–233.
112 Bolognino, A. and Spagnolini, U. (2014) *Consensus based distributed estimation with local-accuracy exchange in dense wireless systems*, IEEE International Conference on Communications (ICC 2014), pp. 4620–4625, Sydney 10-14 June 2014.
113 Bolognino, A. and Spagnolini, U. (2014) *Cooperative estimation for under-determined linear systems*, 48th Annual Conference on Information Sciences and Systems (CISS-2014), pp. 1–5, Princeton 19-21 March 2014.
114 Gallison, P. (2003) *Einstein's clocks, Poincaré's maps: empires of time*, W. W. Norton & Company, Inc.
115 Strogatz, S. (2003) *Sync: the emerging science of spontaneous order*, Hyperion.
116 Winfree, A. T. (1967) Biological rhythms and the behavior of populations of coupled oscillators, *J. Theor. Biol.*, 16, pp. 15–42.
117 Kuramoto, Y. (1984) *Chemical oscillations, waves and turbulence*, Spinger, Berlin.
118 Lindsey, W.C., Ghazvinian, F., Hagmann, W.C. and Desseouky, K. (1985) Network synchronization, *Proc. of the IEEE*, vol. 73, no. 10, pp. 1445–1467.
119 Hong, Y.-W. and Scaglione, A. (2005) A scalable synchronization protocol for large scale sensor networks and its applications, *IEEE Journal on Selected Areas in Communications*, vol. 23, no. 5, pp. 1085–1099.
120 Leng, M. and Wu, Y.C. (2011) Distributed Clock Synchronization for Wireless Sensor Networks Using Belief Propagation, *IEEE Trans. on Signal Processing*, vol. 59, n. 11, pp. 5404–5414.
121 Mills, D.L. (1991) Internet time synchronization: the Network Time Protocol, *IEEE Trans. Communications*, vol. 39 n. 10, pp. 1482–1493.
122 Diestel, R. (2005) *Graph Theory* - Graduate Texts in Mathematics vol.173, Springer-Verlag , 3rd ed.
123 Biggs, N. (1993) *Algebraic Graph Theory*, 2nd ed., Cambridge Univ. Press.
124 Lehmann, E.L. (1959) *Testing Statistical Hypotheses*, John Wiley & Sons, Inc., New York.
125 Davenport, W.B. and Root, W.L. (1958) *An introduction to the theory of random signals and noise*, McGraw-Hill.
126 Middleton, D. (1960) *An introduction to statistical communication theory*, McGraw-Hill.

127 Fukunaga, K. (1972) *Introduction to Statistical Pattern Recognition*, Academic Press.

128 Van Trees, H.L. (2001) *Detection, Estimation, and Modulation theory: radar-sonar signal processing and Gaussian signals in noise*, vol.3, John Wiley & Sons, Ltd.

129 Fukunaga, K. (1990) *Introduction to statistical pattern recognition* (2nd ed.), Academic Press.

130 Samuel, A.L. (1959) *Some studies in machine learning using the game of checkers*, IBM Journal of Res. and Develop. pp. 210–229, vol. 3, no. 3.

131 Jain, A.K., Murty, M.N. and Flynn, P.J. (1999) Data Clustering: a review, *ACM Computing Survey*, pp. 264–323, vol. 31, n. 3.

132 Kay, S. (1998) *Fundamentals of Statistical Signal Processing: detection theory*, Prentice Hall Inc.

133 Fisher, R.A. (1936) The use of multiple measurements in taxonomic problems, *Ann. Eugenic.*, pp. 111–132, vol. 7.

134 McLachlan, G. and Basford, K. (1998) *Mixture models: inference and application to clustering*, Marcel Dekker.

135 Anderson, T.W. and Bahadur, R.R. (1962) Classification into two multivariate normal distributions with different covariance matrices, *The Annals of Mathematical Statistics*, vol. 33, no. 2, pp. 420–431

136 Bishop, C.M. (2006) *Pattern recognition and machine learning*, Springer Ed.

137 Cortes, C. and Vapnik, V. (1995) Support-vector networks, *Machine learning*, vol. 20, no. 3, pp. 273–297.

138 Osuna, E., Freund, R. and Girosi, F. (1997) *Support vector machines: training and applications*, Massachusetts Institute of Technology.

139 Xu, H., Caramanis, C. and Mannor, S. (2009) Robustness and regularization of support vector machines, *Journal of Machine Learning Research*, vol. 10 (Jul), pp. 1485–1510.

140 Wang, L. (2005) *Support vector machines: theory and applications*, Springer Science & Business Media.

141 MacQueen, J. (1967) *Some Methods for classification and analysis of multivariate observations*, Proc. 5th Berkeley Symposium on Mathematical Statistics and Probability, Univ. of California Press, vol.1, pp. 281–297, Berkeley.

142 Tan, P.N., Steinbach, M., and Kumar, V. (2006) *Introduction to data mining*. Pearson Education, 2006

143 Figueiredo, M.A.T. and Jain, A.K. (2002) Unsupervised learning of finite mixture models, *in IEEE Trans. on Pattern Analysis and Machine Intelligence*, vol. 24, no. 3, pp. 381–396.

Index

Pages (bold numbering refers to the most relevant pages)

Statistical Signal Processing in Engineering, First Edition. Umberto Spagnolini.

Companion website: www.wiley.com/go/spagnolini/signalprocessing

h

i

j

k

l

m

Printed and bound by CPI Group (UK) Ltd, Croydon, CR0 4YY
16/04/2025
14658472-0005